Extreme Weather and Climate

Extreme Weather and Climate

UCAR/David Baumhefner

C. Donald Ahrens
Emeritus, Modesto Junior College

Perry Samson
University of Michigan

Australia • Brazil • Japan • Korea •
Mexico • Singapore • Spain •
United Kingdom • United States

Extreme Weather and Climate

C. Donald Ahrens, Perry Samson

Earth Science Editor: Laura Pople

Developmental Editor: Jake Warde

Assistant Editor: Samantha Arvin

Editorial Assistant: Kristina Chiapella

Media Editor: Alexandria Brady

Marketing Manager: Nicole Mollica

Marketing Assistant: Kevin Carroll

Marketing Communications Manager: Belinda Krohmer

Content Project Manager: Hal Humphrey

Art Director: John Walker

Print Buyer: Karen Hunt

Rights Acquisitions Account Manager, Text: Mardell Glinski Schultz

Rights Acquisitions Account Manager, Image: Mandy Groszko

Production Service: Janet Bollow Associates

Text Designer: Janet Bollow Associates and Yvo Riezebos

Art Editor: Janet Bollow Associates

Photo Researcher: Chris Burt and Prepress PMG

Copy Editor: Stuart Kenter

Illustrator: Charles Preppernau

Cover Designer: Robin Terra

Cover Image: © Douglas Keister/Corbis

Photos used with Focus on Extreme Weather, Focus on Instruments, and Extreme Weather Watch are © iStockPhoto. Photo used with Focus on a Special Topic is © J. L. Medeiros.

Compositor: MPS Limited, A Macmillan Company

Student Edition:

ISBN-13: 978-0-495-11857-2

ISBN-10: 0-495-11857-5

Brooks/Cole
20 Davis Drive
Belmont, CA 94002-3098
USA

Cengage Learning is a leading provider of customized learning solutions with office locations around the globe, including Singapore, the United Kingdom, Australia, Mexico, Brazil, and Japan. Locate your local office at:
www.cengage.com/global

Cengage Learning products are represented in Canada by Nelson Education, Ltd.

For your course and learning solutions, visit **www.cengage.com/brookscole.**

Purchase any of our products at your local college store or at our preferred online store **www.CengageBrain.com**

Printed in the United States of America
1 2 3 4 5 6 7 13 12 11 10

Brief Contents

1 The Turbulent Atmosphere 2
2 Energy that Drives the Storms 34
3 Temperature and Humidity Extremes 62
4 Condensation in the Atmosphere 94
5 Clouds and Stability 120
6 Precipitation Extremes 144
7 Atmospheric Motions 174
8 Wind Systems 206
9 Air Masses and Fronts 236
10 Mid-Latitude Cyclonic Storms 268
11 Thunderstorms 296
12 Tornadoes 332
13 Hurricanes 360
14 Weather Forecasting 398
15 The Earth's Changing Climate 428

A Units, Conversions, and Abbreviations 463
B Equations and Constants 466
C Weather Symbols and the Station Model 469
D Humidity and Dew-Point Tables (Psychrometric Tables) 471
E Changing GMT and UTC to Local Time 475
F Average Annual Global Precipitation 476
G Hurricane Tracking Chart 478

Additional Reading Material 479
Glossary 481
Index 495

Contents

Preface xv

1 The Turbulent Atmosphere 2

Threats from the Sky 4
Reasons for Concern 6
The Earth's Atmosphere 7
Properties of the Atmosphere 13
Air Temperature 13
Temperature Scales 14
Vertical Profile of Air Temperature 14
Air Pressure 16

FOCUS ON INSTRUMENTS
Air Pressure and Barometers 18

Wind 19
Moisture 21

FOCUS ON A SPECIAL TOPIC
Rising Air Produces Clouds 23

A Brief Look at Relative Humidity and Dew Point 24
Hydrologic Cycle 24
Extremes of Weather and Climate 25
Summary 31
Key Terms 32
Review Questions 32
Online Learning 33

2 Energy that Drives the Storms 34

Energy and Heat Transfer 36
Latent Heat—The Hidden Warmth 36
Conduction 38
Convection 39

FOCUS ON A SPECIAL TOPIC
Comparing Energy in Storms 39

Radiation 41

FOCUS ON A SPECIAL TOPIC
Is Tanning Healthy? 42

Energy Balancing Act—Absorption, Emission, and Equilibrium 45
Selective Absorbers and the Atmospheric Greenhouse Effect 45
Enhancement of the Greenhouse Effect 47
Warming the Air from Below 48
Shortwave Radiation Streaming from the Sun 49
The Earth's Annual Energy Balance 50
Why the Earth has Seasons 52

FOCUS ON EXTREME WEATHER
Space Weather 54

Seasons in the Northern Hemisphere 56
Seasons in the Southern Hemisphere 58

FOCUS ON EXTREME WEATHER
A Year without a Summer? 59

Summary 60
Key Terms 60
Review Questions 61
Online Learning 61

UCAR/Carlye Calvin

3 Temperature and Humidity Extremes 62

Daily Temperatures 64
Warm Days 64
Extreme Heat 65
Cold Nights 67

FOCUS ON EXTREME WEATHER
Extreme Heat Inside a Car 68

Extreme Cold 69
Daily Temperature Variations 71

FOCUS ON INSTRUMENTS
Measuring Temperature 74

Regional Temperature Variations 75
Weather Extremes and Human Discomfort 79
Wind and Cold 79
Cold, Damp Weather 81
Humidity—A Real Factor in How Hot or Cold We Feel 81
What Is "Relative" in Relative Humidity? 81
High Dew Points Mean Humid Air 82
Dew-Point Extremes 83
Dry Air with a High Relative Humidity 84

FOCUS ON A SPECIAL TOPIC
Humid Air and Dry Air Do Not Weigh the Same 84

Extreme Relative Humidity in the Home 86
It's Not the Heat, It's the Humidity 86
The Heat Index 87
Deadly Heat Waves 88
Beating the Heat—Dealing with Heat Waves 89

FOCUS ON INSTRUMENTS
Measuring Humidity 90

Summary 91
Key Terms 92
Review Questions 92
Online Learning 93

4 Condensation in the Atmosphere 94

The Formation of Dew, Frost, and Haze 96
Fog 97

FOCUS ON A SPECIAL TOPIC
Why Are Headlands Usually Foggier Than Beaches? 100

Foggy Weather 101

FOCUS ON EXTREME WEATHER
When Fog Turns to Smog 103

Clouds: Identification from the Surface 104
High Clouds 105
Middle Clouds 106
Low Clouds 106
Clouds with Vertical Development 109
Some Unusual Clouds 112
Clouds: Observations from Space 114
Summary 118
Key Terms 118
Review Questions 119
Online Learning 119

UCAR

5 Clouds and Stability 120

Atmospheric Stability 122
Types of Atmospheric Stability 124
A Stable Atmosphere 124

FOCUS ON A SPECIAL TOPIC
Stability and Subsidence Inversions 127

An Unstable Atmosphere 128
A Conditionally Unstable Atmosphere 128
Causes of Instability 130
Cloud Development 132
Convection and Clouds 133
Topography and Clouds 135

FOCUS ON EXTREME WEATHER
Lake-Effect Snowstorms and Atmospheric Stability 136

Changing Cloud Forms 139
Summary 141
Key Terms 142
Review Questions 142
Online Learning 143

6 Precipitation Extremes 144

Precipitation Processes 146
Collision and Coalescence Process 146
Ice-Crystal Process 147
Cloud Seeding and Precipitation 149
Precipitation in Clouds 151
Precipitation Types 152
Rain 152
Snow 153
The Effect of a Snowfall 155
Sleet, Freezing Rain, and Ice Storms 157
Hail 159

FOCUS ON EXTREME WEATHER
Aircraft Icing—A Hazard to Flying 160

Precipitation—Extreme Events 162
The Influence of Mountains 162
Wet Regions, Dry Regions, and Precipitation Records 163
Too Much Rain—Floods and Flash Floods 165

FOCUS ON EXTREME WEATHER
The Almost Record Snowfall at Montague, New York 166

FOCUS ON INSTRUMENTS
Doppler Radar and Precipitation 168

When it Doesn't Rain—Drought and the Palmer Index 169
Some Notable Droughts 171
African Drought 171
North American Drought 171
Summary 172
Key Terms 173
Review Questions 173
Online Learning 173

7 Atmospheric Motions 174

Horizontal Pressure Changes and Wind 176
Surface and Upper-Level Charts 178
 The Surface Map 179
 Isobaric (Constant Pressure) Charts 180

FOCUS ON A SPECIAL TOPIC
Flying along a Constant Pressure Surface—When the Air Temperature Drops, Look Out for Mountaintops 182

Why the Wind Blows 183
 Newton's Laws of Motion 183
 Forces that Influence the Wind 184
 Pressure Gradient Force 184
 Coriolis Force 186
How the Wind Blows 188
 Straight-Line Flow Aloft 188
 Curved Winds Around Lows and Highs Aloft 189
 Winds on Upper-Level Charts 191
Surface Winds 192

FOCUS ON INSTRUMENTS
Measuring Winds 192

 Locating the Center of Storms 194
The Influence of Extreme Winds 195
 Strong Winds Blowing over Land 195
 Extreme Winds and Water 198

FOCUS ON EXTREME WEATHER
Winds, Waves, and a Seasick Semester at Sea 200

Windy Places 201
Extreme Winds 201

FOCUS ON A SPECIAL TOPIC
How Chicago Came to Be Known as the "Windy City" 202

Summary 204
Key Terms 204
Review Questions 205
Online Learning 205

8 Wind Systems 206

General Circulation of the Atmosphere 208
 Single-Cell Model 208
 Three-cell Model 209
 Average Surface Winds and Pressure: The Real World 211
 The General Circulation and Precipitation Patterns 213
 Average Wind Flow and Pressure Patterns Aloft 214
Jet Streams 215
Extreme (and Not So Extreme) Local Wind Systems 219
 Monsoon Winds 219

FOCUS ON EXTREME WEATHER
Aircraft Turbulence—Fasten Your Seat Belts 220

 Sea and Land Breezes 223
 Mountain and Valley Breezes 224

FOCUS ON A SPECIAL TOPIC
Windy Afternoons 225

 Katabatic Winds 226
 Chinook Winds 227
 Santa Ana Winds 229
 Other Extreme Winds of Interest 231
Summary 234
Key Terms 234
Review Questions 235
Online Learning 235

UCAR/Carlye Calvin

9 Air Masses and Fronts 236

Air Masses 238
Source Regions 238
Classification 239
Air Masses of North America 239
Continental Polar (cP) and Continental Arctic (cA) Air Masses 239
Extremely Cold Outbreaks 241
Extremely Cold Air Masses Produce a Record Cold Winter 242
Modification of Cold Air Masses 243
Maritime Polar (mP) Air Masses 244
Maritime Tropical (mT) Air Masses 245
Continental Tropical (cT) Air Masses 249
FOCUS ON EXTREME WEATHER
A Scorcher by Noon in June 250
Fronts 251
Stationary Fronts 252
Cold Fronts 253
Typical Cold Fronts 253
"Back Door" Cold Fronts 255
A Strong Cold Front 256
Warm Fronts 257
FOCUS ON EXTREME WEATHER
The Extraordinary Cold Front of 1836 257
The Dryline 259
FOCUS ON EXTREME WEATHER
Warm Fronts and Ice Storms 260
Occluded Fronts 261
Upper-Level Fronts 264
Summary 265
Key Terms 265
Review Questions 266
Online Learning 267

10 Mid-Latitude Cyclonic Storms 268

Polar Front Theory 270
Stages of a Developing Storm 270
A Notorious Storm Follows the Model 272
Where Do Mid-Latitude Cyclones Tend to Form? 273
FOCUS ON EXTREME WEATHER
Northeasters 274
Developing Mid-Latitude Cyclones 275
The Role of Convergence and Divergence 276
FOCUS ON A SPECIAL TOPIC
A Closer Look at Convergence and Divergence 276
Waves in the Westerlies 277
Upper-Air Support for the Developing Storm 278
The Role of the Jet Stream 279
FOCUS ON A SPECIAL TOPIC
Jet Streaks and Storms 280
Conveyor Belt Model of Mid-Latitude Cyclones 282
"Storm of the Century"—The March Storm of 1993 283
FOCUS ON A SPECIAL TOPIC
Ranking East Coast Storms 284
The Columbus Day Storm—October 12, 1962 288
Mid-Latitude Cyclones and Great Plains Blizzards 289
Polar Lows 291
FOCUS ON EXTREME WEATHER
Blizzard Safety Tips 293
Summary 294
Key Terms 294
Review Questions 295
Online Learning 295

UCAR

11 Thunderstorms 296

Thunderstorm Development 298
Ordinary Cell Thunderstorms 298
Multicell Thunderstorms 301
The Gust Front 301
Microbursts 303
Heat Bursts 306
Squall-Line Thunderstorms 306
Mesoscale Convective Complexes 310
Supercell Thunderstorms 311
Thunderstorms and the Dryline 315
Thunderstorms and Flooding 316
Distribution of Thunderstorms 316

FOCUS ON EXTREME WEATHER
A Terrifying Flash Flood 317

Lightning and Thunder 320
How Far Away is the Lightning? — Start Counting 320
How does Lightning Form? 320
Electrification of Clouds 321
The Lightning Stroke 322
The Different Forms of Lightning 324
Forked Lightning 324
Heat Lightning 324
Sheet Lightning 324
Ribbon Lightning 325
Bead Lightning 325
Ball Lightning 325
Dry Lightning 325
St. Elmo's Fire 325
When Lightning Strikes 325

FOCUS ON A SPECIAL TOPIC
Strange Lightning in the Upper Atmosphere 326

Lightning Detection and Suppression 329
Summary 330
Key Terms 330
Review Questions 331
Online Learning 331

12 Tornadoes 332

What is a Tornado? 334
Tornado Life Cycle 336
Tornado Occurrence and Distribution 336
Tornado Winds 339
Seeking Shelter 340
The Fujita Scale 342

FOCUS ON A SPECIAL TOPIC
Flying Cows or "The Herd Shot Around the World" 342

Tornado Outbreaks 345
Tornado Formation 347
Supercell Tornadoes 347
Nonsupercell Tornadoes 350

FOCUS ON EXTREME WEATHER
Takin' a Chase on the Wild Side 351

Tornadic Winds and Doppler Radar 353

FOCUS ON A SPECIAL TOPIC
A Field Study to Explore Tornadoes 355

Waterspouts 356
Summary 357
Key Terms 358
Review Questions 358
Online Learning 359

Harald Richter/NOAA/NSSL

13 Hurricanes 360

Tropical Weather 362
Anatomy of a Hurricane 362
Hurricane Formation and Dissipation 365
  The Right Environment 365
  The Developing Storm 367
  The Storm Dies Out 368
  Investigating the Storm 369
  Hurricane Stages of Development 369
  Hurricane Movement 370
  Eastern Pacific Hurricanes 371
  North Atlantic Hurricanes 374
Naming Hurricanes and Tropical Storms 374

FOCUS ON A SPECIAL TOPIC
How Do Hurricanes Compare with Middle-Latitude Cyclones? 375

Devastating Winds, Flooding, and the Storm Surge 376
Extreme Flooding with a Tropical Storm 380
Some Notable Hurricanes 381
  Camille, 1969 381
  Hugo, 1989 381
  Andrew, 1992 382
  Ivan, 2004 384

FOCUS ON EXTREME WEATHER
Hunting Hugo 384

  Katrina, 2005 386
Other Devastating Hurricanes 387

FOCUS ON A SPECIAL TOPIC
The Record-Setting Atlantic Hurricane Seasons of 2004 and 2005 389

Hurricane Watches, Warnings, and Forecasts 391
Modifying Hurricanes 393
Hurricanes in a Warmer World 393
Summary 395
Key Terms 396
Review Questions 396
Online Learning 397

14 Weather Forecasting 398

Acquisition of Weather Information 400

FOCUS ON EXTREME WEATHER
Watches, Warnings, and Advisories 401

Weather Forecasting Tools 402
Weather Forecasting Methods 404
  The Computer and Weather Forecasting: Numerical Weather Prediction 404
  Why NWS Forecasts go Awry and Steps to Improve Them 406
  Other Forecasting Methods 408
  Types of Forecasts 410
  Accuracy and Skill in Forecasting 411
Predicting Short-Term Hazardous and Severe Weather Events 412
  Forecasting Weather in the West 412
  Forecasting Weather for the Great Plains 414
  Forecasting Southern Storms 415

FOCUS ON A SPECIAL TOPIC
Forecasting Rain or Snow with the Thickness Chart 416

  A Forecast of Southeastern Ice 417
  A Forecast of Hot and Dry for Florida 417
  Weather Forecast for the Northeast 417
Predicting Long-Term Weather and Climate Patterns 418
  El Niño, La Niña, and the Southern Oscillation 418
  Pacific Decadal Oscillation 423
  North Atlantic Oscillation 424
  Arctic Oscillation 424
Summary 425
Key Terms 426
Review Questions 426
Online Learning 427

NOAA/NSSL

15 The Earth's Changing Climate 428

Reconstructing Past Climates 430
Climate Throughout the Ages 432
Temperature Trends during the Past 1000 Years 433

FOCUS ON EXTREME WEATHER
The Ocean's Influence on Rapid Climate Change 434

Temperature Trend during the Past 100-Plus Years 436
Climate Change Caused by Natural Events 437
Climate Change: Feedback Mechanisms 437
Climate Change: Plate Tectonics and Mountain Building 438
Climate Change: Variations in the Earth's Orbit 439
Climate Change: Variations in Solar Output 442
Climate Change: Atmospheric Particles 442
Climate Change Caused by Human (Anthropogenic) Activities 445
Climate Change: Aerosols Injected into the Lower Atmosphere 445
Climate Change: Increasing Levels of Greenhouse Gases 446
Climate Change: Land Use Changes 446

FOCUS ON EXTREME WEATHER
Catastrophic Climate Change Brought on by Nuclear War 447

Global Warming 448
Recent Global Warming: Perspective 448
Radiative Forcing Agents 448
Climate Models and Recent Temperature Trends 449
Future Global Warming: Projections 450
Uncertainties about Greenhouse Gases 451
The Question of Clouds 452

FOCUS ON A SPECIAL TOPIC
Contrails and Climate Change 453

The Ocean's Impact 454
Consequences of Global Warming: The Possibilities 454
Global Warming: Possible Impacts on Extreme Weather 456
Temperature and Humidity 456
Precipitation—Drought and Floods 456
Mid-Latitude Cyclonic Storms 457
Lake-Effect Snowstorms 457
Thunderstorms and Tornadoes 457
Hurricanes 458
Summary of Extreme Weather Events and Global Warming 459
Global Warming: Efforts to Curb 459
Summary 460
Key Terms 460
Review Questions 461
Online Learning 461

A Units, Conversions, and Abbreviations 463

B Equations and Constants 466

C Weather Symbols and the Station Model 469

D Humidity and Dew-Point Tables (Psychrometric Tables) 471

E Changing GMT and UTC to Local Time 475

F Average Annual Global Precipitation 476

G Hurricane Tracking Chart 478

Additional Reading Material 479
Glossary 481
Index 495

UCAR/Carlye Calvin

Preface

Most everyone is curious and concerned about extreme weather events. This curiosity may come from an actual encounter with extreme weather or from news headlines that describe the power and devastation wrought by a powerful storm, such as Hurricane Katrina. Because of the interest generated by extreme weather, a growing number of colleges and universities are teaching meteorology in a way that is exciting and fascinating for students by offering introductory courses aimed at the science of extreme weather. Some schools have modified existing introductory courses to focus more on extreme and unusual weather, while others have created new courses. Many science departments have experienced increased enrollments through these efforts.

This book is intended for students taking a course in atmospheric science that emphasizes extreme, hazardous, or unusual weather. The extreme weather events, such as heat waves, cold spells, floods, tornadoes, hurricanes, lightning and so on are the focus used to teach the concepts in every chapter. Students will come away from this book knowing more about where, when, and how extremes of temperatures, humidity, thunderstorms, tornadoes, and floods happen. In addition, the fundamentals that a student should know about meteorology are found in this book.

Explanations of complex atmospheric processes are offered in a language that is accessible to a wide range of students, from nonscience majors needing to fulfill a science requirement to potential meteorology majors wishing to gain an overview of the field. Exceptional exciting photos and colorful graphics round out the learning experience, clarifying key concepts and deepening the student's understanding of them.

Extreme Weather and Climate blends coverage found in C. Donald Ahrens' market-leading meteorology texts with insights from Perry Samson, teacher and creator of the popular extreme weather course at the University of Michigan in Ann Arbor. Vividly illustrated and full of technology support, this text, written in Don Ahrens' hallmark down-to-earth style, will offer students a solid background in the science of meteorology and will help them develop a new appreciation for the power of nature. For a number of years, Professor Samson has organized expeditions for University of Michigan undergraduate students to help researchers from Texas Tech University explore the dynamic nature of supercell thunderstorms that form over the Great Plains. Here, his insights and experiences are brought to bear on Don Ahrens' accessible and engaging explanations of the science of meteorology. The result is a unique text for introductory courses in extreme weather.

Extreme Weather and Climate is organized into fifteen chapters that offer a flexible progression. Each chapter builds from atmospheric basics through the fundamental processes that are needed to understand the development of extreme weather events. Moreover, each chapter is self-contained and can be covered in any desired order.

Chapter 1, "The Turbulent Atmosphere," provides an overview of the atmosphere. Here, we are introduced to extreme weather events and the risks they pose compared with other risks in life. Chapter 2, "Energy that Drives the Storms," examines how energy is transferred and distributed over the earth, and how this energy warms the planet, causes seasonal variations, and can become concentrated in a relatively small area for storm development. Chapter 3, "Temperature and Humidity Extremes," not only looks at average daily temperature variations, but it also looks at the coldest, hottest, and most humid places found on earth. Chapter 4, "Condensation in the Atmosphere," examines clouds and other forms of condensation, such as fog that poses hazards to ground transportation and aviation.

Chapter 5, "Clouds and Stability," examines how changes in temperature and moisture, both at and above the earth's surface, influence the development of clouds, and why some clouds are able to grow into huge thunderstorms whereas others are not. In addition, this chapter contains a Focus section that details the role that atmospheric stability plays in the development of lake-effect snowstorms. Another Focus section in this chapter describes how stability and inversions are related to air pollution. Chapter 6, "Precipitation Extremes," explains how the various types of precipitation form, and where the wettest, driest, and snowiest regions are found on earth. This chapter also includes information on extreme weather events such as blizzards, ice storms, and drought.

Chapter 7, "Atmospheric Motions," describes how horizontal changes in air pressure produce atmospheric circulations. Here we learn about why and how the wind blows and about the effect that strong winds have on the surface of the earth. This chapter contains a Focus section on how Chicago became known as "The Windy City." Chapter 8, "Wind Systems," looks at the large-scale general circulation of air around the earth, as well as smaller-scale winds, such as the Chinook and Santa Ana wind. A number of extreme wind events are described toward the end of this chapter.

Chapter 9, "Air Masses and Fronts," describes the typical weather that occurs with air masses and fronts, as well as the temperature extremes that can accompany them. The chapter on mid-latitude cyclonic storms (Chapter 10) examines how these storms form and the variety of extreme weather they are capable of producing. Several major storms of the past are detailed toward the end of the chapter, including the March storm of 1993 and the October, 1962, storm along the west coast of the United States.

Chapter 11, "Thunderstorms," describes the many aspects of thunderstorms, including the different types, and the variety of extreme weather they can produce, such as large hail, lightning, high winds, and flash floods. The chapter on tornadoes (Chapter 12) covers the atmospheric conditions necessary to produce tornadoes, as well as some of the most infamous, destructive, and deadly tornadoes ever to occur. A Focus section in this chapter describes the harrowing experience that a team of undergraduate students had while encountering a supercell thunderstorm near Oberlin, Kansas. Another Focus section details the experiments conducted by over 100 scientists, students, and technicians in the summer of 2009 during the VORTEX 2 research project. Chapter 13, "Hurricanes," looks at how and where hurricanes form and why they are so destructive. The beginning of the chapter on forecasting (Chapter 14) covers basic forecasting methods. Toward the end of the chapter, an entire section is dedicated to the forecasting of extreme weather events. The final chapter (Chapter 15) presents an overview to the science of climate change with an emphasis on how a warmer climate might affect changes in the frequency and intensity of extreme weather events.

Additional features of *Extreme Weather and Climate* include:

- *Focus Sections* Each chapter contains at least two Focus sections that either expand on material in the main text or explore a subject closely related to what is being discussed. Focus sections fall into one of three categories: Focus on Extreme Weather, Focus on a Special Topic, and Focus on Instruments.
- *Extreme Weather Watch Boxes* At least three Extreme Weather Watch boxes in each chapter highlight extreme weather events and support the chapter content by illustrating the full range of weather events that impact our lives.
- *Keywords* Important terms are boldfaced in the text, with their definitions appearing in a glossary toward the end of the book.
- *Key phrases* Important phrases are italicized.
- *Units* Metric equivalents of English units in most cases are immediately provided in parenthesis.
- *Embedded Content Review* A Brief Review is placed toward the middle of most chapters.
- *Reinforcement* Summaries at the end of each chapter review the chapter's main points.
- *Key Terms* A list of key terms following each chapter allows the student to review and reinforce their knowledge of the chief concepts they have encountered. Each key term is followed by the number of the page on which the term appears in the text.
- *Review Questions* Questions at the end of chapters act to check how well students have assimilated the material.
- *Geophysical Map* At the back of the book is a geophysical map of North America that serves as a quick reference for locating states, provinces, and geographical features, such as mountain ranges and large bodies of water.

Supplemental Material and Technology Support

Technology for the Instructor

- *PowerLecture* This DVD-ROM, free to adopters, includes art, photos, and tables from the text, as well as prepared lecture outlines in PowerPoint to get you started. Stepped art figures, zoom art, video library, and an instructor's manual and test bank are also included, to help create dynamic presentations.
- *Online Instructor's Manual* Free to adopters. Also available on PowerLecture.
- *Online Test Bank* Free to adopters. Also available on PowerLecture.
- *ExamView* Allows an instructor to easily create and customize tests, see them on the screen exactly as they will print, and print them out.

- *WebTutor Toolbox for WebCT or Blackboard* Jump-start your course with customizable, rich, text-specific content within your Course Management System. WebTutor offers a wide array of web quizzes, activities, exercises, and web links. Robust communication tools—such as a course calendar, asynchronous discussion, real-time chat, a whiteboard, and an integrated e-mail system—make it easy to stay connected to the course.

Technology for the Student

- *Meteorology Resource Center* This password-protected website features an array of resources to complement students' experience with meteorology—including animations, videos on climate change and natural disasters, current news feeds, and a virtual field trip on tornadoes.
- *Virtual Field Trips in Meteorology: Tornadoes* Bring the field of storm chasing into the classroom! Students interact with media to explore the dynamic nature of tornadoes, including the formation and components of supercell thunderstorms, pressure changes within a tornado, the classification of tornadoes, and more. Questions are posed to students along the way to ensure they grasp the material.

Acknowledgements

First and foremost Don Ahrens would like to thank his wife, Lita, for her patience, understanding, and valuable assistance with many aspects of this book, including proofreading and indexing. Thanks to Chris Burt for obtaining beautiful photos and providing the Extreme Weather Watch boxes. A special thanks goes to Charles Preppernau for his beautiful rendering of the art. Thanks goes to Robert S. Robinson and Jan Null for their input, and to Mabel Labiak for her careful proofreading. Thanks to Janet Alleyn, who turned the manuscript into a beautiful book, and to Stuart Kenter for his conscientious editing. My special thanks goes to all the people at Cengage Learning who worked on this book and especially Laura Pople and Jake Warde who saw this book through its many stages.

Perry Samson first thanks Don Ahrens for the opportunity to help construct this text, and is likewise indebted to Prof. Chris Weiss and John Schroeder of Texas Tech University for including the University of Michigan student team in their supercell chases. Perry also acknowledges the careful proofreading provided by his daughter, Karis.

Acknowledge reviewers

Jeff Robertson / Arkansas Tech. University
Colleen Garrity / SUNY-Geneseo
John Cassano / University of Colorado
Scott Curtis / East Carolina University
Kevin Law / Marshall University
Matthew Huber / Purdue University
Dorothy Freidel / Sonoma State University
Bill Gustin / Lansing Community College
Segun Ogunjemiyo / CA State, Fresno
Robert S. Robinson/Clear Water, Inc.
Richard Bagby/Embry Riddle Aeronautical University, Daytona Beach, Florida
Paul Nutter/ University of N. Colorado
Robert Vaughn/Graceland University

To the Student

The earth's atmosphere is generally a tranquil, safe environment shielding us from the harsh cold and intense radiation of space. But occasionally events occur that represent the extremes of atmospheric behavior. Tornadoes, hurricanes, floods, lightning, hail, heat waves, snowstorms and more hold the potential to change our lives in a moment. The purpose of this book is to teach you how these extreme events form, grow, and dissipate so you will more fully understand the risks and what can be done to minimize those risks in your life. Moreover, the book intends to challenge you to think critically about if and how these events may change in frequency and/or intensity in the future as the earth becomes warmer from climatic changes.

Don Ahrens and Perry Samson

The Turbulent Atmosphere

CONTENTS

THREATS FROM THE SKY
REASONS FOR CONCERN
THE EARTH'S ATMOSPHERE
PROPERTIES OF THE ATMOSPHERE
EXTREMES OF WEATHER AND CLIMATE
SUMMARY
KEY TERMS
REVIEW QUESTIONS

Lightning strikes from a severe thunderstorm as it looms over the Great Plains near Grand Rapids, Nebraska, on May 5, 2005.

Normally, our atmosphere is a sheltering and stable environment that has sustained and nurtured life for millions of years. It protects us from the bitter cold of space and from most of the sun's dangerous radiant energy. Its gases trap a portion of the radiant energy emitted by the earth, and this energy warms the lower atmosphere. Precipitation that falls over large areas helps support a wide variety of life, which over time, has adapted to the normal range of weather and climate we find on earth.

Over most of the earth, the atmosphere is quiescent, with a mix of blue skies and white clouds enveloping the globe (see ⟫ Fig. 1.1). However, there are times when the sky turns ominous and the power of the atmosphere becomes focused on specific areas in events of wild fury that often last for short periods of time. These days of "extreme and unusual weather" are the focus of public fears, and are often the impetus behind our quest for knowledge about the atmosphere. While atmospheric scientists must understand the processes that govern the atmosphere even on the relatively quiet days, it is the ability to foretell the extreme and unusual weather events that ultimately prove the greatest challenge, as well as the greatest value, to society. It is these extreme weather events that attract the largest portion of our attention.

Threats from the Sky

Extreme and severe weather captures our imagination because of its sheer power and potential to cause personal injury, destruction, and death. In the United States, the risk of death due to a weather event is relatively small (about two per million people) when compared to other risks (see ⟫ Table 1.1). Violent weather, however, deserves and demands study because it delivers its damage both unexpectedly and in catastrophic ways.

Even though the risk of losing one's life due to a weather event is low, such deaths, and the fact that weather-related events cause an estimated $10 billion

NASA Goddard Space Flight Center Image by Reto Stöckli

⟫ FIGURE 1.1 A composite view of earth from space shows an atmosphere enveloped with white clouds of all shapes and sizes.

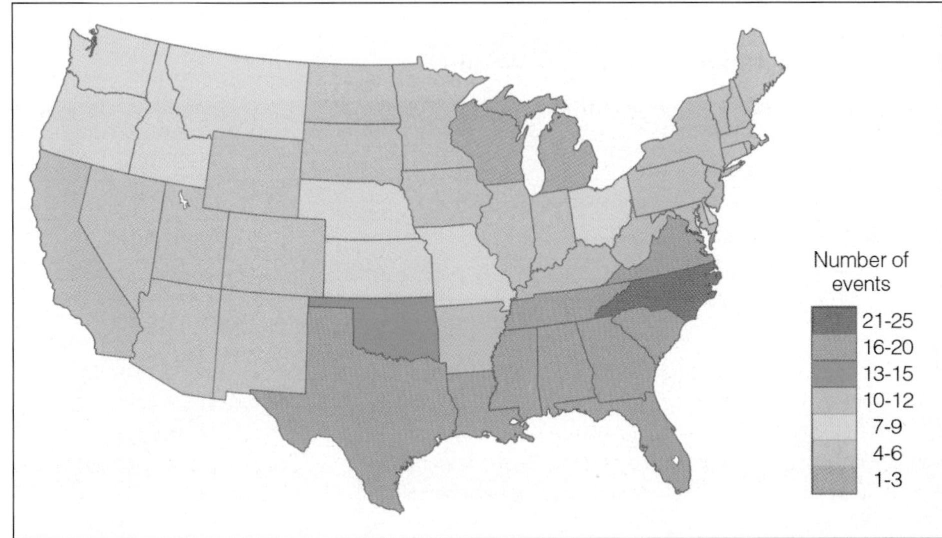

FIGURE 1.2

The total number of billion dollar weather and climate disasters from 1980 through 2004.

in property damage annually (see Fig. 1.2), drive atmospheric scientists to constantly search for a better understanding of how the atmosphere behaves and, in particular, how severe weather events form. Consequently, understanding the laws of nature that govern these extreme weather events will hopefully lead to better forecast accuracy so that, in the future, even fewer lives will be lost.

On average, weather events take about 500 lives annually in the United States. Of this total, almost half are due to excessive heat (see Fig. 1.3). Typically, people over the age of 65 are most vulnerable to heat and account for nearly half of these deaths. Tornadoes, lightning, and flooding account for about 36 percent of weather-related deaths, although many hundreds or even thousands may die from a single extreme weather event (see Table 1.2). For example, the Galveston hurricane of 1900 took more than 8000 lives and the tri-state tornado of 1925 killed over 690 people. In the rest of the world, weather events have wrought havoc in many locales; for instance, more than two million people died in China during a horrific flood in 1931.

Table 1.1

Actual Causes of Death per Million Population in the United States

ACTUAL CAUSES OF DEATHS PER 1,000,000	
Tobacco	1,450
Obesity	1,333
Alcohol	283
Microbial agents*	250
Toxic agents**	183
Motor vehicle crashes	143
Firearm accidents	97
Sexual behaviors†	67
Drug use	57
Weather-related events	2

*Includes deaths due to influenza, pneumonia, septicemia, tuberculosis, and other infectious and parasitic diseases.

**Includes deaths due to environmental exposure to pollutants in air and water, including second-hand tobacco smoke.

†Includes deaths due to HIV; other contributors were hepatitis B and C viruses and cervical cancer.

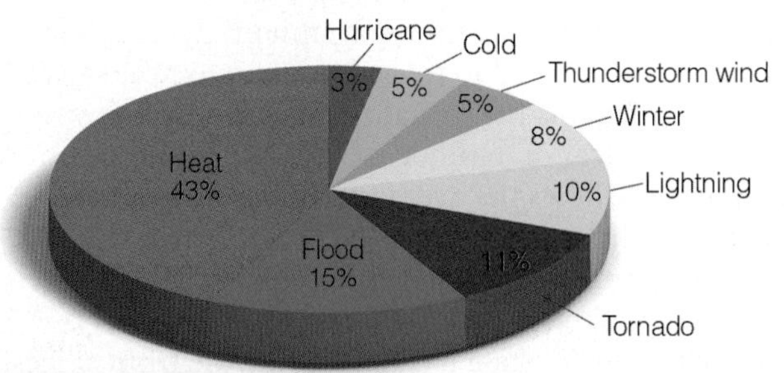

FIGURE 1.3 **Weather-related deaths in the United States broken down by event type. (*Source:* NOAA and Mokdad, A.H., et al., "Actual Causes of Death in the United States, 2000." *Journal of the American Medical Association*, 2004. 291(10): p. 1238–1245.)**

Table 1.2

Deadliest Weather-Related Events (a) in the United States and (b) in the World

(a) UNITED STATES	DEATHS	LOCATION	DATE
DROUGHT/FAMINE	5,000	Central USA	1933-1937
FLOOD	2,200	Johnstown, PA	5/31/1889
TROPICAL STORM	8-12,000	Galveston, TX	9/8-9/1900
HEAT WAVE	739	Chicago, IL	7/13-27/1995
BLIZZARD	400	NY and New England	3/11-13/1888
TORNADO	695	MO, IL, IN	3/18/1925
HAILSTORM	1	several locations and dates	
SMOG	22	Donora, PA	10/26-30/1948
(b) WORLD	**DEATHS**	**LOCATION**	**DATE**
DROUGHT/FAMINE	9,000,000	China	1876-79
FLOOD	2,000,000	Yellow River, China	1931
TROPICAL STORM	500,000	Coastal Bangladesh	11/12-13/1970
HEAT WAVE	35-52,000	Western Europe	8/2003
BLIZZARD	4,000	Iran	2/1972
TORNADO	1,300	Manikganj District, Bangladesh	4/26/1989
HAILSTORM	246	Moradabad, India	4/30/1888
SMOG	4,000	London, England	12/5 10/1952

There is virtually no part of the globe free of the threat of extreme weather. In North America, for example, lightning occurrences are most pronounced in the southeastern United States, notably in Florida, where lightning typically occurs on about a quarter of the days in a given year. Moreover, lightning is a significant threat in mountainous regions where afternoon thunderstorms are common. Although fewer deaths are attributed to tornadoes than to heat waves and flooding, it is the tornadoes that most often strike the greatest fear of weather-related injury and death. Figure 1.4 shows the average number of weather-related fatalities per year per one million people for each state due to flooding (top number), lightning (middle number), and tornadoes (lower number). Notice that every state has had people die from at least one of these weather-related events.

Reasons for Concern

There are a number of concerns about the atmosphere that are not related to isolated extreme weather events. For example, the earth is presently undergoing a warming trend—a warming that is primarily related to human activity. Will this global warming cause an increase in extreme weather events, such as more flooding, a higher incidence of drought, and stronger hurricanes? There is evidence that over the past several decades climate change may have caused an increase in the number of weather-related natural disasters. In addition, the atmosphere serves as a conduit for the transport of toxic materials added, sometimes inadvertently, by the output of industrial systems, or intentionally by bioterrorists. The unthinkable threat of delivering chemical, biological, and radioactive agents through the atmosphere requires our careful understanding of how air moves from one region to another and how the overall atmosphere behaves. Therefore, the remainder of this chapter provides you with a basic understanding of the atmosphere. It is this information that sets the stage for understanding the acute impacts caused by extreme and

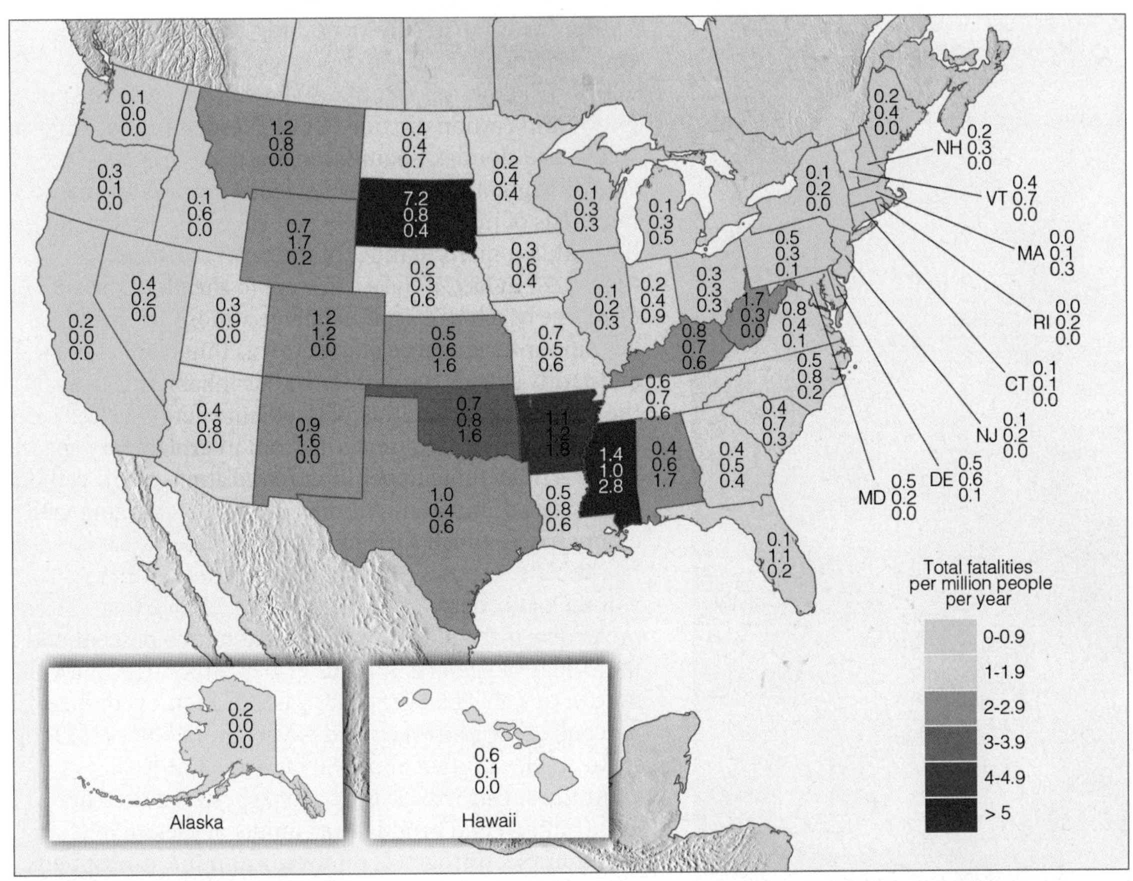

FIGURE 1.4 Average annual weather fatalities per one million people per state due to flooding (top number), lightning (middle number), and tornadoes (lower number) from 1959 through 2005. Color shows combined averages for each state. (Fatalities due to Hurricane Katrina in New Orleans in 2005 are not included.) (Sources include: NOAA, NCDC, and Storm Data Publication.)

unusual weather—the worst, most notorious moments of fury and disaster, which we will explore in subsequent chapters.

The Earth's Atmosphere

Our *atmosphere* is a delicate life-giving blanket of air that surrounds the earth. Living on the surface of the earth, we have adapted so completely to our environment of air that we sometimes forget how truly remarkable this substance is. Even though air is tasteless, odorless, and (most of the time) invisible, it protects us from the scorching rays of the sun and provides us with a mixture of gases that allows life to flourish. Because we cannot see, smell, or taste air, it may seem surprising that between your eyes and the pages of this book are trillions of air molecules. Some of these may have been in a cloud only yesterday, or over another continent last week, or perhaps part of the life-giving breath of a person who lived hundreds of years ago.

Warmth for our planet is provided primarily by the sun's energy. At an average distance from the sun of nearly 150 million kilometers (km), or 93 million miles (mi), the earth intercepts only a very small fraction of the sun's total energy output. However, it is this radiant energy* that drives the atmosphere into the patterns of everyday wind and weather, and allows life to flourish.

At its surface, the earth maintains an average temperature of about 59°F (15°C).** Although this temperature is mild, the earth experiences a wide range of temperatures, as readings can drop below −121°F (−85°C) during a frigid Antarctic night and climb during the day to above 122°F (50°C) on the oppressively hot, subtropical desert.

*Radiant energy, or radiation, is energy transferred in the form of waves that have electrical and magnetic properties. The light that we see is radiation, as is ultraviolet light. More on this important topic is given in Chapter 2.

**The abbreviation °F is used when measuring temperature in degrees Fahrenheit and °C is the abbreviation for degrees Celsius. More information about temperature scales is given later in this chapter.

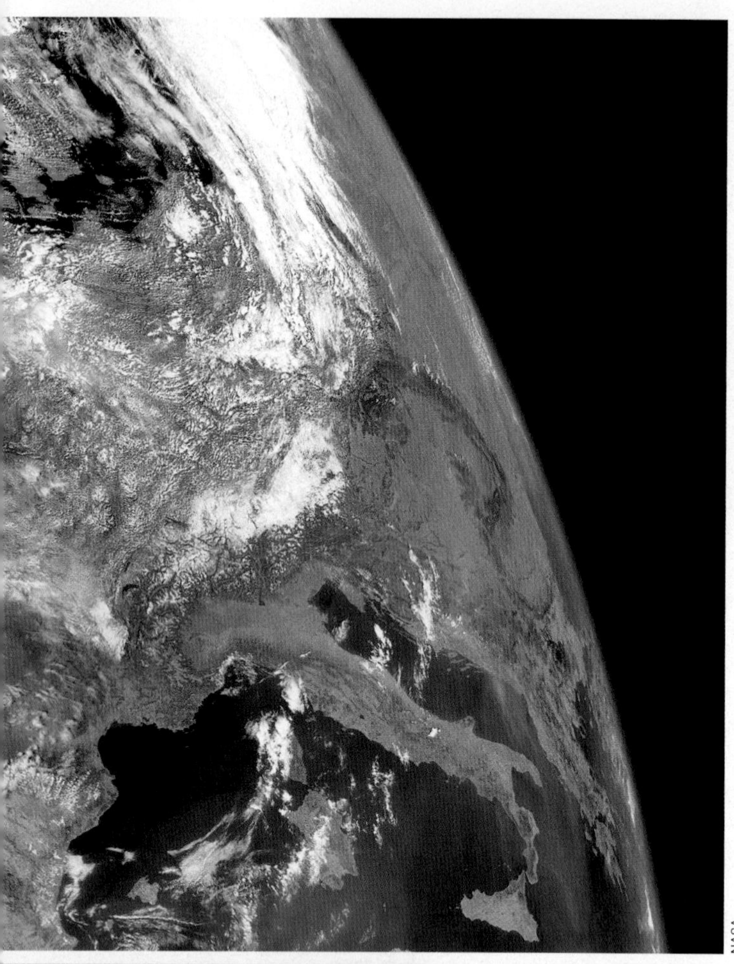
NASA

FIGURE 1.5 The earth's atmosphere as viewed from space. The atmosphere is the thin blue region along the edge of the earth.

The earth's **atmosphere** is a thin, gaseous envelope comprised mostly of nitrogen (N_2) and oxygen (O_2), with small amounts of other gases, such as water vapor (H_2O) and carbon dioxide (CO_2). Nestled in the atmosphere are clouds of liquid water and ice crystals.

Although our atmosphere extends upward for many hundreds of miles, almost 99 percent of the atmosphere lies within a mere 20 miles of the earth's surface (see Fig. 1.5). In fact, if the earth were to shrink to the size of a large beach ball, its inhabitable atmosphere would be thinner than a piece of paper. This thin blanket of air constantly shields the surface and its inhabitants from the sun's dangerous ultraviolet radiant energy, as well as from the onslaught of material from interplanetary space. There is no definite upper limit to the atmosphere; rather, it becomes thinner and thinner, eventually merging with empty space, which surrounds all the planets.

Table 1.3 shows the various gases present in a volume of air near the earth's surface. Notice that molecular **nitrogen** (N_2) occupies about 78 percent and molecular **oxygen** (O_2) about 21 percent of the total volume of dry air. If all the other gases are removed, these percentages for nitrogen and oxygen hold fairly constant up to an elevation of about 80 km (or 50 mi).

At the surface, there is a balance between destruction (output) and production (input) of these gases. For example, nitrogen is removed from the atmosphere primarily by biological processes that involve soil bacteria. In addition, nitrogen is taken from the air by tiny ocean-dwelling plankton that convert it into nutrients that help fortify the ocean's food chain. It is returned to the atmosphere mainly through the decaying of plant and animal matter. Oxygen, on the other hand, is removed from the atmosphere when organic matter

Table 1.3

Composition of the Atmosphere near the Earth's Surface

PERMANENT GASES			VARIABLE GASES			
Gas	Symbol	Percent (by Volume) Dry Air	Gas (and Particles)	Symbol	Percent (by Volume)	Parts per Million (ppm)*
Nitrogen	N_2	78.08	Water vapor	H_2O	0 to 4	
Oxygen	O_2	20.95	Carbon dioxide	CO_2	0.038	385*
Argon	Ar	0.93	Methane	CH_4	0.00017	1.7
Neon	N_e	0.0018	Nitrous oxide	N_2O	0.00003	0.3
Helium	He	0.0005	Ozone	O_3	0.000004	0.04†
Hydrogen	H_2	0.00006	Particles (dust, soot, etc.)		0.000001	0.01–0.15
Xenon	Xe	0.000009	Chlorofluorocarbons (CFCs)		0.00000002	0.0002

*For CO_2, 385 parts per million means that out of every million air molecules 385 are CO_2 molecules.

†Stratospheric values at altitudes between 11 km and 50 km are about 5 to 12 ppm.

FIGURE 1.6 The earth's atmosphere is a rich mixture of many gases, with clouds of condensed water vapor and ice crystals. Here, water evaporates from the ocean's surface. Rising air currents then transform the invisible water vapor into many billions of tiny liquid droplets that appear as puffy cumulus clouds. If the rising air in the cloud should extend to greater heights, where air temperatures are quite low, some of the liquid droplets would freeze into minute ice crystals.

decays and when oxygen combines with other substances, producing oxides. It is also taken from the atmosphere during breathing, as the lungs take in oxygen and release carbon dioxide. The addition of oxygen to the atmosphere occurs during photosynthesis, as plants, in the presence of sunlight, combine carbon dioxide and water to produce sugar and oxygen.

The concentration of the invisible gas **water vapor**, however, varies greatly from place to place, and from time to time. Close to the surface in warm, steamy, tropical locations, water vapor may account for up to 4 percent of the atmospheric gases, whereas in colder arctic areas, its concentration may dwindle to a mere fraction of a percent. Water vapor molecules are, of course, invisible. They become visible only when they transform into larger liquid or solid particles, such as cloud droplets and ice crystals, which may grow in size and eventually fall as rain or snow. The changing of water vapor into liquid water is called *condensation*, whereas the process of liquid water becoming water vapor is called *evaporation*. In the lower atmosphere, water is everywhere. It is the only substance that exists as a gas, a liquid, and a solid at those temperatures and pressures normally found near the earth's surface (see Fig. 1.6).

Water vapor is an extremely important gas in our atmosphere. Not only does it form into both liquid and solid cloud particles that grow in size and fall to earth as precipitation, but it also releases large amounts of heat—called *latent heat*—when it changes from vapor into liquid water or ice. Latent heat is an important source of atmospheric energy, especially for storms, such as thunderstorms and hurricanes. Moreover, water vapor is a potent *greenhouse gas* because it strongly absorbs a portion of the earth's outgoing radiant energy (somewhat like the glass of a greenhouse prevents the heat inside from escaping and mixing with the outside air). Thus, water vapor plays a significant role in the earth's heat-energy balance.

Carbon dioxide (CO_2), a natural component of the atmosphere, occupies a small (but important) percent of a volume of air, about 0.039 percent. Carbon dioxide enters the atmosphere mainly from the decay of

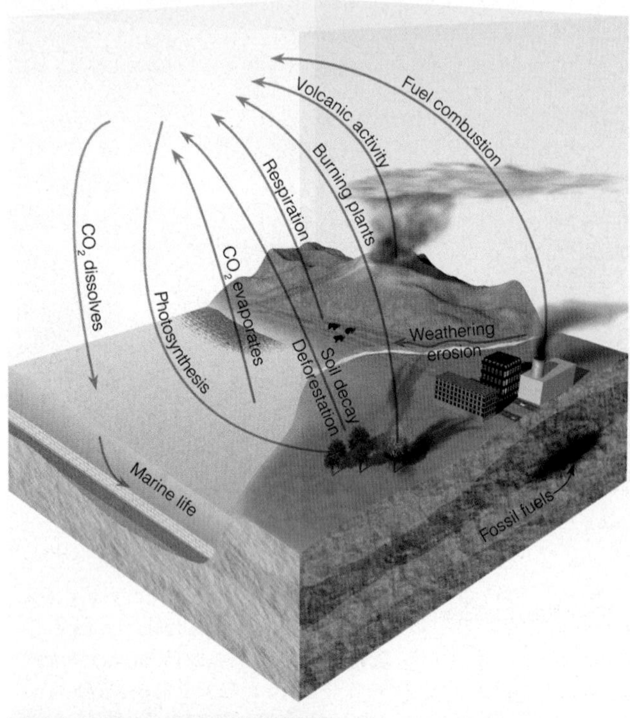

FIGURE 1.7 The main components of the atmospheric carbon dioxide cycle. The gray lines show processes that put carbon dioxide into the atmosphere, whereas the red lines show processes that remove carbon dioxide from the atmosphere.

vegetation, but it also comes from volcanic eruptions, the exhalations of animal life, from the burning of fossil fuels (such as coal, oil, and natural gas), and from deforestation. The removal of CO_2 from the atmosphere takes place during *photosynthesis*, as plants consume CO_2 to produce green matter. The CO_2 is then stored in roots, branches, and leaves. The oceans act as a huge reservoir for CO_2, as phytoplankton (tiny drifting plants) in surface water fix CO_2 into organic tissues. Carbon dioxide that dissolves directly into surface water, mixes downward and circulates through greater depths. Estimates are that the oceans hold more than 50 times the total atmospheric CO_2 content. Figure 1.7 illustrates important ways carbon dioxide enters and leaves the atmosphere.

Figure 1.8 reveals that the atmospheric concentration of CO_2 has risen more than 23 percent since 1958, when it was first measured at Mauna Loa Observatory in Hawaii. This increase means that CO_2 is entering the atmosphere at a greater rate than it is being removed. The increase appears to be due mainly to the burning of fossil fuels, such as coal and oil; however, deforestation also plays a role as cut timber, burned or left to rot, releases CO_2 directly into the air, perhaps accounting for about 20 percent of the observed increase. Measurements of CO_2 also come from ice cores. In Greenland and Antarctica, for example, tiny bubbles of air trapped within the ice sheets reveal that before the industrial revolution, CO_2 levels were stable at about 280 parts per million (ppm). Since the early 1800s, however, CO_2 levels have increased by more than 37 percent. With CO_2 levels presently increasing by about 0.4 percent annually

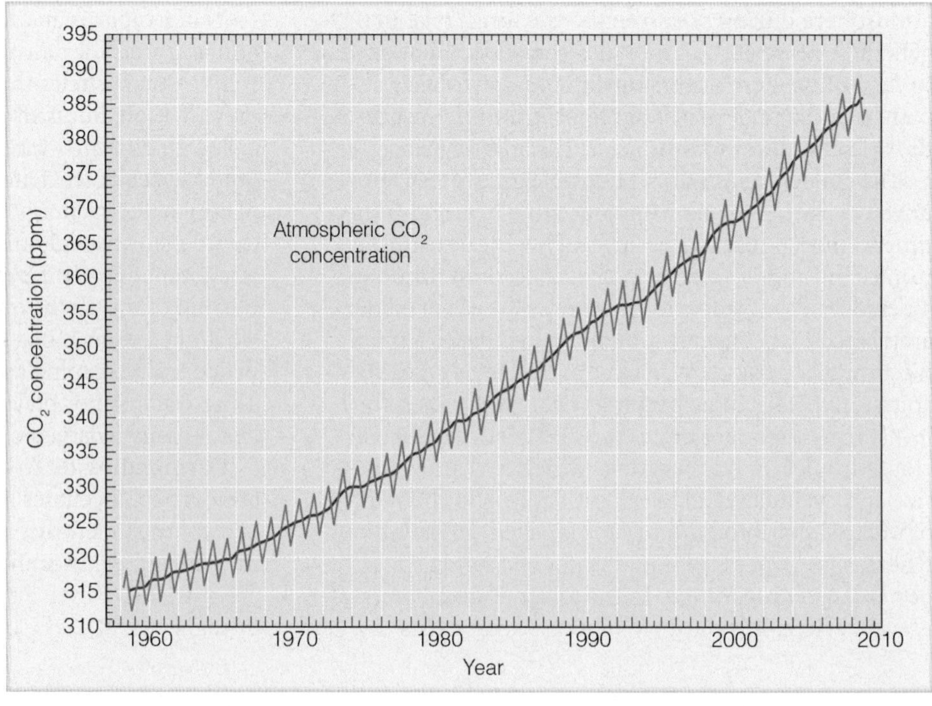

FIGURE 1.8 Measurements of CO_2 in parts per million (ppm) at Mauna Loa Observatory, Hawaii, since 1958. Higher readings occur in winter when plants die and release CO_2 to the atmosphere. The solid line is the average yearly value. Notice that the concentration of CO_2 has increased by more than 23 percent.

(1.9 ppm/year), scientists now estimate that the concentration of CO_2 will likely rise from its current value of about 385 ppm to a value perhaps greater than 750 ppm by the end of this century.

Carbon dioxide is an important greenhouse gas because, like water vapor, it traps a portion of the earth's outgoing energy. Consequently, with everything else being equal, as the atmospheric concentration of CO_2 increases, so should the average global surface air temperature. In fact, over the last hundred years or so, the earth's average surface temperature has warmed by more than 1.4°F (0.8°C). Mathematical climate models that predict future atmospheric conditions estimate that if CO_2 (and other greenhouse gases) continue to increase at their present rates, the earth's surface could warm by an additional 5°F by the end of this century with consequences such as rising sea levels and the rapid melting of polar ice.

Carbon dioxide and water vapor are not the only greenhouse gases. Others include *methane* (CH_4), *nitrous oxide* (N_2O), and *chlorofluorocarbons* (CFCs). Levels of methane have been rising over the past hundred years, as have levels of nitrous oxide.

Chlorofluorocarbons (CFCs) represent a group of greenhouse gases that, up until recently, had been increasing in concentration. At one time, they were the most widely used propellants in spray cans. Today, however, they are mainly used as refrigerants, as propellants for the blowing of plastic-foam insulation, and as solvents for cleaning electronic microcircuits. Although their average concentration in a volume of air is quite small (see Table 1.3, p. 8), they have an important effect on our atmosphere as they not only have the potential for raising global temperatures, they also play a part in destroying the gas ozone in the stratosphere, a region in the atmosphere located between about 11 km and 50 km above the earth's surface.

At the surface, **ozone** (O_3) is the primary ingredient of *photochemical smog,** which irritates the eyes and throat and damages vegetation. But the majority of atmospheric ozone (about 97 percent) is found in the upper atmosphere—in the stratosphere—where it is formed naturally, as oxygen atoms combine with oxygen molecules. Here, the concentration of ozone averages less than 0.002 percent by volume. This small quantity is important, however, because it shields plants, animals, and humans from the sun's harmful ultraviolet rays. It is ironic that ozone, which damages plant life in a polluted

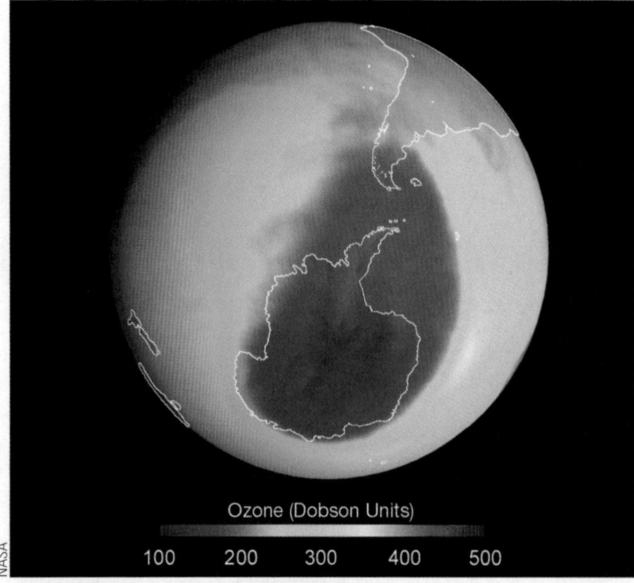

FIGURE 1.9 The darkest color represents the area of lowest ozone concentration, or ozone hole, over the Southern Hemisphere on September 22, 2004. Notice that the hole is larger than the continent of Antarctica. A Dobson unit (DU) is the physical thickness of the ozone layer if it were brought to the earth's surface, where 500 DU equals 5 millimeters.

environment, provides a natural protective shield in the upper atmosphere so that plants on the surface may survive.

When CFCs enter the stratosphere, ultraviolet rays break them apart, and the CFCs release ozone-destroying chlorine. Because of this effect, ozone concentration in the stratosphere has been decreasing over parts of the Northern and Southern Hemispheres. The reduction in stratospheric ozone levels over springtime Antarctica has plummeted at such an alarming rate that during September and October, there is an **ozone hole** over the region. Figure 1.9 illustrates the extent of the ozone hole above Antarctica during September, 2004.

Impurities from both natural and human sources are also present in the atmosphere: Wind picks up dust and soil from the earth's surface and carries it aloft; small saltwater drops from ocean waves are swept into the air (upon evaporating, these drops leave microscopic salt particles suspended in the atmosphere); smoke from forest fires is often carried high above the earth; and volcanoes spew many tons of fine ash particles and gases into the air (see Fig. 1.10). Collectively, these tiny solid or liquid suspended particles of various composition are called **aerosols**.

Some natural impurities found in the atmosphere are quite beneficial. Small, floating particles, for instance, act as surfaces on which water vapor

*Originally the word *smog* meant the combining of smoke and fog. Today, however, the word usually refers to the type of smog that forms in large cities, such as Los Angeles, California. Because this type of smog forms when chemical reactions take place in the presence of sunlight, it is termed *photochemical smog.*

FIGURE 1.10 Erupting volcanoes can send tons of particles into the atmosphere, along with vast amounts of water vapor, carbon dioxide, and sulfur dioxide.

condenses to form clouds. However, most human-made impurities (and some natural ones) are a nuisance, as well as a health hazard. These we call **pollutants**. For example, automobile engines emit copious amounts of *nitrogen dioxide* (NO_2), *carbon monoxide* (CO), and *hydrocarbons*. In sunlight, nitrogen dioxide reacts with hydrocarbons and other gases to produce ozone. Carbon monoxide is a major pollutant of city air. Colorless and odorless, this poisonous gas forms during the incomplete combustion of carbon-containing fuel. Hence, over 75 percent of carbon monoxide in urban areas comes from road vehicles.

The burning of sulfur-containing fuels (such as coal and oil) releases the colorless gas *sulfur dioxide* (SO_2) into the air. When the atmosphere is sufficiently moist, the SO_2 may transform into tiny dilute drops of sulfuric acid. Rain containing sulfuric acid corrodes metals and painted surfaces, and turns freshwater lakes acidic. **Acid rain** is a major environmental problem, especially downwind from major industrial areas. In addition, high concentrations of SO_2 produce serious respiratory problems in humans, such as bronchitis and emphysema, and have an adverse effect on plant life.

EXTREME WEATHER WATCH

The ozone hole reached its greatest extent on record in September, 2006, when it expanded to 10.6 million square miles, an area larger than the size of North America. When the ozone hole expands to this size it encompasses the Patagonia region of South America, the only permanently inhabited region of the world to occasionally be in an ozone layer-free atmosphere.

BRIEF REVIEW

Before going on to the next several sections, here is a review of some of the important concepts presented so far:

- **The risk of dying from an extreme weather event in the United States is low, about 2 per million people or about 500 annually.**
- **Extreme weather captures our attention by the power unleashed by the atmosphere, which can cause destruction and personal injury.**
- **The earth's atmosphere is a mixture of many gases. In a volume of dry air near the surface, nitrogen (N_2) occupies about 78 percent and oxygen (O_2) about 21 percent.**
- **Water vapor, which normally occupies less than 4 percent in a volume of air near the surface, can condense into liquid cloud droplets or transform into delicate ice crystals. Water is the only substance in our atmosphere that is found naturally as a gas (water vapor), as a liquid (water), and as a solid (ice).**
- **Both water vapor and carbon dioxide (CO_2) are important greenhouse gases.**
- **Ozone (O_3) in the stratosphere protects life from harmful ultraviolet (UV) radiation. At the surface, ozone is the main ingredient of photochemical smog.**

Properties of the Atmosphere

When we talk about the **weather**, we are talking about the condition of the atmosphere at any particular time and place.* Weather—which is always changing—is comprised of these elements:

1. *air temperature*
2. *air pressure*
3. *wind*
4. *humidity*
5. *clouds*
6. *precipitation*
7. *visibility*

If we measure and observe these *weather elements* over a specified interval of time, say, for many years, we would obtain the "average weather" or the **climate** of a particular region. Climate, therefore, represents the accumulation of daily and seasonal weather events (the average range of weather) over a long period of time. The concept of climate is much more than this, for it also includes the extremes of weather—the heat waves of summer and the cold spells of winter—that occur in a particular region. The *frequency* of these extremes is what helps us distinguish among climates that have similar averages.

If we were able to watch the earth for many thousands of years, even the climate would change. We would see rivers of ice moving down stream-cut valleys and huge glaciers—sheets of moving snow and ice—spreading their icy fingers over large portions of North America. Advancing slowly from Canada, a single glacier might extend as far south as Kansas and Illinois, with ice several thousands of meters thick covering the region now occupied by Chicago. Over an interval of 2 million years or so, we would see the ice advance and retreat several times. Of course, for this phenomenon to happen, the average temperature of North America would have to decrease, and then rise in a cyclic manner.

Suppose we could photograph the earth once every thousand years for many hundreds of millions of years. In time-lapse film sequence, these photos would show that not only is the climate altering, but the whole earth itself is changing as well: Mountains would rise up only to be torn down by erosion; isolated puffs of smoke and steam would appear as volcanoes spew hot gases and fine dust into the atmosphere; and the entire surface of the earth would undergo a gradual transformation as some ocean basins widen and others shrink.*

In summary, the earth and its atmosphere are dynamic systems that are constantly changing. While major transformations of the earth's surface are completed only after long spans of time, the state of the atmosphere can change in a matter of minutes, just as a beautiful summer afternoon can turn quickly into a raging windstorm.

AIR TEMPERATURE Air is a mixture of countless billions of atoms and molecules. If they could be seen, they would appear to be moving about in all directions, freely darting, twisting, spinning, and colliding with one another like an angry swarm of bees. Close to the earth's surface, each individual molecule will travel only about a thousand times its diameter before colliding with another molecule. Moreover, we would see that all the atoms and molecules are not moving at the same speed, as some are moving faster than others. The **temperature** of the air (or any substance) is *a measure of the average speed of the atoms and molecules*, where higher temperatures correspond to faster average speeds and lower temperatures to slower average speeds.

If we were to cool the air, its atoms and molecules would move more slowly. If we had the ability to cool

*While *weather* is the present state of the atmosphere, *meteorology* is the *study* of the atmosphere and its phenomena.

*The movement of the ocean floor and continents is explained in the widely acclaimed theory of *plate tectonics*.

the air to extremely low temperatures, its atoms and molecules would move slower and slower until the air reached a temperature of −273°C (−459°F), which is the lowest temperature possible. At this theoretical temperature, called *absolute zero,* all molecular motion ceases.

Temperature Scales At absolute zero, we can begin a temperature scale called the *absolute scale*, or *Kelvin scale*, after Lord Kelvin (1824–1907), a famous British scientist who first introduced it. Since the Kelvin scale contains no negative numbers, it is quite convenient for scientific calculations. Two other temperature scales commonly used today are the Fahrenheit and Celsius (formerly centigrade). The **Fahrenheit scale** was developed in the early 1700s by the physicist G. Daniel Fahrenheit, who assigned the number 32 to the temperature at which water freezes, and the number 212 to the temperature at which water boils. The zero point was simply the lowest temperature that he obtained with a mixture of ice, water, and salt. Between the freezing and boiling points are 180 equal divisions, each of which is called a degree. A thermometer calibrated with this scale is referred to as a Fahrenheit thermometer, for it measures an object's temperature in degrees Fahrenheit (°F).

The **Celsius scale** was introduced later in the eighteenth century. The number 0 (zero) on this scale is assigned to the temperature at which pure water freezes, and the number 100 to the temperature at which pure water boils at sea level. The space between freezing and boiling is divided into 100 equal degrees. Therefore, each Celsius degree (°C) is 180/100 or 1.8 times bigger than a Fahrenheit degree. Put another way, an increase in temperature of 1°C equals an increase of 1.8°F.

A formula for converting °F to °C is

$$°C = \tfrac{5}{9}(°F - 32).$$

On the **Kelvin scale**, degrees Kelvin are called *Kelvins* (abbreviated K). Each degree on the Kelvin scale is exactly the same size as a degree Celsius, and a temperature of 0 K is equal to –273°C. Converting from °C to K can be done by simply adding 273 to the Celsius temperature, as

$$K = °C + 273.$$

Figure 1.11 compares the Kelvin, Celsius, and Fahrenheit scales. Notice that the highest temperature ever recorded worldwide was 136°F (58°C), which is equivalent to 331K.

Although throughout this book we will concentrate on temperature extremes found at the earth's surface, it is important to see how the average air temperature changes with height above the earth.

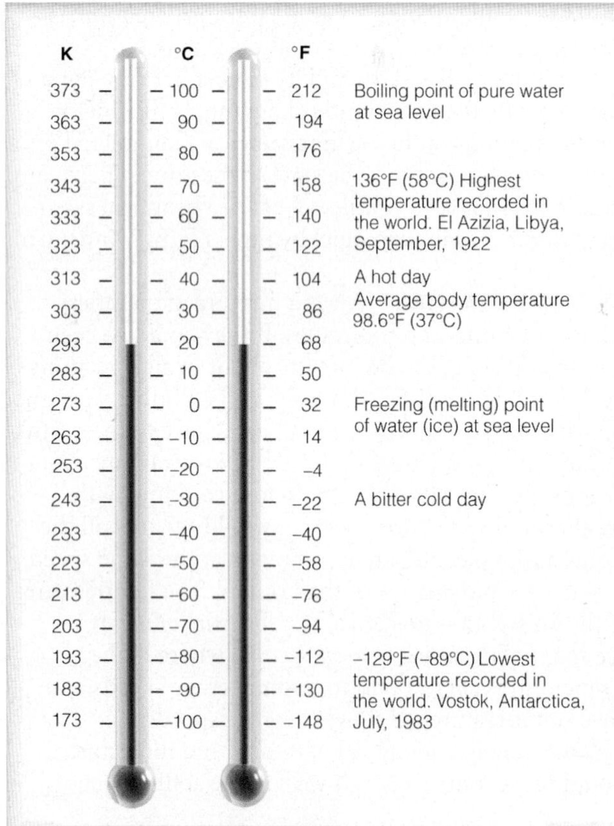

FIGURE 1.11 Comparison of the Kelvin, Celsius, and Fahrenheit scales, along with some world temperature extremes.

Vertical Profile of Air Temperature Look closely at Fig. 1.12 and notice that air temperature normally decreases from the earth's surface up to an altitude of about 11 km, which is nearly 36,000 ft, or 7 mi. This decrease in air temperature with increasing height is due primarily to the fact (investigated further in Chapter 2) that sunlight warms the earth's surface, and the surface, in turn, warms the air above it. The rate at which the air temperature decreases with height is called the temperature **lapse rate**. The *average* (or *standard*) *lapse rate* in this region of the lower atmosphere is about 6.5 degrees Celsius for every 1000 meters (m) or about 3.6°F for every 1000 ft rise in elevation. Keep in mind that these values are only averages. On some days, the air becomes colder more quickly as we move upward, which would increase or steepen the lapse rate. On other days, the air temperature would decrease more slowly with height, and the lapse rate would be less. Occasionally, the air temperature may actually *increase* with height, producing a condition known as

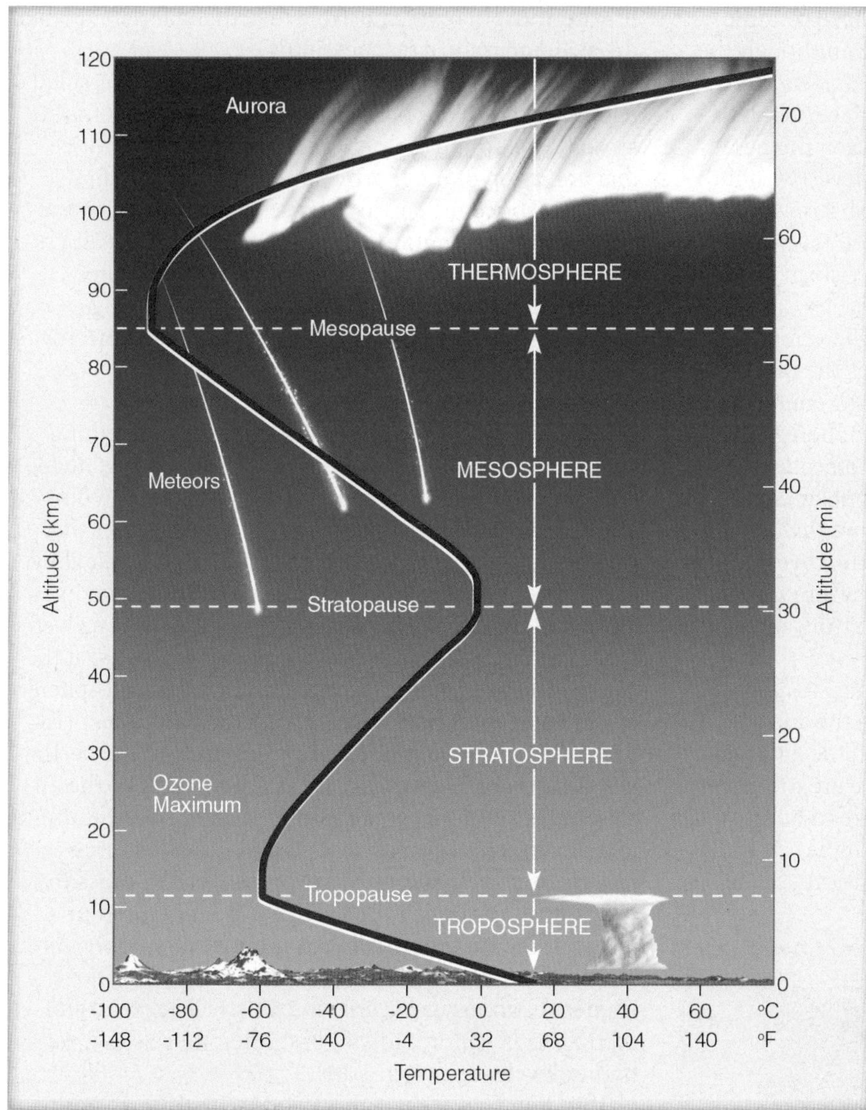

FIGURE 1.12

The average profile of air temperature (heavy red line) above the earth's surface. The changing air temperature with height allows scientists to divide the atmosphere into regions or layers.

a **temperature inversion**. So the lapse rate fluctuates, varying from day to day and season to season. And as we will see later in this book, the development of severe weather, such as thunderstorms and tornadoes, is strongly influenced by the lapse rate.

The region of the atmosphere from the surface up to about 11 km contains all of the weather we are familiar with on earth. Also, this region is kept well stirred by rising and descending air currents. Here, it is common for air molecules to circulate through a depth of more than 10 km (6 mi) in just a few days. This region of circulating air extending upward from the earth's surface to where the air stops becoming colder with height is called the **troposphere**—from the Greek *tropein*, meaning to turn or to change.

Above the troposphere is the **stratosphere**. The boundary separating the troposphere from the stratosphere is the *tropopause*. The height of the tropopause varies. It is normally found at higher elevations over equatorial regions, and it decreases in elevation as we travel poleward. Generally, the tropopause is higher in summer and lower in winter at all latitudes. In some regions, the tropopause "breaks" and is difficult to locate and, here, scientists have observed tropospheric air mixing with stratospheric air and vice versa. These breaks also mark the position of *jet streams*—high winds that meander in a narrow channel like an old river, often at speeds exceeding 115 miles per hour (mi/hr) which is the same speed as 100 knots.*

From Fig. 1.12 we can see that in the stratosphere the air temperature begins to increase with height, producing a *temperature inversion*. The *inversion region* tends to keep the vertical currents of the troposphere from spreading into the stratosphere. The inversion also tends to reduce the amount of vertical motion in the stratosphere itself; hence, it is a stratified layer. Even though the air temperature is increasing with height, the air at an altitude of 30 km is extremely cold, averaging less than −46°C.

The reason for the inversion in the stratosphere is that the gas ozone plays a major part in heating the air at this altitude. Recall that ozone is important because it absorbs energetic ultraviolet (UV) solar energy. Some of this absorbed energy warms the stratosphere, which explains why there is an inversion. If ozone were not present, the air probably would become colder with height as it does in the troposphere.**

*A knot is a nautical mile per hour. One knot is equal to 1.15 miles per hour (mi/hr) or 1.9 kilometers per hour (km/hr).

**Recall from an earlier discussion that the concentration of stratospheric ozone is decreasing over portions of the globe as chlorofluorocarbons break apart and release ozone-destroying chlorine in the process.

Above the stratosphere is the **mesosphere** (middle sphere). The air here is extremely thin. Even though the percentage of nitrogen and oxygen in the mesosphere is about the same as it was at the earth's surface, a breath of mesospheric air contains far fewer oxygen molecules than a breath of tropospheric air. At this level, without proper oxygen-breathing equipment, the brain would soon become oxygen-starved—a condition known as *hypoxia*—and suffocation would result. With an average temperature of −90°C, the top of the mesosphere represents the coldest part of our atmosphere.

The "hot layer" above the mesosphere is the **thermosphere**. Here, oxygen molecules (O_2) absorb energetic solar rays, warming the air. In the thermosphere, there are relatively few atoms and molecules. Consequently, the absorption of a small amount of energetic solar energy can cause a large increase in air temperature that may exceed 500°C, or 900°F (see Fig. 1.13). Moreover it is in the thermosphere where charged particles from the sun interact with air molecules to produce dazzling aurora displays.

Even though the temperature in the thermosphere is exceedingly high, a person shielded from the sun would not necessarily feel hot in this extreme environment. The reason for this fact is that there are too few molecules in this region of the atmosphere to bump against something (exposed skin, for example) and transfer enough heat to it to make it feel warm. So, in the thermosphere it is possible to actually feel cold when the air temperature reaches 500°C.

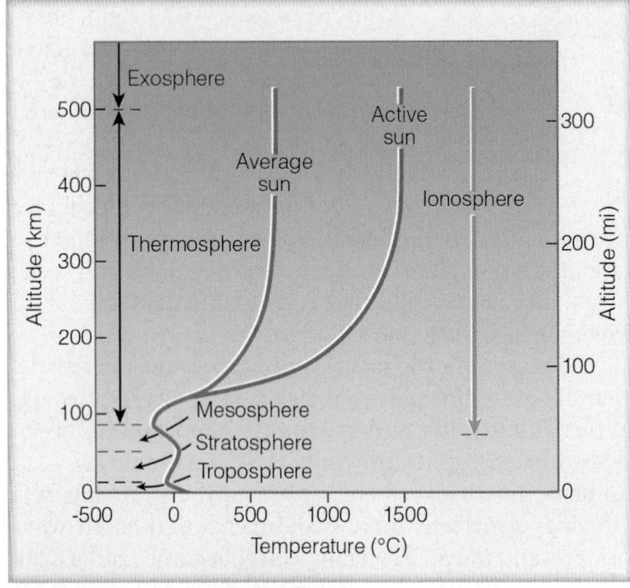

FIGURE 1.13 The hottest place in our atmosphere is in the thermosphere, where air temperatures can reach 1500°C.

Notice in Fig. 1.13 that the *ionosphere* (an electrified region within the upper atmosphere) extends from the mesosphere to the top of the atmosphere. Also notice that at the top of the thermosphere, about 500 km (300 mi) above the earth's surface, is the *exosphere*, the upper limit of our atmosphere. Here molecules can move great distances before they collide with other molecules, and many of the lighter, faster-moving molecules traveling in the right direction actually escape the earth's gravitational pull and shoot off into space.

AIR PRESSURE Air molecules (as well as everything else) are held near the earth by *gravity.* This strong, invisible force pulling down on the air above squeezes (compresses) air molecules closer together, which causes their number in a given volume to increase. The more air above a level, the greater the squeezing effect or compression. Since *air density* is the number of air molecules in a given space (volume), it follows that air density is greatest at the surface and decreases as we move up into the atmosphere.*

Air molecules have weight.** In fact, air is surprisingly heavy. The weight of all the air around the earth is a staggering 5600 trillion tons. The weight of the air molecules acts as a force upon the earth. The amount of force exerted over an area of surface is called *atmospheric pressure* or, simply, **air pressure**. The pressure at any level in the atmosphere may be measured in terms of the total mass of the air above any point. As we climb in elevation, fewer air molecules are above us; hence, *atmospheric pressure always decreases with increasing height*, rapidly at first, then more slowly at higher levels (see Fig. 1.14).

If in Fig. 1.14 we weigh a column of air one square inch in cross section, extending from the average height of the ocean surface (sea level) to the "top" of the atmosphere, it would weigh very nearly 14.7 pounds. Thus, normal atmospheric pressure near sea level is close to 14.7 pounds per square inch. If more molecules are packed into the column, it becomes more dense, the air weighs more, and the surface pressure goes up. On the other hand, when fewer molecules are in the column, the air weighs less, and the surface

*Density is defined as the mass of air in a given volume of air. Density = mass/volume.

**The *weight* of an object, including air, is the force acting on the object due to gravity. In fact, weight is defined as the mass of an object times the acceleration of gravity. An object's *mass* is the quantity of matter in the object. Consequently, the mass of air in a rigid container is the same everywhere in the universe. However, if you were to instantly travel to the moon, where the acceleration of gravity is much less than that of earth, the mass of air in the container would be the same, but its weight would decrease.

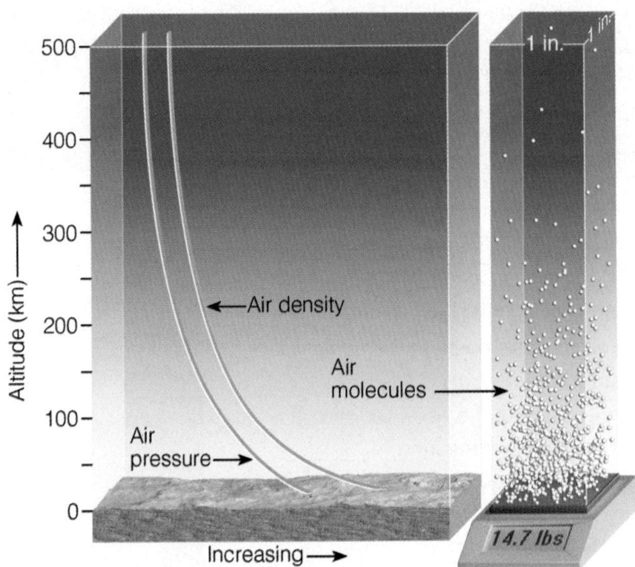

FIGURE 1.14 Both air pressure and air density decrease with increasing altitude. The weight of all the air molecules above the earth's surface produces an average pressure near 14.7 lbs/in.^{2}

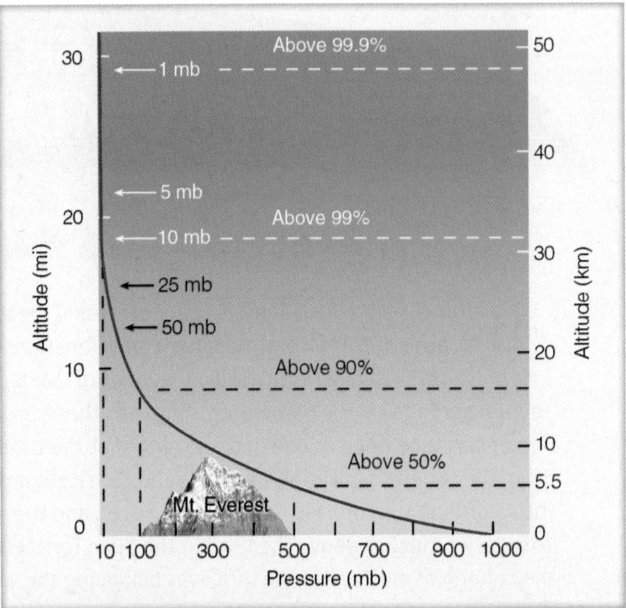

FIGURE 1.15 Atmospheric pressure decreases rapidly with height. Climbing to an altitude of only 5.5 km (3.5 mi) where the pressure is 500 mb, would put you above one-half of the atmosphere's molecules.

pressure goes down. So, a change in air density can bring about a change in air pressure.

Pounds per square inch is, of course, just one way to express air pressure. Presently, the most common unit for air pressure found on surface weather maps is the *millibar* (mb), although the *hectopascal** (hPa) is gradually replacing the millibar as the preferred unit of pressure on surface maps. Another unit of pressure is *inches of mercury* (Hg), which is commonly used both in the field of aviation and in television and radio weather broadcasts. At sea level, the *standard value* for atmospheric pressure is

$$1013.25 \text{ mb} = 1013.25 \text{ hPa} = 29.92 \text{ in. Hg.}$$

Figure 1.14 (and Fig. 1.15) illustrates how rapidly air pressure decreases with height. Near sea level, atmospheric pressure decreases rapidly, whereas at high levels it decreases more slowly. With a sea-level pressure near 1000 mb, we can see in Fig. 1.15 that, at an altitude of only 5.5 km (3.5 mi), the air pressure is about 500 mb, or half of the sea-level pressure. This situation means that, if you were at a mere 5.5 km (which is about 18,000 feet) above the surface, you would be above one-half of all the molecules in the atmosphere.

At an elevation approaching the summit of Mount Everest (about 9 km or 29,000 ft), the air pressure would be about 300 mb. The summit is above nearly 70 percent of all the molecules in the atmosphere. At an altitude of about 50 km (30 mi), the air pressure is about 1 mb, which means that 99.9 percent of all the air molecules are below this level. Yet the atmosphere extends upwards for many hundreds of kilometers, gradually becoming thinner and thinner until it ultimately merges with outer space.

Atmospheric pressure normally changes much more quickly when we move upward than it does when we move sideways. Because air pressure changes so rapidly with increasing height (about 10 mb for every 100-meter increase in elevation near sea level), cities located at

EXTREME WEATHER WATCH

On December 31, 1989, the barometric pressure peaked at 31.55 in. (1078 mb) at Northway, Alaska, establishing the record for the highest barometric pressure reading ever measured in the United States. Some aircraft were actually grounded because their altimeters were unable to calibrate for such a high pressure.

*One hectopascal equals 1 millibar.

FOCUS ON
INSTRUMENTS

Air Pressure and Barometers

Barometers are instruments that detect and measure atmospheric pressure. Because of this fact, atmospheric pressure is also referred to as *barometric pressure*. Evangelista Torricelli, a student of Galileo's, invented the *mercury barometer* in 1643. His barometer, similar to those used today, consisted of a long glass tube open at one end and closed at the other (see Fig. 1). Removing air from the tube and covering the open end, Torricelli immersed the lower portion into a dish of mercury. He removed the cover, and the mercury rose up the tube to nearly 30 inches above the level in the dish. Torricelli correctly concluded that the column of mercury in the tube was balancing the weight of the air above the dish, and, hence, its height was a measure of atmospheric pressure.

The most common type of home barometer—the *aneroid barometer*—contains no fluid. Inside this instrument is a small, flexible metal box called an *aneroid cell*. Before the cell is tightly sealed, air is partially removed, so that small changes in external air pressure cause the cell to expand or contract. The size of the cell is calibrated to represent different pressures, and any change in its size is amplified by levers and transmitted to an indicating arm, which points to the current atmospheric pressure (see Fig. 2).

Notice that the aneroid barometer often has descriptive weather-related words printed above specific pressure values. These descriptions indicate the most likely weather conditions when the needle is pointing to that particular pressure reading. Generally, the higher the readings, the more likely clear weather will occur, and the lower the reading, the better are the chances for inclement weather. This situation occurs because surface high-pressure areas are associated with sinking air and normally fair weather, whereas surface low-pressure areas are associated with rising air and usually cloudy, wet weather. A steady rise in atmospheric pressure (a rising barometer) usually indicates clearing weather or fair weather, whereas a steady drop in atmospheric pressure (a falling barometer) often signals the approach of a storm with inclement weather.

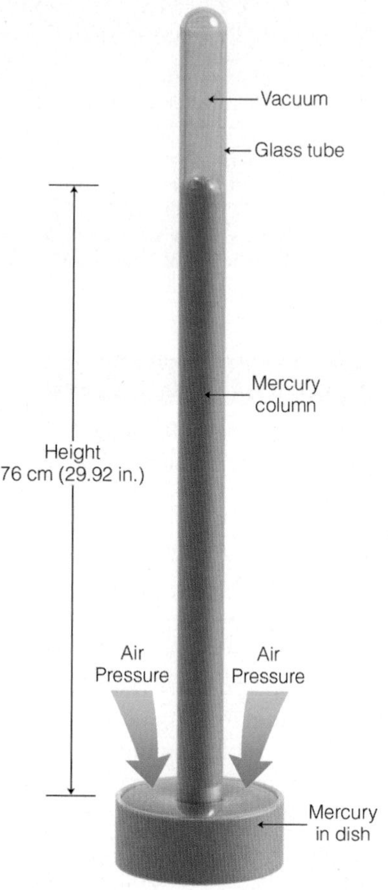

Figure 1 The mercury barometer. The height of the mercury column is a measure of atmospheric pressure.

The *altimeter* and *barograph* are two types of aneroid barometers. Altimeters are aneroid barometers that measure pressure, but are calibrated to indicate altitude. Barographs are recording aneroid barometers. Basically, the barograph consists of a pen attached to an indicating arm that marks a continuous record of pressure on chart paper. The chart paper is attached to a drum rotated slowly by an internal mechanical clock (see Fig. 3).

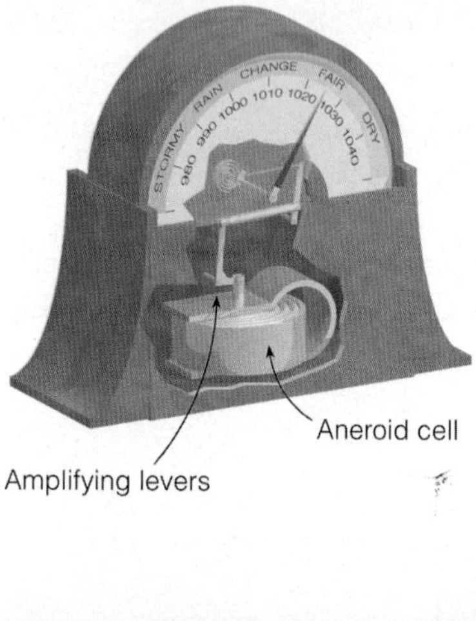

Figure 2 An aneroid barometer.

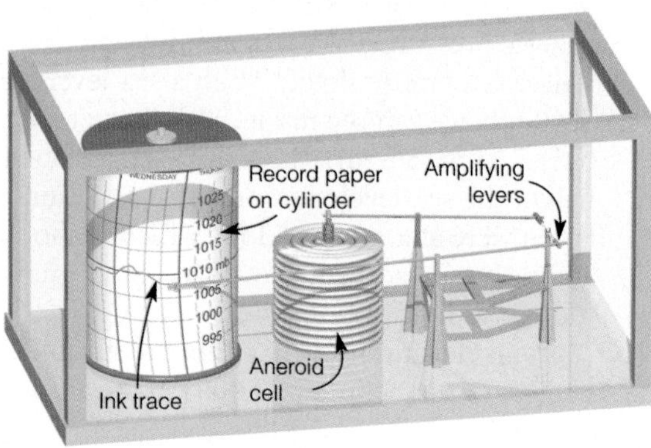

Figure 3 A recording barograph.

different elevations will have much different pressure readings. Consequently, pressure observations are normally adjusted to a level representing the average level of the ocean, called *mean sea level*, and the adjusted pressure reading is referred to as *sea-level pressure*. Figure 1.16 compares sea-level pressure readings in millibars and in inches of mercury, and shows some extreme sea-level pressure readings.

If you are wondering how inches of mercury (a unit of length) can be a measure of atmospheric pressure (which is a force over a given area), read the Focus section concerning barometers on p. 18.

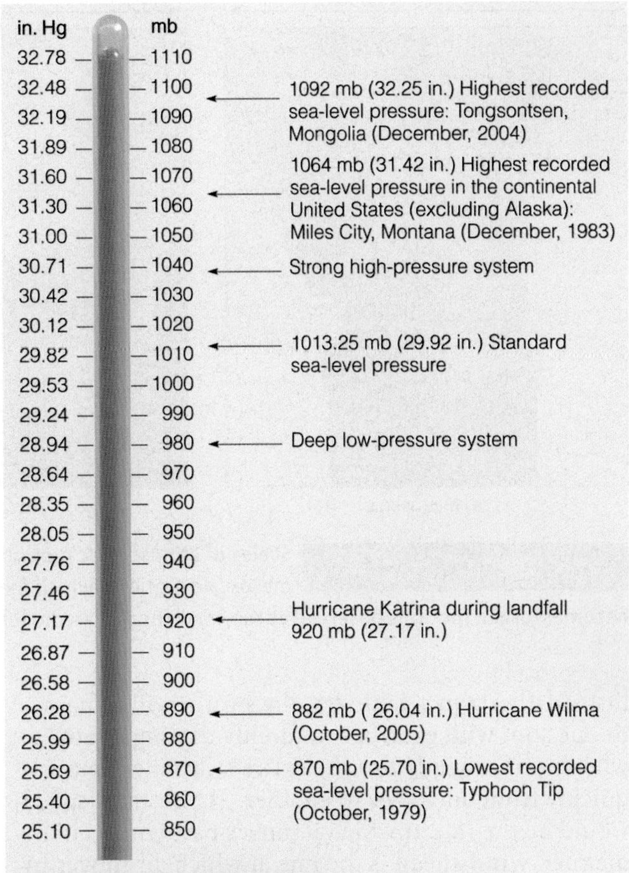

FIGURE 1.16 Atmospheric pressure in inches of mercury and in millibars.

WIND The air in motion—what we commonly call **wind**—is invisible, yet we see evidence of it nearly everywhere we look. It sculptures rocks, moves leaves, blows smoke, and lifts water vapor upward to where it can condense into clouds. The wind is with us wherever we go. On a hot day, it can cool us off; on a cold day, it can make us shiver. A breeze can sharpen our appetite when it blows the aroma from the local bakery in our direction. The wind is a powerful element. The work horse of weather, it moves storms and large fair-weather systems around the globe. It transports heat, moisture, dust, insects, bacteria, and pollens from one area to another.

Circulations of all sizes exist within the atmosphere. Little whirls form inside bigger whirls, which encompass even larger whirls—one huge mass of turbulent, twisting *eddies*.* For clarity, scientists arrange circulations according to their size. This hierarchy of motion from tiny gusts to giant storms is called the **scales of motion**.

Consider smoke rising from a chimney into the otherwise clean air in the industrial section of a large city (see Fig. 1.17a). Within the smoke, small chaotic motions—tiny eddies—cause it to tumble and turn. These eddies constitute the smallest scale of motion—the *microscale*. At the microscale, eddies with diameters of a few meters or less not only disperse smoke, they also sway branches and swirl dust and papers into the air.

In Fig. 1.17b observe that, as the smoke rises, it drifts toward the center of town. Here the smoke rises even higher and is carried many kilometers downwind. This circulation of city air constitutes the next larger scale—the *mesoscale* (meaning middle scale). Typical mesoscale winds range from a few kilometers to about a hundred kilometers in diameter. Generally, they last longer than microscale motions, often many minutes, hours, or in some cases as long as a day. Thunderstorms and tornadoes fall under the heading of mesoscale winds.

When we look at the smokestack on a surface weather map (see Fig. 1.17c), neither the smokestack nor the circulation of city air shows up. All that we see are the circulations around high- and low-pressure areas. We are now looking at the largest scale, or *macroscale*. Circulations of this magnitude dominate regions of hundreds to even thousands of square kilometers and, although the life spans of these features vary, they typically last for days and sometimes weeks.

Wind is characterized by its direction, speed, and gustiness. If we imagine air molecules as being a swarm of bees, the wind may be seen as the movement of the entire swarm. This analogy can be carried a

*Eddies are spinning globs of air that have a life history of their own.

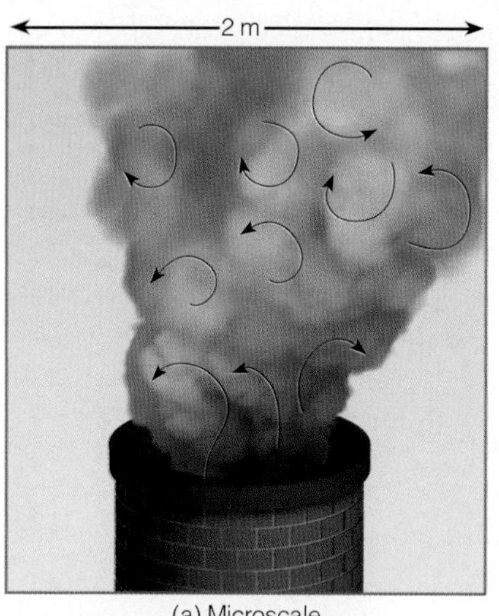

(a) Microscale

(b) Mesoscale

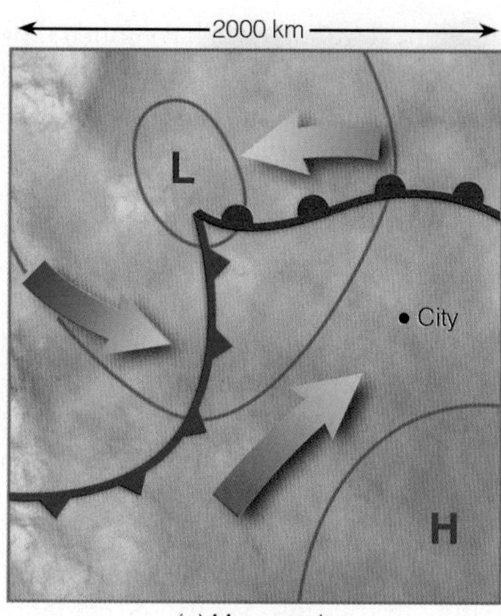

(c) Macroscale

FIGURE 1.17 Scales of atmospheric motion. The tiny microscale motions constitute a part of the larger mesoscale motions, which in turn are part of the much larger macroscale. Notice that as the scale becomes larger, motions observed at the smaller scale are no longer visible.

little further: On a calm day, the swarm will remain in one spot with each bee randomly darting about; whereas on a windy day the entire swarm will move quickly from one place to another. The swarm's speed would be the rate at which it moves past you. In like manner, **wind speed** is the rate at which air moves by a stationary observer. This movement can be expressed as the distance in nautical miles traveled in one hour (knots) or as the number of meters traveled in one second (m/sec).

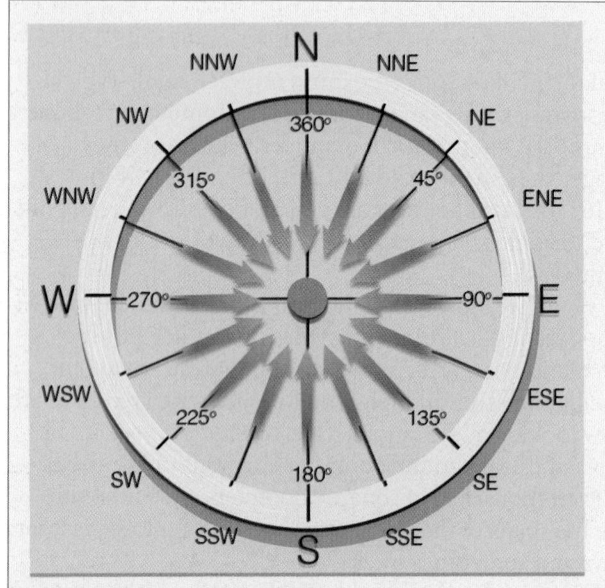

FIGURE 1.18 Wind direction can be expressed in degrees about a circle or as compass points. A north wind blows from the north and is reported as a wind direction of 360°.

Unlike a swarm of bees, air is invisible; we cannot really see it. Rather, we see things being moved by it. Therefore, we can determine **wind direction**—the direction *from which* the wind blows—by watching the movement of objects as air passes them. For example, the rustling of small leaves, smoke drifting near the ground, and flags waving on a pole all indicate wind direction. In a light breeze, a tried and true method of determining wind direction is to raise a wet finger into the air. The dampness quickly evaporates on the wind-facing side, cooling the skin. Wind direction is often given as degrees about a 360° circle or as compass points (see Fig. 1.18).

The wind blows due to horizontal differences in atmospheric pressure. The greater the difference in pressure, the stronger the wind. We can obtain a fairly good idea as to how differences in atmospheric pressure develop by looking at Fig. 1.19. Notice in Fig. 1.19a that there are more air molecules in the air column above the letter "H" than there are in the air column above the letter "L." Because atmospheric pressure is related to the total mass of air above any level, the atmospheric pressure at the letter "H" must be higher than that at the letter "L." If we remove the air columns in Fig. 1.19a and simply look at the areas of surface high pressure (marked "H") and surface low pressure (marked "L") in Fig. 1.19b, we see that the *wind tends to blow from higher pressure toward lower pressure.* It does so because the difference

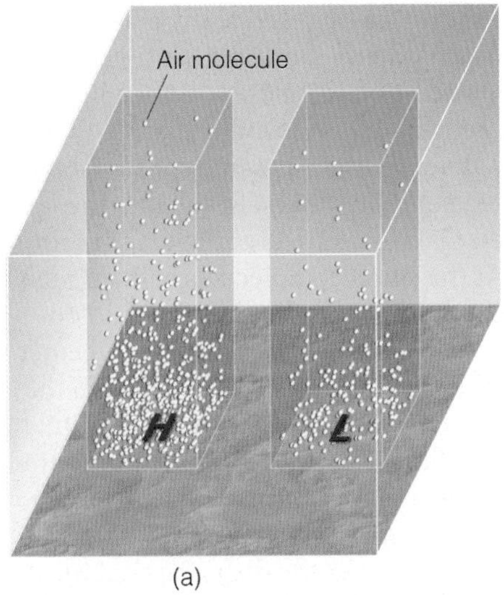

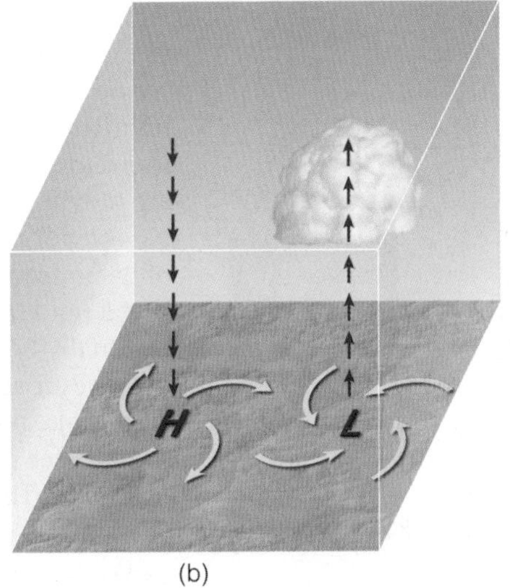

FIGURE 1.19

(a) There are more air molecules above an area of surface high pressure than above an area of surface low pressure. (b) In the Northern Hemisphere, surface winds blow counterclockwise and in toward an area of low pressure; they blow clockwise and outward around an area of high pressure. The air normally rises above an area of surface low pressure and sinks above an area of surface high pressure.

in horizontal air pressure creates a force that starts the air moving from higher pressure toward lower pressure. However, because the earth rotates, the wind does not move in a straight-line path, but is deflected from its path toward the *right* in the Northern Hemisphere.* This deflection causes surface winds in the Northern Hemisphere to blow *clockwise* and *outward* around areas of high pressure and *counterclockwise* and *inward* around areas of low pressure, as shown in Fig. 1.19b.

As the surface air spins inward, toward the area of low pressure, it flows together and rises much like toothpaste does when its open tube is squeezed. The rising air cools and water vapor in the air condenses into clouds. Hence, above areas of surface low pressure, we often find clouds and precipitation. As the surface air flows outward away from the area of surface high pressure, sinking air from above gradually replaces the laterally spreading surface air. Since sinking air does not usually produce clouds, we find generally clear skies and fair weather associated with regions of high atmospheric pressure.

MOISTURE We learned earlier in this chapter that, in a given volume of air, water vapor molecules only account for a small percentage of the total molecules. Water vapor molecules are invisible, yet if we could see them, we would find that in the lower atmosphere they are everywhere. If we could observe just one single water molecule by magnifying it billions of times, we would see an H_2O molecule in the shape of a tiny head that somewhat resembles Mickey Mouse (see Fig 1.20).

* As we will see later when we look more closely at the subject of winds, the deflecting force called the *coriolis force* causes objects to deflect toward the left of their path in the Southern Hemisphere.

The bulk of the "head" of the molecule is the oxygen atom. The "mouth" is a region of excess negative charge. The "ears" are partially exposed protons of the hydrogen atom, which are regions of excess positive charge.

When we look at many H_2O molecules, we see that, as a gas, water vapor molecules move about quite freely, mixing well with neighboring atoms and molecules (see Fig. 1.21). The higher the temperature of the gas, the faster the molecules move. In the liquid state, the water molecules are closer together, constantly jostling and bumping into one another. If we lower the temperature of the liquid, water molecules would move slower and slower until, when cold enough, they arrange themselves into an orderly pattern with each molecule more or less locked into a rigid position, able to vibrate but not able to move about freely. In this solid state, called *ice*, the

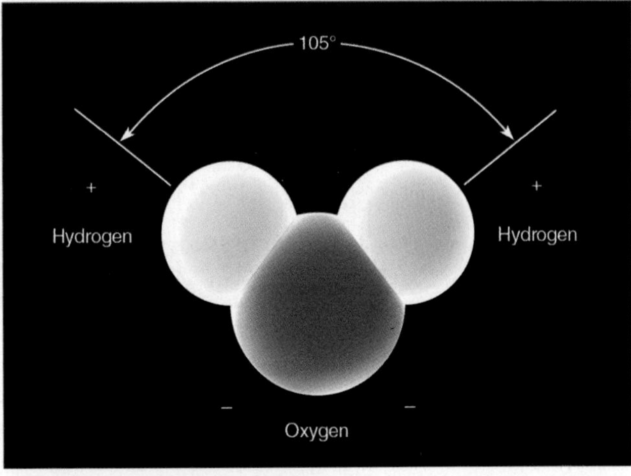

FIGURE 1.20 **The water molecule.**

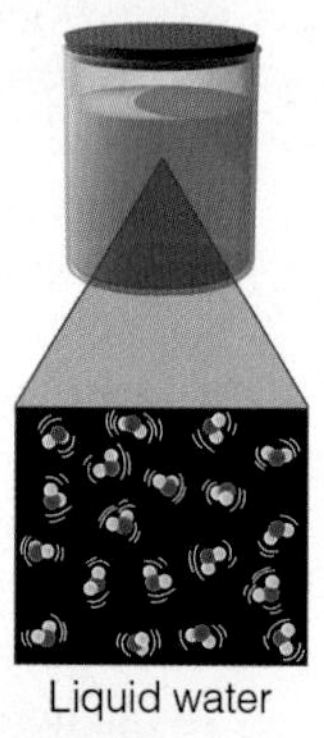
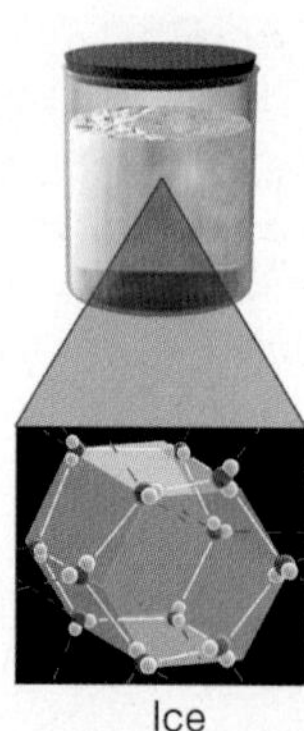

FIGURE 1.21

The three states of matter. Water as a gas, as a liquid, and as a solid.

shape and charge of the water molecule helps arrange the molecules into six-sided (hexagonal) crystals. If we apply warmth to the ice, its molecules would vibrate faster. In fact, some of the molecules would actually vibrate out of their rigid crystal pattern into a disorderly condition—that is, the ice melts.

And so water vapor is a gas that becomes visible to us only when millions of molecules join together to form tiny cloud droplets or ice crystals. In this process water only changes its disguise, not its identity.

To obtain a slightly different picture of water in the atmosphere, suppose we examine water in a beaker similar to the one shown in Fig. 1.22a. If we were able to magnify the surface water about a billion times, we would see water molecules fairly close together, jiggling, bouncing, and moving about. We would also see that the molecules are not all moving at the same speed—some are moving much faster than others. Recall that the *temperature* of the water is a measure of the average motion of its molecules. At the surface, molecules with enough speed (and traveling in the right direction) would occasionally break away from the liquid surface and enter into the air above. These molecules, changing from the *liquid state into the vapor state*, are **evaporating**. While some water molecules are leaving the liquid, others are returning. Those returning are **condensing** as they are changing from a *vapor state to a liquid state.*

When a cover is placed over the dish (Fig. 1.22b), after awhile the total number of molecules escaping from the liquid (evaporating) would be balanced by the number returning (condensing). When this condition exists, the air is said to be **saturated** with water vapor. For every molecule that evaporates, one must condense, and no net loss of liquid or water vapor molecules results.

If we could examine the air above the water in Fig. 1.22a, we would observe the water vapor molecules freely darting about and bumping into each other as well as neighboring molecules of oxygen and nitrogen. We would also observe that mixed in with all of the air molecules are microscopic bits of dust, smoke, and salt from ocean spray. Since many of these serve as surfaces on which water vapor may condense, they are called *condensation nuclei.* In the warm air above the water, fast-moving vapor molecules strike the nuclei with such impact that they simply bounce away. However, if the air is chilled, the molecules move more slowly and are more apt to stick and condense to the nuclei. When many billions of these vapor molecules condense onto the nuclei, tiny liquid cloud droplets form.

We can see then that condensation is more likely to happen as the air cools and the speed of the vapor molecules decreases. As the air temperature increases, condensation is less likely because most of the molecules have sufficient speed (sufficient energy) to remain as a vapor. As we will see in this and other chapters, *condensation occurs primarily when the air is cooled.*

Even though condensation is more likely to occur when the air cools, it is important to note that no matter how cold the air becomes, there will always be a few molecules with sufficient speed (sufficient energy) to remain as a vapor. It should be apparent, then, that with the same number of water vapor molecules in the air, saturation is more likely to occur in cool air than in warm air. This idea often leads to the statement that

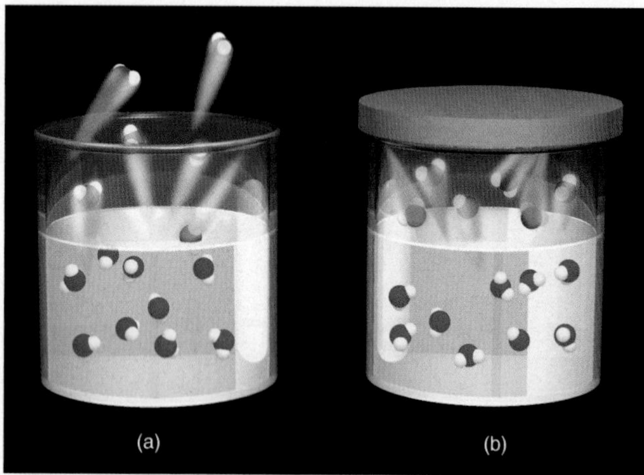

FIGURE 1.22 **(a) Water molecules at the surface of the water are evaporating (changing from liquid into vapor) and condensing (changing from vapor into liquid). Since more molecules are evaporating than condensing, net evaporation is occurring. (b) When the number of water molecules escaping from the liquid (evaporating) balances those returning (condensing), the air above the liquid is saturated with water vapor. (For clarity, only water molecules are illustrated.)**

"warm air can hold more water vapor molecules before becoming saturated than can cold air" or, simply, "warm air has a greater capacity for water vapor than does cold air." It is important to realize that although these statements are correct, the use of such words as "hold" and "capacity" are misleading when describing water vapor content, as air does not really "hold" water vapor in the sense of making "room" for it.

At this point, it is important to realize that most clouds form when air is rising, because rising air always cools. Clouds have weight because they are composed of tiny liquid droplets and ice crystals. But clouds do not fall to the earth because small, rising air currents keep the tiny cloud particles suspended in the atmosphere. (If you are curious as to why rising air always cools, read the Focus section below.)

FOCUS ON

A SPECIAL TOPIC

Rising Air Produces Clouds

To understand why rising air cools and sinking air warms, we need to examine some air. Suppose we place air in an imaginary thin, elastic wrap about the size of a large balloon (see Fig. 4). This invisible balloonlike "blob" is called an *air parcel*. The air parcel can expand and contract freely, but neither external air nor heat is able to mix with the air inside. By the same token, as the parcel moves, it does not break apart, but remains as a single unit.

At the earth's surface, the parcel has the same temperature and pressure as the air surrounding it. Suppose we lift the parcel. Recall from earlier in this chapter that air pressure always decreases as we move up into the atmosphere. Consequently, as the parcel rises, it enters a region where the surrounding air pressure is lower. To equalize the pressure, the parcel molecules inside push the parcel walls outward, expanding it. Because there is no other energy source, the air molecules inside use some of their own energy to expand the parcel. This energy loss shows up as slower molecular speeds, which represent a lower parcel temperature. Hence, *any air that rises always expands and cools.*

If the rising air is sufficiently moist, the air will cool to its saturation point (dew point). At the dew point, invisible water vapor transforms into billions of tiny liquid or ice particles, producing a visible cloud.

If the parcel is lowered to the earth (as shown in Fig. 4), it returns to a region where the air pressure is higher. The higher outside pressure squeezes (compresses) the parcel back to its original (smaller) size. Because air molecules have a faster rebound velocity after striking the sides of a collapsing parcel, the average speed of the molecules inside goes up. (A Ping-Pong ball moves faster after striking a paddle that is moving toward it.) This increase in molecular speed represents a warmer parcel temperature. Therefore, *any air that sinks (subsides), warms by compression.*

As the air sinks and warms, the air temperature and dew point move farther apart and the relative humidity decreases. As a consequence, sinking air does not generally produce clouds.

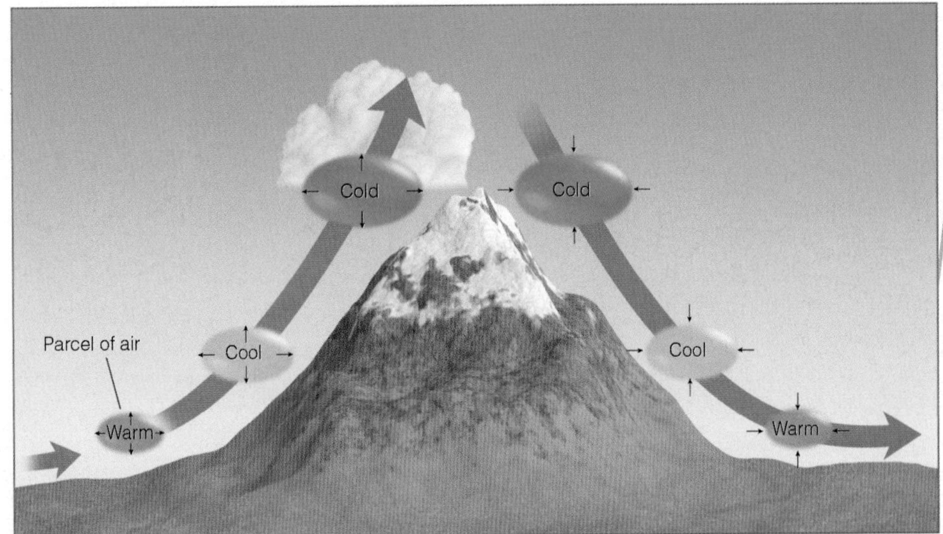

Figure 4 Rising air expands, cools, and often produces clouds. Sinking air is compressed, warms, and typically does not produce clouds.

A Brief Look at Relative Humidity and Dew Point

Although there are many ways of specifying the amount of water vapor in the air, *relative humidity* is the most common way of describing atmospheric moisture. The concept of relative humidity does not indicate the actual amount of water vapor in the air. Rather, it tells us how close the air is to being saturated. The **relative humidity** (RH) is the ratio of the amount of water vapor actually in the air to the maximum amount of water vapor required for saturation at that particular temperature (and pressure). It is the ratio of the air's water vapor *content* to its *capacity*; thus

$$RH = \frac{\text{water vapor content}}{\text{water vapor capacity}} \times 100 \text{ percent.}$$

Relative humidity is always given as a percent. Air with a 50 percent relative humidity is holding half the water in the vapor state that it can hold at that temperature. As the air approaches saturation, the relative humidity increases. When the air is saturated with water vapor, the relative humidity is 100 percent.

The **dew-point temperature** (or simply *dew point*) represents the temperature to which air must be cooled (with no change in moisture content or air pressure) in order for saturation to occur. When the air temperature and dew point are close together, the air is nearly saturated and the relative humidity is high. When the air temperature and dew point are far apart, the air is far from being saturated, and the relative humidity is low. When the air temperature equals the dew point, the air is saturated and the relative humidity is 100 percent.

Hydrologic Cycle We can sum up some of the concepts covered so far by looking at the circulation of water in our atmosphere. Since the oceans occupy over 70 percent of the earth's surface, we can think of this circulation as beginning over the ocean, where the sun's energy transforms enormous quantities of liquid water into water vapor. Winds then transport this invisible moist air to other regions, where the water vapor condenses back into liquid, forming clouds. Under certain conditions, the liquid (or solid) cloud particles may grow in size and fall to the surface as **precipitation**—rain, snow, or hail. If the precipitation falls into an ocean, the water is ready to begin its cycle again. If, on the other hand, the precipitation falls on a continent, a great deal of the water returns to the ocean in a complex journey. This cycle of moving and transforming water molecules from liquid to vapor and back to liquid again is called the **hydrologic** (water) **cycle**. In the form with which we are most concerned, water molecules travel from ocean to atmosphere to land and then back to the ocean.

》Figure 1.23 illustrates the complexities of the hydrologic cycle. For example, before falling rain ever reaches the ground, a portion of it evaporates back into

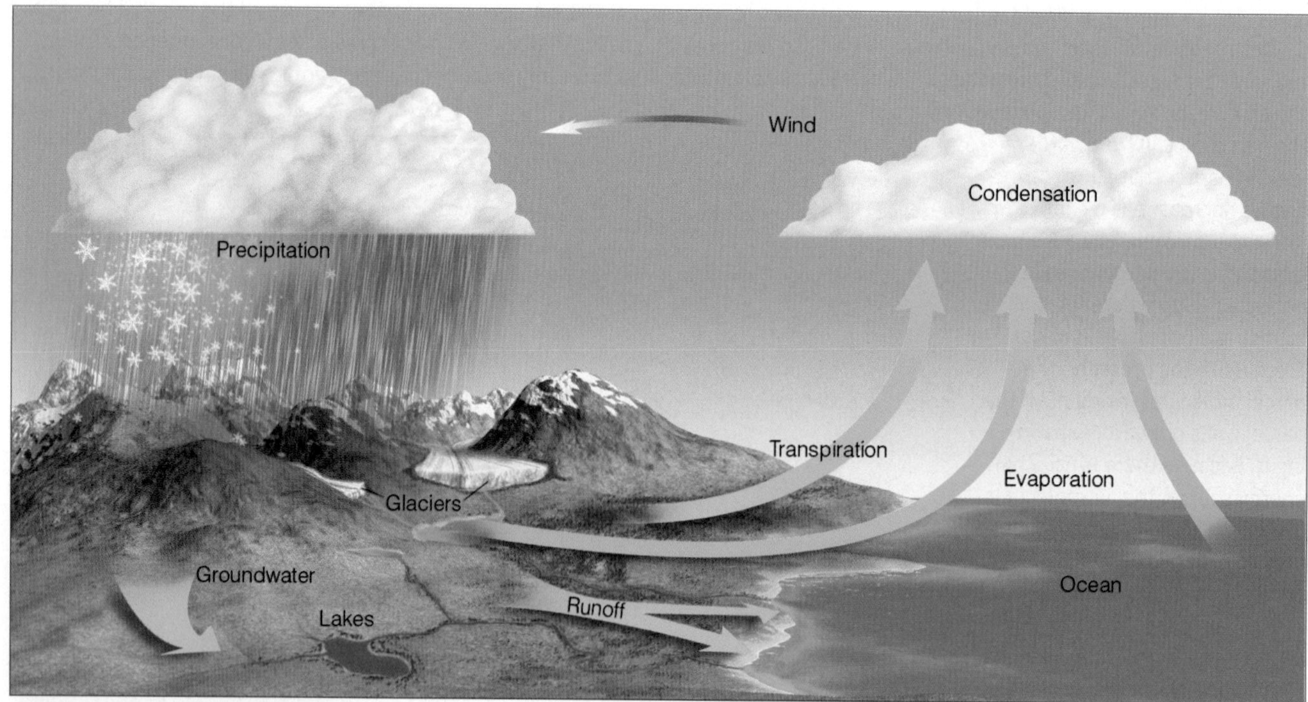

》FIGURE 1.23 The hydrologic cycle—the circulation of water within the atmosphere.

the air. Some of the precipitation may be intercepted by vegetation, where it evaporates or drips to the ground long after a storm has ended. Once on the surface, a portion of the water soaks into the ground by percolating downward through small openings in the soil and rock, forming groundwater that can be tapped by wells. What does not soak in collects in puddles of standing water or runs off into streams and rivers, which find their way back to the ocean. Even the underground water moves slowly and eventually surfaces, only to evaporate or be carried seaward by rivers.

Over land, a considerable amount of water vapor is added to the atmosphere through evaporation from the soil, lakes, and streams. Even plants give up moisture by a process called *transpiration*. The water absorbed by a plant's root system moves upward through the stem and emerges from the plant through numerous small openings on the underside of the leaf. In all, evaporation and transpiration from continental areas amount to only about 15 percent of the hundreds of trillions of gallons of water vapor that annually evaporate into the atmosphere; the remaining 85 percent evaporates from the oceans. If all of this water vapor were to suddenly condense and fall as rain, it would be enough to cover the entire globe with 2.5 centimeters (or 1 inch) of water. The total mass of water vapor stored in the atmosphere at any moment adds up to only a little over a week's supply of the world's precipitation. Since this amount varies only slightly from day to day, the hydrologic cycle is exceedingly efficient in circulating water in the atmosphere.

Up to this point, we have looked at some of the important concepts and principles of the atmosphere that we will use throughout this book. We will now turn our attention to some of the extreme weather events that take place within our atmosphere. As you read this section, keep in mind that the content serves as a broad overview of material that will be covered in more detail in later chapters.

BRIEF REVIEW

Before going on to the next section, here is a review of some of the important concepts presented in the last several sections.

- **Temperature of the air (or any substance) is a measure of the average speed (motion) of its atoms and molecules.**
- **The rate at which the air temperature decreases with height is called the *lapse rate*. A measured increase in air temperature with height is called an *inversion*.**
- **We live at the bottom of the troposphere, which is an atmospheric layer where the air temperature normally decreases with height. The troposphere is a region that contains all of the weather we are familiar with.**
- **The hottest atmospheric layer is the thermosphere, where temperatures can exceed 900°F (500°C). The coldest layer is the mesosphere, where temperatures can drop to −130°F (−90°C). Most of the gas ozone is found in the stratosphere.**
- **Atmospheric (air) pressure at any level represents the total mass of air above that level, and atmospheric pressure always decreases with increasing height above the surface.**
- **Wind is the horizontal movement of air. Differences in atmospheric pressure cause the wind to blow.**
- **The wind direction is the direction from which the wind blows—a north wind blows from the north. Wind speed is the rate at which wind moves by a stationary object.**
- **Clouds and precipitation are often associated with regions of surface low pressure; clear skies with regions of surface high pressure.**
- **Relative humidity does not tell us how much water vapor is actually in the air; rather, it tells us how close the air is to being saturated.**
- **The dew point temperature is the temperature to which air would have to be cooled in order for saturation to occur.**

Extremes of Weather and Climate

Weather and climate play a major role in our lives. Weather, for example, often dictates the type of clothing we wear, while climate influences the type of clothing we buy. Climate determines when to plant crops as well as what type of crops can be planted. Weather determines if these same crops will grow to maturity. Although weather and climate affect our lives in many ways, perhaps their most immediate effect is on our comfort. In order to survive the cold of winter and heat of summer, we build homes, heat them, air condition them, insulate them—only to find that when we leave our shelter, we are at the mercy of the weather elements.

Even when we are dressed for the weather properly, wind, humidity, and precipitation can change our perception of how cold or warm it feels. On a cold, windy day the effects of *wind chill* tell us that it feels much colder than it really is, and, if not properly dressed, we

EXTREME WEATHER WATCH

What a jump! On August 27, 1960, U. S. Air Force Captain Joseph Kittinger parachuted from a manned balloon into the stratosphere at an altitude of 102,800 ft (about 19.5 mi or 31.4 km) where the air temperature is near −70°F (−56°C) and the air pressure is close to 10 mb. When he engaged his parachute at 18,000 ft, he established the world record for the longest parachute freefall (84,800 ft) in 4 minutes and 36 seconds.

run the risk of *frostbite* or even *hypothermia* (the rapid, progressive mental and physical collapse that accompanies the lowering of human body temperature). On a hot, humid day we normally feel uncomfortably warm and blame it on the humidity. If we become too warm, our bodies overheat and *heat exhaustion* or *heat stroke* may result. Those most likely to suffer these maladies are the elderly with impaired circulatory systems and infants, whose heat regulatory mechanisms are not yet fully developed.

For some people, a warm dry wind blowing downslope (a *chinook wind*) adversely affects their behavior (they often become irritable and depressed). Hot, dry, downslope *Santa Ana* winds in Southern California can turn burning dry vegetation into a huge firestorm.

When the weather turns much colder or warmer than normal, it impacts directly on the lives and pocketbooks of many people. For example, the exceptionally warm January of 2006 over the United States saved people millions of dollars in heating costs. On the other side of the coin, the colder than normal winter of 2000–2001 over much of North America sent heating costs soaring as demand for heating fuel escalated.

Major cold spells accompanied by heavy snow and ice can play havoc by snarling commuter traffic, curtailing airport services, closing schools, and downing power lines, thereby cutting off electricity to thousands of customers (see ❱ Fig. 1.24). For example, a huge ice storm during January, 1998, in northern New England and Canada left millions of people without power and caused over a billion dollars in damages, and a devastating snow storm during March, 1993, buried parts of the East Coast with 14-foot snow drifts and left Syracuse, New York, paralyzed with a snow depth of 36 inches. When the frigid air settles into the Deep South, many millions of dollars worth of temperature-sensitive fruits and vegetables may be ruined, the eventual consequence being higher produce prices in the supermarket.

Prolonged drought, especially when accompanied by high temperatures, can lead to a shortage of food and, in some places, widespread starvation. Parts of Africa, for example, have periodically suffered through major droughts and famine. During the summer of 2007, the southeastern section of the United States experienced

❱ FIGURE 1.24

Ice storm near Oswego, New York, caused utility poles and power lines to be weighed down, forcing road closure.

a terrible drought as searing summer temperatures wilted crops, causing losses in excess of a billion dollars. When the climate turns hot and dry, animals suffer too. In 1986, over 500,000 chickens perished in Georgia during a two-day period at the peak of a summer heat wave. Severe drought also has an effect on water reserves, often forcing communities to ration water and restrict its use. During periods of extended drought, vegetation often becomes tinder-dry and, sparked by lightning or a careless human, such a dried-up region can quickly become a raging inferno. During the winter of 2005–2006, hundreds of thousands of acres in drought-stricken Oklahoma and northern Texas were ravaged by wildfires.

Every summer, scorching *heat waves* take many lives. During the past 20 years, an annual average of more than 300 deaths in the United States were attributed to excessive heat exposure. In one particularly devastating heat wave that hit Chicago, Illinois, during July, 1995, high temperatures coupled with high humidity claimed the lives of more than 700 people. And Europe suffered through a devastating heat wave during the summer of 2003 when many people died, including 14,000 in France alone. In California during July, 2006, more than 100 people died as air temperatures climbed to over 115°F (46°C).

Every year, the violent side of weather influences the lives of millions. It is amazing how many people whose family roots are in the Midwest know a story about someone who was severely injured or killed by a tornado. **Tornadoes** are intense rotating columns of air that extend downward from the base of a thunderstorm. Sometimes called *twisters* or *cyclones*, they may appear as ropes or as a large circular cylinder (see ❯ Fig. 1.25). The majority are less than one-quarter of a mile wide and many are smaller than a football field. Tornado winds may exceed 200 mi/hr but most probably peak at less than 135 mi/hr. Tornadoes annually cause damage to buildings and property totaling in the hundreds of millions of dollars, as a single large powerful tornado can level an entire section of a town.

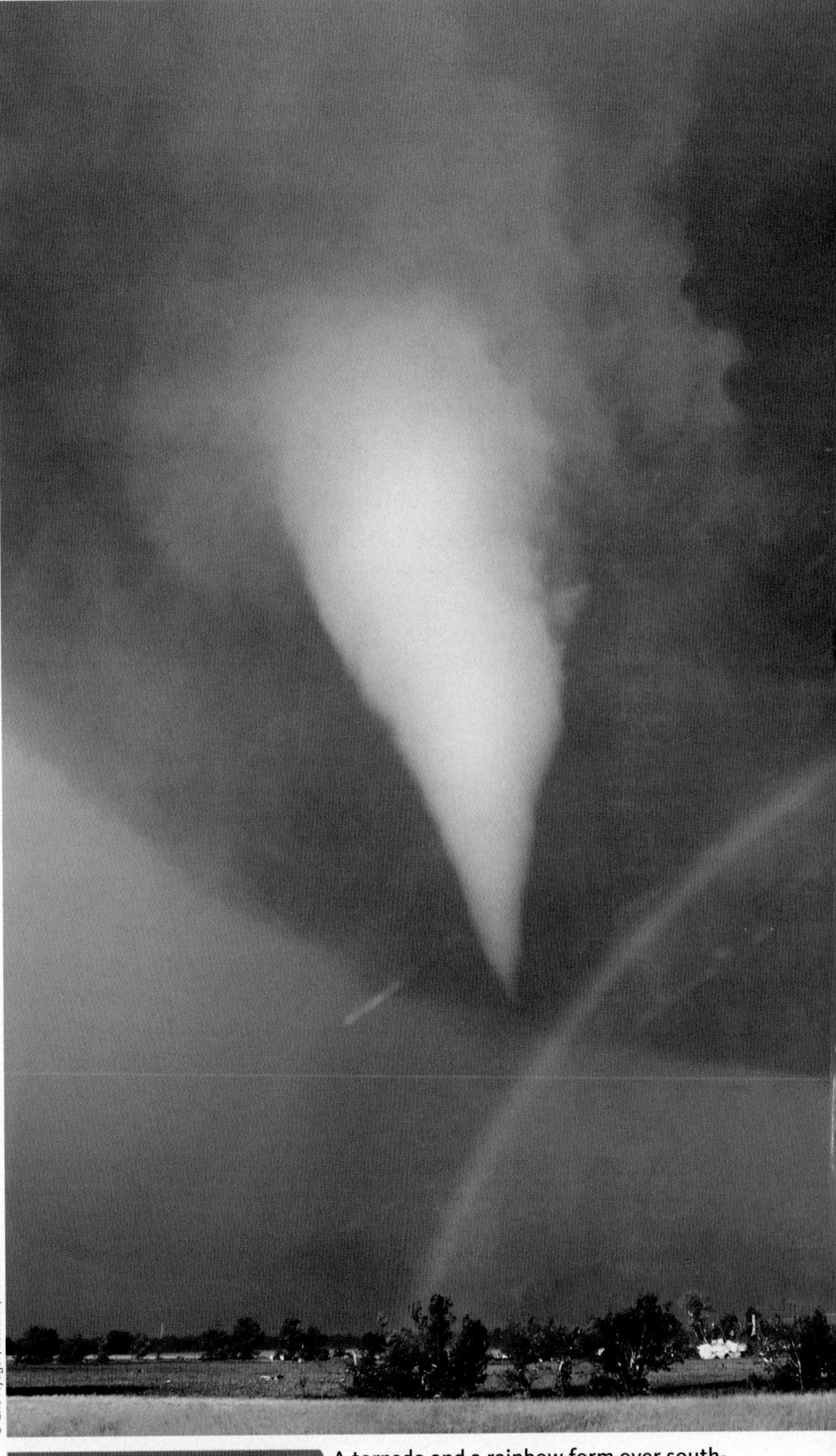

❯ FIGURE 1.25 A tornado and a rainbow form over south-central Kansas during June, 2004. White streaks in the sky are descending hailstones.

Although the gentle rains of a typical summer *thunderstorm* are welcome over much of North America, the heavy downpours, high winds, and hail of the **severe thunderstorm** are not (see ❯ Fig. 1.26). Cloudbursts from slowly moving, intense thunderstorms can provide too much rain too quickly, creating *flash floods* as small streams become raging rivers composed of mud and sand entangled with uprooted plants and trees (see ❯ Fig. 1.27). On the average, more people die in the United States from floods and flash floods than from either lightning or tornadoes. Strong downdrafts originating inside an intense thunderstorm (a *downburst*) create turbulent winds that are capable of destroying crops and inflicting damage upon surface structures. Several airline crashes have been attributed to the turbulent *wind shear* zone within the downburst. Annually, hail damages crops worth millions of dollars, and lightning takes the lives of about eighty people in the United States and starts fires that destroy many thousands of acres of valuable timber (see ❯ Fig. 1.28).

Often thunderstorms are part of a much larger storm system. In middle latitudes, for example, thunderstorms may form with huge **middle-latitude cyclonic storms** that develop, move eastward, and often bring a variety of weather (sometimes extreme) with them. ❯Figure 1.29 shows two mid-latitude cyclones over the North Pacific Ocean. The center of each storm (which is the center of lowest pressure) is marked by the red letter "L." Because these storms are in the Northern Hemisphere, their winds blow counterclockwise and in toward their centers. The colored lines extending away from the center of each storm are weather **fronts**. Each front represents a boundary where there is often a change in wind direction, temperature, and humidity. Along the fronts, the air usually rises and condenses into clouds, from which rain or snow may fall.

The solid red line in Fig. 1.29 represents a *warm front*, with half circles showing its general direction of movement. Along the front, warm air to the south is replacing cooler air to the north. Solid blue lines are *cold fronts*, where cooler air is replacing warmer air, and arrowheads show the front's general direction of movement. It is along a cold front where extreme weather often forms, such as severe thunderstorms with strong, gusty winds and heavy precipitation. Where the cold

❯ FIGURE 1.26 A huge, severe thunderstorm produces large hail and 70 mi/hr winds as it moves across the plains near Sutherland, Nebraska, on June 10, 2006.

FIGURE 1.27

Flood waters race through Florence, Italy, during November, 1966, damaging many historical sites.

FIGURE 1.28

Estimates are that lightning strikes the earth about 100 times every second. About 25 million lightning strikes hit the United States each year.

front has caught up to the warm front, and cold air is now replacing cool air, an *occluded front* is drawn in purple with alternating arrowheads and half circles showing how it is moving.

A usually smaller, but often more vigorous storm is the *tropical cyclone* that forms over warm tropical oceans. Called **hurricanes** in the Atlantic and eastern north Pacific, these storms can generate huge waves, heavy rain, and high winds that may exceed 150 mi/hr. (See Fig. 1.30.) In the Northern Hemisphere, surface winds blow counterclockwise around a central *eye* that marks the center of the storm. With its high winds, heavy rain, and huge surge of water that moves inland as the storm makes landfall, hurricanes are capable of inflicting great destruction and taking many lives. Tragically, such an event occurred in New Orleans where about 1,300 people perished after Hurricane Katrina slammed into the Mississippi-Louisiana coastline on the morning of August 29, 2005.

Even the quiet side of weather has its influence. When winds die down and humid air becomes more tranquil, fog may form. Dense fog can restrict visibility at airports causing flight delays and cancellations. Every winter, deadly fog-related auto accidents

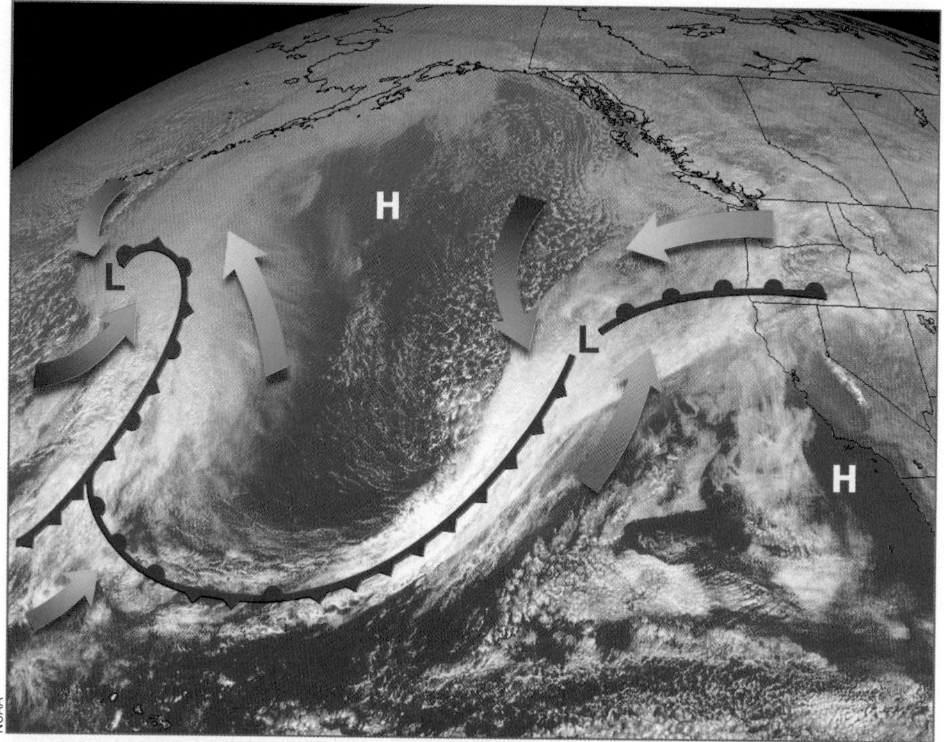

FIGURE 1.29

Visible satellite image of the North Pacific Ocean with two mid-latitude cyclones in different stages of development during February, 2000. Superimposed on the image are surface wind-flow arrows and fronts. In the middle and high latitudes, these large cyclonic storm systems tend to move in a west-to-east direction.

FIGURE 1.30

Visible satellite image of Hurricane Rita over the Gulf of Mexico on September 21, 2005. A major hurricane with winds exceeding 155 mi/hr, Rita was a large, powerful storm capable of inflicting great damage.

occur along our busy highways and turnpikes. But fog has a positive side, too, especially during a dry spell, as fog moisture collects on tree branches and drips to the ground, where it provides water for the tree's root system.

Weather and climate have become so much a part of our lives that the first thing many of us do in the morning is to listen to the local weather forecast. Moreover, when severe weather is imminent, people tend to turn on their radio or television to find out if bad weather is heading their way. For this reason, many radio and television newscasts have their own "weatherperson" to present weather information and give daily forecasts. More and more of these people are professionally trained in meteorology, and many stations require that the weathercaster obtain a seal of approval from the American Meteorological Society (AMS), or a certificate from the National Weather Association (NWA). To make their weather presentation as up-to-the-minute as possible, an increasing number of stations are taking advantage of the information provided by the National Weather Service (NWS), such as computerized weather forecasts, time-lapse satellite images, and color Doppler radar displays.

For more than twenty-five years, a staff of trained professionals at "The Weather Channel" have provided weather information twenty-four hours a day on cable television. And finally, the National Oceanic and Atmospheric Administration (NOAA), in cooperation with the National Weather Service, sponsors weather radio broadcasts at selected locations across the United States. Known as *NOAA weather radio* (and transmitted at VHF—FM frequencies), this service provides continuous weather information and regional forecasts (as well as special weather advisories, including watches and warnings of severe and unusual weather) for over 90 percent of the United States.

Summary

This chapter provides an overview of the earth's atmosphere and a preview of extreme weather events we will investigate in later chapters. The chapter begins by examining the risk of dying from an extreme weather event. Although the risk is low, extreme weather garners a great deal of attention due to the atmosphere's awesome power and potential to cause destruction and personal injury.

We examined the earth's atmosphere and found it to be a mixture of many gases, the most abundant being nitrogen and oxygen. We found that water vapor normally occupies a small percentage of a volume of air, and that increasing levels of carbon dioxide and other greenhouse gases are responsible for a warming of the planet. We looked at the properties of the atmosphere and found that air temperature is related to the average speed of the molecules. We investigated the layers of the atmosphere and looked at the troposphere (the lowest layer), where almost all weather events occur; the stratosphere, where ozone protects us from a portion of the sun's harmful rays; the mesosphere, the coldest layer; and the thermosphere, the warmest layer.

We examined the concept of air pressure and found that surface air pressure is related to the total mass of air above the surface. We also learned that air pressure always decreases with increasing height above the surface and that horizontal differences in air pressure cause the wind to blow. We looked at atmospheric moisture and found that the relative humidity represents how close the air is to being saturated and that the dew point represents the temperature to which air must be cooled in order for saturation to occur. We also learned that condensation—the transformation of vapor into liquid—is more likely to happen where the air cools and that the circulation of water within the atmosphere is called the hydrologic cycle.

We looked at the atmosphere and found it to contain a variety of storms of varying sizes. The movement, intensification, and weakening of these systems, as well as the dynamic nature of the air itself, provides a variety of weather events, some of which are extreme and life-threatening. The sum total of weather and its extremes over a long period of time is what we call climate. Finally, we discussed some of the many ways extreme weather can influence our lives.

Key Terms

The following terms are listed (with page number) in the order they appear in the text. Define each. Doing so will aid you in reviewing the material covered in this chapter.

atmosphere, 8
nitrogen, 8
oxygen, 8
water vapor, 9
carbon dioxide, 9
ozone, 11
ozone hole, 11
aerosols, 11
pollutants, 12
acid rain, 12
weather, 13
climate, 13
temperature, 13
Fahrenheit scale, 14
Celsius scale, 14
Kelvin scale, 14
lapse rate, 14
temperature inversion, 14
troposphere, 15
stratosphere, 15
mesosphere, 16
thermosphere, 16
air pressure, 16
wind, 19
scales of motion, 19
wind speed, 20
wind direction, 20
evaporation, 22
condensation, 22
saturation, 22
relative humidity, 24
dew-point temperature, 24
precipitation, 24
hydrologic cycle, 24
tornadoes, 27
severe thunderstorms, 28
middle-latitude cyclonic storms, 28
fronts, 28
hurricanes, 29

Review Questions

1. (a) How does the risk of death from weather events compare with other risks in life? (Hint: See Table 1.1.)
(b) On average, in the United States, which weather-related event takes the most lives annually? (Hint: See Fig. 1.3.)
2. According to Fig. 1.4, p. 7, in the state where you presently live which weather event (flooding, lightning, or tornadoes) poses the greatest risk for a weather-related fatality?
3. List the two most abundant gases in today's atmosphere. What percentage does each one occupy in a volume of dry air near the earth's surface?
4. List the most abundant greenhouse gases in the earth's atmosphere. What makes them greenhouse gases?
5. What are some of the important roles that water vapor plays in our atmosphere?
6. Briefly explain the production and natural destruction of carbon dioxide near the earth's surface. Give two reasons for the increase in carbon dioxide over the past 100 years or so.
7. Explain how ozone is a needed gas in the stratosphere, but an unwanted gas near the earth's surface.
8. Above what continent would you find the ozone hole?
9. How does weather differ from climate?
10. How does the average speed of air molecules relate to the air temperature?
11. How does the Kelvin temperature scale differ from the Celsius scale? On a day when the outside air temperature is 273 K, would this air be considered warm or cold?
12. On the basis of temperature, list the layers of the atmosphere from the lowest layer to the highest. Which layer is the coldest? The warmest? Which layer contains all of our weather?
13. What is the average (standard) temperature lapse rate in the troposphere?
14. (a) Explain the concept of air pressure in terms of mass of air above some level.
(b) Why does air pressure always decrease with increasing height above the surface?
15. What is the standard atmospheric pressure at sea level in
(a) inches of mercury?
(b) millibars?
(c) hectopascals?
16. Describe the three scales of motion when discussing wind and give an example of each.

17. What causes the wind to blow?
18. Does a north wind blow from the north toward the south, or from the south toward the north?
19. Describe how the wind blows around areas of surface low- and high-pressure in the Northern Hemisphere.
20. Why are clouds more common with areas of low pressure?
21. Explain why condensation is more likely when the air temperature lowers.
22. Define relative humidity and dew point.
23. If the air temperature is close to the dew point, would the relative humidity be high or low? Explain.
24. Briefly describe the movement of water in the hydrologic cycle.
25. List some of the ways extreme weather and climate can influence our lives.

Online Learning

STUDENT COMPANION WEBSITE: Visit this book's companion website at: www.cengage.com/ahrens/extreme1e and choose Chapter 1 for many study aids and ideas for further reading and research. These include flashcards, practice quizzing, and web links.

METEOROLOGY RESOURCE CENTER: For students with access, log on at www.cengage.com/login for more assets, including animations, videos, and more. If your textbook did not come with access, visit www.CengageBrain.com

Energy That Drives the Storms

2

CONTENTS

ENERGY AND HEAT TRANSFER
ENERGY BALANCING ACT—ABSORPTION, EMISSION, AND EQUILIBRIUM
WHY THE EARTH HAS SEASONS
SUMMARY
KEY TERMS
REVIEW QUESTIONS

As water evaporates from the ocean into the atmosphere, it carries a hidden form of energy that is released into the atmosphere when the water vapor condenses into clouds.

Each day the earth as a whole receives the same amount of energy from the sun. But this energy is not distributed evenly over the earth. Annually, tropical regions receive more energy than do polar regions. It is this energy imbalance that drives the atmosphere into the dynamic pattern we experience as wind and weather. Moreover, the energy that drives severe and extreme weather events is the same energy that produces calm, quiescent weather. In severe weather, however, energy becomes concentrated in a relatively small area. Therefore, in this chapter we will examine how energy is distributed over the earth, how energy is transferred from one region to another, and how energy can be concentrated in a relatively small region.

Energy and Heat Transfer

By definition **energy** is the ability or capacity to do work on some form of matter. (Matter is anything that has mass and occupies space.) Work is done on matter when matter is either pushed, pulled, or lifted over some distance. When we lift a brick, for example, we exert a force against the pull of gravity—we "do work" on the brick. The higher we lift the brick, the more work we do. So, by doing work on something, we give it "energy," which it can, in turn, use to do work on other things. The brick that we lifted, for instance, can now do work on your toe—by falling on it.

The total amount of energy stored in any object (internal energy) determines how much work that object is capable of doing. A lake behind a dam contains energy by virtue of its position. This situation typifies what is called *gravitational potential energy* or simply **potential energy** because it represents the potential to do work—a great deal of destructive work if the dam were to break.

Any moving substance possesses energy of motion or **kinetic energy**. The kinetic energy (KE) of an object is equal to half its mass multiplied by its velocity squared; thus

$$KE = \frac{1}{2} mv^2.$$

The faster something moves, the greater its kinetic energy; hence, a strong wind possesses more kinetic energy than a light breeze. Since kinetic energy also depends on the object's mass, a volume of water and an equal volume of air may be moving at the same speed, but, because the water has greater mass, it has more kinetic energy, which is why water moving at a speed of 40 miles per hour can do a great deal more damage than a wind moving at the same speed. The atoms and molecules that comprise all matter have kinetic energy due to their motion. This form of kinetic energy is often referred to as *heat energy*. Probably the most important form of energy in terms of weather and climate is the energy we receive from the sun—*radiant energy*.

Energy, therefore, takes on many forms, and it can change from one form into another. But the total amount of energy in the universe remains constant. *Energy cannot be created nor can it be destroyed*. It merely changes from one form to another in any ordinary physical or chemical process. In other words, the energy lost during one process must equal the energy gained during another. This balance is what we mean when we say that energy is conserved.

How is energy related to temperature? We know that air is a mixture of countless billions of atoms and molecules. As previously mentioned, if they could be seen, they would appear to be moving about in all directions, freely darting, twisting, spinning, and colliding with one another like an angry swarm of bees. Close to the earth's surface, each individual molecule would travel about a thousand times its diameter before colliding with another molecule. Moreover, we would see that all the atoms and molecules are not moving at the same speed, as some are moving faster than others. The energy associated with this motion is kinetic energy, and the *temperature* of the air is a measure of this average kinetic energy. Or as we learned in Chapter 1, the temperature of the air (or any substance) is a measure of the average motion of its atoms and molecules.

Suppose we examine a volume of surface air about the size of a large flexible balloon as shown in ❱ Fig. 2.1a. If we warm the air inside, the molecules would move faster, but they also would move slightly farther apart—the air becomes less dense, as illustrated in Fig. 2.1b. Conversely, if we cool the air back to its original temperature, the molecules would slow down, crowd closer together, and the air would become more dense. This molecular behavior is why, in many places throughout the book, we refer to surface air as either *warm, less-dense air* or as *cold, more-dense air*.

The atmosphere contains internal energy, which is the total energy stored in its molecules. **Heat**, on the other hand, *is energy in the process of being transferred from one object to another because of the temperature difference between them*. After heat is transferred, it is stored as internal energy. In the atmosphere, heat is transferred by *conduction, convection*, and *radiation*. We will examine these mechanisms of energy transfer after we look at the important concept of latent heat.

LATENT HEAT—THE HIDDEN WARMTH We know from Chapter 1 that water vapor is an invisible gas that becomes visible when it changes into larger liquid or solid (ice) particles. This process of transformation is known as a *change of state* or, simply, a *phase change*. The heat energy required to change a substance, such as water, from one state to another is called **latent heat**. But why is

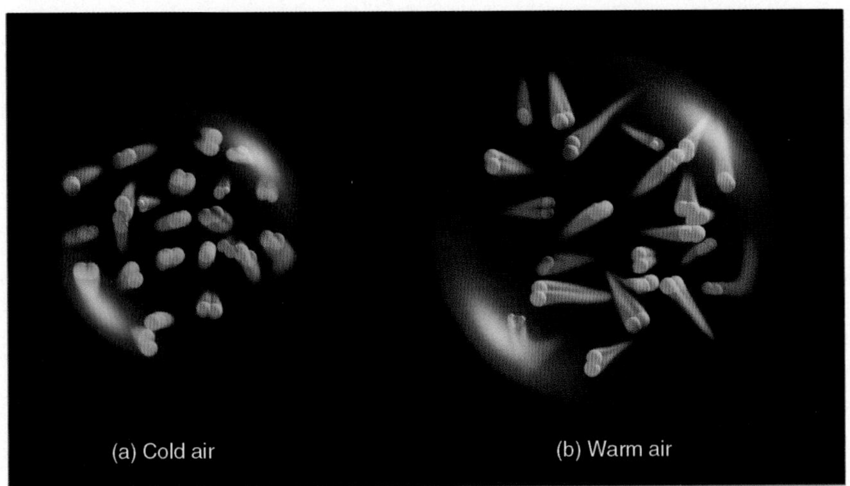

FIGURE 2.1

Air temperature is a measure of the average speed of the molecules. In the cold volume of air, the molecules move more slowly and crowd closer together. In the warm volume, they move faster and farther apart.

this heat referred to as "latent"? To answer this question, we will begin with something familiar to most of us—the cooling produced by evaporating water.

Suppose we microscopically examine a small drop of pure water. At the drop's surface, molecules are constantly escaping (evaporating). Because the more energetic, faster-moving molecules escape most easily, the average motion of all the molecules left behind decreases as each additional molecule evaporates. Since temperature is a measure of average molecular motion, the slower motion suggests a lower water temperature. *Evaporation is, therefore, a cooling process.* Stated another way, evaporation is a cooling process because the energy needed to evaporate the water—that is, to change its phase from a liquid to a gas—may come from the water or other sources, including the air.

The energy lost by liquid water during evaporation can be thought of as carried away by, and "locked up" within, the water vapor molecule. The energy is thus in a "stored" or "hidden" condition and is, therefore, called *latent heat.* It is latent (hidden) in that the temperature of the substance changing from liquid to vapor is still the same. However, the heat energy will reappear as **sensible heat** (the heat we can feel and measure with a thermometer) when the vapor condenses back into liquid water. Therefore, *condensation* (the opposite of evaporation) *is a warming process.*

The heat energy released when water vapor condenses to form liquid droplets is called *latent heat of condensation.* Conversely, the heat energy used to change liquid into vapor at the same temperature is called *latent heat of evaporation* (vaporization). Nearly 600 calories* are required to evaporate a single gram of water at room temperature. With many hundreds of grams of water evaporating from the body, it is no wonder that after a shower we feel cold before drying off. Figure 2.2 summarizes the concepts examined so far. When the change of state

*By definition, a calorie is the amount of heat required to raise the temperature of 1 gram of water from 14.5°C to 15.5°C. In the International System (SI), the unit of energy is the joule (J), where 1 calorie = 4.186 J. (For pronunciation: joule rhymes with pool.)

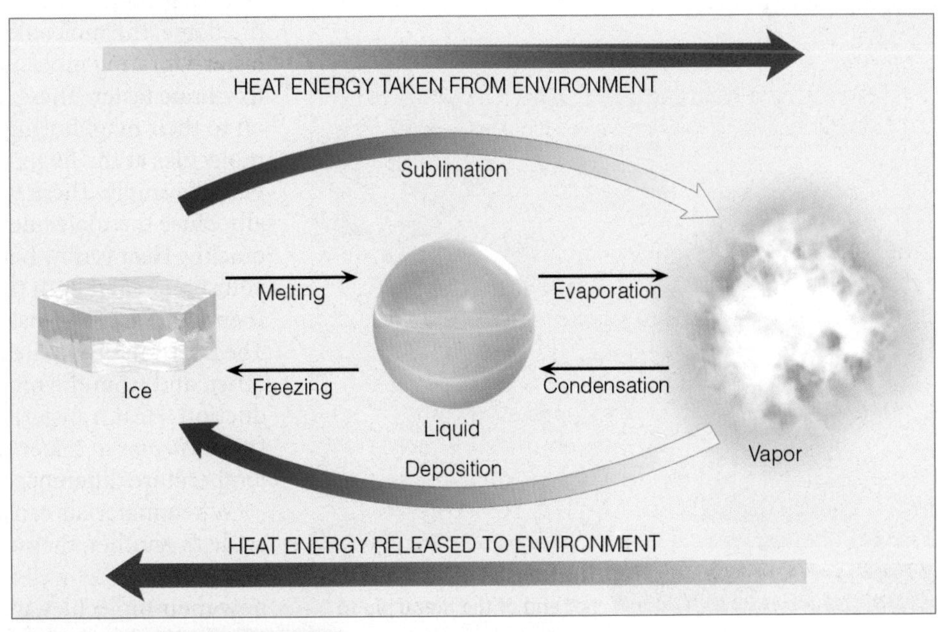

FIGURE 2.2 **Heat energy absorbed and released.**

© C.Donald Ahrens

FIGURE 2.3 Every time a cloud forms, it warms the atmosphere. Inside this developing thunderstorm, a vast amount of stored heat energy (latent heat) is given up to the air, as invisible water vapor becomes countless billions of water droplets and ice crystals. In fact, for the duration of this storm alone, more heat energy is released inside this cloud than is unleashed by a small nuclear bomb.

is from left to right, heat is absorbed by the substance and taken away from the environment. The processes of melting, evaporation, and sublimation (ice to vapor) all cool the environment. When the change of state is from right to left, heat energy is given up by the substance and added to the environment. The processes of freezing, condensation, and deposition (vapor to ice) all warm their surroundings.

Latent heat is an important source of atmospheric energy. Once vapor molecules become separated from the earth's surface, they are swept away by the wind, like dust before a broom. Rising to high altitudes where the air is cold, the vapor changes into liquid and ice cloud particles. During these processes, a tremendous amount of heat energy is released into the environment. This heat provides the fuel for storms such as hurricanes, middle-latitude cyclones, and thunderstorms (see Fig. 2.3). (Additional information on the energy for these storms is provided in the Focus section on p. 39.)

Water vapor evaporated from warm, tropical water can be carried into polar regions, where it condenses and gives up its heat energy. Thus, as we will see, evaporation-transportation-condensation is an extremely important mechanism for the relocation of heat energy (as well as water) in the atmosphere.

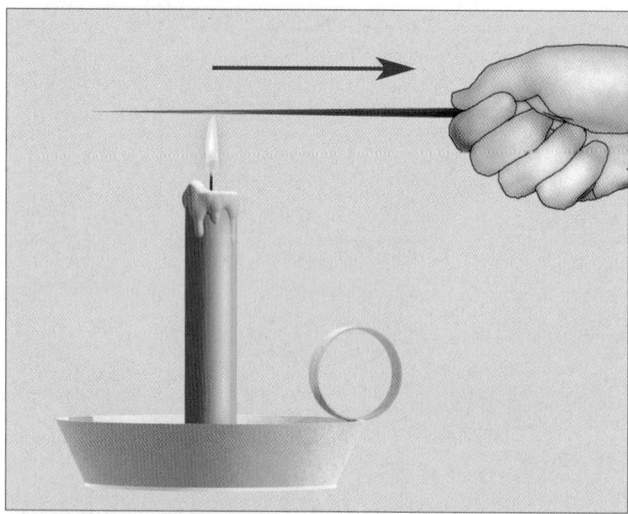

FIGURE 2.4 The transfer of heat from the hot end of the metal pin to the cool end by molecular contact is called *conduction*.

CONDUCTION The transfer of heat from molecule to molecule within a substance is called **conduction.** Hold one end of a metal straight pin between your fingers and place a flaming candle under the other end (see Fig. 2.4). Because of the energy they absorb from the flame, the molecules in the pin vibrate faster. The faster-vibrating molecules cause adjoining molecules to vibrate faster. These, in turn, pass vibrational energy on to their neighboring molecules, and so on, until the molecules at the finger-held end of the pin begin to vibrate rapidly. These fast-moving molecules eventually cause the molecules of your finger to vibrate more quickly. Heat is now being transferred from the pin to your finger, and both the pin and your finger feel hot. If enough heat is transferred, you will drop the pin. The transmission of heat from one end of the pin to the other, and from the pin to your finger, occurs by conduction. Heat transferred in this fashion always flows from *warmer to colder* regions. Generally, the greater the temperature difference, the more rapid the heat transfer.

When materials can easily pass energy from one molecule to another, they are considered to be good conductors of heat. How well they conduct heat depends upon how their molecules are structurally bonded together. Table 2.1 shows that solids, such as metals, are good heat conductors. It is often difficult, therefore, to judge the temperature of metal objects. For example, if you grab a metal pipe at room temperature, it will seem to be much colder than it actually is because the metal conducts heat away from the hand quite rapidly. Conversely, *air is an extremely poor conductor of heat,* which is why most insulating materials have a large number of air spaces trapped within them. Air is such a poor heat conductor that, in calm weather, the hot ground only warms a shallow layer of air a few centimeters thick by conduction. Yet, air can carry this energy rapidly from one region to another. How, then, does this phenomenon happen?

CONVECTION The transfer of heat by the mass movement of a fluid (such as water and air) is called **convection.** This type of heat transfer takes place in liquids and gases because they can move freely, and it is possible to set up currents within them.

Convection happens naturally in the atmosphere. On a warm, sunny day certain areas of the earth's surface absorb more heat from the sun than others; as a result, the air near the earth's surface is heated somewhat unevenly. Air molecules adjacent to these hot surfaces bounce against them, thereby gaining some extra energy by conduction. The heated air expands and becomes less dense than the surrounding cooler air. The expanded warm air is buoyed upward and rises. In

EXTREME WEATHER WATCH

Flying into an area of convection can produce a bumpy plane ride, as passengers are jostled around by the turbulence brought on by the rising and descending air.

FOCUS ON

A SPECIAL TOPIC

Comparing Energy in Storms

Extreme weather events produce damage through the release of kinetic energy (high winds) and electrical energy (lightning). They obtain some of this energy from the conversion of latent heat to potential energy, and from potential energy to kinetic energy. Estimates are that a typical hurricane releases more than a trillion joules of kinetic energy per day over a relatively large area. If we include the release of latent heat in a hurricane, its *total energy* released averages more than a thousand trillion joules per day (see Fig. 1), which is roughly half of the present electrical needs of the world. Notice in Fig. 1 that even an average-size thunderstorm can release tremendous amounts of energy. While it is not likely that the energy of hurricanes and thunderstorms will ever be harnessed for electricity, these numbers point out the huge magnitude of energy contained in storms.

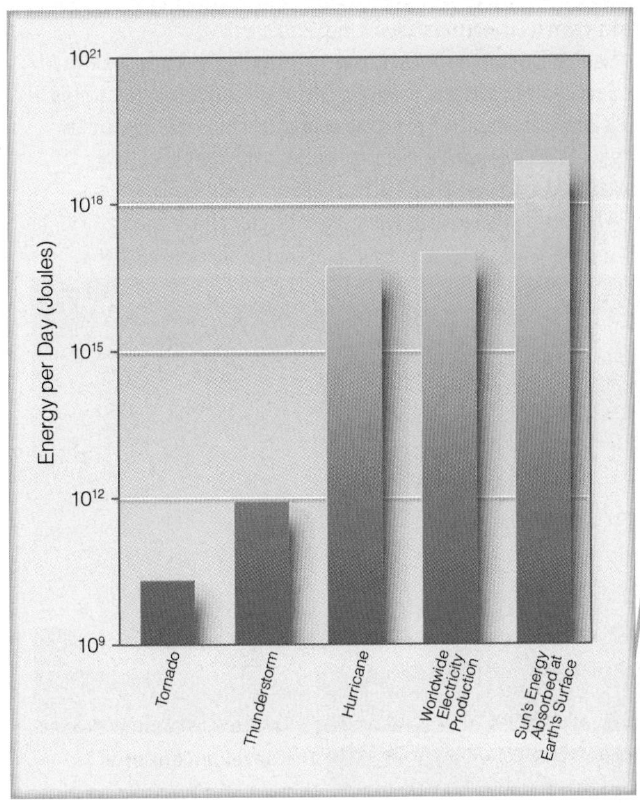

Figure 1
The estimated energy released per day in various storms compared to worldwide electricity production and to the sun's energy absorbed at the earth's surface.

Table 2.1

Heat Conductivity* of Various Substances

SUBSTANCE	HEAT CONDUCTIVITY (Watts† Per Meter Per °C)
Still air	0.023 (at 20°C)
Wood	0.08
Dry soil	0.25
Water	0.60 (at 20°C)
Snow	0.63
Wet soil	2.1
Ice	2.1
Sandstone	2.6
Granite	2.7
Iron	80
Silver	427

*Heat (thermal) conductivity describes a substance's ability to conduct heat as a consequence of molecular motion.

†A watt (W) is a unit of power where one watt equals one joule (J) per second (J/s). One joule equals 0.24 calories.

this manner, large bubbles of warm air rise and transfer heat energy upward. Cooler, heavier air flows toward the surface to replace the rising air. This cooler air becomes heated in turn, rises, and the cycle is repeated. In meteorology, this vertical exchange of heat is called *convection,* and the rising air bubbles are known as **thermals** (see Fig. 2.5).

The rising air expands and gradually spreads outward. It then slowly begins to sink. Near the surface, it moves back into the heated region, replacing the rising air. In this way, a *convective circulation,* or thermal "cell," is produced in the atmosphere (see Fig. 2.6).

Although the entire process of heated air rising, spreading out, sinking, and finally flowing back toward its original location is known as a convective circulation, meteorologists usually restrict the term *convection* to the process of the rising and sinking part of the circulation.

The horizontally moving part of the circulation (called *wind)* carries properties of the air in that particular area with it. The transfer of these properties by horizontally moving air is called **advection**. For example, wind blowing across a body of water will "pick up" water vapor from the evaporating surface and transport it elsewhere in the atmosphere. If the air cools, the water vapor may condense into cloud droplets and release latent heat. In a sense, then, heat is advected (carried) by the water vapor as it is swept along with the wind. Earlier, we saw that this is an important way to redistribute heat energy in the atmosphere.

FIGURE 2.5 The development of a thermal. A thermal is a rising bubble of air that carries heat energy upward by *convection.*

BRIEF REVIEW

Before moving on to the next section, here is a summary of some of the important concepts and facts we have covered:

- **The temperature of a substance is a measure of the average kinetic energy (average speed) of its atoms and molecules.**
- **Evaporation (the transformation of liquid into vapor) is a cooling process that can cool the air, whereas condensation (the transformation of vapor into liquid) is a warming process that can warm the air.**
- **Heat is energy in the process of being transferred from one object to another because of the temperature difference between them.**
- **In conduction, which is the transfer of heat by**

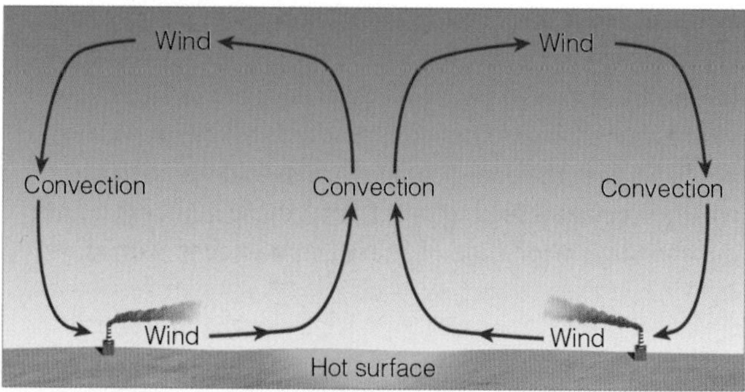

FIGURE 2.6 The rising of hot air and the sinking of cool air sets up a convective circulation. Normally, the vertical part of the circulation is called *convection,* while the horizontal part is called *wind.* Near the surface the wind is advecting smoke from one region to another.

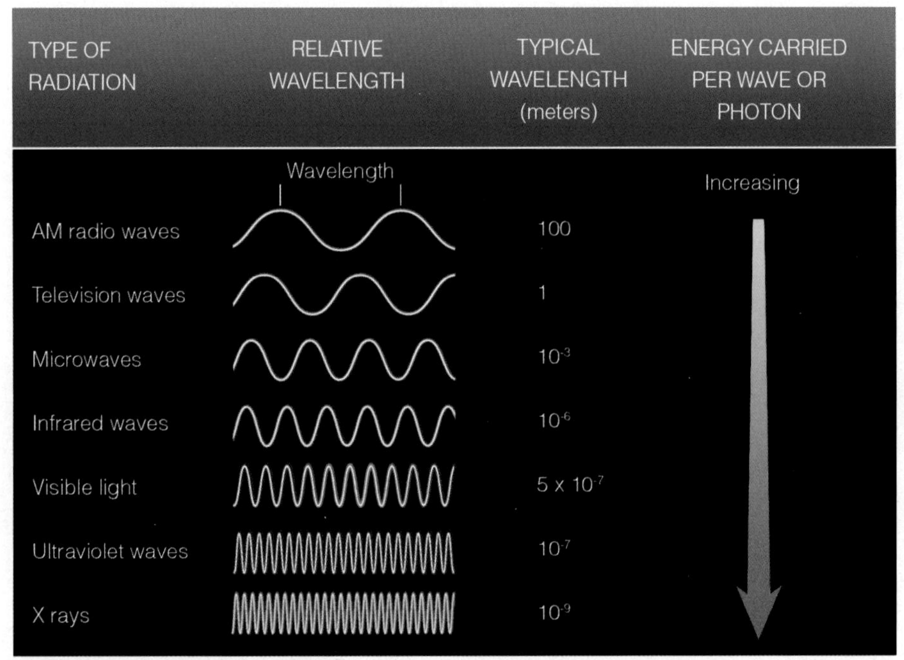

FIGURE 2.7 Radiation characterized according to wavelength. As the wavelength decreases, the energy carried per wave increases.

molecule-to-molecule contact, heat always flows from warmer to colder regions.

- **Air is a poor conductor of heat.**
- **Convection is an important mechanism of heat transfer, as it represents the vertical movement of warmer air upward and cooler air downward.**

There is yet another mechanism for the transfer of energy—radiation, or *radiant energy*, which is what we receive from the sun. In this method, energy may be transferred from one object to another without the space between them necessarily being heated.

RADIATION On a summer day, you may have noticed how warm and flushed your face feels as you stand facing the sun. Sunlight travels through the surrounding air with little effect upon the air itself. Your face, however, absorbs this energy and converts it to thermal energy. Thus, sunlight warms your face without actually warming the air. The energy transferred from the sun to your face is called **radiant energy**, or **radiation**. It travels in the form of waves that release energy when they are absorbed by an object. Because these waves have magnetic and electrical properties, we call them **electromagnetic waves**. Electromagnetic waves do not need molecules to propagate them. In a vacuum, they travel at a constant speed of nearly 300,000 km (186,000 mi) per second—the speed of light.

Figure 2.7 shows some of the different wavelengths of radiation. Notice that the *wavelength* (which is often expressed by the Greek letter lambda, λ) is the distance measured along a wave from one crest to another. Also notice that some of the waves have exceedingly short lengths. For example, radiation that we can see (visible light) has an average wavelength of less than one-millionth of a meter—a distance nearly one-hundredth the diameter of a human hair. To measure these short lengths, we introduce a new unit of measurement called a **micrometer** (abbreviated μm), which is equal to one-millionth of a meter (m); thus

$$1 \text{ micrometer } (\mu\text{m}) = 0.000001 \text{ m} = 10^{-6} \text{ m}.$$

In Fig. 2.7, we can see that the average wavelength of visible light is about 0.0000005 meters, which is the same as 0.5 μm. To give you a common object for comparison, the average height of a letter on this page is about 2000 μm, or 2 millimeters (2 mm), whereas the thickness of this page is about 100 μm.

We can also see in Fig. 2.7 that the longer waves carry less energy than do the shorter waves. When comparing the energy carried by various waves, it is useful to give electromagnetic radiation characteristics of particles in order to explain some of the wave's behavior. We can actually think of radiation as streams of particles, or **photons**, that are discrete packets of energy.*

An ultraviolet (UV) photon carries more energy than a photon of visible light. In fact, certain ultraviolet photons have enough energy to produce sunburns and penetrate skin tissue, sometimes causing skin cancer. (Additional information on radiant energy and its effect on humans is given in the Focus section on p. 42.)

To better understand the concept of radiation, here are a few important concepts and facts to remember:

1. All things (whose temperature is above absolute zero), no matter how big or small, emit radiation. The air, your body, flowers, trees, the earth, the stars are all radiating a wide range of electromagnetic waves. The

*Packets of photons make up waves, and groups of waves make up a beam of radiation.

energy originates from rapidly vibrating electrons, billions of which exist in every object.

2. The wavelengths of radiation that an object emits depend primarily on the object's temperature. *The higher the object's temperature, the shorter are the wavelengths of emitted radiation.* By the same token, as an object's temperature increases, its peak emission of radiation shifts toward shorter wavelengths. This relationship between temperature and wavelength is called *Wien's law** (*or Wien's displacement law*) after the German physicist Wilhelm Wien (pronounced *Ween*, 1864–1928) who discovered it.

*Wien's law:

$$\lambda_{max} = \frac{\text{constant}}{T}.$$

Where λ_{max} is the wavelength at which maximum radiation emission occurs, T is the object's temperature in Kelvins (K) and the constant is 2897 μmK. More information on Wien's law is given in Appendix B.

FOCUS ON
A SPECIAL TOPIC

Is Tanning Healthy?

Earlier, we learned that shorter waves of radiation carry much more energy than longer waves, and that a photon of ultraviolet light carries more energy than a photon of visible light. In fact, ultraviolet (UV) wavelengths in the range of 0.20 and 0.29 μm (known as *UV-C radiation*) are harmful to living things, as certain waves can cause chromosome mutations, kill single-celled organisms, and damage the cornea of the eye. Fortunately, virtually all the ultraviolet radiation at wavelengths in the UV-C range is absorbed by ozone in the stratosphere.

Ultraviolet wavelengths between about 0.29 and 0.32 μm (known as *UV-B radiation*) reach the earth in small amounts. Photons in this wavelength range have enough energy to produce sunburns and penetrate skin tissues, sometimes causing skin cancer. About 90 percent of all skin cancers are linked to sun exposure and UV-B radiation. Oddly enough, these same wavelengths activate provitamin D in the skin and convert it into vitamin D, which is essential to health.

Longer ultraviolet waves with lengths of about 0.32 to 0.40 μm (called *UV-A radiation*) are less energetic, but can still tan the skin. Although UV-B is mainly responsible for burning the skin, UV-A can cause skin redness. It can also interfere with the skin's immune system and cause long-term skin damage that shows up years later as accelerated aging and skin wrinkling. Moreover, recent studies indicate that the longer UV-A exposures needed to create a tan pose about the same cancer risk as a UV-B tanning dose.

Upon striking the human body, ultraviolet radiation is absorbed beneath the outer layer of skin. To protect the skin from these harmful rays, the body's defense mechanism kicks in. Certain cells (when exposed to UV radiation) produce a dark pigment (*melanin*) that begins to absorb some of the UV radiation. (It is the production of melanin that produces a tan.) Consequently, a body that produces little melanin—one with pale skin—has little natural protection from UV-B.

Additional protection can come from a sunscreen. Unlike the old lotions that simply moisturized the skin before it baked in the sun, sunscreens today block UV rays from ever reaching the skin. Some contain chemicals (such as zinc oxide) that reflect UV radiation. (These are the white

©Ben Visbeek

Figure 2 Is sun tanning healthy? Every year too much ultraviolet radiation from the sun results in many thousands of new cases of malignant melanoma, the deadliest form of skin cancer.

3. Objects that have a high temperature emit radiation at a greater rate or intensity than objects with a lower temperature. Thus, as the temperature of an object increases, more total radiation (over a given surface area) is emitted each second. This relationship between temperature and emitted radiation is known as the *Stefan-Boltzmann law** after Josef Stefan (1835–1893) and Ludwig Boltzmann (1844–1906), who devised it.

*Stefan-Boltzmann law:

$$E = \sigma T^4$$

Where E is the maximum rate of radiation emitted by each square meter of surface of an object, σ (the Greek letter sigma) is a constant, and T is the object's surface temperature in Kelvins (K). Additional information on the Stefan-Boltzmann law is given in Appendix B.

EXPOSURE CATEGORY	UV INDEX	PROTECTIVE MEASURES
Minimal	0–2	Apply SPF 15 sunscreen
Low	3–4	Wear a hat and apply SPF 15 sunscreen
Moderate	5–6	Wear a hat, protective clothing, and sunglasses with UV-A and UV-B protection; apply SPF 15+ sunscreen
High	7–9	Wear a hat, protective clothing, and sunglasses; stay in shady areas; apply SPF 15+ sunscreen
Very high	10+	Wear a hat, protective clothing, and sunglasses; use SPF 15+ sunscreen; avoid being in sun between 11 A.M. and 3 P.M.

Figure 3 The UV index.

pastes once seen on the noses of lifeguards.) Others consist of a mixture of chemicals that actually absorb ultraviolet radiation, usually UV-B, although new products with UV-A–absorbing qualities are now on the market. The *Sun Protection Factor* (*SPF*) number on every container of sunscreen dictates how effective the product is in protecting from UV-B— the higher the number, the better the protection.

Protecting oneself from excessive exposure to the sun's energetic UV rays is certainly wise. Estimates are that, in a single year, over 60,000 Americans will be diagnosed with malignant melanoma, the most deadly form of skin cancer. And if the protective ozone shield continues to diminish, there is an ever-increasing risk of problems associated with UV-B. Using a good sunscreen and proper clothing can certainly help. The best way to protect yourself from too much sun, however, is to limit your time in direct sunlight, especially between the hours of 11 A.M. and 3 P.M., when the sun is highest in the sky and its rays are most direct.

Presently, the National Weather Service makes a daily prediction of UV radiation levels for selected cities throughout the United States. The forecast, known as the *UV Index,* gives the UV level at its peak, around noon standard time or 1 P.M. daylight savings time. The 15-point index corresponds to five exposure categories set by the Environmental Protection Agency (EPA). An index value of between 0 and 2 is considered "minimal," whereas a value of 10 or greater is deemed "very high" (see Fig. 3). Depending on skin type, a UV index of 10 means that in direct sunlight (without sunscreen protection) a person's skin will likely begin to burn in about 6 to 30 minutes.

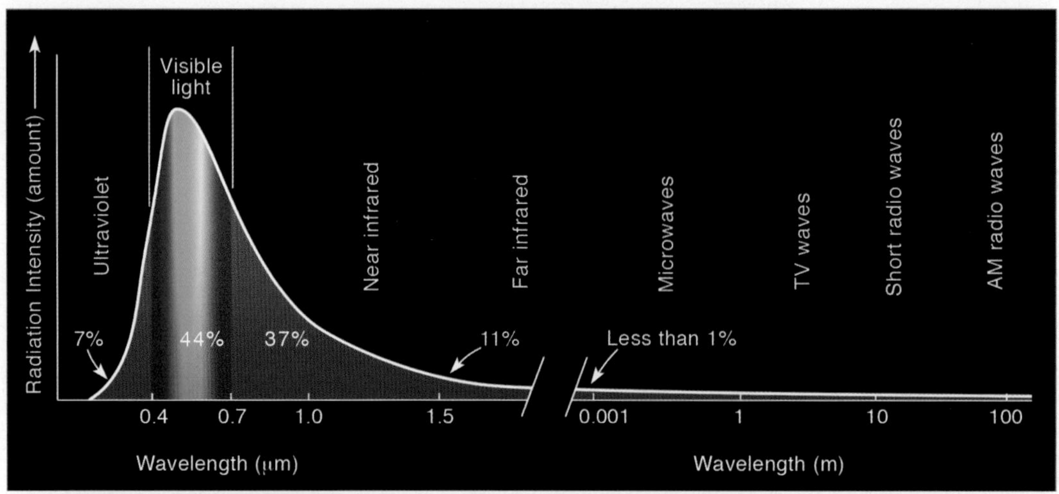

FIGURE 2.8 The sun's electromagnetic spectrum and some of the descriptive names of each region. The numbers underneath the curve approximate the percent of energy the sun radiates in various regions.

Objects at a high temperature (above about 500°C) radiate waves with many lengths, but some of them are short enough to stimulate the sensation of color. We actually see these objects glow red. Objects cooler than this radiate at wavelengths that are too long for us to see. The page of this book, for example, is radiating electromagnetic waves. But because its temperature is only about 68°F (20°C), the waves emitted are much too long to stimulate vision. We are able to see the page, however, because light waves from other sources (such as lightbulbs or the sun) are being *reflected* (bounced) off the paper. If this book were carried into a completely dark room, it would continue to radiate, but the pages would appear black because there are no visible light waves in the room to reflect off the page.

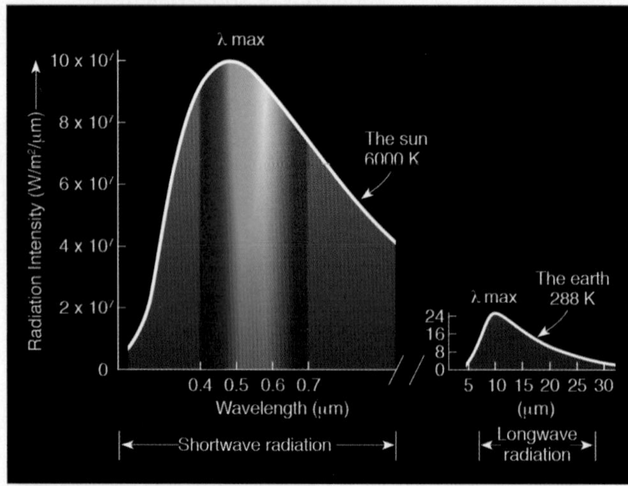

FIGURE 2.9 The hotter sun not only radiates more energy than that of the cooler earth (the area under the curve), but it also radiates the majority of its energy at much shorter wavelengths. (The area under the curves is equal to the total energy emitted, and the scales for the two curves differ by a factor of 100,000.)

The sun emits radiation at almost all wavelengths, but because its surface is hot—6000 K (10,500°F)—it radiates the majority of its energy at relatively short wavelengths. If we look at the amount of radiation given off by the sun at each wavelength, we obtain the sun's *electromagnetic spectrum.* A portion of this spectrum is shown in Fig. 2.8.

Notice that the sun emits a maximum amount of radiation at wavelengths near 0.5 μm. Since our eyes are sensitive to radiation between 0.4 and 0.7 μm, these waves reach the eye and stimulate the sensation of color. This portion of the spectrum is therefore referred to as the **visible region**, and the radiant energy that reaches our eye is called *visible light.* The color violet is the shortest wavelength of visible light. Wavelengths shorter than violet (0.4 μm) are **ultraviolet (UV).** The longest wavelengths of visible light correspond to the color red. Wavelengths longer than red (0.7 μm) are called **infrared (IR).**

Whereas the hot sun emits only a part of its energy in the infrared portion of the spectrum, the relatively cool earth emits almost all of its energy at infrared wavelengths. In fact, the earth, with an average surface temperature near 288 K (59°F, or 15°C) radiates nearly all its energy between 5 and 20 μm, with a peak intensity in the infrared region near 10 μm (see Fig. 2.9). Since the sun radiates the majority of its energy at much shorter wavelengths than does the earth, solar radiation is often called **shortwave radiation**, whereas the earth's radiation is referred to as **longwave** (or *terrestrial*) **radiation**.

Energy Balancing Act—Absorption, Emission, and Equilibrium

If the earth and all things on it are continually radiating energy, why doesn't everything get progressively colder? The answer is that all objects not only radiate energy, they absorb it as well. If an object radiates more energy than it absorbs, it becomes colder; if it absorbs more energy than it emits, it becomes warmer. On a sunny day, the earth's surface warms by absorbing more energy from the sun and the atmosphere than it radiates, whereas at night the earth cools by radiating more energy than it absorbs from its surroundings. When an object emits and absorbs energy at equal rates, its temperature remains constant.

The rate at which something radiates and absorbs energy depends strongly on its surface characteristics, such as color, texture, and moisture, as well as temperature. For example, a black object in direct sunlight is a good absorber of solar radiation. It converts energy from the sun into internal energy, and its temperature ordinarily increases. You need only walk barefoot on a black asphalt road on a summer afternoon to experience this. At night, the blacktop road will cool quickly by emitting infrared energy and, by early morning, it may be cooler than surrounding surfaces.

Any object that is a perfect absorber (that is, absorbs all the radiation that strikes it) and a perfect emitter (emits the maximum radiation possible at its given temperature) is called a **blackbody.** Blackbodies do not have to be colored black, they simply must absorb and emit all possible radiation. Since the earth's surface and the sun absorb and radiate with nearly 100 percent efficiency for their respective temperatures, they both behave as blackbodies.

When we look at the earth from space, we see that half of it is in sunlight, the other half is in darkness. The outpouring of solar energy constantly bathes the earth with radiation, while the earth, in turn, constantly emits infrared radiation. If we assume that there is no other method of transferring heat, then, when the rate of absorption of solar radiation equals the rate of emission of infrared earth radiation, a state *of radiative equilibrium* is achieved. The average temperature at which this occurs is called the **radiative equilibrium temperature**. At this temperature, the earth (behaving as a blackbody) is absorbing solar radiation and emitting infrared radiation at equal rates, and its average temperature does not change. As the earth is about 150 million km (93 million mi) from the sun, the earth's radiative equilibrium temperature is about 255 K (0°F, –18°C). But this temperature is *much* lower than the earth's observed average surface temperature of 288 K (59°F, 15°C). Why is there such a large difference?

The answer lies in the fact that *the earth's atmosphere absorbs and emits infrared radiation.* Unlike the earth, the atmosphere does *not* behave like a blackbody, as it absorbs some wavelengths of radiation and is transparent to others. Objects that selectively absorb and emit radiation, such as gases in our atmosphere, are known as **selective absorbers.**

SELECTIVE ABSORBERS AND THE ATMOSPHERIC GREENHOUSE EFFECT There are many selective absorbers in our environment. Snow, for example, is a good absorber of infrared radiation but a poor absorber of sunlight. Objects that selectively absorb radiation usually selectively emit radiation at the same wavelength. Snow is therefore a good emitter of infrared energy. At night, a snow surface usually emits much more infrared energy than it absorbs from its surroundings. This large loss of infrared radiation (coupled with the insulating qualities of snow) causes the air above a snow surface on a clear, winter night to become extremely cold.

› Figure 2.10 shows some of the most important selectively absorbing gases in our atmosphere (the shaded area represents the absorption characteristics of each gas at various wavelengths). Notice that both water vapor (H_2O) and carbon dioxide (CO_2) are strong absorbers of infrared radiation and poor absorbers of visible solar radiation. Other, less important, selective absorbers include nitrous oxide (N_2O), methane (CH_4), and ozone (O_3), which is most abundant in the stratosphere. As these gases absorb infrared radiation emitted from the earth's surface, they gain kinetic energy (energy of motion). The gas molecules share this energy by colliding with neighboring air molecules, such as oxygen and nitrogen (both of which are poor absorbers of infrared energy). These collisions increase the average kinetic energy of the air, which results in an increase in air temperature. Thus, most of the infrared energy emitted from the earth's surface keeps the lower atmosphere warm.

Besides being selective absorbers, water vapor and CO_2 selectively emit radiation at infrared wavelengths.* This radiation travels away from these gases in all directions. A portion of this energy is radiated toward the earth's surface and absorbed, thus heating the ground. The earth, in turn, radiates infrared energy upward, where it is absorbed and warms the lower atmosphere. In this way, water vapor and CO_2 absorb and radiate infrared energy and act as an insulating layer around the earth, keeping part of the earth's infrared radiation from escaping rapidly into space. Consequently, the earth's surface and the lower atmosphere are much warmer than they would be if these selectively absorbing gases were not present. In fact, as we saw earlier, the earth's mean radiative equilibrium temperature

*Nitrous oxide, methane, and ozone also emit infrared radiation, but their concentration in the atmosphere is much smaller than water vapor and carbon dioxide (see Table 1.3, p. 8).

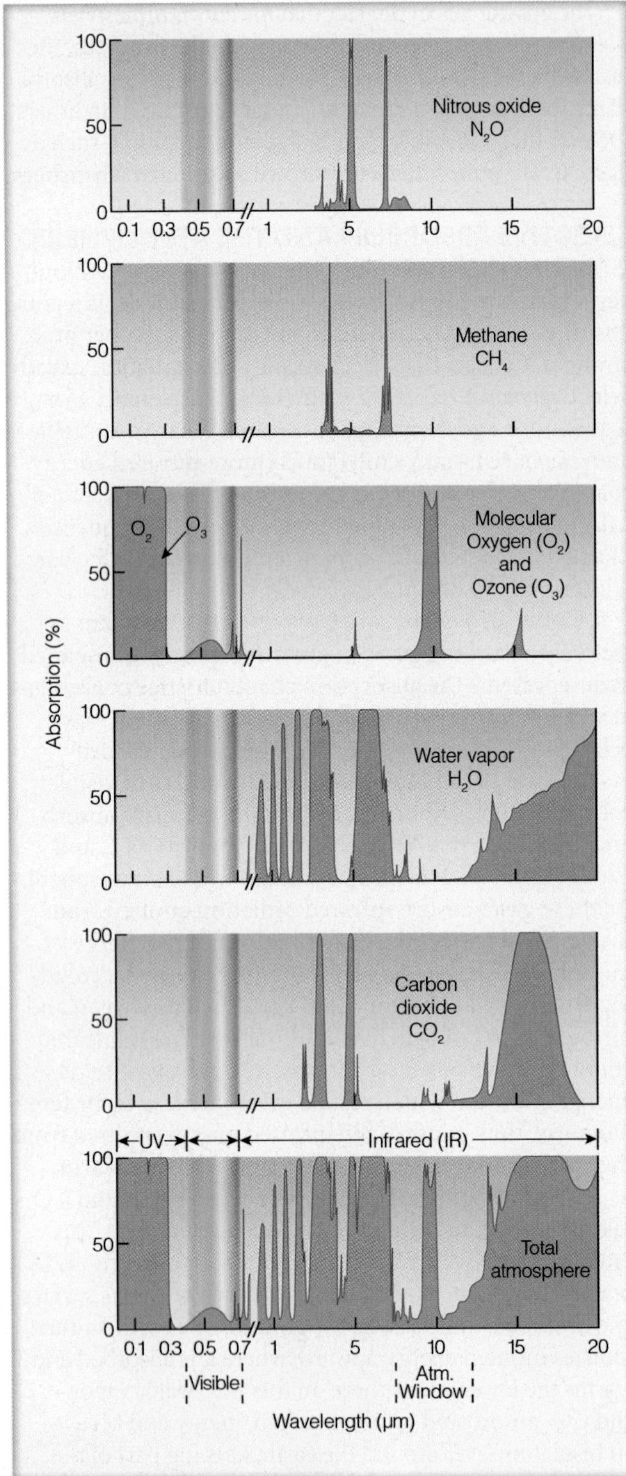

Active › FIGURE 2.10 Absorption of radiation by gases in the atmosphere. The shaded area represents the percent of radiation absorbed by each gas. The strongest absorbers of infrared radiation are water vapor and carbon dioxide. The bottom figure represents the percent of radiation absorbed by all of the atmospheric gases. Visit the Meteorology Resource Center to view this and other Active figures at www.cengage.com/login.

without CO_2 and water vapor would be around 0°F (–18°C) or about 59°F (33°C) lower than at present.

The absorption characteristics of water vapor, CO_2, and other gases (such as methane and nitrous oxide depicted in Fig. 2.10 were at one time thought to be similar to the glass of a florist's greenhouse. In a greenhouse, the glass allows visible radiation to come in, but inhibits to some degree the passage of outgoing infrared radiation. For this reason, the absorption of infrared radiation from the earth by water vapor and CO_2 is popularly called the **greenhouse effect**. However, studies have shown that the warm air inside a greenhouse is probably caused more by the air's inability to circulate and mix with the cooler outside air, rather than by the entrapment of infrared energy. Because of these findings, some scientists suggest that the greenhouse effect should be called the *atmosphere effect*. To accommodate everyone, we will usually use the term *atmospheric greenhouse effect* when describing the role that water vapor, CO_2, and other **greenhouse gases*** play in keeping the earth's mean surface temperature higher than it otherwise would be.

Look again at Fig. 2.10 and observe that, in the bottom diagram, there is a region between about 8 and 11 μm where neither water vapor nor CO_2 readily absorbs infrared radiation. Because these wavelengths of emitted energy pass upward through the atmosphere and out into space, the wavelength range (between 8 and 11 μm) is known as the **atmospheric window**. Clouds can enhance the atmospheric greenhouse effect. Tiny liquid cloud droplets are selective absorbers in that they are good absorbers of infrared radiation but poor absorbers of visible solar radiation. Clouds even absorb the wavelengths between 8 and 11 μm, which are otherwise "passed up" by water vapor and CO_2. Thus, they have the effect of enhancing the atmospheric greenhouse effect by closing the atmospheric window.

Clouds—especially low, thick ones—are excellent emitters of infrared radiation. Their tops radiate infrared energy upward and their bases radiate energy back to the earth's surface where it is absorbed and, in a sense, radiated back to the clouds. This process keeps calm, cloudy nights warmer than calm, clear ones. If the clouds remain into the next day, they prevent much of the sunlight from reaching the ground by reflecting it back to space. Since the ground does not heat up as much as it would in full sunshine, cloudy, calm days are normally cooler than clear, calm days. Hence, the presence of clouds tends to keep nighttime temperatures higher and daytime temperatures lower.

In summary, the atmospheric greenhouse effect occurs because water vapor, CO_2, and other greenhouse

*The term "greenhouse gases" derives from the standard use of "greenhouse effect." Greenhouse gases include, among others, water vapor, carbon dioxide, methane, nitrous oxide, and ozone.

EXTREME WEATHER WATCH

Earth is not the only planet with a greenhouse effect as the thick, dense carbon dioxide atmosphere of Venus produces an exceptionally strong greenhouse effect, and a surface temperature near 950°F, or 500°C.

gases are selective absorbers. They allow most of the sun's radiation to reach the surface, but they absorb a good portion of the earth's outgoing infrared radiation, preventing it from escaping into space. It is the atmospheric greenhouse then, that keeps the temperature of our planet at a level where life can survive. The greenhouse effect is not just a "good thing"—it is essential to life on earth for, without it, air at the surface would be extremely cold (see Fig. 2.11).

ENHANCEMENT OF THE GREENHOUSE EFFECT

In spite of the inaccuracies that have plagued temperature measurements, studies suggest that, during the past century, the earth's surface air temperature has been undergoing a warming of about 0.6°C (about 1°F). In recent years, this *global warming* trend has not only continued, but has increased. In fact, scientific computer models, that mathematically simulate the physical processes of the atmosphere, oceans, and ice, predict that if such a warming should continue unabated, we would be irrevocably committed to the negative effects of climate change, such as a rise in sea level and a shift in global precipitation patterns.

The main cause of this *climate change* appears to be the greenhouse gas CO_2 whose concentration has been increasing primarily due to the burning of fossil fuels and to deforestation. However, increasing concentration of other greenhouse gases, such as methane (CH_4), nitrous oxide (N_2O), and chlorofluorocarbons (CFCs), has collectively been shown to have an effect almost equal to that of CO_2. Look back at Fig. 2.10 and notice that both CH_4 and N_2O absorb strongly at infrared wavelengths. Moreover, a particular CFC (CFC-12) absorbs in the region of the atmospheric window between 8 and 11 μm. Thus, in terms of its absorption impact on infrared radiation, the addition of a single CFC-12 molecule to the atmosphere is the equivalent of adding 10,000 molecules of CO_2. Overall, water vapor accounts for about 60 percent of the atmospheric greenhouse effect, CO_2 accounts for about 26 percent, and the remaining greenhouse gases contribute about 14 percent.

Presently, the concentration of CO_2 in a volume of air near the surface is about 0.039 percent. Climate models predict that a continuing increase of CO_2 to an

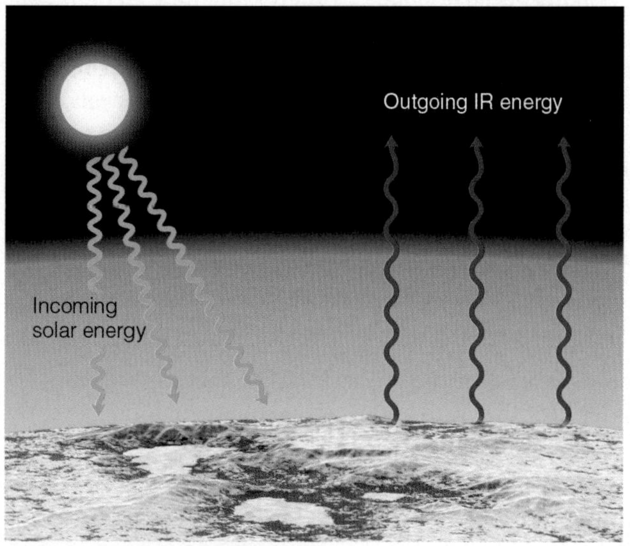

(a) Without greenhouse gases

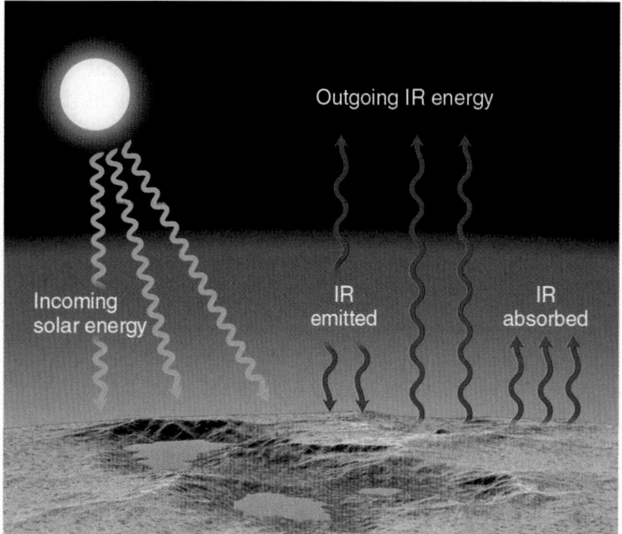

(b) With greenhouse gases

FIGURE 2.11 (a) Near the surface in an atmosphere with little or no greenhouse gases, the earth's surface would constantly emit infrared (IR) radiation upward, both during the day and at night. Incoming energy from the sun would equal outgoing energy from the surface, but the surface would receive virtually no IR radiation from its lower atmosphere. (No atmospheric greenhouse effect.) The earth's surface air temperature would be quite low, and small amounts of water found on the planet would be in the form of ice. (b) In an atmosphere with greenhouse gases, the earth's surface not only receives energy from the sun but also infrared energy from the atmosphere. Incoming energy still equals outgoing energy, but the added IR energy from the greenhouse gases raises the earth's average surface temperature to a more habitable level.

amount more than double its pre-industrial value of 0.028 percent, along with the continued increase of other greenhouse gases, will cause the earth's current average surface air temperature to possibly rise by an additional 5°F (about 3°C) by the end of this century. How can increasing such a small quantity of CO_2 and adding miniscule amounts of other greenhouse gases bring about such a large temperature increase?

Mathematical climate models predict that rising ocean temperatures will cause an increase in evaporation rates. The added *water vapor*—the primary greenhouse gas—will enhance the atmospheric greenhouse effect and double the temperature rise in what is known as a *positive feedback.* But there are other feedbacks to consider.*

The two potentially largest and least understood feedbacks in the climate system are the clouds and the oceans. Clouds can change area, depth, and radiation properties simultaneously with climatic changes. The net effect of all these changes is not totally clear at this time. Oceans, on the other hand, cover 70 percent of the planet. The response of ocean circulations, ocean temperatures, and sea ice to global warming will determine the global pattern and speed of climate change. Unfortunately, it is not now known how quickly each of these will respond.

Satellite data from the *Earth Radiation Budget Experiment* (ERBE) suggest that clouds overall appear to *cool* the earth's climate, as they reflect and radiate away more energy than they retain. (The earth would be warmer if clouds were not present.) So an increase in global cloudiness (if it were to occur) might offset some of the global warming brought on by an enhanced atmospheric greenhouse effect. Therefore, if clouds were to act on the climate system in this manner, they would provide a *negative feedback* on climate change.**

Uncertainties unquestionably exist about the impact that increasing levels of CO_2 and other greenhouse gases will have on enhancing the atmospheric greenhouse effect. Nonetheless, the most recent studies on climate change say that climate change is presently occurring worldwide due primarily to increasing levels of greenhouse gases. The evidence for this conclusion comes from increases in global average air and ocean temperatures, as well as from the widespread melting of snow and ice, and rising sea levels.

*A feedback is a process whereby an initial change in a process will tend to either reinforce the process (positive feedback) or weaken the process (negative feedback). The *water vapor-greenhouse* feedback is a positive feedback because the initial increase in temperature is reinforced by the addition of more water vapor, which absorbs more of the earth's infrared energy, thus strengthening the greenhouse effect and enhancing the warming.

**Overall, the most recent climate models tend to show that changes in clouds would provide a small positive feedback on climate change.

BRIEF REVIEW

In the last several sections, we have explored examples of some of the ways radiation is absorbed and emitted by various objects. Before reading the next several sections, let's review a few important facts and principles:

- ***All* objects with a temperature above absolute zero emit radiation.**
- **The higher an object's temperature, the greater the amount of radiation emitted per unit surface area and the shorter the wavelength of maximum emission.**
- **The earth absorbs solar radiation only during the daylight hours; however, it emits infrared radiation continuously, both during the day and at night.**
- **The earth's surface behaves as a blackbody, making it a much better absorber and emitter of radiation than the atmosphere.**
- **Water vapor and carbon dioxide are important atmospheric greenhouse gases that selectively absorb and emit infrared radiation, thereby keeping the earth's average surface temperature warmer than it otherwise would be.**
- **Cloudy, calm nights are often warmer than clear, calm nights because clouds strongly emit infrared radiation back to the earth's surface.**
- **It is *not* the greenhouse effect itself that is of concern, but the *enhancement* of it due to increasing levels of greenhouse gases.**
- **As greenhouse gases continue to increase in concentration, the average surface air temperature is projected to rise substantially by the end of the twenty-first century.**

With these concepts in mind, we will first examine how the air near the ground warms; then we will consider how the earth and its atmosphere maintain a yearly energy balance.

WARMING THE AIR FROM BELOW If you look back at Fig. 2.10 on p. 46, you'll notice that the atmosphere does not readily absorb radiation with wavelengths between 0.3 and 1.0 μm, the region where the sun emits most of its energy. Consequently, on a clear day, solar energy passes through the lower atmosphere with little effect upon the air. Ultimately it reaches the surface, warming it (see Fig. 2.12). Air molecules in contact with the heated surface bounce against it, gain energy by *conduction,* then shoot upward like freshly popped kernels of corn, carrying their energy with them. Because the air near the ground is very dense, these molecules only travel a short distance before they collide with other molecules. During the collision, these more rapidly moving molecules share their energy with less energetic molecules, raising the average temperature of the air. But air is such a poor heat conductor that this process is only important within a few centimeters of the ground.

As the surface air warms, it actually becomes less dense than the air directly above it. The warmer air rises and the cooler air sinks, setting up thermals, or *free convection cells,* that transfer heat upward and distribute it through a deeper layer of air. The rising air expands and cools, and, if sufficiently moist, the water vapor condenses into cloud droplets, releasing latent heat that warms the air. Meanwhile, the earth constantly emits infrared energy. Some of this energy is absorbed by greenhouse gases (such as water vapor and carbon dioxide) that emit infrared energy upward and downward, back to the surface. Since the concentration of water vapor decreases rapidly above the earth, most of the absorption occurs in a layer near the surface. Hence, the lower atmosphere is mainly heated from the ground upward.

SHORTWAVE RADIATION STREAMING FROM THE SUN As the sun's radiant energy travels through space, essentially nothing interferes with it until it reaches the atmosphere. At the top of the atmosphere, solar energy received on a surface perpendicular to the sun's rays appears to remain fairly constant at nearly two calories on each square centimeter each minute, or 1367 W/m^2—a value called the *solar constant.**

*By definition, the solar constant (which, in actuality, is *not* "constant") is the rate at which radiant energy from the sun is received on a surface at the outer edge of the atmosphere perpendicular to the sun's rays when the earth is at an average distance from the sun. Satellite measurements from the *Earth Radiation Budget Satellite* suggest the solar constant varies slightly as the sun's radiant output varies. The average is about 1.96 cal/cm^2/min, or between 1365 W/m^2 and 1372 W/m^2 in the SI system of measurement.

When solar radiation enters the atmosphere, a number of interactions take place. For example, some of the energy is absorbed by gases, such as ozone, in the upper atmosphere. Moreover, when sunlight strikes very small objects, such as air molecules and dust particles, the light itself is deflected in all directions—forward, sideways, and backwards. The distribution of light in this manner is called **scattering**. (Scattered light is also called *diffuse light.*) Because air molecules are much smaller than the wavelengths of visible light, they are more effective scatterers of the shorter (blue) wavelengths than the longer (red) wavelengths. Hence, when we look away from the direct beam of sunlight, blue light strikes our eyes from all directions, turning the daytime sky blue.

Sunlight can be **reflected** from objects. Generally, reflection differs from scattering in that during the process of reflection more light is sent *backwards.* **Albedo** is the percent of radiation returning from a given surface compared to the amount of radiation initially striking that surface. Albedo, then, represents the *reflectivity* of the surface. In ❯Table 2.2 notice that thick clouds have a higher albedo than thin clouds. On the average, the albedo of clouds is near 60 percent. When solar energy strikes a surface covered with snow, up to 95 percent of the sunlight may be reflected. Most of this energy is in the visible and ultraviolet wavelengths. Consequently, reflected radiation, coupled with direct sunlight, can produce severe sunburns on the exposed skin of unwary snow skiers, and unprotected eyes can suffer the agony of snow blindness.

Water surfaces, on the other hand, reflect only a small amount of solar energy. For an entire day, a smooth water surface will have an average albedo of about 10 percent. Averaged for an entire year, the earth and its atmosphere (including its clouds), will redirect about 30 percent of the sun's incoming radiation back to space, which gives the earth and its atmosphere a combined albedo of 30 percent (see ❯Fig. 2.13).

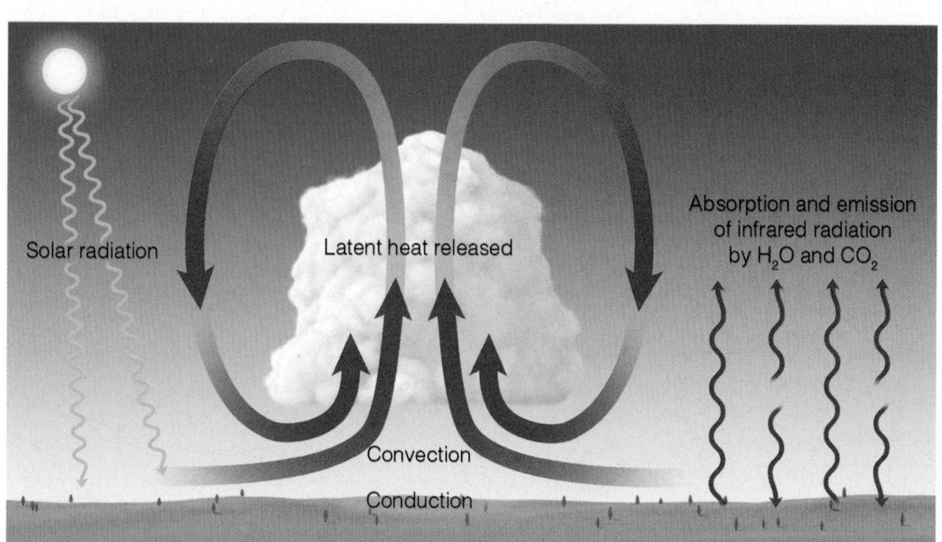

Active ❯ FIGURE 2.12

Air in the lower atmosphere is heated from the ground upward. Sunlight warms the ground, and the air above is warmed by conduction, convection, and infrared radiation. Further warming occurs during condensation as latent heat is given up to the air inside the cloud. Visit the Meteorology Resource Center to view this and other active figures at www.cengage.com/login.

Table 2.2

Typical Albedo of Various Surfaces

SURFACE	ALBEDO (PERCENT)
Fresh snow	75 to 95
Clouds (thick)	60 to 90
Clouds (thin)	30 to 50
Venus	78
Ice	30 to 40
Sand	15 to 45
Earth and atmosphere	30
Mars	17
Grassy field	10 to 30
Dry, plowed field	5 to 20
Water	10*
Forest	3 to 10
Moon	7

* Daily average.

THE EARTH'S ANNUAL ENERGY BALANCE

Although the average temperature at any one place may vary considerably from year to year, the earth's overall average equilibrium temperature changes only slightly from one year to the next. This fact indicates that, each year, the earth and its atmosphere combined must send off into space just as much energy as they receive from the sun. The same type of energy balance must exist between the earth's surface and the atmosphere. That is, each year, the earth's surface must return to the atmosphere the same amount of energy that it absorbs. If this did not occur, the earth's average surface temperature would change. How do the earth and its atmosphere maintain this yearly energy balance?

Suppose 100 units of solar energy reach the top of the earth's atmosphere. We already know from Fig. 2.13 that, on the average, clouds, the earth, and the atmosphere reflect and scatter 30 units back to space, and that the atmosphere and clouds together absorb 19 units, which leaves 51 units of direct and indirect (diffuse) solar radiation to be absorbed at the earth's surface.

Figure 2.14 shows approximately what happens to the solar radiation that is absorbed by the surface and the atmosphere. Out of 51 units reaching the surface, a large amount (23 units) is used to evaporate water, and about 7 units are lost through conduction and convection, which leaves 21 units to be radiated away as infrared energy. Look closely at Fig. 2.14 and notice that the earth's surface actually radiates upward a whopping 117 units. It does so because, although it receives solar radiation only during the day, it constantly emits infrared energy both

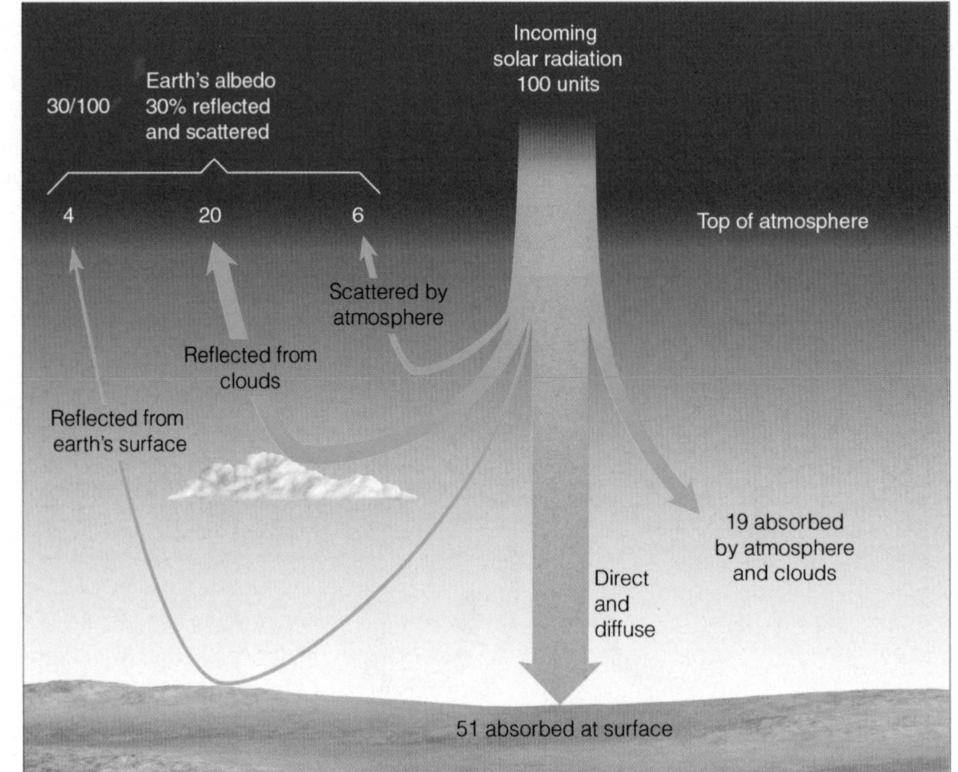

FIGURE 2.13

On the average, of all the solar energy that reaches the earth's atmosphere annually, about 30 percent (30/100) is reflected and scattered back to space, giving the earth and its atmosphere an albedo of 30 percent. Of the remaining solar energy, about 19 percent is absorbed by the atmosphere and clouds, and 51 percent is absorbed at the surface.

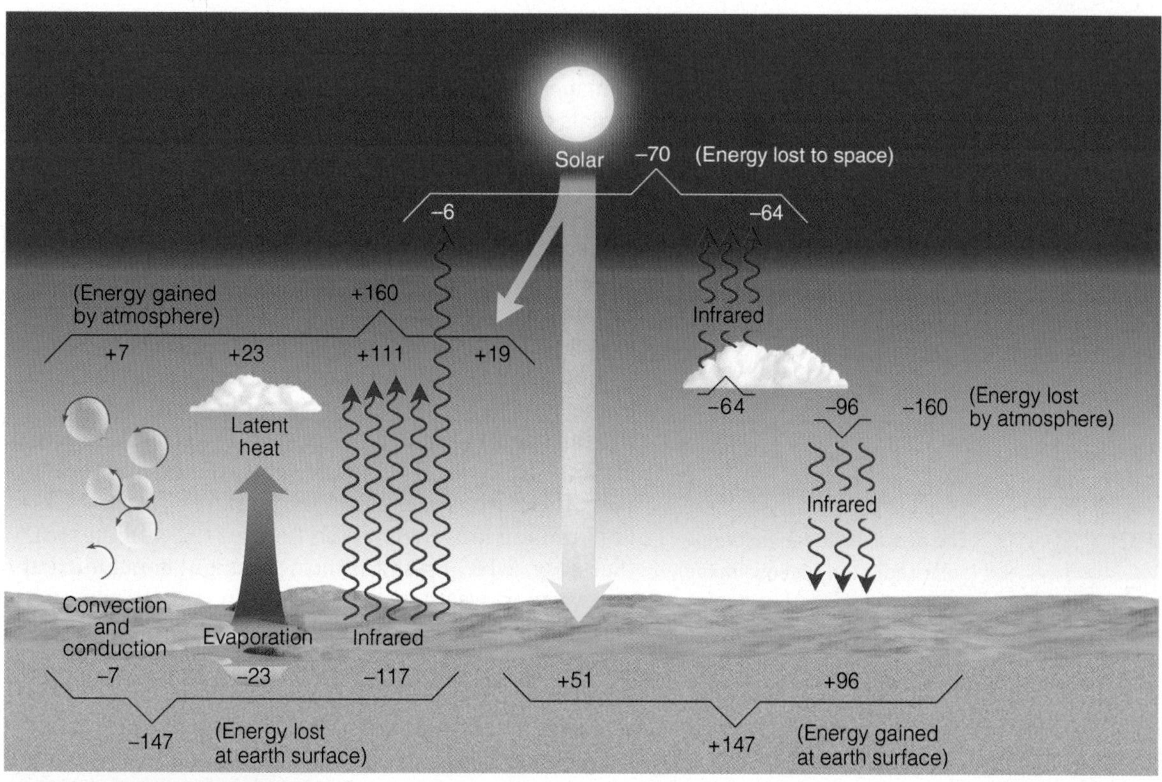

FIGURE 2.14 The earth-atmosphere energy balance. Numbers represent approximations based on surface observations and satellite data. While the actual value of each process may vary by several percent, it is the relative size of the numbers that is important.

during the day and at night. Additionally, the atmosphere above only allows a small fraction of this energy (6 units) to pass through into space. The majority of it (111 units) is absorbed mainly by the greenhouse gases water vapor and CO_2, and by clouds. Much of this energy (96 units) is then radiated back to earth, producing the atmospheric greenhouse effect. Hence, the earth's surface receives nearly twice as much longwave infrared energy from the atmosphere as it does shortwave radiation from the sun. In all these exchanges, notice that the energy lost at the earth's surface (147 units) is exactly balanced by the energy gained there (147 units).

A similar balance exists between the earth's surface and its atmosphere. Again in Fig. 2.14 observe that the energy gained by the atmosphere (160 units) balances the energy lost. Moreover, averaged for an entire year, the solar energy received at the earth's surface (51 units) and that absorbed by the earth's atmosphere (19 units) balances the infrared energy lost to space by the earth's surface (6 units) and its atmosphere (64 units).

We can see the effect that conduction, convection, and latent heat play in the warming of the atmosphere if we look at the energy balance only in radiative terms. The earth's surface receives 147 units of radiant energy from the sun and its own atmosphere, while it radiates away 117 units, producing a *surplus* of 30 units. The atmosphere, on the other hand, receives 130 units (19 units from the sun and 111 from the earth), while it loses 160 units, producing a *deficit* of 30 units. The balance (30 units) is the warming of the atmosphere produced by the heat transfer processes of conduction and convection (7 units) and by the release of latent heat (23 units).

And so, the earth and the atmosphere absorb energy from the sun, as well as from each other. In all of the energy exchanges, a delicate balance is maintained. Essentially, there is no yearly gain or loss of total energy, and the average temperature of the earth and the atmosphere remains fairly constant from one year to the next. This equilibrium does not imply that the earth's average temperature does not change, but that the changes are small from year to year (usually less than one-tenth of a degree Celsius), and become significant only when measured over many years.

Even though the earth and the atmosphere together maintain an annual energy balance, such a balance is not maintained at each latitude. High latitudes tend to lose more energy to space each year than they receive from the sun, while low latitudes tend to gain more energy during the course of a year than they lose. From

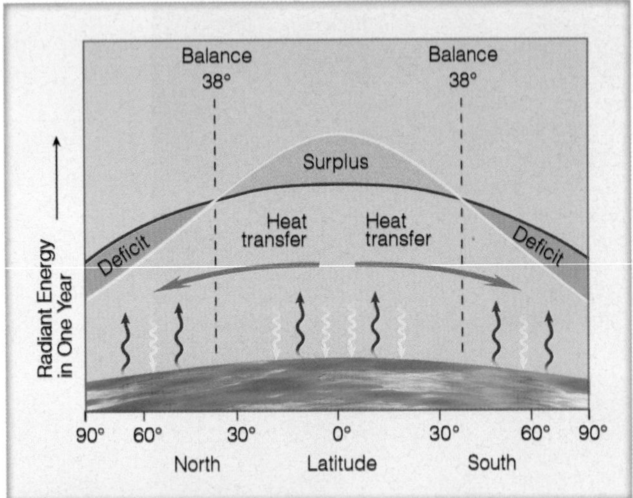

FIGURE 2.15 The average annual incoming solar radiation (yellow lines) absorbed by the earth and the atmosphere along with the average annual infrared radiation (red lines) emitted by the earth and the atmosphere.

Fig. 2.15 we can see that only at middle latitudes near 38° does the amount of energy received each year balance the amount lost. From this situation, we might conclude that polar regions are growing colder each year, while tropical regions are becoming warmer. But this does not happen. To compensate for these gains and losses of energy, winds in the atmosphere and currents in the oceans circulate warm air and water toward the poles, and cold air and water toward the equator. Thus, the transfer of heat energy by atmospheric and oceanic circulations prevents low latitudes from steadily becoming warmer and high latitudes from steadily growing colder. These circulations are extremely important to weather and climate, and will be treated more completely when we examine atmospheric circulations.

We now turn our attention to how incoming solar energy produces the earth's seasons. Before doing so, you may wish to read the Focus section on p. 54 that explains how space weather can disturb airplane routes, damage satellites, and cause a widespread loss of GPS signals on earth.

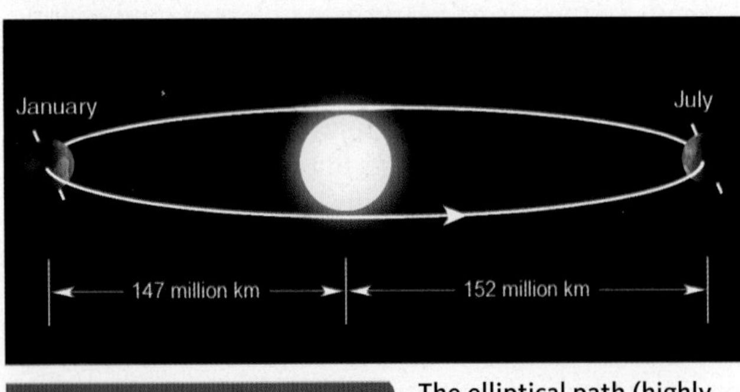

FIGURE 2.16 The elliptical path (highly exaggerated) of the earth about the sun brings the earth slightly closer to the sun in January than in July.

Why the Earth Has Seasons

The earth revolves completely around the sun in an elliptical path (not quite a circle) in slightly longer than 365 days (one year). As the earth revolves around the sun, it spins on its own axis, completing one spin in 24 hours (one day). The average distance from the earth to the sun is 150 million km (93 million mi). Because the earth's orbit is an ellipse instead of a circle, the actual distance from the earth to the sun varies during the year. The earth comes closer to the sun in January (147 million km) than it does in July (152 million km).* (See Fig. 2.16.) From this fact, we might conclude that our warmest weather should occur in January and our coldest weather in July. But, in the Northern Hemisphere, we normally experience cold weather in January when we are closer to the sun and warm weather in July when we are farther away. If nearness to the sun were the primary cause of the seasons then, indeed, January would be warmer than July. However, nearness to the sun is only a small part of the story.

Our seasons are regulated by the amount of solar energy received at the earth's surface. This amount is determined primarily by the angle at which sunlight strikes the surface, and by how long the sun shines on any latitude (daylight hours). Let's look more closely at these factors.

Solar energy that strikes the earth's surface perpendicularly (directly) is much more intense than solar energy that strikes the same surface at an angle. Think of shining a flashlight straight at a wall—you get a small circular spot of light (see Fig. 2.17). Now, tip the flashlight and notice how the spot of light spreads over a larger area. The same principle holds for sunlight. Sunlight striking the earth at an angle spreads out and must heat a larger region than sunlight impinging directly on the earth. Everything else being equal, an area experiencing more direct solar rays will receive more heat than the same size area being struck by sunlight at an angle. In addition, the more the sun's rays are slanted from the perpendicular, the more atmosphere they must penetrate. And the more atmosphere they penetrate, the more they can be scattered and absorbed (attenuated). As a consequence,

*The time around January 3rd, when the earth is closest to the sun, is called *perihelion* (from the Greek *peri*, meaning "near" and *helios*, meaning "sun"). The time when the earth is farthest from the sun (around July 4th) is called *aphelion* (from the Greek *ap*, meaning "away from").

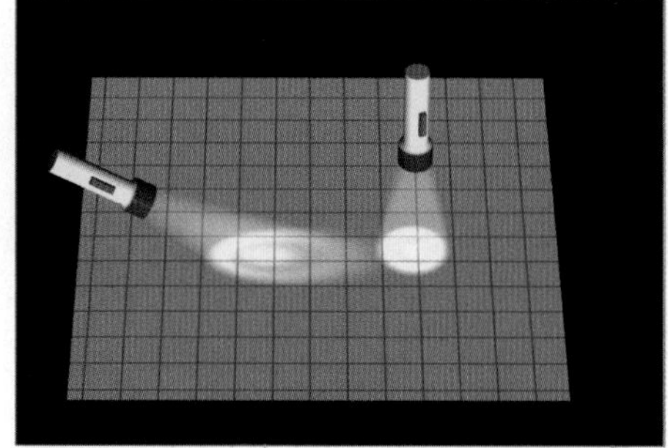

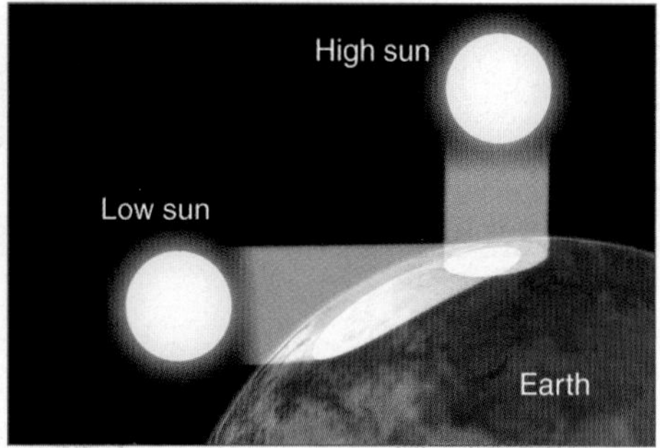

Active › FIGURE 2.17
Sunlight that strikes a surface at an angle is spread over a larger area than sunlight that strikes the surface directly. Oblique sun rays deliver less energy (are less intense) to a surface than direct sun rays. Visit the Meteorology Resource Center to view this and other active figures at www.cengage.com/login.

when the sun is high in the sky, it can heat the ground to a much higher temperature than when it is low on the horizon.

The second important factor determining how warm the earth's surface becomes is the length of time the sun shines each day. Longer daylight hours, of course, mean that more energy is available from sunlight. In a given location, more solar energy reaches the earth's surface on a clear, long day than on a day that is clear but much shorter. Hence, more surface heating takes place.

From a casual observation, we know that summer days have more daylight hours than winter days. Also, the noontime summer sun is higher in the sky than is the noontime winter sun. Both of these events occur because our spinning planet is inclined on its axis (tilted) as it revolves around the sun. As › Fig. 2.18 illustrates, the angle of tilt is 23½° from the perpendicular drawn to the plane of the earth's orbit. The earth's axis points to the same direction in space all year long; thus, the Northern Hemisphere is tilted toward the sun in summer (June), and away from the sun in winter (December).

(*text continues on p. 56*)

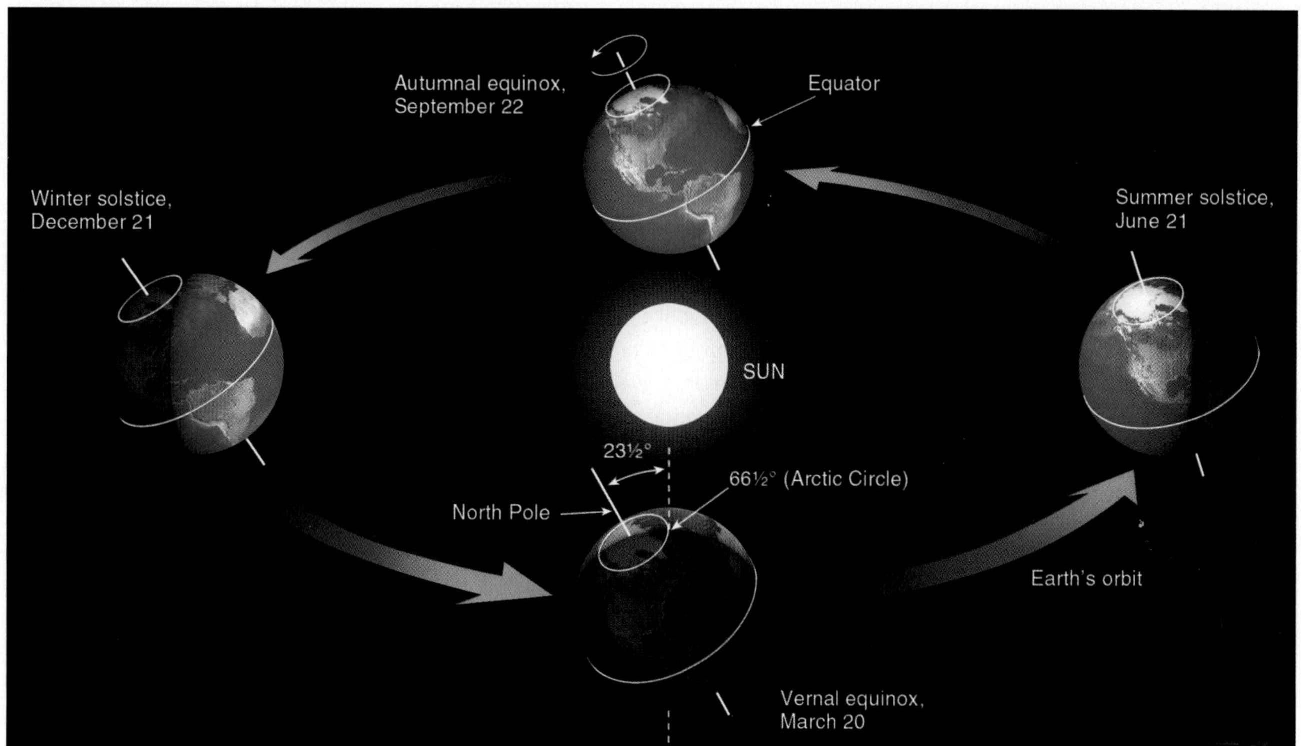

Active › FIGURE 2.18 As the earth revolves about the sun, it is tilted on its axis by an angle of 23½°. The earth's axis always points to the same area in space (as viewed from a distant star). Thus, in June, when the Northern Hemisphere is tipped toward the sun, more direct sunlight and long hours of daylight cause warmer weather than in December, when the Northern Hemisphere is tipped away from the sun. Visit the meteorology Resource Center to view this and other active figures at www.cengage.com/login.

FOCUS ON

EXTREME WEATHER

Space Weather

We observe our weather on earth in terms of temperature, pressure, humidity, moisture, and so on. In space, these elements don't exist. But there is a set of ever-changing conditions that do exist, and have come to be known as *space weather.* These conditions include electromagnetic radiation, powerful magnetic fields, and solar particles that zip through space. Since space weather is driven by activity on the sun, we will examine the sun's interior first.

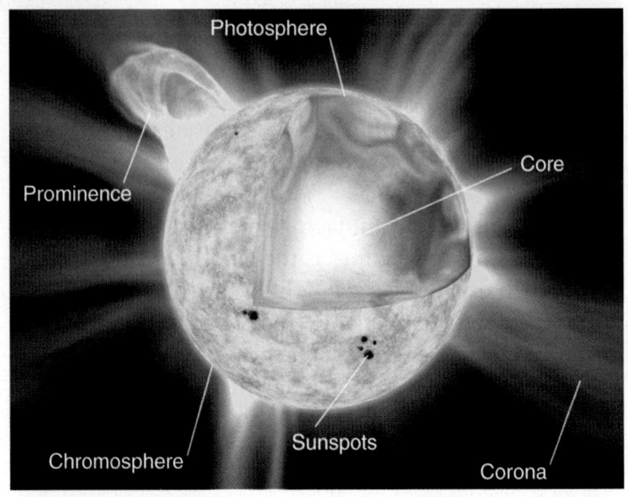

Figure 4 Various regions of the sun.

The sun is a giant celestial furnace. Its core is extremely hot, with a temperature estimated to be near 15 million degrees Celsius. In the core, hydrogen nuclei (protons) collide at such fantastically high speeds that they fuse together to form helium nuclei. This thermonuclear process generates an enormous amount of energy, which gradually works its way to the sun's outer luminous surface—*the photosphere,* "sphere of light" (see Fig. 4). Temperatures here are much cooler than in the interior, generally near 6000°C.

Dark blemishes on the photosphere called *sunspots* are huge, cooler regions that typically average more than five times the diameter of the earth. Sunspots are known to be regions of strong magnetic fields. They are cyclic, with the maximum number of spots occurring approximately every eleven years.

Above the photosphere are the *chromosphere* and the *corona.* The chromosphere ("color sphere") acts as a boundary between the relatively cool photosphere and the much hotter corona, the outermost envelope of the solar atmosphere.

Violent solar activity occasionally occurs in the regions of sunspots. The most dramatic of these events are *prominences* and *flares.* Prominences are huge cloudlike jets of gas that often shoot up into the corona in the form of an arch. Solar flares are tremendous, but brief, eruptions. They emit large quantities of high-energy ultraviolet radiation, powerful magnetic fields, as well as energized charged particles, mainly protons and electrons, which stream outward away from the sun at extremely high speeds.

Even in the absence of violent solar activity there is a constant discharge of charged particles from the sun's surface and tenuous atmosphere. As these charged particles travel through space, they are called *plasma* or *solar wind.* When the solar wind moves close enough to the earth, it interacts with the earth's magnetic field, severely deforming it into a tear-shaped cavity known as the *magnetosphere* (see Fig. 5). The interaction of the solar wind with gases inside the magnetosphere causes high-energy particles to excite gases in the upper atmosphere. These gases then emit visible light—the *aurora*—a beautiful display of color known as the northern or southern lights (see Fig. 6).

Normally, the solar wind travels through space at an average speed of about 250 mi/sec. However, during periods of high solar activity (many sunspots and flares), the solar wind is more dense, travels much faster, and carries more energy.

A number of adverse effects can occur during these periods of active space weather. The high energy radiation can endanger astronauts as well as pilots and passengers in high-flying aircraft. Moreover, an intense solar flare can disturb the earth's magnetic field, producing a so-called *magnetic*

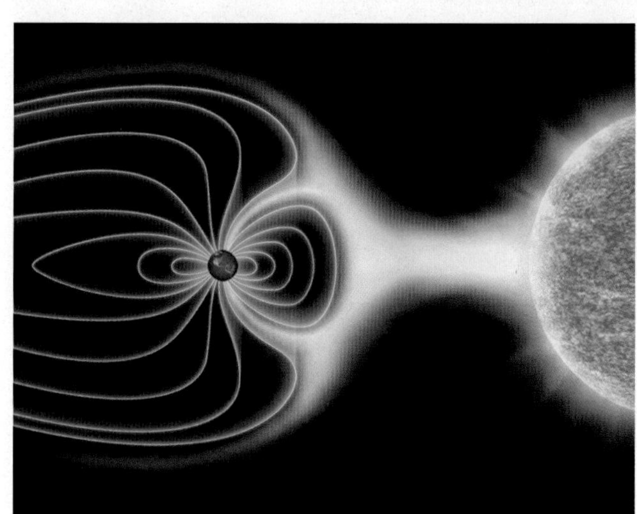

Figure 5 **The stream of charged particles from the sun—called the *solar wind*—distorts earth's magnetic field into a teardrop shape known as the *magnetosphere*.**

storm. Because these storms can intensify the electrical properties of the upper atmosphere, they are often responsible for interruptions in radio, television, and satellite communications. One such storm knocked out electricity throughout the province of Quebec, Canada, during March, 1989. And in May, 1998, after a period of intense solar activity, a communications satellite failed, causing 45 million pagers to suddenly go dead. Moreover, on Halloween, 2003, violent solar activity produced the largest flare ever recorded. The particles emitted by this event lashed the earth at nearly 6 million mi/hr and caused power outages in Sweden, disturbed airplane routes around the world, and damaged 28 satellites, ending the service life of two.

The effect that space weather has on our weather here on earth is not well understood at this time. However, as more information becomes available through ground-based and satellite measurements, hopefully we will have a clearer picture as to how small changes in energy emitted from the sun can influence weather and climate on earth.

Figure 6 **The northern lights (aurora borealis) is a phenomenon that forms as energetic particles from the sun interact with the earth's atmosphere.**

FIGURE 2.19

Land of the Midnight Sun. A series of exposures of the sun taken before, during, and after midnight in northern Alaska during July.

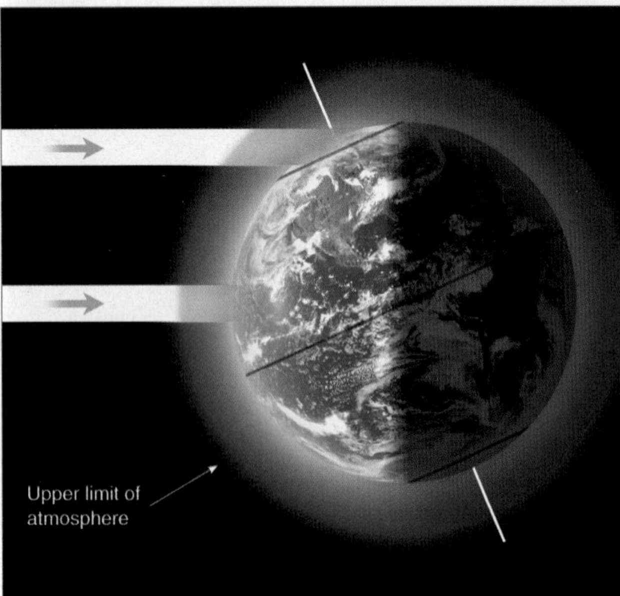

FIGURE 2.20 **During the Northern Hemisphere summer, sunlight that reaches the earth's surface in far northern latitudes has passed through a thicker layer of absorbing, scattering, and reflecting atmosphere than sunlight that reaches the earth's surface farther south. Sunlight is lost through both the thickness of the pure atmosphere and by impurities in the atmosphere. As the sun's rays become more oblique, these effects become more pronounced.**

SEASONS IN THE NORTHERN HEMISPHERE

Notice in Fig. 2.18 that on June 21, the northern half of the world is directed toward the sun. At noon on this day, solar rays beat down upon the Northern Hemisphere more directly than during any other time of year. The sun is at its highest position in the noonday sky, directly above 23½° north (N) latitude (Tropic of Cancer). If you were standing at this latitude on June 21, the sun at noon would be directly overhead. This day, called the **summer solstice**, is the astronomical first day of summer in the Northern Hemisphere.*

Study Fig. 2.18 closely and notice that, as the earth spins on its axis, the side facing the sun is in sunshine and the other side is in darkness. Thus, half of the globe is always illuminated. If the earth's axis were not tilted, the noonday sun would always be directly overhead at the equator, and there would be 12 hours of daylight and 12 hours of darkness at each latitude every day of the year. However, the earth is tilted. Since the Northern Hemisphere faces towards the sun on June 21, each latitude in the Northern Hemisphere will have more than 12 hours of daylight. The farther north we go, the longer are the daylight hours. When we reach the Arctic Circle (66½°N), daylight lasts for 24 hours, as the sun does not set. Notice in Fig. 2.18 how the region above 66½°N never gets into the "shadow" zone as the earth spins. At the North Pole, the sun actually rises above the horizon on March 20 and has six months until it sets on September 22. No wonder this region is called the "Land of the Midnight Sun"! (See Fig. 2.19.)

Even though in the far north the sun is above the horizon for many hours during the summer (see Table 2.3), the surface air there is not warmer than the air farther south, where days are appreciably shorter. The reason for this fact is shown in Fig. 2.20. When *in*coming *sol*ar r*ation* (called *insolation)* enters the atmosphere, fine dust, air molecules, and clouds

*As we will see later in this chapter, the seasons are reversed in the Southern Hemisphere. Hence, in the Southern Hemisphere, this same day is the winter solstice, or the astronomical first day of winter.

reflect and scatter it, and some of it is absorbed by atmospheric gases. Generally, the greater the thickness of atmosphere that sunlight must penetrate, the greater are the chances that it will be either reflected or absorbed by the atmosphere. During the summer in far northern latitudes, the sun is never very high above the horizon, so its radiant energy must pass through a thick portion of atmosphere before it reaches the earth's surface. Some of the solar energy that does reach the surface melts frozen soil or is reflected by snow or ice. And, that which is absorbed is spread over a large area. So, even though northern cities may experience long hours of sunlight they are not warmer than cities farther south. Overall, they receive less radiation at the surface, and what radiation they do receive does not effectively heat the surface.

Look at Fig. 2.18 again and notice that, by September 22, the earth will have moved so that the sun is directly above the equator. Except at the poles, the days and nights throughout the world are of equal length. This day is called the **autumnal** (fall) **equinox**, and it marks the astronomical beginning of fall in the Northern Hemisphere. At the North Pole, the sun appears on the horizon for 24 hours, due to the bending of light by the atmosphere. The following day (or at least within several days), the sun disappears from view, not to rise again for a long, cold six months. Throughout the northern half of the world on each successive day, there are fewer hours of daylight, and the noon sun is slightly lower in the sky. Less direct sunlight and shorter hours of daylight spell cooler weather for the Northern Hemisphere. Reduced sunlight, lower air temperatures, and cooling breezes stimulate the beautiful pageantry of fall colors (see Fig. 2.21).

On December 21 (three months after the autumnal equinox), the Northern Hemisphere is tilted as far away from the sun as it will be all year (see Fig. 2.18, p. 53). Nights are long and days are short. Notice in Table 2.3 that daylight decreases from 12 hours at the equator to 0 (zero) at latitudes above 66½°N. This is the shortest day of the year, called the **winter solstice**—the astronomical beginning of winter in the northern world. On this day, the sun shines directly above latitude 23½°S (Tropic of Capricorn). In the northern half of the world, the sun is at its lowest position in the noon sky. Its rays pass through a thick section of atmosphere and spread over a large area on the surface.

With so little incident sunlight, the earth's surface cools quickly. A blanket of

Table 2.3

Length of Time from Sunrise to Sunset for Various Latitudes on Different Dates

	NORTHERN HEMISPHERE			
Latitude	March 20	June 21	Sept. 22	Dec. 21
0°	12 hr	12.0 hr	12 hr	12.0 hr
10°	12 hr	12.6 hr	12 hr	11.4 hr
20°	12 hr	13.2 hr	12 hr	10.8 hr
30°	12 hr	13.9 hr	12 hr	10.1 hr
40°	12 hr	14.9 hr	12 hr	9.1 hr
50°	12 hr	16.3 hr	12 hr	7.7 hr
60°	12 hr	18.4 hr	12 hr	5.6 hr
70°	12 hr	2 months	12 hr	0 hr
80°	12 hr	4 months	12 hr	0 hr
90°	12 hr	6 months	12 hr	0 hr

FIGURE 2.21 The pageantry of fall colors in New England. The weather most suitable for an impressive display of fall colors is warm, sunny days followed by clear, cool nights with temperatures dropping below 45°F (7°C), but remaining above freezing.

EXTREME WEATHER WATCH

If the tilt of the earth were to suddenly increase to 45°, seasons in the middle latitudes of the Northern Hemisphere would become much more extreme, as summer would be much warmer and winters much colder than at present.

clean snow covering the ground aids in the cooling. In northern Canada and Alaska, arctic air rapidly becomes extremely cold as it lies poised, ready to do battle with the milder air to the south. Periodically, this cold arctic air pushes down into the northern United States, producing a rapid drop in temperature called a *cold wave,* which occasionally reaches far into the south.

Three months past the winter solstice marks the astronomical arrival of spring, which is called the **vernal** (spring) **equinox**. The date is March 20 and, once again, the noonday sun is shining directly on the equator, days and nights throughout the world are of equal length, and, at the North Pole, the sun rises above the horizon after a long six-month absence.

At this point it is interesting to note that although sunlight is most intense in the Northern Hemisphere on June 21, the warmest weather in middle latitudes normally occurs weeks later, usually in July or August. This situation (called the *lag in seasonal temperature)* arises because although incoming energy from the sun is greatest in June, it still exceeds outgoing energy from the earth for a period of at least several weeks. When incoming solar energy and outgoing earth energy are in balance, the highest average temperature is attained. When outgoing energy exceeds incoming energy, the average temperature drops. Because outgoing earth energy exceeds incoming solar energy well past the winter solstice (December 21), we normally find our coldest weather occurring in January or February.

Up to now, we have seen that the seasons are controlled by solar energy striking our tilted planet, as it makes its annual voyage around the sun. This tilt of the earth causes a seasonal variation in both the length of daylight and the intensity of sunlight that reaches the surface. These facts are summarized in ❯ Fig. 2.22, which shows how the sun would appear in the sky to an observer at various latitudes at different times of the year. (Before we look at seasons in the Southern Hemisphere, you may wish to read the Focus section on p. 59 that looks at the year 1816—a year that many claim "had no summer.")

SEASONS IN THE SOUTHERN HEMISPHERE On June 21, the Southern Hemisphere is adjusting to an entirely different season. Again, look back at Fig. 2.18,

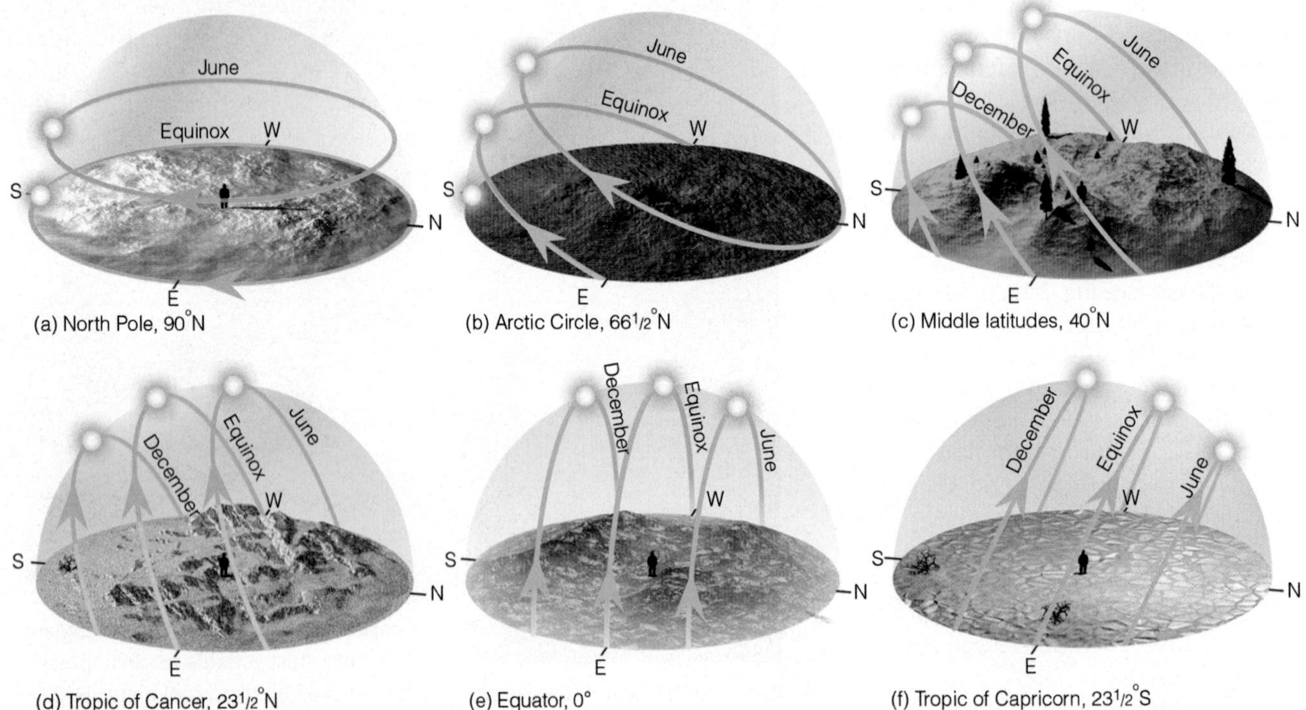

❯ FIGURE 2.22 **The apparent path of the sun across the sky as observed at different latitudes on the June solstice (June 21), the December solstice (December 21), and the equinox (March 20 and September 22).**

p. 53, and notice that this part of the world is now tilted away from the sun. Nights are long, days are short, and solar rays come in at a low angle (see Fig. 2.22). All of these factors keep air temperatures fairly low. The June solstice marks the astronomical beginning of winter in the Southern Hemisphere. In this part of the world, summer will not "officially" begin until the sun is over the Tropic of Capricorn (23½°S)—remember that this occurs on December 21. So, when it is winter and June in the Southern Hemisphere, it is summer and June in the Northern Hemisphere. Conversely, when it is summer and December in the Southern Hemisphere, it is winter and December in the Northern Hemisphere. So,

FOCUS ON
EXTREME WEATHER

A Year without a Summer?

The year 1816 is often referred to as the "year without a summer." Although there was a summer that year, it was a cold one. In early May, an unusually cold blast of arctic air swept through Canada and into the northeastern United States, bringing killing frost to much of the region. Between May and September, one cold outbreak followed another, keeping summer average temperatures well below normal. In early June, heavy snow fell over portions of Pennsylvania, New York, and Vermont, with Williamstown, Vermont, reporting twelve inches of snow on June 7. Additional cold spells in July and August brought more killing frosts. During the warmer days that followed each cold snap, farmers replanted, only to have a cold outbreak damage the planting. Estimates are that killing frosts and storm damage (that occurred with the passage of strong cold fronts) resulted in some counties in the northeast losing more than 90 percent of their harvest.

Over Western Europe, cold rainy weather contributed to a poor wheat crop, and famine spread. In this part of the world, the year without a summer even made its mark on literature as the cold, gloomy summer weather along the shores of Lake Geneva in Switzerland, inspired 18-year-old Mary Shelley to write the novel *Frankenstein*.

This unseasonably cold summer is linked to volcanic activity in the western Pacific. In 1815, Mount Tambora in Indonesia erupted with ferocity, ejecting into the upper atmosphere many tons of sulfur-rich volcanic particles. Some of the particles combined with water vapor to produce tiny, reflecting sulfuric acid particles that grew in size and formed a dense layer of haze that probably remained in the stratosphere for several years. The haze absorbed and reflected back to space a portion of the sun's energy (see Fig. 7). The reflection of incoming sunlight meant less sunlight was able to reach the earth's surface. This situation cooled the surface air, especially in the Northern Hemisphere. In addition, this cooling of the atmosphere in 1816 apparently caused a shift in the upper-level wind pattern that set up a persistent upper-level flow that directed very cold air from northern Canada into the eastern half of the United States.

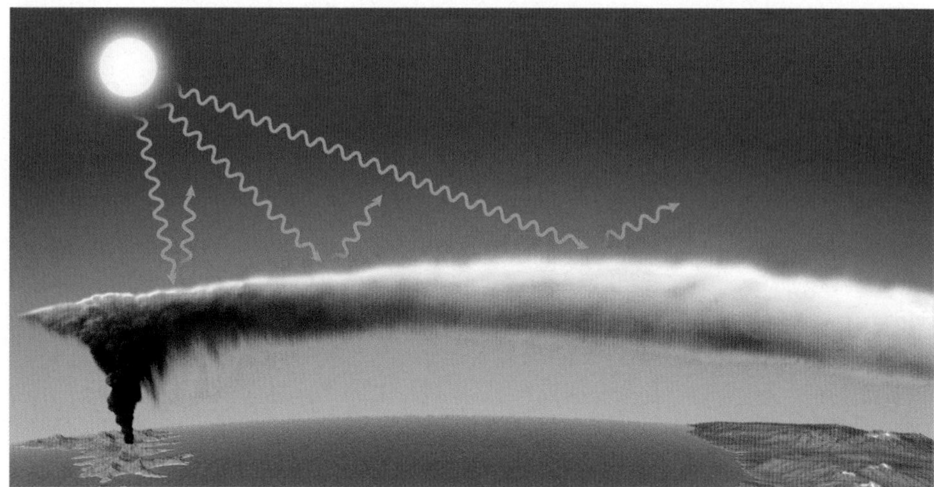

Figure 7 During the eruption of Mount Tambora in 1815, tons of sulfur-rich volcanic gases were lofted into the stratosphere, where they transformed into tiny sulfuric acid particles. These particles drifted eastward with the wind and reflected a portion of the sun's incoming energy. Less sunlight reaching the surface lowered the average temperature over portions of the Northern Hemisphere.

if you are tired of the cold, December weather in your Northern Hemisphere city, travel to the summer half of the world and enjoy the warmer weather. The tilt of the earth as it revolves around the sun makes all this possible.

We know the earth comes nearer to the sun in January than in July. Even though this difference in distance amounts to only about 3 percent, the energy that strikes the top of the earth's atmosphere is almost 7 percent greater on January 3 than on July 4. These statistics might lead us to believe that summer should be warmer in the Southern Hemisphere than in the Northern Hemisphere, which, however, is not the case. A close examination of the Southern Hemisphere reveals that nearly 81 percent of the surface is water compared to 61 percent in the Northern Hemisphere. The added solar energy due to the closeness of the sun is absorbed by large bodies of water, becoming well mixed and circulated within them. This process keeps the average summer (January) temperatures in the Southern Hemisphere cooler than the average summer (July) temperatures in the Northern Hemisphere. Because of water's large heat capacity,* it also tends to keep winters in the Southern Hemisphere warmer than we might expect.

*Heat capacity of a substance is the ratio of the heat energy absorbed (or released) to the corresponding temperature rise (or fall). It takes more heat energy to raise the temperature of a given amount of water by 1°C than it does to raise the same amount of soil by 1°C.

Summary

In this chapter we looked at the concept of energy. Here we learned that energy can take on many forms, and that energy lost during one process must equal energy gained during another. We saw that heat is energy in the process of being transferred from one object to another. We also learned that latent heat is an important source of atmospheric heat energy, especially for storms such as hurricanes and thunderstorms. We examined how the transfer of heat takes place in our atmosphere by conduction, convection, and radiation.

We examined how the hot sun emits the majority of its radiation as shortwave radiation. A portion of this energy heats the earth, and the earth in turn warms the air above. The cool earth emits most of its radiation as longwave infrared energy. Selective absorbing gases in the atmosphere, such as water vapor and carbon dioxide, absorb a portion of this energy, which warms the lower atmosphere. These same gases emit infrared energy back to the earth's surface. The warming of the earth and its atmosphere by the trapping of the earth's outgoing infrared radiation is called the "greenhouse effect." The greenhouse effect keeps the earth's average surface temperature much warmer than it otherwise would be.

We looked at the earth's average equilibrium temperature and found that it remains fairly constant from one year to the next because the amount of energy that the earth and its atmosphere absorb each year is equal to the amount of energy they lose.

Finally, we looked at the seasons and found that the earth has seasons because it is tilted on its axis as it revolves around the sun. The tilt causes seasonal variations in both the length of daylight and the intensity of sunlight that reaches the earth's surface.

Key Terms

The following terms are listed (with page number) in the order they appear in the text. Define each. Doing so will aid you in reviewing the material covered in this chapter.

energy, 36
potential energy, 36
kinetic energy, 36
heat, 36
latent heat, 36
sensible heat, 37
conduction, 38
convection, 39
thermals, 40
advection, 40
radiant energy (radiation), 41
electromagnetic waves, 41
micrometer, 41
photons, 41
visible region, 44
ultraviolet (UV) radiation, 44
infrared (IR) radiation, 44
shortwave radiation, 44
longwave radiation, 44
blackbody, 45
radiative equilibrium temperature, 45
selective absorbers, 45
greenhouse effect, 46
greenhouse gases, 46
atmospheric window, 46
scattering, 49
reflected (light), 49
albedo, 49
summer solstice, 56
autumnal equinox, 57
winter solstice, 57
vernal equinox, 58

Review Questions

1. Distinguish between temperature and heat.
2. What is kinetic energy?
3. Explain how heat is transferred in our atmosphere by (a) conduction, (b) convection, (c) radiation.
4. In the atmosphere, how does advection differ from convection?
5. How does the temperature of an object influence the radiation it emits?
6. How does the amount of radiation emitted by the earth differ from that emitted by the sun?
7. How do the wavelengths of most of the radiation emitted by the sun differ from those emitted by the surface of the earth?
8. When a body reaches a radiative equilibrium temperature, what is taking place?
9. If the earth's surface continually radiates energy, why doesn't it become colder and colder?
10. What are the most abundant selectively absorbing greenhouse gases in the earth's atmosphere?
11. Explain how the earth's atmospheric greenhouse effect works.
12. What gases are responsible for the enhancement of the earth's greenhouse effect?
13. If the earth had no greenhouse effect, would the earth's average surface temperature be higher or lower than it is presently? Explain.
14. How is the lower atmosphere warmed from the surface upward?
15. Why does the albedo of the earth and its atmosphere average about 30 percent?
16. Explain how the earth and its atmosphere balance incoming energy with outgoing energy.
17. What are the main factors that determine seasonal temperature variations?
18. In the Northern Hemisphere, why are summers warmer than winters even though the earth is actually closer to the sun in January?
19. If it is winter and January in New York City, what is the season and month in Sydney, Australia?
20. During the Northern Hemisphere's summer, the daylight hours in far northern latitudes are longer than in middle latitudes. Explain why far northern latitudes are not warmer.
21. In the middle latitudes of the Northern Hemisphere, explain why even though the earth's surface receives maximum sunlight on June 21 (summer solstice) the warmest time of year typically occurs in July.
22. Explain why average summer (January) temperatures in the Southern Hemisphere are cooler than average summer (July) temperatures in the Northern Hemisphere, even though the earth is closer to the sun in January.

Online Learning

STUDENT COMPANION WEBSITE: Visit this book's companion website at: www.cengage.com/ahrens/extreme1e and choose Chapter 2 for many study aids and ideas for further reading and research. These include flashcards, practice quizzing, and web links.

METEOROLOGY RESOURCE CENTER: For students with access, log on at www.cengage.com/login for more assets, including animations, videos, and more. If your textbook did not come with access, visit www.CengageBrain.com

Temperature and Humidity Extremes

Patrick J. Endres/AlaskaPhotoGraphics.com

3

CONTENTS

DAILY TEMPERATURES

WEATHER EXTREMES AND HUMAN DISCOMFORT

BEATING THE HEAT—DEALING WITH HEAT WAVES

SUMMARY

KEY TERMS

REVIEW QUESTIONS

Snow and frost cover spruce trees near the Yukon River, Alaska, one of the coldest wintertime places in North America.

Every summer, scorching heat takes many lives. During the past 20 years, an annual average of more than 300 fatalities in the United States was attributed to excessive heat exposure. On one tragic day in Chicago, Illinois—July 15, 1995—high temperatures coupled with high humidity claimed the lives of more than 270 people. Even the cold side of weather takes its toll. Although extreme cold produces fewer direct deaths than does extreme heat, nonetheless, cold weather can be lethal to those exposed, especially the homeless.

In this chapter, we will not only look at how temperature and humidity vary on a "typical" day, we will also look at the extremes of these two weather elements—the hottest, coldest, and most humid places on this planet.

Daily Temperatures

In Chapter 2, we learned how the sun's energy coupled with the motions of the earth produce the seasons. In a way, each sunny day is like a tiny season as the air goes through a daily cycle of warming and cooling. The air warms during the morning hours, as the sun gradually rises higher in the sky, spreading a blanket of heat energy over the ground. The sun reaches its highest point around noon, after which it begins its slow journey toward the western horizon. It is around noon when the earth's surface receives the most intense solar rays. However, somewhat surprisingly, noontime is usually not the warmest part of the day. Rather, the air continues to be heated, often reaching a maximum temperature later in the afternoon. To find out why this *lag in temperature* occurs, we need to examine a shallow layer of air in contact with the ground.

WARM DAYS As the sun rises in the morning, sunlight warms the ground, and the ground warms the air in contact with it by ground conduction. However, air is such a poor heat conductor that this process only takes place within a few centimeters of the ground. As the sun rises higher in the sky, the air in contact with the ground becomes even warmer, and, on a windless day, a substantial temperature difference usually exists just above the ground. This explains why joggers on a clear, windless, hot summer afternoon may experience an extreme air temperature of over 120°F at their feet and only 95°F at their waists (see ❯ Fig. 3.1).

Near the surface, convection begins, and rising air bubbles (thermals) help to redistribute heat. In calm weather, these thermals are small and do not effectively mix the air near the surface. Thus, large vertical temperature differences are able to exist. On windy days, however, turbulent eddies are able to mix hot, surface air with the cooler air above. This form of mechanical stirring, sometimes called *forced convection,* helps the thermals to transfer heat away from the surface more efficiently. Therefore, on sunny, windy days the temperature difference between the surface air and the air directly above is not as great as it is on sunny, calm days (see ❯ Fig. 3.2).

We can now see why the warmest part of the day is usually in the afternoon. Around noon, the sun's rays are most intense. However, even though incoming solar radiation decreases in intensity after noon, it still exceeds outgoing heat energy from the surface for a time. This situation yields an energy surplus for two to four hours

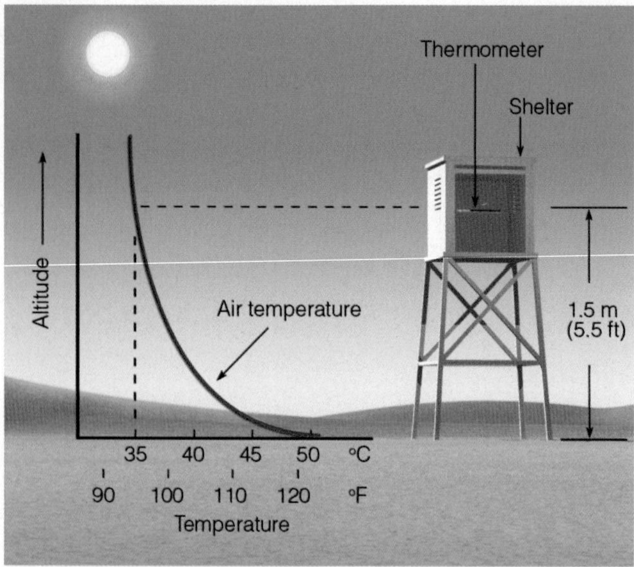

❯ **FIGURE 3.1** On a sunny, calm day, the air near the surface can be substantially warmer than the air a meter or so above the surface.

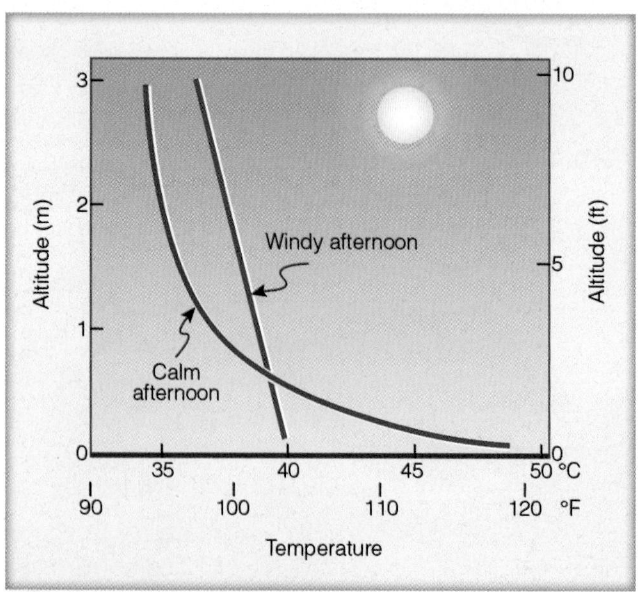

❯ **FIGURE 3.2** Vertical temperature profiles above an asphalt surface for a windy and a calm summer afternoon.

after noon and substantially contributes to a lag between the time of maximum solar heating and the time of maximum air temperature several feet above the surface (see ❯ Fig. 3.3).

The exact time of the highest temperature reading varies somewhat. Where the summer sky remains cloud-free all afternoon, the maximum temperature may occur sometime between 3:00 and 5:00 P.M. Where there is afternoon cloudiness or haze, the temperature maximum usually occurs an hour or two earlier. If clouds persist throughout the day, the overall daytime temperatures are usually lower, as clouds reflect a great deal of incoming sunlight.

Adjacent to large bodies of water, cool air moving inland may modify the rhythm of temperature change such that the warmest part of the day occurs at noon or before. In winter, atmospheric storms circulating warm air northward can even cause the highest temperature to occur at night.

Just how warm the air becomes depends on such factors as the type of soil, its moisture content, and vegetation cover. When the soil is a poor heat conductor (as loosely packed sand is), heat energy does not readily transfer into the ground. This allows the surface layer to reach a higher temperature, availing more energy to warm the air above. On the other hand, if the soil is moist or covered with vegetation, much of the available energy evaporates water, leaving less to heat the air. As you might expect, the highest summer temperatures usually occur over desert regions, where clear skies coupled with low humidities and meager vegetation permit the surface and the air above to warm up rapidly.

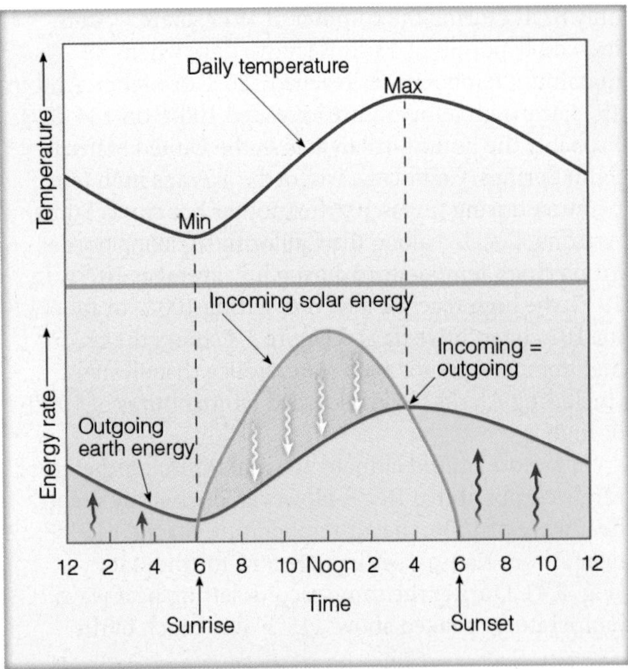

❯ **FIGURE 3.3** The daily variation in air temperature is controlled by incoming energy (primarily from the sun) and outgoing energy from the earth's surface. Where incoming energy exceeds outgoing energy (orange shade), the air temperature rises. Where outgoing energy exceeds incoming energy (blue shade), the air temperature falls.

❯ Table 3.1

Capitalize High Temperature for Some of the Hottest Cities in the United States

CITY	AVERAGE DAILY HIGH TEMPERATURE
Bullhead City, AZ*	112.0°
Palm Springs, CA**	108.2°
Yuma, AZ	107.0°
Phoenix, AZ	106.0°
Las Vegas, NV	104.1°
Tucson, AZ	101.0°
Laredo, TX	100.5°
Redding, CA	99.5°
Bakersfield, CA	98.2°
Fresno, CA	98.1°
Wichita Falls, TX	97.6°
Waco, TX	96.7°
Dallas-Ft. Worth, TX	96.3°
Del Rio, TX	96.2°
El Paso, TX	95.5°

*Hottest inhabited place in the United States (with a population of at least 45,000).
**Hottest major metro area in the United States.

Where the air is humid, haze and cloudiness lower the maximum temperature by preventing some of the sun's rays from reaching the ground. In humid Atlanta, Georgia, the average maximum temperature for July is 87°F. In contrast, Phoenix, Arizona—in the desert southwest at the same latitude as Atlanta—experiences an average July maximum of 106°F.

EXTREME HEAT The hottest region in the United States is the desert southwest, where cities such as Phoenix, Arizona; Palm Springs, California; and Las Vegas, Nevada; all have maximum July temperatures above 100°F (see ❯ Table 3.1). The highest temperature recorded in the United States (134°F) occurred at Greenland Ranch in Death Valley, California (at 282 ft below sea level) on July 10, 1913. Here, air temperatures are persistently hot throughout the summer, with the average maximum for July being 116°F and the average low temperature only 87°F. During the summer of 1917, there was an incredible period of 43 consecutive days when the maximum temperatures reached 120°F or higher. And in 1974, the high temperature exceeded 100°F on 134 days. Probably the hottest urban area in the United States is Palm Springs, California, where the average high temperature during July is 108°F. Another hot city is Yuma, Arizona. Located along the California-Arizona border, Yuma's high temperature during July averages 107°F. In 1937, the high reached a record-setting 100°F or more for 101 consecutive days. ❯ Figure 3.4 shows the record high temperature for each state. Notice that all states (including Alaska) have recorded temperatures of 100°F or higher.

In a more humid climate, the maximum temperature rarely climbs above 105°F. However, during the record heat wave of 1936, the air temperature reached 121°F near Alton, Kansas, setting a record for the state (Fig. 3.4). During the same record-setting heat wave, temperatures peaked above 115°F over much of the Great Plains, with Steele, North Dakota, reporting a record high 121°F. During the heat wave of 1983, which destroyed about $7 billion in crops and increased the nation's air-conditioning bill by an estimated $1 billion, Fayetteville reported North Carolina's all-time record high temperature when the mercury hit 110°F.

These readings, however, do not hold a candle to the hottest place in the world. That distinction probably belongs to Dallol, Ethiopia. Dallol is located in the desert near latitude 12°N, in the hot-dry Danakil Depression (see ❯ Fig. 3.5). A prospecting company kept weather records at Dallol from 1960 to 1966. During this time, the average daily maximum temperature exceeded 100°F every month of the year, except during December and January, when the average maximum lowered to 98°F and 97°F, respectively. On many days, the air temperature exceeded 120°F. The average annual temperature for the six years at Dallol was 94°F. In comparison, the average annual temperature in Yuma is 74°F and at Death Valley, 76°F. The highest temperature reading on earth (under standard conditions) occurred northwest of Dallol at El Azizia, Libya (32°N), when, on September

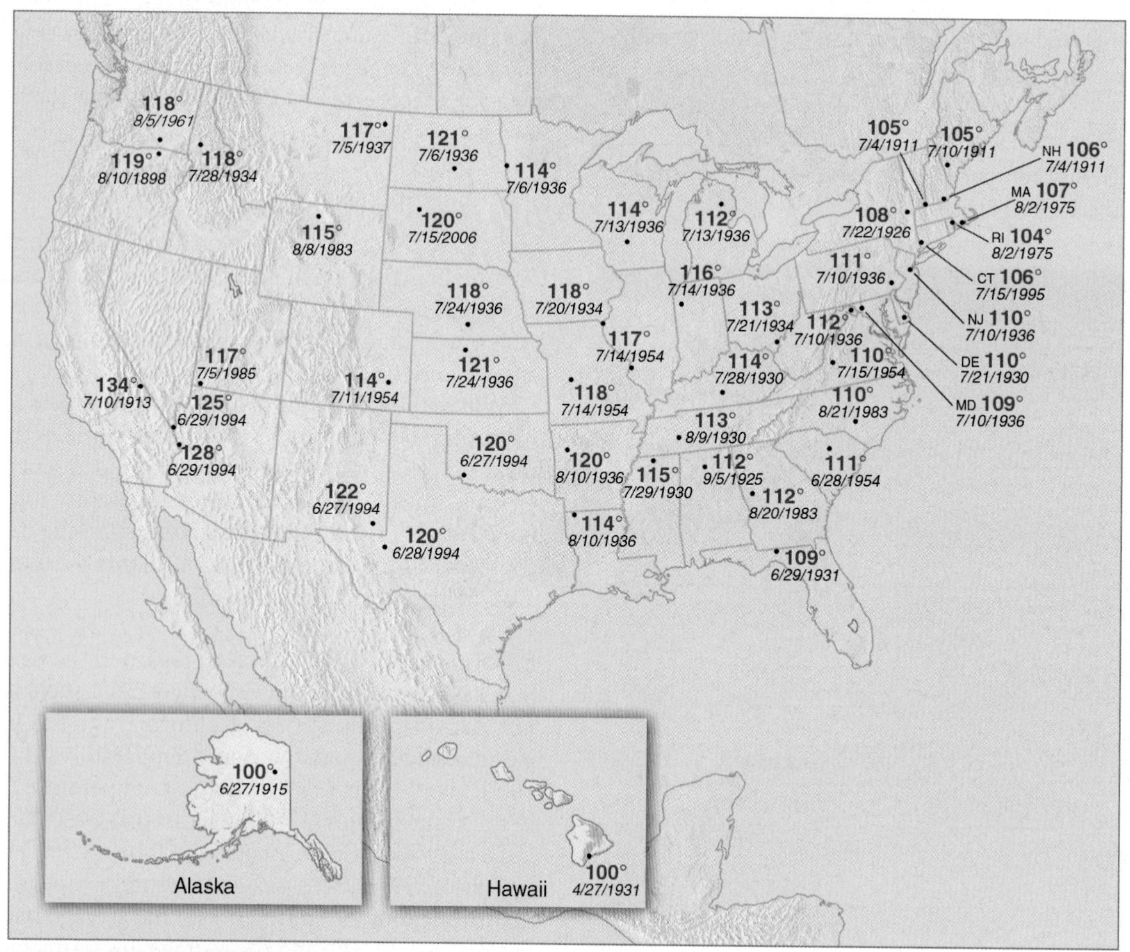

❯ FIGURE 3.4 Record high temperatures (°F) for each state.

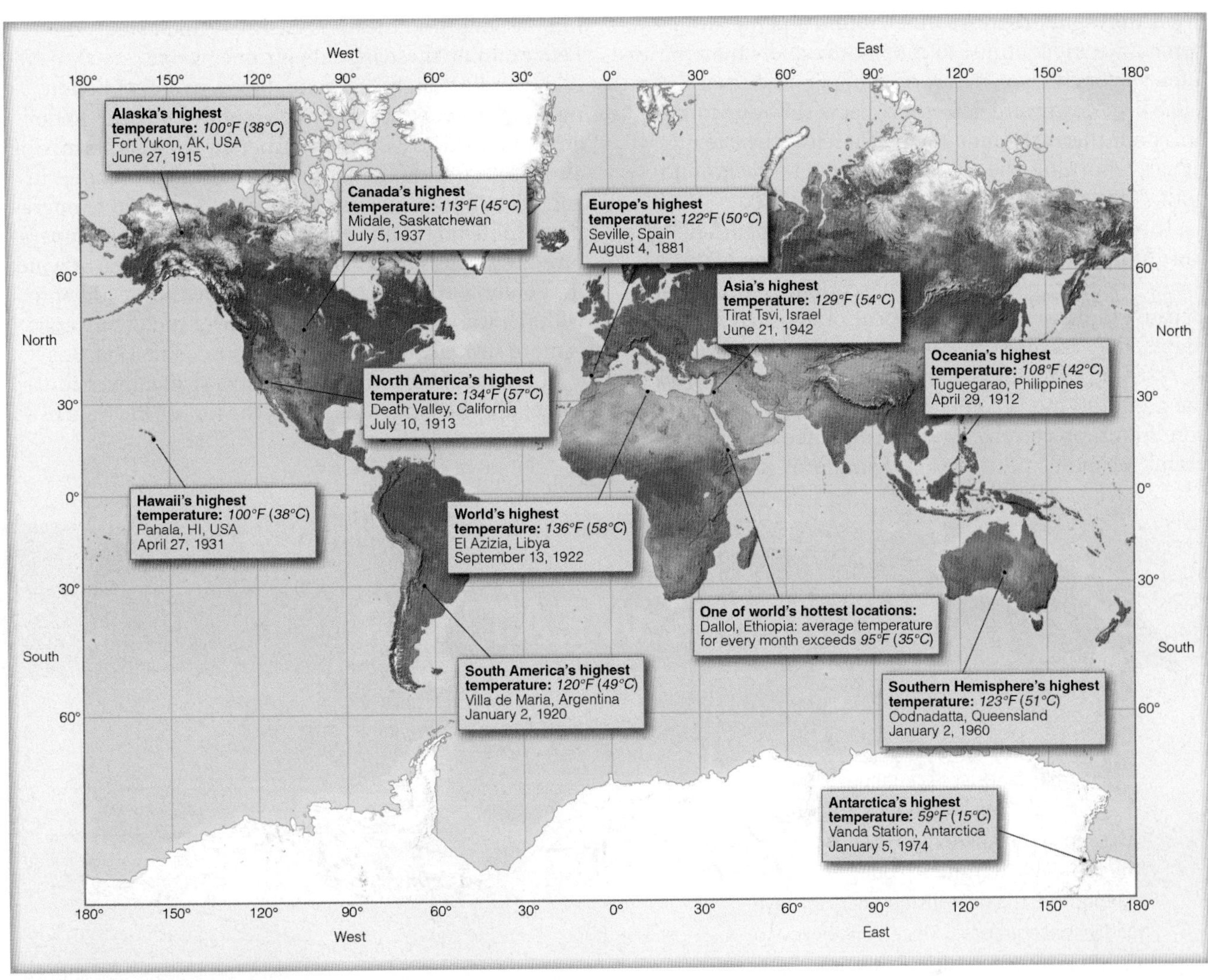

FIGURE 3.5 Record high temperatures throughout the world.

13, 1922, the temperature reached a scorching 136°F. In fact, as we can see in Fig. 3.5, temperatures exceeding 120°F have occurred on all continents except Antarctica.

Up to this point we've looked at the extreme heat in the open air. Inside a car in direct sunlight with windows rolled up, the air temperature can climb above 160°F in a short period of time. More on this topic is provided in the Focus section on p. 68.

COLD NIGHTS We know that nights are typically much cooler than days. The reason for this fact is that, as the afternoon sun lowers, its energy is spread over a larger area, which reduces the heat available to warm the ground. Look back at Fig. 3.3 and observe that sometime in late afternoon or early evening, the earth's surface and air above begin to lose more energy than they receive; hence, they start to cool.

Both the ground and air above cool by radiating infrared energy, a process called **radiational cooling**. The ground, being a much better radiator than air, is able to cool more quickly. Consequently, shortly after sunset, the earth's surface is slightly cooler than the air directly above it. The surface air transfers some energy to the ground by conduction, which the ground, in turn, quickly radiates away.

EXTREME WEATHER WATCH

The world's hottest major city with a population of at least one million people is Bangkok, Thailand, where the daily maximum temperature exceeds 90°F on average 300 days a year, and its average annual temperature is 84.5°F.

As the night progresses, the ground and the air in contact with it continue to cool more rapidly than the air a few meters higher. The warmer upper air does transfer *some* heat downward, a process that is slow due to the air's poor thermal conductivity. Therefore, by late night or early morning, the coldest air is next to the ground, with slightly warmer air above (see ❯ Fig. 3.6).

This measured increase in air temperature just above the ground is known as a **radiation inversion** because it forms mainly through radiational cooling of the surface. Because radiation inversions occur on most clear, calm nights, they are also called *nocturnal inversions.*

A strong radiation inversion occurs when the air near the ground is much colder than the air higher up. Ideal conditions for a strong inversion and, hence, very low nighttime temperatures exist when the air is calm, the night is long, and the air is fairly dry and cloud-free. Let's examine these ingredients one by one.

A windless night is essential for a strong radiation inversion because a stiff breeze tends to mix the colder air at the surface with the warmer air above. This mixing, along with the cooling of the warmer air as it comes in contact with the cold ground, causes a vertical temperature profile that is almost isothermal (constant temperature) in a layer several feet thick. In the absence of wind, the cooler, more-dense surface air does not readily mix with the warmer, less-dense air above, and the inversion is more strongly developed as illustrated in ❯ Fig. 3.7.

A long night also contributes to a strong inversion. Generally, the longer the night, the longer the time of

FOCUS ON

EXTREME WEATHER

Extreme Heat Inside a Car

Just like a florist's greenhouse, the interior of an automobile is warmed by the sun's radiant energy. However, unlike the life-giving warmth to young plants in a greenhouse, the heat inside a car can have deadly consequences. Since 1998, over 450 infants and children, along with untold thousands of pets, have died of heatstroke after being left inside vehicles.*

Not only can the air temperature inside a vehicle reach very high temperatures in excess of 140°F, but the rate of temperature rise is extremely rapid. In a study published by the American Academy of Pediatrics it was found that inside a car, in direct sunlight with windows rolled up, the average temperature can rise 19°F in just 10 minutes and 35°F in 30 minutes (see Fig. 1). Hence, an outside air temperature of 90°F can approach 125°F inside a car in half an hour.

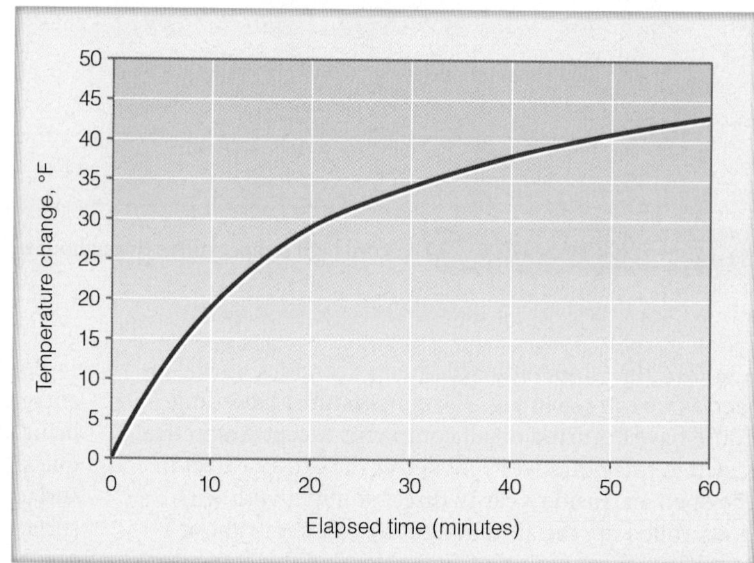

Figure 1 Average temperature rise inside a closed car in direct sunlight for 16 days between May 16 and August 8, 2002, with an outside temperature between 72°F and 96°F. (Jan Null)

This extreme heating is the result of the sun's energy being absorbed by objects within a car that can reach temperatures exceeding 180°F. This heat energy inside the enclosed vehicle effectively warms the trapped air inside. Even on a relatively mild day, when the air temperature is only 70°F, interior vehicle temperatures can exceed deadly levels, reaching 115°F after only an hour of exposure.

* For further information please visit http://ggweather.com/heat/

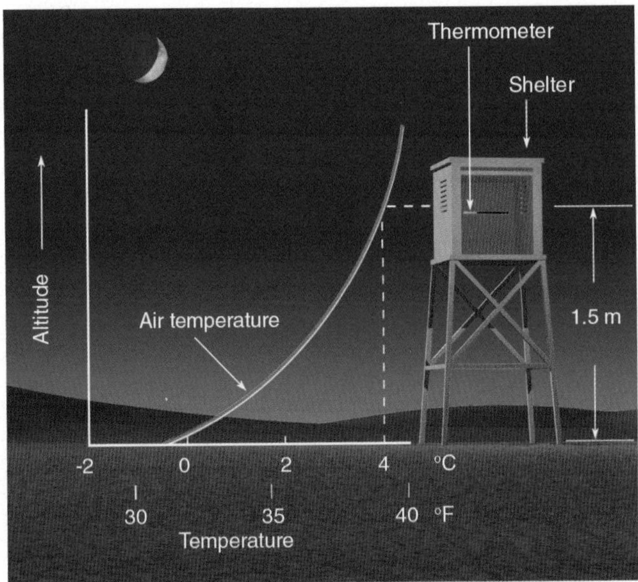

FIGURE 3.6 On a clear, calm night, the air near the surface can be much colder than the air above. The increase in air temperature with increasing height above the surface is called a radiation temperature inversion.

radiational cooling and the better are the chances that the air near the ground will be much colder than the air above. Consequently, winter nights provide the best conditions for a strong radiation inversion, other factors being equal.

Finally, radiation inversions are more likely with a clear sky and dry air. Under these conditions, the ground is able to radiate its energy to outer space and thereby cool rapidly. However, with cloudy weather and moist air, much of the outgoing infrared energy is absorbed and radiated to the surface, retarding the rate of cooling. Also, on humid nights, condensation in the form of fog or dew will release latent heat, which warms the air. So, radiation inversions may occur on any night. But, during long winter nights, when the air is still, cloud-free, and relatively dry, these inversions can become strong and deep. As a consequence, on a cold, dry winter night, it is common to experience below-freezing temperatures near the ground, and air more than 10°F warmer at your waist.

On cold nights, plants and certain crops may be damaged by the low temperatures. Fruit trees are particularly vulnerable to cold weather in the spring when they are blossoming. If the cold occurs over a widespread area for a long enough time to damage certain crops, the extreme cold is called a **freeze**. A single freeze in California, Texas, or Florida can cause crop losses in the millions or even billions of dollars. In fact, citrus crop losses in Florida during the hard freeze of January, 1977, exceeded $2 billion. In California, several freezes during the spring of 2001 caused millions of dollars in damages to California's north coast vineyards, which resulted in higher wine prices.

The coldest air and lowest temperatures are frequently found in low-lying areas. The reason for this situation is that cold, heavy surface air slowly drains downhill during the night and eventually settles in low-lying basins and valleys. In middle latitudes, the warmer hillsides, called **thermal belts,** are less likely to experience freezing temperatures than the valley below (see Fig. 3.8). This phenomenon encourages farmers to plant on hillsides those trees and sensitive crops that are unable to survive the valley's low temperatures. Moreover, on the valley floor, the cold, dense air is unable to rise, so smoke and other pollutants trapped in this heavy air can restrict visibility. Therefore, valley bottoms are not only colder, but are also more frequently polluted than nearby hillsides.

EXTREME COLD

Figure 3.9 shows the lowest temperature reported for each state. Notice that all states (excluding Hawaii) have experienced temperatures of 0°F or below. The lowest official temperature for Alaska is –80°F, whereas the official record low for the forty-eight adjacent states belongs to Rogers Pass, Montana, where on the morning of January 20, 1954, the temperature

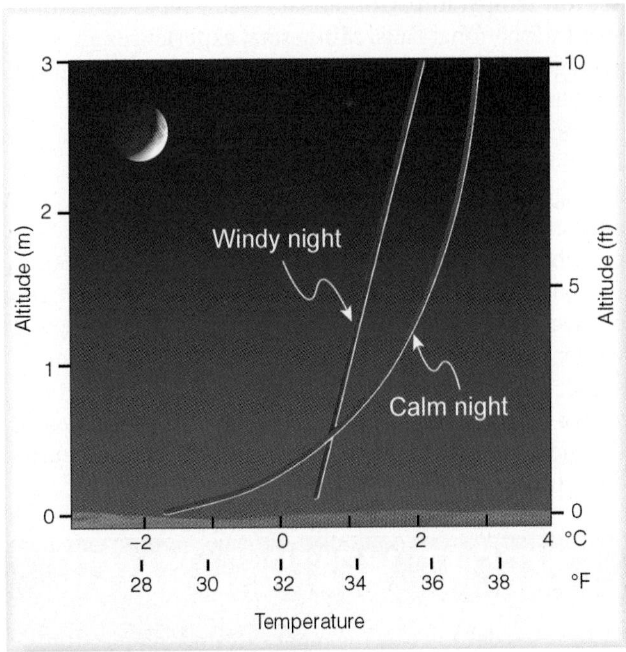

FIGURE 3.7 Vertical temperature profiles just above the ground on a windy night and on a calm night. Notice that the radiation inversion develops better on the calm night.

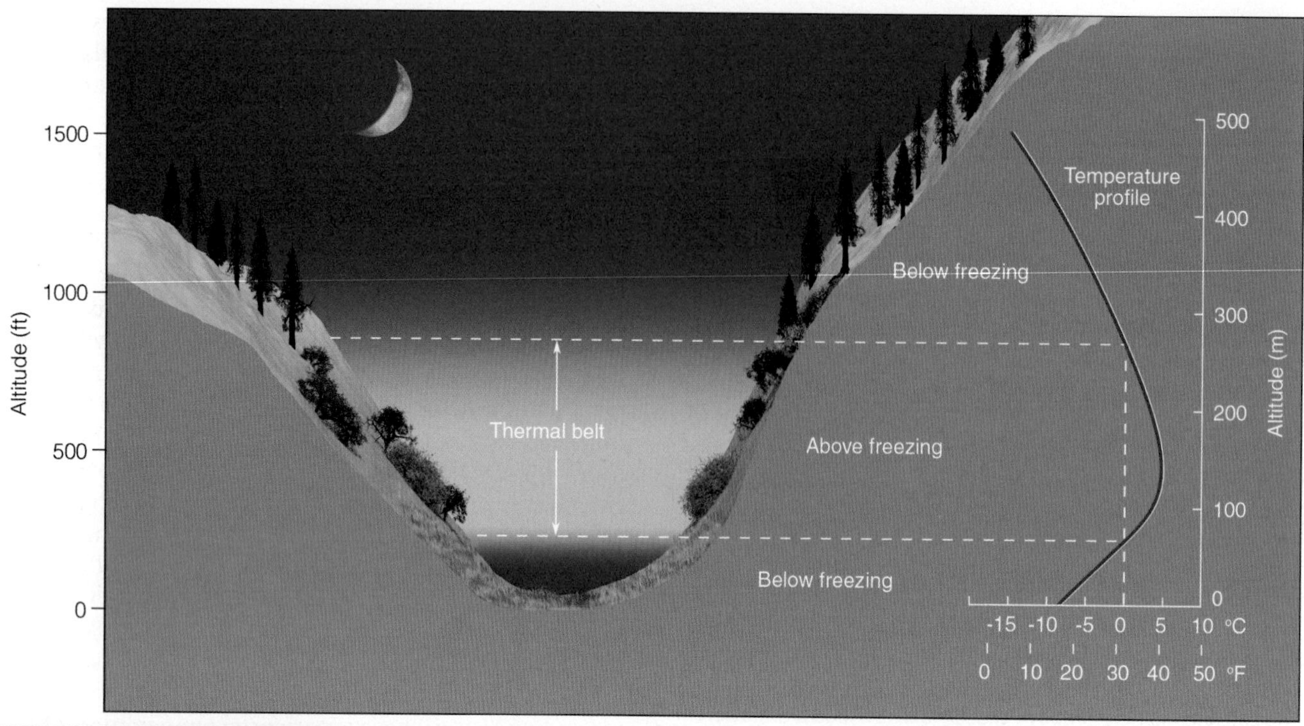

FIGURE 3.8 On cold, clear nights, the settling of cold air into valleys makes them colder than surrounding hillsides. The region along the side of the hill where the air temperature is above freezing is known as a *thermal belt.*

dropped to –70°F. The coldest region of the lower forty-eight states is the northern Great Plains and northern Maine. Here we find some of the coldest average low temperatures for January (see Table 3.2). Notice that International Falls, Minnesota, experiences an average January low temperature of –8°F and an average temperature for the entire month of –3°F.

Located several hundred miles to the south of International Falls, Minneapolis-St. Paul, with an average January minimum temperature of 4°F and an average temperature of 16°F for the three winter months, is the coldest major urban area in the nation. For duration of extreme cold, Minneapolis reported 186 consecutive hours of temperatures below 0°F during the winter of 1911–1912. Within the forty-eight adjacent states, however, the record for the longest duration of severe cold belongs to Langdon, North Dakota, where the thermometer remained below 0°F for 41 consecutive days during the winter of 1936.

EXTREME WEATHER WATCH

The 1930s decade was the most extreme temperature-wise in the United States since record-keeping began. Of the 100 possible state record maximum temperatures and record minimum temperatures, 35 occurred during this decade.

Table 3.2

Average January Low Temperatures (°F) for some of the Coldest Cities in the United States

CITY	AVERAGE JANUARY LOW TEMPERATURE (°F)
Fairbanks, AK	–19.0°
Tower, MN*	–13.6°
Int'l Falls, MN	–8.4°
Gunnison, CO	–6.1°
Grand Forks, ND	–4.3°
Bemidji, MN	–4.3°
Alamosa, CO	–3.7°
Williston, ND	–3.3°
Fargo, ND	–2.3°
St. Cloud, MN	–1.2°
Duluth, MN	–1.2°
Bismarck, ND	–0.6°
Caribou, ME	–0.3°
Aberdeen, SD	0.6°
Minneapolis, MN**	4.3°

*Coldest inhabited place in lower 48 states.
**Coldest major metro area in the United States.

The largest cold wave in modern recorded history in the United States occurred in February of 1899. Temperatures during this cold spell fell below 0°F in every existing state, including Florida. This particular event was the first and only of its kind in recorded history. Record temperatures set during this outbreak still stand today in many cities of the United States.

The coldest areas in North America are found in the Yukon and Northwest Territories of Canada. Resolute, Canada (latitude 75°N), has an average temperature of −26°F for the month of January. The lowest temperatures and coldest winters in the Northern Hemisphere are found in the interior of Siberia and Greenland. For example, the average January temperature in Yakutsk, Siberia (latitude 62°N), is −46°F. There, the average temperature for the entire year is a bitter cold 12°F. At Eismitte, Greenland, the average temperature for February (the coldest month) is −53°F, with the average annual temperature being a frigid −22°F.

Even though these temperatures are extremely low, they do not come close to the coldest area of the world: the Antarctic. At the geographical South Pole, over 9000 feet above sea level, where the Amundsen-Scott scientific station has been keeping records for more than fifty years, the average temperature for the month of July (winter) is −74°F, and the mean annual temperature is −57°F. The lowest temperature ever recorded there (−117°F) occurred under clear skies with a light wind on the morning of June 23, 1983. Cold as it was, it was not the record low for the world. That distinction belongs to the Russian station at Vostok, Antarctica (latitude 78°S), where the temperature plummeted to −129°F on July 21, 1983. ❱ Figure 3.10 provides more information on worldwide record low temperatures. (Before going on to the next section, you may wish to read the Focus section on p. 74 that details where and how air temperature is measured.)

DAILY TEMPERATURE VARIATIONS The greatest variation in daily temperature occurs at the earth's surface. In fact, the difference between the daily maximum and minimum temperature—called the **daily (diurnal)**

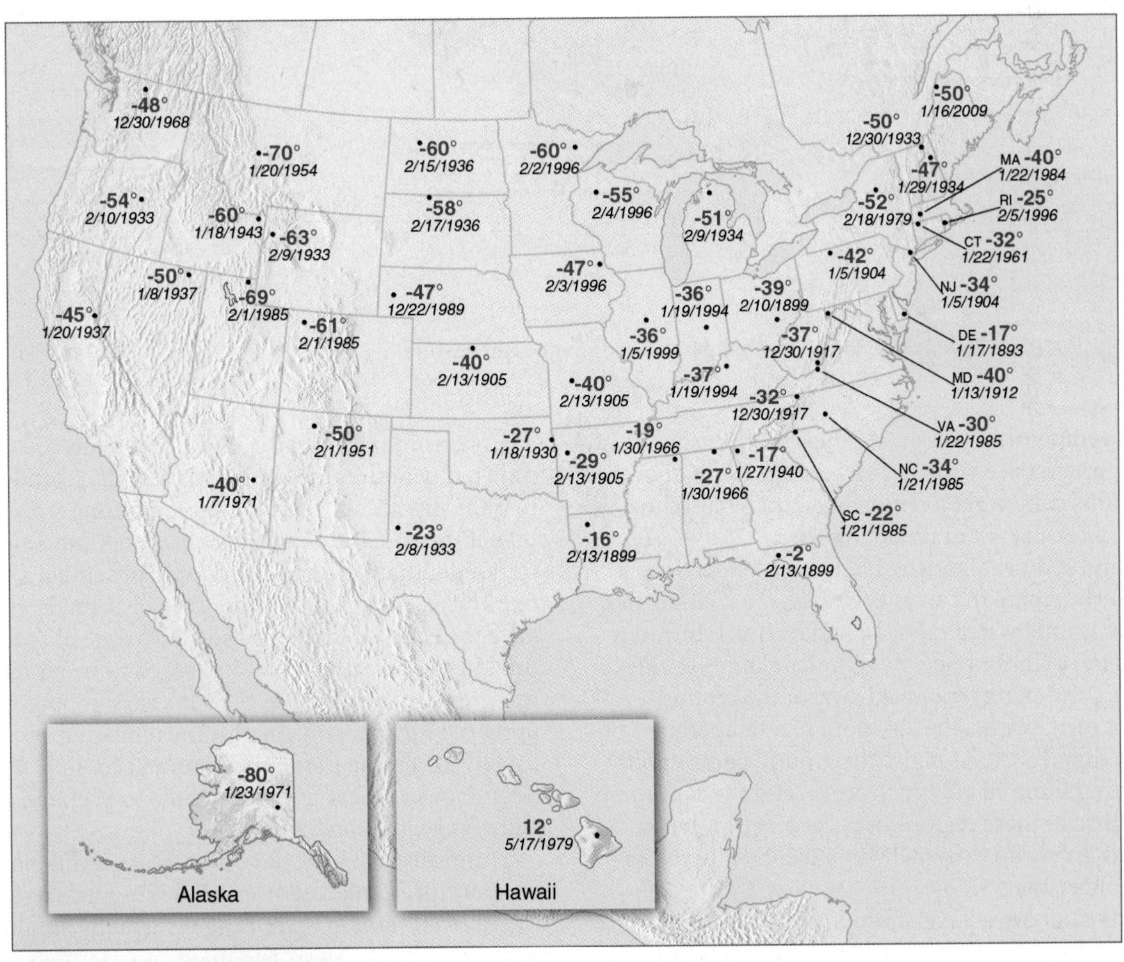

❱ **FIGURE 3.9** Record low temperatures (°F) for each state.

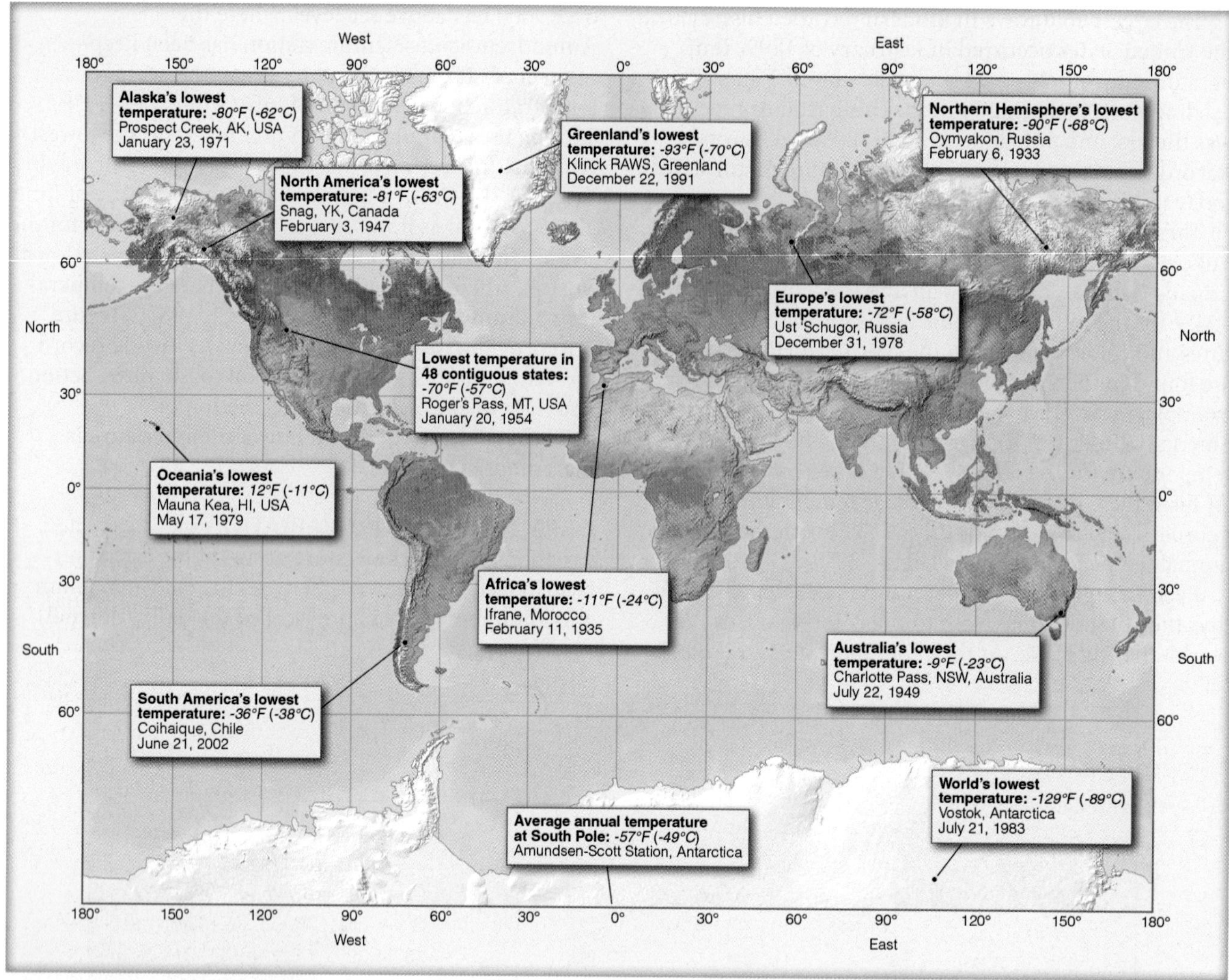

FIGURE 3.10 Record low temperatures throughout the world.

range of temperature—is greatest next to the ground and becomes progressively smaller as we move away from the surface. This daily variation in temperature is also much larger on clear days than on cloudy ones.

The largest diurnal range of temperature occurs on high deserts, where the air is fairly dry, often cloud-free, and there is little water vapor to radiate much infrared energy back to the surface. By day, clear summer skies allow the sun's energy to quickly warm the ground which, in turn, warms the air above to a temperature often exceeding 160°F. At night, the ground cools rapidly by radiating infrared energy to space, and the minimum temperature in these regions occasionally dips below 45°F, thus giving an extremely high daily temperature range of more than 55°F.

Clouds can have a large affect on the daily range in temperature. As we saw in Chapter 2, clouds (especially low, thick ones) are good reflectors of incoming solar radiation, and so they prevent much of the sun's energy from reaching the surface. This effect tends to lower daytime temperatures (see Fig. 3.11a). If the clouds persist into the night, they tend to keep nighttime temperatures higher, as clouds are excellent absorbers and emitters of infrared radiation—the clouds actually emit a great deal of infrared energy back to the surface. Clouds, therefore, have the effect of lowering the daily range of temperature. In clear weather (Fig. 3.11b), daytime air temperatures tend to be higher as the sun's rays impinge directly upon the surface, while nighttime temperatures are usually lower due to rapid radiational cooling. Therefore, clear days and clear nights combine to promote a large daily range in temperature.

Humidity can also have an effect on diurnal temperature ranges. For example, in humid regions, the diurnal temperature range is usually small. Here, haze and clouds lower the maximum temperature by preventing some of the sun's energy from reaching the surface. At night, the moist air keeps the minimum temperature high by absorbing the earth's infrared radiation and radiating a portion of it to the ground. An example of a humid city with a small summer diurnal temperature range is Charleston, South Carolina, where the average July maximum temperature is 90°F, the average minimum is 72°F, and the diurnal range is only 18°F.

Cities near large bodies of water typically have smaller diurnal temperature ranges than cities farther inland. This phenomenon is caused in part by the additional water vapor in the air and by the fact that water warms and cools much more slowly than land.

Moreover, cities whose temperature readings are obtained at airports often have larger diurnal temperature ranges than those whose readings are obtained in downtown areas. The reason for this fact is that nighttime temperatures in cities tend to be warmer than those in outlying rural areas. This nighttime city warmth—called the *urban heat island*—forms as the sun's energy is absorbed by urban structures and concrete; then, during the night, this heat energy is slowly released into the city air.

The average of the highest and lowest temperature for a 24-hour period is known as the **mean (average) daily temperature**. Most newspapers list the mean daily temperature along with the highest and lowest temperatures for the preceding day. The average of the mean daily temperatures for a particular date averaged for a 30-year period gives the average (or "*normal*") temperatures for that date.

Brief Review

Up to this point we have examined daily temperatures along with temperature extremes. Before going on, here is a review of some of the important concepts and facts we have covered:

- **During the day, the earth's surface and air above will continue to warm as long as incoming energy (mainly sunlight) exceeds outgoing energy from the surface.**
- **Record-high temperatures tend to occur in summer over subtropical deserts.**
- **At night, the earth's surface cools, mainly by giving up more infrared radiation than it receives—a process called "radiational cooling."**
- **Radiation inversions exist usually at night, when the air near the ground is colder than the air above.**
- **The coldest nights of winter normally occur when the air is calm, fairly dry (low water-vapor content), and cloud-free.**
- **Record low temperatures tend to occur at high latitudes in the middle of large continents.**
- **The highest temperatures during the day and the lowest temperatures at night are normally observed at the earth's surface.**
- **At most locations, cloudy days and cloudy nights tend to reduce the daily (diurnal) range of temperature. Humid regions tend to have smaller diurnal ranges than "dry" regions. Extremely high diurnal ranges often occur in summer over high deserts.**

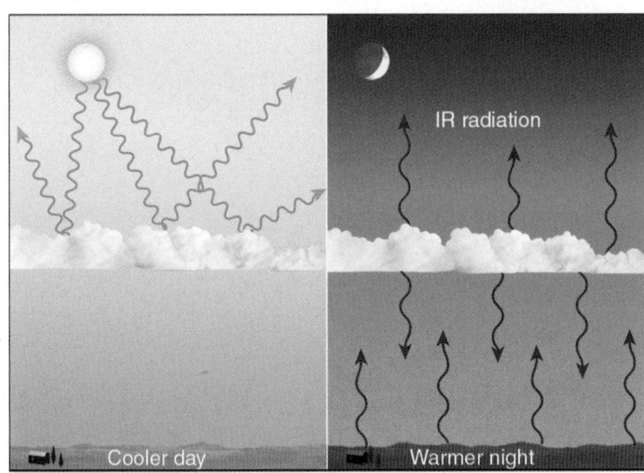

(a) Small daily temperature range

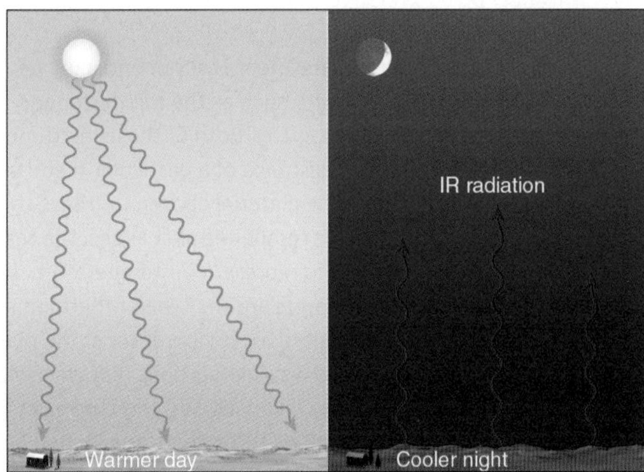

(b) Large daily temperature range

FIGURE 3.11 (a) Clouds tend to keep daytime temperatures lower and nighttime temperatures higher, producing a small daily range in temperature. (b) In the absence of clouds, days tend to be warmer and nights cooler, producing a larger daily range in temperature.

EXTREME WEATHER WATCH

What to wear on this day? On August 17, 2002, Rapid City, South Dakota, set a record low temperature of 39°F. By late afternoon, the air temperature reached 101°F, a record high temperature for that date.

So far we've looked at the hottest and coldest places on earth. We've also looked at how the air temperature changes during a typical day. We will now turn our attention to average seasonal temperatures and why they differ so much from one region to another. Why, for example, are winters in Grand Forks, North Dakota, so much colder than winters in Seattle, Washington, when both cities are situated at nearly the same latitude?

FOCUS ON INSTRUMENTS

Measuring Temperature

A very common thermometer that measures surface air temperature is the *liquid-in-glass thermometer* (see Fig. 2). These thermometers have a glass bulb attached to a sealed, graduated tube about 25 cm (10 in.) long. A very small opening, or bore, extends from the bulb to the end of the tube. A liquid in the bulb (usually mercury or red-colored alcohol) is free to move from the bulb up through the bore and into the tube. When the air temperature increases, the liquid in the bulb expands, and rises up the tube. When the air temperature decreases, the liquid contracts, and moves down the tube. Hence, the length of the liquid in the tube represents the air temperature. Because the bore is very narrow, a small temperature change will show up as a relatively large change in the length of the liquid column.

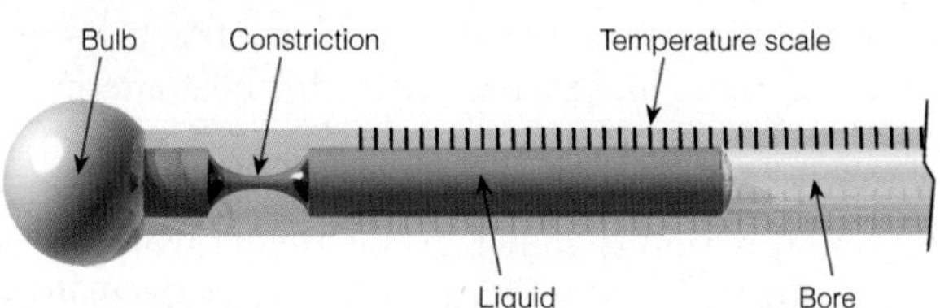

Figure 2 A section of a special type of liquid-in-glass thermometer called a *maximum thermometer* measures the maximum temperature for the day. Once the maximum temperature is reached, the constriction prevents the mercury from moving back into the bulb.

Highly accurate temperature measurements may be made with *electrical thermometers*, such as the *thermistor* and the *electrical resistance thermometer.* Both of these instruments measure the electrical resistance of a particular material. Since the resistance of the material chosen for these thermometers changes as the temperature changes, the resistance can be calibrated to represent air temperature. Electrical resistance thermometers are the type of thermometers used in the measurement of air temperature at the over 900 fully automated surface weather stations (known as *ASOS* for *A*utomated *S*urface *O*bserving *S*ystem) that exist at airports and military facilities throughout the United States.

Thermometers and other instruments are usually housed in an *instrument shelter*. The shelter completely encloses the instruments, protecting them from rain, snow, and the sun's direct rays. It is painted white to reflect sunlight, faces north to avoid direct exposure

Figure 3 The instruments that comprise the ASOS system. The max-min temperature shelter is the middle box.

REGIONAL TEMPERATURE VARIATIONS The main factors that cause variations in temperature from one place to another are called the **controls of temperature.** In the previous chapter, we saw that the greatest factor in determining temperature is the amount of solar radiation that reaches the surface. This amount, of course, is determined by the length of daylight hours and the intensity of incoming solar radiation. Both of these factors are a function of latitude; hence, latitude is considered an important control of temperature. The main controls are:

1. latitude
2. land and water distribution
3. ocean currents
4. elevation

We can obtain a better picture of these controls by examining ❱ Fig. 3.12 and ❱ Fig. 3.13, which show the average monthly temperatures throughout the world for January and July. The lines on the map are **isotherms**—lines connecting places that have the same temperature. Because air temperature normally decreases with height, cities at very high elevations are much colder than their

to sunlight, and usually has louvered sides, so that air is free to flow through it. This construction helps to keep the air inside the shelter at the same temperature as the air outside. Thermometers inside a standard shelter are mounted about 5 to 6 ft above the ground.

The older instrument shelters are gradually being replaced by the *Max-Min Temperature Shelter* (see Fig. 3). The shelter is mounted on a pipe, and wires from the electrical temperature sensor inside are run to a building. A readout inside the building displays the current air temperature and stores the maximum and minimum temperatures for later retrieval. This type of shelter is now used with the automated (ASOS) system.

Because air temperatures vary considerably above different types of surfaces, shelters are usually placed over grass to ensure that the air temperature is measured at the same elevation over the same type of surface. Unfortunately, some shelters are placed on asphalt, others sit on concrete, while others are located on the tops of tall buildings, making it difficult to compare air temperature measurements from different locations. In fact, if either the maximum or minimum air temperature in your area seems suspiciously different from those of nearby towns, find out where the instrument shelter is situated.

A vertical profile of temperature (as well as pressure and humidity) up to an altitude of about 30 km (100,000 ft) can be obtained with an instrument called a *radiosonde.* The radiosonde is a small, lightweight box equipped with weather instruments and a radio transmitter. It is attached to a cord that has a parachute and a gas-filled balloon tied tightly at the end (see Fig. 4). As the balloon rises, the attached radiosonde measures air temperature with a small electrical thermometer located just outside the box.

Air temperature, moisture, and pressure information are transmitted to the surface where a computer reconverts the various radio frequencies into values of temperature and humidity. When plotted on a graph, the vertical distribution of temperature is called a *sounding.* Eventually, the balloon bursts and the radiosonde returns to earth, its descent being slowed by its parachute.

Figure 4 The radiosonde with parachute and balloon.

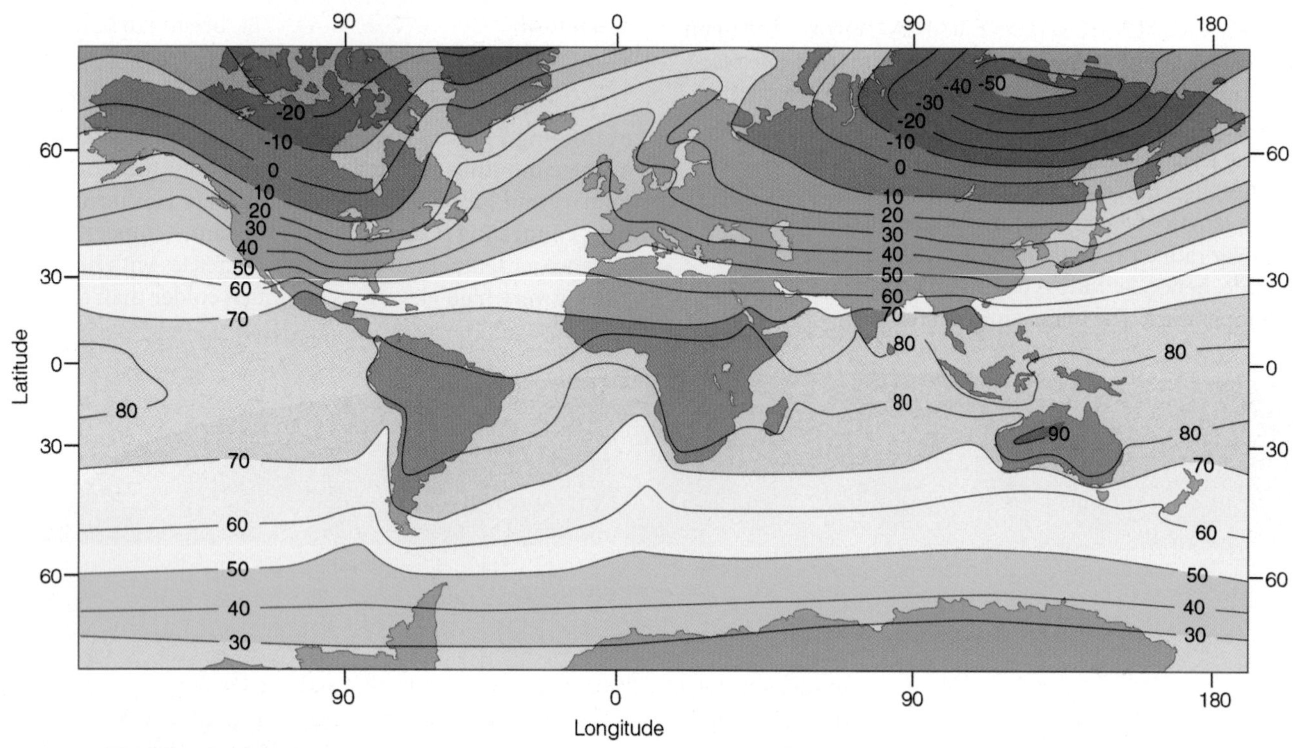

FIGURE 3.12 Average air temperature near sea level in January (°F).

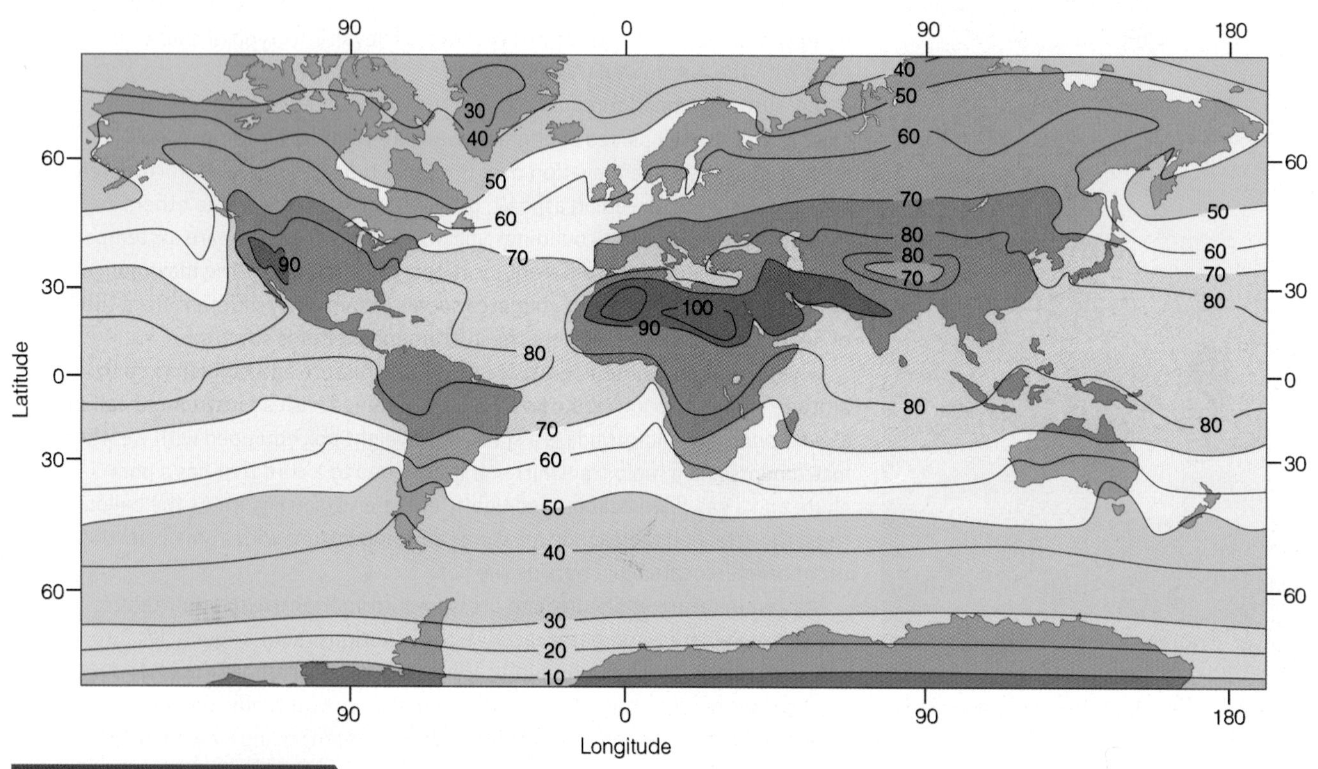

FIGURE 3.13 Average air temperature near sea level in July (°F).

sea-level counterparts. Consequently, the isotherms in Fig. 3.12 and Fig. 3.13 are corrected to read at the same horizontal level (sea level) by adding to each station above sea level an amount of temperature that would correspond to an average temperature change with height.

Figures 3.12 and 3.13 show the importance of latitude on temperature. Notice that on both maps and in both hemispheres the isotherms are oriented east-west, indicating that locations at the same latitude receive nearly the same amount of solar energy. In addition, the annual solar heat that each latitude receives decreases from low-to-high latitudes; hence, average temperatures in January and July tend to decrease from low-to-high latitudes. However, because there is a greater variation in solar radiation between low and high latitudes in winter than in summer, notice that the isotherms in January (during the Northern Hemisphere winter) are closer together (a tighter gradient*) than they are in July. This circumstance means that if you travel from New Orleans to Detroit in January, you are more likely to experience greater temperature variations than if you make the same trip in July.

Even though average temperatures tend to decrease from low latitudes toward high latitudes, notice on the July map (Fig. 3.13) that the highest average temperatures do not occur in the tropics, but rather (as we saw earlier) in the subtropical deserts of the Northern Hemisphere. Here, sinking air associated with high-pressure areas generally produces clear skies and low humidity. These conditions, along with a high sun beating down upon a relatively barren landscape, produce scorching heat.

For extreme cold, notice on the January map (Fig. 3.12) that the lowest average temperatures are found in the interior of Siberia, where the average January temperature dips below –50°F. As cold as this region is, it is even colder over the Antarctic. Extremely cold surface air forms as relatively dry air, high elevations, and as snow-covered surfaces allow for rapid radiational cooling during the Antarctic's dark winter months. Although not shown in Fig. 3.13, the average temperature for the coldest month at the South Pole is below –70°F. And for absolute cold, the lowest average temperature for any month (–100°F) was recorded at the Plateau Station during July, 1968.

So far we've seen that January temperatures in the Northern Hemisphere are much lower in the middle of continents than they are at the same latitude near the oceans. Notice on the July map (Fig. 3.13) that the reverse is true. One reason for these temperature differences can be attributed to the unequal heating and cooling properties of land and water. For one thing, solar energy reaching land is absorbed in a thin layer of soil; reaching water, it penetrates deeply. Because water is able to circulate, it distributes its heat through a much deeper layer. Also, some of the solar energy striking the water is used to evaporate it rather than heat it. It takes a great deal more heat to raise the temperature of a given amount of water by one degree than it does to raise the temperature of the same amount of land by one degree. Water not only heats more slowly than land, it cools more slowly as well, and so the oceans act like huge heat reservoirs. Thus, mid-ocean surface temperatures change relatively little from summer to winter compared to the much larger annual temperature changes over the middle of continents.

At any location, the difference in average temperature between the warmest month (often July in the Northern Hemisphere) and coldest month (often January) is called the **annual range of temperature**. Notice in Fig. 3.12 and Fig. 3.13 that, near the equator, annual temperature ranges are small, whereas, in the middle of high latitude landmasses, annual ranges are quite large. As an example, Yakutsk, located in northeastern Siberia near the Arctic Circle, has an extremely large annual temperature range of 112°F. The average annual temperature of this region is close to 0°F. In North America, Winnipeg, Canada, is a city in the middle of a continent with a large annual temperature range of 68°F (see Fig. 3.14).

*Gradient represents the rate of change of some quantity (in this case temperature) over a given distance.

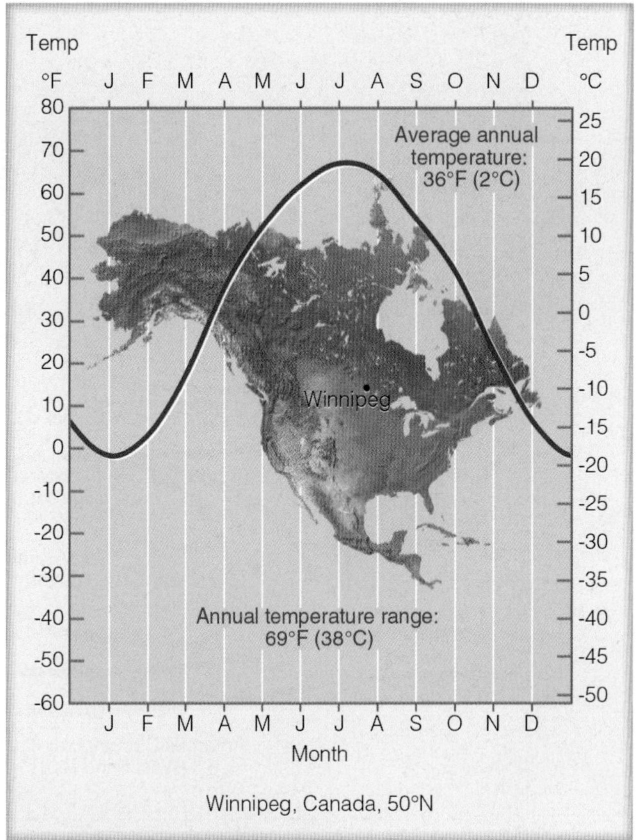

FIGURE 3.14 The red line shows the average monthly temperatures for Winnipeg, Canada. Located at latitude 50°N, in the middle of the continent, Winnipeg has a large annual temperature range of 68°F.

EXTREME WEATHER WATCH

The place with the world's largest range of temperature is the Siberian city of Verkhoyansk, Russia. Here, the lowest temperature ever measured was –90°F and the highest was 98°F, giving the city an extreme range in temperature of 188°F.

Look closely at Figs. 3.12 and 3.13 on p. 76 and notice that the isotherms on both maps tend to bend when they approach an ocean–continent boundary. Such bending is due in part to the unequal heating and cooling properties of land and water that we discussed earlier, and in part to *ocean currents.* Figure 3.15 shows the distribution of the major ocean currents throughout the world. Notice that along the eastern margins of continents, warm ocean currents transport warm water poleward whereas, along the western continental margins, cold currents transport cold water equatorward. Consequently, ocean currents play a vital role in transporting heat from one region of the world to another. For example, the warm Gulf Stream, which forms northward parallel to the coast of North America, gradually widens and slows as it merges with the broader North Atlantic Drift. As this current moves northeastward along the coast of Great Britain and Norway, it brings with it warm water that helps keep winter temperatures much warmer than one would expect this far north (see Fig. 3.16).

Along the west coast of North America, the cool California current flows southward, bringing cool water with it. The cooling of the water is enhanced along the coast of Northern California, as cold water from below rises to the surface in a process called *upwelling.* The cold water helps keep summer temperatures directly along the coast quite low, as the average daily high temperature in downtown

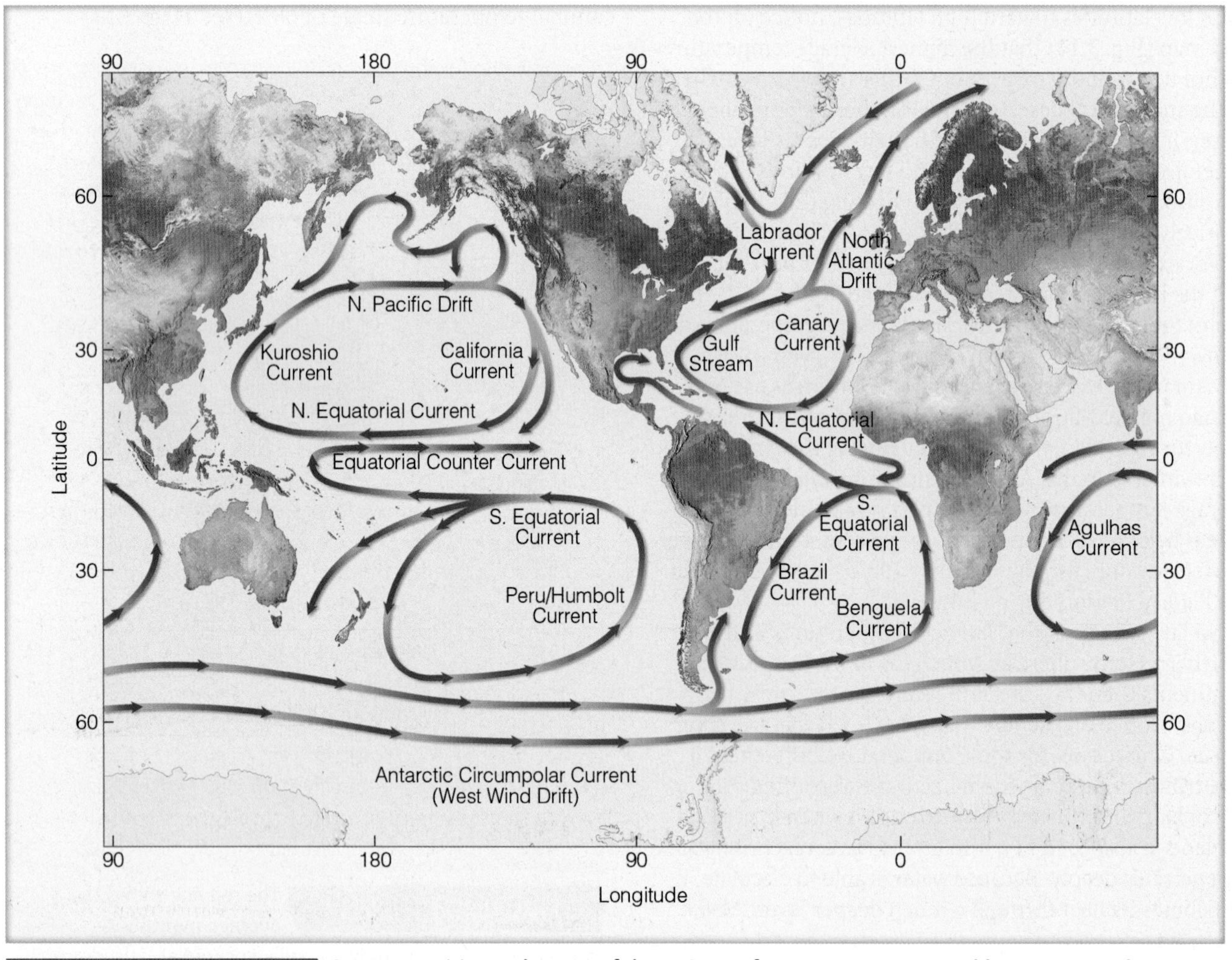

FIGURE 3.15 Average position and extent of the major surface ocean currents. Cold currents are shown in blue; warm currents are shown in red.

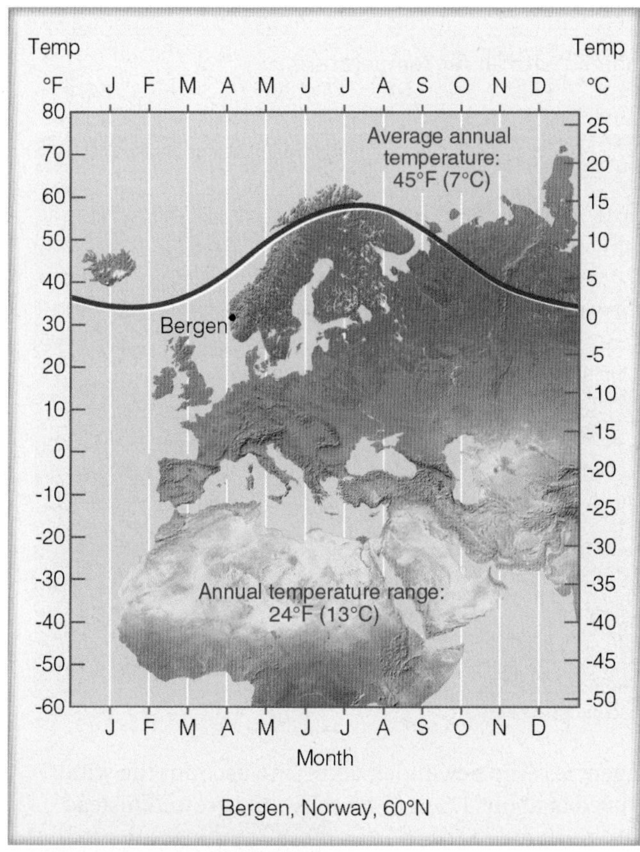

FIGURE 3.16 Monthly temperatures for Bergen, Norway. Located at 60°N latitude, Bergen's average winter temperature (December, January, February) is much higher than one would expect for a city located at this latitude. (Compare Bergen's temperature with that of Winnipeg, Canada, Fig. 3.14.)

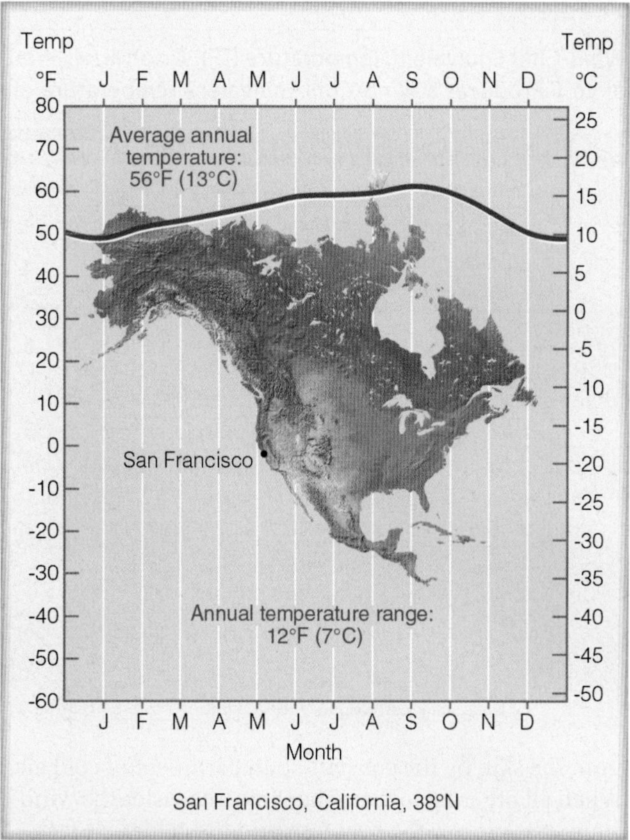

FIGURE 3.17 Monthly temperatures for San Francisco, California. Located along the coast at latitude 37°N, San Francisco experiences a cool ocean current and upwelling that combine to modify the city's climate and make its summers quite cool.

San Francisco during July is only 62°F. These low summer coastal temperatures and a chilling breeze blowing off the ocean led Mark Twain to say, "The coldest winter I ever experienced was a summer in San Francisco." Figure 3.17 shows the effect that upwelling has on average monthly temperatures in San Francisco.

Weather Extremes and Human Discomfort

On an overcast summer afternoon in San Francisco, the air will feel much colder with a wind of 20 mi/hr than when the air is calm. The human body's perception of temperature—called *sensible temperature*—obviously changes with varying atmospheric conditions. The reason for these changes is related to how we exchange heat energy with our environment. Let's look at this concept in more detail.

WIND AND COLD Why do we usually feel colder when the wind blows? The human body stabilizes its temperature, primarily by converting food into heat (*metabolism*). To maintain a constant temperature, the heat produced and absorbed by the body must be equal to the heat it loses to its surroundings. There is, therefore, a constant exchange of heat—especially at the surface of the skin—between the body and the environment.

One way the body loses heat is by emitting infrared energy. But we not only emit radiant energy, we absorb it as well. Another way the body loses and gains heat is by conduction and convection, which transfer heat to and from the body by air motions. On a cold day, a thin layer of warm air molecules forms close to the skin, protecting it from the surrounding cooler air and from the rapid transfer of heat. Thus, in cold weather, when the air is calm, the temperature we perceive is often higher than a thermometer might indicate.

Once the wind starts to blow, the insulating layer of warm air is swept away, and heat is rapidly removed

Table 3.3

Wind-Chill Equivalent Temperature (°F). A 20-Mi/Hr Wind Combined with an Air Temperature of 20°F Produces a Wind-Chill Equivalent Temperature of 4°F*

WIND SPEED (MI/HR)	AIR TEMPERATURE (°F)																
Calm	40	35	30	25	20	15	10	5	0	-5	-10	-15	-20	-25	-30	-35	-40
5	36	31	25	19	13	7	1	-5	-11	-16	-22	-28	-34	-40	-46	-52	-57
10	34	27	21	15	9	3	-4	-10	-16	-22	-28	-35	-41	-47	-53	-59	-66
15	32	25	19	13	6	0	-7	-13	-19	-26	-32	-39	-45	-51	-58	-64	-71
20	30	24	17	11	4	-2	-9	-15	-22	-29	-35	-42	-48	-55	-61	-68	-74
25	29	23	16	9	3	-4	-11	-17	-24	-31	-37	-44	-51	-58	-64	-71	-78
30	28	22	15	8	1	-5	-12	-19	-26	-33	-39	-46	-53	-60	-67	-73	-80
35	28	21	14	7	0	-7	-14	-21	-27	-34	-41	-48	-55	-62	-69	-76	-82
40	27	20	13	6	-1	-8	-15	-22	-29	-36	-43	-50	-57	-64	-71	-78	-84
45	26	19	12	5	-2	-9	-16	-23	-30	-37	-44	-51	-58	-65	-72	-79	-86
50	26	19	12	4	-3	-10	-17	-24	-31	-38	-45	-52	-60	-67	-74	-81	-88
55	25	18	11	4	-3	-11	-18	-25	-32	-39	-46	-54	-61	-68	-75	-82	-81
60	25	17	10	3	-4	-11	-19	-26	-33	-40	-48	-55	-62	-69	-76	-84	-91

*Dark shaded areas represent conditions where frostbite occurs in 30 minutes or less.

from the skin by the constant, bombardment of cold air. When all other factors are the same, the faster the wind blows, the greater the heat loss, and the colder we feel. How cold the wind makes us feel is usually expressed as a **wind-chill index (WCI).**

The modern wind-chill index (see Table 3.3 and Table 3.4) was formulated in 2001 by a joint action group of the National Weather Service and other agencies. The new index takes into account the wind speed at about 1.5 m (5 ft) above the ground instead of the 10 m (33 ft) where "official" readings are usually taken. In addition, it translates the ability of the air to take heat away from a person's face (the air's cooling power) into a wind-chill equivalent temperature. For example, notice in Table 3.3 that an air temperature of 10°F with a wind speed of 10 mi/hr produces a wind-chill equivalent temperature of –4°F. Under these conditions,

Table 3.4

Wind-Chill Equivalent Temperature (°C)*

WIND SPEED (KM/HR)	AIR TEMPERATURE (°C)												
Calm	10	5	0	-5	-10	-15	-20	-25	-30	-35	-40	-45	-50
10	8.6	2.7	-3.3	-9.3	-15.3	-21.1	-27.2	-33.2	-39.2	-45.1	-51.1	-57.1	-63.0
15	7.9	1.7	-4.4	-10.6	-16.7	-22.9	-29.1	-35.2	-41.4	-47.6	-51.6	-59.9	-66.1
20	7.4	1.1	-5.2	-11.6	-17.9	-24.2	-30.5	-36.8	-43.1	-49.4	-55.7	-62.0	-68.3
25	6.9	0.5	-5.9	-12.3	-18.8	-25.2	-31.6	-38.0	-44.5	-50.9	-57.3	-63.7	-70.2
30	6.6	0.1	-6.5	-13.0	-19.5	-26.0	-32.6	-39.1	-45.6	-52.1	-58.7	-65.2	-71.7
35	6.3	-0.4	-7.0	-13.6	-20.2	-26.8	-33.4	-40.0	-46.6	-53.2	-59.8	-66.4	-73.1
40	6.0	-0.7	-7.4	-14.1	-20.8	-27.4	-34.1	-40.8	-47.5	-54.2	-60.9	-67.6	-74.2
45	5.7	-1.0	-7.8	-14.5	-21.3	-28.0	-34.8	-41.5	-48.3	-55.1	-61.8	-68.6	-75.3
50	5.5	-1.3	-8.1	-15.0	-21.8	-28.6	-35.4	-42.2	-49.0	-55.8	-62.7	-69.5	-76.3
55	5.3	-1.6	-8.5	-15.3	-22.2	-29.1	-36.0	-42.8	-49.7	-56.6	-63.4	-70.3	-77.2
60	5.1	-1.8	-8.8	-15.7	-22.6	-29.5	-36.5	-43.4	-50.3	-57.2	-64.2	-71.1	-78.0

*Dark shaded areas represent conditions where frostbite occurs in 30 minutes or less.

the skin of a person's exposed face would lose as much heat in one minute in air with a temperature of 10°F and a wind speed of 10 mi/hr as it would in calm air with a temperature of –4°F. Of course, how cold we feel actually depends on a number of factors, including the fit and type of clothing we wear, the amount of sunshine striking the body, and the actual amount of exposed skin.

High winds, in below-freezing air, can remove heat from exposed skin so quickly that the skin may actually freeze and discolor. The freezing of skin, called **frostbite,** usually occurs on the body extremities first because they are the greatest distance from the source of body heat.

Possibly the lowest wind-chill ever occurred in Antarctica. On August 25, 2005, the Russian Antarctic station of Vostok recorded an air temperature of –99°F and a wind speed of 113 mi/hr, resulting in a wind-chill equivalent temperature below –99°F. The exact wind-chill is not known because the readings of wind-chill do not go that low (see Table 3.3). However, under these extreme conditions, any exposed skin would freeze in a few seconds and any moisture on exposed lips would instantly turn to ice.

COLD, DAMP WEATHER In cold weather, wet skin can be a factor in how cold we feel. A cold rainy day (drizzly, or even foggy) often feels colder than a "dry" one because water on exposed skin conducts heat away from the body better than air does. In fact, in cold, wet, and windy weather a person may actually lose body heat faster than the body can produce it. This process may even occur in relatively mild weather with air temperatures as high as 50°F. The rapid loss of body heat may lower the body temperature below its normal level and bring on a condition known as **hypothermia**—the rapid, progressive mental and physical collapse that accompanies the lowering of human body temperature.

The first symptom of hypothermia is exhaustion. If exposure continues, judgment and reasoning power begin to disappear. Prolonged exposure, especially at temperatures near or below freezing, produces stupor, collapse, and death when the internal body temperature drops to 79°F. Most cases of hypothermia occur when the air temperature is between freezing and 50°F, probably because many people apparently do not realize that wet clothing in windy weather greatly enhances the loss of body heat, even when the temperature is well above freezing.

HUMIDITY—A REAL FACTOR IN HOW HOT OR COLD WE FEEL In cold weather, heat is more easily dissipated through the skin. To counteract this rapid heat loss, the peripheral blood vessels of the body constrict, cutting off the flow of blood to the outer layers of the skin. On the other hand, in hot weather, the blood vessels enlarge, allowing a greater loss of heat energy to the surroundings. In addition to this, we perspire. As evaporation occurs, the skin cools. When the air contains a great deal of water vapor and it is close to being saturated, perspiration does not readily evaporate from the skin. Less evaporational cooling causes most people to feel hotter than it really is, and a number of people start to complain about the "heat and humidity." But what type of humidity are people complaining about? Since there are a number of ways of specifying the amount of water vapor in the air, there are several meanings for the concept of humidity.

What Is "Relative" in Relative Humidity? We already know from Chapter 1 that the term **humidity** often refers to the amount of water vapor in the air. To most of us, a moist day suggests a high humidity. However, it is interesting to note that in hot, "dry" desert air there is often more water vapor than in cold, "damp" polar air. Does this observation mean that the desert air has a higher humidity? The answer is both yes and no, depending on the type of humidity we mean.

To understand the different meanings of humidity, imagine that we enclose a volume of air (about the size of a large balloon) in a thin elastic container—a *parcel*—as illustrated in Fig. 3.18. If we extract the water vapor from the parcel, we would specify the humidity in the following ways:

1. We could compare the weight (mass) of the water vapor with the volume of air in the parcel and obtain the *water vapor density,* or *absolute humidity.*

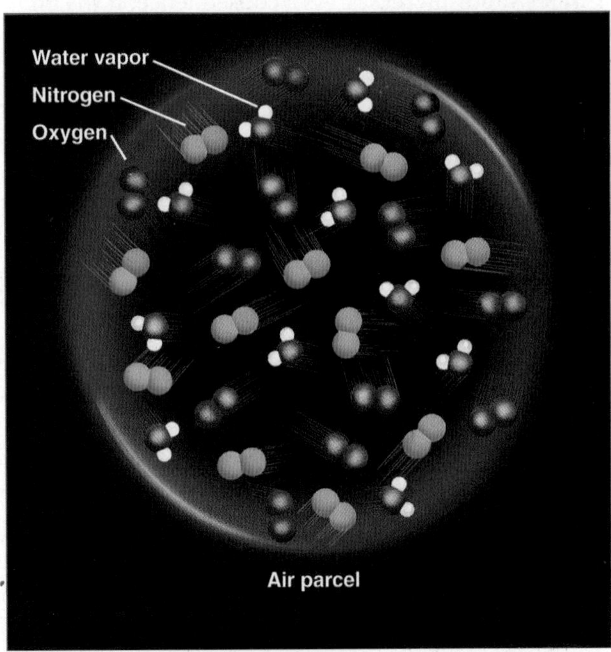

FIGURE 3.18 The water vapor content (humidity) inside this air parcel can be expressed in a number of ways.

2. We could compare the weight (mass) of the water vapor in the parcel with the total weight (mass) of all the air in the parcel (including vapor) and obtain the *specific humidity.*
3. Or, we could compare the weight (mass) of the water vapor in the parcel with the weight (mass) of the remaining dry air and obtain the *mixing ratio.*
4. If we leave the water vapor inside the parcel, we could express humidity of the air in terms of the pressure the water vapor molecules are exerting on the sides of the parcel. This form of humidity, which indicates the air's total water vapor content, is called *vapor pressure.*
5. If the air inside the parcel is saturated with water vapor, the vapor pressure at that temperature becomes the *saturation vapor pressure.*

Again, in Chapter 1 we learned that the most common way of describing atmospheric moisture is *relative humidity.* Recall that the **relative humidity (RH)** *is the ratio of the amount of water vapor actually in the air to the maximum amount of water vapor required for saturation at that particular temperature (and pressure),* as:

$$\text{RH} = \frac{\text{water vapor content}}{\text{water vapor capacity}} \times 100 \text{ percent.}$$

What can bring about a change in the air's relative humidity? We know from Chapter 1 that, as the air temperature rises, the air can hold more water molecules in the vapor state; that is, the air's capacity for water vapor increases. As a consequence, if the air temperature increases (and the content of water vapor in the air remains the same), the relative humidity will drop. Conversely, if the air temperature decreases (again, with no change in water vapor content), the relative humidity will rise. So a change in air temperature will bring about a change in relative humidity because the air's capacity for water vapor changes *relative* to the air's actual water vapor content. Figure 3.19 shows the effect air temperature has on the relative humidity during a 24-hour day.

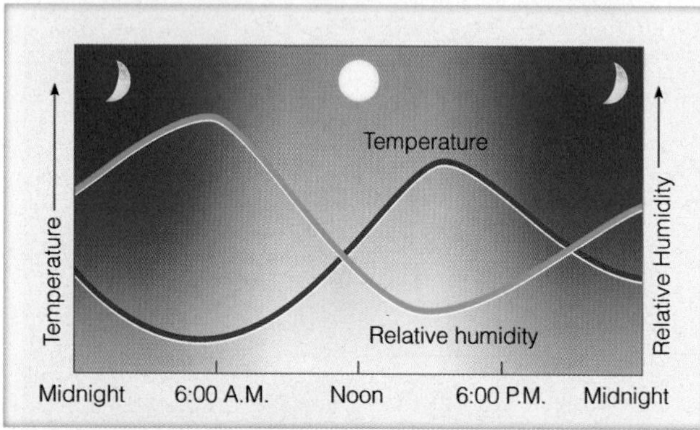

FIGURE 3.19 Due to changes in air temperature, the relative humidity usually changes during the course of a 24-hour day. When the air is cool (morning), the relative humidity is high. When the air is warm (afternoon), the relative humidity is low. These conditions exist in clear weather when the air is calm or of constant wind speed and there is little change in the air's water vapor content.

Relative humidity is always given as a percent. Air with a 100 percent relative humidity is saturated because the air is holding as much water in the vapor state that it can hold (at that temperature). Air with a 50-percent relative humidity is only holding 50 percent of its capacity for water vapor.

Another way to change the air's relative humidity is to increase or decrease the amount of water vapor in the air. At a constant temperature, adding water vapor to the air raises the relative humidity; removing water vapor from the air lowers it.

High Dew Points Mean Humid Air If relative humidity does not tell us how much water vapor is in the air, what does? A good indicator of the amount of water vapor in the air is the *dew-point temperature* or, simply, the **dew point.** Again, from Chapter 1, recall that the dew point *is the temperature to which air would have to be cooled (with no change in air pressure or moisture content) for saturation to occur. High dew points indicate high water vapor content; low dew points, low water vapor content.*

The difference between air temperature and dew point can indicate whether the relative humidity is low or high. When the air temperature and dew point are far apart, the relative humidity is low; when they are close to the same value, the relative humidity is high. When the air temperature and dew point are equal, the air is *saturated* and the relative humidity is 100 percent.

Figure 3.20a shows the average dew-point temperatures across the United States and southern Canada for January. Notice that the dew points are highest (the greatest amount of water vapor in the air) over the Gulf Coast states and lowest over the interior. Compare New Orleans with Fargo. Cold, dry winds from northern Canada flow relentlessly into the Center Plains during the winter, keeping this area dry. But warm, moist air from the Gulf of Mexico helps maintain a higher dew-point temperature in the southern states.

Figure 3.20b is a similar diagram showing the average dew-point temperatures for July. Again, the highest dew points are observed along the Gulf Coast, with some

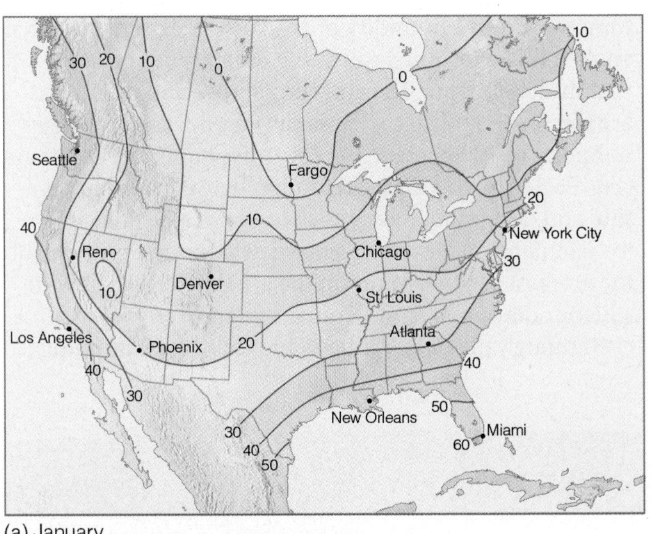

(a) January

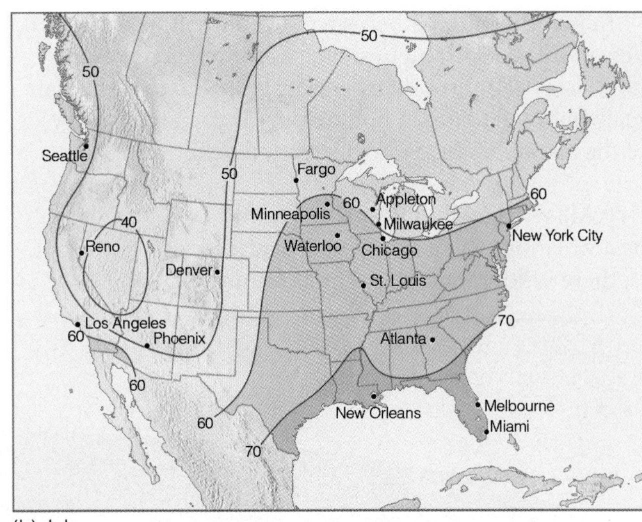

(b) July

FIGURE 3.20 Average surface dew-point temperatures (°F) for (a) January and for (b) July.

areas experiencing average dew-point temperatures near 75°F. Note, too, that the dew points over the eastern and central portion of the United States are much higher in July, meaning that the July air contains between 3 and 6 times more water vapor than the January air. The reason for the high dew points is that this region is almost constantly receiving humid air from the warm Gulf of Mexico. The lowest dew point, and, hence, the driest air, is found in the West, with Nevada experiencing the lowest values—a region surrounded by mountains that effectively shield it from significant amounts of moisture moving in from the southwest and northwest.

Dew-Point Extremes Record high dew points in the United States usually occur with the sweltering heat waves of summer. Dew points exceeding 80°F are rare in the United States; however, during the heat wave of July, 1995, many cities reported extremely high dew points. For example, Waterloo, Iowa, reported a dew point of 84°F and Philadelphia, Pennsylvania, a record high dew point of 82°F. The highest dew point ever in Chicago (83°F), Milwaukee (82°F), and Minneapolis (81°F) all occurred on the same date: July 30, 1999. The highest dew point in the United States (90°F) occurred at three locations: at New Orleans Naval Air Station on July 30, 1987; at Melbourne, Florida, on July 12, 1987; and at Appleton, Wisconsin, on July 14, 1995.

The highest dew points in the world occur in the Middle East near large bodies of extremely warm water that often have sea-surface temperatures in the upper 80s (°F) and low-to-mid 90s (°F). In fact, an extreme sea-surface temperature of 98°F was once measured on the Red Sea by a British ship. From off these extremely warm bodies of water flows hot, muggy air with exceptionally high dew points. As a consequence, Sharjah, in the United Arab Emirates, once reported a dew point of 93°F, and the highest dew point on record occurred at Dhahran, Saudi Arabia, on July 8, 2003, when the dew point reached 95°F (see Fig. 3.21).

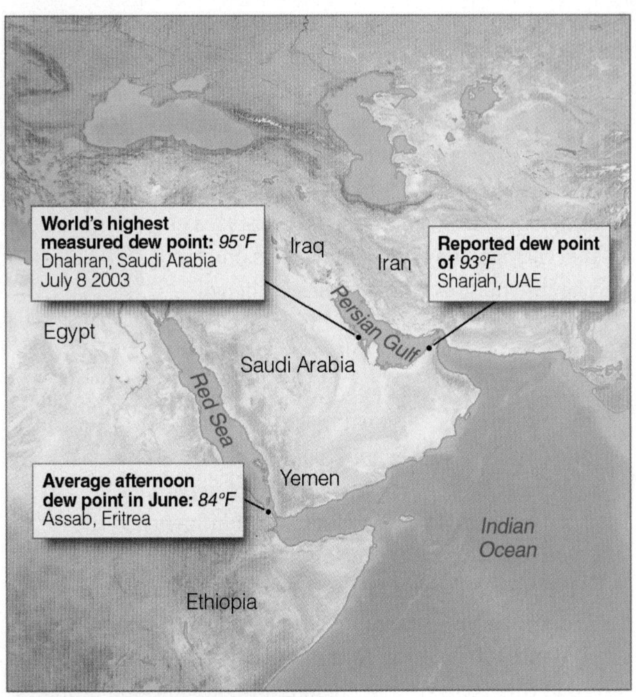

FIGURE 3.21

Record high dew-point temperatures (°F).

In hot, muggy weather, when the air temperature and dew point are both high, some people remark about how "heavy" or how dense the air feels. Is hot, humid air really more dense than hot, dry air? If you are not sure of the answer, read the Focus section below.

Dry Air with a High Relative Humidity Can there be a situation where the air is considered "dry"—that is, there is little water vapor in the air—and at the same time the relative humidity of this air is 100 percent? To answer this question, look at Fig. 3.22a and observe that in the polar air, the air temperature and dew point are the same, and so the air is saturated with a relative humidity of 100 percent. On the other hand, the desert air (Fig. 3.22b), with a large separation between air temperature and dew point, has a much lower relative humidity—21 percent. However, since dew point is a measure of the amount of water vapor in the air, the desert air (with a higher dew point) must contain *more* water vapor. So even though the polar air has a higher relative humidity,

FOCUS ON
A SPECIAL TOPIC

Humid Air and Dry Air Do Not Weigh the Same

Does a volume of hot, humid air really weigh more than a similar size volume of hot, dry air? The answer is no! At the same temperature and at the same level in the atmosphere, hot, humid air is lighter (less dense) than hot, dry air. The reason for this fact is that a molecule of water vapor (H_2O) weighs appreciably less than a molecule of either nitrogen (N_2) or oxygen (O_2). (Keep in mind that we are referring strictly to water vapor—a gas—and not suspended liquid droplets.)

Consequently, in a given volume of air, as lighter water vapor molecules replace either nitrogen or oxygen molecules one for one, the number of molecules in the volume does not change, but the total weight of the air becomes slightly less. Since air density is the mass of air in a volume, the more humid air must be lighter than the drier air. Hence, *hot humid air at the surface is lighter (less dense) than hot dry air.*

Figure 5 On this summer afternoon in Maryland, lighter (less-dense) hot, humid air rises and condenses into towering cumulus clouds.

This fact can have an important influence in the weather. The lighter the air becomes, the more likely it is to rise. All other factors being equal, hot, humid (less-dense) air will rise more readily than hot, dry (more-dense) air (see Fig. 5). It is, of course, the water vapor in the rising air that changes into liquid cloud droplets and ice crystals, which, in turn, grow large enough to fall to the earth as precipitation.

Of lesser importance to weather but of greater importance to sports is the fact that a baseball will "carry" farther in less-dense air. Consequently, without the influence of wind, a ball will travel slightly farther on a hot, humid day than it will on a hot, dry day. So when the sports announcer proclaims "the air today is heavy because of the high humidity" remember that this statement is not true and, in fact, a 404-foot home run on this humid day might simply be a 400-foot out on a very dry day.

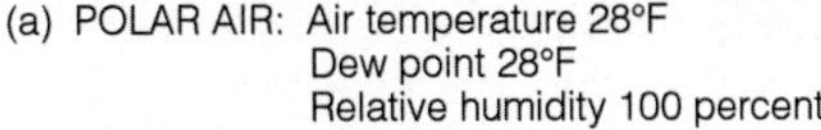
(a) POLAR AIR: Air temperature 28°F
Dew point 28°F
Relative humidity 100 percent

(b) DESERT AIR: Air temperature 95°F
Dew point 50°F
Relative humidity 21 percent

FIGURE 3.22 The polar air has the higher relative humidity; whereas, the desert air, with the higher dew point, contains more water vapor.

the desert air that contains more water vapor has a higher water vapor density, or *absolute humidity,* and a higher specific humidity and mixing ratio as well.

Now we can see why cold polar air is often described as being "dry" when the relative humidity is high (often close to 100 percent). In cold, polar air, the dew point and air temperature are normally close together. But the low dew point means that there is little water vapor in the air. Consequently, the air is said to be "dry" even though the relative humidity is quite high.

BRIEF REVIEW

Up to this point we've looked at the extremes of temperature and humidity. Before going on, here are some of the important concepts and facts we've covered so far:

- The largest annual temperature ranges are found at high latitudes in the middle of continents.
- Ocean currents play a role in the average worldwide distribution of temperature by transporting warm and cold water from one region to another. Warm water flowing northward along the coastline of Eastern Europe helps to keep winters in coastal cities much warmer than cities located in the middle of the continent.
- In cold weather, the stronger the wind blows, the greater the heat loss from exposed skin, and the colder we feel. How cold we feel is often expressed as "wind chill."
- When the amount of water vapor in the air remains constant, cooling the air raises the relative humidity and warming the air lowers it.
- In hot, muggy weather, we often feel warmer than it actually is because when the relative humidity is high, less perspiration is able to evaporate from our skin and cool us off.
- The dew point is a good indicator of the amount of water vapor in the air: High dew points indicate high water-vapor content. The highest measured dew point in the world was 95°F. Dew points in the range of 80°F seldom occur in the United States.
- In very cold weather, low dew points indicate dry air. If the dew point is close to the air temperature, the air is nearly saturated and the relative humidity is close to 100 percent.

FIGURE 3.23 When outside air with an air temperature and a dew point of 5°F is brought indoors and heated to a temperature of 68°F (without adding water vapor to the air), the relative humidity drops to 8 percent, placing adverse stress on plants, animals, and humans living inside. (*T* represents temperature; T_d, dew point; and *RH*, relative humidity.)

Extreme Relative Humidity in the Home During the winter, the relative humidity inside a home can drop to an extremely low value and the inhabitants are usually unaware of it. When cold polar air is brought indoors and heated, its relative humidity decreases dramatically. Notice in Fig. 3.23 that when outside air with a temperature and dew point of 5°F is brought indoors and heated to 68°F, the relative humidity of the heated air drops to 8 percent—a value lower than what you would normally experience in a desert during the hottest time of the day.

Very low relative humidities in a house can have an adverse effect on things living inside. For example, house plants have a difficult time surviving because the moisture from their leaves and the soil evaporates rapidly. Hence, house plants usually need watering more frequently in winter than in summer. People suffer, too, when the relative humidity is quite low. The rapid evaporation of moisture from exposed flesh causes skin to crack, dry, flake, or itch. These low humidities also irritate the mucous membranes in the nose and throat, producing an "itchy" throat. Similarly, dry nasal passages permit inhaled bacteria to incubate, causing persistent infections. The remedy for most of these problems is simply to increase the relative humidity. But how?

The relative humidity in a home can be increased just by heating water and allowing it to evaporate into the air. The added water vapor raises the relative humidity to a more comfortable level. In modern homes, a humidifier, installed near the furnace, adds moisture to the air. The air, with its increased water vapor, is circulated throughout the home by a forced air heating system.

It's Not the Heat, It's the Humidity We are now in a position to examine why people on a hot, muggy day often exclaim "It's not so much the heat, it's the humidity." In warm weather, the main source of body cooling is through evaporation of perspiration. Recall from Chapter 2 that evaporation is a cooling process, so when the air temperature is high and the relative humidity low, perspiration on the skin evaporates quickly, often making us feel that the air temperature is lower than it really is. However, when both the air temperature and relative humidity are high and the air is nearly saturated with water vapor, body moisture does not readily evaporate; instead, it collects on the skin as beads of perspiration. Less evaporation means less cooling, and so we usually feel warmer than we did with a similar air temperature, but a lower relative humidity. Consequently, people normally feel cooler when the air temperature is 90°F and the relative humidity is 20 percent than when the air temperature is 90°F and the relative humidity is 60 percent.

A good measure of how cool the skin can become is the **wet-bulb temperature**—*the lowest temperature that can be reached by evaporating water into the air.** On a hot day when the wet-bulb temperature is low, rapid evaporation (and, hence, cooling) takes place at the skin's surface. As the wet-bulb temperature approaches the air temperature, less cooling occurs, and the skin temperature may begin to rise. When the wet-bulb temperature exceeds the skin's temperature, no net evaporation occurs, and the body temperature can rise quite rapidly. Fortunately, most of the time, the wet-bulb temperature is considerably below the temperature of the skin.

When the weather is hot and muggy, a number of heat-related problems may occur. For example, in hot weather when the human body temperature rises, the *hypothalamus* gland (a gland in the brain that regulates body temperature) activates the body's heat-regulating mechanism, and over ten million sweat glands wet the body with as much as two liters of liquid per hour. As this perspiration evaporates, rapid loss of water and salt can result in a chemical imbalance that may lead to painful *heat cramps.* Excessive water loss through perspiring coupled with an

*Notice that the wet-bulb temperature and the dew-point temperature are different. The wet-bulb temperature is attained by *evaporating water* into the air, whereas the dew-point temperature is reached by *cooling* the air.

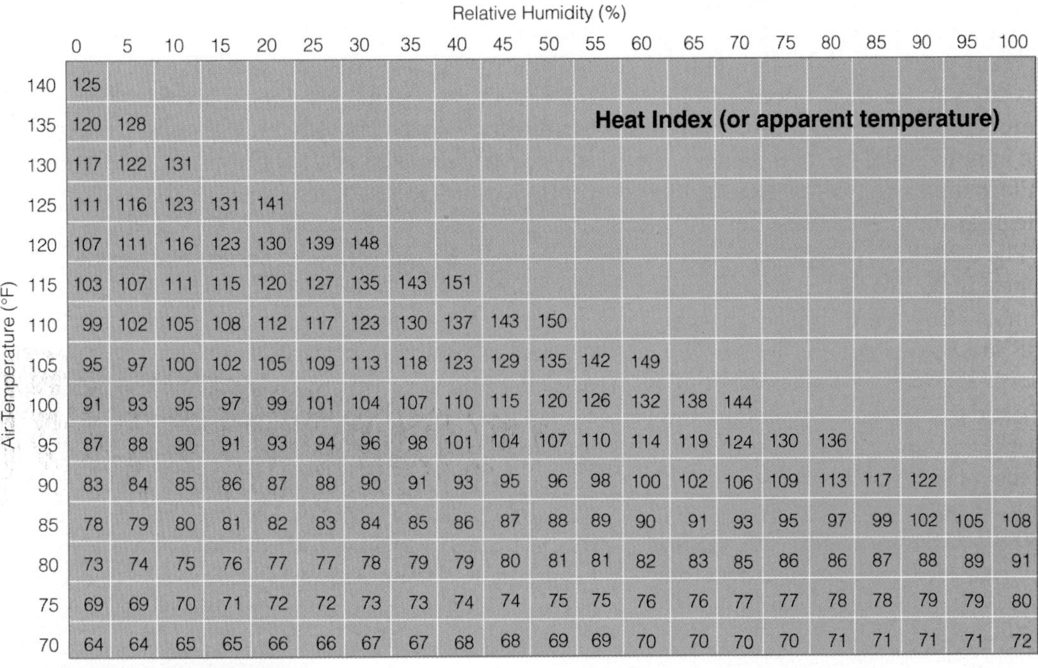

Heat Index (or apparent temperature)

Air Temperature (°F) / Relative Humidity (%)	0	5	10	15	20	25	30	35	40	45	50	55	60	65	70	75	80	85	90	95	100
140	125																				
135	120	128																			
130	117	122	131																		
125	111	116	123	131	141																
120	107	111	116	123	130	139	148														
115	103	107	111	115	120	127	135	143	151												
110	99	102	105	108	112	117	123	130	137	143	150										
105	95	97	100	102	105	109	113	118	123	129	135	142	149								
100	91	93	95	97	99	101	104	107	110	115	120	126	132	138	144						
95	87	88	90	91	93	94	96	98	101	104	107	110	114	119	124	130	136				
90	83	84	85	86	87	88	90	91	93	95	96	98	100	102	106	109	113	117	122		
85	78	79	80	81	82	83	84	85	86	87	88	89	90	91	93	95	97	99	102	105	108
80	73	74	75	76	77	77	78	79	79	80	81	81	82	83	85	86	86	87	88	89	91
75	69	69	70	71	72	72	73	73	74	74	75	75	76	76	77	77	78	78	79	79	80
70	64	64	65	65	66	66	67	67	68	68	69	69	70	70	70	70	71	71	71	71	72

FIGURE 3.24 Air temperature (°F) and relative humidity are combined to determine an apparent temperature or heat index (HI). An air temperature of 95°F with a relative humidity of 55 percent produces an apparent temperature (HI) of 110°F.

increasing body temperature may result in *heat exhaustion*—fatigue, headache, nausea, and even fainting. If one's body temperature rises above about 106°F, **heatstroke** can occur, resulting in complete failure of the circulatory functions. If the body temperature continues to rise, death may result. In fact, as we saw earlier, each year across North America, hundreds of people die from heat-related maladies. Even strong, healthy individuals can succumb to heatstroke, as did the Minnesota Vikings' all-pro offensive lineman, Korey Stringer, who collapsed after practice on July 31, 2001, and died 15 hours later. Before Korey fainted, temperatures on the practice field were in the 90s (°F) with the relative humidity above 55 percent.

The Heat Index In an effort to draw attention to this serious weather-related health hazard, an index called the **heat index (HI)** is used by the National Weather Service. The index combines air temperature with relative humidity to determine an **apparent temperature**—what the air temperature "feels like" to the average person for various combinations of air temperature and relative humidity. For example, in Fig. 3.24 an air temperature of 100°F and a relative humidity of 60 percent produce an apparent temperature of 132°F. As we can see in Table 3.5, heatstroke or sunstroke is imminent when the index reaches this level. However, as we saw in the preceding paragraph, heatstroke related deaths can occur when the heat index value is considerably lower than 130°F.

An extremely high heat index can occur with a high dew-point temperature. For example, when the dew point is 80°F and the air temperature is 95°F, the relative humidity is 63 percent. This combination of heat and moisture produces a heat index of 117°F. On July 8, 2003, when Dhahran, Saudi Arabia, recorded the world's highest dew point of 95°F, the air temperature was 108°F, the

Table 3.5

The Heat Index (HI) and Related Syndrome

CATEGORY	APPARENT TEMPERATURE (°F)	HEAT SYNDROME
I	130° or higher	Heatstroke or sunstroke *imminent*
II	105° – 130°	Sunstroke, heat cramps, or heat exhaustion *likely*, heatstroke *possible* with prolonged exposure and physical activity
III	90°–105°	Sunstroke, heat cramps, and heat exhaustion *possible* with prolonged exposure and physical activity
IV	80°–90°	Fatigue *possible* with prolonged exposure and physical activity

relative humidity 67 percent, and the heat index reached an incredible 176°F! ❱ Table 3.6 shows the average afternoon heat index, relative humidity, and dew point during July for various cities across the United States.

At this point it is important to dispel a common myth that seems to circulate in hot, humid weather. After being outside for awhile, people will say that the air temperature today is 90 degrees and the relative humidity is 90 percent, We see in Fig. 3.24 that this weather condition would produce a heat index of 122°F. Although this weather situation is remotely possible, it is *highly unlikely,* as a temperature of 90°F and a relative humidity of 90 percent can occur only if the dew-point temperature is very high (nearly 87°F), and a dew-point temperature this high, as we saw in an earlier section, rarely occurs in the United States, even on the muggiest of days.

DEADLY HEAT WAVES Heat waves kill.* On average, more Americans every year die from excessive heat than from any other weather-related disaster, including hurricanes and tornadoes. Heat is a silent killer. On average, over 300 people die each year from heat-related maladies. It is especially dangerous for the elderly living in urban areas. Often, it is not apparent that hundreds of victims have been claimed until days after the worst of the heat has abated and city officials are able to appraise mortality rates in relation to what might be heat-related or simply "normal" death rates. This was the case in Chicago, Illinois, during the torrid **heat wave** of July, 1995.

During the week of July 12, 1995, more than 700 deaths in Chicago were attributed to the heat. Of these, over 450 were directly heat-related, and the remainder as "excess than normal" fatalities. ❱ Table 3.7 shows the maximum and minimum temperatures reported at Chicago's Midway Airport on the city's south side, where many of the deaths occurred. Notice that the number of deaths began to soar about three days after the heat wave began. In fact, the highest heat index (119°F) occurred on July 13, when the air temperature reached

*A heat wave is a period of abnormal and uncomfortable hot and often humid weather.

❱ Table 3.6

Average July Afternoon Heat Index (HI) for Selected Cities in the United States along with Average Maximum July Temperatures, Average July Dew-Point Temperatures, and Average Afternoon July Relative Humidities

CITY	AVERAGE MAX TEMP (°F)	AVERAGE DEW POINT (°F)	AVERAGE AFTERNOON RH (%)	AVERAGE AFTERNOON HEAT INDEX (°F)
Yuma, AZ	107	59	21	108
Orlando, FL	92	72	64	107
New Orleans, LA	91	73	66	106
Phoenix, AZ	106	56	20	106
Corpus Christi, TX	93	74	56	106
Houston, TX	94	72	54	105
Tampa, FL	91	73	65	105
Lake Charles, LA	91	73	64	104
Charleston, SC	90	71	66	104
Dallas, TX	96	68	42	103
San Antonio, TX	94	69	43	100
Washington, DC	88	66	57	95
St. Louis, MO	89	67	51	94
Atlanta, GA	88	68	57	94
Sacramento, CA	93	53	28	91
New York, NY	86	64	53	88
Chicago, IL	84	62	54	87
Minneapolis, MN	84	60	50	85

Table 3.7

Maximum and Minimum Temperature Recorded at Chicago's Midway Airport along with Heat-Related Fatalities in the Chicago Area during the Period July 12, 1995, through July 19, 1995

	TEMPERATURE (°F*)		
Date	Max	Min	Fatalities
July 12	97°	75°	0
July 13	105°	81°	12
July 14	102°	84°	120
July 15	99°	77°	275
July 16	93°	76°	160
July 17	88°	73°	120
July 18	86°	69°	30
July 19	89°	69°	22

*The normal max and min temperatures for this period are 85°F and 65°F.

105°F and the dew point climbed to 76°F. In addition, extremely warm nighttime temperatures were as much to blame for the high mortality rates as were daytime maximums.

In the United States, another deadly heat wave took place during the summer of 1936. The exact number of deaths attributed to this event is unknown (due to the lack of accurate records), but probably more than 5000 deaths occurred due to the extreme heat. The intense and widespread heat wave resulted in temperatures soaring above 120°F in ten states, including North Dakota and South Dakota. Along the East Coast temperatures climbed above 110°F in New Jersey and Pennsylvania. In New York City the temperature even reached 106°F in Central Park. The extreme heat across the United States made this the hottest July on record. During August, the center of the heat wave moved south, where cities in Oklahoma, Texas, and Arkansas experienced extreme heat. Altus, Oklahoma, for example, had an average high temperature of 110°F for the month, with a maximum temperature of 120°F on August 12. If you look back at Fig. 3.4, p. 66, you will see that the record high temperature for many states took place during the summer of 1936.

EXTREME WEATHER WATCH

A ninety-ninety kind of day. At 9:00 A.M. on April 26, 2005, the air temperature in Bangkok, Thailand, stood at 91.4°F. With a dew point of 89.6°F and a relative humidity of 94 percent, the heat index exceeded 130°—high enough to cause a person to suffer heat stroke while walking to work.

By far the world's deadliest heat wave in modern times occurred in Western Europe during the first two weeks of August, 2003. Estimates are that between 35,000 and 52,000 people died during this horrific event, with 15,000 deaths reported in France alone. All-time record high temperatures were recorded in many countries, including Portugal (117°F), France (111°F), Switzerland (107°F), Germany (105°F), United Kingdom (101°F), and Luxembourg (100°F). Again, as in Chicago, exceptionally warm nighttime temperatures in the upper 70s and lower 80s (°F) were as much to blame for the deaths as the daily maximum temperatures.

Up to this point, we've looked at extreme temperature and humidity readings. Instruments that measure humidity are presented in the Focus section on p. 90.

Beating the Heat—Dealing with Heat Waves

Studies of the 2003 heat wave in Europe show that the elderly and people in lower socioeconomic groups had the highest risk of death. Many of these fatalities occurred when people were confined to their homes with inadequate options for mitigating the high temperatures.

There are strategies that one can take to help reduce the temperatures inside homes. They include:

- Install window air conditioners that fit snugly and are well insulated.
- Install temporary window reflectors (for use between windows and drapes), such as aluminum foil-covered cardboard, to reflect heat back outside.
- Add weather stripping to doors and sills to keep cool air in.

FOCUS ON
INSTRUMENTS

Measuring Humidity

The common instrument used to obtain dew point and relative humidity is a *psychrometer*, which consists of two liquid-in-glass thermometers mounted side by side and attached to a piece of metal that has either a handle or chain at one end (see Fig. 6). The thermometers are exactly alike except that one has a piece of cloth (wick) covering the bulb. The wick-covered thermometer—called the *wet bulb*—is dipped in clean water, whereas the other thermometer is kept dry. Both thermometers are ventilated for a few minutes, either by whirling the instrument (*sling psychrometer*), or by drawing air past it with an electric fan (*aspirated psychrometer*). Water evaporates from the wick and that thermometer cools. The drier the air, the greater the amount of evaporation and cooling. After a few minutes, the wick-covered thermometer will cool to the lowest value possible. Recall from an earlier section that this is the *wet-bulb temperature*—the lowest temperature that can be attained by evaporating water into the air.

The dry thermometer (commonly called the *dry bulb*) gives the current air temperature, or *dry-bulb temperature*. The temperature difference between the dry bulb and the wet bulb is known as the *wet-bulb depression*. A large depression indicates that a great deal of water can evaporate into the air and that the relative humidity is low. A small depression indicates that little evaporation of water vapor is possible, so the air is close to saturation and the relative humidity is high. If there is no depression, the dry bulb, the wet bulb, and the dew point are the same; the air is saturated and the relative humidity is 100 percent. (Tables used to compute relative humidity and dew point are given in Appendix D.

Instruments that measure humidity are commonly called *hygrometers*. One type—called the *hair hygrometer*—uses human (or horse) hair to measure relative humidity. It is constructed on the principle that, as the relative humidity increases, the length of hair increases and, as the relative humidity decreases, so does the hair length. A number of strands of hair (with oils removed) are attached to a system of levers. A small change in hair length is magnified by a linkage system and transmitted to a dial (Fig. 7) calibrated to show relative humidity, which can then be read directly or recorded on a chart. Often, the chart is attached to a clock-driven rotating drum that gives a continuous record of relative humidity.

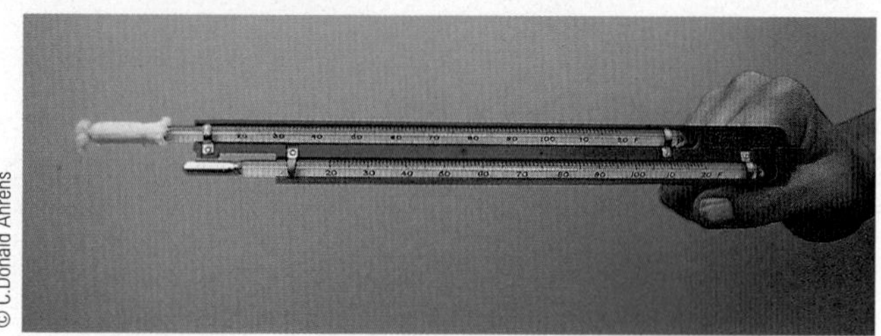

Figure 6 **The sling psychrometer.**

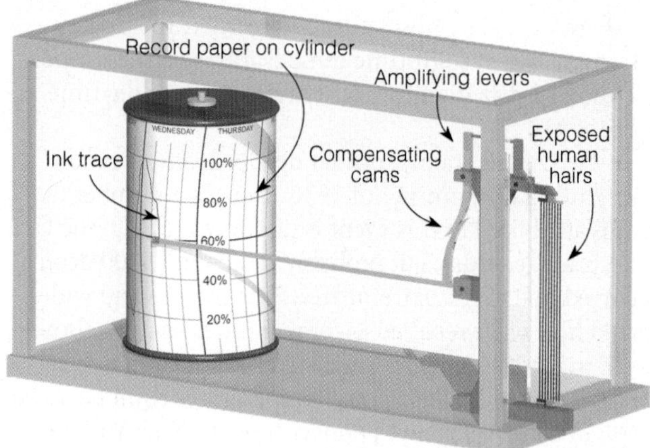

Figure 7 **The hair hygrometer measures relative humidity by amplifying and measuring changes in the length of human (or horse) hair.**

The *electrical hygrometer* is another instrument that measures humidity. It consists of a flat plate coated with a film of carbon. An electric current is sent across the plate. As water vapor is absorbed, the electrical resistance of the carbon coating changes. These changes are translated into relative humidity. This instrument is commonly used in the radiosonde, which gathers atmospheric data at various levels above the earth. The *dew-point hygrometer* measures the dew-point temperature by cooling the surface of a mirror until condensation (dew) forms. This sensor is the type that measures dew-point temperature in the hundreds of fully automated weather stations—Automated Surface Observing System (ASOS)—that exist throughout the United States.*

*A picture of ASOS is shown in Fig. 3 on p. 74.

- Cover windows that receive morning or afternoon sun with drapes, shades, awnings, or louvers. (Outdoor awnings or louvers can reduce the heat that enters a home by up to 80 percent.)
- Keep storm windows up all year.

Some of these options are, unfortunately, not affordable to the very populations at highest risk. For those who cannot access air conditioning or who do not have a temporary location to move to, the best options may be:

- Stay indoors as much as possible (preferably on the lowest floor) and limit exposure to the sun.
- Consider spending the warmest part of the day in air-conditioned public buildings, such as libraries, schools, movie theaters, shopping malls, and other community facilities.
- Use fans. Circulation of air can cool the body by increasing the rate of evaporating perspiration.
- Eat well-balanced, light, and regular meals. Avoid using salt tablets unless directed to do so by a physician.
- Drink plenty of water and limit intake of alcoholic beverages. Persons on fluid-restricted diets or who have a problem with fluid retention should consult a doctor before increasing liquid intake.
- Dress in loose-fitting, lightweight, and light-colored clothes. If you go outside, protect your face and head by wearing a wide-brimmed hat.
- Check on family, friends, and neighbors who do not have air conditioning and who spend much of their time alone.
- Avoid strenuous work during the warmest part of the day. Use a buddy system when working in extreme heat, and take frequent breaks.

Warm air rises. Consequently, the upper floor of an apartment building is often the warmest floor during a heat wave. For this reason, insulation is usually added to the building's attic. Often the roof of a building is painted a light color to increase the roof's reflectivity (albedo). On some roof's, vegetation is planted to reduce the rooftop temperature, so that less heat will transfer into the building (see ❱Fig. 3.25).

❱ FIGURE 3.25 **Vegetation planted on a rooftop in Buenos Aires, Argentina, reduces solar heating of the roof, which reduces the transfer of heat into the upper floor.**

Summary

This chapter examines the daily variation in air temperature and humidity, along with the hottest, coldest, and most humid places found on earth. The chapter begins by looking at the daily variation in air temperature near the surface. Here we find that the air temperature is controlled mainly by the input of energy from the sun and the output of energy from the surface. On a clear, calm day, the surface air warms, as long as heat input (mainly sunlight) exceeds heat output (mainly convection and radiated infrared energy). The warmest part of the day is usually in the afternoon because, until this time, incoming solar energy usually exceeds outgoing energy from the surface. The surface air cools at night as long as heat output exceeds input. Because the ground at night cools more quickly than the air above, the coldest air is normally found at the surface, where a radiation inversion usually forms. The coldest nights of winter tend to occur on clear, calm, dry nights.

The hottest region in the United States is the desert southwest. The hottest region in the world occurs in summer over the subtropical deserts of the Northern Hemisphere, where El Azizia in Libya reported a record high temperature of 136°F. The coldest regions in the United States are found in Alaska, in the northern Great Plains, and in northern Maine. The coldest region in the world is Antarctica, where the air temperature dropped to −129°F. The most humid region in the world—that with the highest dew points—occurs in the Middle East near large bodies of extremely warm water.

The diurnal and annual ranges of temperature are greater in dry climates than in humid ones. Near the equator, the annual range of temperature is small. In the middle of high-latitude landmasses, annual temperature ranges are usually quite large.

The wind and humidity play a role in how cold or hot we feel. In cold weather, the faster the wind blows, the greater the heat loss from our skin, and the colder we feel. How cold we feel can be expressed as wind chill. In hot, humid weather, when the dew point and relative humidity are both high, less perspiration is able to evaporate from the skin and cool us off, so we feel hotter than it actually is. When cold, dry outside air is brought indoors and heated to room temperature, the relative humidity can drop to extremely low values.

The heat index is a measure of how hot it feels to an average person for various combinations of air temperature and relative humidity. On average, more people die from heat-related maladies than from any other weather-related event.

Key Terms

The following terms are listed (with page number) in the order they appear in the text. Define each. Doing so will aid you in reviewing the material covered in this chapter.

radiational cooling, 67
radiation inversion, 68
freeze, 69
thermal belts, 69
daily (diurnal) range of temperature, 71
mean (average) daily temperature, 73
controls of temperature, 75
isotherms, 75
annual range of temperature, 77
wind-chill index (WCI), 80
frostbite, 81
hypothermia, 81
humidity, 81
relative humidity (RH), 82
dew point, 82
wet-bulb temperature, 86
heatstroke, 87
heat index (HI), 87
apparent temperature, 87
heat wave, 88

Review Questions

1. What are some of the factors that determine the daily fluctuation of air temperature just above the ground?
2. Explain how incoming energy and outgoing energy regulate the daily variation in air temperature.
3. On a calm, sunny day, why is the air next to the ground normally much warmer than the air just above?
4. Explain why the warmest time of the day is usually in the afternoon, even though the sun's rays are most direct at noon.
5. Explain how radiational cooling at night produces a radiation temperature inversion.
6. What weather conditions are best suited for the formation of a cold night and a strong radiation inversion?
7. Explain why the daily range of temperature is normally greater (a) in dry regions than in humid regions and (b) on clear days than on cloudy days.
8. Where is the hottest region in the United States? Where is the coldest?
9. Why is the hottest region of the world in the subtropical deserts of the Northern Hemisphere and not at the equator?
10. Why do the first freeze in autumn and the last freeze in spring occur in low-lying areas?
11. Why is the largest annual range of temperature normally observed over continents away from large bodies of water?
12. List the controls of temperature and then explain the influence each has on average global temperatures in January and July?

13. In cold weather, how do increasing wind speeds make us feel colder than it actually is?
14. What conditions can bring on hypothermia?
15. (a) What does the relative humidity represent?
(b) When the relative humidity is given, why is it also important to know the air temperature?
(c) Explain two ways the relative humidity may be changed.
16. Explain why, during a summer day, the relative humidity will change as shown in Fig. 3.19, p. 82.
17. Why do hot, humid summer days usually feel hotter than hot, dry summer days?
18. Why is it very unlikely that you will ever experience a temperature of 90°F with a relative humidity of 90 percent?
19. On average, where do we usually find the highest dew-point temperatures in the United States?
20. Why is cold polar air described as "dry" when the relative humidity of the air is very high?
21. Why is the wet-bulb temperature a good measure of how cool human skin can become?
22. When outside air is brought indoors on an extremely cold winter day, the relative humidity of the heated air inside often drops below 10 percent. Explain why this situation occurs.
23. Which set of conditions do you feel produces the highest heat index: (a) temperature, 100°F; relative humidity, 30 percent; (b) temperature, 90°F; relative humidity, 60 percent. (Hint: Look at Fig. 3.24, p. 87.)

Online Learning

STUDENT COMPANION WEBSITE: Visit this book's companion website at: www.cengage.com/ahrens/extreme1e and choose Chapter 3 for many study aids and ideas for further reading and research. These include flashcards, practice quizzing, and web links.

METEOROLOGY RESOURCE CENTER: For students with access, log on at www.cengage.com/login for more assets, including animations, videos, and more. If your textbook did not come with access, visit www.CengageBrain.com

Condensation in the Atmosphere

4

CONTENTS

THE FORMATION OF DEW, FROST, AND HAZE

FOG

FOGGY WEATHER

CLOUDS: IDENTIFICATION FROM THE SURFACE

CLOUDS: OBSERVATIONS FROM SPACE

SUMMARY

KEY TERMS

REVIEW QUESTIONS

A shallow blanket of fog covers the surface in western Iowa on a cool summer morning during August, 2007.

Extreme or unusual weather may appear suddenly and last for only a few minutes. For example, during the winter while driving your car on a clear, sunny morning, you may suddenly encounter fog—fog that limits visibility to only a few feet in front of your vehicle. As you struggle to see the road, several questions may run through your mind: Why did the fog form here, and will I be out of it soon? In order to answer questions such as these, we will look at fog and other forms of condensation, examining both how and where they form and the hazardous conditions they produce. Since extreme weather is often associated with clouds, we will also look at the different cloud types, noting which are the most dangerous to aviation and which tend to produce severe and unusual weather events.

The Formation of Dew, Frost, and Haze

On clear, calm nights, objects near the earth's surface cool rapidly by emitting infrared radiation. The ground and objects on it often become much colder than the surrounding air. Air that comes in contact with these cold surfaces cools by conduction. Eventually, the air cools to the dew point. As surfaces (such as twigs, leaves, and blades of grass) cool below this temperature, water vapor begins to condense upon them, forming tiny visible specks of water called **dew** (see ▶ Fig. 4.1). If the air temperature should drop to freezing or below, the dew will freeze, becoming tiny beads of ice called *frozen dew*. Because the coolest air is usually at ground level, dew is more likely to form on blades of grass than on objects several feet above the surface. This thin coating of dew not only dampens bare feet, but it also is a valuable source of moisture for many plants during periods of low rainfall.

Dew is more likely to form on nights that are clear and calm than on nights that are cloudy and windy. Clear nights allow objects near the ground to cool rapidly, and calm winds mean that the coldest air will be located at ground level. These atmospheric conditions are usually associated with large fair-weather, high-pressure systems. On the other hand, the cloudy, windy weather that inhibits rapid cooling near the ground and the forming of dew often signifies the approach of a rain-producing storm system. These observations inspired the following folk-rhyme:

> When the dew is on the grass,
> rain will never come to pass.
> When grass is dry at morning light,
> look for rain before the night!

Visible white frost forms on cold, clear, calm mornings when the dew-point temperature is at or below freezing. When the air temperature cools to the dew point (now called the *frost point*) and further cooling occurs, water vapor can change directly to ice without becoming a liquid first—a process called *deposition*.* The delicate, white crystals of ice that form in this manner are called *hoarfrost, white frost*, or simply **frost**. Frost has a treelike branching pattern that easily distinguishes it from the nearly spherical beads of frozen dew (see ▶ Fig. 4.2).

*When the ice changes back into vapor without melting, the process is called *sublimation*.

▶ FIGURE 4.1 Dew forms on clear nights when objects on the surface cool to a temperature below the dew point. If these beads of water should freeze, they would become frozen dew.

▶ FIGURE 4.2 These are the delicate ice-crystal patterns that frost exhibits on a window during a cold winter morning.

FIGURE 4.3

The high relative humidity of the cold air above the lake is causing a layer of haze to form on a still winter morning.

© C. Donald Ahrens

In very dry weather, the air may become quite cold and drop below freezing without ever reaching the frost point, and no visible frost forms. *Freeze* and *black frost* are words denoting this situation—a situation that can severely damage certain crops (see Chapter 3, p. 69).

As a deep layer of air cools during the night, its relative humidity increases. When the air's relative humidity reaches about 75 percent, some of its water vapor may begin to condense onto tiny floating particles of sea salt and other substances—*condensation nuclei*—that are *hygroscopic* ("water seeking") in that they allow water vapor to condense onto them when the relative humidity is considerably below 100 percent. As water collects onto these nuclei, their size increases and the particles, although still small, are now large enough to scatter visible light in all directions, becoming **haze**—a layer of particles dispersed through a portion of the atmosphere (see Fig. 4.3).

As the relative humidity gradually approaches 100 percent, the haze particles grow larger, and condensation begins on the less-active nuclei. Now a large fraction of the available nuclei have water condensing onto them, causing the droplets to grow even bigger, until eventually they become visible to the naked eye. The increasing size and concentration of droplets further restrict visibility. When the visibility lowers to less than 1 km (or 0.62 mi), and the air is wet with millions of tiny floating water droplets, the haze becomes a cloud resting near the ground, which we call **fog.***

Fog

Fog, like any cloud, usually forms in one of two ways: (1) by cooling—air is cooled below its saturation point (dew point); and (2) by evaporation and mixing—water vapor is added to the air by evaporation, and the moist air mixes with relatively dry air. Once fog forms it is maintained by new fog droplets, which constantly form on available nuclei. In other words, the air must maintain its degree of saturation either by continual cooling or by evaporation and mixing of vapor into the air.

Fog produced by the earth's radiational cooling is called **radiation fog,** or *ground fog.* It forms best on clear nights when a shallow layer of moist air near the ground is overlain by drier air. Under these conditions, the ground cools rapidly since the shallow, moist layer does not absorb much of the earth's outgoing infrared radiation. As the ground cools, so does the air directly above it, and a surface inversion forms, with colder air at the surface and warmer air above. The moist, lower layer (chilled rapidly by the cold ground) quickly becomes saturated, and fog forms. The longer the night, the longer the time of cooling and the greater the likelihood of fog. Hence, radiation fogs are most common over land in late fall and winter.

Another factor promoting the formation of radiation fog is a light breeze of less than five knots. Although radiation fog may form in calm air, slight air movement brings more of the moist air in direct contact with the cold ground and the transfer of heat occurs more rapidly. A strong breeze tends to prevent a radiation fog from forming by mixing the air near the surface with the drier air above. The ingredients of clear skies and light winds are associated with large high-pressure areas (anticyclones). Consequently, during the winter, when a high becomes stagnant over an area, radiation fog may form on consecutive days.

*This is the official international definition of *fog.* The United States Weather Service reports fog as a restriction to visibility when fog restricts the visibility to 6 miles or less and the spread between the air temperature and dew point is 5°F or less. When the visibility is less than one-quarter of a mile, the fog is considered *dense.*

© C. Donald Ahrens

FIGURE 4.4

Radiation fog nestled in a valley in central Oregon.

Because cold, heavy air drains downhill and collects in valley bottoms, we normally see radiation fog forming in low-lying areas. Hence, radiation fog is frequently called *valley fog.* The cold air and high moisture content in river valleys make them susceptible to radiation fog. Since radiation fog normally forms in lowlands, hills may be clear all day long, while adjacent valleys are fogged in (see Fig. 4.4).

Radiation fogs are normally deepest around sunrise. Usually, however, a shallow fog layer will dissipate or *burn off* by afternoon. Of course, the fog does not "burn"; rather, sunlight penetrates the fog and warms the ground, causing the temperature of the air in contact with the ground to increase. The warm air rises and mixes with the foggy air above, which increases the temperature of the foggy air. In the slightly warmer air, some of the fog droplets *evaporate,* allowing more sunlight to reach the ground, which produces more heating, and soon the fog completely evaporates and disappears. If the fog layer is quite thick, it may not completely dissipate and a layer of low clouds (called *stratus*) covers the region. This type of fog is sometimes called *high fog.*

When warm, moist air moves over a sufficiently colder surface, the moist air may cool to its saturation point, forming **advection fog**. A good example of advection fog may be observed along the Pacific Coast during summer. The main reason fog forms in this region is that the surface water near the coast is much colder than the surface water farther offshore. Warm, moist air from the Pacific Ocean is carried (advected) by westerly winds over the cold coastal waters. Chilled from below, the air temperature drops to the dew point, and fog is produced. Advection fog, unlike radiation fog, always involves the movement of air, so when there is a stiff summer breeze in San Francisco, it's common to watch advection fog roll in past the Golden Gate Bridge (see Fig. 4.5). It is also more common to see advection fog forming at headlands that protrude seaward than in the mouths of bays. If you are curious as to why, read the Focus section on p. 100.

As summer winds carry the fog inland over the warmer land, the fog near the ground dissipates, leaving a sheet of low-lying gray clouds that block out the sun. Farther inland, the air is sufficiently warm, so that even these low clouds evaporate and disappear.

Because they provide moisture to the coastal redwood trees, advection fogs are important to the scenic beauty of the Pacific Coast. Much of the fog moisture collected by the needles and branches of the redwoods drips to the ground (*fog drip*), where it is utilized by the tree's shallow root system. Without the summer fog, the coast's redwood trees would have trouble surviving the dry California summers. Hence, we find them nestled in the fog belt along the coast. It's important to keep in mind that advection fog forms when wind blows moist air over a cooler surface, whereas radiation fog forms under relatively calm conditions. Figure 4.6 visually summarizes the formation of these two types of fog.

Advection fogs also prevail where two ocean currents with different temperatures flow next to one another. Such is the case in the Atlantic Ocean off the coast of Newfoundland, where the cold southward-flowing Labrador Current lies almost parallel to the warm northward-flowing Gulf Stream. Warm southerly air moving over the cold water produces fog in that region—so frequently that fog occurs on about two out of three days during summer.

Advection fog also forms over land. In winter, warm, moist air from the Gulf of Mexico moves northward over progressively colder and slightly elevated land. As the air cools to its saturation point, a fog forms in the southern or central United States. Because the cold ground is often the result of radiational cooling, fog that forms in this manner is sometimes called *advection-radiation fog.* During this same time of year, air moving across the warm Gulf Stream encounters the colder land of

© Herbert Spichtinger/Corbis Edge/CORBIS

FIGURE 4.5 Advection fog rolling in past the Golden Gate Bridge in San Francisco. As fog moves inland, the air warms and the fog lifts above the surface. Eventually, the air becomes warm enough to totally evaporate the fog.

the British Isles and produces the thick fogs of England. Similarly, fog forms as marine air moves over an ice or snow surface. In extremely cold arctic air, ice crystals form instead of water droplets, producing an *ice fog*.

Fog that forms as moist air flows up along an elevated plain, hill, or mountain is called **upslope fog**. Typically, upslope fog forms during the winter and spring on the eastern side of the Rockies, where the eastward-sloping plains are nearly a kilometer higher than the land farther east. Occasionally, cold air moves from the lower eastern plains westward. The air gradually rises, expands, becomes cooler, and—if sufficiently moist—a fog forms (see Fig. 4.7). Upslope fogs that form over an extensive area may last for many days.

So far, we have seen how the cooling of air produces fog. But remember that fog may also form by the mixing of two unsaturated masses of air. Fog that forms in this manner is usually called *evaporation fog* because evaporation initially enriches the air with water vapor. Probably, a more appropriate name for the fog is **evaporation (mixing) fog**. On a cold day, you may have unknowingly

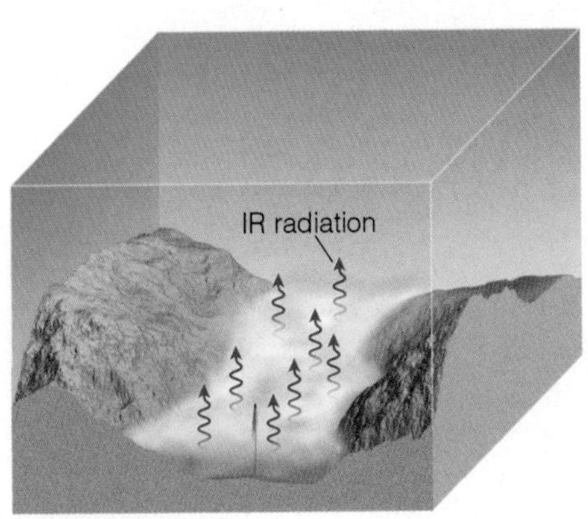

(a) Radiation fog

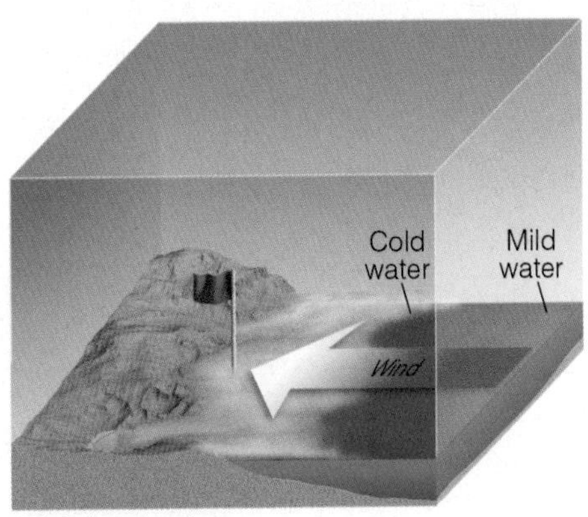

(b) Advection fog

FIGURE 4.6

(a) Radiation fog tends to form on clear, relatively calm nights when cool, moist surface air is overlain by drier air and rapid radiational cooling occurs. (b) Advection fog forms when the wind moves moist air over a cold surface and the moist air cools to its dew point.

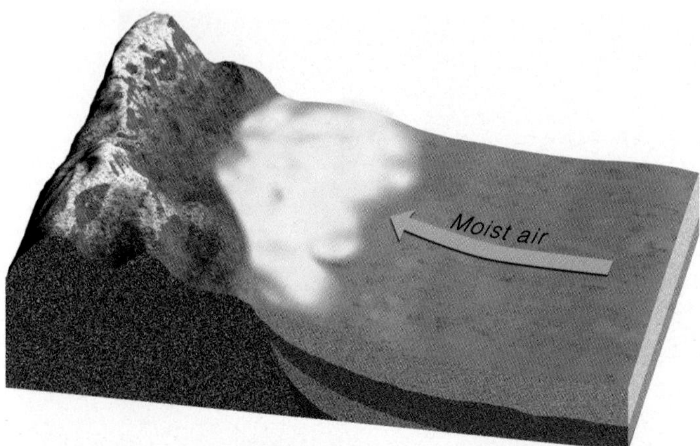

FIGURE 4.7 Upslope fog forms as moist air slowly rises, cools, and condenses over elevated terrain.

produced evaporation (mixing) fog. When moist air from your mouth or nose meets the cold air and mixes with it, the air becomes saturated, and a tiny cloud forms with each exhaled breath.

A common form of evaporation (mixing) fog is the *steam fog,* which forms when cold air moves over warm water. This type of fog forms above a heated outside swimming pool in winter. As long as the water is warmer than the unsaturated air above, water will evaporate from the pool into the air. The increase in water vapor raises the dew point, and, if mixing is sufficient, the air above becomes saturated. The colder air directly above the water is heated from below and becomes warmer than the air directly above it. This warmer air rises and, from a distance, the rising condensing vapor appears as "steam."

It is common to see steam fog forming over lakes on autumn mornings, as cold air settles over water still warm from the long summer. On occasion, over the Great

FOCUS ON

A SPECIAL TOPIC

Why Are Headlands Usually Foggier Than Beaches?

If you drive along a highway that parallels an irregular coastline, you may have observed that advection fog is more likely to form in certain regions. For example, headlands that protrude seaward usually experience more fog than do beaches that are nestled in the mouths of bays. Why?

As air moves onshore, it crosses the coastline at nearly a right angle. This causes the air to flow together or converge in the vicinity of the headlands (see Fig. 1). This area of weak convergence causes the surface air to rise and cool just a little. If the rising air is close to being saturated, it will cool to its dew point, and fog will form.

Meanwhile, near the beach area, the surface air spreads apart or diverges as it crosses the coastline. This area of weak divergence creates sinking and slightly warmer air. Because the sinking of air increases the separation between air temperature and dew point, fog is less likely to form in this region. Hence, the headlands can be shrouded in fog while the beaches are basking in sunshine.

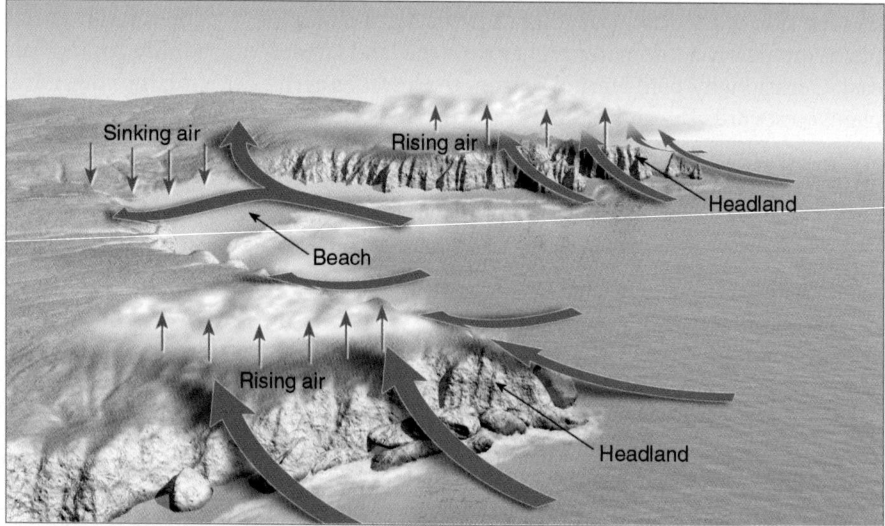

Figure 1 Along an irregular coastline, advection fog is more likely to form at the headland (the region of land extending seaward) where moist surface air converges and rises than at the beach where air diverges and sinks.

EXTREME WEATHER WATCH

Ever hear of Caribou fog? No, it's not the fog that forms in Caribou, Maine, but the fog that forms around herds of caribou. In very cold weather, just a little water vapor added to the air will saturate it. Consequently, the perspiration and breath from large herds of caribou add enough water vapor to the air to create a blanket of fog that hovers around the herd.

Lakes and other warm bodies of water, columns of condensed vapor rise from the fog layer, forming whirling *steam devils,* which appear similar to the dust devils on land. If you travel to Yellowstone National Park, you will see steam fog forming above thermal ponds all year long (see Fig. 4.8). Over the ocean in polar regions, steam fog is referred to as *arctic sea smoke.*

Steam fog may form above a wet surface on a sunny day. This type of fog is commonly observed after a rain shower as sunlight shines on a wet road, heats the asphalt, and quickly evaporates the water. This added vapor mixes with the air above, producing steam fog. Fog that forms in this manner is short-lived and disappears as the road surface dries. However, this type of fog can produce hazardous driving conditions (if just for an instant), especially when sunlight reflects off the fog into the driver's eyes.

A warm rain falling through a layer of cold, moist air can produce fog. As a warm raindrop falls into a cold layer of air, some of the water evaporates from the raindrop into the air. This process may saturate the air, and if mixing occurs, fog forms. Fog of this type is often associated with warm air riding up and over a mass of colder surface air. The fog usually develops in the shallow layer of cold air just ahead of an approaching warm front or behind a cold front, which is why this type of evaporation fog is also known as *precipitation fog,* or *frontal fog.*

FIGURE 4.8 Even in summer, warm air rising above thermal pools in Yellowstone National Park condenses into a type of steam fog.

Foggy Weather

The foggiest regions in the United States are shown in Fig. 4.9. Notice that dense fog is more prevalent in coastal margins (especially those regions lapped by cold ocean currents) than in the center of the continent. In fact, one of the foggiest spots near sea level in

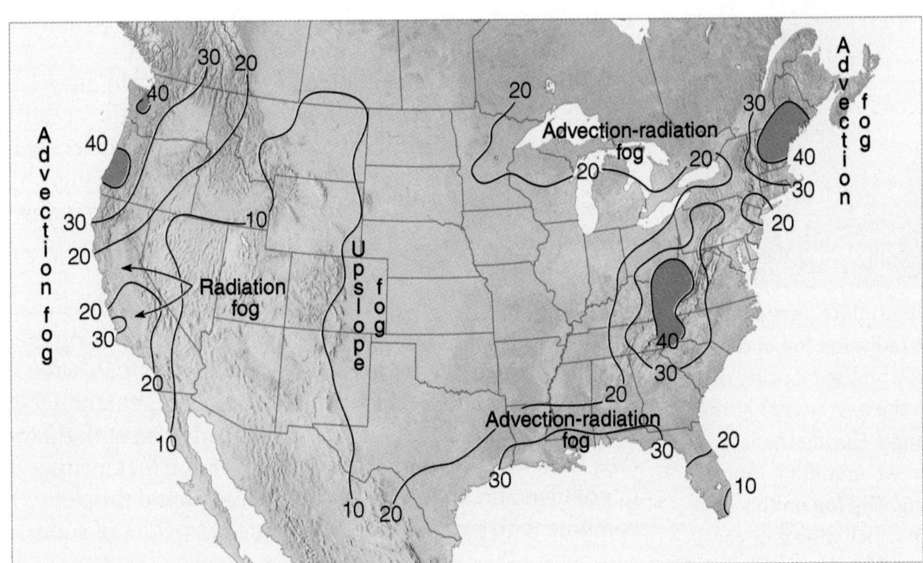

FIGURE 4.9 Average annual number of days with dense fog (visibility less than 0.25 miles) throughout the United States. (NOAA)

EXTREME WEATHER WATCH

One of the foggiest places in the world is also one of the driest—the Skeleton Coast of Namibia in southern Africa. Even though this region experiences fog on more than 100 days a year, its annual average rainfall is less than an inch.

the United States is Cape Disappointment, Washington. Located at the mouth of the Columbia River, it averages 2556 hours (or the equivalent of 106.5 twenty-four-hour days) of dense fog each year.* Anyone who travels to this spot hoping to enjoy the sun during August and September would find its name appropriate indeed. Along the Washington coast, Willapa also experiences extensive fog. During a four-year period, Willapa reported an average of 3,863 hours (161 equivalent days) of dense fog. During one of those years, the station reported 7,613 hours (317 days) of fog.

*Dense fog means that the visibility is restricted by fog to one-quarter mile or less.

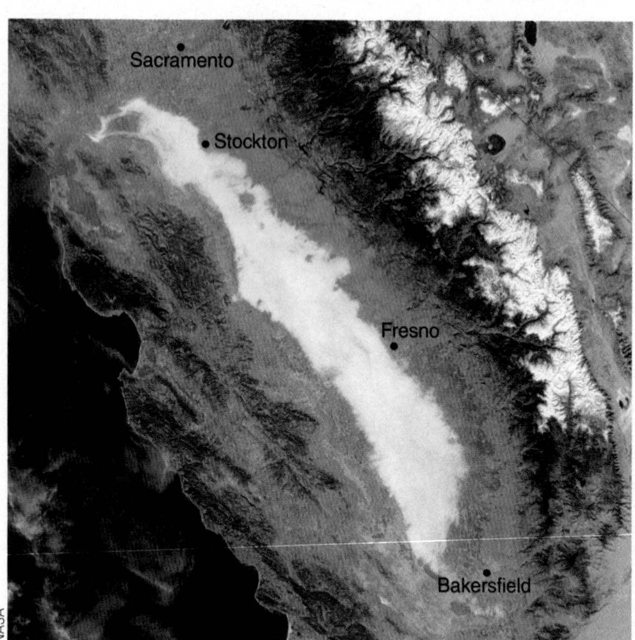

FIGURE 4.10 Visible satellite image of dense radiation fog in the southern half of California's Central Valley on the morning of November 20, 2002. The white region to the east (right) of the fog is the snow-capped Sierra Nevada range. During the late fall and winter, the fog, nestled between two mountain ranges, can last for many days without dissipating. The fog on this day was responsible for several auto accidents, including a 14-car pileup near Fresno, California.

Notice in Fig. 4.9 that the coast of Maine is also foggy. In fact, Moose Peak Lighthouse on Mistake Island averages 1,580 hours (65 equivalent days) of dense fog. To the south, Nantucket Island has on average 2,040 hours (85 equivalent days) of dense fog.

Dense fog also has many negative aspects. Along a gently sloping highway, the elevated sections may have excellent visibility, while in lower regions—only a few kilometers away—fog may cause poor visibility. Driving from the clear area into the fog on a major freeway can be extremely dangerous. In fact, every winter many people are involved in fog-related auto accidents. These usually occur when a car enters the fog and, because of the reduced visibility, the driver puts on the brakes to slow down. The car behind then slams into the slowed vehicle, causing a chain-reaction accident with many cars involved. One such accident actually occurred near Fresno, California, in February, 2002, when 87 vehicles smashed into each other along a stretch of foggy Highway 99. The accident left dozens of people injured, three people dead, and a landscape strewn with cars and trucks twisted into heaps of jagged steel. A similar accident occurred during November of the same year (see Fig. 4.10). Statistics show that more than 40,000 fog-related crashes can occur in a single year, causing over 600 deaths and injuring more than 19,000 people.

Fog-related problems are not confined to land. Even with sophisticated electronic equipment, dense fog in the open sea hampers navigation. A Swedish liner rammed the luxury liner *Andrea Doria* in thick fog off Nantucket Island on July 25, 1956, causing 52 casualties.

Reduction in visibility caused by dense fog has caused about 22 percent of weather-related fatalities in aviation. For example, during March, 1977, on a fog-covered runway in the Canary Islands, two 747 jet airliners collided, taking the lives of over 570 people. And on a foggy runway in Detroit, Michigan, eight people died in December, 1990, when a jet aircraft mistakenly taxied down an active runway and was struck by another jet taking off.

With the same water content, fog that forms in dirty city air often is thicker than fog that forms over the ocean. Normally, the smaller number of condensation nuclei over the middle of the ocean produces fewer, but larger, fog droplets. City air with its abundant nuclei produces many tiny fog droplets, which greatly increase the thickness (or opaqueness) of the fog and reduce visibility. A dramatic example of a thick fog forming in air with abundant nuclei occurred in London, England, during the early 1950s. The fog became so thick, and the air so laden with smoke particles, that sunlight could not penetrate the smoggy air, requiring that streetlights be left on at midday. (See Focus section on p. 103.) Moreover, fog that forms in polluted air can turn acidic as the tiny liquid droplets combine with gaseous impurities, such as oxides of sulfur

FOCUS ON
EXTREME WEATHER

When Fog Turns to Smog

London, like San Francisco, is forever associated in the popular imagination with fog. Although London does average about 30 days a year of dense fog (visibility below one-fourth of a mile), San Francisco's foggy days are determined by how close you are to the ocean. Along the city's Ocean Beach, for example, there are about 60 foggy days a year (mostly during the months of June through August); whereas, at the San Francisco airport, on the bay side of the peninsula and sheltered from the onshore fog-bearing winds by a range of hills, the number of foggy days averages only 17 per year.

The foggiest place in the world (excluding mountaintops in the clouds) is probably the Grand Banks of Maritime Canada where Cape Race, Newfoundland, averages 158 foggy days each year. Other places averaging over 100 days of fog a year (aside from those mentioned in the text) include the southern coast of Chile, the Skeleton Coast of the southern African nation of Namibia, and the Severnaya Zemlya Islands in northern Russia's Kara Sea. London's reputation for fog is actually based upon the smog (smoke and fog) of its past.

Beginning with the Industrial Revolution in the mid-nineteenth century until the early 1960s, Londoners heated their homes with soft-coal–burning stoves and furnaces. Winter temperature inversions sometimes trapped these effluents at ground level for days on end. The mixing of the smoke and fog created the so-called "pea soup" fogs.

The week of December 5 through 12, 1952, was unusually cold by London standards. Because of the cold weather, the city's residents were burning an unusual amount of coal. A temperature inversion trapped the coal smoke at a low elevation and the smoke mixed with radiation fog near the surface. This deadly combination of smoke and fog reduced visibilities to less than a few feet and produced the worst smog-related pollution disaster ever. The smog was so dense that the visibility actually remained below 165 feet for two days, and, at Heathrow Airport, the visibility at one point dropped to only 30 feet (see Fig. 2). Sulfur dioxide (SO_2) levels peaked at 700 parts per billion (ppb). Any reading above 300 ppb is considered hazardous. Health authorities were unaware a catastrophe was developing until undertakers realized they were running short of coffins and hospitals gradually became crammed with patients suffering from pulmonary ailments. In the end, it became apparent that at least 4,000 people had died in one week as a result of the smog. For weeks following this event, deaths remained at a higher than normal level. And, in fact, it is now estimated that perhaps as many as 12,000 Londoners perished as a direct result of this deadly mix of fog and smoke.

Figure 2 Smog was so dense in London during December, 1952, that visibilities were often restricted to less than 100 feet and streetlights had to be turned on during the middle of the day.

and nitrogen. **Acid fog** poses a threat to human health, especially to people with preexisting respiratory problems.

Up to this point, we have looked at the different forms of condensation that occur on or near the earth's surface. In particular, we learned that fog is simply many millions of tiny liquid droplets (or ice crystals) that form near the ground. In the following sections, we will see how these same particles, forming well above the ground, are classified and identified as clouds.

BRIEF REVIEW

However, before going on to the section on clouds, here is a brief review of some of the facts and concepts we covered so far:

- Dew, frost, and frozen dew generally form on clear nights when the temperature of objects on the surface cools below the air's dew-point temperature.
- Visible white frost forms in saturated air when the air temperature is at or below freezing. Under these conditions, water vapor can change directly to ice, in a process called *deposition*.
- Condensation nuclei act as surfaces on which water vapor condenses. Those nuclei that have an affinity for water vapor are called *hygroscopic*.
- Fog is a cloud resting on the ground. It can be composed of water droplets, ice crystals, or a combination of both.
- Radiation fog, advection fog, and upslope fog all form as the air cools. The cooling for radiation fog is mainly radiational cooling at the earth's surface; for advection fog, the cooling is mainly warmer air moving over a colder surface; for upslope fog, the cooling occurs as moist air gradually rises and expands along sloping terrain.
- Evaporation (mixing) fog, such as steam fog and frontal fog, forms as water evaporates and mixes with drier air.

Clouds: Identification from the Surface

Although ancient astronomers named the major stellar constellations about 2000 years ago, clouds were not formally identified and classified until the early nineteenth century. The French naturalist Lamarck (1744–1829) proposed the first system for classifying clouds in 1802; however, his work did not receive wide acclaim. One year later, Luke Howard, an English naturalist, developed a cloud classification system that found general acceptance. In essence, Howard's innovative system employed Latin words to describe clouds as they appear to a ground observer. He named a sheetlike cloud *stratus* (Latin for "layer"); a puffy cloud *cumulus* ("heap"); a wispy cloud *cirrus* ("curl of hair"); and a rain cloud *nimbus* ("violent rain"). In Howard's system, these were the four basic cloud forms. Other clouds could be described by combining the basic types. For example, nimbostratus is a rain cloud that shows layering, whereas cumulonimbus is a rain cloud having pronounced vertical development.

In 1887, Abercromby and Hildebrandsson expanded Howard's original system and published a classification system that, with only slight modification, is still in use today. Ten principal cloud forms are divided into four primary cloud groups. Each group is identified by the height of the cloud's base above the surface: high clouds, middle clouds, and low clouds. The fourth group contains clouds showing more vertical than horizontal development. Within each group, cloud types are identified by their appearance. Table 4.1 lists these four groups and their cloud types.

The approximate base height of each cloud group is given in Table 4.2. Note that the altitude separating the high and middle cloud groups overlaps and varies with latitude. Large temperature changes cause most of this latitudinal variation. For example, high cirriform clouds are composed almost entirely of ice crystals. In subtropical regions, air temperatures low enough to freeze all liquid water usually occur only above about 20,000 feet. In polar regions, however, these same temperatures may be found at altitudes as low as 10,000 feet. Hence, while you may observe cirrus clouds at 12,000 feet over northern Alaska, you will not see them at that elevation above southern Florida.

Clouds cannot be accurately identified strictly on the basis of elevation. Other visual clues are necessary. Some of these are explained in the following sections.

Table 4.1

The Four Major Cloud Groups and Their Types

1. High clouds Cirrus (Ci) Cirrostratus (Cs) Cirrocumulus (Cc)	3. Low clouds Stratus (St) Stratocumulus (Sc) Nimbostratus (Ns)
2. Middle clouds Altostratus (As) Altocumulus (Ac)	4. Clouds with vertical development Cumulus (Cu) Cumulonimbus (Cb)

Table 4.2

Approximate Height of Cloud Bases above the Surface for Various Locations

CLOUD GROUP	TROPICAL REGION	MIDDLE-LATITUDE REGION	POLAR REGION
High Ci, Cs, Cc	20,000 to 60,000 ft (6000 to 18,000 m)	16,000 to 43,000 ft (5000 to 13,000 m)	10,000 to 26,000 ft (3000 to 8000 m)
Middle As, Ac	6500 to 26,000 ft (2000 to 8000 m)	6500 to 23,000 ft (2000 to 7000 m)	6500 to 13,000 ft (2000 to 4000 m)
Low St, Sc, Ns	surface to 6500 ft (0 to 2000 m)	surface to 6500 ft (0 to 2000 m)	surface to 6500 ft (0 to 2000 m)

HIGH CLOUDS High clouds in middle and low latitudes generally form above 20,000 ft (or 6000 m). Because the air at these elevations is quite cold and "dry," high clouds are composed almost exclusively of ice crystals and are also rather thin.* High clouds usually appear white, except near sunrise and sunset, when the unscattered (red, orange, and yellow) components of sunlight are reflected from the underside of the clouds.

The most common high clouds are the **cirrus** (Ci), which are thin, wispy clouds blown by high winds into long streamers called *mares' tails.* Notice in Fig. 4.11 that they can look like a white, feathery patch with a faint wisp of a tail at one end. Cirrus clouds usually move across the sky from west to east, indicating the prevailing winds at their elevation.

Cirrocumulus (Cc) clouds, seen less frequently than cirrus, appear as small, rounded, white puffs that may occur individually, or in long rows (see Fig. 4.12). When in rows, the cirrocumulus cloud has a rippling appearance that distinguishes it from the silky look of the cirrus and the sheetlike cirrostratus. Cirrocumulus seldom cover more than a small portion of the sky. The dappled cloud elements that reflect the red or yellow light of a setting sun make this one of the most beautiful of all clouds. The small ripples in the cirrocumulus strongly resemble the scales of a fish; hence, the expression "*mackerel sky*" commonly describes a sky full of cirrocumulus clouds.

The thin, sheetlike, high clouds that often cover the entire sky are **cirrostratus** (Cs) (Fig. 4.13), which are so thin that the sun and moon can be clearly seen through them. The ice crystals in these clouds bend the light passing through them and will often produce a halo—a ring of light that encircles the sun or moon. In fact, the veil of cirrostratus may be so thin that a halo is the only clue to its presence. Thick cirrostratus clouds give the sky a glary white appearance and frequently form ahead of an advancing mid-latitude cyclonic storm; hence, they can be used to predict rain or snow within twelve to twenty-four hours, especially if they are followed by middle-type clouds. And, as we will see in a later chapter, cirrostratus clouds can form ahead of thunderstorms.

FIGURE 4.11 Cirrus clouds.

*Small quantities of liquid water in cirrus clouds at temperatures as low as –36°C (–33°F) were discovered during research conducted above Boulder, Colorado.

© C. Donald Ahrens

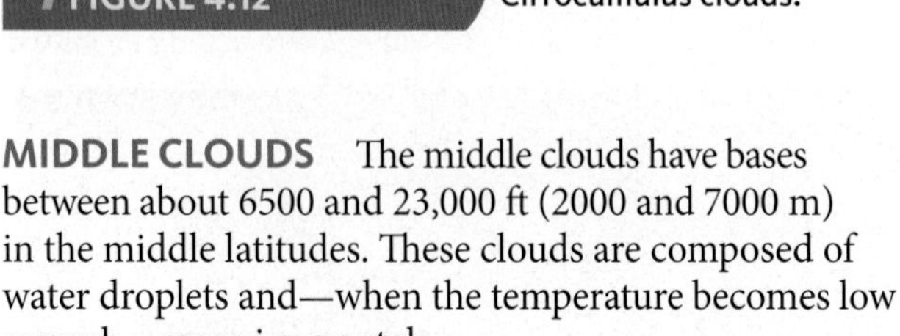
FIGURE 4.12 Cirrocumulus clouds.

© C. Donald Ahrens

FIGURE 4.13 Cirrostratus clouds with a faint halo encircling the sun. The sun is the bright white area in the center of the circle.

MIDDLE CLOUDS The middle clouds have bases between about 6500 and 23,000 ft (2000 and 7000 m) in the middle latitudes. These clouds are composed of water droplets and—when the temperature becomes low enough—some ice crystals.

Altocumulus (Ac) clouds are middle clouds that appear as gray, puffy masses, sometimes rolled out in parallel waves or bands (see Fig. 4.14). Usually, one part of the cloud is darker than another, which helps to separate it from the higher cirrocumulus. Also, the individual puffs of the altocumulus appear larger than those of the cirrocumulus. A layer of altocumulus may sometimes be confused with altostratus; in case of doubt, clouds are called altocumulus if there are rounded masses or rolls present. Altocumulus clouds that look like "little castles" (*castellanus*) in the sky indicate the presence of rising air at cloud level. The appearance of these clouds on a warm, humid summer morning often portends thunderstorms and the possibility of severe weather by late afternoon.

The **altostratus** (As) is a gray or blue-gray cloud that often covers the entire sky over an area that extends over many hundreds of square kilometers. In the thinner section of the cloud, the sun (or moon) may be *dimly visible* as a round disk, which is sometimes referred to as a "watery sun" (see Fig. 4.15). Thick cirrostratus clouds are occasionally confused with thin altostratus clouds. The gray color, height, and dimness of the sun are good clues to identifying an altostratus. The fact that halos only occur with cirriform clouds also helps one distinguish them. Another way to separate the two is to look at the ground for shadows. If there are none, it is a good bet that the cloud is altostratus because cirrostratus are usually transparent enough to produce shadows. Altostratus clouds often form ahead of storms having widespread and relatively continuous precipitation. If precipitation falls from an altostratus, its base usually lowers. If the precipitation reaches the ground, the cloud is then classified as *nimbostratus.*

LOW CLOUDS Low clouds, with their bases lying below 6500 ft (or 2000 m) are almost always composed of water droplets; however, in cold weather, they may contain ice particles and snow.

The **nimbostratus** (Ns) is a dark gray, "wet"-looking cloud layer associated with more or less continuously falling rain or snow (see Fig. 4.16). The intensity of this precipitation is usually light or moderate—it is never of the heavy, showery variety, unless well-developed cumuliform clouds are embedded within the nimbostratus cloud. The base of the nimbostratus cloud is normally impossible to identify clearly and is easily confused with the altostratus. Thin nimbostratus is usually darker gray than thick altostratus, and you cannot see the sun or moon through a layer of nimbostratus. Visibility below a nimbostratus cloud deck is usually quite poor because rain will evaporate and mix with the air in this region. If this air becomes saturated, a lower layer of clouds or fog may form beneath the original cloud base. Since these lower clouds drift rapidly with the wind, they form irregular shreds with a ragged appearance called *stratus fractus,* or *scud.*

A low, lumpy cloud layer is the **stratocumulus** (Sc). It appears in rows, in patches, or as rounded masses with blue sky visible between the individual cloud elements

FIGURE 4.14 Altocumulus clouds.

FIGURE 4.15 Altostratus clouds. The appearance of a dimly visible "watery sun" through a deck of gray clouds is usually a good indication that the clouds are altostratus.

FIGURE 4.16 The nimbostratus is the sheetlike cloud from which light rain is falling. The ragged-appearing cloud beneath the nimbostratus is stratus fractus, or scud.

(see Fig. 4.17). Often they appear near sunset as the spreading remains of a much larger cumulus cloud. The color of stratocumulus ranges from light to dark gray. It differs from altocumulus in that it has a lower base and larger individual cloud elements. (Compare Fig. 4.14 with Fig. 4.17.) To distinguish between the two, hold your hand at arm's length and point toward the cloud. Altocumulus cloud elements will generally be about the size of your thumbnail; stratocumulus cloud elements will usually be about the size of your fist. Rain or snow rarely falls from stratocumulus.

Stratus (St) is a uniform grayish cloud that often covers the entire sky. It resembles a fog that does not reach the ground (see Fig. 4.18). Actually, when a thick fog "lifts," the resulting cloud is a deck of low stratus. Normally, no precipitation falls from the stratus, but sometimes it is accompanied by a light mist or drizzle. This cloud commonly occurs over Pacific and Atlantic coastal waters in summer. A thick layer of stratus might

FIGURE 4.17 Stratocumulus clouds forming along the south coast of Florida. Notice that the rounded masses are larger than those of the altocumulus.

FIGURE 4.18 A layer of low-lying stratus clouds hides the mountains in Iceland.

FIGURE 4.19 Cumulus clouds. Small cumulus clouds such as these are sometimes called *fair weather cumulus*, or *cumulus humilis*.

be confused with nimbostratus, but the distinction between them can be made by observing the base of the cloud. Often, stratus has a more uniform base than does nimbostratus. Also, a deck of stratus may be confused with a layer of altostratus. However, if you remember that stratus clouds are lower and darker gray, the distinction can be made.

CLOUDS WITH VERTICAL DEVELOPMENT Familiar to almost everyone, the puffy **cumulus** (Cu) cloud takes on a variety of shapes, but most often it looks like a piece of floating cotton with sharp outlines and a flat base (see Fig. 4.19). The base appears white to light gray, and, on a humid day, may be only a few thousand feet above the ground and a half a mile or so wide. The top of the cloud—often in the form of rounded towers—denotes the limit of rising air and is usually not very high. These clouds can be distinguished from stratocumulus by the fact that cumulus clouds are detached (usually a great deal of blue sky between each cloud) whereas stratocumulus usually occur in groups or patches. Also, the cumulus has a dome- or tower-shaped top as opposed to the generally flat tops of the stratocumulus. Cumulus clouds that show only slight vertical growth (*cumulus humilis*) are associated with fair weather; therefore, we call these clouds "fair weather cumulus." If the cumulus clouds are small and appear as broken fragments of a cloud with ragged edges, they are called *cumulus fractus.*

Harmless-looking cumulus often develop on warm summer mornings and, by afternoon, become much larger and more vertically developed. When the growing cumulus resembles a head of cauliflower, it becomes a *cumulus congestus,* or *towering cumulus* (Tcu). Most

EXTREME WEATHER WATCH

A true giant in the sky. On June 15, 1996, during a tornado outbreak in Kansas, a cumulonimbus cloud reached an incredible height of 78,000 feet—an altitude of nearly 15 miles above the surface.

FIGURE 4.20 Cumulus congestus. This line of cumulus congestus clouds is building along Maryland's eastern shore.

often, it is a single large cloud, but, occasionally, several grow into each other, forming a line of towering clouds, as shown in Fig. 4.20. Precipitation that falls from a cumulus congestus is always showery.

If a cumulus congestus continues to grow vertically, it develops into a giant **cumulonimbus** (Cb)—a thunderstorm cloud (see Fig. 4.21). While its dark base may be no more than 2000 ft above the earth's surface, its top may extend upward to the tropopause, over 39,000 ft higher. A cumulonimbus can occur as an isolated cloud or as part of a line or "wall" of clouds.

The tremendous amounts of energy released by the condensation of water vapor within a cumulonimbus results in the development of violent up- and down-drafts, which may exceed 90 mi/hr. The lower (warmer) part of the cloud is usually composed of only water droplets. Higher up in the cloud, water droplets and ice crystals both abound, while, toward the cold top, there are only ice crystals. Swift winds at these higher altitudes can reshape the top of the cloud into a huge flattened *anvil.** These great thunderheads may contain all forms of precipitation—large raindrops, snowflakes, snow pellets, and sometimes hailstones—all of which can fall to earth in the form of heavy showers. Lightning, thunder, and even tornadoes are associated with the cumulonimbus. (More information on the violent nature of thunderstorms and tornadoes is given in future chapters.)

Cumulus congestus and cumulonimbus frequently look alike, making it difficult to distinguish between them. However, you can usually distinguish them by looking at the top of the cloud. If the sprouting upper part of the cloud is sharply defined and not fibrous, it is usually a cumulus congestus; conversely, if the top of the cloud loses its sharpness and becomes fibrous in texture, it is usually a cumulonimbus. (Compare Fig. 4.20 with Fig. 4.21.) The weather associated with these clouds also differs: Lightning, thunder, and large hail typically occur with cumulonimbus.

So far, we have discussed the ten primary cloud forms, summarized pictorially in Fig. 4.22. This figure, along with cloud photographs and descriptions, should help you identify the more common cloud forms. Don't worry if you find it hard to estimate cloud heights. This is a difficult procedure, requiring much practice. You can use local objects (hills, mountains, tall buildings) of known height as references on which to base your height estimates.

*An anvil is a heavy block of iron or steel with a smooth, flat top on which metals are shaped by hammering.

FIGURE 4.21 A cumulonimbus cloud. Strong upper-level winds blowing from right to left produce a well-defined anvil. Sunlight scattered by falling ice crystals produces the white (bright) area beneath the anvil. Notice the heavy rain shower falling from the base of the cloud.

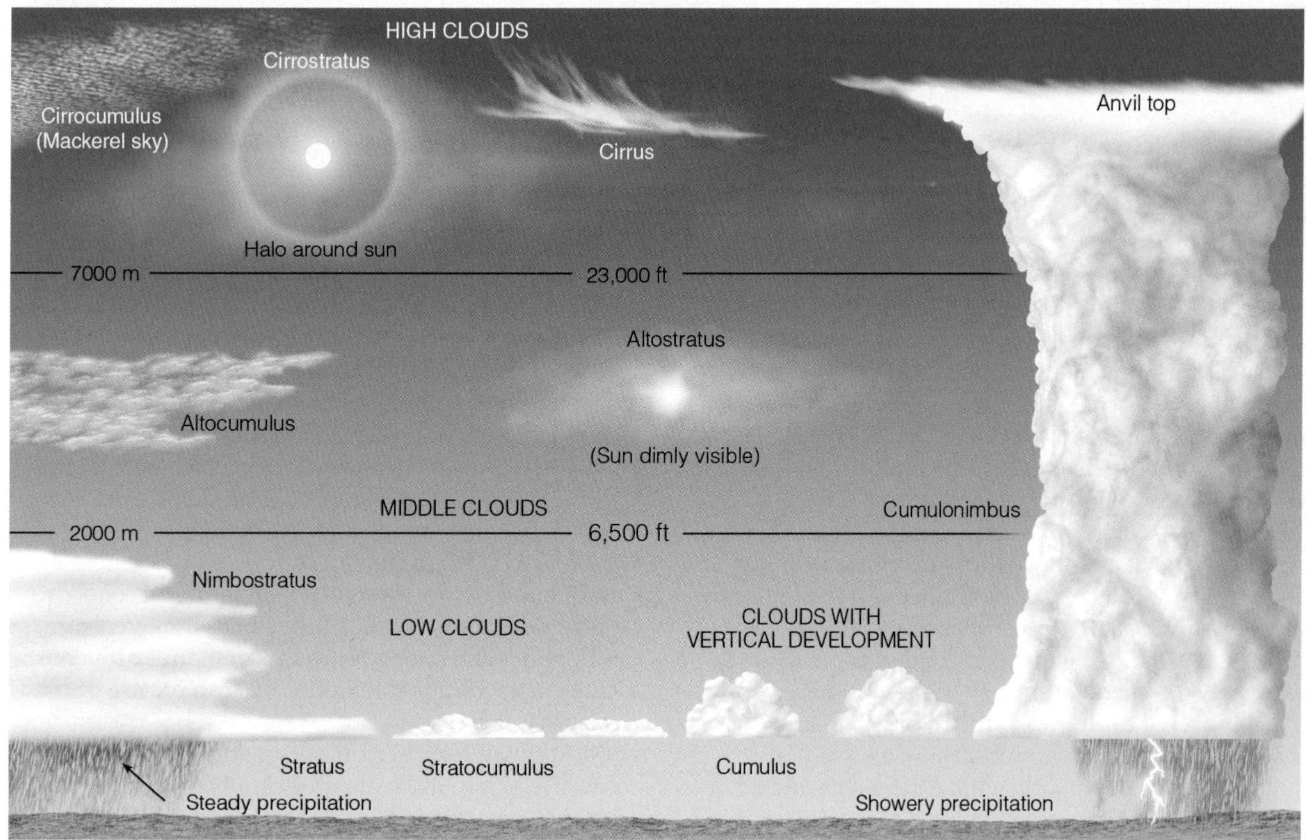

FIGURE 4.22 A generalized illustration of basic cloud types based on height above the surface and vertical development.

FIGURE 4.23 Lenticular clouds forming over a mountain range at sunset.

SOME UNUSUAL CLOUDS Although the ten basic cloud forms are the most frequently seen, there are some unusual clouds that deserve mentioning in the context of extreme weather. For example, moist air crossing a mountain barrier often forms into waves. The clouds that form in the wave crest usually have a lens shape and are, therefore, called **lenticular clouds** (see Fig. 4.23). Frequently, they form one above the other like a stack of pancakes, and at a distance they may resemble hovering spacecraft. Hence, it is no wonder a large number of UFO sightings take place when lenticular clouds are present. Beneath the lenticular cloud, a large swirling, extremely turbulent, rotor may be present that can pose a major hazard to aircraft. If the rotor becomes visible as a cloud, it is called a *rotor cloud.*

When a cloud forms over and extends downwind of an isolated mountain peak, as shown in Fig. 4.24, it is called a **banner cloud.** This cloud often is associated with strong atmospheric turbulence above and downwind of a mountain range. The conditions that lead to the formation of these clouds are particularly dangerous to small aircraft.

Similar to the lenticular cloud is the *cap cloud,* or **pileus,** that usually resembles a silken scarf capping the top of a sprouting cumulus cloud (see Fig. 4.25). Pileus clouds form when moist winds are deflected up and over the top of a building cumulus congestus or cumulonimbus. If the air flowing over the top of the cloud condenses, a pileus often forms. Again, these clouds indicate that strong winds and strong vertical air motions exist in the vicinity of the cloud—conditions that can prove hazardous to small aircraft.

Most clouds form in rising air, but the mammatus forms in sinking air. **Mammatus clouds** derive their name from their appearance—baglike sacs that hang beneath the cloud and resemble a cow's udder (see Fig. 4.26). Although mammatus most frequently form on the underside

© C. Donald Ahrens

FIGURE 4.24 The cloud forming over and downwind of Mt. Rainier is called a banner cloud.

© C. Donald Ahrens

FIGURE 4.25 A pileus cloud forming above a developing cumulus cloud.

of cumulonimbus, they can develop beneath cirrocumulus, altostratus, altocumulus, and stratocumulus as well.* For mammatus to form, the sinking air must be cooler than the air around it and have a high liquid water or ice content. As saturated air sinks, it warms, but the warming is retarded because of the heat taken from the air to evaporate the liquid or melt ice particles. If the sinking air remains saturated and cooler than the air around it, the sinking air can extend below the cloud base appearing as rounded masses we call mammatus clouds.

Up to now we've looked at clouds from the earth's surface. The next section looks at clouds from a different vantage point—from space.

*Mammatus clouds often form with severe thunderstorms. For this reason, people sometimes will call a sky dotted with mammatus clouds "a tornado sky." Mammatus clouds are not funnel clouds, do not rotate, and their appearance, as we will see in Chapter 12, has no relationship to tornadoes.

© Jorn Olson

FIGURE 4.26 Mammatus clouds forming beneath a thunderstorm.

Clouds: Observations from Space

The weather satellite is a cloud-observing platform in earth's orbit. It provides extremely valuable cloud photographs of areas where there are no ground-based observations. Because water covers over 70 percent of the earth's surface, there are vast regions where few (if any) surface cloud observations are made. Before weather satellites were available, tropical storms, such as hurricanes and typhoons, often went undetected until they moved dangerously near inhabited areas. Residents of the regions affected had little advance warning. Today, satellites spot these storms while they are still far out in the ocean and track them accurately.

There are two primary types of weather satellites in use for viewing clouds. The first are called **geostationary satellites** (or *geosynchronous satellites*) because they orbit the equator at the same rate the earth spins and, hence, remain at nearly 36,000 km (22,300 mi) above a fixed spot on the earth's surface (see ❯ Fig. 4.27). This positioning allows continuous monitoring of a specific region.

Geostationary satellites are also important because they use a "real time" data system, meaning that the satellites transmit images to the receiving system on the ground as soon as the camera takes the picture. Successive cloud images from these satellites can be put into a time-lapse movie sequence to show the cloud movement, dissipation, or development associated with weather fronts and storms. This information is a great help in forecasting the progress of large weather systems. Wind directions and speeds at various levels may also be approximated by monitoring cloud movement with the geostationary satellite.

To complement the geostationary satellites, there are **polar-orbiting satellites,** which closely parallel the earth's meridian lines. These satellites pass over the north and south polar regions on each revolution. As the earth rotates to the east beneath the satellite, each pass monitors an area to the west of the previous pass (see ❯ Fig. 4.28). Eventually, the satellite covers the entire earth.

Polar-orbiting satellites have the advantage of photographing clouds directly beneath them. Thus, they provide sharp pictures in polar regions, where photographs from a geostationary satellite are distorted because of the low angle at which the satellite "sees" this region. Polar orbiters also circle the earth at a much lower altitude (about 850 km) than geostationary satellites and provide detailed photographic information about objects, such as violent storms and cloud systems.

Continuously improved detection devices make weather observation by satellites more versatile than ever. Early satellites, such as *TIROS I,* launched on April 1, 1960, used television cameras to photograph clouds. Contemporary satellites use radiometers, which can observe clouds during both day and night by detecting radiation that emanates from the top of the clouds. Additionally, satellites have the capacity to obtain vertical profiles of atmospheric temperature and moisture by detecting emitted radiation from atmospheric gases, such as water vapor. In modern satellites, a special type of advanced radiometer (called an *imager*) provides satellite pictures with much better resolution than did previous imagers. Moreover, another type of special radiometer (called a *sounder*) gives a more accurate profile of temperature and moisture at different levels in the atmosphere than did earlier instruments. In the latest Geostationary Operational Environment Satellite (*GOES*) series, the imager and sounder are able to operate independent of each other.

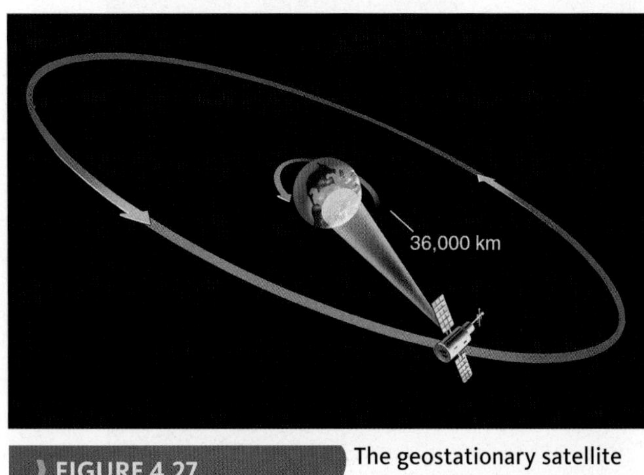

❯ FIGURE 4.27 The geostationary satellite moves through space at the same rate that the earth rotates, so it remains above a fixed spot on the equator and monitors one area constantly.

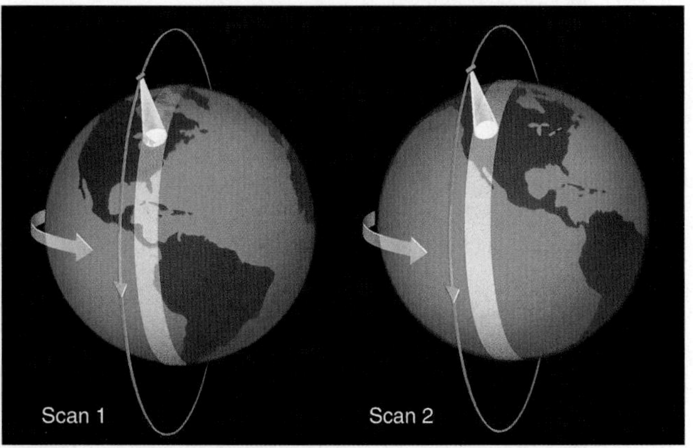

❯ FIGURE 4.28 Polar-orbiting satellites scan from north to south, and on each successive orbit the satellite scans an area farther to the west.

Information on cloud thickness and height can be deduced from satellite images. Visible images show the sunlight reflected from a cloud's upper surface. Because thick clouds have a higher albedo (reflectivity) than thin clouds, they appear brighter on a visible satellite image. However, high, middle, and low clouds have just about the same albedo, so it is difficult to distinguish among them simply by using visible light photographs. To make this distinction, *infrared cloud images* are used. Such pictures produce a better image of the actual radiating surface because they do not show the strong visible reflected light. Since warm objects radiate more energy than cold objects, high temperature regions can be artificially made to appear darker on an infrared photograph. Because the tops of low clouds are warmer than those of high clouds, cloud observations made in the infrared can distinguish between warm low clouds (dark) and cold high clouds (light) (see ❱ Fig. 4.29). Moreover, cloud temperatures can be converted by a computer into a three-dimensional image of the cloud. These are the 3-D cloud photos presented on television by many weathercasters.

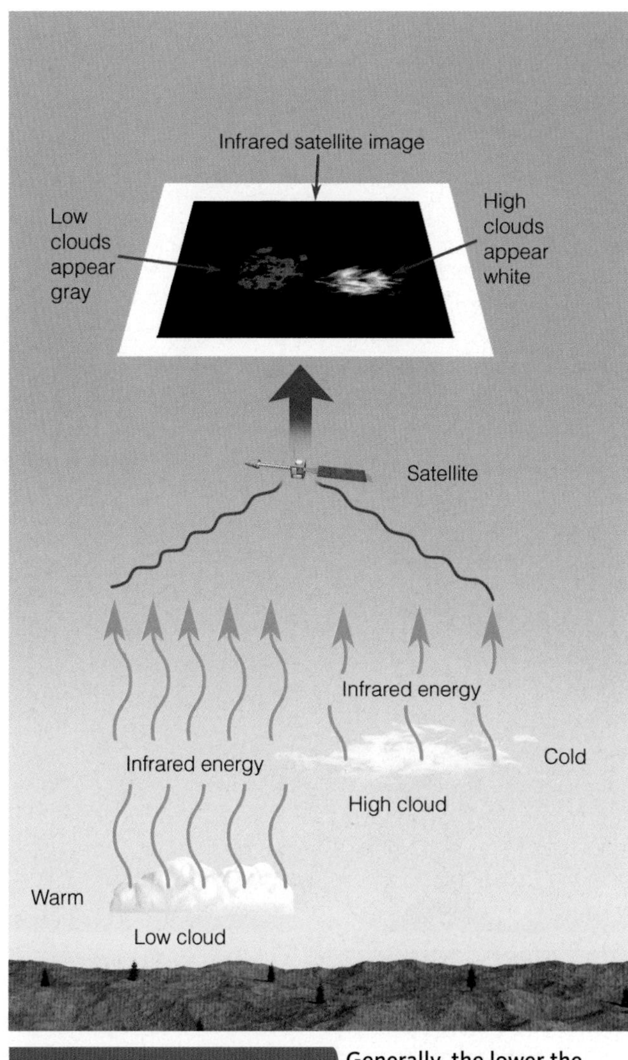

❱ FIGURE 4.29 Generally, the lower the cloud, the warmer its top. Warm objects emit more infrared energy than do cold objects. Thus, an infrared satellite image can distinguish warm, low (gray) clouds from cold, high (white) clouds.

❱ Figure 4.30 shows a visible satellite image (from a geostationary satellite) of a middle-latitude cyclonic storm system in the eastern Pacific. Notice that all of the clouds in the image appear white. However, in the infrared image (see ❱ Fig. 4.31), taken on the same day (and just about the same time), the clouds appear to have many shades of gray. In the visible image, the clouds covering part of Oregon and northern California appear relatively thin compared to the thicker, bright clouds to the west. Furthermore, these thin clouds must be high because they also appear bright in the infrared image. The elongated band of clouds off the coast marks the position of an approaching weather front. Here, the clouds appear white and bright in both pictures, indicating a zone of thick, heavy clouds. Behind the front, the lumpy clouds are probably cumulus because they appear gray in the infrared image, indicating that their tops are low and relatively warm.

When temperature differences are small, it is difficult to directly identify significant cloud and surface features on an infrared image. Some way must be found to increase the contrast between features and their backgrounds. Such increase in contrast can be done by a process called *computer enhancement*. Certain temperature ranges in the infrared image are assigned specific shades of gray—grading from black to white. Normally, clouds with cold tops, and those tops near freezing are assigned the darkest gray color.

❱ Figure 4.32 is an infrared-enhanced image for the same day as shown in Figs. 4.30 and 4.31. Often in this type of image, dark blue or red is assigned to clouds with the coldest (highest) tops. Hence, the dark red areas embedded along the front in Fig. 4.32 represent the region where the coldest and, therefore, highest and thickest clouds are found. It is here where the stormiest weather is probably occurring. Also notice that, near the southern tip of the image, the dark red blotches surrounded by areas of white are thunderstorms that have developed over warm tropical waters. They show up clearly as thick white clouds in both the visible and infrared images. By examining the movement of these clouds on successive satellite images, forecasters can predict the arrival of clouds and storms, as well as the passage of weather fronts.

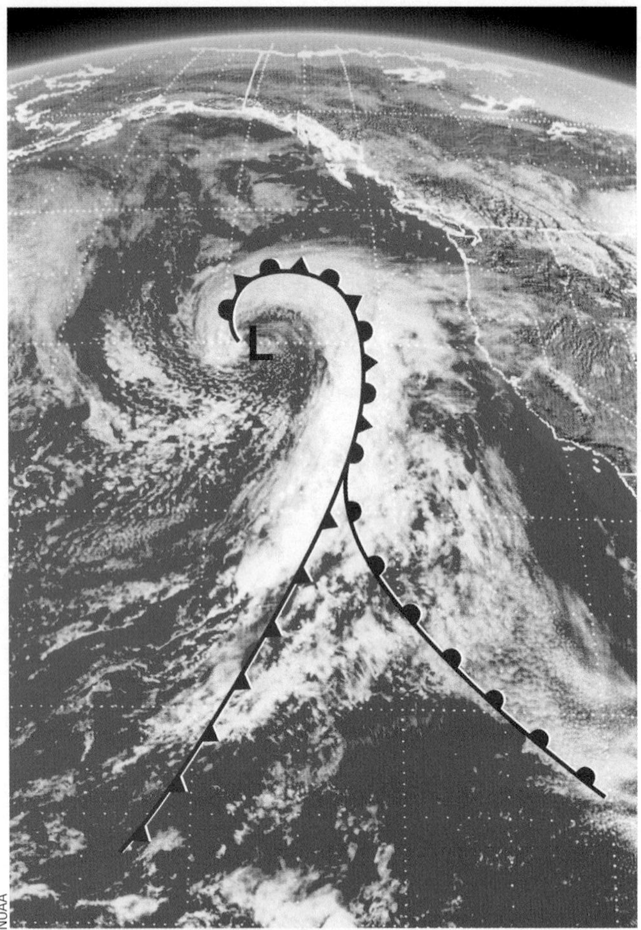

FIGURE 4.30 A visible satellite image of the eastern Pacific taken at just about the same time on the same day as the image in Fig. 4.31. Notice that the clouds in the visible image appear white. Superimposed on the image is the mid-latitude cyclonic storm system with its weather front.

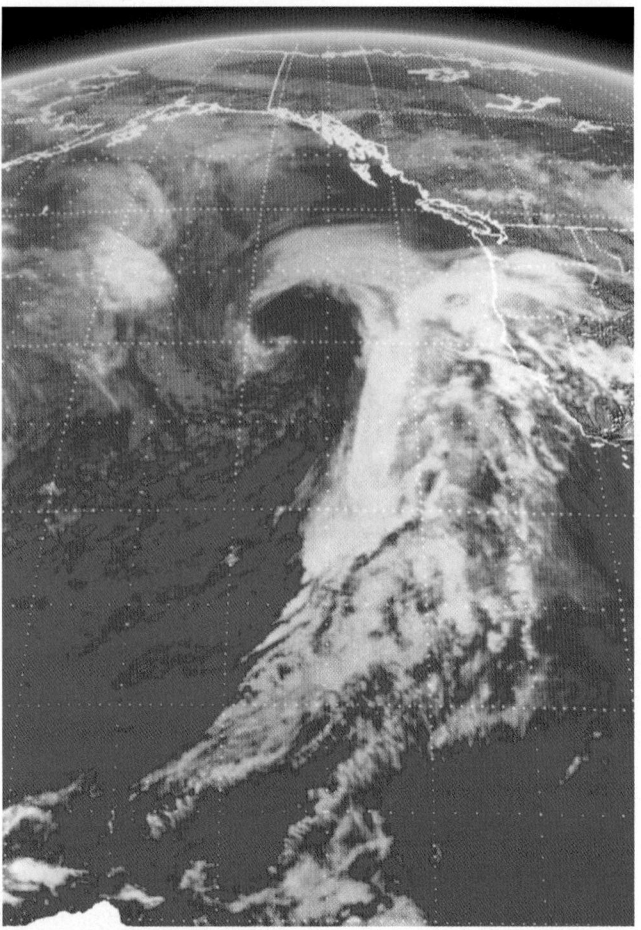

Active FIGURE 4.31 Infrared image of the eastern Pacific taken at just about the same time on the same day as the image in Fig. 4.30. Notice that the low clouds in the infrared image appear in various shades of gray. Visit the Meteorology Resource Center to view this and other active figures at www.cengage.com/login

In regions where there are no clouds, it is difficult to observe the movement of the air. To help with this situation, geostationary satellites are equipped with water-vapor sensors that can profile the distribution of atmospheric water vapor in the middle and upper troposphere (see Fig. 4.33). In time-lapse films, the swirling patterns of moisture clearly show wet regions and dry regions, as well as middle tropospheric swirling wind patterns and jet streams.

The *T*ropical *R*ainfall *M*easuring *M*ission *TRMM* satellite provides information on clouds and precipitation from about 35°N to 35°S. A joint venture of NASA and the National Space Agency of Japan, this satellite orbits the earth at an altitude of about 400 km (250 mi). From this vantage point the satellite, when looking straight down, can pick out individual cloud features as small as 2.4 km (1.5 mi) in diameter. Some of the instruments onboard the *TRMM* satellite include a visible and infrared scanner, a microwave imager, and precipitation radar. These instruments help provide three-dimensional images of clouds and storms, along with the intensity and distribution of precipitation (see Fig. 4.34). Additional onboard instruments send back information concerning the earth's energy budget and lightning discharges in storms.

FIGURE 4.32 An enhanced infrared image of the eastern Pacific taken on the same day as the images shown in Fig. 4.30 and Fig. 4.31.

NOAA

FIGURE 4.33 Infrared water-vapor image. The darker areas represent dry air aloft; the brighter the gray, the more moist the air in the middle or upper troposphere. Bright white areas represent dense cirrus clouds or the tops of thunderstorms. Green areas represent the coldest cloud tops.

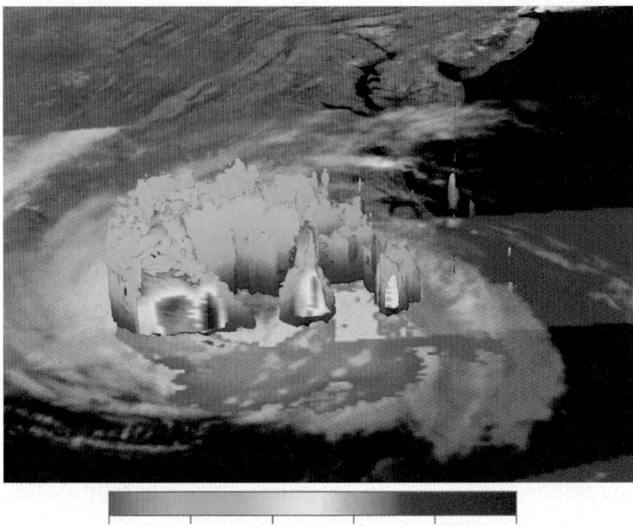

FIGURE 4.34

A three-dimensional *TRMM* satellite image of Hurricane Ophelia along the North Carolina coast on September 14, 2005. The light green areas in the cutaway view represent the region of lightest rainfall, whereas dark red and orange indicate regions of heavy rainfall.

Summary

This chapter focuses on condensation and its many forms. The chapter begins by examining dew and frost. Here we learn that dew forms when the air temperature cools to the dew point in a shallow layer of air near the surface. If the dew should freeze, it becomes frozen dew. Frost forms when the air cools to its dew point that is at freezing or below, and water vapor transforms directly from a vapor into a solid.

As the air cools in a deeper layer near the surface, condensation nuclei in the atmosphere, such as sea salt, begin to attract water vapor when the relative humidity is less than 100 percent. Water condenses onto these hygroscopic nuclei forming haze that restricts visibility. As the relative humidity approaches 100 percent, condensation occurs on most nuclei, and the air becomes filled with tiny liquid droplets (or ice crystals) called fog.

Fog forms in two primary ways: (a) cooling the air and (b) evaporating and mixing water vapor into the air. Radiation fog, advection fog, and upslope fog form by the cooling of air, whereas steam fog and frontal fog are two forms of evaporation (mixing) fog. Although fog has some positive beneficial effects, in many places it is a nuisance, for it disrupts air traffic, and it is the primary cause of a number of auto accidents.

Condensation above the earth's surface produces clouds. When clouds are classified according to their height and physical appearance, they are divided into four main groups: high, middle, low, and clouds with vertical development. Since each cloud has physical characteristics that distinguish it from all the others, careful cloud observations normally lead to correct identification.

Extreme weather, such as strong damaging surface winds, flash floods, and tornadoes, is most often associated with cumulonimbus clouds. An aircraft flying in the vicinity of a lenticular cloud, rotor cloud, or a banner cloud may experience hazardous flying conditions.

Satellites enable scientists to obtain a bird's-eye view of clouds on a global scale. Polar-orbiting satellites obtain data covering the earth from pole to pole, whereas geostationary satellites located above the equator continuously monitor a desired portion of the earth. Both types of satellites use radiometers (imagers) that detect emitted radiation. As a consequence, clouds can be observed both day and night.

Visible satellite images, which show sunlight reflected from a cloud's upper surface, can distinguish thick clouds from thin clouds. Infrared pictures show an image of the cloud's radiating top and can distinguish low clouds from high clouds. To increase the contrast between cloud features, infrared images are enhanced.

Key Terms

The following terms are listed (with page number) in the order they appear in the text. Define each. Doing so will aid you in reviewing the material covered in this chapter.

dew, 96
frost, 96
haze, 97
fog, 97
radiation fog, 97
advection fog, 98
upslope fog, 99
evaporation (mixing) fog, 99
acid fog, 104
cirrus clouds, 105
cirrocumulus clouds, 105
cirrostratus clouds, 105
altocumulus clouds, 106
altostratus clouds, 106
nimbostratus clouds, 106
stratocumulus clouds, 106
stratus clouds, 107
cumulus clouds, 109
cumulonimbus clouds, 110
lenticular clouds, 112
banner cloud, 112
pileus clouds, 112
mammatus clouds, 112
geostationary satellites, 114
polar-orbiting satellites, 114

Review Questions

1. Explain how dew, frozen dew, and visible frost form.
2. How does haze differ from fog?
3. List two primary ways in which fog forms.
4. Describe the atmospheric conditions that are necessary for the formation of
   (a) radiation fog
   (b) advection fog
5. When radiation fog "burns off," what happens to the fog droplets?
6. Why does radiation fog usually "burn off" by afternoon?
7. Explain how evaporation (mixing) fog forms.
8. Describe how steam fog on a road surface can lead to hazardous driving conditions, especially when the sun is up.
9. List some of the adverse effects of fog.
10. Describe the "foggy areas" in the United States.
11. Clouds are most generally classified by height above the earth's surface. List the major height categories and the cloud types associated with each.
12. List at least two distinguishable characteristics of each of the ten basic clouds.
13. Why are high clouds normally thin? Why are they composed almost entirely of ice crystals?
14. How can you distinguish altostratus from cirrostratus?
15. Which cloud is most often associated with severe weather, such as strong surface winds, hail, and tornadoes?
16. Which clouds are associated with each of the following characteristics:
    (a) mackerel sky
    (b) lightning and thunder
    (c) mare's tails
    (d) anvil top
    (e) light continuous rain or snow?
17. A halo around the sun or moon often indicates that a midlatitude cyclonic storm may be approaching from the west. What cloud produces this halo?
18. Explain how a lenticular cloud differs from a pileus cloud.
19. Why would a pilot not want to fly anywhere near a rotor cloud?
20. Explain how mammatus clouds form.
21. How do geostationary satellites differ from polar-orbiting satellites?
22. Explain how visible and infrared satellite images can be used to distinguish:
    (a) high clouds from low clouds;
    (b) thick clouds from thin clouds.
23. In an enhanced infrared satellite image, how would thunderstorms appear?

Online Learning

STUDENT COMPANION WEBSITE: Visit this book's companion website at: www.cengage.com/ahrens/extreme1e and choose chapter 4 for many study aids and ideas for further reading and research. These include flashcards, practice quizzing, and web links.

METEOROLOGY RESOURCE CENTER: For students with access, log on at www.cengage.com/login for more assets, including animations, videos, and more. If your textbook did not come with access, visit www.CengageBrain.com

Clouds and Stability

CONTENTS

ATMOSPHERIC STABILITY
TYPES OF ATMOSPHERIC STABILITY
CLOUD DEVELOPMENT
SUMMARY
KEY TERMS
REVIEW QUESTIONS

Instability generated by cold air aloft causes rapidly rising air and the formation of a thunderstorm near Brewster, Nebraska.

Many severe and extreme weather events occur with rapidly rising air currents called *convection*. Thunderstorms, hail, tornadoes, and hurricanes all form with strong convection. When we think of convection, we usually think of strong surface heating by the sun. Convection does form in this manner, but, as we will see in this chapter, convection may also be initiated by changes in atmospheric temperature and moisture that take place above the earth's surface. Our interest here is to understand how the temperature structure of the atmosphere can influence the development of convection and ultimately the formation of thunderstorms and severe weather.

We know that most clouds form as air rises and cools. Why does air rise on some occasions and not on others? And why do some cumulus clouds grow into huge complex thunderstorms, whereas others only grow into tiny fair-weather cumulus? To answer these questions we need to look at the concept of *atmospheric stability*.

Atmospheric Stability

When we speak of atmospheric stability, we are referring to a condition of equilibrium. For example, rock A resting in the depression in ❱ Fig. 5.1 is in *stable* equilibrium. If the rock is pushed up along either side of the hill and then let go, it will quickly return to its original position. On the other hand, rock B, resting on the top of the hill, is in a state of *unstable* equilibrium, as a slight push will set it moving away from its original position. Applying these concepts to the atmosphere, we can see that air is in stable equilibrium when, after being lifted or lowered, it tends to return to its original position—it resists upward and downward air motions. Air that is in unstable equilibrium will, when given a little push, move farther away from its original position—it favors vertical air currents.

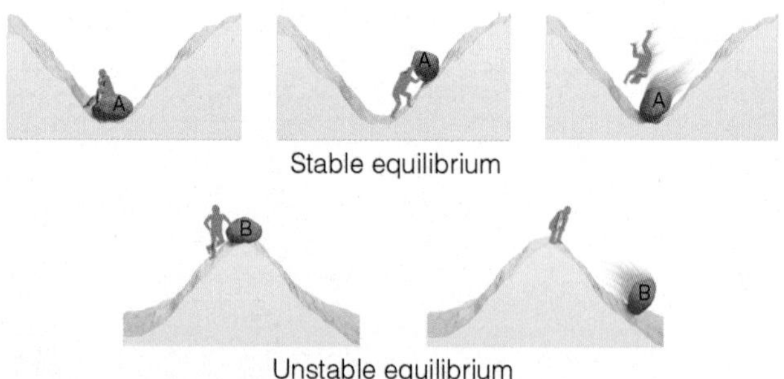

❱ **FIGURE 5.1** **When rock A is disturbed, it will return to its original position; rock B, however, will accelerate away from its original position.**

EXTREME WEATHER WATCH

Though rarely categorized as clouds, tornadoes would qualify. As air is drawn into the low-pressure core of the tornado it expands and cools at the dry adiabatic rate, just as it would if it were rapidly lifted to a region of lower pressure. The cooling proceeds to the point where condensation begins, liquid droplets form, and the funnel becomes visible as a rapidly rotating cloud.

To explore the behavior of rising and sinking air, we must first put some air in an imaginary thin elastic wrap. This small volume of air is referred to as a **parcel of air.***
Although the air parcel can expand and contract freely, it does not break apart but remains as a single unit. At the same time, neither external air nor heat can mix with the air inside the parcel. The space occupied by the air molecules within the parcel defines the air density. The average speed of the molecules is directly related to the air temperature, and the molecules colliding against the parcel walls determine the air pressure inside.

At the earth's surface, the parcel has the same temperature and pressure as the air surrounding it. Suppose we lift the air parcel up into the atmosphere. We know from Chapter 1 that air pressure decreases with height. Consequently, the air pressure surrounding the parcel lowers. The lower pressure outside allows the air molecules inside to push the parcel walls outward, expanding the parcel. Because there is no other energy source, the air molecules inside must use some of their own energy to expand the parcel. This shows up as slower average molecular speeds. Recall from Chapter 2 that temperature is related to the average speed (average kinetic energy) of all the molecules. Consequently, these slower speeds result in a lower parcel temperature. If the parcel is lowered to the surface, it returns to a region where the surrounding air pressure is higher. The higher pressure squeezes (compresses) the parcel back into its original (smaller) volume. This squeezing increases the average speed of the air molecules and the parcel temperature rises. Hence, *a rising parcel of air expands and cools, while a sinking parcel is compressed and warms.***

Another way to look at the temperature change in a rising parcel is to imagine you are playing pool. Suppose you hit the cue ball and it strikes the side of the pool table. The velocity of the ball after rebounding off the

*An air parcel is an imaginary body of air about the size of a large basketball. The concept of an air parcel is illustrated several places in the text, including Fig. 3.18, p. 81.

**Additional information on rising and sinking air parcels is provided in the Focus section in Chapter 1 on p. 23.

side of the table is nearly the same velocity as the ball had before striking it. Suppose that the sides of the table become flexible. Now when you shoot the ball and it strikes the table's side, part of the ball's energy of motion goes into pushing back the side, and the ball bounces away more slowly. This slower speed represents a loss in kinetic energy and a lower temperature.

If a parcel of air expands and cools, or compresses and warms, with no interchange of heat with its surroundings, this situation is called an **adiabatic process.** As long as the air in the parcel is unsaturated (the relative humidity is less than 100 percent), the rate of adiabatic cooling or warming remains constant. This rate of heating or cooling is about 10°C for every 1000 meters (m) of change in elevation (5.5°F per 1000 ft) and applies only to unsaturated air. For this reason, it is called the **dry adiabatic rate*** (see ❯Fig. 5.2).

As the rising air cools, its relative humidity increases as the air temperature approaches the dew-point temperature. If the rising air cools to its dew-point temperature, the relative humidity becomes 100 percent. Further lifting results in condensation, a cloud forms, and latent heat is released inside the rising air parcel. Because the heat added during condensation offsets some of the cooling due to expansion, the air no longer cools at the dry adiabatic rate but at a lesser rate called the **moist adiabatic rate.** (Because latent heat is added to the rising saturated air, the process is not really adiabatic.**) If a saturated parcel containing water droplets were to sink, it would compress and warm at the moist adiabatic rate because evaporation of the liquid droplets would offset the rate of compressional warming. Hence, the rate at which rising or sinking saturated air changes temperature—the moist adiabatic rate—is less than the dry adiabatic rate.

*Using metric units and temperatures in °C make the numbers easy to deal with when describing rising and descending air parcels. Therefore, throughout this chapter, we will use metric units and temperatures in °C.

**If condensed water or ice is removed from the rising saturated parcel, the cooling process is called an *irreversible pseudoadiabatic process.*

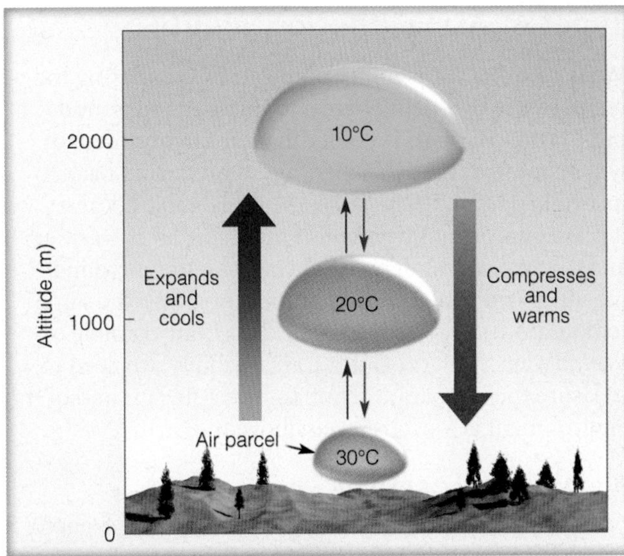

❯ **FIGURE 5.2** The dry adiabatic rate. As long as the air parcel remains unsaturated, it expands and cools by 10°C per 1000 m; the sinking parcel compresses and warms by 10°C per 1000 m.

Unlike the dry adiabatic rate, the moist adiabatic rate is not constant, but varies greatly with temperature and, hence, with moisture content—as warm saturated air produces more liquid water than cold saturated air. The added condensation in warm, saturated air liberates more latent heat. Consequently, the moist adiabatic rate is much less than the dry adiabatic rate when the rising air is warm; however, the two rates are nearly the same when the rising air is very cold (see ❯Table 5.1). Although the moist adiabatic rate does vary, to make the numbers easy to deal with, we will use an average of 6°C per 1000 m (3.3°F per 1000 ft) in most of our examples and calculations.

❯ Table 5.1

The Moist Adiabatic Rate for Different Temperatures and Pressures in °C/1000 m and °F/1000 ft

PRESSURE (MB)	TEMPERATURE (°C)					TEMPERATURE (°F)				
	−40	−20	0	20	40	−40	−5	30	65	100
1000	9.5	8.6	6.4	4.3	3.0	5.2	4.7	3.5	2.4	1.6
800	9.4	8.3	6.0	3.9		5.2	4.6	3.3	2.2	
600	9.3	7.9	5.4			5.1	4.4	3.0		
400	9.1	7.3				5.0	4.0			
200	8.6					4.7				

Types of Atmospheric Stability

We determine the stability of the air by comparing the temperature of a rising parcel to that of its surroundings. If the rising air is colder than its environment, it will be more dense* (heavier) and tend to sink back to its original level. In this case, the air is *stable* because it resists upward movement. If the rising air is warmer and, therefore, less dense (lighter) than the surrounding air, it will continue to rise until it reaches the same temperature as its environment. This is an example of *unstable* air. To figure out the air's stability, we need to measure the temperature both of the rising air and of its environment at various levels above the earth.

A STABLE ATMOSPHERE Suppose we release a balloon-borne instrument called a *radiosonde.* (A photo of a radiosonde is found in Fig. 4, p. 75). As the balloon carries the radiosonde up into the atmosphere, the radiosonde sends back temperature data as shown in Fig. 5.3. (Such a vertical profile of temperature is called *a sounding.*) We measure the air temperature in the vertical and find that it decreases by 4°C for every 1000 m (2°F per 1000 ft). Remember from Chapter 1 that the rate at which the air temperature changes with elevation is called the *lapse rate.* Because this is the rate at which the air temperature surrounding us will be changing if we were to climb upward into the atmosphere, we will refer to it as the **environmental lapse rate**. Now suppose in Fig. 5.3a that a parcel of unsaturated air with a temperature of 30°C is lifted from the surface. As it rises, it cools at the dry adiabatic rate (10°C per 1000 m), and the temperature inside the parcel at 1000 meters would be 20°C, or 6°C lower than the air surrounding it. Look at Fig. 5.3a closely and notice that, as the air parcel rises higher, the temperature difference between it and the surrounding air becomes even greater. Even if the parcel is initially saturated (see Fig. 5.3b), it will cool at the moist rate—6°C per 1000 m—and will be colder than its environment at all levels. In both cases, the rising air is colder and heavier than the air surrounding it. In this example, the atmosphere is **absolutely stable**. *The*

*When, at the same level in the atmosphere, we compare parcels of air that are equal in size but vary in temperature, we find that cold air parcels are more dense than warm air parcels; that is, in the cold parcel, there are more molecules that are crowded closer together.

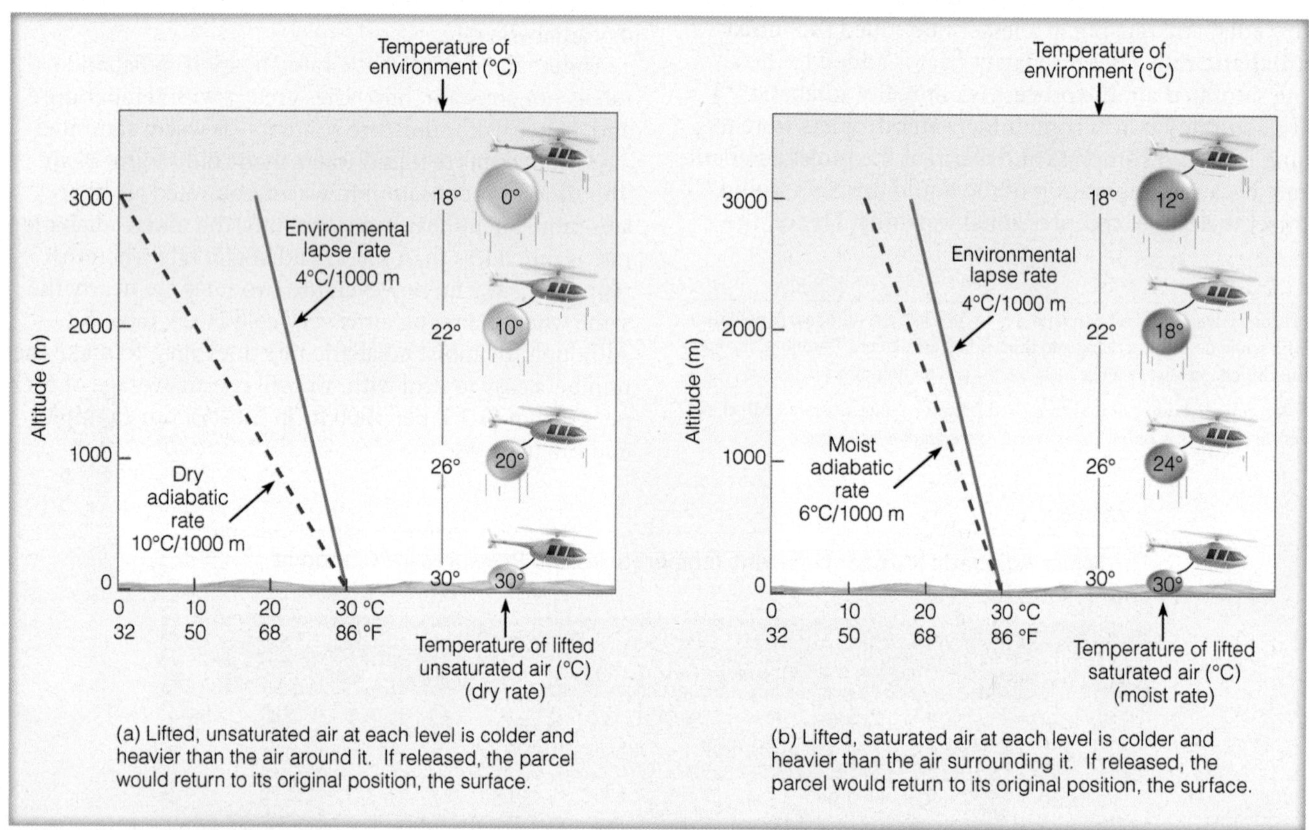

FIGURE 5.3 An absolutely stable atmosphere occurs when the environmental lapse rate is less than the moist adiabatic rate. In a stable atmosphere, a rising air parcel is colder and more dense than the air surrounding it, and, if given the chance (that is, released), it will return to its original position, the surface. (In both situations, the helicopter lifts the air parcel. In the real world, this type of parcel lifting would be impossible.)

atmosphere is always absolutely stable when the environmental lapse rate is less than the moist adiabatic rate.

Since air in an absolutely stable atmosphere strongly resists upward vertical motion, it will, *if forced to rise*, tend to spread out horizontally. If clouds form in this rising air, they, too, will spread horizontally in relatively thin layers and usually have flat tops and bases. We might expect to see clouds—such as cirrostratus, altostratus, nimbostratus, or stratus—forming in stable air.

What conditions are necessary to bring about a stable atmosphere? As we have just seen, the atmosphere is stable when the environmental lapse rate is small, that is, when the difference in temperature between the surface air and the air aloft is relatively small. Consequently, the atmosphere tends to become more stable—that is, it stabilizes—as the air aloft warms or the surface air cools. If the air aloft is being replaced by warmer air (warm advection), and the surface air is not changing appreciably, the environmental lapse rate decreases and the atmosphere becomes more stable. Similarly, the environmental lapse rate decreases and the atmosphere becomes more stable when the lower layer cools (see ❱ Fig. 5.4). The *cooling* of the *surface air* may be due to:

1. nighttime radiational cooling of the surface
2. an influx of cold surface air brought in by the wind (cold advection)
3. air moving over a cold surface

Consequently, on any given day, the atmosphere is most stable in the early morning around sunrise, when the lowest surface air temperature is recorded. If the surface air becomes saturated in a stable atmosphere, a persistent layer of haze or fog may form (see ❱ Fig. 5.5). This situation helps us understand why radiation fog—a common type of fog described in the previous chapter—tends to be most dense in the early morning, and why many fog-related auto accidents occur during this time of day.

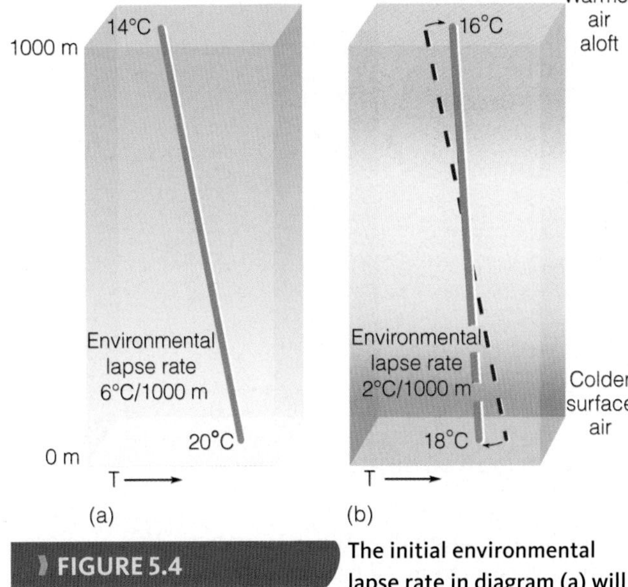

❱ FIGURE 5.4 The initial environmental lapse rate in diagram (a) will become more stable (stabilize) as the air aloft warms and the surface air cools, as illustrated in diagram (b).

Another way the atmosphere becomes more stable is when an entire layer of air sinks. For example, if a layer of unsaturated air over 1000 meters thick and covering a large area subsides, the entire layer will warm by adiabatic compression. As the layer subsides, it becomes compressed by the weight of the atmosphere and shrinks vertically. The upper part of the layer sinks farther, and,

❱ FIGURE 5.5 Cold surface air, on this morning, produces a stable atmosphere that inhibits vertical air motions and allows the fog and haze to linger close to the ground.

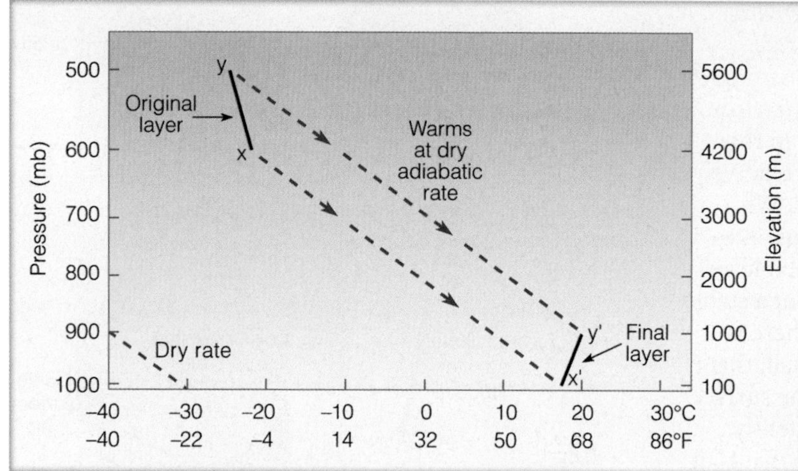

FIGURE 5.6

The layer x–y is initially 1400 m thick. If the entire layer slowly subsides, it shrinks in the more-dense air near the surface. As a result of the shrinking, the top of the layer warms more than the bottom, the entire layer (x′–y′) becomes more stable, and in this example forms an inversion.

hence, warms more than the bottom part. This phenomenon is illustrated in Fig. 5.6. After subsiding, the top of the layer is actually warmer than the bottom, and an inversion* is formed. Inversions that form as air slowly sinks over a large area are called **subsidence inversions**. They sometimes occur at the surface, but more frequently, they are observed aloft and are often associated with large high-pressure areas because of the sinking air motions associated with these systems.

An inversion represents an atmosphere that is absolutely stable. Why? Within the inversion, warm air overlies cold air, and, if air rises into the inversion, it is becoming colder, while the air around it is getting warmer. Obviously, the colder air would tend to sink. Inversions, therefore, act as lids on vertical air motion. When an inversion exists near the ground, stratus clouds, fog, haze, and pollutants are all kept close to the surface. In fact, most air pollution episodes occur with subsidence inversions. (For additional information on subsidence inversions, read the Focus section on p. 127.)

At this point, we can see why extremely high surface air temperatures can occur when the air aloft is sinking. Notice in Fig. 5.7 that the air inside an air parcel at an altitude of about 3000 m (10,000 ft), where the air pressure is 700 mb, has a moderately cool temperature of 10°C (50°F). If this air parcel sinks all the way to the surface (0 meters), where the pressure is 1000 mb, the final temperature inside the parcel having warmed at the dry adiabatic rate will be a whopping 40°C (104°F)! Thus, the air parcel at 700 mb has the potential of being very warm when brought to the surface.**

Before we turn our attention to unstable air, let's first examine a condition known as **neutral stability**. If the lapse rate is exactly equal to the dry adiabatic rate, rising or sinking unsaturated air will cool or warm at the same rate as the air around it. At each level, it would have the same temperature and density as the surrounding air. Because this air tends neither to continue rising nor sinking, the atmosphere is said to be neutrally stable. For saturated air, *neutral stability* exists when the environmental lapse rate is equal to the moist adiabatic rate.

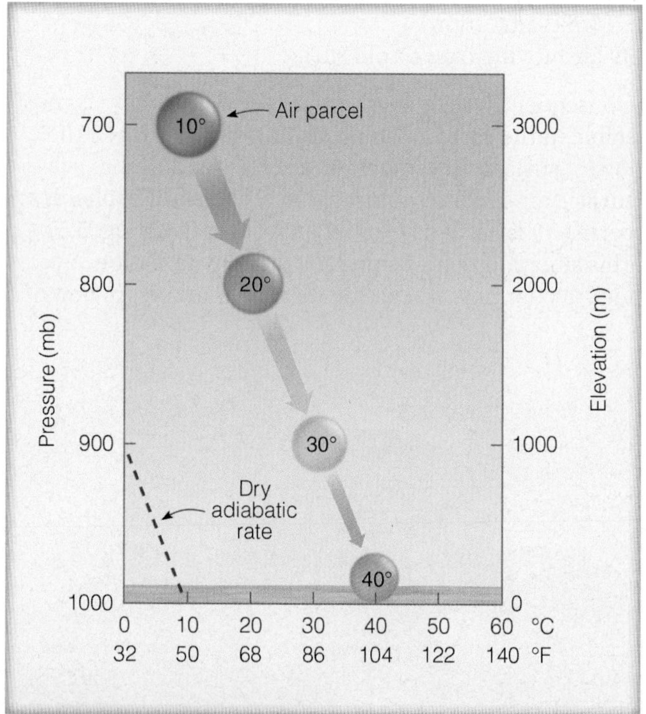

FIGURE 5.7 An air parcel initially with a temperature of 10°C (50°F), at an altitude where the pressure is 700 mb, has the potential to warm to 40°C (104°F) if it were to sink to the surface.

*Recall from Chapter 1 that an inversion represents an atmospheric condition where the air becomes warmer with height.

**The temperature an air parcel would have if lowered at the dry adiabatic rate to a pressure of 1000 millibars is called *potential temperature*. Moving parcels to the same level allows them to be observed under identical conditions so it can be determined which parcels are potentially warmer than others. In this example, the potential temperature of the parcel is 40°C, or 313K.

FOCUS ON
A SPECIAL TOPIC

Stability and Subsidence Inversions

Figure 1 shows a typical summertime vertical profile of air temperature (solid red line) and dew point (purple dashed line) measured with a radiosonde near the coast of California. Notice that the air temperature decreases from the surface up to an altitude of about 300 m (1000 ft). Notice also that, where the air temperature reaches the dew point, a cloud forms.

Above about 300 m the air temperature increases rapidly up to an altitude near 900 m (about 3000 ft). This region of increasing air temperature with increasing height marks the region of the subsidence inversion. Within the inversion, air from aloft slowly sinks and warms by compression.

The sinking air at the top of the inversion is not only warm (about 24°C or 75°F) but also dry with a low relative humidity, as indicated by the large spread between air temperature and dew point. The subsiding air, which does not reach the surface, is associated with a large high-pressure area, located to the west of California.

Immediately below the base of the inversion lies cool, moist air. The cool air is unable to penetrate the inversion because a lifted parcel of cool, marine air within the inversion would be much colder and heavier than the air surrounding it. Since the colder air parcel would fall back to its original position, the atmosphere is absolutely stable within the inversion. The subsidence inversion, therefore, acts as a lid on the air below, preventing the air from mixing vertically into the inversion. And so the marine air with its pollution and clouds is confined to a relatively shallow region near the earth's surface.

If the base of the inversion were to lower, the air pollutants below would be concentrated in a smaller space, producing a potentially dangerous situation. It is this trapping of air near the surface, associated with a strong subsidence inversion, that helps to make West Coast cities such as Los Angeles very polluted.

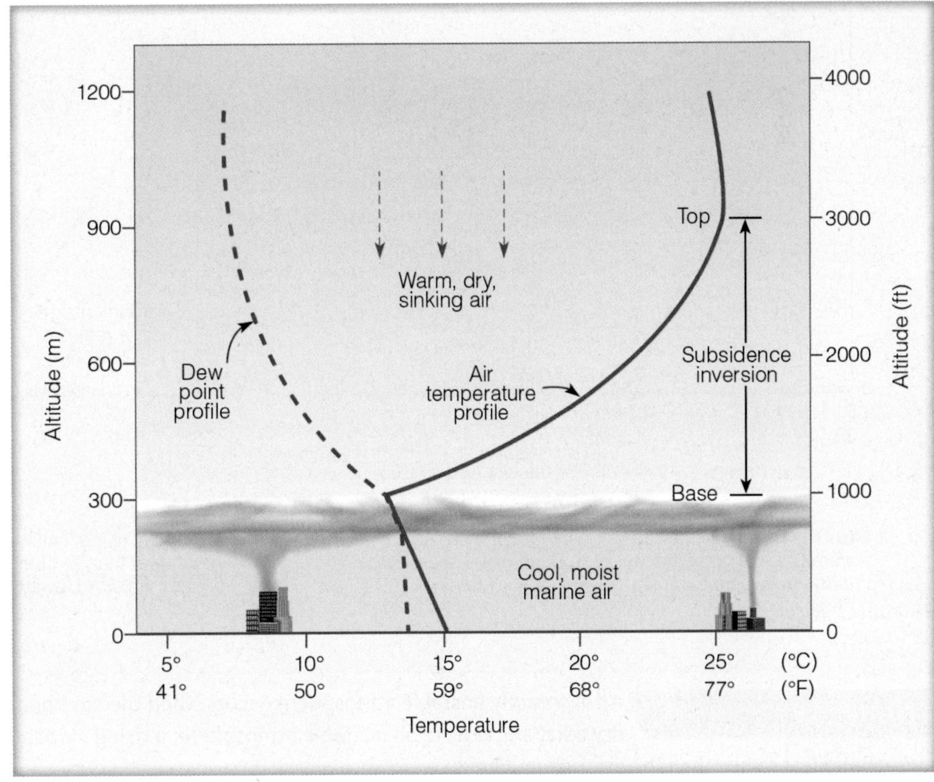

Figure 1 A strong subsidence inversion along the coast of California. The base of the stable inversion acts as a cap or lid on the cool, marine air below. An air parcel rising into the inversion layer would sink back to its original level because the rising air parcel would be colder and more dense than the air surrounding it.

AN UNSTABLE ATMOSPHERE Suppose a radiosonde sends back the temperatures above the earth as plotted in Fig. 5.8a. Once again, we determine the atmosphere's stability by comparing the environmental lapse rate to the moist and dry adiabatic rates. In this case, the environmental lapse rate is 11°C per 1000 m (6°F per 1000 ft). A rising parcel of unsaturated surface air will cool at the dry adiabatic rate. Because the dry adiabatic rate is less than the environmental lapse rate, the parcel will be warmer than the surrounding air and will continue to rise, constantly moving upward, away from its original position. The atmosphere is unstable. Of course, a parcel of saturated air cooling at the lower moist adiabatic rate will be even warmer than the air around it (see Fig. 5.8b). In both cases, the air parcels, once they start upward, will continue to rise on their own because the rising air parcels are warmer and less dense than the air around them. The atmosphere in this example is said to be **absolutely unstable**. *Absolute instability results when the environmental lapse rate is greater than the dry adiabatic rate.*

It should be noted, however, that deep layers in the atmosphere are seldom, if ever, absolutely unstable. Absolute instability is usually limited to a very shallow layer near the ground on hot, sunny days. Here, the environmental lapse rate can exceed the dry adiabatic rate, and the lapse rate is called *superadiabatic*.

So far, we have seen that the atmosphere is absolutely stable when the environmental lapse rate is less than the moist adiabatic rate and absolutely unstable when the environmental lapse rate is greater than the dry adiabatic rate. However, a typical type of atmospheric instability exists when the lapse rate lies between the moist and dry adiabatic rates.

A CONDITIONALLY UNSTABLE ATMOSPHERE The environmental lapse rate in Fig. 5.9 is 7°C per 1000 m (4°F per 1000 ft). When a parcel of unsaturated air rises, it cools dry adiabatically and is colder at each level than the air around it (see Fig. 5.9a). It will, therefore, tend to sink back to its original level because it is in a stable atmosphere. Now, suppose the rising parcel is saturated. As we can see in Fig. 5.9b, the rising air is warmer than its environment at each level. Once the parcel is given a

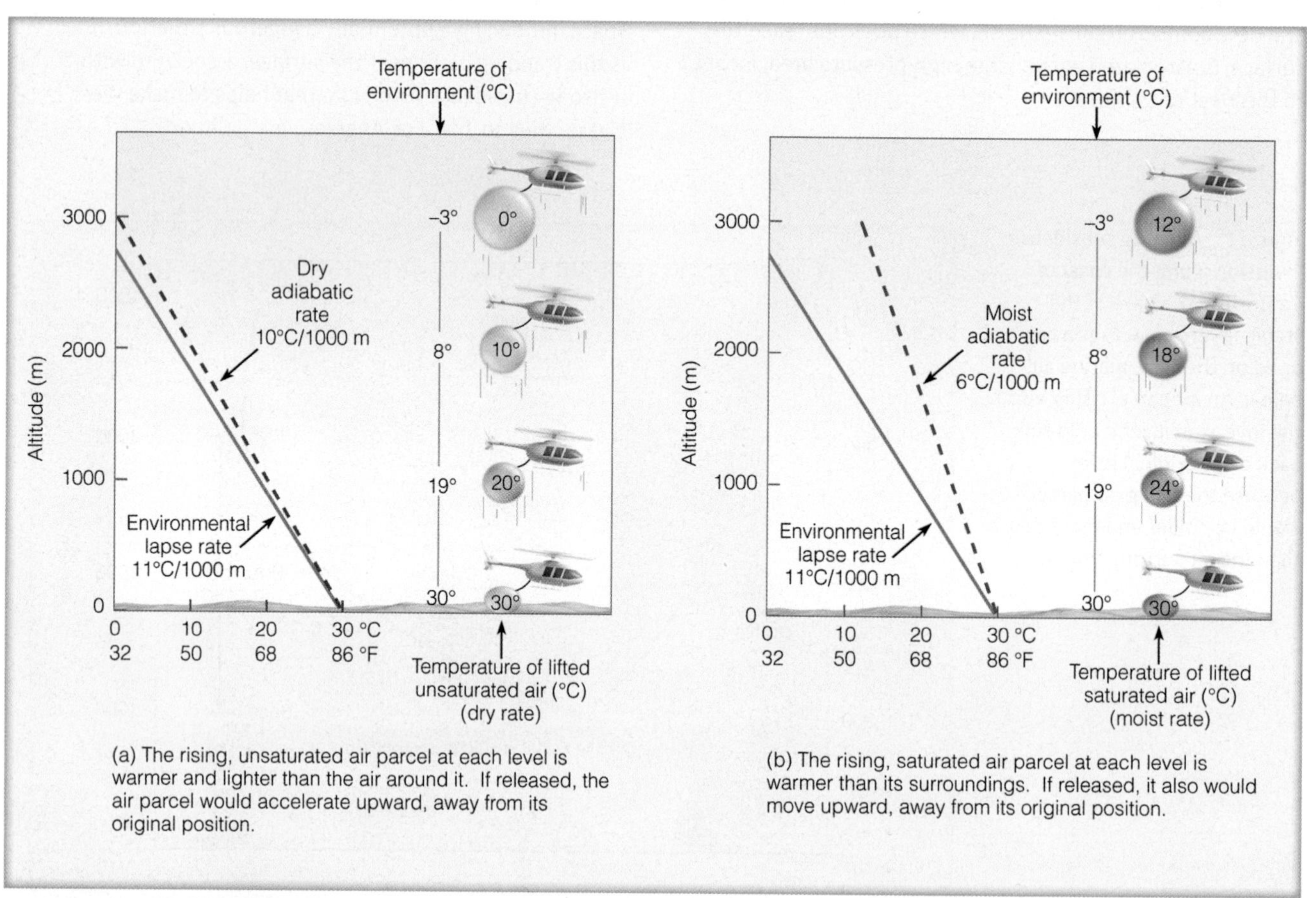

FIGURE 5.8 An absolutely unstable atmosphere occurs when the environmental lapse rate is greater than the dry adiabatic rate. In an unstable atmosphere, a rising air parcel will continue to rise because it is warmer and less dense than the air surrounding it.

push upward, it will tend to move in that direction; the atmosphere is unstable for the saturated parcel. In this example, the atmosphere is said to be **conditionally unstable.** This type of stability depends upon whether or not the rising air is saturated. When the rising parcel of air is unsaturated, the atmosphere is stable; when the parcel of air is saturated, the atmosphere is unstable. Conditional instability means that, if unsaturated air could be lifted to a level where it becomes saturated, instability would result.

We can obtain a more visual concept of this idea by examining the rising parcel of air in ❯ Fig. 5.10. The initial air parcel at the surface is unsaturated (but humid) and is somehow forced to rise. (What causes an air parcel to rise will be discussed in a later section.) As the parcel rises, it expands, and cools at the *dry adiabatic rate* until its air temperature cools to its dew point. At this level, the air is saturated, the relative humidity is 100 percent, and further lifting results in condensation and the formation of a cloud. The elevation above the surface where the cloud first forms (in this example, 1000 meters) is called the **condensation level.**

In Fig. 5.10 notice that above the condensation level, the rising saturated air now cools at the *moist adiabatic rate*. Notice also that from the surface up to a level near 2000 meters, the rising, lifted air is colder than the air surrounding it. The atmosphere up to this level is *stable*. However, due to the release of latent heat, the rising air near 2000 meters has actually become warmer than the air around it. Since the lifted air can rise on its own accord, the atmosphere is now *unstable*. The level in the atmosphere where the air parcel, after being lifted, becomes warmer than the air surrounding it, is called the *level of free convection*. Here, where the air parcel is warmer and less dense than the air surrounding it, there is an upward-directed force (called *buoyant force*) acting on it. The warmer the air parcel compared to its surroundings, the greater the buoyant force, and the more rapidly the air rises.

The atmospheric layer from the surface up to 4000 meters in Fig. 5.10 has gone from stable to unstable because the rising air was humid enough to become

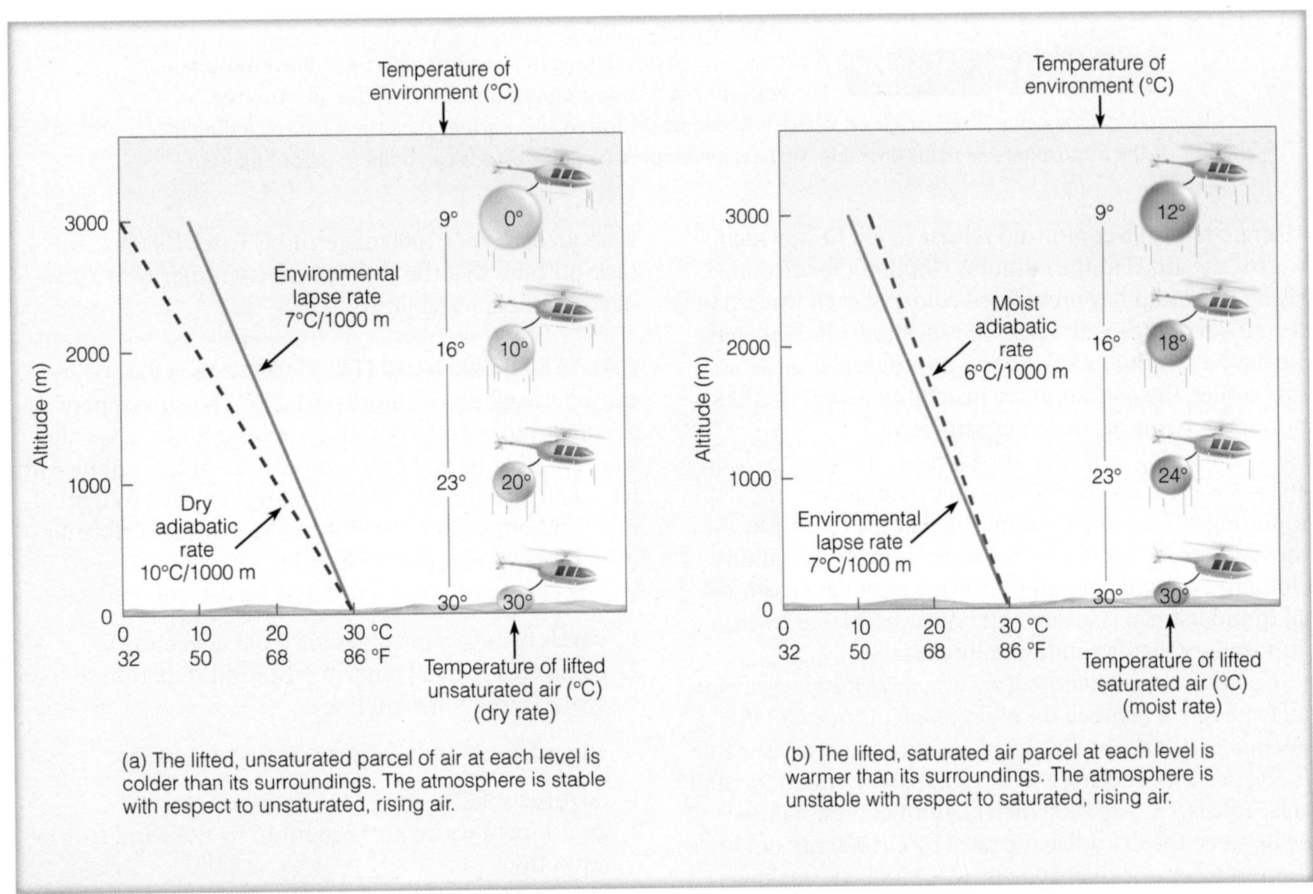

(a) The lifted, unsaturated parcel of air at each level is colder than its surroundings. The atmosphere is stable with respect to unsaturated, rising air.

(b) The lifted, saturated air parcel at each level is warmer than its surroundings. The atmosphere is unstable with respect to saturated, rising air.

Active ❯ FIGURE 5.9 Conditionally unstable atmosphere. The atmosphere is *stable* if the rising air is *unsaturated* (a), but *unstable* if the rising air is *saturated* (b). A conditionally unstable atmosphere occurs when the environmental lapse rate is between the moist adiabatic rate and the dry adiabatic rate. Visit the Meteorology Resource Center to view this and other active figures at www.cengage.com/login

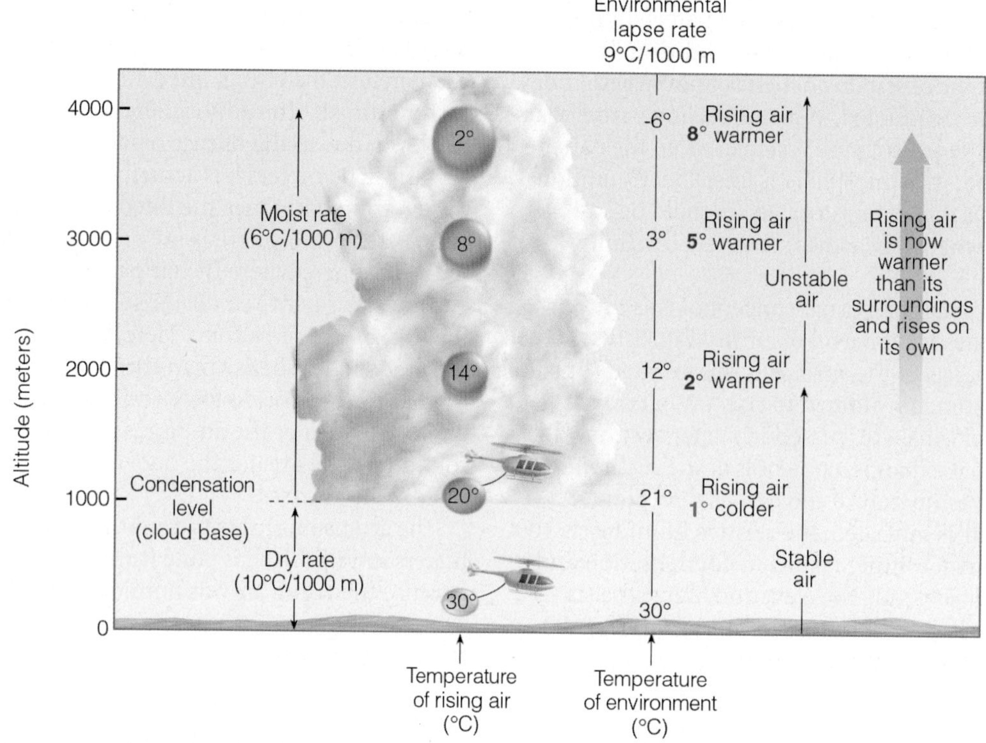

FIGURE 5.10 A cumulus cloud developing in a conditionally unstable atmosphere. The atmosphere is conditionally unstable because unsaturated, stable air is being lifted to a level where it becomes saturated and warmer than the air surrounding it. If the atmosphere remains unstable, vertical developing cumulus clouds can build to great heights.

saturated, form a cloud, and release latent heat, which warms the air. Had the cumulus cloud not formed, the rising air would have remained colder at each level than the air surrounding it. From the surface to 4000 meters, we have a *conditionally unstable atmosphere* and, as we saw earlier, the condition for instability being whether or not the rising air becomes saturated.

Notice in Fig. 5.10 that the rising air inside the cloud at 4000 meters is now 8°C warmer than the air surrounding it. This situation means that the air inside the cloud could continue to rise, building the cumulus cloud to even greater heights. As a consequence, almost all thunderstorms (severe and nonsevere) form in an atmosphere that is conditionally unstable.

Conditional instability occurs whenever the environmental lapse rate is between the moist adiabatic rate and the dry adiabatic rate. In Fig. 5.9, the environmental lapse rate is 7°C per 1000 meters, and in Fig. 5.10 the environmental lapse rate is 9°C per 1000 meters. Both of these values lie between the dry adiabatic rate (10°C/1000 m) and the average moist adiabatic rate (6°C/1000 m), making both atmospheres conditionally unstable. Looking at the atmosphere as a whole, you may recall from Chapter 1 that the average (standard) lapse rate in the troposphere is about 6.5°C per 1000 meters (3.6°F per 1000 ft). This fact indicates that the atmosphere is ordinarily in a state of conditional instability.

CAUSES OF INSTABILITY What causes the atmosphere to become more unstable? The atmosphere becomes more unstable as the environmental lapse rate steepens, that is, as the air temperature drops rapidly with increasing height. This circumstance may be brought on by either air aloft becoming colder or the surface air becoming warmer (see Fig. 5.11).

The *cooling of the air aloft* may be due to:

1. winds bringing in colder air (cold advection)
2. clouds (or the air) emitting infrared radiation to space (radiational cooling)

The *warming of the surface air* may be due to:

1. daytime solar heating of the surface
2. an influx of warm air brought in by the wind (warm advection)
3. air moving over a warm surface

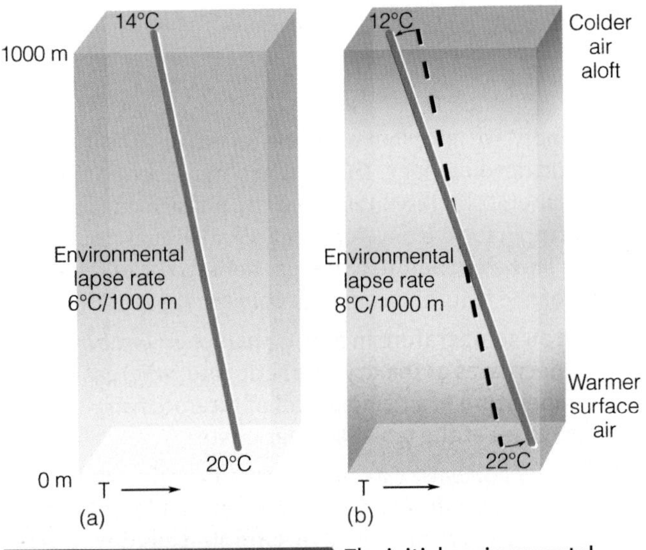

FIGURE 5.11 The initial environmental lapse rate in diagram (a) will become more unstable (that is, destabilize) as the air aloft cools and the surface air warms, as illustrated in diagram (b).

The combination of cold air aloft and warm surface air can produce a steep lapse rate and atmospheric instability.

At this point, we can see that the stability of the atmosphere changes during the course of a day. In clear, calm weather around sunrise, surface air is normally colder than the air above it, a radiation inversion exists, and the atmosphere is quite stable as indicated by smoke or haze lingering close to the ground. As the day progresses, sunlight warms the surface and the surface warms the air above. As the air temperature near the ground increases, the lower atmosphere gradually becomes more unstable—that is, it *destabilizes*—with maximum instability usually occurring during the hottest part of the day. On a humid summer afternoon, this phenomenon can be observed by watching cumulus clouds develop into thunderstorms.

Up to now, we have seen that a layer of air may become more unstable by either cooling the air aloft or warming the air at the surface. A layer of air may also be made more unstable by either mixing or lifting. Let's look at mixing first. In Fig. 5.12, the environmental lapse rate before mixing is less than the moist rate, and the layer is stable (*A*). Now, suppose the air in the layer is mixed either by convection or by wind-induced turbulent eddies. Air is cooled adiabatically as it is brought up from below and heated adiabatically as it is mixed downward. The up and down motion in the layer redistributes the air in such a way that the temperature at the top of the layer decreases, while, at the base, it increases. This steepens the environmental lapse rate and makes the layer more unstable. If this mixing continues for some time, and the air remains unsaturated, the vertical temperature distribution will eventually be equal to the dry adiabatic rate (*B*).

Just as lowering an entire layer of air makes it more stable, the lifting of a layer makes it more unstable. In Fig. 5.13, the air lying between 1000 and 900 mb is initially absolutely stable since the environmental lapse rate of layer x–y is less than the moist adiabatic rate. The layer is lifted, and, as it rises, the rapid decrease in air density aloft causes the layer to stretch out vertically. If

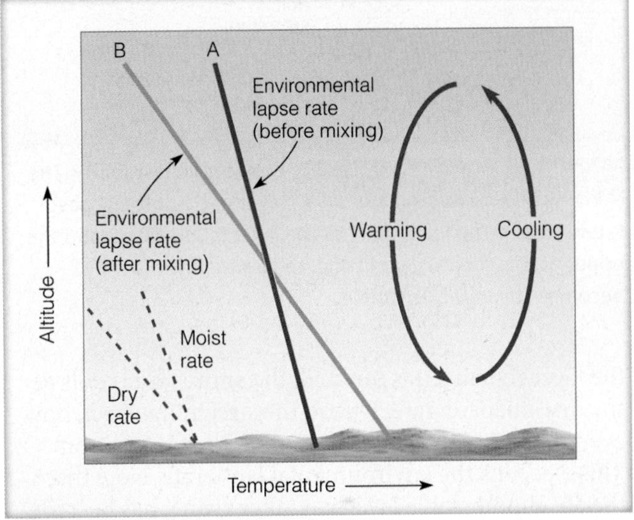

FIGURE 5.12 Mixing tends to steepen the lapse rate. Rising, cooling air lowers the temperature toward the top of the layer, while sinking, warming air increases the temperature near the bottom.

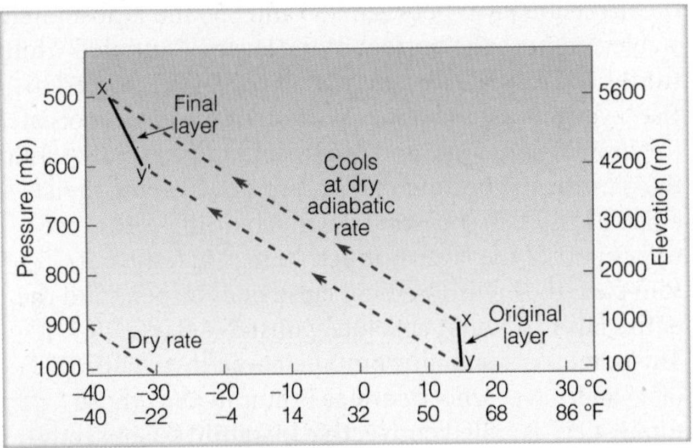

FIGURE 5.13 The lifting of an entire layer of air tends to increase the instability of the layer. The initial stable layer (x–y) after lifting is now a conditionally unstable layer (x′–y′).

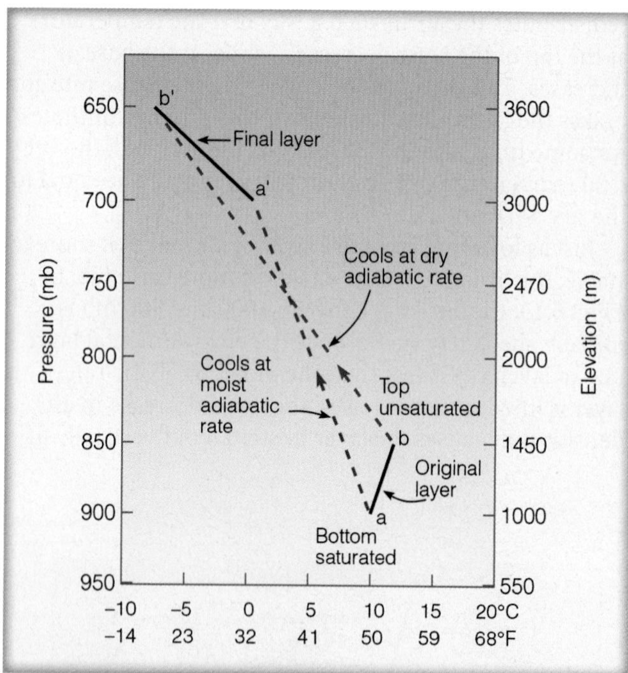

FIGURE 5.14 Convective instability. The layer a–b is initially absolutely stable. The lower part of the layer is saturated, and the upper part is "dry." After lifting, the entire layer (a′–b′) becomes absolutely unstable.

the layer remains unsaturated, the entire layer cools at the dry adiabatic rate. Due to the stretching effect, however, the top of the layer cools more than the bottom. This steepens the environmental lapse rate. Note that the absolutely stable layer *x–y*, after rising, has become conditionally unstable between 500 and 600 mb (layer *x′–y′*).

A very stable air layer may be converted into an absolutely unstable layer when the lower portion of a layer is moist and the upper portion is quite dry. In Fig. 5.14, the inversion layer between 900 and 850 mb is absolutely stable. Suppose the bottom of the layer is saturated while the air at the top is unsaturated. If the layer is forced to rise, even a little, the upper portion of the layer cools at the dry adiabatic rate and grows cold quite rapidly, while the air near the bottom cools more slowly at the moist adiabatic rate. It does not take much lifting before the upper part of the layer is much colder than the bottom part; the environmental lapse rate steepens and the entire layer becomes absolutely unstable (layer *a′–b′*). The potential instability, brought about by the lifting of a stable layer whose surface is humid and whose top is "dry," is called **convective instability.** Convective instability is associated with the development of severe storms, such as thunderstorms and tornadoes, which are investigated more thoroughly in later chapters.

BRIEF REVIEW

Up to now, we have looked briefly at stability as it relates to cloud development. The next section describes how atmospheric stability influences the physical mechanisms responsible for the development of individual cloud types. However, before going on, here is a brief review of some of the facts and concepts concerning stability.

- The air temperature in a rising parcel of *unsaturated* air decreases at the dry adiabatic rate, whereas the air temperature in a rising parcel of *saturated* air decreases at the moist adiabatic rate.
- The dry adiabatic rate and moist adiabatic rate of cooling are different due to the fact that latent heat is released in a rising parcel of saturated air.
- In a *stable atmosphere*, a lifted parcel of air will be colder (heavier) than the air surrounding it. Because of this fact, the lifted parcel will tend to sink back to its original position.
- In an *unstable atmosphere*, a lifted parcel of air will be warmer (lighter) than the air surrounding it, and thus will continue to rise upward, away from its original position.
- The atmosphere becomes more stable (stabilizes) as the surface air cools, the air aloft warms, or a layer of air sinks (subsides) over a vast area.
- The atmosphere becomes more unstable (destabilizes) as the surface air warms, the air aloft cools, or a layer of air is either mixed or lifted.
- A conditionally unstable atmosphere exists when the environmental lapse rate is between the moist adiabatic rate and the dry adiabatic rate.
- The atmosphere is normally most stable in the early morning and most unstable in the afternoon.
- Layered clouds tend to form in a stable atmosphere, whereas cumuliform clouds tend to form in a conditionally unstable atmosphere.

Cloud Development

We know that most clouds form as air rises, cools, and condenses. Since air normally needs a "trigger" to start it moving upward, what is it that causes the air to rise so that clouds are able to form? Basically, the following mechanisms are responsible for the development of the majority of clouds we observe:

1. surface heating and free convection
2. uplift along topography
3. widespread ascent due to convergence of surface air
4. uplift along weather fronts (see Fig. 5.15).

The first mechanism that can cause the air to rise is convection. Although we briefly looked at convection in Chapter 2 when we examined rising thermals and how they transfer heat upward into the atmosphere, we will now look at convection from a slightly different perspective—how rising thermals are able to form into cumulus clouds.

CONVECTION AND CLOUDS Some areas of the earth's surface are better absorbers of sunlight than others and, therefore, heat up more quickly. The air in contact with these "hot spots" becomes warmer than its surroundings. A hot "bubble" of air—a *thermal*—breaks away from the warm surface and rises, expanding and cooling as it ascends. As the thermal rises, it mixes with the cooler, drier air around it and gradually loses its identity. Its upward movement now slows. Frequently, before it is completely diluted, subsequent rising thermals penetrate it and help the air rise a little higher. If the rising air cools to its saturation point, the moisture will condense, and the thermal becomes visible to us as a cumulus cloud.

Observe in ❯ Fig. 5.16 that the air motions are downward on the outside of the cumulus cloud. The

EXTREME WEATHER WATCH

The smell of a thermal. On sunny days, when the atmosphere is conditionally unstable, rising thermals break away from the surface and rapidly rise up into the atmosphere, carrying surface air with them. If the thermal originated over a field of alfalfa, the smell of alfalfa may be carried thousands of feet above the field.

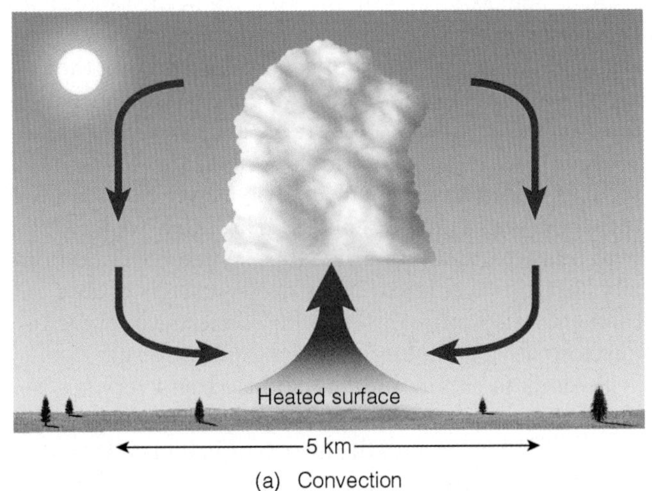

(a) Convection

(b) Lifting along topography

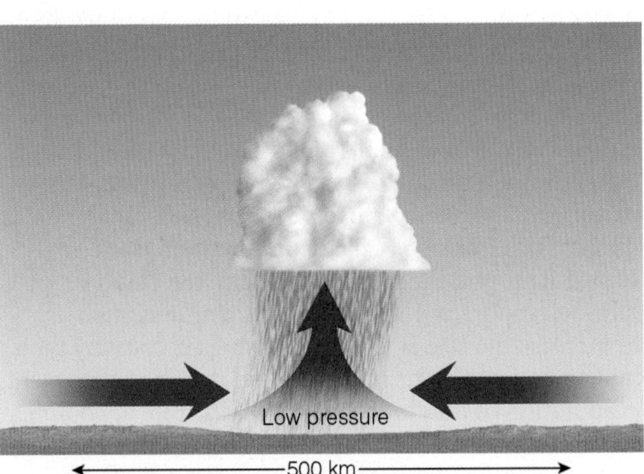

(c) Convergence of air

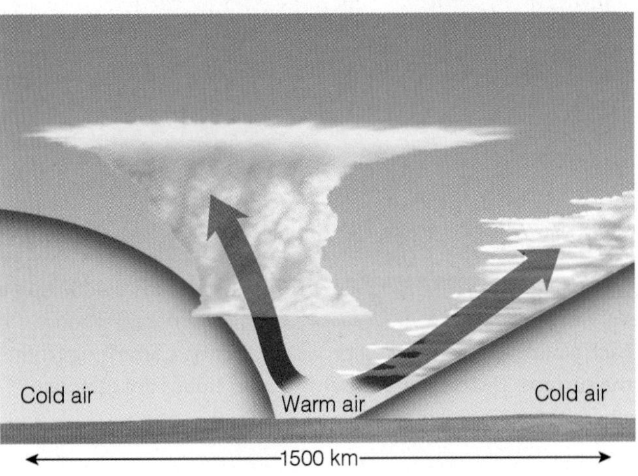

(d) Lifting along weather fronts

❯ FIGURE 5.15 The primary ways clouds form: (a) surface heating and convection; (b) forced lifting along topographic barriers; (c) convergence of surface air; (d) forced lifting along weather fronts.

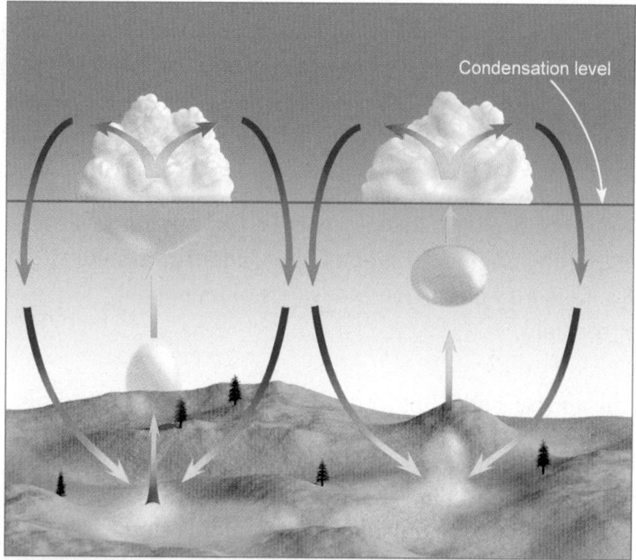

FIGURE 5.16 Cumulus clouds form as hot, invisible air bubbles detach themselves from the surface, then rise and cool to the condensation level. Below and within the cumulus clouds, the air is rising. Around the cloud, the air is sinking.

FIGURE 5.17 Cumulus clouds building on a warm summer afternoon. Each cloud represents a region where thermals are rising from the surface. The clear areas between the clouds are regions where the air is sinking.

downward motions are caused in part by evaporation around the outer edge of the cloud, which cools the air, making it heavy. Another reason for the downward motion is the completion of the convection current started by the thermal. Cool air slowly descends to replace the rising warm air. Therefore, we have rising air in the cloud and sinking air around it. Since subsiding air greatly inhibits the growth of thermals beneath it, small cumulus clouds usually have a great deal of blue sky between them (see Fig. 5.17).

As the cumulus clouds grow, they shade the ground from the sun. This, of course, cuts off surface heating and upward convection. Without the continual supply of rising air, the cloud begins to erode as its droplets evaporate. Unlike the sharp outline of a growing cumulus, the cloud now has indistinct edges, with cloud fragments extending from its sides. As the cloud dissipates (or moves along with the wind), surface heating begins again and regenerates another thermal, which becomes a new cumulus. This is why you often see cumulus clouds form, gradually disappear, then reform in the same spot.

The stability of the atmosphere above the base of the cumulus cloud plays a major role in determining the growth of the cloud. Notice in Fig. 5.18 that, when a deep stable layer begins a short distance above the cloud base, only fair weather cumulus humilis are able to form. If a deep conditionally unstable layer exists above the cloud base, cumulus congestus are likely to grow, with billowing cauliflowerlike tops. When the conditionally unstable layer is extremely deep—usually greater than 4 km (2.5 mi)—the cumulus congestus may even develop into a cumulonimbus—a thunderstorm.

Seldom do cumulonimbus clouds extend very far above the tropopause. The stratosphere is quite stable, so once a cloud penetrates the tropopause, it usually stops growing vertically and spreads horizontally. The low temperature at this altitude produces ice crystals in the upper section of the cloud. In the middle latitudes, high winds near the tropopause blow the ice crystals laterally, producing the flat anvil-shaped top so characteristic of cumulonimbus clouds (see Fig. 5.19).

The vertical development of a convective cloud also depends upon the mixing that takes place around its periphery. The rising, churning cloud mixes cooler air into it. Such mixing is called **entrainment**. If the environment around the cloud is very dry, the cloud droplets quickly evaporate. The effect of entrainment, then, is to increase the rate at which the rising air cools by the injection of cooler air into the cloud and the subsequent evaporation of the cloud droplets. If the rate of cooling approaches the dry adiabatic rate, the air stops rising and the cloud no longer builds, even though the lapse rate may indicate a conditionally unstable atmosphere.

Up to now, we have looked at convection over land. Convection and the development of cumulus clouds

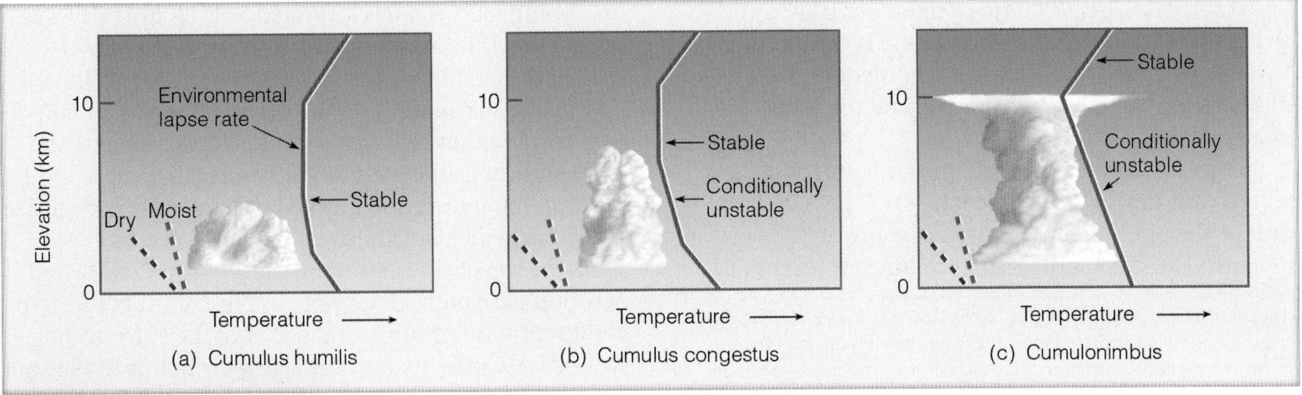

FIGURE 5.18 The air's stability greatly influences the growth of cumulus clouds.

also occur over large bodies of water. As cool air flows over a body of relatively warm water, the lowest layer of the atmosphere becomes warm and moist. This induces instability—convection begins and cumulus clouds form. If the air moves over progressively warmer water, as is sometimes the case over the open ocean, more active convection occurs and a cumulus cloud can build into cumulus congestus and finally into cumulonimbus. This sequence of cloud development is observed from satellites as cold northerly winds move southward over the northern portions of the Atlantic and Pacific oceans (see Fig. 5.20 on p. 138).

Once a convective cloud forms, stability, humidity, and entrainment all play a part in its vertical development. The level at which the cloud initially forms, however, is determined primarily by the surface temperature and moisture content of the original thermals.

Before going on to the next section, you may wish to read the Focus section on p. 136 that details the role that atmospheric stability and convection play in the formation of some extraordinary snowstorms called *lake-effect snows.*

TOPOGRAPHY AND CLOUDS Horizontally moving air obviously cannot go through a large obstacle, such as a mountain, so the air must go over it. Forced lifting along a topographic barrier is called **orographic uplift.** Often, large masses of air rise when they approach a long chain of mountains like the Sierra Nevada or Rockies. This lifting produces cooling, and, if the air is humid, clouds form. Clouds produced in this manner are called *orographic clouds.* The type of cloud that forms will depend on the air's stability and moisture content. On the leeward (downwind) side of the mountain, as the air moves downhill, it warms. This sinking air is now drier, since much of its moisture was removed in the form of clouds and precipitation on the windward side. This region on the leeward side, where precipitation is noticeably less, is called a **rain shadow.**

An example of orographic uplift, cloud development, and the formation of a rain-shadow is given in Fig. 5.21. Before rising up and over the barrier, the air at the base of the mountain (0 m) on the windward side has an air temperature of 20°C (68°F) and a dew-point temperature of 12°C (54°F). Notice that the atmosphere is conditionally unstable, as indicated by the environmental

FIGURE 5.19 Cumulus clouds developing into thunderstorms in a conditionally unstable atmosphere over the Great Plains. Notice that, in the distance, the cumulonimbus with the anvil top has reached the stable part of the atmosphere.

lapse rate of 8°C per 1000 m. (Remember from our earlier discussion that the atmosphere is conditionally unstable when the environmental lapse rate falls between the dry adiabatic rate and the moist adiabatic rate.)

As the unsaturated air rises, the air temperature decreases at the dry adiabatic rate (10°C per 1000 m) and the dew-point temperature decreases at 2°C per 1000 m.* Notice that the rising, cooling air reaches its dew point and becomes saturated at 1000 m. This level (called the *lifting condensation level,* or *LCL*) marks the base of the cloud that has formed as air is lifted (in this case by the mountain). As the rising saturated air condenses into many billions of liquid cloud droplets, and as latent heat is liberated by the condensing vapor, both the air temperature and dew-point temperature decrease at the moist adiabatic rate.

At the top of the mountain, the air temperature and dew point are both –2°C. Note in Fig. 5.21 that this temperature (–2°C) is higher than that of the surrounding air (–4°C). Consequently, the rising air at this level is not

*The decrease in dew-point temperature is caused by the rapid decrease in air pressure within the rising air. Since the dew point is directly related to the actual vapor pressure of the rising air, a decrease in total air pressure causes a corresponding decrease in vapor pressure and, hence, a lowering of the dew-point temperature.

FOCUS ON
EXTREME WEATHER

Lake-Effect Snowstorms and Atmospheric Stability

During the winter when the weather in the Midwest is dominated by clear and cold air, people living on the eastern shores of the Great Lakes brace themselves for heavy snow showers. Snowstorms that form on the downwind side of one of these lakes are known as *lake-effect snows*. These storms serve as a good example of how the atmosphere can destabilize as cold air flows over a relatively warm body of water.

Lake-effect snowstorms are highly localized, extending from just a few miles to more than 100 miles inland. The snow usually falls as a heavy shower or squall in a concentrated zone. So centralized is the region of snowfall, that one part of a city may accumulate many inches of snow while in another part, the ground is bare.

Lake-effect snows are most numerous from November to January. During these months, cold air moves over the lakes when they are relatively warm and not quite frozen. The contrast in temperature between water and air can be as much as 25°C. Studies show that the greater the contrast in temperature, the greater the potential for snow showers. In Fig. 2 we can see that, as the cold air moves over the warmer water, the air near the surface is quickly warmed from below, making it more buoyant and less stable. Rapidly, the air sweeps up moisture, soon becoming saturated. Out over the water, the vapor condenses into steam fog. As the air continues to warm, it rises and forms billowing cumuliform clouds, which continue to grow as the air becomes more unstable. Eventually, these clouds produce heavy showers of snow (see Fig. 2), which make the lake seem like a snow factory.

Once the air and clouds reach the downwind side of the lake, additional lifting is

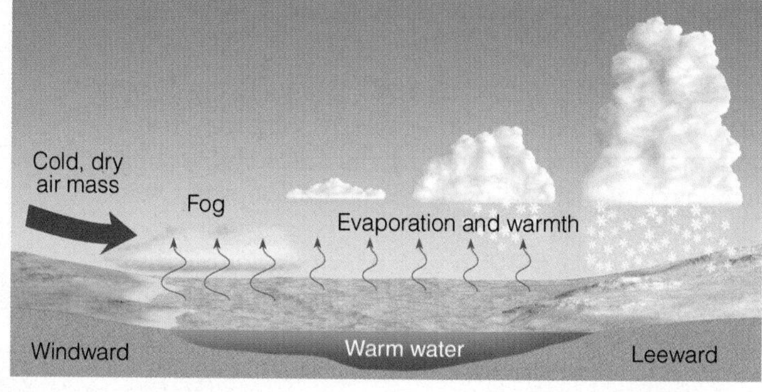

Figure 2 The formation of lake-effect snows. Cold, dry air crossing the lake gains moisture and warmth from the water, and the atmosphere destabilizes. The more buoyant air now rises, forming clouds that deposit large quantities of snow on the lake's leeward (downwind) shores.

only warmer, but unstable with respect to its surroundings. Therefore, the rising air should continue to rise and build into a much larger cumuliform cloud.

Suppose, however, that the air at the top of the mountain (temperature and dew point of −2°C) is forced to descend to the base of the mountain (0 m) on the leeward side. If we assume that the cloud remains on the windward side and does not extend beyond the mountaintop, the temperature of the sinking air will increase at the dry adiabatic rate (10°C per 1000 m) all the way down to the base of the mountain. (The dew-point temperature increases at a much lower rate of 2°C per 1000 m.)

EXTREME WEATHER WATCH

The most extreme rain shadow in the United States is that provided by the 5140-foot-high Mt. Waialeale on the Hawaiian island of Kauai. During March, 1982, a United States monthly rainfall record of 148.83 inches was reported there. Just ten miles to the west, in the mountain's rain shadow, the town of Mana recorded only 3.80 inches during that same month.

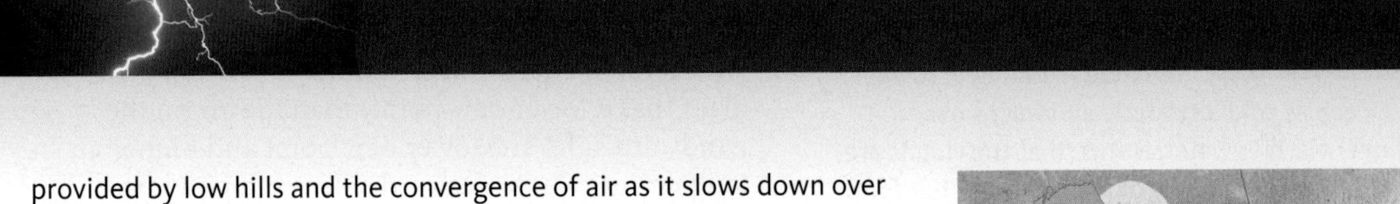

provided by low hills and the convergence of air as it slows down over the rougher terrain. In late winter, the frequency and intensity of lake-effect snows often taper off as the temperature contrast between water and air diminishes and larger portions of the lakes freeze.

Generally, the longer the stretch of water over which the air mass travels (the longer the fetch), the greater the amount of warmth and moisture derived from the lake, and the greater the potential for heavy snow showers. In fact, studies show that, for significant snowfall to occur, the air must move across 80 km (50 mi) of open water. Consequently, forecasting lake-effect snowfalls depends to a large degree on determining the trajectory of the air as it flows over the lake. Regions that experience heavy lake-effect snowfalls are shown in Fig. 3.

Figure 3 Areas shaded white show regions that experience heavy lake-effect snows.

As the cold air moves farther east, the heavy snow showers usually taper off; however, the western slope of the Appalachian Mountains produces further lifting, enhancing the possibility of more and heavier showers. The heat given off during condensation warms the air and, as the air descends the eastern slope, compressional heating warms it even more. Snowfall ceases, and by the time the air arrives in Philadelphia, New York, or Boston, the only remaining trace of the snow showers occurring on the other side of the mountains are the puffy cumulus clouds drifting overhead.

Heavy lake-effect snows can literally bury a city in a relatively short period of time. For example, in less than four days Buffalo, New York,* received nearly seven feet of snow during December, 2001. And Montague, New York (which lies on the eastern side of Lake Ontario), received over seven feet of snow in less than forty-eight hours during January, 1997.

Lake-effect snows are not confined to the Great Lakes. In fact, any large unfrozen lake (such as the Great Salt Lake) can enhance snowfall when cold, relatively dry air sweeps over it. Moreover, a type of lake-effect snow occurs when cold air moves over a relatively warm ocean, then lifts slightly as it moves over a landmass. Such *ocean-effect snows* are common over Cape Cod, Massachusetts, in winter.

*Buffalo, New York is a city that experiences heavy lake-effect snows. Visit the National Weather Service website in Buffalo at http://www.erh.noaa.gov/buf/lakeffect/indexlk.html and read about lake-effect snowstorms measured in feet, as well as other interesting weather stories.

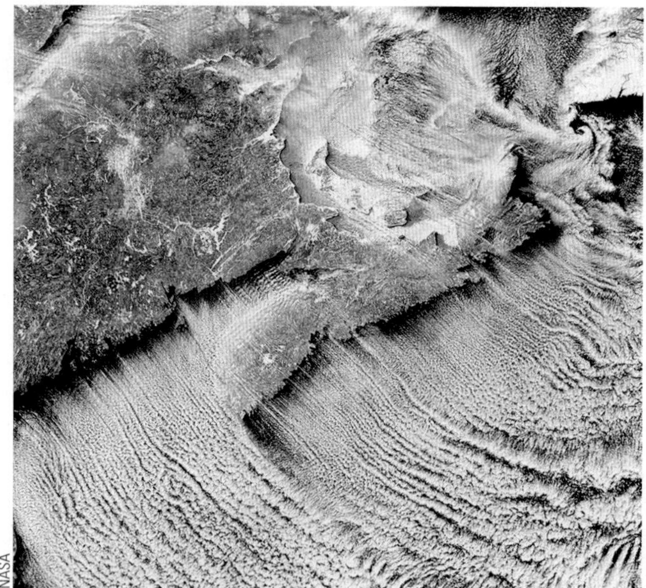

FIGURE 5.20 Satellite view of stratocumulus clouds forming in rows over the Atlantic Ocean as cold, dry arctic air sweeps over Canada, then out over warmer water. Notice that the clouds are absent over the landmass and directly along the coast, but form and gradually thicken as the surface air warms and destabilizes farther offshore.

We can see in Fig. 5.21 that on the leeward side, after descending 3000 m (about 10,000 ft), the air temperature is 28°C (82°F) and the dew-point temperature is 4°C (39°F). The air is now 8°C (14°F) warmer than it was before being lifted over the barrier. The higher air temperature on the leeward side is the result of latent heat being converted into sensible heat during condensation on the windward side. (In fact, the rising air at the *top* of the mountain is considerably warmer than it would have been had condensation not occurred.) The lower dew-point temperature and, hence, drier air on the leeward side are the result of water vapor condensing and then remaining as liquid cloud droplets and precipitation on the windward side.

We now have two important concepts to remember:

1. Air descending a mountain warms by compressional heating and, upon reaching the surface, can be much warmer than the air at the same level on the upwind side.
2. Air on the leeward side of a mountain is normally drier (has a lower dew point) than the air on the windward side. The lower dew point and higher air temperature on the leeward side produce a lower relative humidity, a greater potential for evaporation of water, and a rain shadow desert.

Although clouds are more prevalent on the windward side of mountains, they may, under certain atmospheric

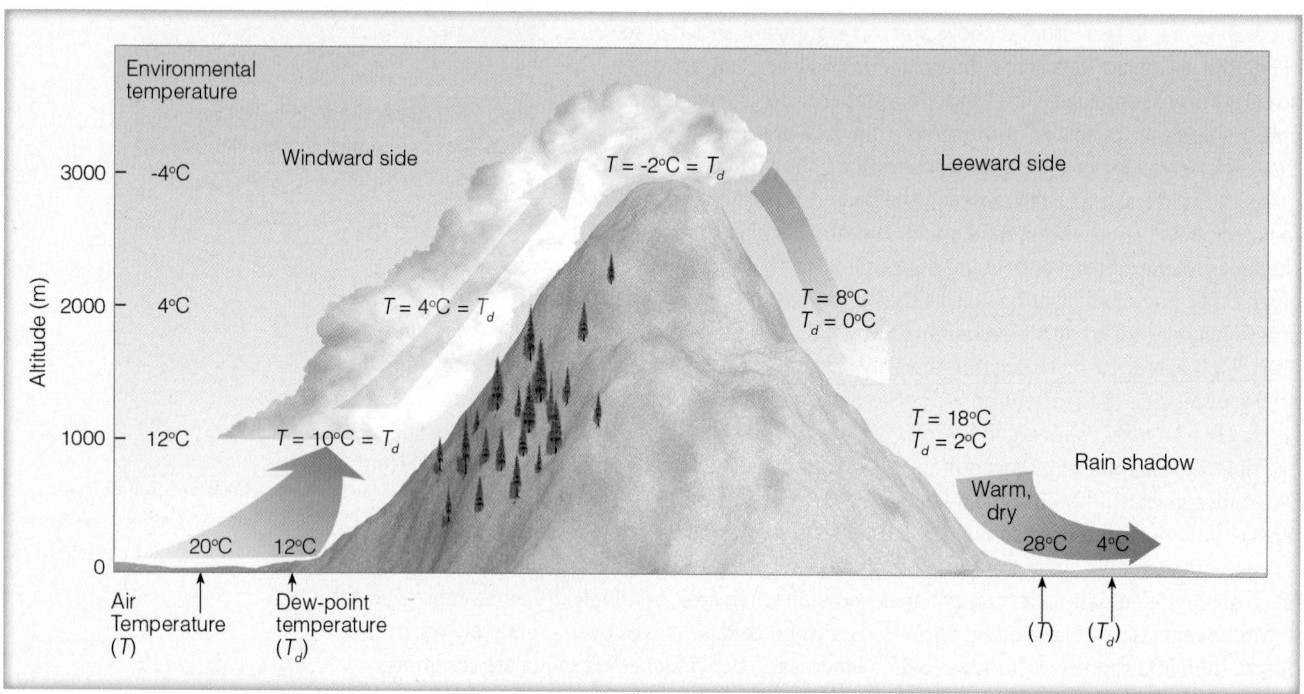

Active FIGURE 5.21 Orographic uplift, cloud development, and the formation of a rain shadow. Visit the Meteorology Resource Center to view this and other active figures at www.cengage.com/login

conditions, form on the leeward side as well. For example, stable air flowing over a mountain often moves in a series of waves that may extend for several hundred kilometers on the leeward side (see Fig. 5.22). These waves resemble the waves that form in a river downstream from a large boulder. Recall from Chapter 4 that *wave clouds* often have a characteristic lens shape and are commonly called *lenticular clouds.*

The formation of lenticular clouds is shown in Fig. 5.23. As moist air rises on the upwind side of the wave, it cools and condenses, producing a cloud. On the downwind side of the wave, the air sinks and warms; the cloud evaporates. Viewed from the ground, the clouds appear motionless as the air rushes through them; hence, they are often referred to as *standing wave clouds.*

When the air between the cloud-forming layers is too dry to produce clouds, lenticular clouds will form one above the other. Actually, when a strong wind blows almost perpendicular to a high mountain range, mountain waves may extend into the stratosphere, producing a spectacular display, sometimes resembling a fleet of hovering spacecraft (see Fig. 5.24).

Notice in Fig. 5.23 that beneath the lenticular cloud downwind of the mountain range, a large swirling eddy forms. The rising part of the eddy may cool enough to produce **rotor clouds**. The air in the rotor is extremely turbulent and presents a major hazard to aircraft in the vicinity. Dangerous flying conditions also exist near the leeside of the mountain, where strong downwind air motions are present.

CHANGING CLOUD FORMS When a cloud changes its form, it sometimes indicates that thunderstorms (and perhaps severe weather) may develop later in the day. For example, occasionally altocumulus clouds will begin to grow vertically and show tower-like extensions that often resemble castles, and for this reason, they are called *altocumulus castellanus* (see Fig. 5.25). They form when

FIGURE 5.22 Satellite view of wave clouds forming many kilometers downwind of the mountains in Scotland and Ireland.

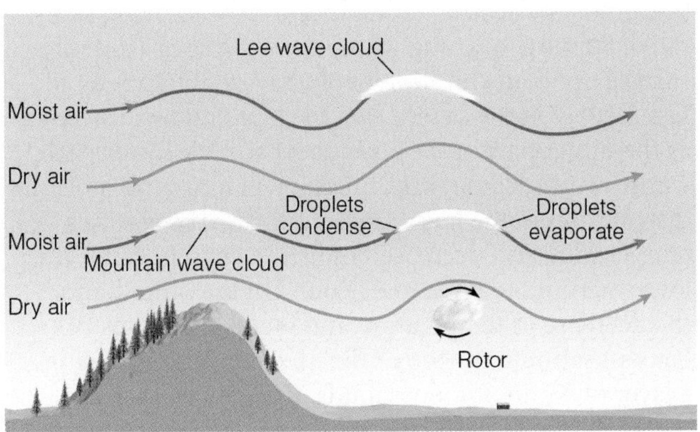

FIGURE 5.23 Clouds that form in the wave directly over the mountains are called *mountain wave clouds*, whereas those that form downwind of the mountain are called *lee wave clouds*.

FIGURE 5.24 A mass of moist stable air gliding up and over Mikeno Volcano in the Republic of the Congo condenses into a spectacular lenticular cloud.

FIGURE 5.25 An example of altocumulus castellanus that portend the formation of thunderstorms in the afternoon.

rising currents within the cloud extend into conditionally unstable air above the cloud. Apparently, the buoyancy for the rising air comes from the latent heat released during condensation within the cloud. When altocumulus castellanus appear, they indicate that the mid-level of the troposphere is becoming more unstable (destabilizing). This destabilization is often the precursor to shower activity. So a morning sky full of altocumulus castellanus will likely become afternoon showers and even thunderstorms.

Occasionally, the stirring of a moist layer of stable air will change a clear day into a cloudy one. In Fig. 5.26a, the atmospheric layer is stable and close to being saturated. Suppose a strong wind mixes the layer from the surface up to an elevation of 600 m (2000 ft). (See Fig. 5.26b.) As we saw earlier, the lapse rate will steepen as the upper part of the layer cools and the lower part warms. At the same time, mixing will make the moisture distribution in the layer more uniform. The warmer temperature and decreased moisture content cause the lower part of the layer to dry out. On the other hand, the decrease in temperature and increase in moisture content saturate the top of the mixed layer, producing a layer of stratocumulus clouds. In Fig. 5.26b, notice that the air above the region of mixing is still stable and inhibits further mixing. In some cases, an inversion may

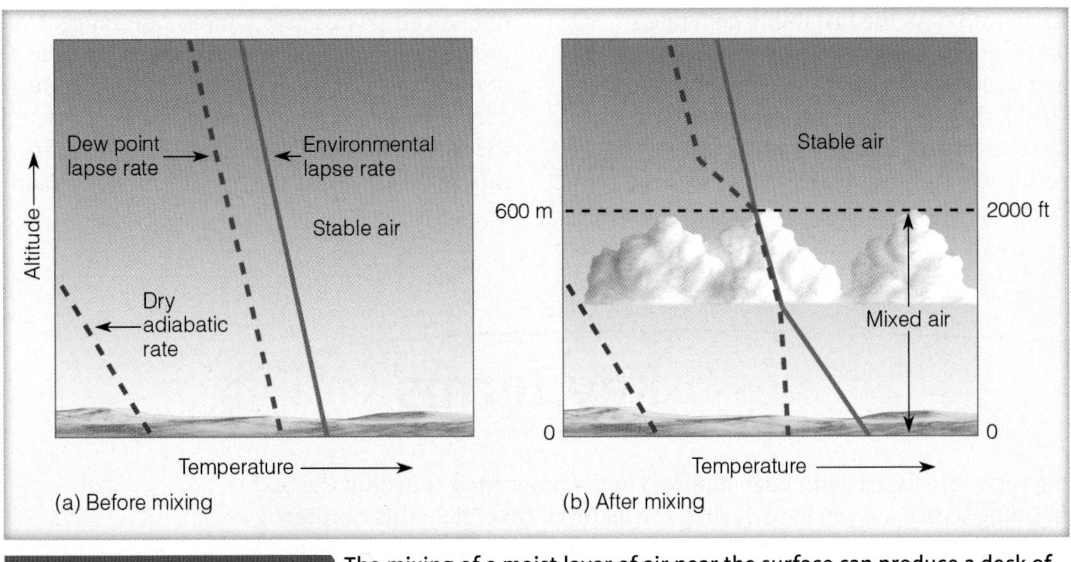

FIGURE 5.26 The mixing of a moist layer of air near the surface can produce a deck of stratocumulus clouds.

actually form above the clouds. However, if the surface warms substantially, rising thermals may penetrate the stable region and the stratocumulus clouds may change into more widely separated clouds, such as cumulus or cumulus congestus. A stratocumulus layer changing to a sky dotted with growing cumulus clouds often occurs as surface heating increases on a warm, humid summer day.

Summary

In this chapter, we tied together the concepts of stability and the formation of clouds. We learned that rising unsaturated air cools at the dry adiabatic rate. Due to the release of latent heat, rising saturated air cools at the moist adiabatic rate. The environmental lapse rate is the rate at which the air temperature changes with increasing height. A rapid decrease in air temperature with height indicates a steep environmental lapse rate and a more unstable atmosphere. A slow decrease in air temperature with height indicates a gentle environmental lapse rate and a more stable atmosphere.

In a stable atmosphere, a lifted parcel of air will be colder (heavier) than the air surrounding it at each new level, and it will sink back to its original position. Because stable air tends to resist upward vertical motions, clouds forming in a stable atmosphere often spread horizontally and have a stratified appearance, such as cirrostratus and altostratus. Either cooling the surface air or warming the air aloft, or both, will cause the atmosphere to become more stable. Also, the sinking (subsidence) of an entire layer of air will cause the layer to become more stable. When the sinking air is warmer than the air beneath it, a subsidence inversion will form.

In an unstable atmosphere, a lifted parcel of air will be warmer (lighter) than the air surrounding it at each new level, and it will continue to raise upward away from its original position. In a conditionally unstable atmosphere, an unsaturated parcel of air can be lifted to a level where condensation begins, latent heat is released, and instability results, as the temperature inside the rising parcel becomes warmer than the air surrounding it. In a conditionally unstable atmosphere, rising air tends to form clouds that develop vertically, such as cumulus congestus and cumulonimbus. Instability may be caused by warming the surface air, cooling the air aloft, or by the lifting or mixing of an entire layer of air. Lifting an entire layer of air where the base of the layer is humid and the top "dry," produces convective instability, an important mechanism for the formation of intense thunderstorms.

On warm, humid days the instability generated by surface heating produces convection, often in the form of rising thermals that may develop into cumulus clouds. Instability may also cause change in existing clouds, as convection changes an altocumulus into a more towering altocumulus castellanus. Near the surface, mixing can change a clear day into a cloudy one. As air passes over a mountain range, clouds and precipitation often form on its upward side. The removal of moisture and the release of latent heat during condensation produce warmer and drier descending air on the mountain's downwind side. The drier region on the mountain's leeward side is called a rain shadow.

Key Terms

The following terms are listed (with page number) in the order they appear in the text. Define each. Doing so will aid you in reviewing the material covered in this chapter.

parcel of air, 122
adiabatic process, 123
dry adiabatic rate, 123
moist adiabatic rate, 123
environmental lapse rate, 124
absolutely stable atmosphere, 124
subsidence inversion, 126
neutral stability, 126
absolutely unstable atmosphere, 128
conditionally unstable atmosphere, 129
condensation level, 129
convective instability, 132
entrainment, 134
orographic uplift, 135
rain shadow, 135
rotor clouds, 139

Review Questions

1. What is an adiabatic process?
2. Why are moist and dry adiabatic rates of cooling different?
3. Under what conditions would the moist adiabatic rate of cooling be almost equal to the dry adiabatic rate?
4. Explain the difference between environmental lapse rate and dry adiabatic rate.
5. How would one normally obtain the environmental lapse rate?
6. What is a stable atmosphere, and how can it form?
7. Describe the general characteristics of clouds associated with stable and unstable atmospheres.
8. List and explain several processes by which a stable atmosphere can be made unstable.
9. If the atmosphere is conditionally unstable, what condition is necessary to bring on instability?
10. Explain why cumulus clouds are conspicuously absent over a cool water surface.
11. Why are cumulus clouds more frequently observed during the afternoon than at night?
12. Explain why an inversion represents an absolutely stable atmosphere.
13. How and why does lifting or lowering a layer of air change its stability?
14. How does atmospheric stability influence the formation of lake-effect snowstorms?
15. List and explain several processes by which an unstable atmosphere can be made stable.
16. Why do cumulonimbus clouds (thunderstorms) often have flat tops?
17. Is strong convection more likely in a stable or unstable atmosphere? Explain your reasoning.
18. On a warm, sunny day the sky is dotted with small developing cumulus clouds. How does the stability of the atmosphere determine whether these clouds will remain small or grow into cumulonimbus clouds (thunderstorms)?

19. Why are there usually large spaces of blue sky between cumulus clouds?
20. List four primary ways clouds form and describe the formation of one cloud type by each method.
21. Explain why rain shadows form on the leeward side of mountains.
22. Describe the necessary conditions for the formation of convective instability. How do you feel convective instability may play a role in the formation of intense thunderstorms?
23. Describe the conditions necessary to produce stratocumulus clouds by mixing.
24. Briefly describe how each of the following clouds forms:
    (a) lenticular
    (b) rotor
    (c) castellanus
25. If you see a sky full of altocumulus castellanus in the late afternoon, what does this observation tell you about the stability of the atmosphere where the clouds have formed?
26. Why are downslope winds on the leeward side of a high mountain range usually much warmer and drier than the upslope winds on the windward side?

Online Learning

STUDENT COMPANION WEBSITE: Visit this book's companion website at: www.cengage.com/ahrens/extreme1e and choose chapter 5 for many study aids and ideas for further reading and research. These include flash cards, practice quizzing, and web links.

METEOROLOGY RESOURCE CENTER: For students with access, log on at www.cengage.com/login for more assets, including animations, videos, and more. If your textbook did not come with access, visit www.CengageBrain.com to purchase.

Precipitation Extremes

CONTENTS

PRECIPITATION PROCESSES

PRECIPITATION TYPES

PRECIPITATION—EXTREME EVENTS

SUMMARY

KEY TERMS

REVIEW QUESTIONS

Summer thunderstorms over the otherwise dry Painted Desert of Arizona, create heavy rain and flash floods.

As we all know, cloudy weather does not necessarily mean that it will rain or snow. In fact, clouds may form, linger for many days, and never produce **precipitation.*** In Eureka, California, the August daytime sky is overcast more than 50 percent of the time, yet the average precipitation there for August is merely one-tenth of an inch. How, then, do cloud droplets grow large enough to produce rain? Why do some clouds produce rain, but not others? And where are the wettest and driest places in the world located?

To answer these and other questions, we will begin this chapter by examining how tiny cloud droplets are able to grow into much larger raindrops. We will then look at the formation of other types of precipitation, such as sleet, freezing rain, and hail. Toward the end of the chapter, we will look at where one would find the wettest, driest, and snowiest places on this planet.

Precipitation Processes

In ❱ Fig. 6.1, we can see that an ordinary cloud droplet is extremely small, having an average diameter of 0.02 millimeters (mm), which is less than one-thousandth of an inch. Also, notice in Fig. 6.1 that the diameter of a typical cloud droplet is 100 times smaller than a typical raindrop. Clouds, then, are composed of many small droplets—too small to fall as rain. These minute droplets require only slight upward air currents to keep them suspended. Those droplets that do fall descend slowly and evaporate in the drier air beneath the cloud.

*Recall from Chapter 1 that precipitation is any form of water (liquid or solid) that falls from a cloud and reaches the ground.

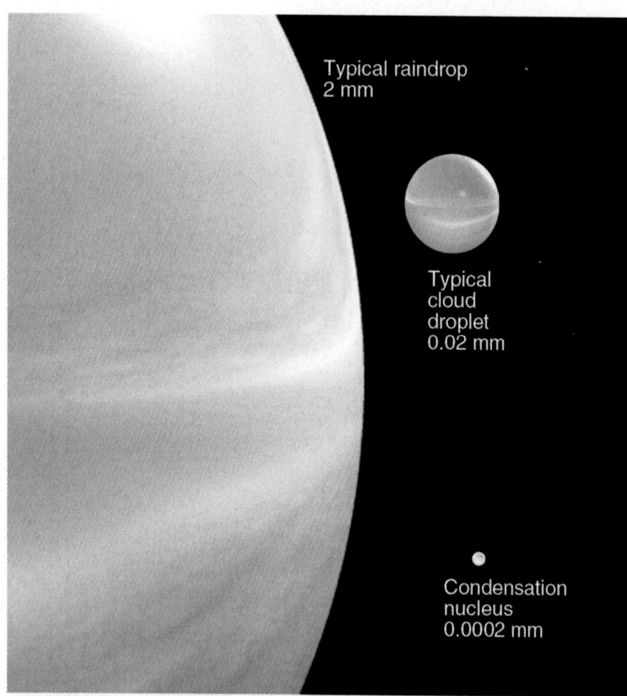

❱ FIGURE 6.1 Relative sizes of raindrops, cloud droplets, and condensation nuclei in millimeters (mm).

In Chapter 4 we learned that condensation begins on tiny particles called *condensation nuclei*. The growth of cloud droplets by condensation is slow and, even under ideal conditions, it would take several days for this process alone to create a raindrop. It is evident, then, that the condensation process by itself is entirely too slow to produce rain. Yet, observations show that clouds can develop and begin to produce rain in less than an hour. Since it takes about 1 million average size cloud droplets to make an average size raindrop, there must be some other process by which cloud droplets grow large and heavy enough to fall as precipitation.

Even though all the intricacies of how rain is produced are not yet fully understood, two important processes stand out: (1) the collision-coalescence process and (2) the ice-crystal (or Bergeron) process.

COLLISION AND COALESCENCE PROCESS In clouds with tops warmer than about 5°F (–15°C), the **collision-coalescence process** can play a significant role in producing precipitation. To produce the many collisions necessary to form a raindrop, some cloud droplets must be larger than others. Larger drops may form on large condensation nuclei, such as salt particles, or through random collisions of droplets. Studies also suggest that turbulent mixing between the cloud and its drier environment may play a role in producing larger droplets.

Larger cloud droplets fall faster than smaller droplets. This fact means that large droplets can grow even larger by overtaking and colliding with smaller droplets in their path (see ❱ Fig. 6.2). This merging of cloud droplets by collision is called **coalescence**. Coalescence appears to be enhanced if colliding droplets have opposite (and, hence, attractive) electrical charges.

An important factor influencing cloud droplet growth by the collision process is the amount of time the droplet spends in the cloud. Since rising air currents slow the rate at which droplets fall, a thick cloud with strong updrafts will maximize the time cloud droplets spend in a cloud and, hence, the size to which they grow.

Clouds that have above-freezing temperatures at all levels are called *warm clouds*. In tropical regions, where warm cumulus clouds build to great heights, strong convective updrafts frequently occur. In ❱ Fig. 6.3, suppose a cloud droplet is caught in a strong updraft. As the droplet rises, it collides with and captures smaller drops in its path, and grows until it reaches a size of about 1 mm. At this point, the updraft in the cloud is just able to balance the pull of gravity on the drop. Here, the drop remains suspended until it grows just a little bigger. Once the fall velocity of the drop is greater than the updraft velocity in the cloud, the drop slowly descends. As the drop falls, some of the smaller droplets get caught in the airstream around it, and are swept aside. Larger cloud droplets are captured by the falling drop, which then grows larger. By the time this drop reaches the bottom of the cloud, it will be a large raindrop with a diameter of over 5 mm. Because raindrops of this size fall faster and reach the ground first, they typically occur at the beginning of a rain shower originating in these warm, convective cumulus clouds.

So far, we have examined the way cloud droplets in warm clouds (that is, those clouds with temperatures above freezing) grow large enough by the collision-coalescence process to fall as raindrops. The most important factor in the production of raindrops is the cloud's liquid water content. In a cloud with sufficient water, other significant factors are:

1. the range of droplet sizes
2. the cloud thickness
3. the updrafts of the cloud
4. the electric charge of the droplets and the electric field in the cloud

Relatively thin stratus clouds with slow, upward air currents are, at best, only able to produce drizzle (the lightest form of rain), whereas the towering cumulus clouds associated with rapidly rising air can cause heavy showers. Now, let's turn our attention to how clouds with temperatures below freezing are able to produce precipitation.

ICE-CRYSTAL PROCESS The **ice-crystal** (or **Bergeron***) **process** of rain formation proposes that both ice crystals and liquid cloud droplets must co-exist in clouds at temperatures below freezing. Consequently, this process of rain formation is extremely important in middle and high latitudes, where clouds are able to extend upwards into regions where air temperatures are below freezing. Such clouds are called *cold clouds*. Figure 6.4 illustrates a typical cumulonimbus cloud that has formed over the Great Plains of North America.

In the warm region of the cloud (below the freezing level) where only water droplets exist, we might expect to observe cloud droplets growing larger by the collision

*The ice-crystal process is also known as the *Bergeron process* after the Swedish meteorologist Tor Bergeron, who proposed that essentially all raindrops begin as ice crystals.

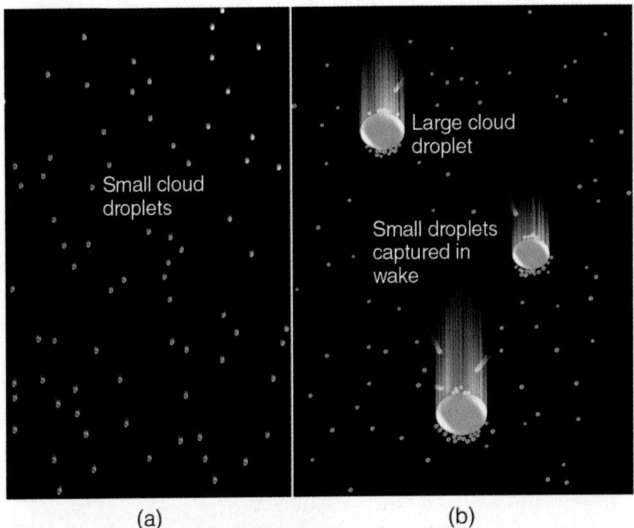

FIGURE 6.2 Collision and coalescence. (a) In a relatively warm cloud composed only of small cloud droplets of uniform size, the droplets are less likely to collide as they all fall very slowly at about the same speed. Those droplets that do collide, frequently do not coalesce because of the strong surface tension that holds together each tiny droplet. (b) In a cloud composed of different size droplets. larger droplets fall faster than smaller droplets. Although some tiny droplets are swept aside, some collect on the larger droplet's forward edge, while others (captured in the wake of the larger droplet) coalesce on the droplet's backside.

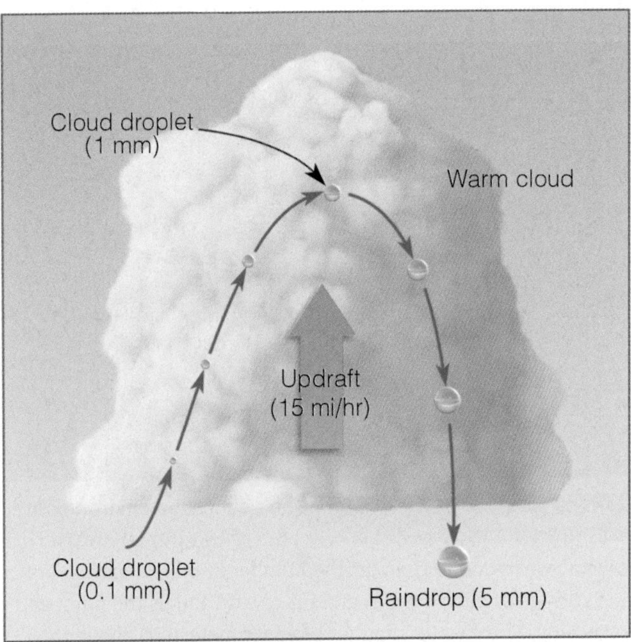

FIGURE 6.3 A cloud droplet rising then falling through a warm cumulus cloud can grow by collision and coalescence and emerge from the cloud as a large raindrop.

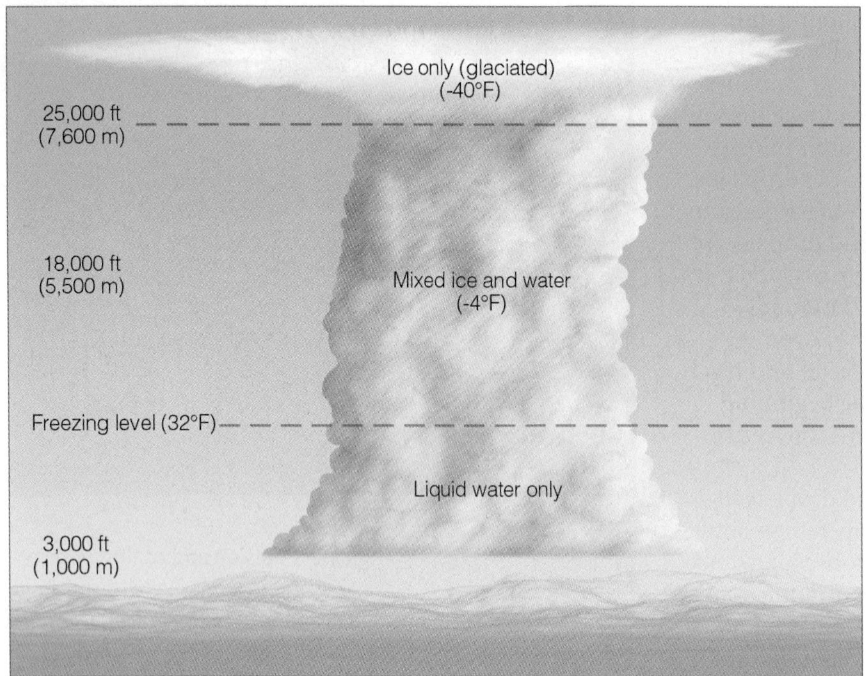

FIGURE 6.4

The distribution of ice and water in a cumulonimbus cloud.

and coalescence process described in the previous section. Surprisingly, in the cold air just above the freezing level, almost all of the cloud droplets are still composed of liquid water. Water droplets existing at temperatures below freezing are referred to as **supercooled**. At higher levels, ice crystals become more numerous, but are still outnumbered by water droplets. Ice crystals exist overwhelmingly in the upper part of the cloud, where air temperatures drop to well below freezing. Why are there so few ice crystals in the middle of the cloud, even though temperatures there, too, are below freezing? Laboratory studies reveal that the smaller the amount of pure water, the lower the temperature at which water freezes. Since cloud droplets are extremely small, it takes very low temperatures to turn them into ice.

Just as liquid cloud droplets form on condensation nuclei, ice crystals may form in subfreezing air if there are ice-forming particles present called *ice nuclei*. The number of ice-forming nuclei available in the atmosphere is small, especially at temperatures above 14°F (–10°C). Although some uncertainty exists regarding the principal source of ice nuclei, it is known that certain clay minerals, bacteria in decaying plant leaf material, and ice crystals themselves are excellent ice nuclei. Moreover, particles serve as excellent ice-forming nuclei if their geometry resembles that of an ice crystal.

We can now understand why there are so few ice crystals in the subfreezing region of some clouds. Liquid cloud droplets may freeze, but only at very low temperatures. Ice nuclei may initiate the growth of ice crystals, but they do not abound in nature. Therefore, we are left with a cold cloud that contains many more liquid droplets than ice particles, even at low temperatures. Neither the tiny liquid nor solid particles are large enough to fall as precipitation. How, then, does the ice-crystal process produce rain and snow?

In the subfreezing air of a cloud, many supercooled liquid droplets will surround each ice crystal. Suppose that the ice crystal and liquid droplet in Fig. 6.5 are part of a cold (5°F), supercooled, saturated cloud. Since the air is saturated, both the liquid droplet and the ice crystal are in equilibrium, meaning that the number of molecules leaving the surface of both the droplet and the ice crystal must equal the number of molecules returning. Observe, however, that there are more vapor

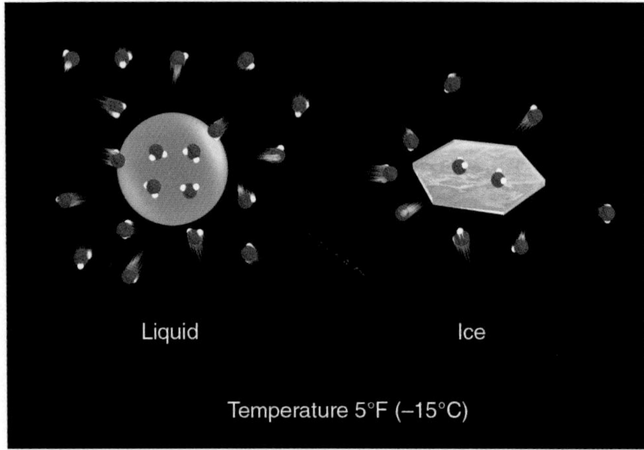

FIGURE 6.5 In a saturated environment, the water droplet and the ice crystal are in equilibrium, as the number of molecules leaving the surface of each droplet and ice crystal equals the number returning. The greater number of vapor molecules above the liquid produces a greater vapor pressure above the droplet. This situation means that, at saturation, the pressure exerted by the water molecules is greater over the water droplet than above the ice crystal.

molecules above the liquid. The reason for this fact is that molecules escape the surface of water much easier than they escape the surface of ice. Consequently, more molecules escape the water surface at a given temperature, requiring more in the vapor phase to maintain saturation. Therefore, it takes more vapor molecules to saturate the air directly above the water droplet than it does to saturate the air directly above the ice crystal. At saturation, the greater number of water vapor molecules above the liquid droplet produces a higher vapor pressure. As a consequence, the pressure exerted by the vapor molecule above the liquid droplet is greater than the vapor pressure exerted by the vapor molecules above the ice crystal.

This difference in vapor pressure causes water vapor molecules to move (diffuse) from the droplet toward the ice crystal. The removal of vapor molecules reduces the vapor pressure above the droplet. Since the droplet is now out of equilibrium with its surroundings, it evaporates to replenish the diminished supply of water vapor above it. This process provides a continuous source of moisture for the ice crystal, which absorbs the water vapor and grows rapidly (see ❱ Fig. 6.6). Hence, during the *ice-crystal (Bergeron) process, ice crystals grow larger at the expense of the surrounding water droplets.*

The ice crystals may now grow even larger. For example, in some clouds, ice crystals might collide with supercooled liquid droplets. Upon contact, the liquid droplets freeze into ice and stick together. This process of ice crystals growing larger as they collide with supercooled cloud droplets is called *accretion*. The icy matter that forms is called *graupel* (or *snow pellets*). As the graupel falls, it may fracture or splinter into tiny ice particles when it collides with cloud droplets. These splinters may then go on themselves to become new graupel, which, in turn, may produce more splinters. In colder clouds, the delicate ice crystals may collide with other crystals and fracture into smaller ice particles, or tiny seeds, which freeze hundreds of supercooled droplets on contact. In both cases, a chain reaction may develop, producing many ice crystals (see ❱ Fig. 6.7). As they fall, they collide and stick to one another, forming an aggregate of ice crystals called a *snowflake*. If the snowflake melts before reaching the ground, it continues its fall as a raindrop. Therefore, much of the rain falling in middle and northern latitudes—even in summer—begins as snow.

CLOUD SEEDING AND PRECIPITATION The primary goal in many experiments concerning **cloud seeding** is to inject (or seed) a cloud with small particles that will act as nuclei, so that the cloud particles will grow large enough to fall to the surface as precipitation.

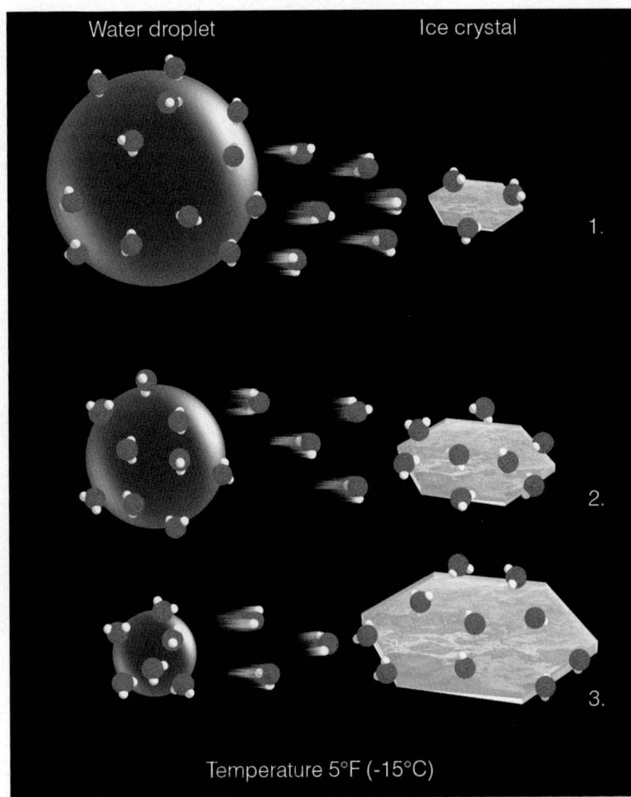

❱ FIGURE 6.6 The ice-crystal process. The greater number of water vapor molecules around the liquid droplets causes water molecules to diffuse from the liquid drops toward the ice crystals. The ice crystals absorb the water vapor and grow larger, while the water droplets grow smaller.

The first ingredient in any seeding project is, of course, the presence of clouds, as seeding does not generate clouds. However, at least a portion of the cloud (preferably the upper part) must be supercooled because cloud seeding uses the ice-crystal process to cause the cloud particles to grow. The idea is to find clouds that have too low a ratio of ice crystals to droplets and then to add enough artificial ice nuclei so that the ratio of crystals to droplets is optimal (about 1:100,000) for producing precipitation.

Some of the first experiments in cloud seeding were conducted by Vincent Schaefer and Irving Langmuir during the late 1940s. To seed a cloud, they dropped crushed pellets of *dry ice* (solid carbon dioxide) from a plane. Because dry ice has a temperature of −108°F (−78°C), it acts as a cooling agent. As the extremely cold, dry ice pellets fall through the cloud, they quickly cool the air around them. This cooling causes the air around the pellet to become supersaturated. In this supersaturated air, water vapor forms directly into many tiny cloud droplets. In the very cold air created by the

(a) Falling ice crystals may freeze supercooled droplets on contact (accretion), producing larger ice particles.

(b) Falling ice particles may collide and fracture into many tiny (secondary) ice particles.

(c) Falling ice crystals may collide and stick to other ice crystals (aggregation), producing snowflakes.

FIGURE 6.7 Ice particles in clouds.

falling pellets (below −40°F), the tiny droplets instantly freeze into tiny ice crystals. The newly formed ice crystals then grow larger by deposition as the water vapor molecules attach themselves to the ice crystals at the expense of the nearby liquid droplets and, upon reaching a sufficiently large size, they fall as precipitation.

In 1947, Bernard Vonnegut demonstrated that silver iodide (AgI) could be used as a cloud-seeding agent. Because silver iodide has a crystalline structure similar to an ice crystal, it acts as an effective ice nucleus at temperatures of 25°F (−4°C) and lower. Silver iodide causes ice crystals to form in two primary ways:

1. Ice crystals form when silver iodide crystals come in contact with supercooled liquid droplets.
2. Ice crystals grow in size as water vapor deposits onto the silver iodide crystal.

Silver iodide is much easier to handle than dry ice, since it can be supplied to the cloud from burners located either on the ground or on the wing of a small aircraft. Although other substances, such as lead iodide and cupric sulfide, are also effective ice nuclei, silver iodide still remains the most commonly used substance in cloud-seeding projects.

Just how effective is artificial seeding with silver iodide in increasing precipitation? This is a much-debated question among meteorologists. First of all, it is difficult to evaluate the results of a cloud-seeding experiment. When a seeded cloud produces precipitation, the question always remains as to how much precipitation would have fallen had the cloud not been seeded. Other factors must be considered when evaluating cloud-seeding experiments: the type of cloud, its temperature, moisture content, droplet size distribution, and updraft velocities in the cloud.

The business of cloud seeding can be a bit tricky, since overseeding can produce too many ice crystals. When this phenomenon occurs, the cloud becomes glaciated (all liquid droplets become ice) and the ice particles, being very small, do not fall as precipitation. Since few liquid droplets exist, the ice crystals cannot grow by the ice-crystal (Bergeron) process; rather, they evaporate, leaving a clear area in a thin, stratified cloud.

Although some experiments suggest that cloud seeding does not increase precipitation, others seem to indicate that seeding *under the right conditions* may enhance precipitation between 5 percent and 20 percent. And so the controversy continues.

Under certain conditions, clouds may be seeded naturally. For example, when cirriform clouds lie directly above a lower cloud deck, ice crystals may descend from the higher cloud and seed the cloud below (see Fig. 6.8). As the ice crystals mix into the lower cloud, supercooled droplets are converted to ice crystals, and the precipitation process is enhanced. Sometimes the ice crystals in the lower cloud may settle out, leaving a clear area or "hole" in the cloud. When the cirrus clouds form waves downwind from a mountain chain, bands of precipitation often form—producing heavy precipitation in some areas and practically no precipitation in others (see Fig. 6.9).

There are even conditions where cloud seeding may be inadvertent. For example, some industries emit large concentrations of condensation nuclei and ice nuclei into the air. Studies have shown that these particles are at least partly responsible for increasing precipitation in, and downwind of, cities. On the other hand, studies have also indicated that the burning of certain types of agricultural waste may produce smoke containing many condensation nuclei. These particles produce clouds that yield less precipitation because they contain numerous, but very small, droplets.

In summary, cloud seeding in certain instances may lead to more precipitation; in others, to less precipitation, and, in still others, to no change in precipitation amounts. Many of the questions about cloud seeding have yet to be resolved.

PRECIPITATION IN CLOUDS In cold, strongly convective clouds, precipitation may begin only minutes after the cloud forms, and may be initiated by either the collision-coalescence or the ice-crystal (Bergeron) process. Once either process begins, most precipitation growth is by accretion, as supercooled liquid droplets freeze on impact with snowflakes and ice crystals. Although precipitation is commonly absent in warm-layered clouds, such as stratus, it is often associated with such cold-layered clouds as nimbostratus and altostratus. This precipitation is thought to form principally by the ice-crystal (Bergeron) process because the liquid water content of these clouds is generally lower than that in convective clouds, thus making the collision-coalescence process much less effective. Nimbostratus clouds are normally thick enough to extend to levels where air temperatures are quite low, and they usually last long enough for the ice-crystal process to initiate precipitation.

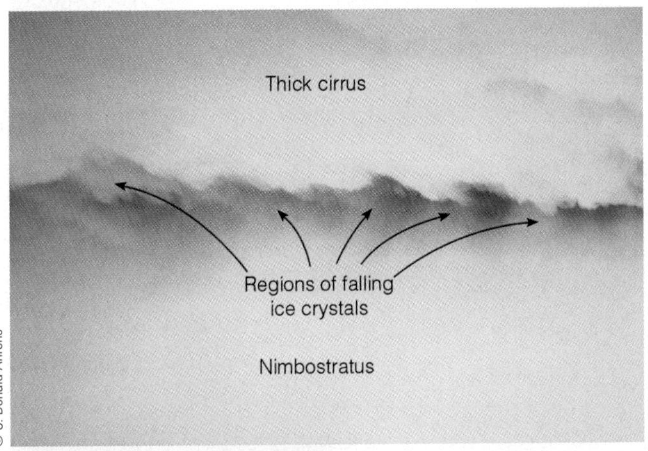

FIGURE 6.8 Ice crystals falling from a dense cirriform cloud into a lower nimbostratus cloud. This photo was taken at an altitude near 6 km (19,700 ft) above western Pennsylvania. At the surface, moderate rain was falling over the region.

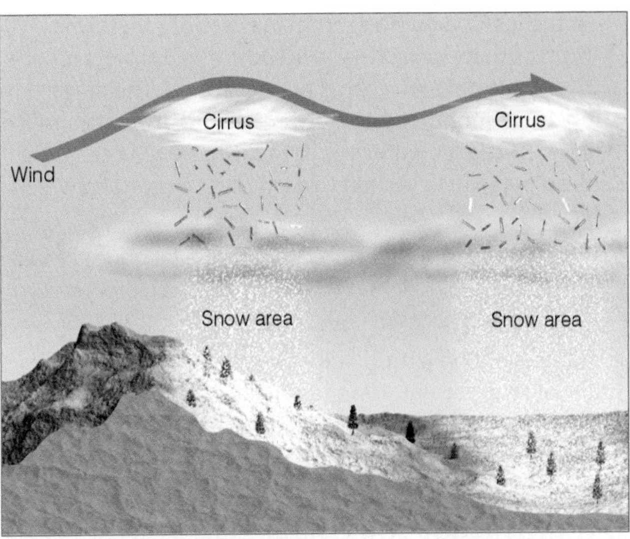

FIGURE 6.9 Natural seeding by cirrus clouds may form bands of precipitation downwind of a mountain chain. Notice that heavy snow is falling only in the seeded areas.

BRIEF REVIEW

In the last few sections we encountered a number of important concepts and ideas about how cloud droplets can grow large enough to fall as precipitation. Before examining the various types of precipitation, here is a summary of some of the important ideas presented so far:

- **Cloud droplets are very small, much too small to fall as rain.**
- **Cloud droplets form on cloud condensation nuclei.**
- **Cloud droplets, in above-freezing air, can grow larger as faster-falling, bigger droplets collide and coalesce with smaller droplets in their path.**

- In the ice-crystal (Bergeron) process of rain formation, both ice crystals and liquid cloud droplets must coexist at below-freezing temperatures. The difference in vapor pressure between liquid droplets and ice crystals causes water vapor to diffuse from the liquid droplets (which shrink) toward the ice crystals (which grow).
- Most of the rain that falls over middle latitudes results from melted snow that formed from the ice-crystal (Bergeron) process.
- Cloud seeding with silver iodide can only be effective in coaxing precipitation from clouds if the cloud is supercooled and the proper ratio of cloud droplets to ice crystals exists.

Precipitation Types

Up to now, we have seen how cloud droplets are able to grow large enough to fall to the ground as rain or snow. While falling, raindrops and snowflakes may be altered by atmospheric conditions encountered beneath the cloud and transformed into other forms of precipitation that can profoundly influence our environment by creating extremely hazardous weather conditions.

RAIN Most people consider **rain** to be any falling drop of liquid water. To the meteorologist, however, that falling drop must have a diameter equal to, or greater than, 0.5 mm (0.02 in.) to be considered rain. Fine uniform drops of water whose diameters are smaller than 0.5 mm (which is a diameter about one-half the width of the letter "o" on this page) are called **drizzle**. Most drizzle falls from stratus clouds; however, small raindrops may fall through air that is unsaturated, partially evaporate, and reach the ground as drizzle. Occasionally, the rain falling from a cloud never reaches the surface because the low humidity causes rapid evaporation. As the drops become smaller, their rate of fall decreases, and they appear to hang in the air as a rain streamer. These evaporating streaks of precipitation are called **virga*** (see Fig. 6.10).

Raindrops may also fall from a cloud and not reach the ground if they encounter the rapidly rising air of an

*Studies suggest that the "rain streamer" is actually caused by ice (which is more reflective) changing to water (which is less reflective). Apparently, most evaporation occurs below the virga line.

FIGURE 6.10 The streaks of falling precipitation that evaporate before reaching the ground are called *virga*.

updraft. If the updraft weakens or changes direction and becomes a downdraft, the suspended drops will fall to the ground as a sudden rain **shower**. The showers falling from cumuliform clouds are usually brief and sporadic, as the cloud moves overhead and then drifts on by. If the shower is excessively heavy, it is termed a *cloudburst.* Beneath a cumulonimbus cloud, which normally contains large up and down convection currents, it is entirely possible that one side of a street may be dry (updraft side), while a heavy shower is occurring across the street (downdraft side). Continuous rain, on the other hand, usually falls from a layered cloud that covers a large area and has smaller vertical air currents. These are the conditions normally associated with nimbostratus clouds.

Raindrops that reach the earth's surface are seldom larger than about 6 mm (0.2 in.), the reason being that the collisions (whether glancing or head-on) between raindrops tend to break them up into many smaller drops. Additionally, when raindrops grow too large they become unstable and break apart.

Rain is usually transparent, but occasionally it takes on various colors. For example, during November, 1933, black rain fell over New York State after a tremendous dust storm spread dirt from Montana to Maine. When many tons of dust or volcanic ash are carried aloft, tiny mud or ash balls have actually fallen from clouds. Such was the case on April 12, 1902, when mud showers reportedly fell over a large area of the Mid-Atlantic states. Apparently, an intense dust storm over Illinois lifted dirt into the air, and strong upper-level winds then carried the dirt eastward.

After a rainstorm, visibility usually improves primarily because precipitation removes (scavenges) many of the suspended particles. When rain combines with gaseous pollutants, such as oxides of sulfur and nitrogen, it becomes acidic. *Acid rain*, which has an adverse effect on plants and water resources, is becoming a major problem in many industrialized regions of the world.

SNOW We have learned that much of the precipitation reaching the ground actually begins as **snow**. In summer, the freezing level is usually high and the snowflakes falling from a cloud melt before reaching the surface. In winter, however, the freezing level is much lower, and falling snowflakes have a better chance of survival. In fact, snowflakes can generally fall about 1000 ft (300 m) below the freezing level before completely melting. When the warmer air beneath the cloud is relatively dry, the snowflakes partially melt. As the liquid water evaporates, it chills the snowflake, which retards its rate of melting. Consequently, in air that is relatively dry, snowflakes may reach the ground even when the air temperature is considerably above freezing.

EXTREME WEATHER WATCH

Maybe it has never rained cats and dogs, but it has rained maggots. In Acapulco, Mexico, during October, 1968, swarms of maggots (about an inch in length) fell from the sky during a heavy rain shower, covering everything, even people who had gathered there to witness a yachting event. Apparently, the maggots were swept into a thunderstorm by strong vertical air currents.

Is it ever "too cold to snow"? Although many believe this expression, the fact remains that it is *never* too cold to snow. True, more water vapor will condense from warm saturated air than from cold saturated air. But, no matter how cold the air becomes, it always contains some water vapor that could produce snow. In fact, tiny ice crystals have been observed falling at temperatures as low as −53°F. We usually associate extremely cold air with "no snow" because the coldest winter weather occurs on clear, calm nights—conditions that normally prevail with strong high pressure areas that have few if any clouds.

Snowflakes falling through moist air that is slightly above freezing slowly melt as they descend. A thin film of water forms on the edge of the flakes, which acts like glue when other snowflakes come in contact with it. In this way, several flakes join to produce giant snowflakes that often measure an inch or more in diameter. In fact, snowflakes up to 5.5 inches wide were reported falling in Nashville, Tennessee, on January 24, 1894. These large, soggy snowflakes are associated with moist air and temperatures near freezing. However, when snowflakes fall through extremely cold air with a low moisture content, they do not readily stick together and small, powdery flakes of "dry" snow accumulate on the ground.

If you catch falling snowflakes on a dark object and examine them closely, you will see that the most common snowflake form is a fernlike branching shape called *dendrite* (see ❱ Fig. 6.11). As ice crystals fall through a cloud, they are constantly exposed to changing temperatures and moisture conditions. Since many ice crystals can join together *(aggregate)* to form a much larger snowflake, ice crystals may assume many complex patterns.

Snow falling from developing cumulus clouds is often in the form of **flurries**. These are usually light showers that fall intermittently for short durations and produce only light accumulations. A more intense snow shower is called a **snow squall**. These brief but heavy falls of snow are comparable to summer rain showers

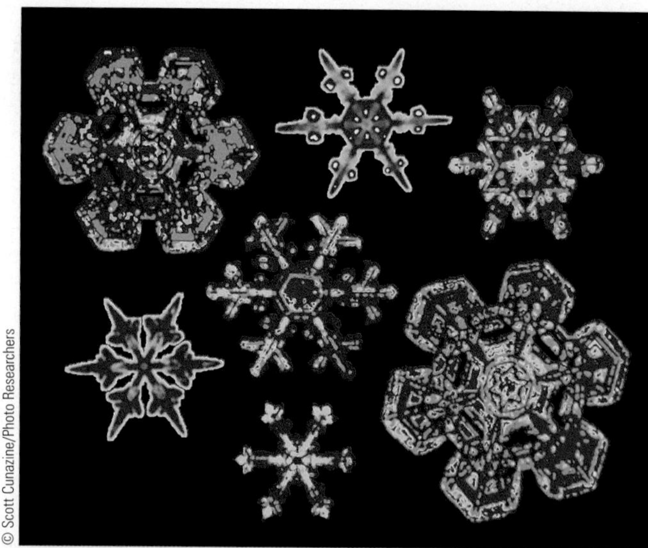

FIGURE 6.11 Computer color-enhanced image of dendrite snowflakes.

and, like snow flurries, usually fall from cumuliform clouds. A more continuous snowfall (sometimes steadily, for several hours) accompanies nimbostratus and altostratus clouds.

When a strong wind is blowing at the surface, snow can be picked up and deposited into huge drifts. Drifting snow is usually accompanied by *blowing snow*; that is, snow lifted from the surface by the wind and blown about in such quantities that horizontal visibility is greatly restricted. The combination of drifting and blowing snow, after falling snow has ended, is called a *ground blizzard*. A true **blizzard** is a weather condition characterized by low temperatures and strong winds (greater than 35 mi/hr) bearing large amounts of fine, dry, powdery particles of snow, which can reduce visibility to only a few feet (see Fig. 6.12). In the United States and Great Britain, however, any heavy snowstorm accompanied by high winds is often referred to as a blizzard.

Small, opaque grains of snow with diameters less than 1 mm (0.04 in.) are called **snow grains**. They tend to be fairly flat or elongated. The solid equivalent of drizzle, snow grains usually fall steadily from stratus clouds. Upon striking a hard surface, they neither bounce nor shatter. **Snow pellets** (or *graupel*), on the other hand, are white grains of ice snow about the size of an average raindrop. They are sometimes confused with snow grains. The distinction is easily made, however, by remembering that, unlike snow grains, snow pellets are brittle, crunchy, and bounce (or break apart) upon hitting a hard surface. They usually fall as showers, especially from cumulus congestus clouds.

Snow pellets form as snow crystals collide with supercooled water droplets that freeze on contact. The icy matter that forms on the snowflake is called *rime*. Notice in Fig. 6.13 that when the snowflake accumulates a light coating of rime, it becomes a *rimed snowflake*. When the snowflake accumulates a heavy coating of rime, it becomes a more rounded aggregate of icy matter containing many air spaces, called, as previously mentioned, a *snow pellet* or *graupel*.

During the winter, when the freezing level is at a low elevation, snow pellets reach the surface as a light, rounded clump of snow-like ice. On the surface, the accumulation of snow pellets sometimes gives the appearance of tapioca pudding; hence, it can be referred

FIGURE 6.12

Winds gusting up to 60 mi/hr produce a ground blizzard with near zero visibility in Western Iowa on January 29, 2009.

to as *tapioca snow*. In a summer thunderstorm, when the freezing level is well above the surface, snow pellets often melt and reach the surface as large raindrops.

Most of the time snowflakes fall as delicate white crystals. But various colored snows have been reported. For instance, pink snow actually fell on Durango, Colorado, during January, 1932, and red snow fell over portions of Italy and Switzerland during February, 1852. Apparently, the reddish snow was due to dust that lifted and mixed into snow-producing clouds. Black snow was reported in New York in April, 1889. The Weather Bureau suggested that the black snow may have been caused by either dark sediment or vegetable mold. During March, 1879, yellow snow fell over portions of Bethlehem, Pennsylvania. The yellow snow contained pollen from pine trees that were in bloom.

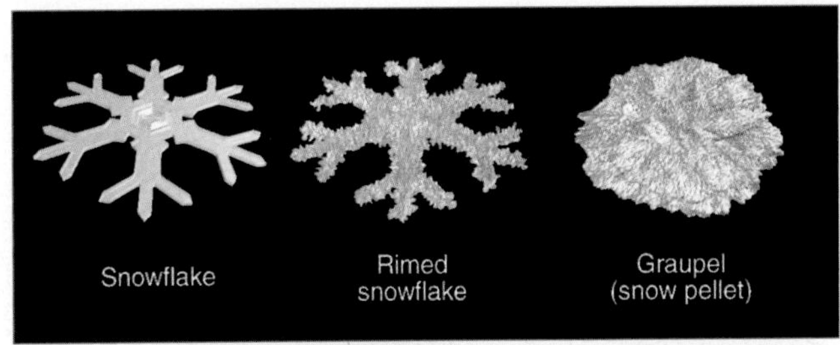

FIGURE 6.13 A snowflake becoming a rimed snowflake, then finally graupel (a snow pellet).

THE EFFECT OF A SNOWFALL A mantle of snow covering the landscape is much more than a beautiful setting—it is a valuable resource provided by nature. A blanket of snow is a good insulator (poor heat conductor). In fact, the more air spaces there are between the individual snowflake crystals, the better insulator they become. A light, fluffy covering of snow protects sensitive plants and their root systems from damaging low temperatures by retarding the loss of ground heat.

On winter nights, items covered with dry, fluffy snow maintain a higher temperature than those exposed to the cold air. For this reason, in extremely cold weather, individuals should leave a covering of fluffy snow on the hood of their cars. The insulating properties of the snow will help to keep the car's engine a bit warmer and make starting the car a bit easier.

In a similar way, snow can prevent the ground from freezing downward to great depths. In cold climates that receive little snow, it is often difficult to grow certain crops because the frozen soil makes spring cultivation almost impossible. Frozen ground also prevents early spring rains from percolating downward into the soil, leading to rapid water runoff and flooding. If subsequent rains do not fall, the soil could even become moisture-deficient.

If you become lost in a cold and windy snowstorm, build a snow cave and climb inside. It not only will protect you from the wind, but it also will protect you from the extreme cold by slowing the escape of heat your body generates.

Winter snows may be beautiful, but they are not without hardships and potential hazards. As spring approaches, rapid melting of the snowpack may flood low-lying areas. Too much snow on the side of a steep hill or mountain may become an **avalanche** (see Fig. 6.14). The added weight of snow on the roof of a building may cause it to collapse, leading to costly repairs and even loss of life. Each winter, heavy snows clog streets and disrupt transportation. To

FIGURE 6.14 An avalanche of powdery snow falls from Gasherbrum, Baltoro, a glacier in Pakistan.

Table 6.1

Snowiest Cities in the United States

CITY	ANNUAL AVERAGE SNOWFALL (inches)
Truckee, CA	120.3
Marquette, MI (airport)	179.8
Marquette, MI (city)	118.2
Steamboat Springs, CO	173.3
Oswego, NY	153.3
Sault St. Marie, MI	131.2
Syracuse, NY	120.2
Meadville, PA	111.2
Flagstaff, AZ	111.1
Watertown, NY	110.8
Muskegon, MI	105.9
Rochester, NY	99.5
Utica, NY	98.5
Montpelier, VT	97.9
Traverse City, MI	96.8
Buffalo, NY	95.7
Juneau, AK	96.3

keep traffic moving, streets must be plowed and sanded, or salted to lower the temperature at which the snow freezes (melts). This effort can be expensive, especially if the snow is heavy and wet. Cities unaccustomed to snow are usually harder hit by a moderate snowstorm than cities that frequently experience snow. A January snowfall of several inches in New Orleans, Louisiana, can bring traffic to a standstill, while a snowfall of several inches in Buffalo, New York, would go practically unnoticed.

Figure 6.15 shows the annual average snowfall across the United States. As you would expect, annual snowfall totals tend to be low in the southern states and higher as you move north. Notice that in areas of the northeast and in the mountainous west, annual snowfall totals exceed 72 inches. In fact, as we will see later in this chapter, Paradise Ranger Station on Mount Rainier, Washington, receives an annual average of 692 inches of snow, making it one of the snowiest places in the world. Although snow is exceedingly rare in south Florida and along the Gulf of Mexico, on Christmas Day, 2004, Brownsville, Texas, received 1.4 inches of snow and on the morning of December 11, 2008, up to an inch of snow covered grassy areas in New Orleans, Louisiana. And on January 19, 1977, for the first time in recorded history, snowflakes were reported falling in Miami, Florida. Table 6.1 shows some of the snowiest cities (populations of at least 10,000) in the United States.

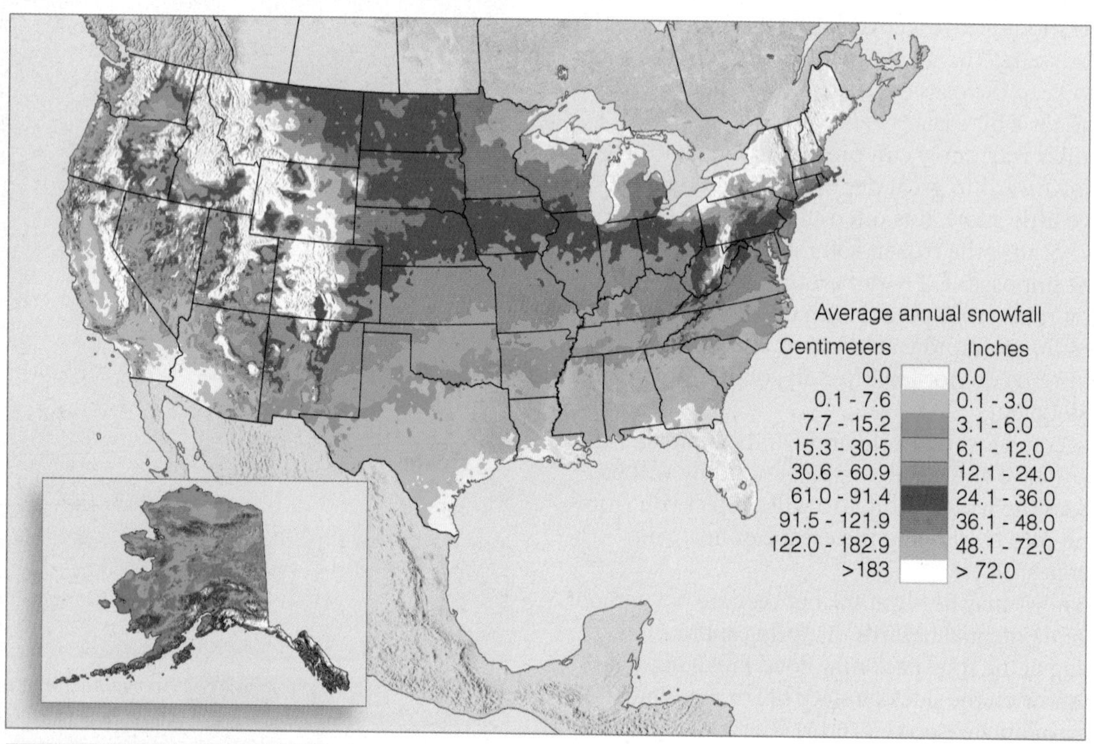

FIGURE 6.15 Average annual snowfall over the United States.

EXTREME WEATHER WATCH

The greatest Gulf Coast snowstorm on record occurred during February, 1895, when 20 inches of snow fell in Houston, Texas, 8 inches in New Orleans, Louisiana, and 6 inches in Brownsville, Texas. During the storm, snow actually fell on the Gulf Coast city of Tampico, Mexico. At a latitude of 22°, this weather event marks the southernmost sea-level snowfall ever in the Northern Hemisphere.

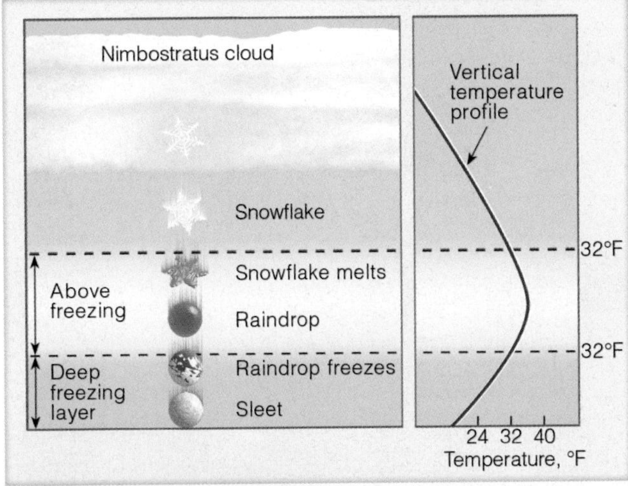

FIGURE 6.16 Sleet forms when a partially melted snowflake or a cold raindrop freezes into a pellet of ice before reaching the ground.

SLEET, FREEZING RAIN, AND ICE STORMS Consider the falling snowflake in Fig. 6.16. As it falls into warmer air, it begins to melt. When it falls through the deep subfreezing surface layer of air, the partially melted snowflake or cold raindrop turns back into ice, not as a snowflake, but as a tiny transparent (or translucent) *ice pellet* called **sleet**.* Generally, these ice pellets bounce when striking the ground and produce a tapping sound when they hit a window or piece of metal.

The cold surface layer beneath a cloud may be too shallow to freeze raindrops as they fall. In this case, they reach the surface as supercooled liquid drops. Upon striking a cold object, the drops spread out and almost immediately freeze, forming a thin veneer of ice. This form of precipitation is called **freezing rain**, or *glaze*. If the drops are quite small, the precipitation is called *freezing drizzle*. When small, supercooled cloud or fog droplets strike an object whose temperature is below freezing, the tiny droplets freeze, forming an accumulation of white or milky granular ice called **rime** (see Fig. 6.17).

FIGURE 6.17 An accumulation of rime forms on tree branches as supercooled fog droplets freeze on contact in the below-freezing air.

Occasionally, light rain, drizzle, or supercooled fog droplets will come in contact with surfaces, such as bridges and overpasses, that have cooled to a temperature below freezing. The tiny liquid droplets freeze on contact to the road surface or pavement, producing a sheet of ice that often appears relatively dark. Such ice is usually called **black ice**. On any road surface, black ice produces extremely hazardous driving conditions.

Freezing rain can create a beautiful winter wonderland by coating everything with silvery, glistening ice. At the same time, highways turn into skating rinks for automobiles, and the destructive weight of the ice—which can be many tons on a single tree—breaks tree branches, power lines, and telephone cables. Where there is a substantial accumulation of freezing rain or freezing drizzle, these storms are called **ice storms** (see Fig. 6.18).

The area most frequently hit by these storms extends over a broad region from Texas into Minnesota, and eastward into the middle Atlantic states and New England. Also subject to ice storms are low-lying valleys and other regions that allow warm air to ride up and over cold surface air. Ice storms are extremely rare in most of California and Florida (see Fig. 6.19).

Ice storms can be very costly. A case in point is the huge ice storm of January, 1998, which left millions of people without power in northern New England and Canada, some with no power for up to two weeks after

*Occasionally, the news media incorrectly use the term *sleet* to represent a mixture of rain and snow. The term used in this manner is, however, the British meaning.

FIGURE 6.18 A huge ice storm during January, 1998, covered Syracuse, New York, with a heavy coating of freezing rain, causing tree limbs to break and power lines to sag.

the storm. The ice storm covered some communities with up to 4 inches of ice and caused over $1 billion in damage, making this storm Canada's costliest natural disaster. In the worst ice storm to hit Kansas and Missouri in 100 years, two inches of ice covered sections of these states in January, 2002, causing over 300,000 people to be without power. Another ice storm made a mess of the Northeast during December, 2008, when a thick coating of ice covering trees and power lines caused more than one million homes and businesses from Pennsylvania to Maine to lose power. Yet, another potent ice storm during January, 2009, left 700,000 people without power across sections of the Ohio River Valley. The storm caused more than $200 million in damages and caused 55 deaths, 25 in Kentucky alone.

In summary, Fig. 6.20 shows various winter temperature profiles and the type of precipitation associated with each. In profile (a), the air temperature is below freezing at all levels, and snowflakes reach the surface. In (b), a zone of above-freezing air causes snowflakes to partially melt; then, in the deep, subfreezing air at the surface, the liquid freezes into sleet. In the shallow subfreezing surface air in (c), the melted snowflakes, now supercooled liquid drops, freeze on contact, producing freezing rain. In (d), the air temperature is above freezing in a sufficiently deep layer so that precipitation reaches the surface as rain. (Before going on to the section on hail, you may wish to read the Focus section on p. 160 that describes the hazards of flying into a region of freezing rain.)

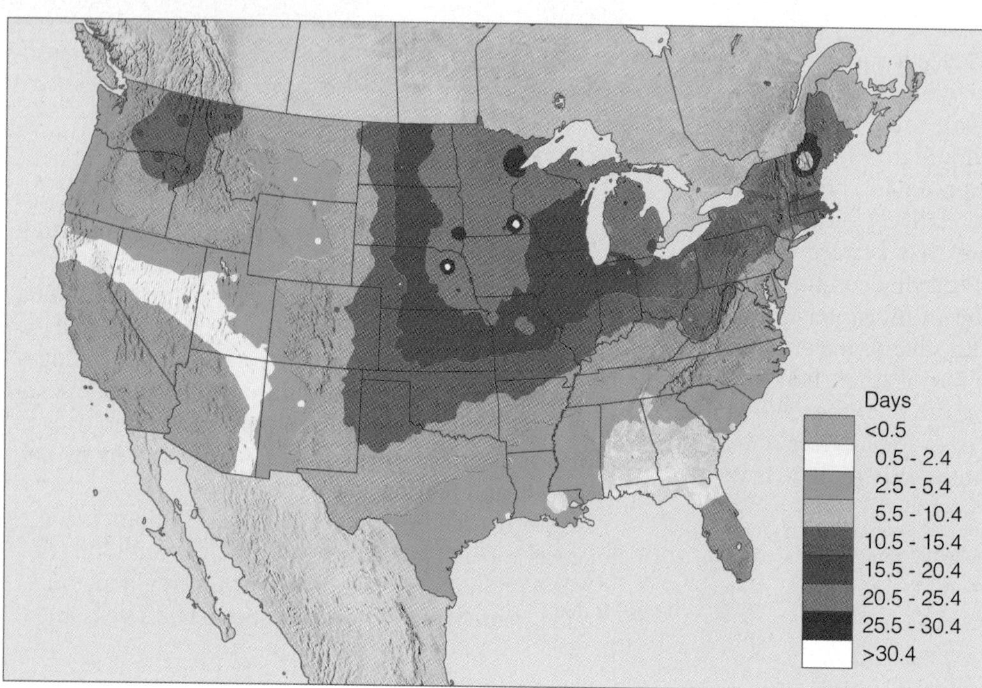

FIGURE 6.19 Average annual number of days with freezing rain and freezing drizzle over the United States. (NOAA)

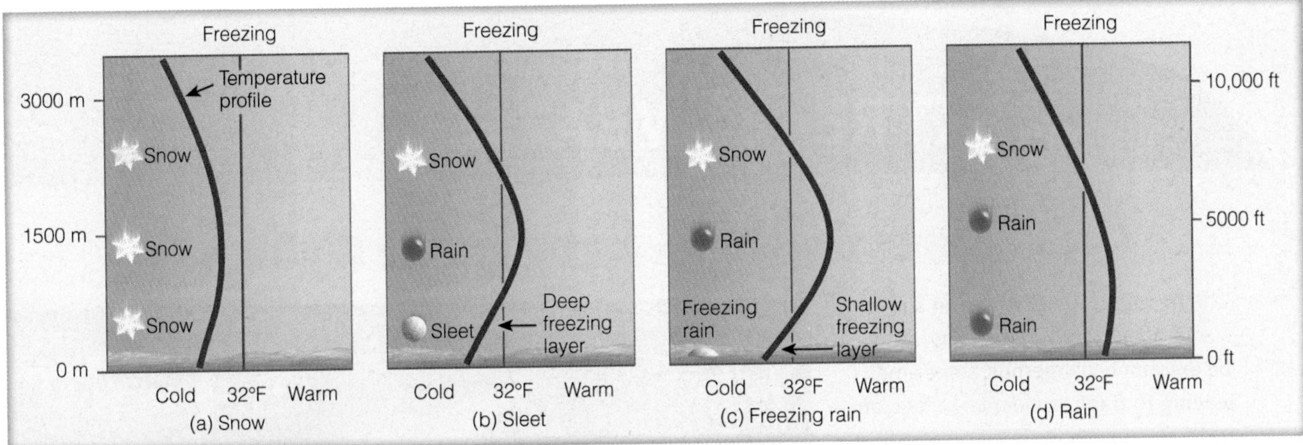

FIGURE 6.20 Vertical temperature profiles (solid red line) associated with different forms of precipitation.

HAIL **Hailstones** are pieces of ice, either transparent or partially opaque, ranging in size from that of small peas to that of golf balls or larger (see Fig. 6.21). Some are round; others take on irregular shapes. The largest authenticated hailstone in the United States fell on Aurora, Nebraska, during June, 2003. This giant hailstone had a measured diameter of 7 inches and a circumference of 18.7 inches (see Fig. 6.22). Although an accurate weight was difficult to obtain, the hailstone (being almost as large as a soccer ball) probably weighed over 1.75 lbs. Canada's record hailstone fell on Cedoux, Saskatchewan, during August, 1973. It weighed about 0.6 pounds and measured about 4 inches in diameter. Needless to say, large hailstones are quite destructive. They can break windows, dent cars, batter roofs of homes, and cause extensive damage to livestock and crops. In fact, a single hailstorm can destroy a farmer's crop in a matter of minutes, which is why farmers sometimes call it "the white plague."

Estimates are that, in the United States alone, hail damage amounts to hundreds of millions of dollars annually. The costliest hailstorm on record in the United States battered the Front Range of the Rocky Mountains in Colorado with golf ball- and baseball-size hail on July 11, 1990. The storm damaged thousands of roofs and tens of thousands of cars, trucks, and streetlights, causing an estimated $625 million in damage. Although hailstones are potentially lethal, only two fatalities due to

FIGURE 6.21 The accumulation of small hail after a thunderstorm. The hail formed as supercooled cloud droplets collected on ice particles called *graupel* inside a cumulonimbus cloud.

FIGURE 6.22 This giant hailstone—the largest ever reported in the United States with a diameter of 7 in. (17.8 cm)—fell on Aurora, Nebraska, during June, 2003.

FOCUS ON
EXTREME WEATHER

Aircraft Icing—A Hazard to Flying

The formation of ice on an aircraft—called *aircraft icing*—can be extremely dangerous, sometimes leading to tragic accidents. In fact, aircraft icing may have been responsible for the downing of a commuter plane that plummeted from the sky onto a house near Niagara International Airport at Buffalo, New York, in February, 2009—killing all 49 on board and one person on the ground.

© Annebique Bernard/CORBIS SYGMA

Figure 1 An aircraft undergoing de-icing during inclement winter weather.

How does aircraft icing form? Consider an aircraft flying through an area of freezing rain or through a region of large supercooled droplets in a cumuliform cloud. As the large, supercooled drops strike the leading edge of the wing, they break apart and form a film of water, which quickly freezes into a solid sheet of ice. This smooth, transparent ice—called *clear ice*—is similar to the freezing rain or glaze that coats trees during ice storms. Clear ice can build up quickly; it is heavy and difficult to remove, even with modern de-icers.

When an aircraft flies through a cloud composed of tiny, supercooled liquid droplets, *rime ice* may form. Rime ice forms when some of the cloud droplets strike the wing and freeze before they have time to spread, thus leaving a rough and brittle coating of ice on the wing. Because the small, frozen droplets trap air between them, rime ice usually appears white (see Fig. 6.17, p. 157). Even though rime ice redistributes the flow of air over the wing more than clear ice does, it is lighter in weight and is more easily removed with de-icers.

Because the raindrops and cloud droplets in most clouds vary in size, a mixture of clear and rime ice usually forms on aircraft. Also, because concentrations of liquid water tend to be greatest in warm air, icing is usually heaviest and most severe when the air temperature is between 32°F and 14°F (0°C and −10°C).

A major hazard to aviation, icing reduces aircraft efficiency by increasing weight. Icing has other adverse effects, depending on where it forms. On a wing or fuselage, ice can disrupt the air flow and decrease the plane's flying capability. When ice forms in the air intake of the engine, it robs the engine of air, causing a reduction in power. Icing may also affect the operation of brakes, landing gear, and instruments. Because of the hazards of ice on an aircraft, its wings are usually sprayed with a type of antifreeze before taking off during cold, inclement weather (see Fig. 1).

falling hail have been documented in the United States during the last 100 years or so. However, hundreds have died in hailstorms in India, Bangladesh, and China. In fact, in the deadliest hailstorm on record, 246 people died in Moradabad District, India, on April 30, 1888.

Hail is produced in a cumulonimbus cloud—usually an intense thunderstorm—when graupel (snow pellets) or large frozen raindrops, or just about any particles (even insects) act as *embryos* that grow by accumulating supercooled liquid droplets—*accretion*. It takes a million cloud droplets to form a single raindrop, but it takes about 10 billion cloud droplets to form a golf ball–size hailstone. For a hailstone to grow to this size, it must remain in the cloud between 5 and 10 minutes. Violent, upsurging air currents within the storm carry small ice particles high above the freezing level where the ice particles grow by colliding with supercooled liquid cloud droplets. Violent rotating updrafts in severe thunderstorms are even capable of sweeping the growing ice particles laterally through the cloud. In fact, it appears that the best trajectory for hailstone growth is one that is nearly horizontal through the storm (see ❱ Fig. 6.23).

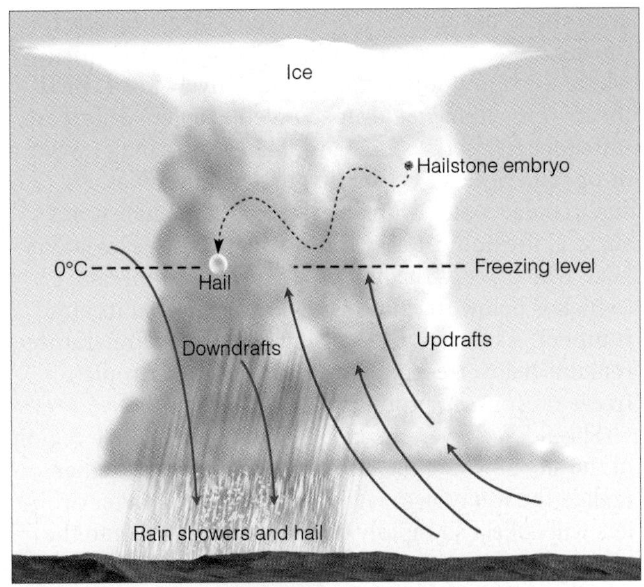

❱ FIGURE 6.23 Hailstones begin as embryos (usually ice particles called *graupel*) that remain suspended in the cloud by violent updrafts. When the updrafts are tilted, the ice particles are swept horizontally through the cloud, producing the optimal trajectory for hailstone growth. Along their path, the ice particles collide with supercooled liquid droplets, which freeze on contact. The ice particles eventually grow large enough and heavy enough to fall toward the ground as hailstones.

As growing ice particles pass through regions of varying liquid water content, a coating of ice forms around them, causing them to grow larger and larger. In a strong updraft, the larger hailstones ascend very slowly, and may appear to "float" in the updraft, where they continue to grow rapidly by colliding with numerous supercooled liquid droplets. When winds aloft carry the large hailstones away from the updraft or when the hailstones reach appreciable size, they become too heavy to be supported by the rising air, and they begin to fall.

In the warmer air below the cloud, the hailstones begin to melt. Small hail often completely melts before reaching the ground, but in the violent thunderstorms of late spring and summer, hailstones often grow large enough to reach the surface before completely melting. Strangely, then, we find the largest form of frozen precipitation occurring during the warmest time of the year.

❱ Figure 6.24 shows a cut section of a very large hailstone. Notice that it has distinct concentric layers of milky white and clear ice. We know that a hailstone

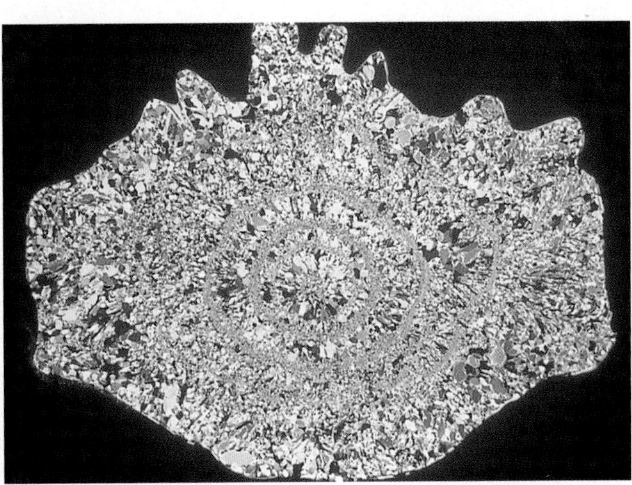

❱ FIGURE 6.24 A large hailstone first cut, then photographed under regular light (left) and polarized light (right). This procedure reveals its layered structure.

grows by accumulating supercooled water droplets. If the growing hailstone enters a region inside the storm where the liquid water content is relatively low (called the *dry growth regime*), supercooled droplets will freeze immediately on the stone, producing a coating of white or opaque rime ice containing many air bubbles. As supercooled water droplets freeze onto the hailstone's surface, the liquid-to-ice transformation releases latent heat, which keeps the hailstone's surface temperature (which is below freezing) warmer than that of its environment. As long as the hailstone's surface temperature remains below freezing, liquid supercooled droplets freeze on contact, producing a coating of rime.

Should, however, the hailstone get swept into a region of the storm where the liquid-water contact is higher (called the *wet growth regime*), supercooled water droplets will collect so rapidly on the stone that, due to the release of latent heat, the stone's surface temperature will remain at 0°C. Now the supercooled droplets no longer freeze on impact; instead, they spread a coating of water around the hailstone, filling in the porous regions. As the water coating the hailstone slowly freezes, air bubbles are able to escape, leaving a layer of clear ice around the stone. Therefore, as a hailstone passes through a thunderstorm of changing liquid-water content (the dry and wet growth regimes) alternating layers of opaque and clear ice form, as illustrated in Fig. 6.24.

As a thunderstorm moves along, it may deposit its hail in a long narrow band (often a mile-and-a-half wide and about 6 miles long) known as a **hailstreak**. If the storm should remain almost stationary for a period of time, substantial accumulation of hail is possible. For example, in June, 1984, a devastating hailstorm lasting over an hour dumped knee-deep hail on the suburbs of Denver, Colorado. In addition to its destructive effect, accumulation of hail on a roadway is a hazard to traffic, as when, for example, four people lost their lives near Soda Springs, California, in a 15-vehicle pile up on a hail-covered freeway during September, 1989. ❱ Figure 6.25 shows that the greatest frequency of hailstorms is over the western Great Plains, a region favorable for the development of hail-producing thunderstorms.

Precipitation—Extreme Events

A generalized picture of global precipitation is shown in ❱ Fig. 6.26.* Notice that certain regions stand out as very wet or very dry. The equatorial regions, for example, are wet, as converging wind belts produce rising air, towering clouds, and heavy precipitation all year long. On the other hand, precipitation is sparse near latitude 30°. Here, sinking air associated with large high-pressure areas produce a "dry belt" around the globe where precipitation tends to be light and highly variable. The Sahara Desert of North Africa is in this region. The polar regions also tend to be dry because, in the cold air, there is little moisture that can be rung out in the form of precipitation.

THE INFLUENCE OF MOUNTAINS Mountain ranges disrupt the generalized pattern of global precipitation (1) by promoting convection (because their slopes are warmer than the surrounding air) and (2) by forcing air

*A more detailed map of global precipitation is in Appendix.

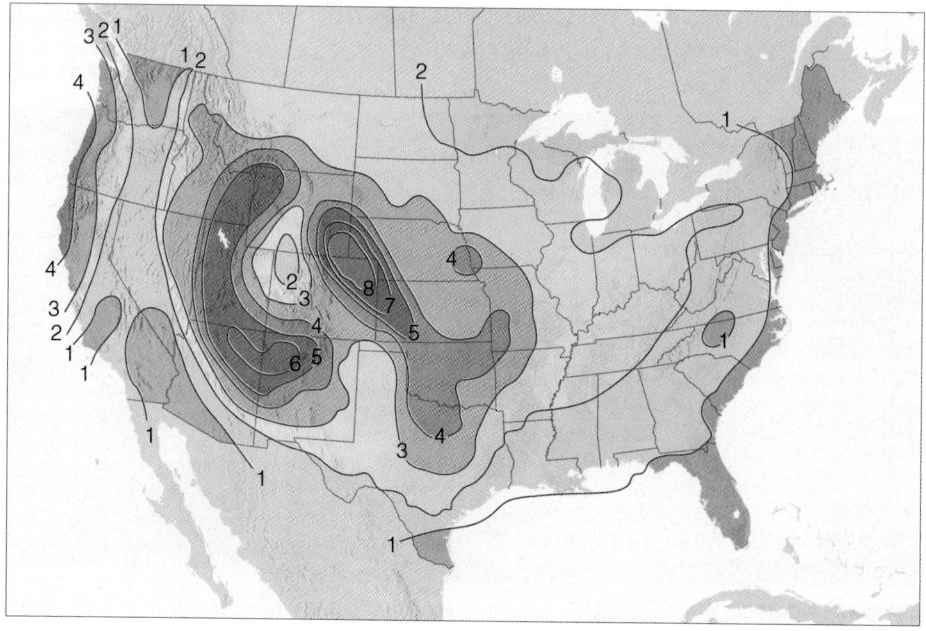

❱ **FIGURE 6.25**

The average number of days each year on which hail is observed throughout the United States from 1950 to 1975. (NOAA)

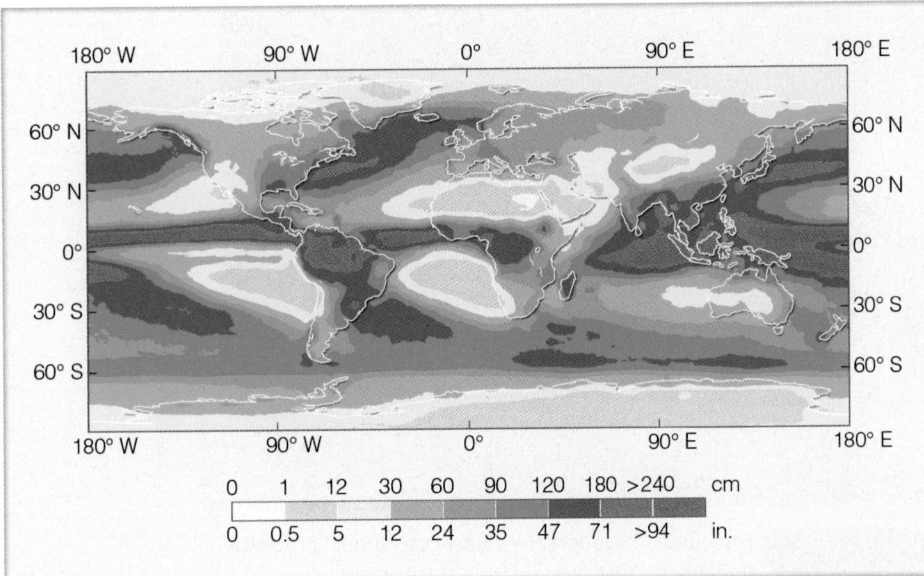

FIGURE 6.26

A generalized picture of annual average global precipitation.

to rise along their windward slopes (*orographic uplift*). Consequently, the windward side of mountains tends to be "wet." As air descends and warms along the leeward side, there is less likelihood of clouds and precipitation. Thus, the leeward (downwind) side of mountains tends to be "dry." As Chapter 5 points out, a region on the leeward side of a mountain where precipitation is noticeably less is called a *rain shadow*.

A good example of the rain shadow effect occurs in the northwestern part of Washington State. Situated on the western side at the base of the Olympic Mountains, the Hoh River Valley annually receives an average 150 inches of precipitation (see Fig. 6.27). On the eastern (leeward) side of this range, only about 60 miles from the Hoh rain forest, the mean annual precipitation is less than 18 inches, and irrigation is necessary to grow certain crops. Figure 6.28 shows a classic example of how topography produces several rain shadow effects.

WET REGIONS, DRY REGIONS, AND PRECIPITATION RECORDS Most of the "rainiest" places in the world are located on the windward side of mountains. For example, Mount Waialeale on the island of Kauai, Hawaii, has the greatest annual average rainfall in the United States: 460 inches (see Fig. 6.29). Mawsynram, on the crest of the southern slopes of the Khasi Hills in northeastern India, is considered the wettest place in the world as it receives an average of 467 inches of rainfall each year, the majority of which falls during the summer monsoon, between April and October. Cherrapunji, which is only about 10 miles from Mawsynram, holds the greatest twelve-month rainfall total of 1,042 inches, and once received 150 inches of rain in just five days.

Record rainfall amounts are often associated with tropical storms. On the island of La Reunion (about 400 miles east of Madagascar in the Indian Ocean), a tropical cyclone dumped 53 inches of rain on Belouve in twelve hours. Heavy rains of short duration often occur with severe thunderstorms that move slowly or stall over a region. On July 4, 1956, 1.2 inches of rain fell from a thunderstorm on Unionville, Maryland, in one minute.

The wettest areas in the United States occur in Hawaii, along the south coast of Alaska, along the coastal Pacific Northwest, and in the southeastern states. The "raininess" of a city can be a measure of its total precipitation over a year or the number of days each year it experiences measurable rain. Table 6.2a (p. 165) shows cities in the United States with the greatest annual average

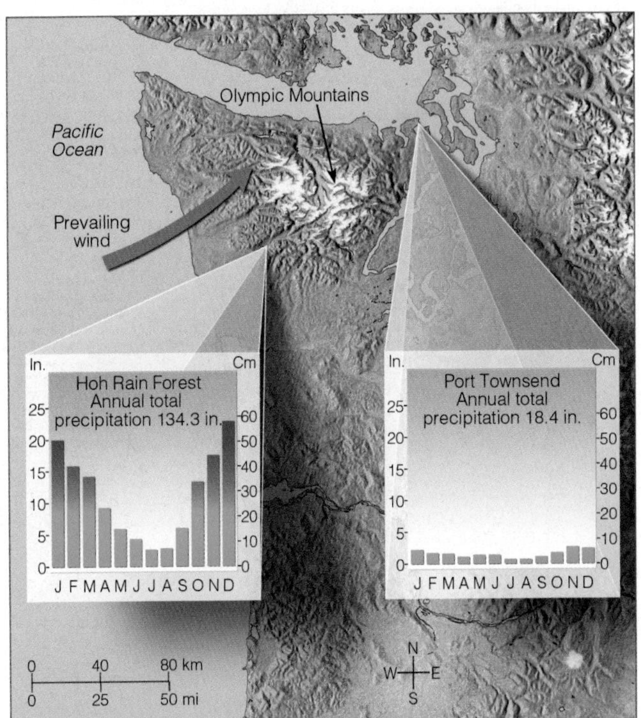

FIGURE 6.27 The effect of the Olympic Mountains in Washington State on average annual precipitation.

WEST
EAST
Sierra Nevada
Coast Range mountains
Rain shadow desert
Santa Cruz
San Jose
Mt. Hamilton
Los Banos
Merced
Mariposa
Yosemite
Ranger Station
Bishop
Tonopah, Nevada
Precipitation (in.)
70
50
30
10
0
Precipitation (cm)
175
125
75
25
0

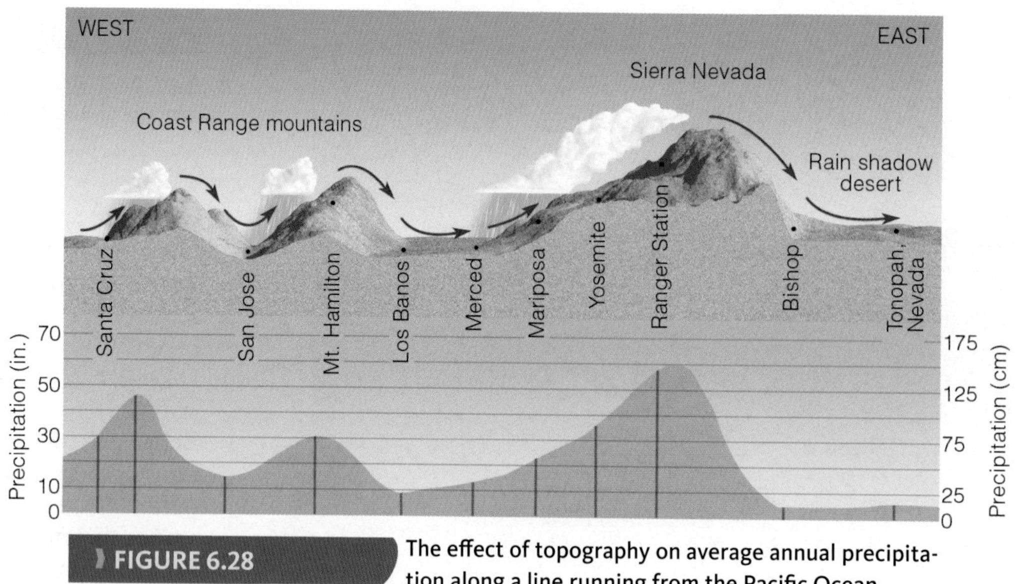

FIGURE 6.28 The effect of topography on average annual precipitation along a line running from the Pacific Ocean through Central California into western Nevada.

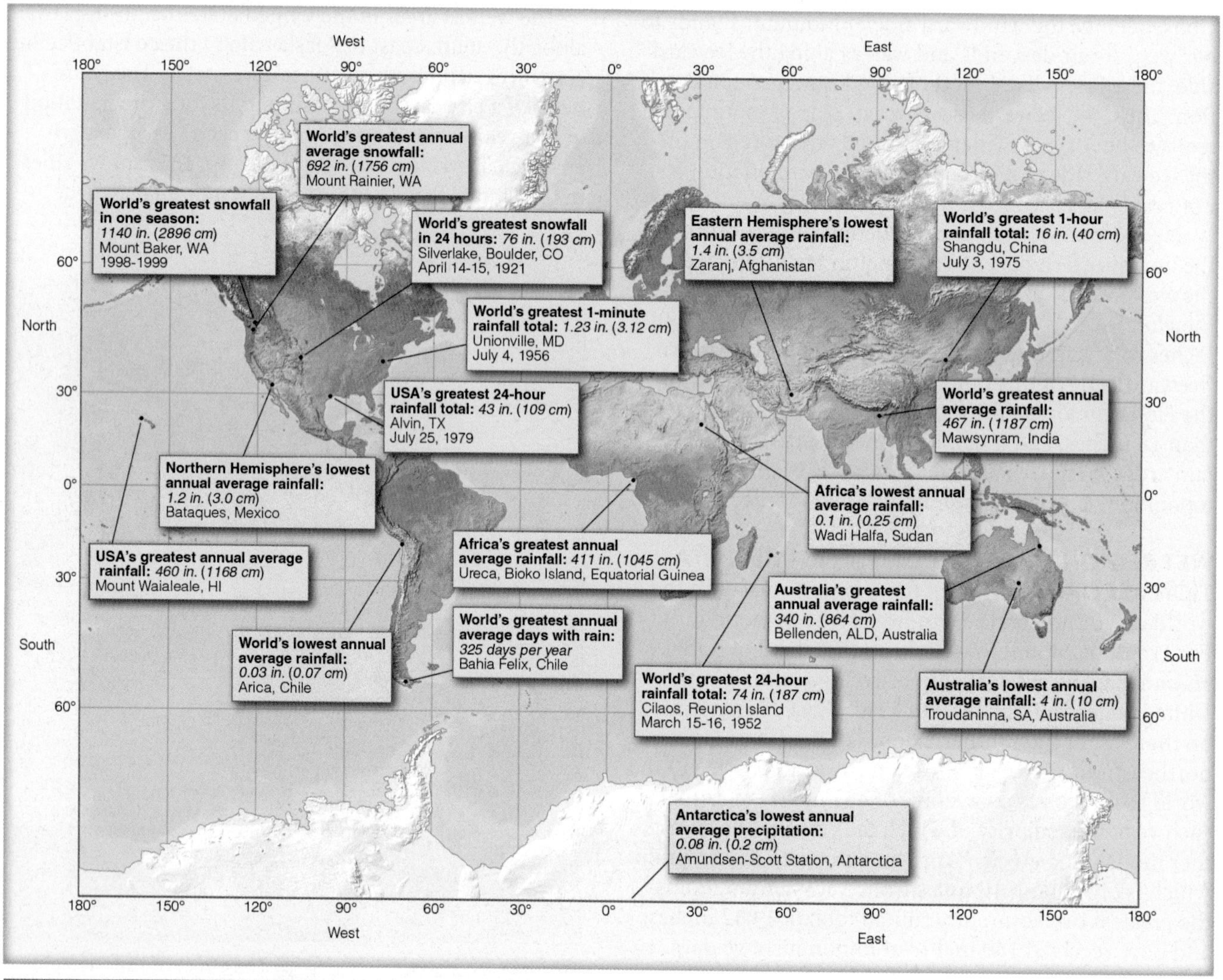

West
East
North
South
180° 150° 120° 90° 60° 30° 0° 30° 60° 90° 120° 150° 180°
60° 30° 0° 30° 60°
World's greatest annual average snowfall: 692 in. (1756 cm) Mount Rainier, WA
World's greatest snowfall in one season: 1140 in. (2896 cm) Mount Baker, WA 1998-1999
World's greatest snowfall in 24 hours: 76 in. (193 cm) Silverlake, Boulder, CO April 14-15, 1921
Eastern Hemisphere's lowest annual average rainfall: 1.4 in. (3.5 cm) Zaranj, Afghanistan
World's greatest 1-hour rainfall total: 16 in. (40 cm) Shangdu, China July 3, 1975
World's greatest 1-minute rainfall total: 1.23 in. (3.12 cm) Unionville, MD July 4, 1956
USA's greatest 24-hour rainfall total: 43 in. (109 cm) Alvin, TX July 25, 1979
World's greatest annual average rainfall: 467 in. (1187 cm) Mawsynram, India
Northern Hemisphere's lowest annual average rainfall: 1.2 in. (3.0 cm) Bataques, Mexico
Africa's lowest annual average rainfall: 0.1 in. (0.25 cm) Wadi Halfa, Sudan
Africa's greatest annual average rainfall: 411 in. (1045 cm) Ureca, Bioko Island, Equatorial Guinea
USA's greatest annual average rainfall: 460 in. (1168 cm) Mount Waialeale, HI
Australia's greatest annual average rainfall: 340 in. (864 cm) Bellenden, ALD, Australia
World's greatest annual average days with rain: 325 days per year Bahia Felix, Chile
World's lowest annual average rainfall: 0.03 in. (0.07 cm) Arica, Chile
World's greatest 24-hour rainfall total: 74 in. (187 cm) Cilaos, Reunion Island March 15-16, 1952
Australia's lowest annual average rainfall: 4 in. (10 cm) Troudaninna, SA, Australia
Antarctica's lowest annual average precipitation: 0.08 in. (0.2 cm) Amundsen-Scott Station, Antarctica

FIGURE 6.29 Some precipitation records throughout the world.

precipitation, and Table 6.2b shows those cities with the greatest number of days with measurable precipitation (0.01 inches or greater). Notice that Seattle, Washington, which is often considered a "rainy city" has many rainy days each year (about 158), but its annual average total precipitation is only 38 inches. This situation tells us that most of Seattle's precipitation falls in light amounts, with relatively few rainy days exceeding one inch. New Orleans, Louisiana, on the other hand, experiences far fewer rainy days than Seattle (about 114), but on about 23 of these days it rains a total of one inch or more. Hence, New Orleans with an annual average precipitation total of about 64 inches, experiences many heavy rainfall days and more extreme rain events than Seattle. Consequently, it is far more likely to experience a deluge in New Orleans than in Seattle.

Snowfalls tend to be heavier where cool, moist air rises along the windward slopes of mountains. One of the snowiest places in North America is located at the Paradise Ranger Station in Mt. Rainier National Park, Washington. Situated at an elevation of 5,400 ft above sea level, this station receives an average 692 inches of snow annually, and holds the world's record annual snowfall total of 1,224 inches, which fell between February, 1971, and February, 1972. A record seasonal snowfall total of 1,140 inches fell at Mt. Baker Ski Area, Washington, during the winter of 1998–1999. The greatest snowfall during a 24-hour period (76 inches) occurred at Silverlake near Boulder, Colorado, on April 14–15, 1921. But this record may have been surpassed during January, 1997, in Montague, New York, when 77 inches of snow fell during a 24-hour period. (More information on this amazing snowfall event, and why it may not be a world record, is given in the Focus section on p. 166.)

As we noted earlier, the driest regions of the world lie in the frigid polar region, the leeward side of mountains, and in the belt of subtropical high pressure, between 15° and 30° latitude. Arica in northern Chile holds the world record for lowest annual rainfall—0.03 inches. In the United States, the driest regions are found in the desert southwest, the southern San Joaquin Valley of California, and Death Valley in Southern California, which averages only 1.78 inches of precipitation annually. Table 6.3 shows the driest cities in the United States.

EXTREME WEATHER WATCH

The deepest accumulation of snow measured in the world was the 465.4 inches (almost 39 ft) recorded at the 5000-foot level of Mt. Ibuki in the Japanese Alps of Honshu Island on February 14, 1927. Cold Siberian air blowing over the Sea of Japan creates tremendous snowfalls on the windward slopes of these mountains.

TOO MUCH RAIN—FLOODS AND FLASH FLOODS

Flood waters that rise rapidly with little or no warning are called **flash floods**. Such flooding often results when thunderstorms stall or move very slowly, causing heavy rainfall over a relatively small area. During June, 2006, flash flooding (and overall flooding) occurred over

Table 6.2

The Wettest Cities in the United States Based on (a) Annual Average Precipitation, and (b) Average Number of Days Each Year with Measurable Precipitation

(a) ANNUAL AVERAGE PRECIPITATION (INCHES)		(b) AVERAGE NUMBER OF DAYS EACH YEAR WITH PRECIPITATION	
City	Precipitation (in.)	City	Days
Yakutat, AK	159.7	Hilo, HI	277
Hilo, HI	126.1	Yakutat, AK	235
Quillayute, WA	102.0	Juneau, AK	223
Sitka, AK	85.9	Quillayute, WA	209
Aberdeen, WA	83.7	Astoria, OR	196
Newport, OR	67.8	Marquette, MI	175
Astoria, OR	67.1	Buffalo, NY	169
Mobile, AL	66.3	Elkins, WV	169
Miami (Hialeah), FL	66.0	Syracuse, NY	169
Biloxi, MS	65.8	Sault St. Marie, MI	165
Baton Rouge, LA	65.1	Olympia, WA	164
North Bend, OR	64.4	Binghamton, NY	162
Pensacola, FL	64.3	Erie, PA	162
New Orleans, LA	64.2	Caribou, ME	160
Fort Lauderdale, FL	64.2	Youngstown, OH	160
		Seattle, WA	158
		Rochester, NY	157
		Cleveland, OH	155

FOCUS ON
EXTREME WEATHER

The Almost Record Snowfall at Montague, New York

In Chapter 5, on p. 136, we learned that areas downwind of the Great Lakes are subject to heavy lake-effect snows. Within this snowbelt region, the areas that receive the heaviest accumulation of snow, averaging more than 200 inches annually are: (1) the Keweenaw Peninsula, a thumb of land that juts into Lake Superior on Michigan's Upper Peninsula, and (2) the upstate New York counties of Oswego, Jefferson, and Lewis—all of which are downwind of Lake Ontario. Both regions have hills that provide uplift necessary to squeeze moisture from the air that has moved over long stretches of lake water.

Figure 2 During the winter, air from off Lake Ontario lifts along the Tug Hill Plateau and forms into clouds, which dump heavy lake-effect snows over the region.

One of the heaviest snowfall events ever during a 24-hour period took place near the township of Montague in Lewis County, New York, on January 11–12, 1997, when 77 inches of lake-effect snow was measured by a National Weather Service snow spotter on the Tug Hill Plateau, about 1600 feet above Lake Ontario (see Fig. 2). During this remarkable snowstorm, the snow fell so fast that snowplow operators claimed the visibility was limited to the distance from the driver to the front of the plow (about 10 feet). They also stated that at the height of the storm, snow was accumulating on the road at an astonishing rate of about one foot per hour! Although 77 inches of snow reportedly fell in just 24 hours, the snowstorm lasted three days with a total reported accumulation of 95 inches—almost 8 feet (see Fig. 3).

Figure 3 Heavy lake-effect snow falling in Montague, New York, on January 12, 1997. The total 24-hour snow accumulation from this storm was 77 inches—a record, but not an "official" one.

The 77-inch snowfall was quickly celebrated as a new world record. (The old record of 76 inches occurred at Silver Lake, Colorado, during April, 1921.) However, under close scrutiny by a review committee of the National Weather Service, it was found that the snow spotter had measured the snow six times during the 24-hour period. For the snowfall to be considered as an "officially accepted record," no more than *four* measurements can be made during a 24-hour period (no more than one every six hours), and so the world-record claim was ultimately dismissed. Record or no record, this whopper of a snowstorm during January, 1997, remains memorable.

parts of New England and the mid-Atlantic states when tropical moist air developed into intense thunderstorms that produced heavy rain and extensive flooding that damaged thousands of homes. Flooding may also occur when thunderstorms move quickly, but keep passing over the same area, a phenomenon called **training**. (Like railroad cars, one after another, passing over the same tracks.) During late winter or spring, large chunks of ice from a frozen river may accumulate in the river's channel. The ice forms a dam that produces local flooding called **ice jam flooding**. In recent years, flash floods in the United States have claimed on average more than 100 lives a year, and have accounted for untold property and crop damage.

In some areas, flooding occurs primarily in the spring when heavy rain and melting snow cause rivers to overflow their banks. During March, 1997, heavy downpours over the Ohio River Valley caused extensive flooding that forced thousands from their homes along rivers and smaller streams in Ohio, Kentucky, Tennessee, and West Virginia. One month later, heavy rain coupled with melting snow caused the Red River to overflow its banks inundating 90 percent of the city of Grand Forks, North Dakota. Flooding also occurs with tropical storms that deposit torrential rains over an extensive area. This situation occurred during June, 2001, when tropical storm Allison dumped more than 20 inches of rain over Houston, Texas, within a twelve-hour period, submerging a vast part of the city. In six days, the storm deposited a staggering 37 inches (over 3 feet) of rain on the Port of Houston.

During the summer of 1993, thunderstorm after thunderstorm rumbled across the upper Midwest, causing the worst flood ever in that part of the United States. What began as a wetter than normal winter and spring for most of the upper Midwest turned into "The Great Flood of 1993" by the end of July. In mid-June, thunderstorms began to form almost daily along a persistent frontal boundary that stretched across the upper Midwest.

Fed by warm, humid air from the Gulf of Mexico, thunderstorms almost daily rolled through an area that stretched eastward from Nebraska and South Dakota into Minnesota and Wisconsin, and southward into Iowa, Illinois, and Missouri (see Fig. 6.30). Torrential rains from these storms quickly saturated the soil, and soon runoff began to raise the water level in creeks and rivers. By the end of June, communities in the northern regions of the Mississippi River Valley were experiencing flooding.

As the thunderstorms continued into July, city after city was claimed by the rising waters (see Fig. 6.31). Between April and July, many areas had received twice their normal rainfall, and rivers continued to crest well

Table 6.3

The Driest Cities in the United States Based on (a) Annual Average Precipitation and (b) Average Number of Days Each Year with Measurable Precipitation

(a) ANNUAL AVERAGE PRECIPITATION (INCHES)		(b) AVERAGE NUMBER OF DAYS EACH YEAR WITH PRECIPITATION	
City	Precipitation (in.)	City	Days
Yuma, AZ	3.0	Yuma, AZ	17
Las Vegas, NV	4.5	Palm Springs, CA	17
Barrow, AK	4.6	Las Vegas, NV	26
Wendover, UT	4.8	Bishop, CA	29
Bishop, CA	5.0	Los Angeles, CA	35
Palm Springs, CA	5.1	Phoenix, AZ	36
Bakersfield, CA	6.5	Santa Barbara, CA	36
Reno, NV	7.5	Bakersfield, CA	37
Winslow, AZ	8.0	San Diego, CA	43
Phoenix, AZ	8.3	Wendover, UT	44
Yakima, WA	8.3	Fresno, CA	45
Winnemucca, NV	8.3	El Paso, TX	46
El Paso, TX	9.4	Reno, NV	51
Albuquerque, NM	9.5	Tucson, AZ	53
San Diego, CA	10.8	Albuquerque, NM	59

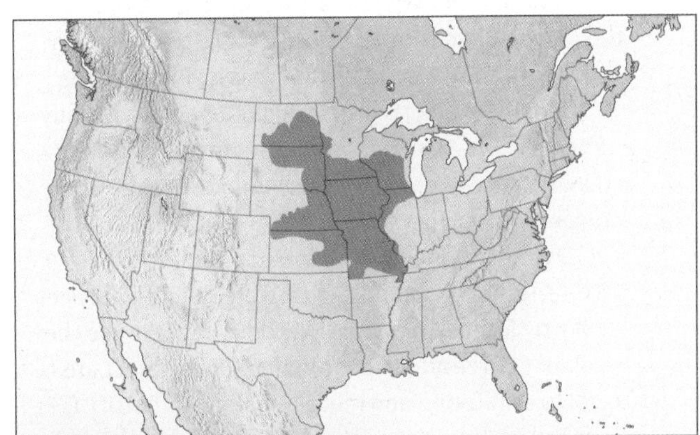

FIGURE 6.30 The blue-shaded area represents counties that experienced flooding during "The Great Flood of 1993." Many of these counties were officially declared disaster areas.

above flood stage through July. By the time the water began to recede in August, more than 60 percent of the levees along the Mississippi River had been destroyed, and an area larger than the state of Texas had been covered with water (the blue-shaded area in Fig. 6.30). Estimates are that $6.5 billion in crops was lost, as millions of acres of valuable farmland were inundated by flood waters. The worst flooding this area had ever seen took 45 human lives, damaged or destroyed 45,000 homes, and forced the evacuation of 74,000 people. And this great flood had an impact on both the living and the dead, as the Hardin Cemetery in Missouri had more than 700 graves opened. Some caskets were swept away by raging flood waters and deposited many miles downstream, and some were never found.

Before turning our attention to drought and its impact on humanity, you may wish to read the Focus section below that explains how Doppler radar measures precipitation.

FOCUS ON INSTRUMENTS

Doppler Radar and Precipitation

Radar (radio detection and ranging) has become an essential tool of the atmospheric scientist, for it gathers information about storms and precipitation in previously inaccessible regions. Atmospheric scientists use radar to examine the inside of a cloud much like physicians use X-rays to examine the inside of a human body. Essentially, the radar unit consists of a transmitter that sends out short, powerful microwave pulses. When this energy encounters a foreign object—called a *target*—a fraction of the energy is scattered back toward the transmitter and is detected by a receiver (see Fig. 4). The returning signal is amplified and displayed on a screen, producing an image or "echo" from the target. The elapsed time between transmission and reception indicates the target's distance.

The brightness of the echo is directly related to the amount (intensity) of rain falling in the cloud. So, the radar screen shows not only where precipitation is occurring, but also how intense it is. Typically the radar image is displayed using various colors, usually ranging from green or blue to dark red, to denote the intensity of precipitation within the range of the radar unit (see Fig. 5).

During the 1990s, *Doppler radar* replaced the conventional radar units that were put into service shortly after World War II. Doppler radar is like the older conventional radar in that it can detect areas of precipitation and measure rainfall intensity (see Fig. 6a). Using special computer programs called *algorithms*, the rainfall intensity, over a given area for a given time, can be computed and displayed as an estimate of total rainfall over that particular area (see Fig. 6b). But the Doppler radar can do more than conventional radar.

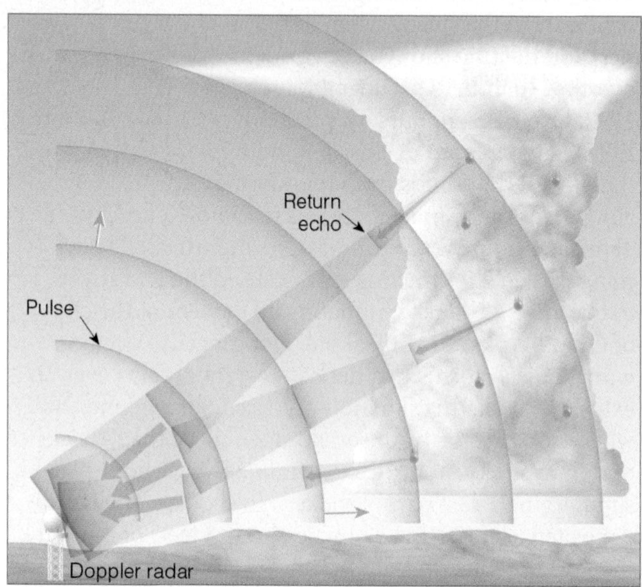

Figure 4 A microwave pulse is sent out from the radar transmitter. The pulse strikes raindrops and a fraction of its energy is reflected back to the radar unit, where it is detected and displayed, as shown in Fig. 5.

Because the Doppler radar uses the principle called *Doppler shift*,* it has the capacity to measure the speed at which

*The *Doppler shift* (or effect) is the change in the frequency of waves that occurs when the emitter or the observer is moving toward or away from the other. As an example, suppose a high-speed train is approaching you. The higher-pitched (higher frequency) whistle you hear as the train approaches will shift to a lower pitch (lower frequency) after the train passes.

WHEN IT DOESN'T RAIN—DROUGHT AND THE PALMER INDEX When a region's average precipitation drops dramatically for an extended period of time, drought may result. The word **drought** refers to a period of abnormally dry weather that produces a number of negative consequences, such as crop damage or an adverse impact on a community's water supply. Prolonged drought, especially when accompanied by high temperatures, can lead to a shortage of food and, in some places, widespread starvation. While many extreme or unusual weather events are short-lived, drought is a more gradual phenomenon, slowly taking hold of an area and tightening its grip with time. In severe cases, drought can last for many years and have devastating effects on people and animals.

Keep in mind that drought is more than a dry spell. For example, during the summer in the Central Valley of California it may not rain from May through September. This dry spell is normal for this region and, therefore,

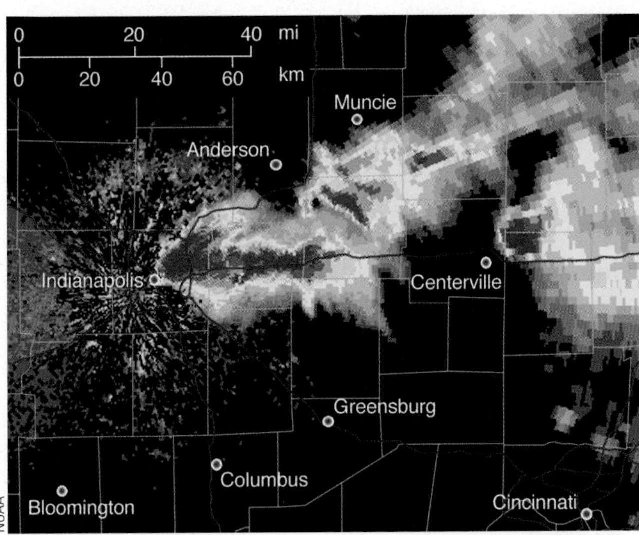

Figure 5 Doppler radar image showing rainfall intensity over Indianapolis, Indiana, on April 14, 2006. The areas shaded green indicate where light-to-moderate rain is falling. Yellow indicates heavier rainfall. The red-shaded areas represent the heaviest rainfall and the presence of severe thunderstorms.

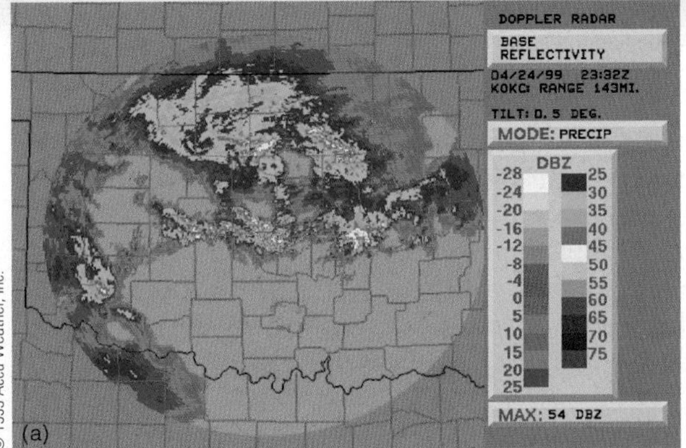

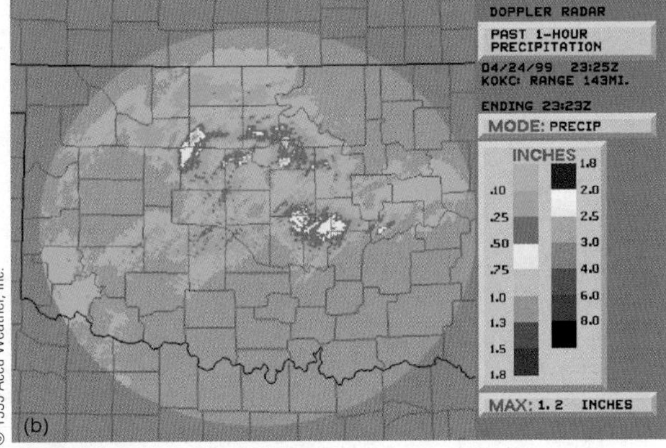

Figure 6 (a) Doppler radar display showing precipitation intensity over Oklahoma for April 24, 1999. The lightest precipitation is shown as blue and green; heavier rainfall is indicated by the color yellow. The numbers under the letters DBZ represent the logarithmic scale for measuring the size and volume of precipitation particles. (b) Doppler radar display showing 1-hour rainfall amounts over Oklahoma for April 24, 1999.

falling rain is moving horizontally toward or away from the radar antenna. Falling rain moves with the wind. Consequently, Doppler radar allows scientists to peer into a tornado-generating thunderstorm and observe its wind. We will investigate these ideas further in later chapters when we consider the formation of severe thunderstorms and tornadoes.

FIGURE 6.31 Flooding during the summer of 1993 covered a vast area of the upper Midwest. Here, the state capital of Missouri is surrounded by flood waters from the swollen Missouri River.

would not be considered a drought. However, if this summer dry spell were to occur in a humid region, such as in the southeastern United States, the lack of rain could be disastrous for many aspects of the area, and a drought would ensue.

In an attempt to measure drought severity, Wayne Palmer, a scientist with the National Weather Service, developed the **Palmer Drought Severity Index** (PDSI). The index takes into account average temperature and precipitation values to define drought severity. The index is most effective in assessing long-term drought that lasts several months or more. Drought conditions are indicated by a set of numbers that range from 0 (normal) to −4 (extreme drought). (See Table 6.4.) The index also assesses wet conditions with numbers that range from +2 (unusually moist) to +4 (extremely moist). The *Palmer Hydrological Drought Index* (PHDI) expands the PDSI by taking into account additional water (hydrological) information, such as a region's groundwater reserves and reservoir levels.

Figure 6.32 shows the PHDI across the United States for December, 2007. Notice that several regions are in an extreme drought (dark red shade), including a large

Table 6.4

Palmer Drought Severity Index

VALUE	DROUGHT	VALUE	MOISTURE
-4.0 or less	Extreme	+4.0 or greater	Extremely Moist
-3.0 to -3.9	Severe	+3.0 to +3.9	Very Moist
-2.0 to -2.9	Moderate	+2.0 to +2.9	Unusually Moist
−1.9 to +1.9	Normal	-1.9 to +1.9	Normal

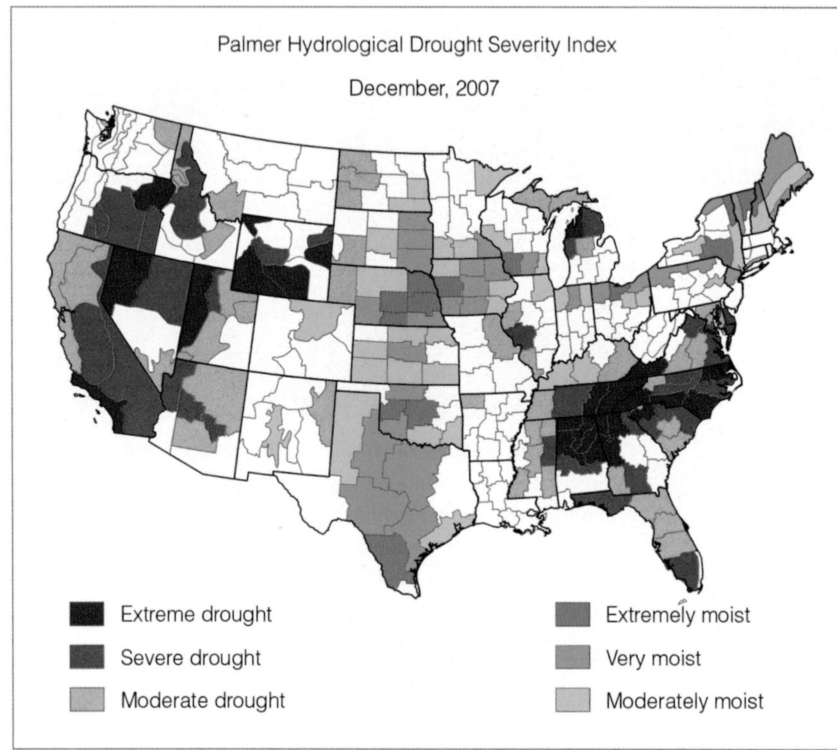

FIGURE 6.32 The Palmer Hydrological Drought Index for December, 2007, showing long-term drought conditions, and regions with sufficient moisture. (National Climatic Data Center, NOAA)

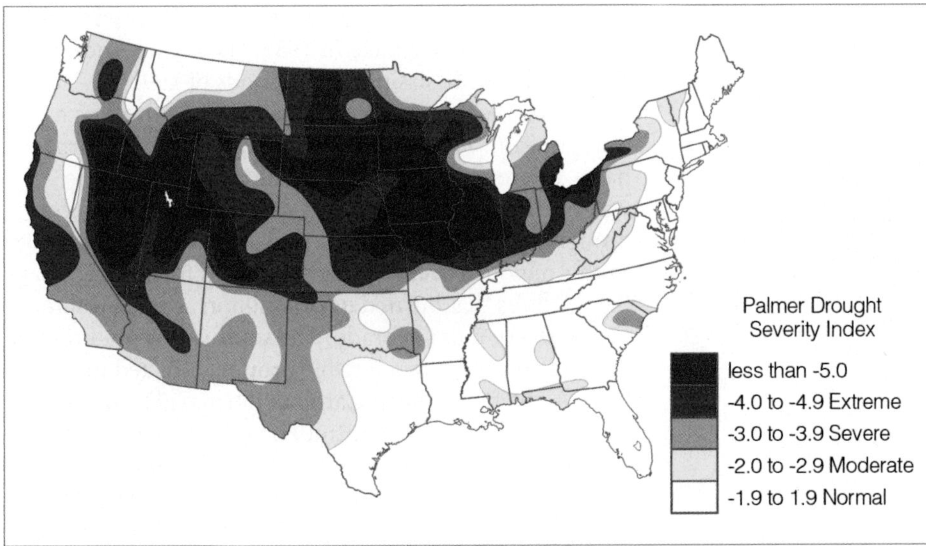

FIGURE 6.33

Drought severity index in the United States on July 1, 1934. (U.S. Department of Agriculture)

portion of the southeastern United States and the western half of the nation. Notice also that there is an extensive area of green from Texas northward into the Dakotas where there is ample moisture.

SOME NOTABLE DROUGHTS Extensive drought has afflicted at some point almost every major continent. In fact, drought has occurred in all areas at one time or another. Drought and its related famine account for more deaths worldwide than any other weather-related disaster. (Look back at Table 1.2, p. 6.) One continent especially hard hit by drought is Africa.

African Drought One region in Africa that has had its share of drought is the Sahel of North Africa. Located between about 14° and 18°N latitude, the Sahel is bounded on the north by the dry Sahara Desert and on the south by the grassland of the Sudan. The Sahel is a semi-arid region of variable rainfall where precipitation totals may exceed 20 inches in the southern portion and less than 5 inches in the north.

During the winter, the Sahel is dry, but as summer approaches, the rains usually begin. The inhabitants of the Sahel are mostly nomadic people who migrate to find grazing land for their cattle and goats. In the late 1960s, the annual rains did not reach as far north as usual, marking the beginning of a series of dry years and severe drought.

The decrease in rainfall, along with over-grazing, turned thousands of square miles of pasture into barren wasteland. By 1973, when the severe drought reached its climax, rainfall totals were 50 percent of the long-term average, and perhaps 50 percent of the cattle and goats had died. The Sahara Desert had migrated southward into the northern fringes of the region, and a great famine had taken the lives of more than 100,000 people. Although low rainfall years have been followed by wetter ones, relatively dry conditions have persisted over the region for the past 40 years or so. The overall dryness of the region has created many problems, including the shrinking of many of the larger lakes, such as Lake Chad. Severe drought has plagued other parts of Africa, including Somalia, Kenya, Ethiopia, and Morocco. In fact, a terrible drought during 1984 and 1988 in the Horn of Africa led to a famine that killed more than 750,000 people.

North American Drought There have been a number of severe droughts in North America, too. In fact, probably the worst weather related disaster to hit the United States during the twentieth century was the great drought of the 1930s. That drought, which tragically coincided with the Great Depression, actually began in the late 1920s and continued into the late 1930s. It not only lasted a long time, but it extended over a vast area (see Fig. 6.33). The drought, coupled with poor farming practices, left the top soil of the Great Plains ripe for wind erosion. As a result, wind storms lifted millions of tons of soil into the air, creating vast dust storms that buried whole farm houses, reducing millions of acres to an unproductive wasteland, and financially ruining thousands of families (see Fig. 6.34). Because of the infamous dust storms, the 1930s are often referred to as "the Dust Bowl years." To worsen an already bad situation, the drought was accompanied by extreme summers heat that was most severe during the summers of 1934 and 1936.

One misconception of this great drought is that "the rains never came." Actually, over most areas it did rain. In fact, some areas experienced above-normal rainfall totals for several months, and in some places for an entire season. But, unfortunately, the wet spells were unable to mitigate the extended dry periods that progressively became more and more severe, and eventually affected the lives of millions of people. Many were left destitute and moved elsewhere, especially to California.

FIGURE 6.34 A truck tries to outrun one of the many devastating dust storms that roared over the Central Plains in the 1930s.

More normal rainfall patterns returned to the Plains in the 1940s, and agriculture began to recover. But severe drought has returned to the region on several occasions—in the 1950s and again in 1988. The worst drought in 50 years affected at least 35 states during the long, hot summer of 1988. Rainfall totals over the Midwest, Northern Plains, and the Rockies were 50 to 85 percent below normal. In some areas, the lack of rainfall dated back to 1984. Crops and livestock died, and the region suffered greatly. Nationwide losses from the drought exceeded $40 billion, exceeding the losses caused by the San Francisco earthquake in 1989, Hurricane Andrew in 1992, and the Mississippi River floods of 1993.

The northeastern United States suffered through a drought from 1963 to 1965, and in the west, California had a severe drought during the mid-1970s, causing water rationing to be imposed statewide. A more recent drought occurred in the west and southwest during 2002. In this region, extremely dry vegetation turned into roaring wildfires that burned over a million acres in California, Arizona, and Colorado. The southeast is not immune to severe drought, as a terrible drought afflicted the area from Alabama to North Carolina during 2007 and 2008.

As global warming continues, will severe droughts become more numerous? Climate models predict that precipitation will decrease over regions of the subtropics, and some models indicate that severe droughts will actually become more prevalent. The decrease in rainfall may already be taking place as recent studies suggest that annual average rainfall totals in the subtropics have been decreasing during the last century. If this situation is true, and the models prove correct, greater stress in the future could be placed on agriculture and on the inhabitants of the region.

Summary

In this chapter, we have seen that cloud droplets are too small and light to reach the ground as rain. Cloud droplets can grow in size as large cloud droplets, falling through clouds, collide and merge with smaller droplets in their path. In clouds where the air temperature is below freezing, ice crystals can grow larger at the expense of the surrounding liquid droplets. As an ice crystal begins to fall, it grows even larger by colliding with supercooled liquid droplets, which freeze on contact. In an attempt to coax more precipitation from them, some clouds are seeded with silver iodide.

Precipitation can reach the surface in a variety of forms. In winter, raindrops may freeze on impact, producing freezing rain that can disrupt electrical service by downing power lines. Raindrops may freeze into tiny pellets of ice above the ground and reach the surface as sleet. Depending on conditions, snow may fall as pellets, grains, or flakes. Strong updrafts in a cumulonimbus cloud may carry ice particles high above the freezing level, where they acquire a further coating of ice and form destructive hailstones. Because the formation of hail requires clouds containing strong convection, hail is more likely to form during the warmer months.

Precipitation amounts can be strongly influenced by topography. Because air rises on a mountain's windward side and descends on its leeward side, rainfall amounts are usually higher on the mountain's windward side. Where precipitation is conspicuously absent on the mountain's leeward side, there is a rain shadow.

Floods occur when heavy rain and sometimes melting snow produce runoff that exceeds the holding capacity of a river or stream. Flash flooding can occur when slow-moving, intense thunderstorms move over an area or when multiple thunderstorms move repeatedly over the same area in a process called training.

When the total precipitation over a region drops dramatically for an extended period of time, drought is possible. The Palmer Drought Severity Index takes into account average temperature and precipitation values to assess drought severity. Almost every continent has at one time or another experienced extensive drought. As global warming continues, severe drought may actually become more prevalent over certain regions of the world.

Key Terms

The following terms are listed (with page number) in the order they appear in the text. Define each. Doing so will aid you in reviewing the material covered in this chapter.

precipitation, 146
collision-coalescence process, 146
coalescence, 146
ice-crystal (Bergeron) process, 147
supercooled (water droplets), 148
cloud seeding, 149
rain, 152
drizzle, 152
virga, 152
shower (rain), 153
snow, 153
flurries (of snow), 153
snow squall, 153
blizzard, 154
snow grains, 154
snow pellets, 154
avalanche, 155
sleet, 157
freezing rain, 157
rime, 157
black ice, 157
ice storm, 157
hailstone, 159
hailstreak, 162
flash flood, 165
training, 167
ice jam flooding, 167
drought, 169
Palmer Drought Severity Index, 170

Review Questions

1. What is the primary difference between a cloud droplet and a raindrop?
2. Describe how the process of collision and coalescence produces precipitation.
3. Would the collision-and-coalescence process work better at producing rain in (a) a warm, thick nimbostratus cloud or (b) a warm, towering cumulus congestus cloud? Explain.
4. How does the ice-crystal (Bergeron) process produce precipitation? What is the main premise behind this process?
5. Inside a cloud, at which of the following temperatures would you expect to see the most supercooled cloud droplet?
   (a) 35°F (2°C)
   (b) 15°F (−9°C)
   (c) −40°F (−40°C)
6. Why do heavy showers usually fall from cumuliform clouds? Why does steady precipitation normally fall from stratiform clouds?
7. Explain the main principle behind cloud seeding.
8. How are clouds naturally seeded?
9. How does rain differ from drizzle?
10. Why is it *never* too cold to snow?
11. Explain how snowflakes in a cloud grow large and heavy enough to fall toward the ground.
12. Describe some of the positive consequence of a snowfall.
13. What is the difference between freezing rain and sleet?
14. How does black ice form?
15. Why are ice storms often responsible for the loss of power to many thousands, and sometimes millions, of people?
16. Describe how icing can be hazardous to aircraft.
17. Describe how hail might form in a cumulonimbus cloud.
18. Why is hail more common in summer than in winter?
19. How do the atmospheric conditions that produce sleet differ from those that produce hail?
20. Where are the wettest and driest regions of the world located?
21. Explain how mountains influence precipitation.
22. Describe how a rain shadow forms.
23. Explain several ways a flash flood might form.
24. What are some of the negative consequences of an extensive drought?
25. The Palmer Drought Severity Index uses what weather elements to assess drought severity?

Online Learning

STUDENT COMPANION WEBSITE: Visit this book's companion website at: www.cengage.com/ahrens/extreme1e and choose Chapter 6 for many study aids and ideas for further reading and research. These include flashcards, practice quizzing, and web links.

METEOROLOGY RESOURCE CENTER: For students with access, log on at www.cengage.com/login for more assets, including animations, videos, and more. If your textbook did not come with access, visit www.CengageBrain.com to purchase.

Atmospheric Motions

7

CONTENTS

HORIZONTAL PRESSURE CHANGES AND WIND
SURFACE AND UPPER-LEVEL CHARTS
WHY THE WIND BLOWS
HOW THE WIND BLOWS
SURFACE WINDS
THE INFLUENCE OF EXTREME WINDS
WINDY PLACES
EXTREME WINDS
SUMMARY
KEY TERMS
REVIEW QUESTIONS

Strong winds blowing over Lake Superior produce large waves and pounding surf along the shoreline of Minnesota.

Many extreme weather events are associated with strong winds. In fact, much of the damage caused by tornadoes and severe thunderstorms is due to brute wind power. Some of the wind damage results from the wind blowing directly against an object. Other damage results from the wind blowing over an object (think of the wind blowing a roof off a building). Wind damage is so intimately related to storms that it is essential that we understand why and how the wind blows.

Although air does move up and down, the majority of atmospheric winds are horizontal, more or less parallel to the earth's surface. From the gentle breeze to the raging tornado, all winds form by the same process: differences in atmospheric pressure. Consequently, in this chapter we will first look at how horizontal changes in atmospheric pressure produce atmospheric circulations. Then we will examine the forces that influence atmospheric motions both aloft and at the surface. Toward the end of the chapter, we will examine a variety of ways the force of the wind exerts its presence over land and water.

Horizontal Pressure Changes and Wind

In Chapter 1, we learned several important concepts about atmospheric pressure. One stated that *air pressure* is simply the mass of air above a given level. As we climb in elevation above the earth's surface, there are fewer air molecules above us; hence, *atmospheric pressure always decreases with increasing height.* Another concept we learned was that most of our atmosphere is crowded close to the earth's surface, which causes *air pressure to decrease with height, rapidly at first, then more slowly at higher altitudes.* So one way to change air pressure is to simply move up or down in the atmosphere. But what causes the air pressure to change in the horizontal? And why does the air pressure change at the surface?

To answer these questions, we eliminate some of the complexities of the atmosphere by constructing *models.* Figure 7.1 shows a simple atmospheric model—a column of air, extending well up into the atmosphere. In the column, the dots represent air molecules. Our model assumes: (1) that the air molecules are the same size and have the same weight; (2) that the air molecules are not crowded close to the surface and, unlike the real atmosphere, the air density remains constant from the surface up to the top of the column; (3) that the width of the column does not change with height; and (4) that the air is unable to freely move into or out of the column.

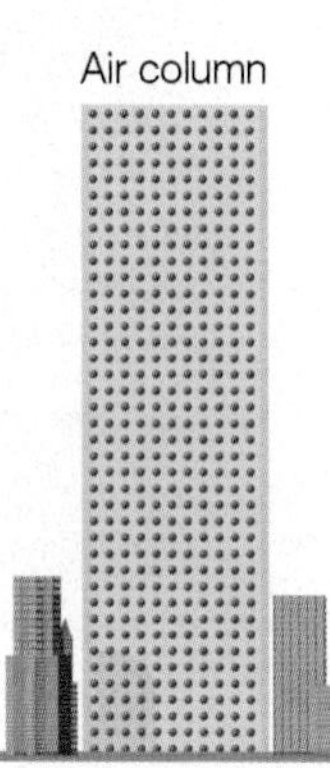

FIGURE 7.1

A model of the atmosphere where air density remains constant with height. The air pressure at the surface is related to the number of molecules above. When air of the same temperature is stuffed into the column, the surface air pressure rises. When air is removed from the column, the surface pressure falls.

Suppose we somehow force more air into the column in Fig. 7.1. What would happen? The added air would make the column more dense, and the added weight of the air in the column would increase the surface air pressure. Likewise, if a great deal of air were removed from the column, the surface air pressure would decrease. Consequently, to change the surface air pressure, we need to change the mass of air (the number of molecules) in the column above the surface. But how can this feat be accomplished?

Look at the air columns in Fig. 7.2a.* Suppose both columns are located at the same elevation, both have the same air temperature, and both have the same surface air pressure. This condition, of course, means that there must be the same number of molecules (same mass of air) in each column above both cities. Further suppose that the surface air pressure for both cities remains the same, while the air above city 1 cools and the air above city 2 warms (see Fig. 7.2b).

As the air in column 1 cools, the molecules move more slowly and crowd closer together—the air becomes more dense. In the warm air above city 2, the molecules move faster and spread farther apart—the air becomes less dense. Since the width of the columns does not change (and if we assume an invisible barrier exists between the columns), the total number of molecules above each city remains the same, and the surface pressure does not change. Therefore, in the more-dense cold air above city 1, the column shrinks, while the column rises in the less-dense, warm air above city 2.

We now have a cold, shorter dense column of air above city 1 and a warm, taller less-dense air column above city 2. From this situation, we can conclude that *it takes a shorter column of cold, more-dense air to exert the same surface pressure as a taller column of warm, less-dense air.* This concept has a great deal of meteorological significance.

Atmospheric pressure decreases more rapidly with height in the cold column of air. In the cold air above city 1 (Fig. 7.2b), move up the column and observe how quickly you pass through the densely packed molecules. This activity indicates a rapid change in pressure. In the

*We will keep our same assumptions as in Fig. 7.1; that is, (1) the air molecules are not crowded close to the surface, (2) the width of the columns does not change, and (3) air is unable to move into or out of the columns.

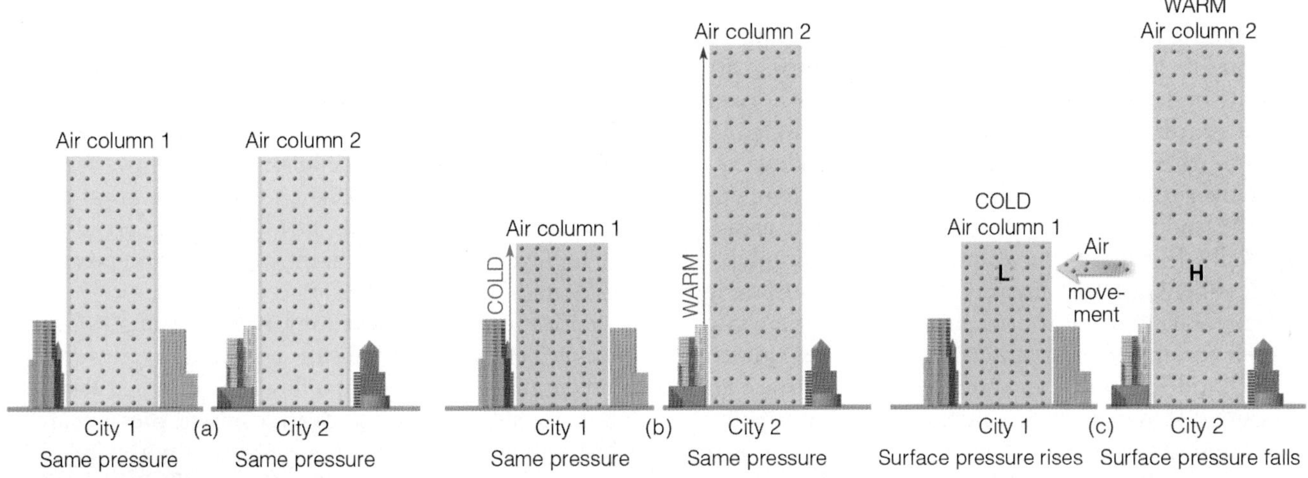

FIGURE 7.2 (a) Two air columns, each with identical mass, have the same surface air pressure. (b) Because it takes a shorter column of cold air to exert the same pressure as a taller column of warm air, as column 1 cools, it must shrink, and as column 2 warms, it must rise. (c) Because at the same level in the atmosphere there is more air above the H in the warm column than above the L in the cold column, warm air aloft is associated with high pressure and cold air aloft with low pressure. The pressure differences aloft create a force that causes the air to move from a region of higher pressure toward a region of lower pressure. The removal of air from column 2 causes its surface pressure to drop, whereas the addition of air into column 1 causes its surface pressure to rise. (The difference in height between the two columns is greatly exaggerated.)

warmer, less-dense air, the pressure does not decrease as rapidly with height, simply because you climb above fewer molecules in the same vertical distance.

In Fig. 7.2c move up the warm, red column until you come to the letter *H.* Now move up the cold, blue column the same distance until you reach the letter *L.* Notice that there are more molecules above the letter *H* in the warm column than above the letter *L* in the cold column. The fact that the number of molecules above any level is a measure of the atmospheric pressure leads to an important concept: *Warm air aloft is normally associated with high atmospheric pressure, and cold air aloft is associated with low atmospheric pressure.*

In Fig. 7.2c, the horizontal difference in temperature creates a horizontal difference in pressure. The pressure difference establishes a force (called the *pressure gradient force*) that causes the air to move from higher pressure toward lower pressure. Consequently, if we remove the invisible barrier between the two columns and allow the air aloft to move horizontally, the air will move from column 2 toward column 1. As the air aloft leaves column 2, the mass of the air in the column decreases, and so does the surface air pressure. Meanwhile, the accumulation of air in column 1 causes the surface air pressure to increase.

Higher air pressure at the surface in column 1 and lower air pressure at the surface in column 2 cause the surface air to move from city 1 towards city 2 (see Fig. 7.3). As the surface air moves out away from city 1, the air aloft slowly sinks to replace this outwardly spreading surface air. As the surface air flows into city 2, it slowly rises to replace the depleted air aloft. In this manner, a complete circulation of air—called a *direct thermal circulation*—is established due to the heating and cooling of air columns. As we will see in Chapter 8, this type of circulation is the basis for a wide range of wind systems around the world.

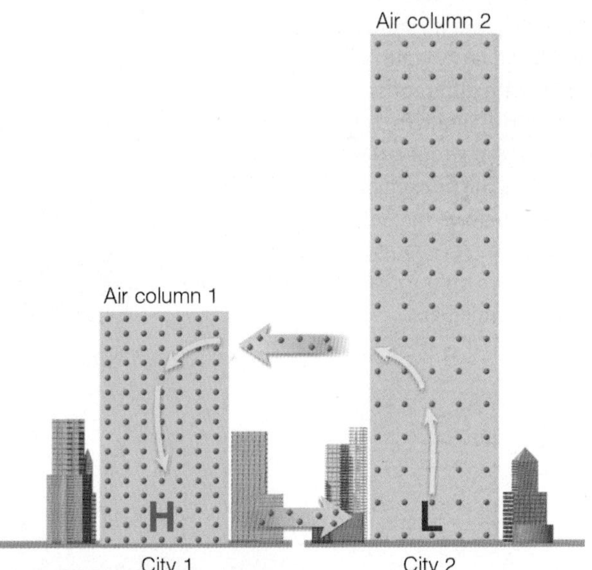

FIGURE 7.3 The heating and cooling of air columns causes horizontal pressure variations aloft and at the surface. These pressure variations force the air to move from areas of higher pressure toward areas of lower pressure. In conjunction with these horizontal air motions, the air slowly sinks above the surface high and rises above the surface low.

In summary, we can see how heating and cooling columns of air can establish horizontal variations in air pressure both aloft and at the surface. It is these horizontal differences in air pressure that cause the wind to blow.

From what we have learned so far, we might expect to see the surface pressure dropping as the air temperature rises, and vice versa. Over large continental areas, especially the southwestern United States in summer, hot, less-dense surface air is accompanied by surface low pressure. Likewise, bitter cold, dense arctic air in winter is often accompanied by surface high pressure. However, in the middle latitudes, any cyclic change in surface air pressure due to daily temperature changes is normally overshadowed by pressure changes brought about by the movement of large-scale high- and low-pressure areas. Regions of high and low pressure normally show up well on surface and upper-level charts. We will carefully examine these charts, keeping in mind that it is on these charts where we can locate large differences in atmospheric pressure and high winds.

Surface and Upper-Level Charts

We begin our study of weather maps by examining the air pressure measured at the four different locations in ❱ Fig. 7.4a. Notice that each city (as indicated by a dot) has a different pressure. The atmospheric pressure measured at each city is called *station pressure* because it represents the air pressure at a station (city) at a particular elevation. Since atmospheric pressure always decreases with increasing height, the higher a city is above sea level (the average level of the ocean's surface), the lower its station pressure. (Compare the lower station pressure of 952 millibars* at city A with the higher station pressure of 979 millibars at city B.)

Since variations in elevation have such a huge influence on a city's station pressure, and because pressure gradient that influences wind flow are examined along a horizontal level, readings of atmospheric pressure are usually adjusted to a common elevation—normally mean sea level. The adjusted reading is called **sea-level pressure**. By having all of the pressure readings at the same level, the influence of altitude on pressure is eliminated. The size of the correction to sea level is based primarily on how high the city is above sea level.

Near the earth's surface, atmospheric pressure decreases on the average by about 10 millibars for every 100 meters' increase in elevation (about 1 in. of mercury for each

*Recall from Chapter 1 that a millibar (mb) is a unit of pressure, where 1000 mb equals 29.53 inches of mercury. Standard atmospheric pressure is 1013.2 mb, which is equal to 29.92 inches of mercury.

❱ FIGURE 7.4

The top diagram (a) shows four cities (A, B, C, and D) at varying elevations above sea level, all with different station pressures. The middle diagram (b) represents sea-level pressures of the four cities plotted on a sea-level chart. The bottom diagram (c) shows sea-level pressure readings of the four cities plus other sea-level pressure readings, with isobars drawn on the chart (gray lines) at intervals of 4 millibars.

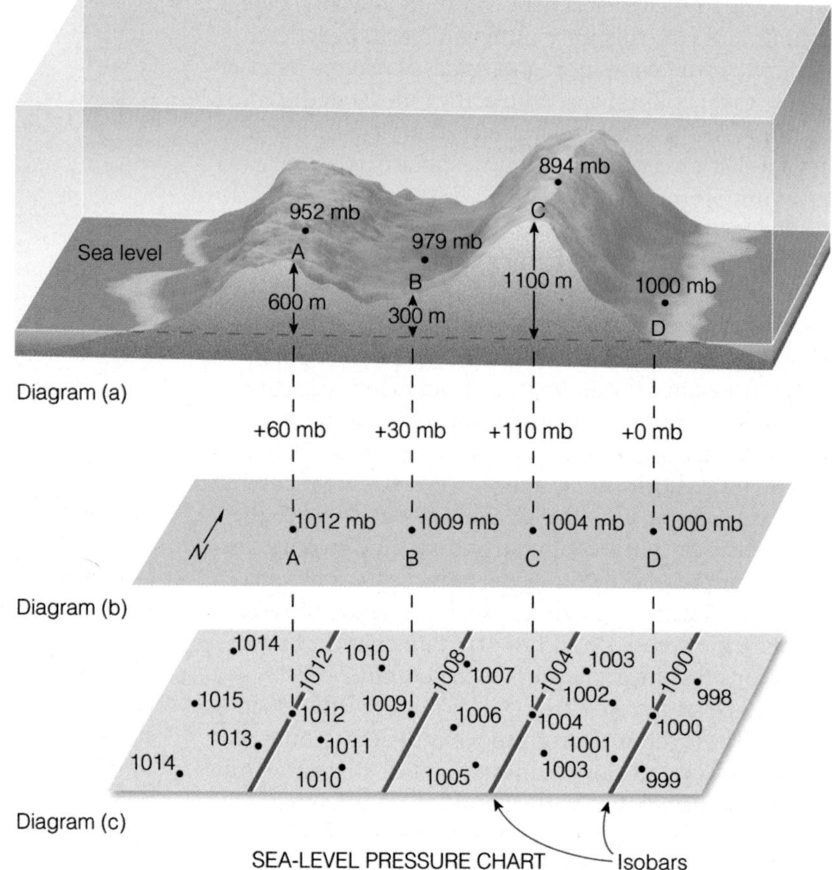

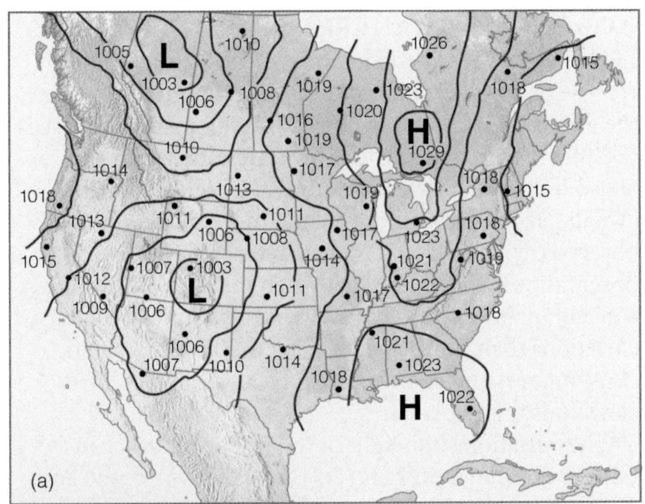

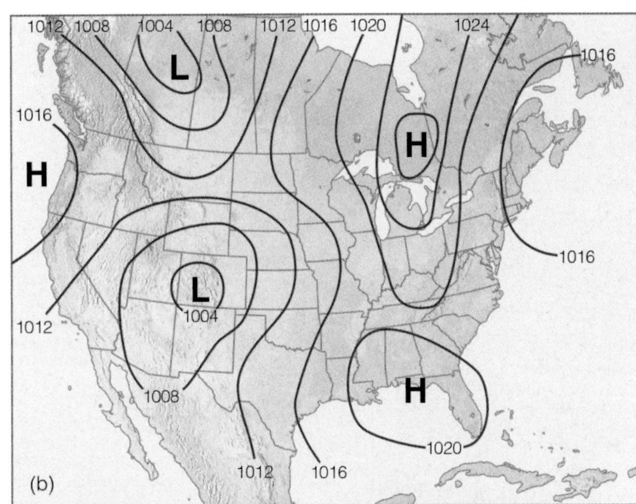

FIGURE 7.5 (a) Sea-level isobars drawn so that each observation is taken into account. Not all observations are plotted. (b) Sea-level isobars after smoothing.

1000-ft rise).* Notice in Fig. 7.4a that city A has a station pressure of 952 mb. Notice also that city A is 600 meters above sea level. Adding 10 mb per 100 m to its station pressure yields a sea-level pressure of 1012 millibars (Fig. 7.4b). After all the station pressures are adjusted to sea level (Fig. 7.4c), we are able to see the horizontal variations in sea-level pressure—something we were not able to see from the station pressures alone in Fig. 7.4a.

When more pressure data are added (Fig. 7.4c), the chart can be analyzed and the pressure pattern visualized. **Isobars** (lines connecting points of equal pressure) are drawn as solid dark lines at intervals of 4 mb, with 1000 millibars being the base value. Note that the isobars do not pass through each point, but, rather, between many of them, with the exact values being interpolated from the data given on the chart. For example, follow the 1008-mb line from the top of the chart southward and observe that there is no plotted pressure of 1008 mb. The 1008-mb isobar, however, comes closer to the station with a sea-level pressure of 1007 mb than it does to the station with a pressure of 1010 mb. Notice also that on the right side of the 1008 isobar all pressure values are less than 1008 mb, whereas on the left side they are greater than 1008 mb.

With its isobars, the bottom chart in Fig. 7.4c shows variations in pressure along a horizontal surface, sea level. This chart is now called a *sea-level pressure chart,* or simply a **surface map**. When weather data are plotted on the map, it becomes a *surface weather map.* Because winds are produced by differences in pressure on a horizontal surface, surface maps are extremely useful tools in the analysis of wind-flow patterns.

THE SURFACE MAP The isobars on the surface map in Fig. 7.5a are drawn precisely, with each individual observation taken into account. Notice that many of the lines are irregular, especially in mountainous regions over the Rockies. The reason for the wiggle is due, in part, to small-scale local variations in pressure and to errors introduced by correcting observations that were taken at high-altitude stations. An extreme case of this type of error occurs at Leadville, Colorado, with an elevation of over 3000 m or 10,000 ft, is the highest city in the United States. Here, the station pressure is typically near 700 mb. This means that nearly 300 mb must be added to obtain a sea-level pressure reading! A mere 1 percent error in estimating the exact correction would result in a 3-mb error in sea-level pressure. For this reason, isobars are smoothed through readings from high-altitude stations and from stations that might have small observational errors. Figure 7.5b shows how the isobars appear on the surface map after they are smoothed.

The sea-level pressure chart described so far represents the atmospheric pressure at a constant level—in this case, sea level. The same type of chart could be drawn to show horizontal variations in pressure at any level in the atmosphere; for example, at 10,000 ft. However, the reality is that measurements above the surface are made using a balloon-borne package of instruments (called a radiosonde) that measures the altitude of a specific pressure at a height above sea-level. For example, a pressure of 500 millibars might be observed at 18,400 ft

*This decrease in atmospheric pressure with height (10 mb/100 m) occurs when the air temperature decreases at the standard lapse rate of 6.5°C/1000 m. Because atmospheric pressure decreases more rapidly with height in cold (more-dense) air than it does in warm (less-dense) air, the vertical rate of pressure change is typically greater than 10 mb per 100 m in cold air and less than that in warm air.

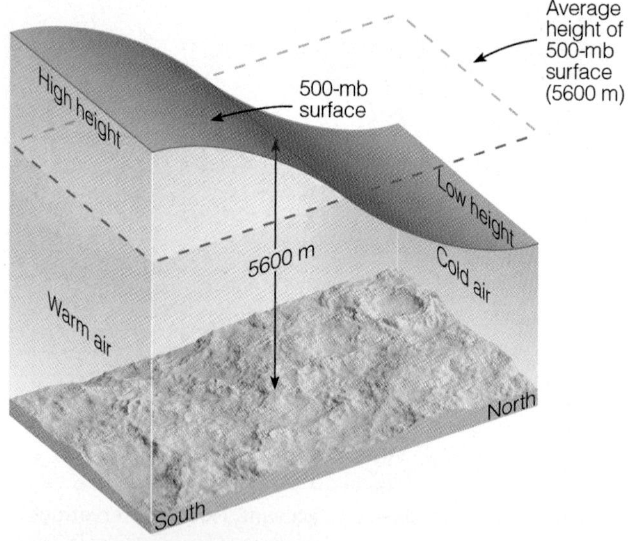

FIGURE 7.6 The area shaded gray in the above diagram represents a surface of constant pressure, or isobaric surface. Because of the changes in air density, the isobaric surface rises in warm, less-dense air and lowers in cold, more-dense air. Where the horizontal temperature changes most quickly, the isobaric surface changes elevation most rapidly. (The relative change in height of the 500-mb surface is exaggerated in this figure.)

(5,600 m) above sea-level. For this reason, the preferred chart commonly in use in the study of weather above the surface is the *constant pressure chart,* or **isobaric chart**. Instead of showing pressure variations at a constant altitude, isobaric charts show height variations along a surface of constant pressure.

ISOBARIC (CONSTANT PRESSURE) CHARTS In Fig. 7.6, the area shaded gray at the top of the column represents a constant pressure (isobaric) surface, where the atmospheric pressure at all points along this surface is 500 mb. Notice that in the warmer, less-dense air the 500-mb pressure surface is found at a higher (than average) level, while in the colder, more-dense air, it is observed at a much lower (than average) level. From these observations, we can see that when the air aloft is warm, constant pressure surfaces are typically found at higher elevations than normal, and when the air aloft is cold, constant pressure surfaces are typically found at lower elevations than normal.

The variations in height of the isobaric surface in Fig. 7.6 are shown in Fig. 7.7. Note that where the constant altitude lines intersect the 500-mb pressure surface, **contour lines** (lines connecting points of equal elevation) are drawn on the 500-mb map. Each contour line, of course, tells us the altitude above sea level at which we can obtain a pressure reading of 500 mb. In the warmer air to the south, the elevations are high, while in the cold air to the north, the elevations are low. The contour lines are crowded together in the middle of the chart, where the pressure surface dips rapidly due to the changing air temperatures. Where there is little horizontal temperature change, there are also few contour lines. Although contour lines are height lines, keep in mind that they illustrate pressure as do isobars in that contour lines of low height represent a region of lower pressure and contour lines of high height represent a region of higher pressure.

Since cold air aloft is normally associated with low heights or low pressures, and warm air aloft with high

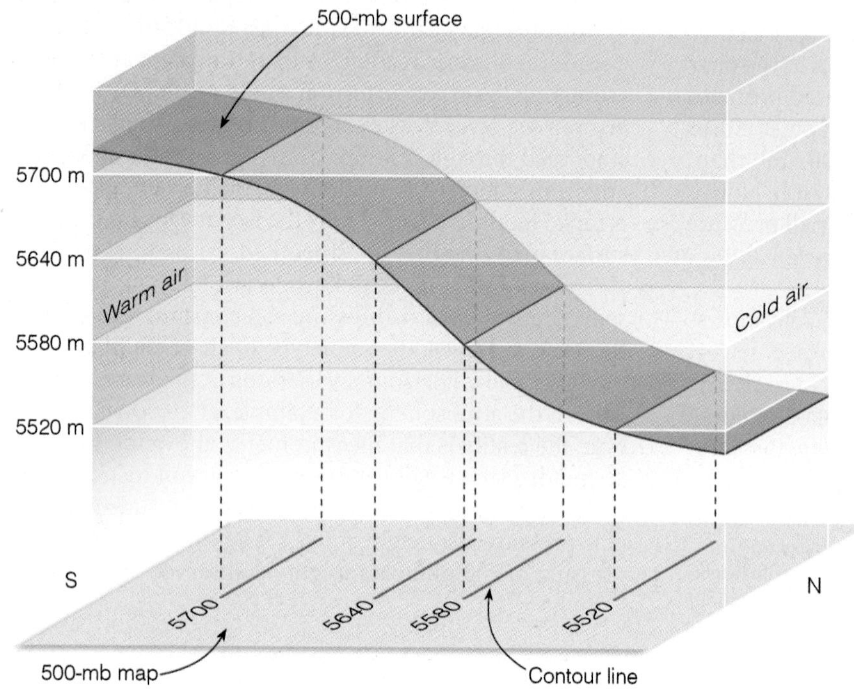

FIGURE 7.7 Changes in elevation of an isobaric surface (500-mb) show up as contour lines on an isobaric (500-mb) map. Where the surface dips most rapidly, the lines are closer together.

Table 7.1

Common Isobaric Charts and Their Approximate Elevation above Sea Level

ISOBARIC SURFACE (MB) CHARTS	APPROXIMATE ELEVATION (M)	(FT)
1000	120	400
850	1,460	4,800
700	3,000	9,800
500	5,600	18,400
300	9,180	30,100
200	11,800	38,700
100	16,200	53,200

heights or high pressures, on upper-air charts representing the Northern Hemisphere, contour lines and isobars usually decrease in value from south to north because the air is typically warmer to the south and colder to the north. The lines, however, are not straight; they bend and turn, indicating **ridges** *(elongated highs)* where the air is warm and indicating depressions, or **troughs** *(elongated lows),* where the air is cold. In Fig. 7.8, we can see how the wavy contours on the map relate to the changes in altitude of the isobaric surface.

Although we have examined only the 500-mb chart, other isobaric charts are commonly used. Table 7.1 lists the charts of interest in meteorology and their approximate heights above sea level.

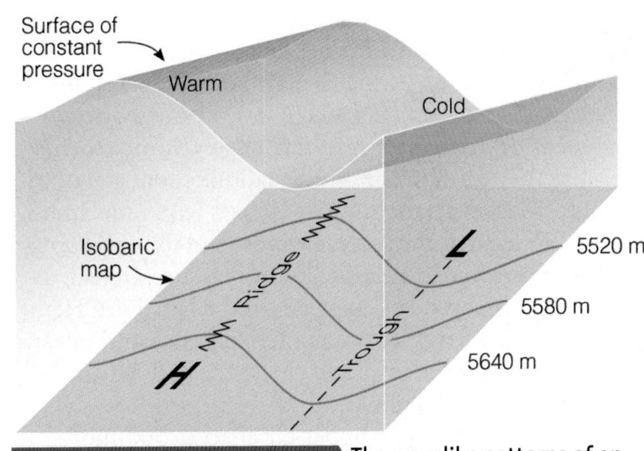

FIGURE 7.8 The wavelike patterns of an isobaric surface reflect the changes in air temperature. An elongated region of warm air aloft shows up on an isobaric map as higher heights and a ridge; the colder air shows as lower heights and a trough.

Upper-level charts are a valuable tool. As we will see, they show wind-flow patterns that are extremely important in forecasting all types of weather, including severe weather events. They can also be used to determine the movement of weather systems and to predict the behavior of surface pressure areas. To the pilot of a small aircraft, a constant pressure chart can help determine whether the plane is flying at an altitude either higher or lower than its altimeter indicates. (For more information on this topic, read the Focus section on pp. 182–183.)

Figure 7.9a is a simplified surface map that shows areas of high and low pressure and arrows that indicate

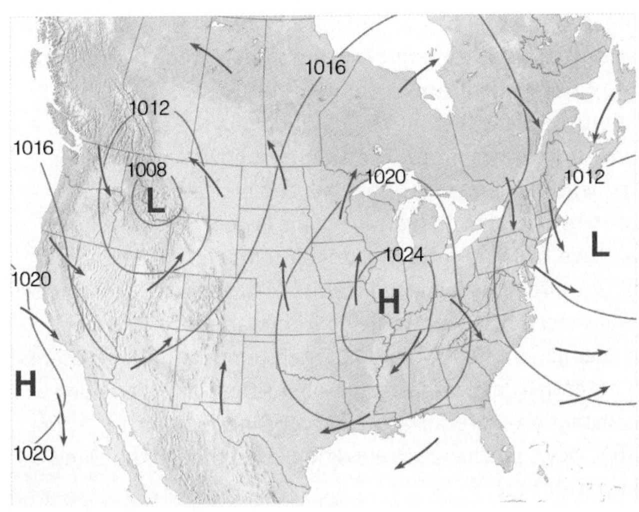

(a) Surface map

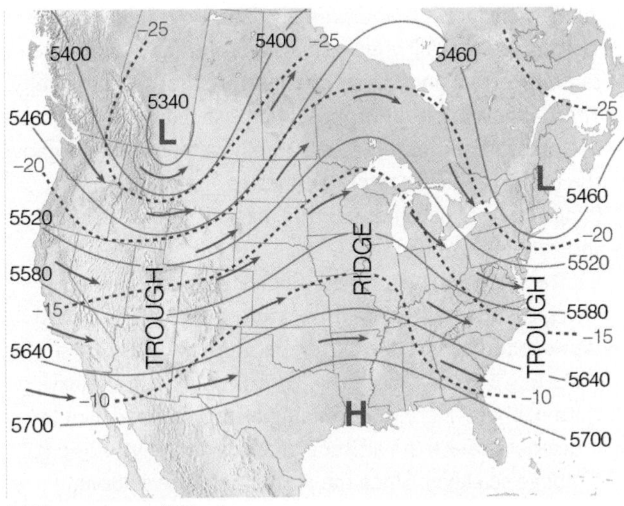

(b) Upper-air map (500-mb)

FIGURE 7.9 (a) Surface map showing areas of high and low pressure. The solid lines are isobars drawn at 4-mb intervals. The arrows represent wind direction. Notice that the wind blows *across* the isobars. (b) The upper-air (500-mb) map fo r the same day as the surface map. Solid lines on the map are contour lines in meters above sea level. Dashed red lines are isotherms in °C. Arrows show wind direction. Notice that, on this upper-air map, the wind blows *parallel* to the contour lines.

wind direction—the direction from which the wind is blowing. The large blue H's on the map indicate the centers of high pressure, which are also called **anticyclones**. The large L's represent centers of low pressure, also known as *depressions* or **mid-latitude cyclonic storms** because they form in the middle latitudes, outside of the tropics. The solid dark lines are isobars with units in millibars. Notice that the surface winds tend to blow across the isobars toward regions of lower pressure. In fact, as we briefly observed in Chapter 1, in the Northern Hemisphere surface winds blow counterclockwise and inward toward the center of lows and clockwise and outward from the center of highs.

Figure 7.9b shows an upper-air chart (a 500-mb isobaric map) for the same day as the surface map in Fig. 7.9a. The solid gray lines on the map are contour lines given in meters above sea level. The difference in elevation between each contour line (called the *contour interval)* is 60 meters. Superimposed on this map are dashed red lines, which represent lines of equal temperature (isotherms). Observe how the contour lines tend to parallel the isotherms. As we would expect, the contour lines tend to decrease in value from south to north.

The arrows on the 500-mb map show the wind direction. Notice that, unlike the surface winds that cross the isobars in Fig. 7.9a, the winds on the 500-mb chart (Fig. 7.9b) tend to flow *parallel* to the contour lines in a wavy west-to-east direction. Why does the wind tend to cross the isobars on a surface map yet blow parallel to the contour lines (or isobars) on an upper-air chart? To answer this question we will now examine the forces that affect winds.

FOCUS ON

A SPECIAL TOPIC

Flying along a Constant Pressure Surface—When the Air Temperature Drops, Look Out for Mountaintops

"It was a cloudy windy day, and the plane seemed to fly right into the mountain." That is a quote from an observer who witnessed a plane crash. In a world with so much technology, how could such a tragic event take place? The fact is that extremely dangerous flying conditions often prevail when a small aircraft flies over mountainous terrain and the air temperature drops dramatically. This dangerous situation exists because many small aircraft use pressure altimeters to indicate how high they are above the terrain. Unfortunately, altimeters are instruments (small aneroid barometers) that measure pressure but only indicate elevation.

The elevation an altimeter indicates is determined using the assumption that the atmosphere is a standard atmosphere where the air temperature decreases at a rate of 3.6°F per 1000 ft (6.5°C/1000 m). In a standard atmosphere, a pressure reading of 700 millibars would be equivalent to an altitude of approximately 10,000 ft above sea level. Since the air temperature seldom, if ever, decreases at exactly the standard rate, an aircraft's pressure altimeter will often indicate an altitude either higher or lower than the aircraft's true elevation.

An aircraft using a pressure altimeter flies along a surface of constant pressure. Recall from an earlier discussion that constant pressure surfaces rise in warm air and descend in cold air. Notice in Fig. 1 that in the warm air the altimeter is measuring a pressure of 700 mb and indicating an altitude of 10,000 ft, but the aircraft is flying at an altitude higher than the altimeter indicates.

As the plane flies into much colder air, the constant pressure surface drops. The altimeter, however, continues to measure a pressure of 700 mb and indicate an altitude of 10,000 ft, even though the plane is now flying at a lower altitude. With no correction for air temperature, in the cold air the aircraft is flying at an altitude lower than the altimeter indicates. Consequently, flying from warmer air into much colder air (with no correction for air temperature) represents a potentially dangerous situation. As the plane flies along a constant pressure surface that drops in elevation, the altimeter shows no change in elevation, even though the plane is losing altitude.

This problem can be serious, especially for planes flying above mountainous terrain with poor visibility, and where high winds and turbulence can reduce the air pressure drastically. To ensure adequate clearance under these conditions, pilots fly their aircraft higher than they normally would,

Why the Wind Blows

Our understanding of why the wind blows stretches back through several centuries, with many scientists contributing to our knowledge. When we think of the movement of air, however, one great scholar stands out—Isaac Newton (1642–1727), who formulated several fundamental laws of motion.

NEWTON'S LAWS OF MOTION Newton's first law of motion states that *an object at rest will remain at rest and an object in motion will remain in motion (and travel at a constant velocity along a straight line) as long as no force is exerted on the object.* For example, a baseball in a pitcher's hand will remain there until a force (a push) acts upon the ball. Once the ball is pushed (thrown), it would continue to move in that direction forever if it were not for the force of air friction (which slows it down), the force of gravity (which pulls it toward the ground), and the catcher's mitt (which exerts an equal but opposite force to bring it to a halt). Similarly, to start air moving, to speed it up, to slow it down, or even to change its direction requires the action of an external force. This brings us to Newton's second law.

Newton's second law states that *the force exerted on an object equals its mass times the acceleration produced.** In symbolic form, this law is written as

$$F = ma.$$

*Newton's second law may also be stated in this way: The acceleration of an object (times its mass) is caused by all of these forces acting on it.

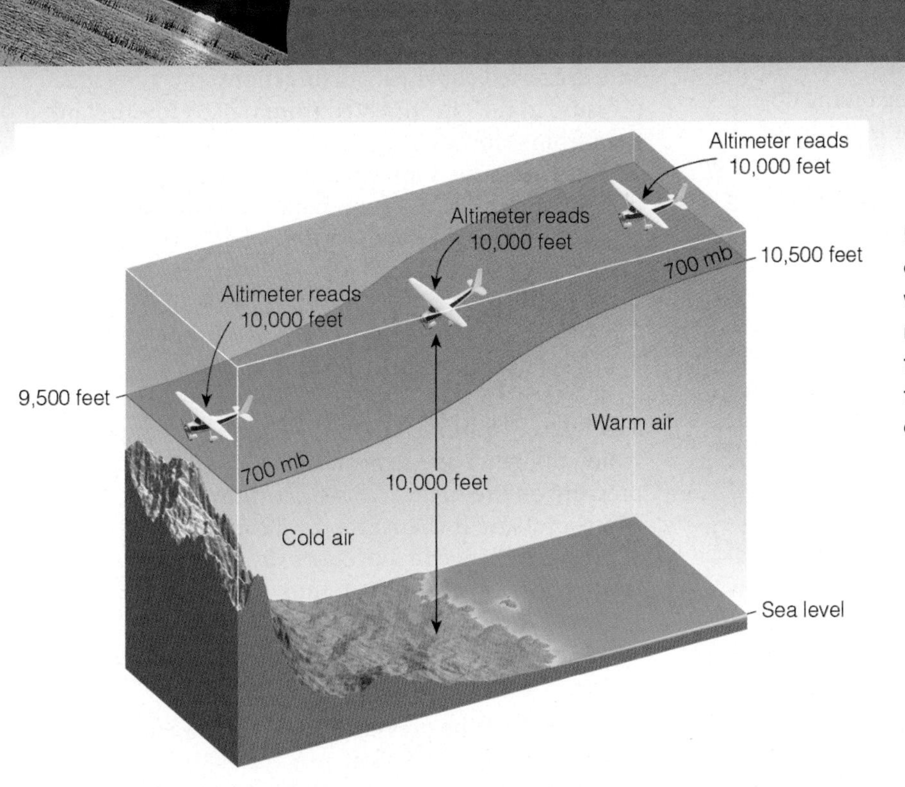

Figure 1 As an aircraft flies along a constant pressure surface from warmer air into colder air, without making temperature corrections the aircraft will lose altitude even though the pressure altimeter continues to read the same altitude.

consider air temperature, and compute a more realistic altitude by resetting their altimeters to reflect these conditions.

Because of the inaccuracies inherent in the pressure altimeter, many high performance and commercial aircraft are equipped with a radio altimeter. This device is like a small radar unit that measures the altitude of the aircraft by sending out radio waves, which bounce off the terrain. The time it takes these waves to reach the surface and return is a measure of the aircraft's altitude. If used in conjunction with a pressure altimeter, a pilot can determine the variations in a constant pressure surface simply by flying along that surface and observing how the true elevation measured by the radio altimeter changes.

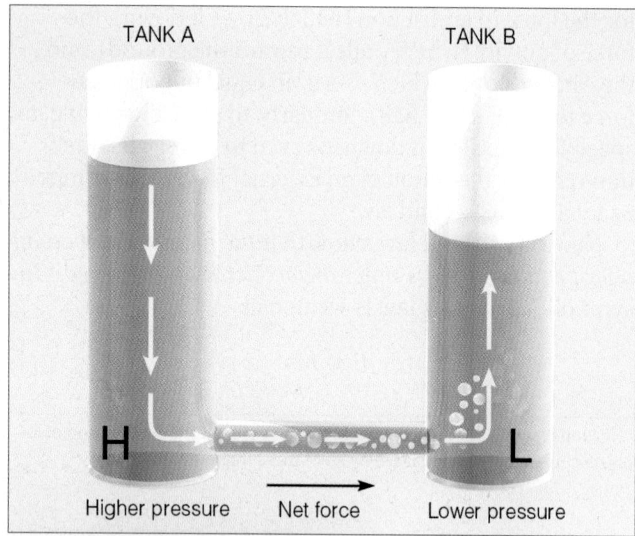

FIGURE 7.10 The higher water level creates higher fluid pressure at the bottom of tank A and a net force directed toward the lower fluid pressure at the bottom of tank B. This net force causes water to move from higher pressure toward lower pressure.

From this relationship we can see that, when the mass of an object is constant, the force acting on the object is directly related to the acceleration that is produced. A force in its simplest form is a push or a pull. *Acceleration is the speeding up, the slowing down, or the changing of direction of an object.* (More precisely, acceleration is the change of velocity* over a period of time.)

Because more than one force may act upon an object, Newton's second law always refers to the *net,* or total, force that results. An object will always accelerate in the direction of the total force acting on it. Therefore, to determine in which direction the wind will blow, we must identify and examine all of the forces that affect the horizontal movement of air. These forces include:

1. pressure gradient force
2. Coriolis force
3. centripetal force
4. friction

*Velocity specifies both the speed of an object and its direction of motion.

We will first study the forces that influence the flow of air aloft. Then we will see which forces modify winds near the ground.

FORCES THAT INFLUENCE THE WIND We have already learned that horizontal differences in atmospheric pressure cause air to move and, hence, the wind to blow. Since air is an invisible gas, it may be easier to see how pressure differences cause motion if we examine a visible fluid, such as water.

In Fig. 7.10, the two large tanks are connected by a pipe. Tank A is two-thirds full and tank B is only one-half full. Since the water pressure at the bottom of each tank is proportional to the weight of water above, the pressure at the bottom of tank A is greater than the pressure at the bottom of tank B. Moreover, since fluid pressure is exerted equally in all directions, there is a greater pressure in the pipe directed from tank A toward tank B than from B toward A.

Since pressure is force per unit area, there must also be a net force directed from tank A toward tank B. This force causes the water to flow from left to right, from higher pressure toward lower pressure. The greater the pressure difference, the stronger the force, and the faster the water moves. In a similar way, horizontal differences in atmospheric pressure cause air to move.

Pressure Gradient Force Figure 7.11 shows a region of higher pressure on the map's left side, lower pressure on the right. The isobars show how the horizontal pressure is changing. If we compute the amount of pressure change that occurs over a given distance, we have the **pressure gradient**; thus

$$\text{Pressure gradient} = \frac{\text{difference in pressure}}{\text{distance}}.$$

In Fig. 7.11, the pressure gradient between points 1 and 2 is 4 mb per 100 km.

Suppose the pressure in Fig. 7.11 were to change, and the isobars become closer together. This condition would produce a rapid change in pressure over a relatively short distance, or what is called a *steep* (or *strong*) *pressure gradient.* However, if the pressure were

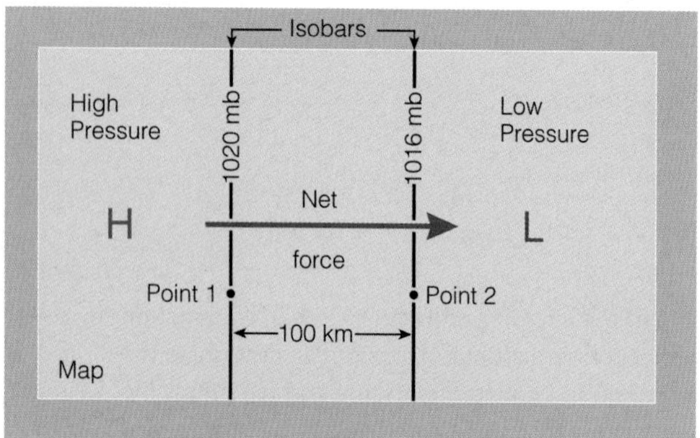

FIGURE 7.11 The pressure gradient between point 1 and point 2 is 4 mb per 100 km. The net force directed from higher toward lower pressure is the *pressure gradient force.*

to change such that the isobars spread farther apart, then the difference in pressure would be small over a relatively large distance. This condition is called a *gentle* (or *weak*) *pressure gradient.*

Notice in Fig. 7.11 that when differences in horizontal air pressure exist there is a net force acting on the air. This force, called the **pressure gradient force** *(PGF), is always directed from higher toward lower pressure at right angles to the isobars.* The magnitude of the force is directly related to the pressure gradient. Steep pressure gradients correspond to strong pressure gradient forces and vice versa. ❯ Figure 7.12 shows the relationship between pressure gradient and pressure gradient force.

The *pressure gradient force is the force that causes the wind to blow.* Because of this fact, closely spaced isobars on a weather chart indicate steep pressure gradients, strong forces, and high winds. On the other hand, widely spaced isobars indicate gentle pressure gradients, weak forces, and light winds. An example of a steep pressure gradient and strong winds is illustrated on the surface weather map in ❯ Fig. 7.13. Notice that the tightly

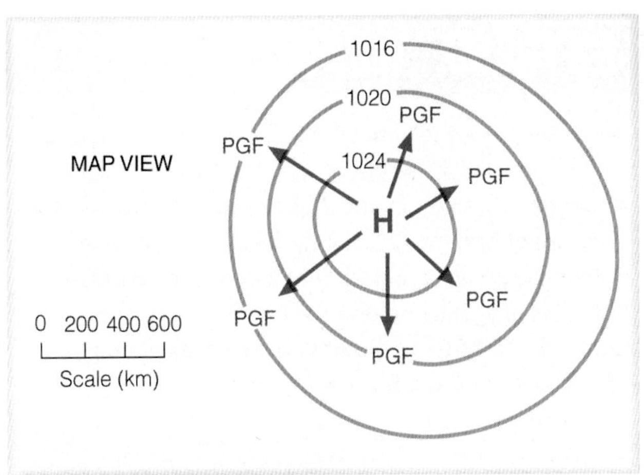

❯ FIGURE 7.12 The closer the spacing of the isobars, the greater the pressure gradient. The greater the pressure gradient, the stronger the pressure gradient force (*PGF*). The stronger the *PGF*, the greater the wind speed. The red arrows represent the relative magnitude of the force, which is always directed from higher toward lower pressure.

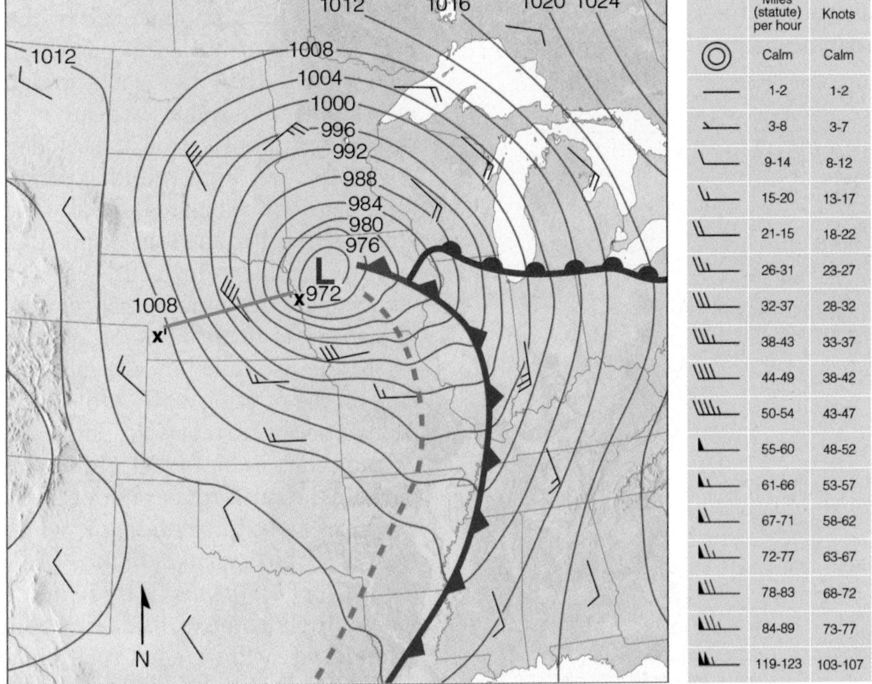

❯ FIGURE 7.13 Surface weather map for 6 A.M. (CST), Tuesday, November 10, 1998. Dark gray lines are isobars with units in millibars. The interval between isobars is 4 mb. A deep low-pressure area with a central pressure of 972 mb (28.70 in.) is moving over northwestern Iowa. The distance along the green line X-X′ is 500 km. The difference in pressure between X and X′ is 32 mb, producing a pressure gradient of 32 mb/500 km. The tightly packed isobars along the green line are associated with strong northwesterly winds of 40 knots, with gusts even higher. Wind directions are given by lines that parallel the wind. Wind speeds are indicated by barbs and flags. (A wind indicated by the symbol ⟍ would be a wind from the northwest at 10 knots. See blue insert.) The solid blue line is a cold front, the solid red line a warm front, and the solid purple line an occluded front. The dashed gray line is a trough.

EXTREME WEATHER WATCH

On October 12, 1962, one of the most powerful storms ever to develop along the Pacific Northwest coast, slammed into Oregon and Washington. With a central pressure of 960 mb (28.35 in.), the storm had such a strong pressure gradient that it produced winds in excess of 75 mi/hr along the Pacific Coast from California to British Columbia with a peak wind gust of 179 mi/hr at Cape Blanco, Oregon.

packed isobars along the green line are producing a steep pressure gradient of 32 mb per 500 km and strong surface winds of 40 knots with much higher gusts.

In fact, the deep, low-pressure area illustrated in Fig. 7.13 was quite a storm, that produced extreme weather over a vast area. The intense low with its tightly packed isobars and strong pressure gradient produced extremely high winds that gusted over 100 mi/hr in Wisconsin. The extreme winds caused blizzard conditions over the Dakotas, closed many Interstate highways, shut down airports, and overturned trucks. The winds pushed a school bus off the road near Albert Lea, Minnesota, injuring two children, and blew the roofs off homes in Wisconsin. This notorious deep storm set an all-time record low pressure of 963 mb (28.43 in.) for Minnesota on November 10, 1998.

If the pressure gradient force were the only force acting upon air, we would always find winds blowing directly from higher toward lower pressure. However, the moment air starts to move, it is deflected in its path by the *Coriolis force.*

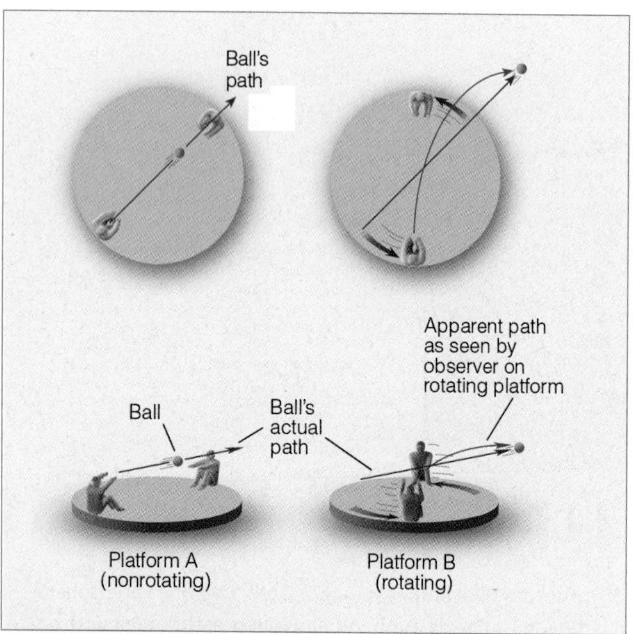

FIGURE 7.14 **On nonrotating platform A, the thrown ball moves in a straight line. On platform B, which rotates counterclockwise, the ball continues to move in a straight line. However, platform B is rotating while the ball is in flight; thus, to anyone on platform B, the ball appears to deflect to the right of its intended path.**

Coriolis Force The **Coriolis force** describes an apparent force that is due to the rotation of the earth. To understand how it works, consider two people playing catch as they sit opposite one another on the rim of a merry-go-round (see Fig. 7.14, platform A). If the merry-go-round is not moving, each time the ball is thrown, it moves in a straight line to the other person.

Suppose the merry-go-round starts turning counterclockwise—the same direction the earth spins as viewed from above the North Pole. If we watch the game of catch from above, we see that the ball moves in a straight-line path just as before. However, to the people playing catch on the merry-go-round, the ball seems to veer to its right each time it is thrown, always landing to the right of the point intended by the thrower (see Fig. 7.14, platform B). This perspective is due to the fact that, while the ball moves in a straight-line path, the merry-go-round rotates beneath it; by the time the ball reaches the opposite side, the catcher has moved. To anyone on the merry-go-round, it seems as if there is some force causing the ball to deflect to the right. This apparent force is called the *Coriolis force* after Gaspard- Gustavede Coriolis, a nineteenth-century French scientist who worked out a mathematical way of describing it. (Because it is an *apparent force* due to the rotation of the earth, it is also called the *Coriolis effect.*) This effect occurs on the rotating earth, too. All free-moving objects, such as ocean currents, aircraft, artillery projectiles, and air molecules seem to deflect from a straight-line path because the earth rotates under them.

The Coriolis force *causes the wind to deflect to the right of its intended path in the Northern Hemisphere and to the left of its intended path in the Southern Hemisphere.* To illustrate, consider a satellite in polar circular orbit. If the earth were not rotating, the path of the satellite would be observed to move directly from north to south, parallel to the earth's meridian lines. However, the earth *does* rotate, carrying us and meridians eastward with it. Because of this rotation, in the Northern Hemisphere we see the satellite moving southwest instead of due south; it seems to veer off its path and move toward *its right.* In the Southern Hemisphere, the earth's direction of rotation is clockwise as viewed from above the South Pole. Consequently, a satellite moving northward from the South Pole would appear to move northwest and, hence, would veer to the *left* of its path.

As the wind speed increases, the Coriolis force increases; hence, *the stronger the wind, the greater the deflection.* Additionally, the Coriolis force increases for all wind speeds from a value of *zero at the equator to a maximum at the poles.* This phenomenon is illustrated in ❱ Fig. 7.15 where three aircraft, each at a different latitude, are flying along a straight-line path, with no external forces acting on them. The destination of each aircraft is due east and is marked on the illustration in Fig. 7.15a. Each plane travels in a straight path relative to an observer positioned at a fixed spot in space. The earth rotates beneath the moving planes, causing the destination points at latitudes 30° and 60° to change direction slightly—to the observer in space (see Fig. 7.15b). To an observer standing on the earth, however, it is the plane that appears to deviate. The amount of deviation is greatest toward the pole and nonexistent at the equator. Therefore, the Coriolis force has a far greater effect on the plane at high latitudes (large deviation) than on the plane at low latitudes (small deviation). On the equator, it has no effect at all. The same is true of its effect on winds.

In summary, to an observer on the earth, objects moving in *any direction* (north, south, east, or west) are deflected to the *right* of their intended path in the Northern Hemisphere and to the *left* of their intended path in the Southern Hemisphere. The amount of deflection depends upon:

1. the rotation of the earth
2. the latitude
3. the object's speed

In addition, *the Coriolis force acts at right angles to the wind, only influencing wind direction and never wind speed.*

The Coriolis "force" behaves as a real force, constantly tending to "pull" the wind to its right in the Northern Hemisphere and to its left in the Southern Hemisphere. Moreover, this effect is present in all motions relative to the earth's surface. However, in most of our everyday experiences, the Coriolis force is so small (compared to other forces involved in those experiences) that it is negligible and, contrary to popular belief, does not cause water to turn clockwise or counterclockwise when draining from a sink.

The Coriolis force is also minimal on small-scale winds, such as those that blow inland along coasts in summer. Here, the Coriolis force might be strong because of high winds, but the force cannot produce much deflection over the relatively short distances. Only where winds blow over vast regions is the effect significant.

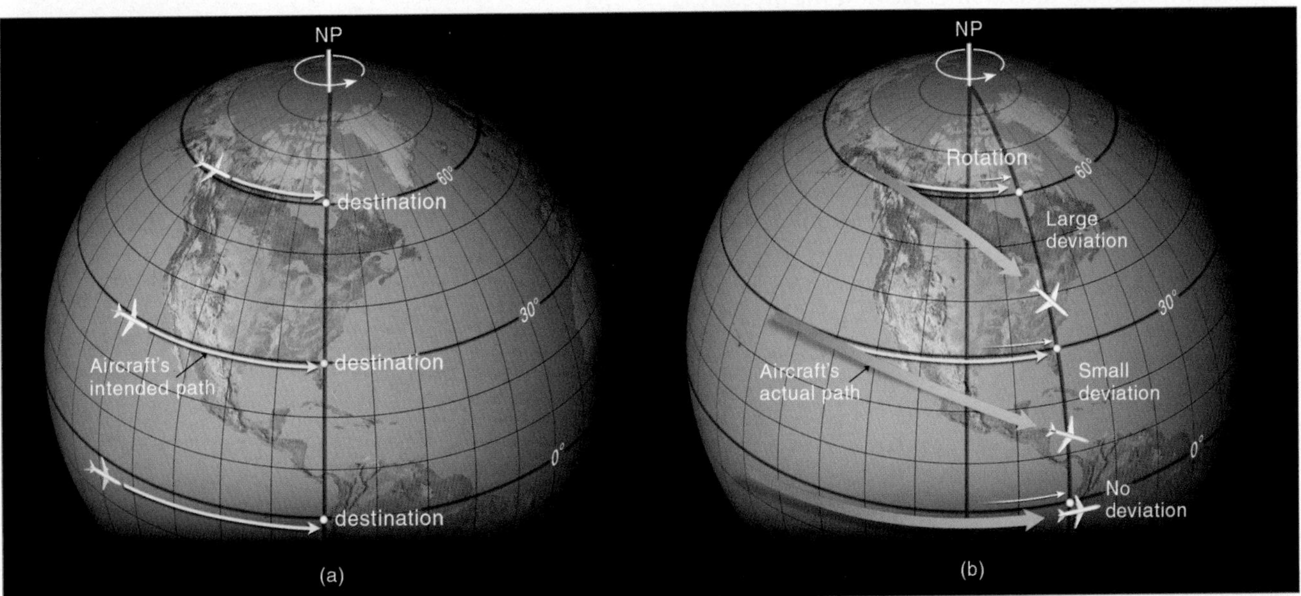

Active ❱ FIGURE 7.15 Except at the equator, a free-moving object heading either east or west (or any other direction) will appear from the earth to deviate from its path as the earth rotates beneath it. The deviation (Coriolis force) is greatest at the poles and decreases to zero at the equator. Visit the Meteorology Resource Center to view this and other active figures at www.cengage.com/login

BRIEF REVIEW

In summary, we know that:

- A change in surface air pressure can be brought about by changing the mass (amount of air) above the surface.
- Heating and cooling columns of air can establish horizontal variations in atmospheric pressure aloft and at the surface.
- A difference in horizontal air pressure produces a horizontal pressure gradient force.
- The pressure gradient force is always directed from higher pressure toward lower pressure, and it is the pressure gradient force that causes the air to move and the wind to blow.
- Steep pressure gradients (tightly packed isobars on a weather map) indicate strong pressure gradient forces and high winds; gentle pressure gradients (widely spaced isobars) indicate weak pressure gradient forces and light winds.
- Once the wind starts to blow, the Coriolis force causes it to bend to the right of its intended path in the Northern Hemisphere and to the left of its intended path in the Southern Hemisphere.

How the Wind Blows

We will now turn our attention to see how the pressure gradient force and the Coriolis force produce winds that blow more or less in a straight-line path and those that blow in a curved path, aloft and at the surface.

STRAIGHT-LINE FLOW ALOFT Earlier in this chapter, we saw that the winds aloft on an upper-level chart blow more or less parallel to the isobars or contour lines. We can see why this phenomenon happens by carefully looking at Fig. 7.16, which shows a map in the Northern Hemisphere, above the earth's frictional influence,* with horizontal pressure variations at an altitude of about 3300 feet above the earth's surface. The evenly spaced isobars indicate a constant pressure gradient force (*PGF*) directed from south toward north as indicated by the red arrow at the left. Why, then, does the map show a wind blowing from the west? We can answer this question by placing a parcel of air at position 1 in the diagram and watching its behavior.

*The friction layer (the layer where the wind is influenced by frictional interaction with objects on the earth's surface) usually extends from the surface up to about 1000 m (3300 ft) above the ground.

At position 1, the *PGF* acts immediately upon the air parcel, accelerating it northward toward lower pressure. However, the instant the air begins to move, the Coriolis force begins to deflect the air toward its right, curving its path. As the parcel of air increases in speed (positions 2, 3, and 4), the magnitude of the Coriolis force increases (as shown by the blue arrows), bending the wind more and more to its right. Eventually, the wind speed increases to a point where the Coriolis force just balances the *PGF.* At this point (position 5), the wind no longer accelerates because the net force is zero. Here the wind flows in a straight path, parallel to the isobars at a constant speed.* This flow of air is called a **geostrophic** (*geo:* earth; *strophic:* turning) **wind**. Notice that the geostrophic wind blows in the Northern Hemisphere with lower pressure to its left and higher pressure to its right.

When the flow of air is purely geostrophic, the isobars (or contour lines) are straight and evenly spaced, and the wind speed is constant. In the atmosphere, isobars are

*At first, it may seem odd that the wind blows at a constant speed with no net force acting on it. But when we remember that the net force is necessary only to accelerate ($F = ma$) the wind, it makes more sense. For example, it takes a considerable net force to push a car and get it rolling from rest. But once the car is moving, it only takes a force large enough to counterbalance friction to keep it going. There is no net force acting on the car, yet it rolls along at a constant speed.

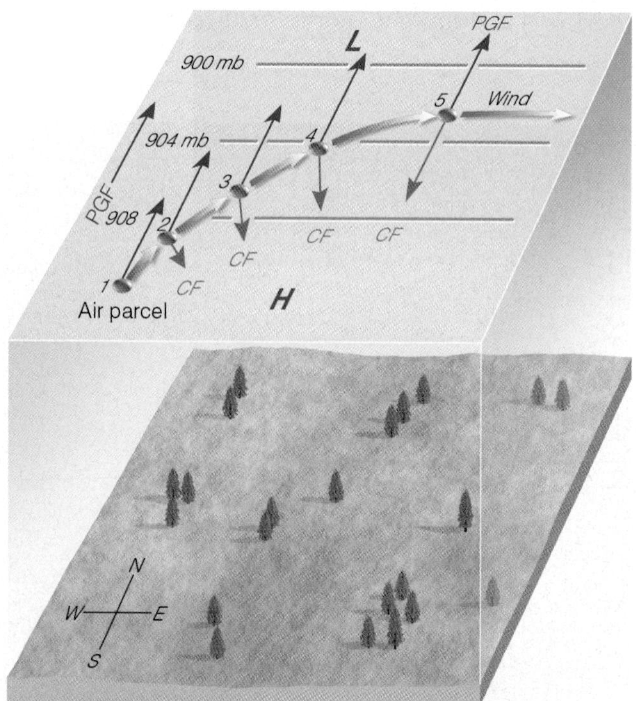

FIGURE 7.16 Above the level of friction, air initially at rest will accelerate until it flows parallel to the isobars at a steady speed with the pressure gradient force (*PGF*) balanced by the coriolis force (*CF*). Wind blowing under these conditions is called *geostrophic.*

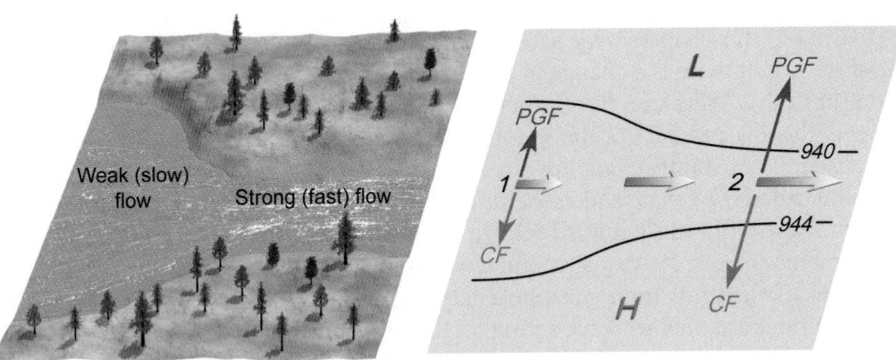

FIGURE 7.17

The isobars and contours on an upper-level chart are like the banks along a flowing stream. When they are widely spaced, the flow is weak; when they are narrowly spaced, the flow is stronger. The increase in winds on the chart results in a stronger Coriolis force (*CF*), which balances a larger pressure gradient force (*PGF*).

rarely straight or evenly spaced, and the wind normally changes speed as it flows along. So, the geostrophic wind is usually only an approximation of the real wind. However, the approximation is generally close enough to help us more clearly understand the behavior of the winds aloft.

As we would expect from our previous discussion of winds, the speed of the geostrophic wind is directly related to the pressure gradient. In Fig. 7.17, we can see that a geostrophic wind flowing parallel to the isobars is similar to water in a stream flowing parallel to its banks. At position 1, the wind is blowing at a low speed; at position 2, the pressure gradient increases and the wind speed picks up. Notice also that at position 2, where the wind speed is greater, the Coriolis force is greater and balances the stronger pressure gradient force.

We know that the winds aloft do not always blow in a straight line; frequently, they curve and bend into meandering loops as they tend to follow the patterns of the isobars. In the Northern Hemisphere, winds blow counterclockwise around lows and clockwise around highs. The next section explains why.

CURVED WINDS AROUND LOWS AND HIGHS ALOFT Because lows are also known as cyclones, the counterclockwise flow of air around them is often called *cyclonic flow.* Likewise, the clockwise flow of air around a high, or anticyclone, is called *anticyclonic flow.* Look at the wind flow around the upper-level low (Northern Hemisphere) in Fig. 7.18a. At first, it appears as though the wind is defying the Coriolis force by bending to the left as it moves counterclockwise around the system. Let's see why the wind blows in this manner.

Suppose we consider a parcel of air initially at rest at position 1 in Fig. 7.18a. The pressure gradient force accelerates the air inward toward the center of the low and the Coriolis force deflects the moving air to its right, until the air is moving parallel to the isobars at position 2. If the wind were geostrophic, at position 3 the air would move northward parallel to straight-line isobars at a constant speed. The wind is blowing at a constant speed, but parallel to curved isobars. A wind that blows at a constant speed parallel to *curved isobars* above the level of frictional influence is termed a **gradient wind**.

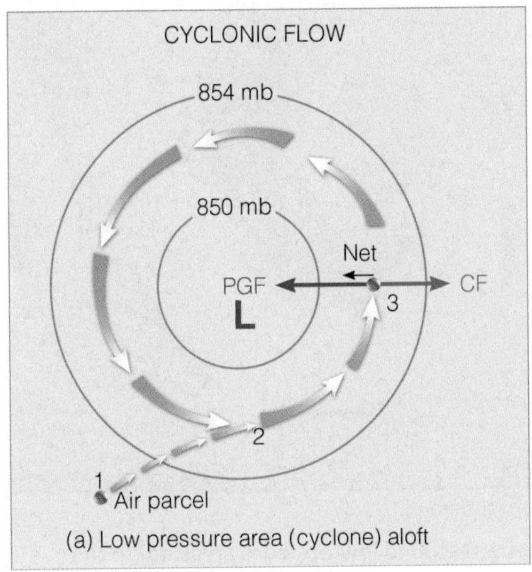

(a) Low pressure area (cyclone) aloft

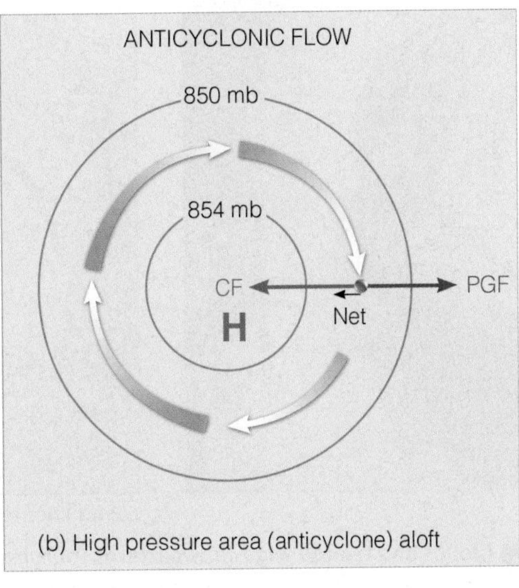

(b) High pressure area (anticyclone) aloft

FIGURE 7.18

Winds and related forces around areas of low and high pressure above the friction level in the Northern Hemisphere. Notice that the pressure gradient force (*PGF*) is in red, while the Coriolis force (*CF*) is in blue.

Earlier in this chapter we learned that an object accelerates when there is a change in its speed or direction (or both). Therefore, the gradient wind blowing *around* the low-pressure center is constantly accelerating because it is constantly changing direction. This acceleration, called the *centripetal acceleration,* is directed at right angles to the wind, inward toward the low center.

Remember from Newton's second law that, if an object is accelerating, there must be a net force acting on it. In this case, the net force acting on the wind must be directed toward the center of the low, so that the air will keep moving in a counterclockwise, circular path. This inward-directed force is called the **centripetal force** *(centri:* center; *petal:* to push toward), and results from an imbalance between the Coriolis force and the pressure gradient force.*

Again, look closely at position 3 (Fig. 7.18a) and observe that the inward-directed pressure gradient force *(PGF)* is greater than the outward-directed Coriolis force *(CF)*. The difference between these forces—the net force—is the inward-directed centripetal force. In Fig. 7.18b, the wind blows clockwise around the center of the high. The spacing of the isobars tells us that the magnitude of the *PGF* is the same as in Fig. 7.18a. However, to keep the wind blowing in a circle, the inward-directed Coriolis force must now be greater in magnitude than the outward-directed pressure gradient force, so that the centripetal force (again, the net force) is directed inward.

In the Southern Hemisphere, the pressure gradient force starts the air moving and the Coriolis force deflects the moving air to the *left,* thereby causing the wind to blow *clockwise around lows* and *counterclockwise around highs.* Figure 7.19 shows a satellite image of clouds and wind (dark arrows) around a strong low-pressure area in the Northern Hemisphere (7.19a) and in the Southern Hemisphere (7.19b).

We know that in the middle and high latitudes of the Northern Hemisphere the winds aloft tend to blow in a west-to-east direction. In the Southern Hemisphere, the winds aloft in middle and high latitudes also tend to blow from the west. The reason for this behavior is due to the fact that as air aloft moves southward from high pressure (warm air) over the equator toward low pressure (cold air) near the poles, the Coriolis force bends the moving air to the left, until it blows from the west.

Near the equator, where the Coriolis force is minimum, winds may blow around intense tropical storms with the centripetal force being almost as large as the pressure gradient force. In this type of flow, the Coriolis force is considered negligible, and the wind is called *cyclostrophic.*

*In some cases, it is more convenient to express the centripetal force (and the centripetal acceleration) as the *centrifugal force,* an apparent force that is equal in magnitude to the centripetal force, but directed outward from the center of rotation. The gradient wind is then described as a balance of forces between the centrifugal force, the pressure gradient force, and the Coriolis force.

(a) Northern Hemisphere

(b) Southern Hemisphere

FIGURE 7.19 Clouds and related wind-flow patterns (purple arrows) around low-pressure areas. (a) In the Northern Hemisphere, winds blow counterclockwise around an area of low pressure. (b) In the Southern Hemisphere, winds blow clockwise around an area of low pressure.

So far we have seen how winds blow in theory, but how do they appear on an actual map?

WINDS ON UPPER-LEVEL CHARTS On the upper-level 500-mb map (Fig. 7.20), notice that, as we would expect, the winds tend to parallel the contour lines in a wavy west-to-east direction. Notice also that the contour lines tend to decrease in elevation from south to north. This situation occurs because the air at this level is warmer to the south and colder to the north. On the map, where horizontal temperature contrasts are large there is also a large height gradient—the contour lines are close together and the winds are strong. Where the horizontal temperature contrasts are small, there is a small height gradient—the contour lines are spaced farther apart and the winds are weaker. In general, on maps such as this we find stronger north-to-south temperature contrasts in winter than in summer, which is why the winds aloft are usually stronger in winter.

In Fig. 7.20, the wind is geostrophic where it blows in a straight path parallel to evenly spaced lines; it is gradient where it blows parallel to curved contour lines. Where the wind flows in large, looping meanders, following a more or less north-south trajectory (such as along the west coast of North America), the wind-flow pattern is called **meridional**. Where the winds are blowing in a west-to-east direction (such as over the eastern third of the United States), the flow is termed **zonal**.

Because the winds aloft in middle and high latitudes generally blow from west to east, planes flying in this direction have a beneficial tail wind, which explains why a flight from San Francisco to New York City takes about thirty minutes less than the return flight. If the flow aloft is zonal, clouds, storms, and surface anticyclones tend to move more rapidly from west to east. However, where the flow aloft is meridional, as we will see in Chapter 10 surface storms tend to move more slowly, often intensifying into major storm systems that produce extreme weather over a large area.

Take a minute and look back at Fig. 7.13 on p. 185. Observe that the winds on this surface map tend to cross the isobars, blowing from higher pressure toward lower pressure. Observe also that along the green line, the tightly packed isobars are producing a steady surface wind of 40 knots. However, this same pressure gradient (with the same air temperature) would, on an upper-level chart, produce a much stronger wind. Why do surface winds normally cross the isobars and why do they blow more slowly than the winds aloft? (Before going on to the section on surface winds, you may wish to read the Focus section on pp. 192–193, which describes the instruments that measure wind speed and direction.)

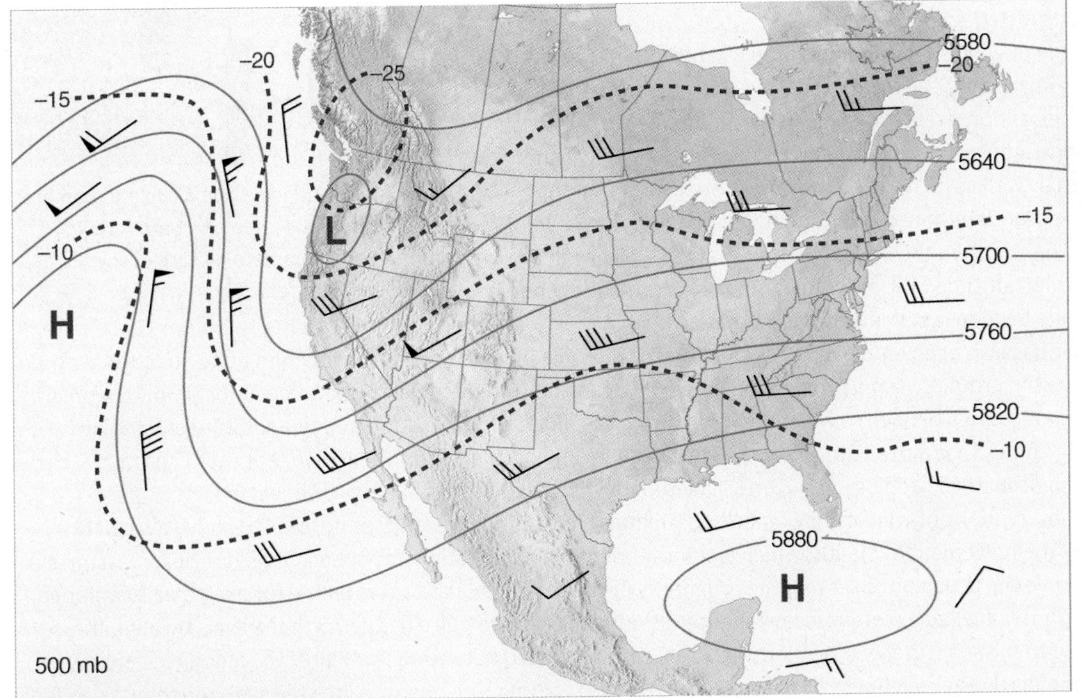

Miles (statute) per hour	Knots
Calm	Calm
1-2	1-2
3-8	3-7
9-14	8-12
15-20	13-17
21-25	18-22
26-31	23-27
32-37	28-32
38-43	33-37
44-49	38-42
50-54	43-47
55-60	48-52
61-66	53-57
67-71	58-62
72-77	63-67
78-83	68-72
84-89	73-77
119-123	103-107

FIGURE 7.20 An upper-level 500-mb map showing wind direction, as indicated by lines that parallel the wind. Wind speeds are indicated by barbs and flags. (See the blue insert.) Solid gray lines are contours in meters above sea level. Dashed red lines are isotherms in °C.

Surface Winds

Winds on a surface weather map do not blow exactly parallel to the isobars; instead, they cross the isobars, moving from higher to lower pressure. The angle at which the wind crosses the isobars varies, but averages about 30°. The reason for this behavior is *friction.*

The frictional drag of the ground slows the wind down. Because the effect of friction decreases as we move away from the earth's surface, wind speeds tend to increase with height above the ground. The atmospheric layer that is influenced by friction, called the **friction layer** (or *boundary layer),* usually extends

FOCUS ON

INSTRUMENTS

Measuring Winds

A very old, yet reliable, weather instrument for determining wind direction is the *wind vane.* Most wind vanes consist of a long arrow with a tail, which is allowed to move freely about a vertical post. (See Fig. 2.) The arrow always points into the wind and, hence, always gives the wind direction.

The instrument that measures wind speed is the *anemometer.* Most anemometers today consist of three (or more) hemispherical cups *(cup anemometer)* mounted on a vertical shaft as shown in Fig. 2. The difference in wind pressure from one side of a cup to the other causes the cups to spin about the shaft. The rate at which they rotate is directly proportional to the speed of the wind. The spinning of the cups is usually translated into wind speed through a system of gears, and may be read from a dial or transmitted to a recorder.

A vertical profile of wind speed and direction can be obtained with the balloon-borne instrument package called a *radiosonde.* (See Chapter 3, Fig. 4, p. 75) As the balloon rises from the surface, equipment located on the ground constantly tracks the balloon, measuring its vertical and horizontal angles, as well as its height above the ground. From this information, a computer determines and prints the vertical profile of wind from the surface up to where the balloon normally pops, typically in the stratosphere near an altitude of 100,000 feet. The observation of winds using a radiosonde balloon is called a *rawinsonde observation.*

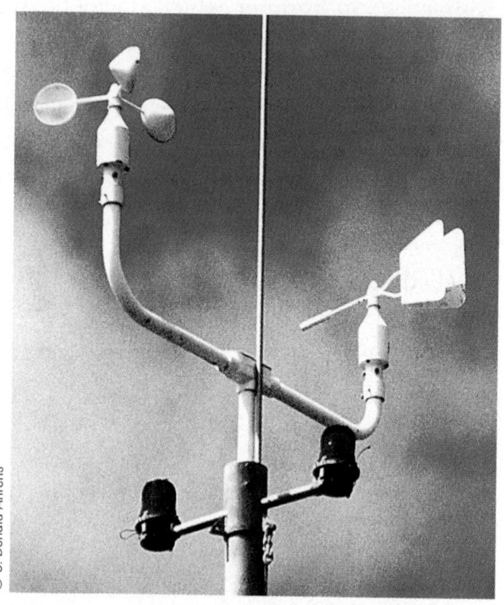

Figure 2 A wind vane and a cup anemometer. These instruments are part of the ASOS system. (For a complete picture of the system, see Chapter 3, Fig. 3, p. 74.)

Doppler radar has also been employed to obtain a vertical profile of wind speed and direction up to an altitude of 10 miles or so above the ground. Such a profile is called a *wind sounding,* and the radar, a *wind profiler* (or simply a *profiler).* Doppler radar emits pulses of microwave energy that strike irregularities in moisture and temperature created by turbulent, twisting eddies that move with the wind. The Doppler wind profilers are so sensitive that they can translate the energy returning from these eddies into a vertical picture of wind speed and direction as shown in Fig. 3.

Upper-air observations of wind can be made by satellites. Geostationary satellites positioned above a particular location show the movement of clouds, which is translated into wind direction and speed. Satellites now measure surface winds above the ocean by observing the roughness of the sea. The *QuickScat* satellite, for example, is equipped with an instrument (called a *scatterometer*) that sends out a microwave pulse of energy that travels through the clouds, down to the sea surface. A portion of this energy is bounced back to the satellite. The amount of energy returning depends on the roughness of the sea—rougher seas return more energy. Since the sea's roughness depends upon the strength of the wind blowing over it, the intensity of the returning energy can be translated into surface wind speed and direction, as illustrated in Fig. 4.

upward to an altitude near 1000 m or 3000 ft above the surface, but this altitude may vary somewhat since both strong winds and rough terrain can extend the region of frictional influence.

In ❯ Fig. 7.21, the wind aloft is blowing at a level above the frictional influence of the ground. At this level, the wind is approximately geostrophic and blows parallel to the isobars with the pressure gradient force (*PGF*) on its left balanced by the Coriolis force (*CF*) on its right. Notice, however, that at the surface the wind speed is slower. Apparently, the same pressure gradient force aloft

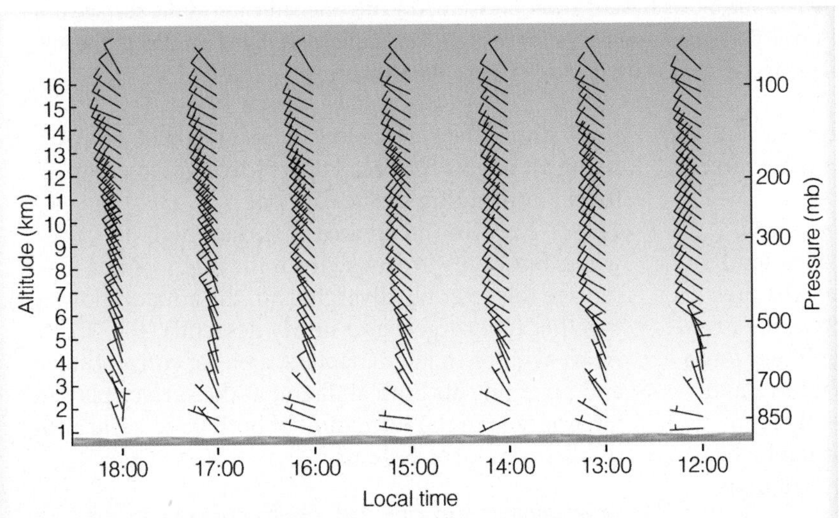

Figure 3 A profile of wind direction and speed above Hillsboro, Kansas, on June 28, 2006.

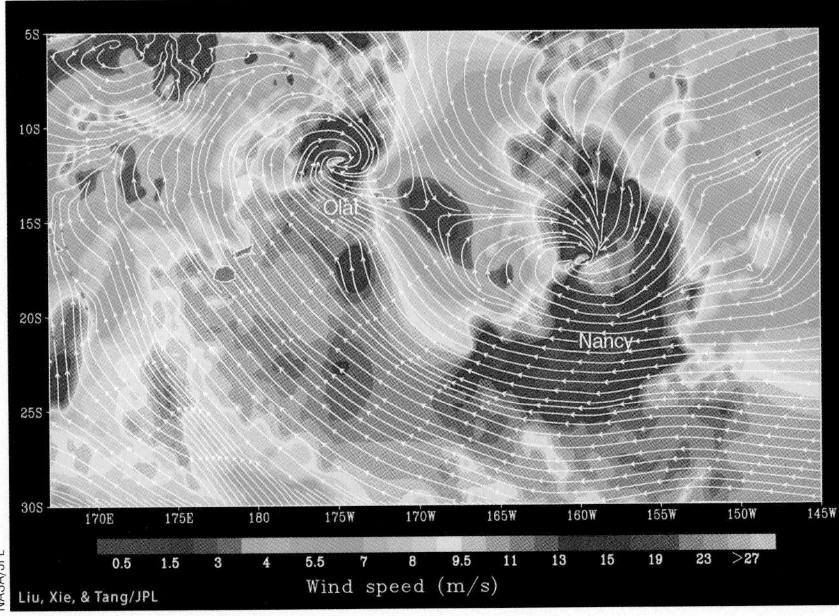

Figure 4 A satellite image of wind direction and wind speed obtained from the *QuickScat* satellite over the South Pacific Ocean on February 15, 2005. Wind direction is shown with arrows. Wind speed is indicated by colors, where purple represents the lightest winds and light pink the strongest winds. Names on the image represent the tropical cyclones Olaf and Nancy.

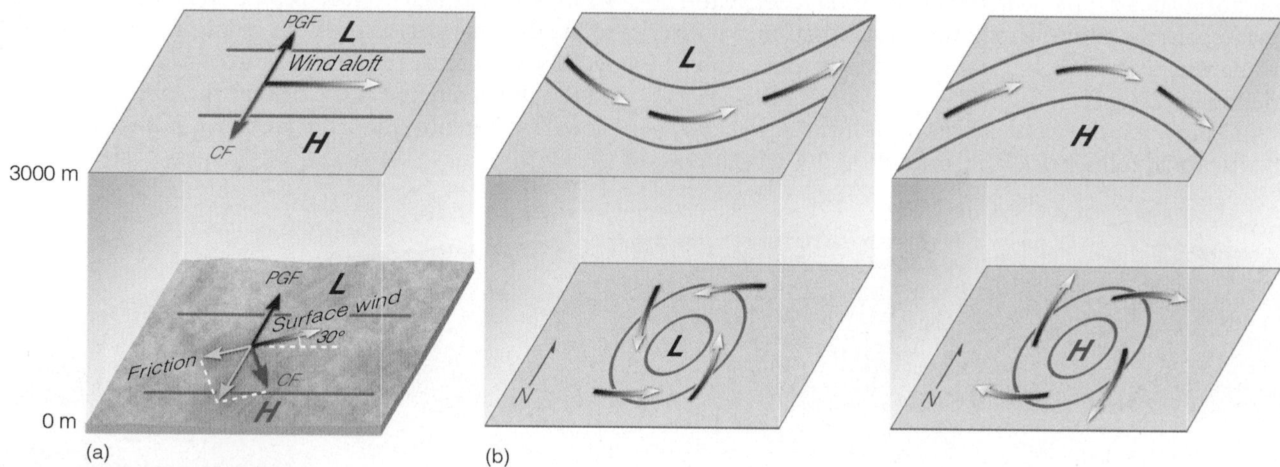

FIGURE 7.21 (a) The effect of surface friction is to slow down the wind so that, near the ground, the wind crosses the isobars and blows toward lower pressure. (b) This phenomenon at the surface produces an inflow of air around a surface low and an outflow of air around a surface high. Aloft, the winds blow parallel to the lines, usually in a wavy west-to-east pattern. (Both diagrams (a) and (b) are in the Northern Hemisphere.)

will not produce the same wind speed at the surface, and the wind at the surface will not blow in the same direction as it does aloft.

Near the surface, *friction reduces the wind speed, which in turn reduces the Coriolis force.* Consequently, the weaker Coriolis force no longer balances the pressure gradient force, and the wind blows across the isobars toward lower pressure. The pressure gradient force is now balanced by the sum of the frictional force and the Coriolis force. Therefore, in the Northern Hemisphere, we find surface winds blowing counterclockwise and *into* a low; they flow clockwise and *out* of a high (see Fig. 7.21b).

In the Southern Hemisphere, winds blow *clockwise* and *inward* around surface lows, *counterclockwise* and *outward* around surface highs.

Because surface winds blow in toward the center of a low-pressure area, the converging surface air must go somewhere. Since it can't go into the ground, it slowly rises. Above the surface low (at about 20,000 ft or so), the rising air begins to spread apart (diverge) to compensate for the converging surface air (see Fig. 7.22). Around an area of surface high pressure, winds blow outward away from the high's center. To replace this laterally diverging air, the air aloft flows together (converges) and slowly descends (Fig. 7.22). As we will see in later chapters, areas of converging and diverging air both aloft and at the surface play an integral role in the formation of both large-scale and small-scale storm systems.

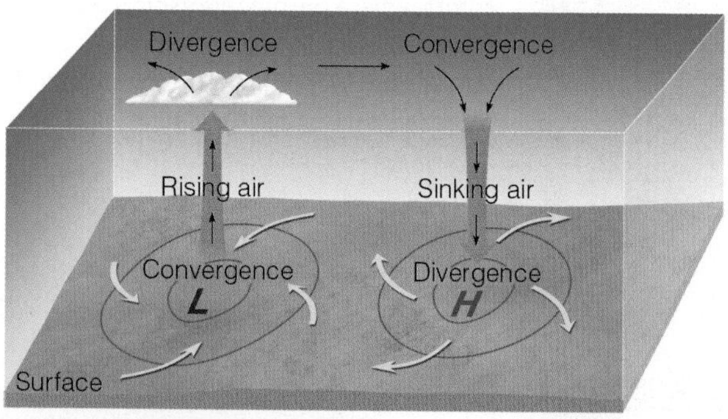

FIGURE 7.22 Winds and air motions associated with surface highs and lows in the Northern Hemisphere.

LOCATING THE CENTER OF STORMS By observing the wind, we can locate the center of storms. The storms we are referring to here are the large-scale, mid-latitude cyclonic storms that cover vast regions, usually on the order of hundreds of square miles. Suppose, for example, you are standing at the surface in the Northern Hemisphere and you notice middle- or high-level clouds drifting overhead from south to north. Since the clouds are blown by the wind, the wind aloft must be a south wind as illustrated in Fig. 7.23. We know from an earlier section that the wind aloft blows almost parallel to the isobars (or contours), with lower pressure to its left and higher pressure to its right.

Therefore, if you stand with your back to the wind aloft, the center of lowest pressure (and the center of the cyclonic storm system) will be to your left, which in Fig. 7.23a is to your west.* Weather in the middle latitudes of the Northern Hemisphere tends to move from west to east. As a consequence, the storm system and

*This statement for wind and pressure aloft in the Northern Hemisphere is often referred to as *Buys-Ballot's law*, after the Dutch meteorologist Christoph Buys-Ballot (1817–1890), who formulated it.

any weather fronts have yet to pass your area, and later on you may be in for extreme weather in the form of thunderstorms and high surface winds.

Suppose there are no clouds above you and the only wind you can observe is a surface wind blowing from the southeast as illustrated in Fig. 7.23b. We can locate the center of the surface mid-latitude cyclonic storm in a similar way we located the upper-level storm center, but because of friction we need to make a slight modification. Now, if you stand with your back to the surface wind and then turn clockwise about 30° (to account for surface friction), the center of lowest pressure will be to your left, or to the west of you, as shown in Fig. 7.23b. And, again, you may be in for extreme weather as the storm system approaches.

The Influence of Extreme Winds

A small increase in wind speed can greatly increase the wind force on an object. The reason for this fact is that the amount of force exerted by the wind over an area increases as the square of the wind velocity. This relationship is shown by

$$F \sim V^2,$$

where F is the force of the wind and V is the wind velocity. From this relationship, we can see that if the wind velocity doubles, the force goes up by a factor of 2^2, or 4, which means that walking into a 20 mi/hr wind requires four times as much effort as walking into a 10 mi/hr wind. This fact also explains why a wind of 80 mi/hr blowing against a building will likely only cause windows to break; whereas, a wind of 120 mi/hr blowing against the same building can produce moderate structural damage.

A convenient scale for estimating the strength of the wind is the **Beaufort wind scale**, invented in the early nineteenth century by Admiral Sir Francis Beaufort of the British Navy. Although modified and modernized, it still uses the movement of visible objects (such as trees and flags) to estimate how strong the wind is blowing (see ❯ Table 7.2).

STRONG WINDS BLOWING OVER LAND Strong winds blow down trees, overturn trucks, and even move railroad cars (see ❯ Fig. 7.24). For example, in February, 1965, the wind presented people in North Dakota with a "ghost train" as it pushed five railroad cars from Portal to Minot (about 80 miles) without a locomotive. And on May 2, 2009, while the Dallas Cowboys rookie football players were going through workouts at the indoor practice facilities near Dallas, Texas, a strong wind—

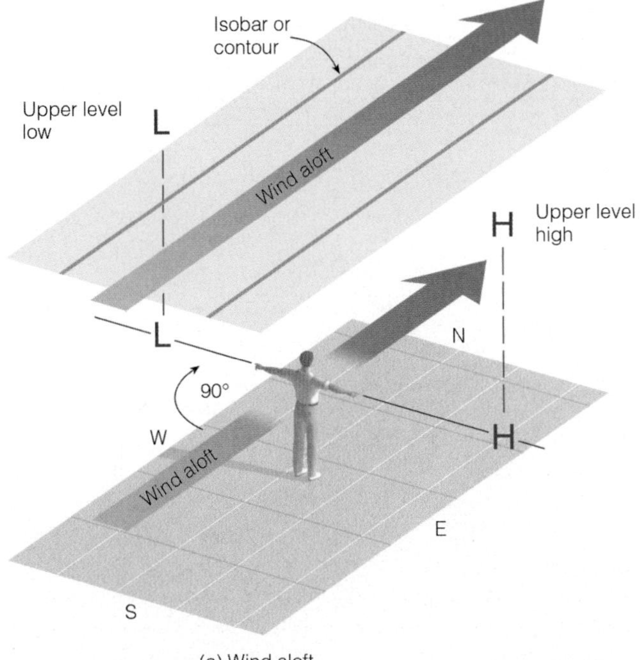

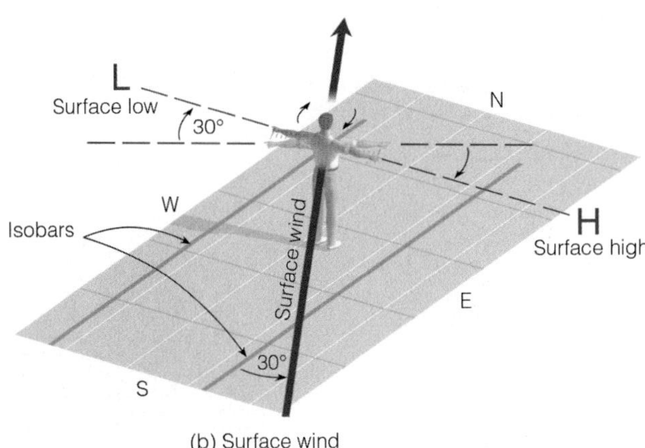

Northern Hemisphere

❯ FIGURE 7.23 (a) In the Northern Hemisphere, if you stand with the wind aloft at your back, lower pressure aloft (and the center of the upper-level, mid-latitude cyclone) will be to your left. (b) At the surface, the center of lowest pressure and the mid-latitude cyclone will be to your left if, with your back to the wind, you turn clockwise about 30°.

Table 7.2

Beaufort Wind Scale (over land)

BEAUFORT NUMBER	DESCRIPTION	WIND SPEED Mi/Hr	WIND SPEED Knots	OBSERVATIONS
0	Calm	0-1	0-1	Smoke rises vertically
1	Light air	1-3	1-3	Direction of wind shown by drifting smoke, but not by wind vanes
2	Light breeze	4-7	4-6	Wind felt on face; leaves rustle; wind vanes moved by wind; flags stir
3	Gentle breeze	8-12	7-10	Leaves and small twigs move; wind will extend light flag
4	Moderate breeze	13-18	11-16	Wind raises dust and loose paper; small branches move; flags flap
5	Fresh breeze	19-24	17-21	Small trees with leaves begin to sway; flags ripple
6	Strong breeze	25-31	22-27	Large tree branches in motion; umbrellas used with difficulty
7	Moderate gale	32-38	28-33	Whole trees in motion; exerts a strong force on people walking into wind; flags extend
8	Fresh gale	39-46	34-40	Wind breaks twigs off trees; walking is difficult
9	Strong gale	47-54	41-47	Slight structural damage; slates and tiles torn from roofs
10	Whole gale	55-63	48-55	Trees are broken and uprooted
11	Storm	64-74	56-64	Trees are uprooted and blown some distance
12	Hurricane	≥ 75	≥ 65	Winds produce extensive damage

FIGURE 7.24

The strong winds of Hurricane Jeanne blew this truck over and onto an SUV at a travel center parking lot in Vero Beach, Florida, on September 26, 2004.

FIGURE 7.25 Strong winds flowing past an obstruction, such as these hills, can produce a reverse flow of air that strikes an object from the side opposite the general wind direction. A car confronted with this situation may actually move into the center lane as the driver adjusts to the changing wind direction.

estimated at over 70 mi/hr—ripped the roof off the facility, injuring 12 people, two seriously.

Wind blowing with sufficient force to rip the roof off a training facility is uncommon. However, wind blowing with enough force to move an automobile is very common, especially when the automobile is exposed to a strong crosswind. On a normal road, the force of a crosswind is usually insufficient to move a car sideways because of the reduced wind flow near the ground. However, when the car crosses a high bridge, where the frictional influence of the ground is reduced, the increased wind speed can be felt by the driver. Near the top of a high bridge, where the wind flow is typically strongest, complicated, turbulent whirls (eddies) pound against the car's side as the air moves past obstructions, such as guard railings and posts. In a strong wind, these eddies may even break into extremely turbulent whirls that buffet the car, causing difficult handling as it moves from side to side. If there is a wall on the bridge, the wind may swirl around and strike the car from the side opposite the wind direction, producing hazardous driving conditions.

A similar effect occurs where the wind moves over low hills paralleling a highway (see Fig. 7.25). When the vehicle moves by the obstruction, a wind gust from the opposite direction can suddenly and without warning push it to the opposite side. This wind hazard is a special problem for trucks, campers, and trailers, and highway signs warning of gusty wind areas are often posted.

Strong winds blowing over the earth's surface have a number of interesting effects. When the wind blows over exposed soil, it takes an active role in shaping the landscape. In the desert, for example, loose particles of sand and dust can be lifted from the surface and carried away by the wind, leaving the surface lower than it once was. These same winds may also move rocks across a wet ground surface as shown in Fig. 7.26.

On a landscape covered with snow, strong winds may lift the snow and blow it horizontally. If the snow-covered ground should cause the winds to break into twisting eddies, and if the snow on the ground is moist and sticky, some of it may be picked up by the wind and sent rolling. As it rolls along, it collects more snow and grows bigger. If the wind is sufficiently strong, the moving clump of snow becomes cylindrical, often with a

FIGURE 7.26 Winds may have helped to push a rock across a wet surface at Death Valley National Park, California.

hole extending through it lengthwise. These **snow rollers** may range from egg size to that of small barrels (see ❯ Fig. 7.27). The tracks they make in the snow are typically less than 1 centimeter deep and several meters long. Snow rollers are rare, but, when they occur, they create a striking winter scene. In populated areas, they may escape notice as they are often mistaken as having been made by children rather than by nature.

At many locations, the wind blows more frequently from one direction than from any other. The **prevailing wind** is the name given to the wind direction most often observed during a given time period. Prevailing winds can greatly affect the climate of a region. For example, where the prevailing winds are upslope, the rising, cooling air makes clouds, fog, and precipitation more likely than where the winds are downslope. Prevailing onshore winds in summer carry moisture, cool air, and fog into coastal regions, whereas prevailing offshore breezes carry warmer and drier air into the same locations. In the high country, strong prevailing winds can bend and twist tree branches toward the downwind side, producing *wind-sculptured "flag" trees* (see ❯ Fig. 7.28).

❯ FIGURE 7.27 Snow rollers—natural cylindrical rolls of snow—grow larger as the wind blows them down a hillside.

❯ FIGURE 7.28 In the high country, trees standing unprotected from strong prevailing winds are often sculpted into "flag" trees, such as these trees in Wyoming.

EXTREME WINDS AND WATER The impact of the wind on the earth's surface is not limited to land; wind also influences water—it makes waves, which sometimes may cause a great deal of destruction. Waves forming by wind blowing over the surface of the water are known as **wind waves**. Just as air blowing over the top of a water-filled pan creates tiny ripples, so waves are created as the frictional drag of the wind transfers energy to the water. In general, the greater the wind speed, the greater the amount of energy added, and the higher the waves will be. Actually, the amount of energy transferred to the water (and thus the height to which a wave can build) depends upon three factors:

1. the wind speed
2. the length of time that the wind blows over the water
3. the *fetch*, or distance, of deep water over which the wind blows

A sustained 55 mi/hr wind blowing steadily for nearly three days over a distance of about 1600 miles can generate waves with an average height near 50 feet. Thus, a huge stationary storm system centered somewhere over the open sea is capable of creating large waves with heights occasionally measuring 100 feet. Even winds blowing over a relatively small area can generate waves high enough to capsize a small boat. This situation apparently happened on February 28, 2009, off the Florida Gulf Coast when a line of thunderstorms produced winds and waves that capsized a 21-foot fishing boat. Three of four friends onboard drowned, including National Football League football players Marquis Cooper, linebacker of the Oakland Raiders, and free-agent defensive lineman Corey Smith.

Traveling in the open ocean, waves represent a form of energy. As they move into a region of weaker winds, they gradually change: Their crests become lower and more rounded, forming what are commonly called *swells*. When waves reach a shoreline they transfer their energy—sometimes catastrophically—to the coast and structures along it. High storm-induced waves can hurl thousands of tons of water against the shore. If this happens during an unusually high tide, resort homes overlooking the ocean can be pounded by the surf into a twisted mass of board and nails. Bear in mind that the storm creating these waves may be thousands of miles away and, in fact, may never reach the shore. Some of the largest, most damaging waves ever to strike the beach communities of Southern California arrived on what was described as "one of the clearest days imaginable." On the more positive side, in the Hawaiian Islands these high waves are excellent for surfboarding.

Strong winds blowing over an open body of water, such as a lake, can have an interesting effect—it can cause the water to slosh back and forth rhythmically. This sloshing causes the water level to periodically rise and fall, much like water does at both ends of a bathtub when the water is disturbed. Such water waves that oscillate back and forth are called **seiches** (pronounced "sayshes"). In addition to strong winds, seiches may also be generated by sudden changes in atmospheric pressure or by earthquakes.* Around the Great Lakes, the term seiche applies to any sudden rise in water level whether or not it oscillates. During December, 1986, seiches generated by strong easterly winds caused extensive coastal flooding along the southwestern shores of Lake Michigan. More recently, in November, 2003, strong westerly winds, gusting to more than 60 mi/hr created a seiche on Lake Erie that caused a 12-foot difference in lake level between Toledo, Ohio (on its western shore), and Buffalo, New York (on its eastern shore). (For an account of how high winds and waves can disrupt a "semester at sea," read the Focus section on p. 200.)

The wind blowing over a large body of deep water can have yet another effect—it can cause cold water from below to rise to the surface in a process called **upwelling.**

*Earthquakes and other disturbances on a lake floor can cause the water to slosh back and forth, producing a seiche. Earthquakes on the ocean basin floor can cause a tsunami, a Japanese word meaning "harbor waves" because these waves build in height as they enter a bay or harbor.

EXTREME WEATHER WATCH

More than 200 supertankers and container ships over 650 feet in length have been lost at sea over the past 20 years due to severe weather and huge waves. On February 8, 2000, the British royal research ship *Discovery* measured a gigantic 98-foot-high wave while on a scientific expedition 155 miles west of Scotland. The wind had been blowing at 50 mi/hr or greater for over 12 hours in the vicinity of the observed wave.

For upwelling to occur, the wind must flow more or less parallel to the coastline. Notice in Fig. 7.29 that summer winds tend to parallel the coastline of California. As the wind blows over the ocean, the surface water beneath it is set in motion. As the surface water moves, it bends slightly to its right due to the Coriolis effect. (Remember, it would bend to the left in the Southern Hemisphere.) The water beneath the surface also moves, and it too bends slightly to its right. The net effect of this phenomenon is that a rather shallow layer of surface water moves at right angles to the surface wind and heads seaward. As the surface water drifts away from the coast, cold, nutrient-rich water from below rises (upwells) to replace it. Upwelling is strongest and surface water is coolest where the wind parallels the coast, such as it does in summer along the coast of northern California.

Because of the cold coastal water, summertime weather along the West Coast often consists of low clouds and fog, as the air over the water is chilled to its saturation point. On the brighter side, upwelling produces good fishing, as higher concentrations of nutrients are brought to the surface. But swimming is only for the hardiest of souls, as the average surface

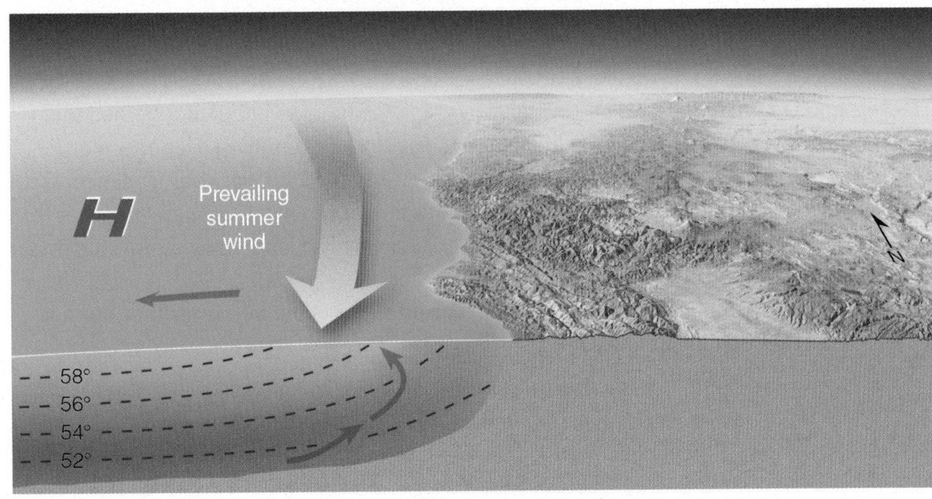

FIGURE 7.29

As winds blow parallel to the west coast of North America, surface water is transported to the right (out to sea). Cold water moves up from below (upwells) to replace the surface water. The large H represents the position of a large high-pressure area (the Pacific High) in summer.

FOCUS ON

EXTREME WEATHER

Winds, Waves, and a Seasick Semester at Sea

Most of us know someone who at one time was "sick of school." But what about someone who was "sick in school?" That's what happened to many of the nearly 700 college students who took off from Vancouver, British Columbia, on January 18, 2005, for a 100-day "semester at sea."

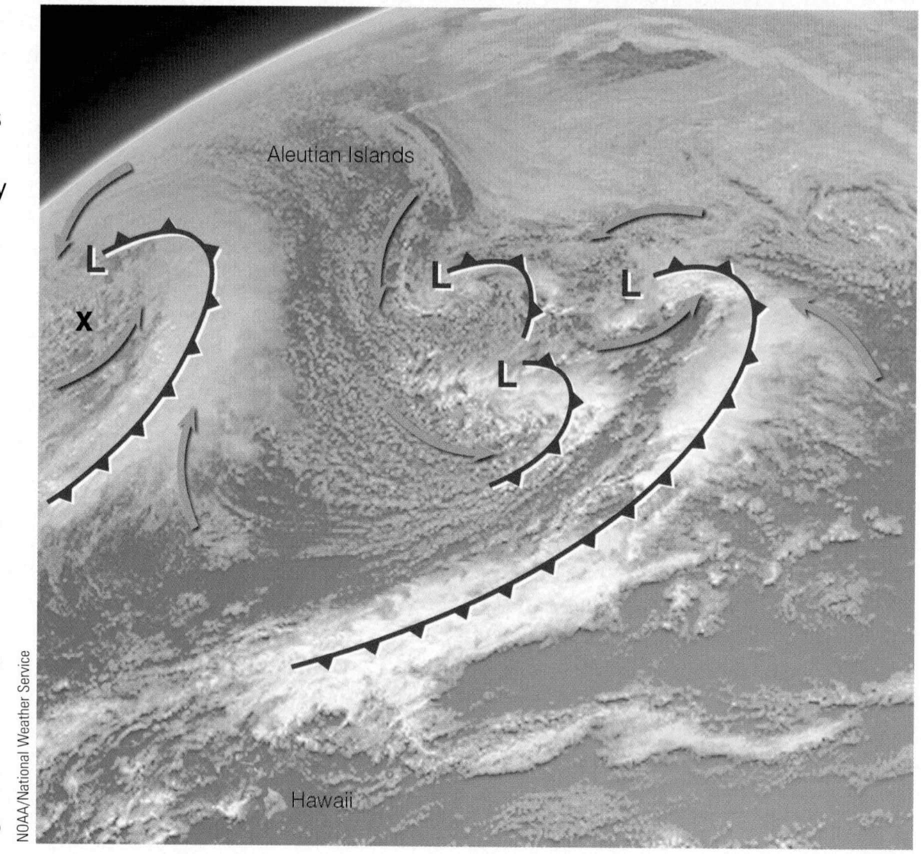

After days of riding the huge waves of the North Pacific, the weather turned real ugly, as a storm with high winds greeted the ship (see Fig. 5). In the wee hours of the morning on January 26, high winds approaching 75 mi/hr created huge waves—one estimated at 50 feet high—that smashed the glass on the bridge and shorted out the ship's electrical and navigational systems. With three of the four engines disabled by the waves, the 590-foot vessel swayed from side to side, knocking students from their bunks onto the floor, where they dodged flying books, television sets, coffee pots, and furniture.

Aleutian Islands

L

L

L

L

X

Hawaii

NOAA/National Weather Service

Figure 5 Visible satellite image of the North Pacific on January 26, 2005, shows a series of cyclonic storms with frontal bands marching across the ocean toward North America. Arrows show surface winds. The large X represents the approximate position where a passenger ship carrying hundreds of college students encountered high winds, huge waves, and rough seas.

The captain ordered everyone to put on their life jackets and get into the ship's narrow hallways. As the students huddled together, the ship continued to roll from side to side as waves battered it. Students found themselves tumbling over one another as they slid across the floor. The storm with its strong winds gradually subsided and the ship was soon under control. The vessel with its cargo of anxious students then limped into Honolulu, Hawaii, for repairs. Fortunately, except for a few bumps and bruises and many nauseated passengers, no one was seriously injured, and many of the students continued their "semester at sea" in the air, flying from one destination to another.

What these students learned firsthand on this adventure was how the interaction between strong winds and the ocean can have a rather exciting, if not violent, outcome.

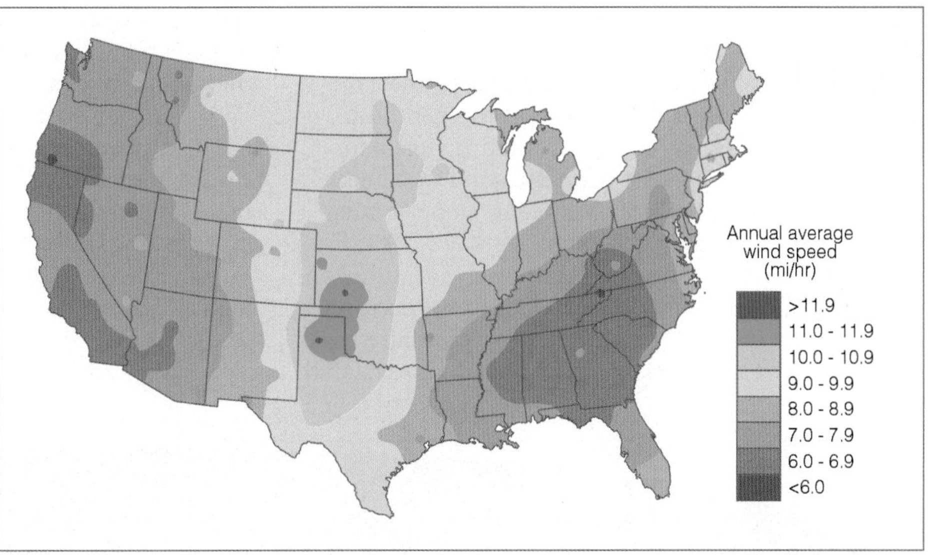

FIGURE 7.30

Annual average wind speeds across the United States.

water temperature along the Pacific Coast in summer is nearly 20°F colder than the average coastal water temperature found at the same latitude along the Atlantic Coast.

Windy Places

Figure 7.30 shows the annual average wind speed across the United States. Notice that the windiest region is in the Central Plains, where several locations have average winds exceeding 12 mi/hr. Other windy spots include Alaska (where Gold Bay's average wind is 16.9 mi/hr), Hawaii, and in a few places along the Atlantic and Pacific coasts.

Mountaintops and passes tend to be windy, especially where air funnels through a narrow gap. In fact, the windiest place in the United States (where winds are measured by instruments) is at the top of Mount Washington, New Hampshire. At an elevation of 6,288 feet above sea level, this site experiences an average wind speed of 35.3 mi/hr, with the month of January averaging gale-force winds of 46.3 mi/hr. In fact, hurricane-force winds in excess of 74 mi/hr occur there on average about 100 days a year. Probably the windiest place closer to sea level is the Blue Hill observatory in Massachusetts. Situated on the top of a hill near Boston, this station has an average wind speed of 15.4 mi/hr.

The windiest cities in the United States are given in Table 7.3. Overall, the windiest city is Dodge City, Kansas, with an annual average wind speed of 14.0 mi/hr. Other windy cities include Amarillo, Texas (13.5 mi/hr), and Rochester, Minnesota (13.1 mi/hr). Windy large cities are Boston (12.5 mi/hr), La Guardia Airport in New York City (12.2 mi/hr), and Oklahoma City (12.3 mi/hr). It is interesting to note that Chicago, which has the label "the wind city," is far down the list. If you are interested in how Chicago got its label, read the Focus section on p.202.

Table 7.3

Annual Average Wind Speed at Selected Cities in the United States

CITY	ANNUAL AVERAGE WIND SPEED (MI/HR)
Dodge City, KS*	14.0
Amarillo, TX	13.5
Rochester, MN	13.1
Casper, WY	12.9
Cheyenne, WY	12.9
Kahului, HI	12.8
Great Falls, MT	12.7
Goodland, KS	12.6
Boston, MA	12.5
Lubbock, TX	12.4
Lihue, HI	12.3
Wichita, KS	12.3
Fargo, ND	12.3
Oklahoma City, OK	12.3
New York, NY (La Guardia Airport)	12.2
San Francisco, CA (Airport)	10.6
Chicago, IL	10.4
Philadelphia, PA	9.6
Washington, DC	9.4
Medford, OR**	4.8

* Windiest city
** Least windy city

Extreme Winds

Estimates are that the maximum speed at which the wind can blow at sea level is in the range of 200 to 225 mi/hr. Above this speed, friction with the earth's surface creates such a drag on the wind that it cannot blow any faster. Indeed, no wind has ever been measured near sea level above 207 mi/hr (see Fig. 7.31).

Wind speeds in excess of 225 mi/hr are possible on mountaintops, in areas that may locally funnel winds into narrow valleys, and in rapidly rotating violent tornadoes. In fact, the greatest estimated wind speed near the surface (301 mi/hr) was obtained by Doppler radar inside a large tornado in Oklahoma on May 3, 1999. The strongest "official" wind speed ever reliably measured on the earth's surface by instruments (231 mi/hr) occurred at the summit of Mt. Washington, New Hampshire, over 6000 ft above sea level on April 12,1934. Moreover, wind speeds above 200 mi/hr may have occurred on Cannon Mountain in New Hampshire on April 2, 1973, and on the summit of Grandfather Mountain in North Carolina on January 26, 2006. At both locations, the anemometer (an instrument that measures wind speed) recorded and stuck for several seconds at 199.5 mi/hr—the maximum capability of the instruments in use at the time. Aircraft investigating strong tropical cyclones have occasionally measured winds above 200 mi/hr, but only at relatively high altitudes.

FOCUS ON

A SPECIAL TOPIC

How Chicago Came to Be Known as the "Windy City"

Is Chicago really windy? According to Table 7.3, Chicago, with an annual average wind speed of 10.4 mi/hr, does not come close to being the windiest city in the United States. In fact, it does not even register as one of the top ten. New York City, with an annual average wind speed of 12.4 mi/hr, is actually windier. So, how is it that Chicago got the Windy City nickname?

Actually, it is not totally clear how Chicago became known as the "Windy City." An early reference to this label dates back to 1871, when a popular New York City magazine published a rhyme that referred to Chicago as "the windy old town of Chicago." In 1884, the *Chicago Tribune* enhanced the windy idea by proclaiming that Chicago made an excellent summer resort due to the cooling breezes that blow into the city from off Lake Michigan. In 1885, Chicago was referred to as the "windy city" in the *Louisville Courier* and the *New York Evening Telegraph.*

Figure 6 **Aerial view of Chicago. The downtown region is next to Lake Michigan, whereas O'Hare Airport, where wind observations are presently taken, is about 16 miles to the west.**

In 1893, several cities were bidding for the World's Fair, two of which were Chicago and New York City. In an attempt to win the bid, the Chicago Fair Committee put on a rather exuberant presentation before Congress. The editor of the *New York Sun,* Charles Dana, wrote an article about the Chicago presenters, stating: "Don't pay attention to the nonsensical claims of the windy city. Its people could not build a world's fair even if they won it." The "windy city" label was now firmly in place and the nickname has stuck ever since.

But hold on, maybe Chicago is windier than the data imply. Wind measurements for Chicago have been taken at O'Hare Airport for the past forty years or so. Earlier wind measurements were taken atop a building in the downtown district, closer to the lake and miles to the east of the airport (see Fig. 6). The annual average wind speed from the downtown site is 15.8 mi/hr, which is windier than Dodge City—the windiest city in Table 7.3.

This increase in wind speed is no doubt due in part to the fact that the measurements were taken at a higher level than those at the airport. Another influence on wind speed could be the summer breezes that blow into the heart of the city from the lake. At street level, it's windy and gusty because the tall downtown buildings help to increase the wind speed by tunneling the air through narrow passages. So, if you really want to know why Chicago is known as the Windy City, ask anyone who is trying to walk against the wind on a blustery day in downtown Chicago.

When measuring high winds, it's interesting to note that there are few locations in the world that have in place anemometers capable of measuring wind speeds over 200 mi/hr. By the same token, many instruments are simply blown away by winds of this magnitude. Furthermore, the odds are slim that a storm such as a tropical cyclone with winds in excess of 200 mi/hr will strike land, and even slimmer odds that such a storm will conveniently track its strongest winds over a weather station with an anemometer capable of measuring such high winds. However, when Hurricane Andrew came ashore south of Miami on August 24, 1992, its strongest winds passed directly over the National Hurricane Center facility in Coral Cables. The anemometer there gave out after measuring winds as strong as 164 mi/hr. Wind speeds during the peak of the storm were estimated to be at least 175 mi/hr. Additional information on extreme wind events is provided in Fig. 7.31.

EXTREME WEATHER WATCH

Although the wind gust of 231 mi/hr atop Mt. Washington, New Hampshire, is considered the official world-record wind speed, several other stronger wind gusts have been measured but never officially accepted. A 3-second wind gust of 253 mi/hr was measured on Barrow Island off the coast of Western Australia on April 10, 1996, when tropical cyclone Olivia passed over a natural gas facility owned by the Chevron Oil Corporation. Although the anemometer in use during the storm was never considered inaccurate, the record has not been accepted as official because the wind speed was so out of range for what might be expected from a cyclone of Olivia's intensity, and no other wind measurements in the vicinity came close to 253 mi/hr.

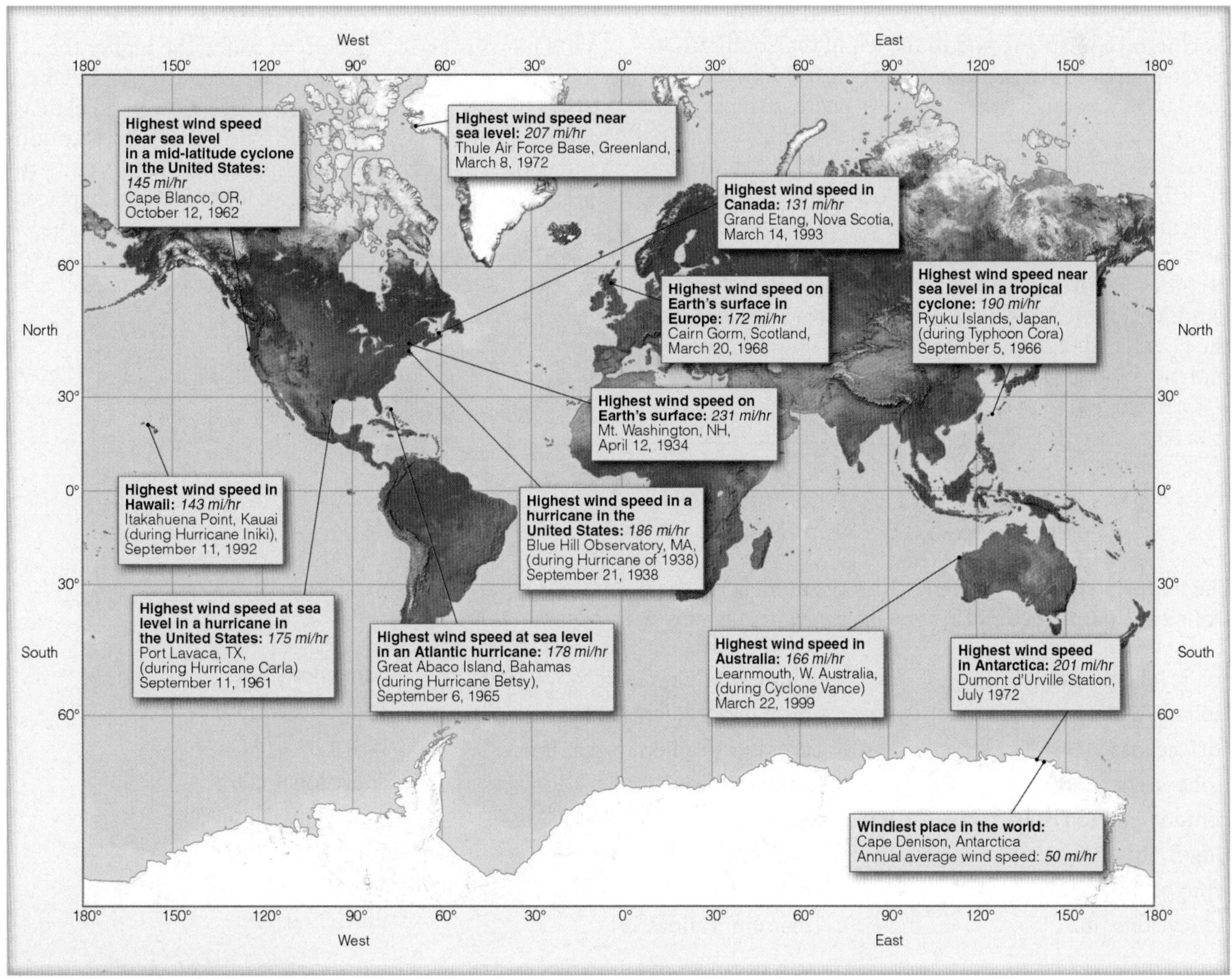

FIGURE 7.31 Record high wind speeds (mi/hr) throughout the world.

Summary

In this chapter we examined why and how the wind blows. We looked at surface and upper-air charts and found that surface weather maps use isobars to show horizontal variations in pressure, whereas constant pressure (isobaric) charts use contour lines to show horizontal changes in height. Height changes on an isobaric chart represent pressure changes, just as isobars do on a surface weather map. Where the air aloft is cold, the air pressure is normally lower than average; where the air aloft is warm, the air pressure is normally higher than average. Where horizontal variations in temperature exist, there is a corresponding horizontal change in pressure. The difference in pressure establishes a force—the pressure gradient force—that starts the air moving from higher toward lower pressure.

Once the air is set in motion, the Coriolis force bends the moving air to the right of its intended path in the Northern Hemisphere and to the left in the Southern Hemisphere. Above the level of surface friction, the wind is bent enough so that it blows nearly parallel to the isobars, or contours. Where the wind blows in a straight-line path, and a balance exists between the pressure gradient force and the Coriolis force, the wind is termed geostrophic. Where the wind blows parallel to curved isobars (or contours), the centripetal acceleration becomes important, and the wind is called a gradient wind. When the wind-flow pattern aloft is west-to-east, the flow is called *zonal;* where the wind flow aloft is more north-south, the flow is called *meridional.*

The interaction of the forces causes the wind in the Northern Hemisphere to blow clockwise around regions of high pressure and counterclockwise around areas of low pressure. The effect of surface friction is to slow down the wind. This causes the surface air to blow across the isobars from higher pressure toward lower pressure. Consequently, in both hemispheres, surface winds blow outward, away from the center of a high, and inward, toward the center of a low.

We examined extreme winds that blow over the earth's surface and observed that as the wind speed doubles, the force of the wind blowing against an object increases by a factor of four. Strong winds can blow down trees, overturn trucks, and move desert rocks. Over a snow surface, winds can produce snow rollers. Where high winds blow over a ridge, trees may be sculpted into flag trees. Winds that blow over water create wind waves. The longer the wind blows over a large body of water, the higher the waves. High winds may also cause water in a large lake to slosh back and forth. Where winds blow parallel to a coastline, surface water beneath the wind is set in motion and cooler, deeper water may rise to the surface in a process called upwelling. Finally, we looked at windy places and found that the highest wind ever measured at the earth's surface was at the observation site atop the 6,288-foot Mount Washington in New Hampshire.

Key Terms

The following terms are listed (with page number) in the order they appear in the text. Define each. Doing so will aid you in reviewing the material covered in this chapter.

sea-level pressure, 178
isobars, 179
surface map, 179
isobaric chart, 180
contour lines (on isobaric charts), 180
ridges, 181
troughs, 181
anticyclone, 182
mid-latitude cyclonic storm, 182
pressure gradient, 184
pressure gradient force, 185
Coriolis force, 186
geostrophic wind, 188
gradient wind, 189
centripetal force, 190
meridional flow, 191
zonal flow, 191
friction layer, 192
Beaufort wind scale, 195
snow rollers, 198
prevailing wind, 198
wind waves, 198
seiches, 199
upwelling, 199

Review Questions

1. What can cause the air pressure to change at the bottom of a column of air?
2. Why does air pressure decrease with height more rapidly in cold air than in warm air?
3. How does sea-level pressure differ from station pressure? Can the two ever be the same?
4. On an upper-level chart, is cold air aloft generally associated with low or high pressure? What about warm air aloft?
5. What do Newton's first and second laws of motion tell us?
6. Explain why, in the Northern Hemisphere, the average height of contour lines on an upper-level isobaric chart tend to decrease northward.
7. What is the force that initially sets the air in motion?
8. What does the Coriolis force do to moving air
   (a) in the Northern Hemisphere?
   (b) in the Southern Hemisphere?
9. How does a steep (or strong) pressure gradient appear on a weather map?
10. Explain why on a map, closely spaced isobars (or contours) indicate strong winds, and widely spaced isobars (or contours) indicate weak winds.
11. What is a geostrophic wind? Why would you *not* expect to observe a geostrophic wind at the equator?
12. Why do upper-level winds in the middle latitudes of both hemispheres generally blow from the west?
13. Describe how the wind blows around highs and lows aloft and near the surface in the Northern Hemisphere.
14. On a surface weather map, why do the winds cross the isobars?
15. Describe the type of vertical air motions associated with surface high- and low-pressure areas.
16. If clouds above you are moving from southwest to northeast, in which general direction would you expect to find an upper-level low pressure area? If the surface wind on this same day is blowing from the southeast, in which general direction would you expect to find a surface mid-latitude cyclonic storm?
17. How are wind-sculpted flag trees able to show the prevailing wind?
18. With the same wind speed, explain why a camper is more easily moved by the wind than a car?
19. What are the necessary conditions for the development of large wind waves?
20. How can a coastal area have huge, damaging waves on a clear non-stormy day?
21. Explain how seiches can form.
22. What conditions are necessary for upwelling to occur along the west coast of North America? How does upwelling influence the formation of fog?
23. It is unlikely that near sea-level winds will ever exceed 225 mi/hr, yet the highest wind speed ever measured at the earth's surface is 231 mi/hr. Explain this apparent paradox.

Online Learning

STUDENT COMPANION WEBSITE: Visit this book's companion website at: www.cengage.com/ahrens/extreme1e and choose Chapter 7 for many study aids and ideas for further reading and research. These include flashcards, practice quizzin g, and web links.

METEOROLOGY RESOURCE CENTER: For students with access, log on at www.cengage.com/login for more assets, including animations, videos, and more. If your textbook did not come with access, visit www.CengageBrain.com to purchase.

Wind Systems

8

CONTENTS

GENERAL CIRCULATION OF THE ATMOSPHERE

JET STREAMS

EXTREME (AND NOT SO EXTREME) LOCAL WIND SYSTEMS

SUMMARY

KEY TERMS

REVIEW QUESTIONS

High winds sweep dust high into the air during May, 2005, near Arnold, Nebraska.

The hottest, coldest, driest, and wettest places on earth are in many ways tied to the general flow of air around our planet. When we look at this worldwide average flow of air, we are examining what scientists call the **general circulation of the atmosphere**. This pattern of global wind flow will help answer many questions about extreme weather and climate: Why, for example, certain regions of the world are extremely dry, whereas others are extremely wet; why some regions have exceptionally wet summers and others, exceptionally wet winters. And why the wind in middle latitudes tends to blow from west to east, whereas in the tropics they blow from east to west. Tied into the general circulation is the jet stream, which plays a major role in extreme weather, especially in the formation of thunderstorms, tornadoes, and huge middle-latitude cyclonic storms. As a consequence, we will first examine the large scale circulation of winds around our planet. We will then look at how these winds influence global participation. Our attention then turns to the jet stream and its movement within the large-scale circulation of air. Near the end of the chapter, we will examine a variety of smaller-scale winds, such as the chinook wind and the Santa Ana wind, describing how they form and the type of extreme weather they bring to a region.

General Circulation of the Atmosphere

Before we study the general circulation, we must remember that it only represents the *average* air flow around the world. Actual winds at any one place and at any given time may vary considerably from this average. Nevertheless, the average can answer why and how the winds blow around the world the way they do—why, for example, prevailing surface winds are northeasterly in Honolulu, Hawaii, and westerly in New York City. The average can also give a picture of the driving mechanism behind these winds, as well as a model of how heat is transported from equatorial regions poleward, keeping the climate in middle latitudes tolerable.

The underlying cause of the general circulation is the unequal heating of the earth's surface. We learned in Chapter 2 that, averaged over the entire earth, incoming solar radiation is roughly equal to outgoing earth radiation. However, we also know that this energy balance is not maintained for each latitude, since the tropics experience a net gain in energy, while polar regions suffer a net loss. To balance these inequities, the atmosphere transports warm air poleward and cool air equatorward. Although seemingly simple, the actual flow of air is complex; certainly not everything is known about it. In order to better understand it, we will first look at some models (that is, artificially constructed simulations) that eliminate some of the complexities of the general circulation.

SINGLE-CELL MODEL The first model is the single-cell model, in which we assume that:

1. The earth's surface is uniformly covered with water (so that differential heating between land and water does not come into play).
2. The sun is always directly over the equator (so that the winds will not shift seasonally).
3. The earth does not rotate (so that the only force we need to deal with is the pressure gradient force).

With these assumptions, the general circulation of the atmosphere on the side of the earth facing the sun would look much like the representation in Fig. 8.1a, a huge thermally driven convection cell in each hemisphere. (For reference, the names of the different regions of the world and their approximate latitudes are given in Figure 8.1b.)

The circulation of air described in Fig. 8.1a is the **Hadley cell** (named after the eighteenth-century English meteorologist George Hadley, who first proposed the idea). It is driven by energy from the sun as warm air rises and cold air sinks. Excessive heating of the equatorial area produces a broad region of surface low pressure, while at the poles excessive cooling creates a region of surface high pressure. In response to the horizontal pressure gradient, cold surface polar air flows equatorward, while at higher levels air flows toward the poles. The entire circulation consists of a closed loop with rising air near the equator, sinking air over the poles, an equatorward flow of air near the surface, and a return flow aloft. In this manner, some of the excess energy of the tropics is transported as sensible and latent heat to the regions of energy deficit at the poles.

Such a simple cellular circulation as this does not actually exist on the earth. For one thing, the earth rotates, so the Coriolis force would deflect the southward-moving surface air in the Northern Hemisphere to the right, producing easterly surface winds at practically all latitudes. These winds would be moving in a direction opposite to that of the earth's rotation and, due to friction with the surface, would slow down the earth's spin. We know that this does not happen and that prevailing winds in middle latitudes actually blow from the west. Therefore, observations alone tell us that a closed circulation of air between the equator and the poles is not the proper model for a rotating earth. But this model does show us how a nonrotating planet would balance an excess of energy at the equator and a deficit at the poles. How, then, does the wind blow on a rotating planet? To answer, we will keep our model simple by retaining our first two assumptions—that is, that the earth is covered with water and that the sun is always directly above the equator.

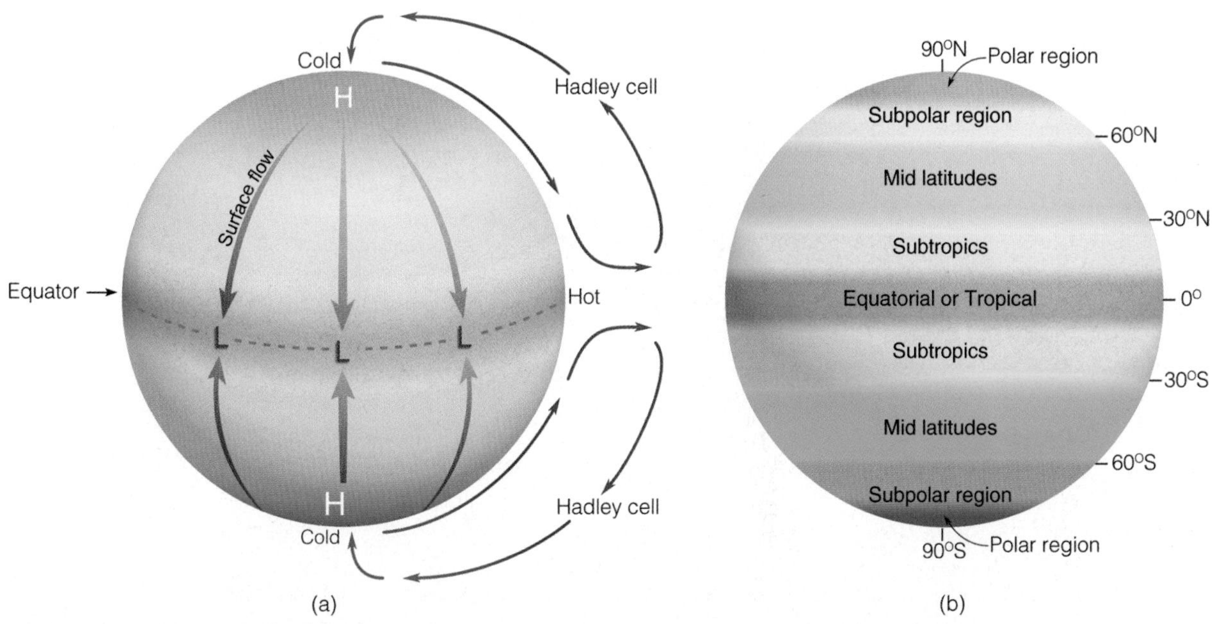

FIGURE 8.1 Diagram (a) shows the general circulation of air on the side of the earth facing the sun on a nonrotating earth uniformly covered with water and with the sun directly above the equator. (Vertical air motions are highly exaggerated in the vertical.) Diagram (b) shows the names that apply to the different regions of the world and their approximate latitudes.

THREE-CELL MODEL If we allow the earth to spin, the simple convection system breaks into a series of cells as shown in Fig. 8.2. Although this model is considerably more complex than the single-cell model, there are some similarities. The tropical regions still receive an excess of heat and the poles a deficit. In each hemisphere, three cells instead of one have the task of energy redistribution. A surface high-pressure area is located at the poles, and a broad trough of surface low pressure still exists at the equator. From the equator to latitude 30°, the circulation is the *Hadley cell.* Let's look at this model more closely by examining what happens to the air above the equator. (Refer to Fig. 8.2 as you read the following section.)

Over equatorial waters, the air is warm, horizontal pressure gradients are weak, and winds are light. This region is referred to as the **doldrums**. (The monotony of the weather in this area has given rise to the expression "down in the doldrums.") Here, warm air rises, often condensing into huge cumulus clouds and thunderstorms that liberate enormous amounts of latent heat that makes the air more buoyant and provides energy to drive the Hadley cell. The rising air reaches the tropopause, which acts like a barrier, causing the air to move laterally toward the poles. The Coriolis force deflects this poleward flow toward the right in the Northern Hemisphere and to the left in the Southern Hemisphere, providing westerly winds aloft in both hemispheres. These westerly winds then direct storms, such as mid-latitude cyclones and thunderstorms, along a more or less west-to-east path. (We will see later that these westerly winds also reach maximum velocity and produce jet streams near 30° and 60° latitudes.)

As air moves poleward from the tropics, it constantly cools by giving up infrared radiation, and at the same time it also begins to converge, especially as it approaches the middle latitudes.* This convergence (piling up) of air aloft increases the mass of air above the surface, which in turn causes the air pressure at the surface to increase. Hence, at latitudes near 30°, the convergence of air aloft produces belts of high pressure called **subtropical highs** (or anticyclones). As the converging, relatively dry air above the highs slowly descends, it warms by compression. This subsiding air produces generally clear skies and warm surface temperatures; hence, it is here that we find the major deserts of the world, such as the Sahara. It is here where we also find some of the highest temperatures ever measured on earth. (Look back at Fig. 3.5, p. 67.)

Over the oceans, the weak pressure gradients in the center of a subtropical high produce only weak winds. According to legend, sailing ships traveling to the New World were frequently becalmed in this region,

*You can see why the air converges if you have a globe of the world. Put your fingers on meridian lines at the equator and then follow the meridians poleward. Notice how the lines and your fingers bunch together in the middle latitudes.

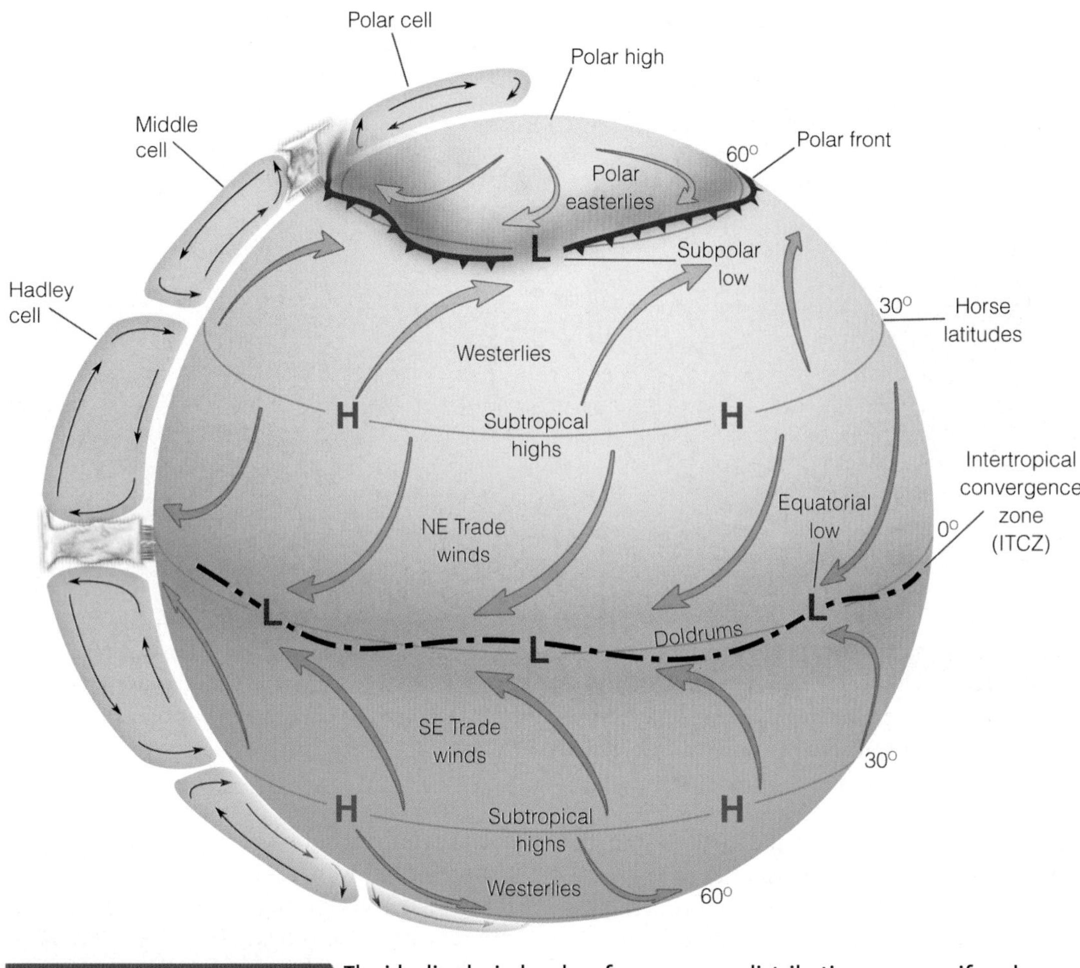

FIGURE 8.2 The idealized wind and surface-pressure distribution over a uniformly water-covered rotating earth.

and, as food and supplies dwindled, horses were either thrown overboard or eaten. As a consequence, this region is sometimes called the **horse latitudes**.

From the horse latitudes, some of the surface air moves back toward the equator. It does not flow straight back, however, because the Coriolis force deflects the air, causing it to blow from the northeast in the Northern Hemisphere and from the southeast in the Southern Hemisphere. These steady winds provided sailing ships with an ocean route to the New World; hence, these winds are called the **trade winds**. Near the equator, the *northeast trades* converge with the *southeast trades* along a boundary called the **intertropical convergence zone (ITCZ)**. In this region of surface convergence, air rises and continues its cellular journey. It is along the ITCZ where rising air develops into huge thunderstorms that drop copious amounts of rain in the form of heavy showers.

Meanwhile, at latitude 30°, not all of the surface air moves equatorward. Some air moves toward the poles and deflects toward the east, resulting in a more or less westerly air flow—called the *prevailing westerlies,* or, simply, **westerlies**—in both hemispheres. Consequently, from Texas northward into Canada, it is much more common to experience winds blowing out of the west than from the east. The westerly flow in the real world is not constant as migrating areas of high and low pressure break up the surface flow pattern from time to time. In the middle latitudes of the Southern Hemisphere, where the surface is mostly water, winds blow more steadily from the west.

As this mild air travels poleward, it encounters cold air moving down from the poles. These two air masses of contrasting temperature do not readily mix. They are separated by a boundary called the **polar front**, a zone of low pressure—the **subpolar low**—where surface air converges and rises, and storms and clouds develop. Some of the rising air returns at high levels to the horse latitudes, where it sinks back to the surface in the vicinity of the subtropical high. The middle cell is completed when surface air from the horse latitudes flows poleward toward the polar front.

Notice in Fig. 8.2 that, in the Northern Hemisphere, behind the polar front, the cold air from the poles is deflected by the Coriolis force, so that the general flow of air is from the northeast. Hence, this is the region of the **polar easterlies**. In winter, the polar front with its cold air can move into middle and subtropical latitudes, producing a cold polar outbreak. Along the front, a portion of the rising air moves poleward, and the Coriolis force deflects the air into a westerly wind at high levels. Air aloft eventually reaches the poles, slowly sinks to the surface, and flows back toward the polar front, completing the weak *polar cell.*

We can summarize all of this by referring back to Fig. 8.2 and noting that, at the surface, there are two major areas of high pressure and two major areas of low pressure. Areas of high pressure exist near latitude 30° and the poles; areas of low pressure exist over the equator and near 60° latitude in the vicinity of the polar front. By knowing the way the winds blow around these systems, we have a generalized picture of surface winds throughout the world. The trade winds extend from the subtropical high to the equator, the westerlies from the subtropical high to the polar front, and the polar easterlies from the poles to the polar front.

How does this three-cell model compare with actual observations of winds and pressure? We know, for example, that upper-level winds at middle latitudes generally blow from the west. The middle cell, however, suggests an east wind aloft as air flows equatorward. Hence, discrepancies exist between this model and atmospheric observations. This model does, however, agree closely with the winds and pressure distribution at the *surface,* and so we will examine this next.

AVERAGE SURFACE WINDS AND PRESSURE: THE REAL WORLD When we examine the real world with its continents and oceans, mountains and ice fields, we obtain an average distribution of sea-level pressure and winds for January and July, as shown in ❱ Figs. 8.3a and 8.3b. Look closely at both maps and observe that there are regions where pressure systems appear to persist throughout the year. These systems are referred to as **semipermanent highs and lows** because they move only slightly during the course of a year.

In Fig. 8.3a, we can see that there are four semipermanent pressure systems in the Northern Hemisphere during January. In the eastern Atlantic, between latitudes 25° and 35°N is the *Bermuda–Azores high,* often called the **Bermuda high**, and, in the Pacific Ocean, its counterpart, the **Pacific high**. These are the subtropical anticyclones that develop in response to the convergence of air aloft near an upper-level jet stream. Since surface winds blow clockwise around these systems, we find the trade winds to the south and the prevailing westerlies to the north. In the Southern Hemisphere, where there is relatively less land area, there is less contrast between land and water, and the subtropical highs show up as well-developed systems with a clearly defined circulation.

Where we would expect to observe the polar front (between latitudes 40° and 65°), there are two semipermanent subpolar lows. In the North Atlantic, there is the *Greenland-Icelandic low,* or simply **Icelandic low**, which covers Iceland and southern Greenland, while the **Aleutian low** sits over the Gulf of Alaska and Bering Sea near the Aleutian Islands in the North Pacific. These zones of cyclonic activity actually represent regions where numerous storms, having traveled eastward, tend to converge, especially in winter. In the Southern Hemisphere, the subpolar low forms a continuous trough that completely encircles the globe.

On the January map (Fig. 8.3a), there are other pressure systems, which are not semipermanent in nature. Over Asia, for example, there is a huge (but shallow) anticyclone called the **Siberian high**, which forms because of the intense cooling of the land. A similar (but less intense) anticyclone (called the *Canadian high*) is evident over North America.

As summer approaches, the land warms and the cold shallow highs disappear. In some regions, areas of surface low pressure replace areas of high pressure. The lows that form over the warm land are shallow *thermal lows.* On the July map (Fig. 8.3b), warm thermal lows are found over the desert southwest of the United States, over the plateau of Iran, and north of India.

When we compare the January and July maps, we can see several changes in the semipermanent pressure systems. The strong subpolar lows so well developed in January over the Northern Hemisphere are hardly discernible on the July map. The subtropical highs, however, remain dominant in both seasons. Because the sun is

EXTREME WEATHER WATCH

The semipermanent Aleutian and Icelandic low-pressure zones have experienced storms with the lowest barometric pressures ever recorded in a nontropical storm environment. On January 11, 1993, a pressure of 913 mb (26.96 in.)—the equivalent of a Category 5 hurricane—was measured at the center of an Icelandic storm near the Shetland Islands north of Scotland. An intense Aleutian cyclone deepened to 926 mb (27.35 in.) at Dutch Harbor, Alaska, on October 25, 1977.

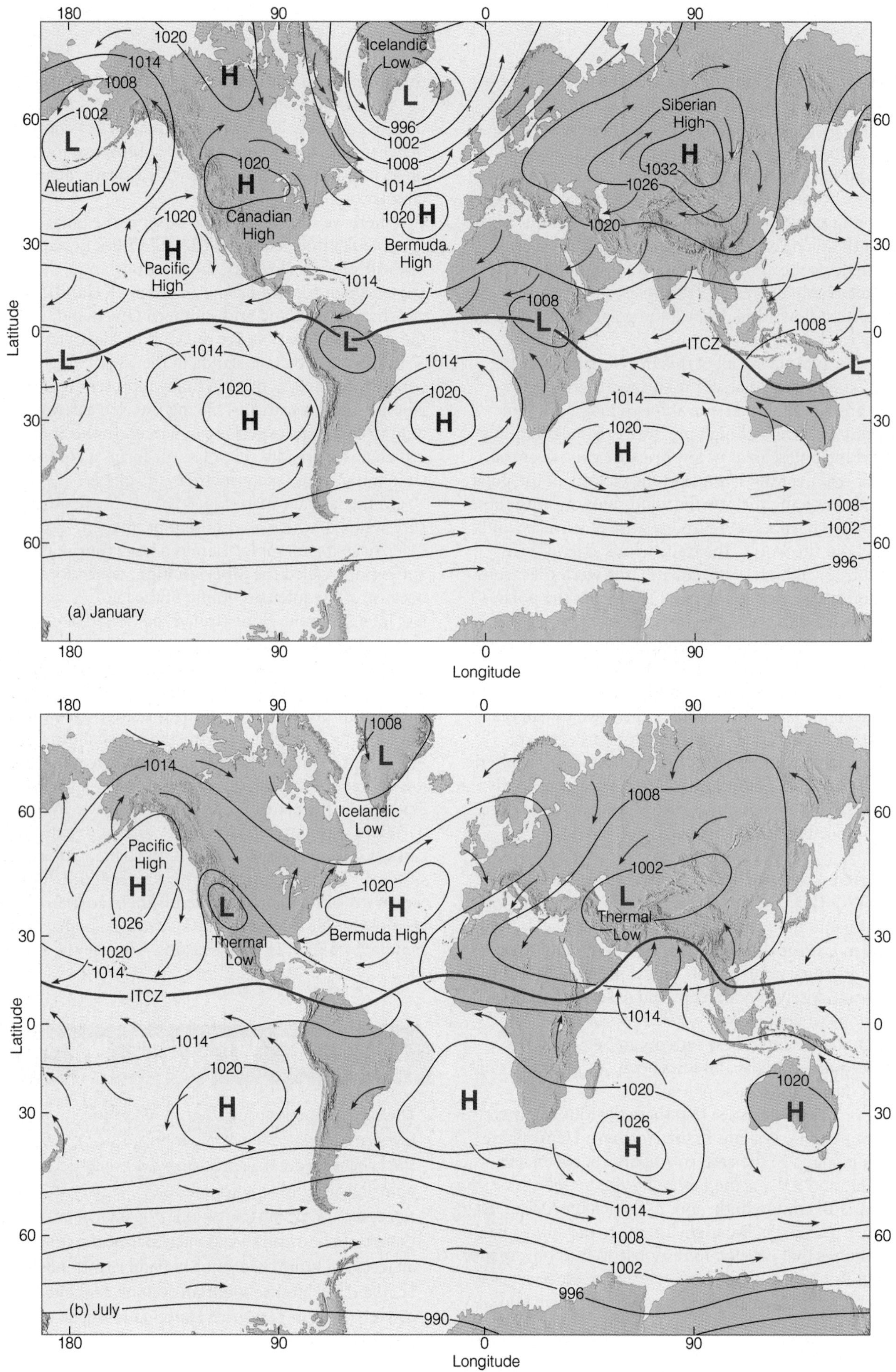

Icelandic Low
Aleutian Low
Canadian High
Pacific High
Bermuda High
Siberian High
ITCZ
(a) January
Latitude
Longitude
Thermal Low
Bermuda High
(b) July

FIGURE 8.3 (*On facing page*) Average sea-level pressure distribution and surface wind-flow patterns for January (a) and for July (b). The solid red line represents the position of the ITCZ.

overhead in the Northern Hemisphere in July and overhead in the Southern Hemisphere in January, the zone of maximum surface heating shifts seasonally. In response to this shift, the major pressure systems, wind belts, and ITCZ (heavy red line in Fig. 8.3) *shift toward the north in July and toward the south in January.**

Figure 8.4 illustrates a winter weather map where the main features of the general circulation have been displaced southward.

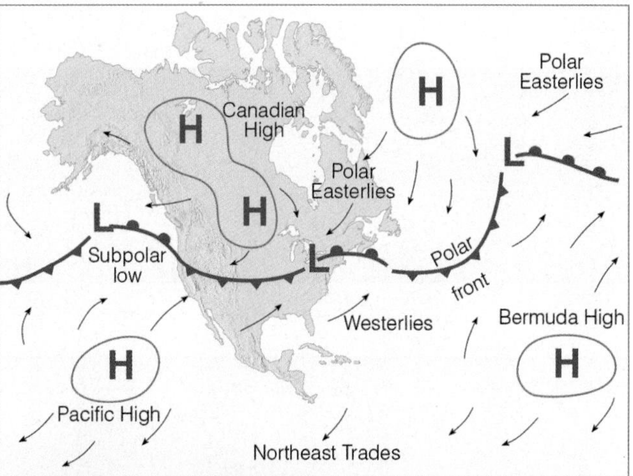

FIGURE 8.4 A winter weather map depicting the main features of the general circulation over North America. Notice that the Canadian high, polar front, and subpolar lows have all moved southward into the United States, and that the prevailing westerlies exist south of the polar front. The arrows on the map illustrate wind direction.

THE GENERAL CIRCULATION AND PRECIPITATION PATTERNS The position of the major features of the general circulation and their latitudinal displacement (which annually averages about 10° to 15°) strongly influence the precipitation of many areas. For example, on the global scale, we would expect abundant rainfall where the air rises and very little where the air sinks. Consequently, areas of high rainfall exist in the tropics, where humid air rises in conjunction with the ITCZ, and between 40° and 55° latitude, where middle-latitude storms and the polar front force air upward. Areas of low rainfall are found near 30° latitude in the vicinity of the subtropical highs and in polar regions where the air is cold and dry (see Fig. 8.5).

Poleward of the equator, between the doldrums and the horse latitudes (between 0° and 30°) the area is influenced by both the ITCZ and the subtropical high. In summer (high sun period), the subtropical high moves poleward and the ITCZ invades this area, bringing with it ample rainfall. In winter (low sun period), the subtropical high moves equatorward, bringing with it clear, dry weather. (See Fig. 8.6.)

Notice in Fig. 8.6 that in the Northern Hemisphere just north of the subtropical high there is a region where summers are dry. In some areas, such as along the coast of Southern California, the dry summer weather may last for more than three months, and it may not rain from May through September. What causes this extremely dry weather? Again, we look at the general circulation for the answer.

In Fig. 8.7, notice that during the summer the Pacific high drifts northward to a position off the California coast. Sinking air on the high's eastern side produces a strong subsidence inversion, which tends to keep sum-

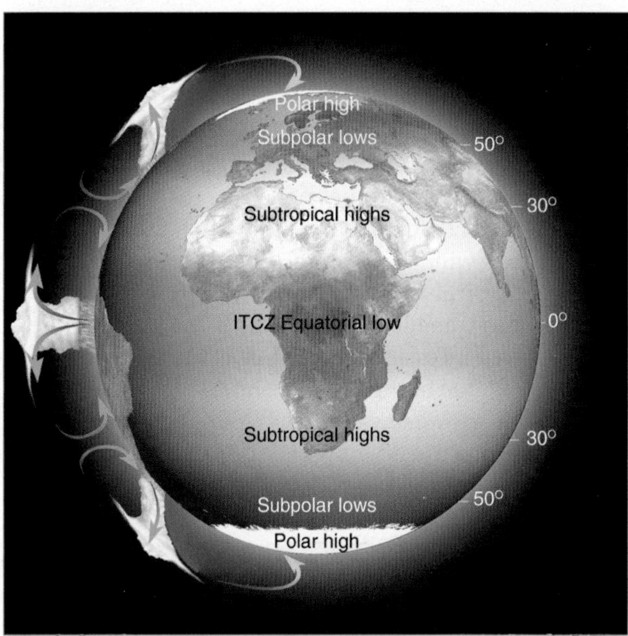

FIGURE 8.5 Rising and sinking air associated with the major pressure systems of the earth's general circulation. Where the air rises, precipitation tends to be abundant (blue shade); where the air sinks, drier regions prevail (tan shade). Note that the sinking air of the subtropical highs produces the major desert regions of the world.

*An easy way to remember the seasonal shift of pressure systems is to think of birds — in the Northern Hemisphere, they migrate south in the winter and north in the summer.

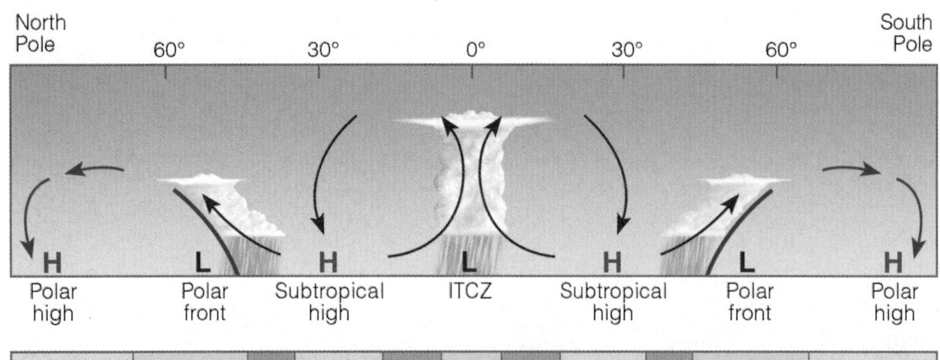

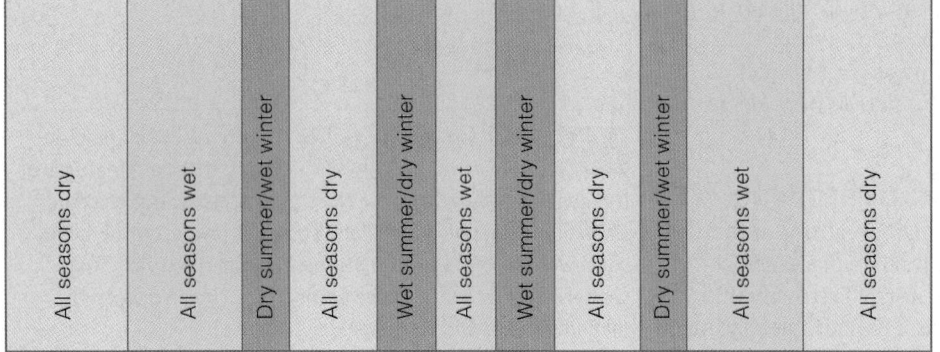

FIGURE 8.6

A vertical cross section along a line running north to south illustrates the main global regions of rising and sinking air associated with the earth's general circulation. The annual shifting of the main features of the general circulation causes wet and dry seasonal changes in different regions of the world.

mer weather along the West Coast quite dry. The rainy season typically occurs in winter when the high moves south and the polar front and storms are able to penetrate the region (look back at Fig. 8.4). Observe in Fig. 8.7 that along the East Coast of the United States this extremely dry summer does not exist. The reason for this situation is that along the East Coast, the clockwise circulation of winds around the Bermuda high brings warm subtropical air northward into the United States and southern Canada from the Gulf of Mexico and the Atlantic Ocean. Because sinking air is not as well developed on this side of the high, the humid air can rise and condense into towering cumulus clouds and thunderstorms. So, in part, it is the air motions associated with the subtropical highs that keep summer weather dry in California and moist in Georgia. (Compare the rainfall patterns for Los Angeles, California, and Atlanta, Georgia, in Fig. 8.8.)

AVERAGE WIND FLOW AND PRESSURE PATTERNS ALOFT Figures 8.9a and 8.9b are average global 500-mb charts for the months of January and July, respectively. Look at both charts carefully and observe that some of the surface features of the general circulation are reflected on these upper-air charts. On the January map, for example, both the Icelandic low and Aleutian

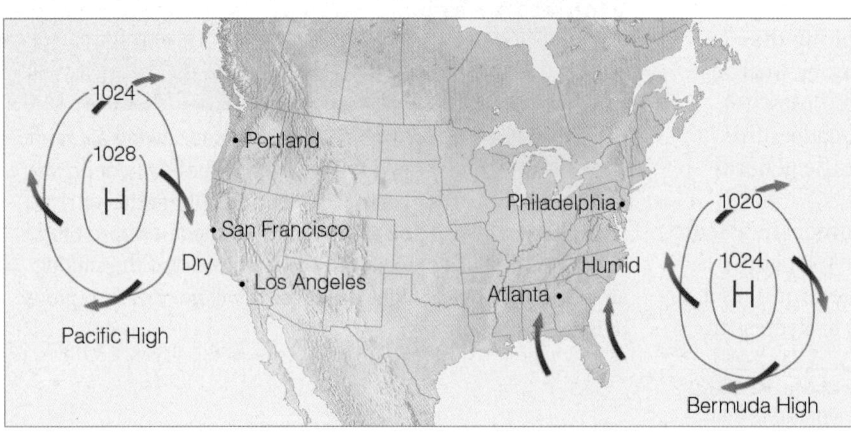

FIGURE 8.7

During the summer, the Pacific high moves northward. Sinking air along its eastern margin (over California) produces a strong subsidence inversion, which causes relatively dry weather to prevail. Along the western margin of the Bermuda high, southerly winds bring in humid air, which rises, condenses, and produces abundant rainfall.

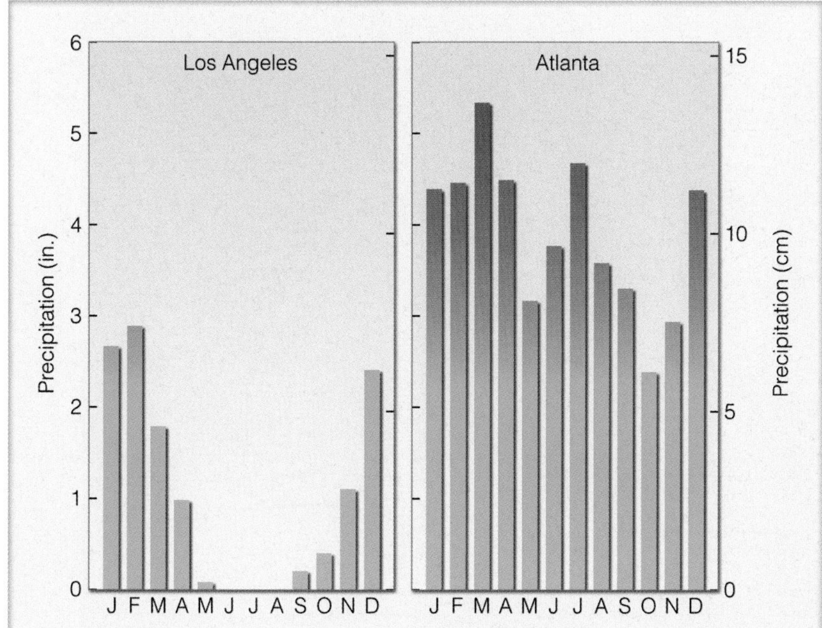

FIGURE 8.8

Average annual precipitation for Los Angeles, California, and Atlanta, Georgia.

low are located to the west of their surface counterparts. On the July map, the subtropical high-pressure areas of the Northern Hemisphere appear as belts of high height (high pressure) that tend to circle the globe south of 30°N. In both hemispheres, the air is warmer over low latitudes and colder over high latitudes. This horizontal temperature gradient establishes a horizontal pressure (contour) gradient that causes the winds to blow from the west especially in middle and high latitudes.* Notice that the temperature gradients and the contour gradients are steeper in January than July. Consequently, the winds aloft are stronger in winter than in summer. The westerly winds, however, do not extend all the way to the equator, as easterly winds appear on the equatorward side of the upper-level subtropical highs. As we will see in later chapters, westerly winds aloft in the middle latitudes tend to steer cyclonic storm systems eastward; whereas, easterly winds aloft in the tropics tend to steer tropical storm systems westward.

In middle and high latitudes, the westerly winds continue to increase in speed above the 500-mb level. An important reason for this increase in wind speed is that the north-to-south temperature gradient causes the horizontal pressure (contour) gradient to increase with height up to the tropopause. As a result, the winds increase in speed up to the tropopause. Above the tropopause, the temperature gradients reverse. This changes the pressure gradients and reduces the strength of the westerly winds. Where strong winds tend to concentrate into narrow bands at the tropopause, we find rivers of fast-flowing air—*jet streams.*

*Remember that, at this level (about 5600 m or 18,000 ft above sea level), the winds are approximately geostrophic, and tend to blow more or less parallel to the contour lines.

Jet Streams

Atmospheric **jet streams** are swiftly flowing air currents thousands of miles long, a few hundred miles wide, and less than a mile thick. Wind speeds in the central core of a jet stream often exceed 100 mi/hr and occasionally exceed 200 mi/hr. Jet streams are usually found at the tropopause at elevations between 10 and 15 km (6 and 9 mi), although they may occur at both higher and lower altitudes.

Jet streams were first encountered by high-flying military aircraft during World War II, but their existence was suspected before the war. Ground-based observations of fast-moving cirrus clouds had revealed that westerly winds aloft must be moving rapidly.

Figure 8.10 illustrates the average position of the jet streams, tropopause, and general circulation of air for the Northern Hemisphere in winter. From this diagram, we can see that there are two main jet streams, both located in tropopause gaps, where mixing between tropospheric and stratospheric air takes place. The jet stream situated near 30° latitude at about 13 km (43,000 ft) above the subtropical high is the **subtropical jet stream.*** The jet stream situated at about 10 km (33,000 ft) near the polar

*The subtropical jet stream is normally found between 20° and 30° latitude.

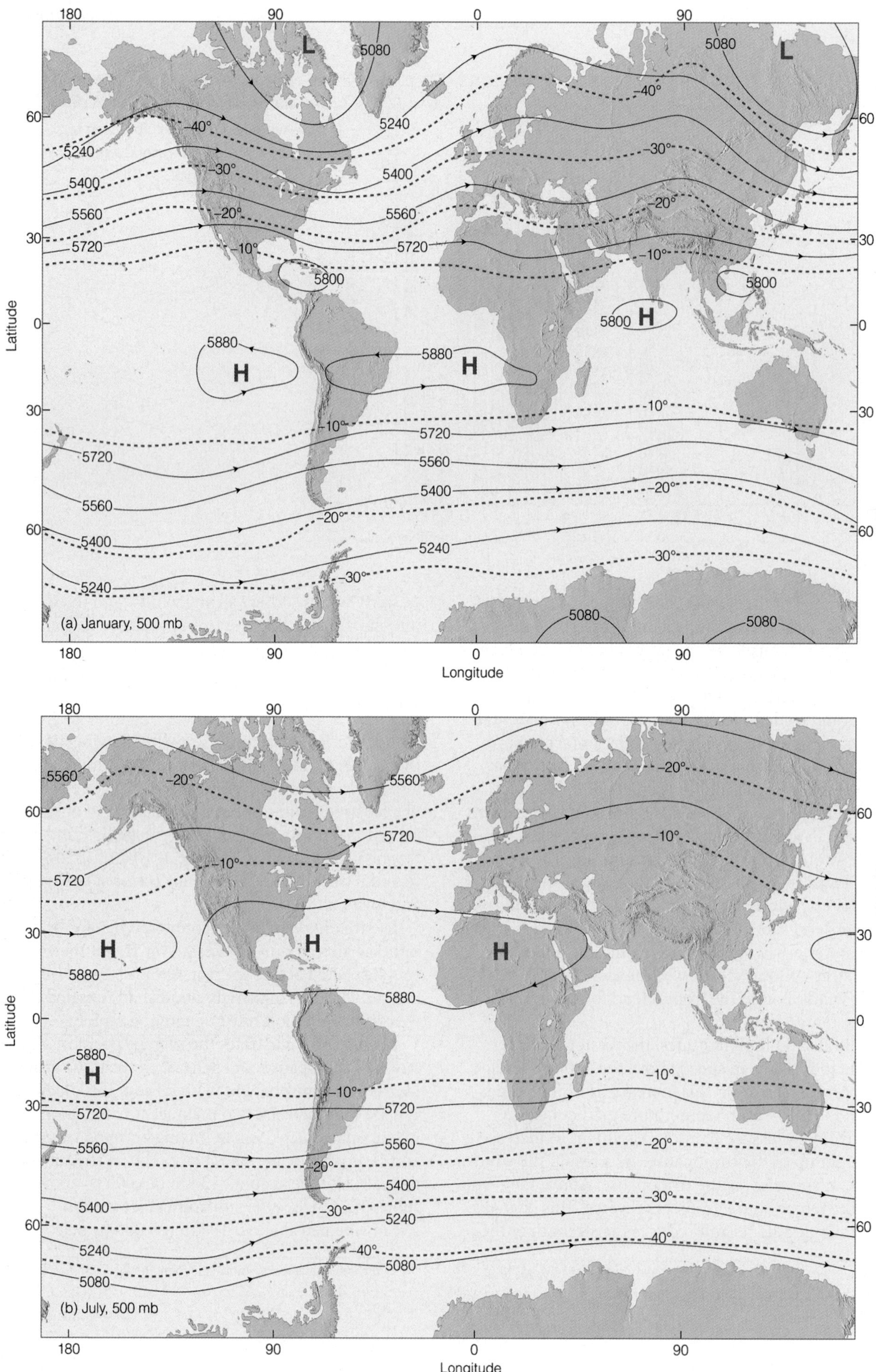

(a) January, 500 mb
Latitude
Longitude
180
90
0
90
60
30
0
30
60
L
L
H
H
H
H
5080
5240
5400
5560
5720
5800
5880
-40°
-30°
-20°
-10°
(b) July, 500 mb
5560
5720
5880
5400
5240
5080
-20°
-10°
-30°
-40°

FIGURE 8.9 (*On facing page*) Average 500-mb chart for the month of January (a) and for July (b). Solid lines are contour lines in meters above sea level. Dashed red lines are isotherms in °C. Arrowheads illustrate wind direction.

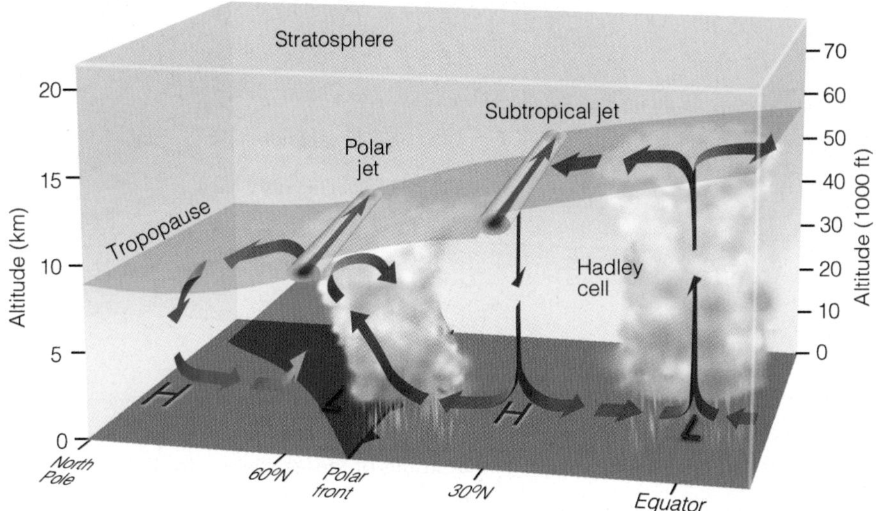

FIGURE 8.10 Average position of the polar jet stream and the subtropical jet stream, with respect to a model of the general circulation in winter. Both jet streams are flowing from west to east.

front is known as the *polar front jet stream* or, simply, the **polar jet stream**.

In Fig. 8.10, the wind in the center of the jet stream would be flowing as a westerly wind away from the viewer. This direction, of course, is only an average, as jet streams often flow in a wavy west-to-east pattern. When the polar jet stream flows in broad loops that sweep north and south, it may even merge with the subtropical jet. Occasionally, the polar jet splits into two jet streams. The jet stream to the north is often called the *northern branch* of the polar jet, whereas the one to the south is called the *southern branch.*

Since jet streams are bands of strong winds, they form in the same manner as all winds do—from horizontal differences in air pressure. In Fig. 8.10, notice that the polar jet stream forms along the polar front where sharp contrasts in temperature produce rapid horizontal pressure changes and strong winds. Due to the fact that the north-to-south temperature contrasts along the front are greater in winter than they are in summer, the polar jet stream shows seasonal variations. In winter, the polar jet stream winds are stronger and the jet moves farther south, sometimes as far south as Florida and Southern California. In summer, the polar jet stream is weaker and forms over higher latitudes.

Observe in Fig. 8.10 that the subtropical jet stream forms on the poleward (north) side of the Hadley cell, at a higher altitude than the polar jet stream. Here, warm air aloft carried poleward by the Hadley cell produces sharp temperature differences, strong pressure gradients, and high winds. Figure 8.11 illustrates how the polar jet stream and the subtropical jet stream might appear as they sweep around the earth in winter.

We can better see the looping pattern of the jet by studying Fig. 8.12a, which shows the position of the

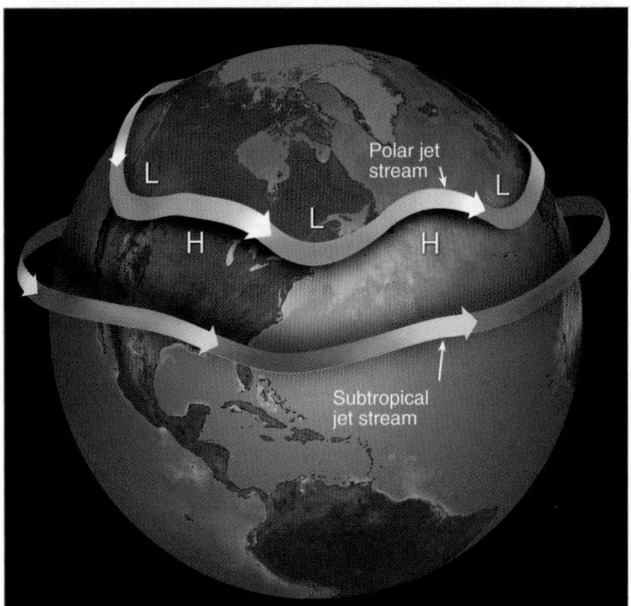

FIGURE 8.11 A jet stream is a swiftly flowing current of air that moves in a wavy west-to-east direction. The figure shows the position of the polar jet stream and subtropical jet stream in winter. Although jet streams are shown as one continuous river of air, in reality they are discontinuous, with their position varying from one day to the next.

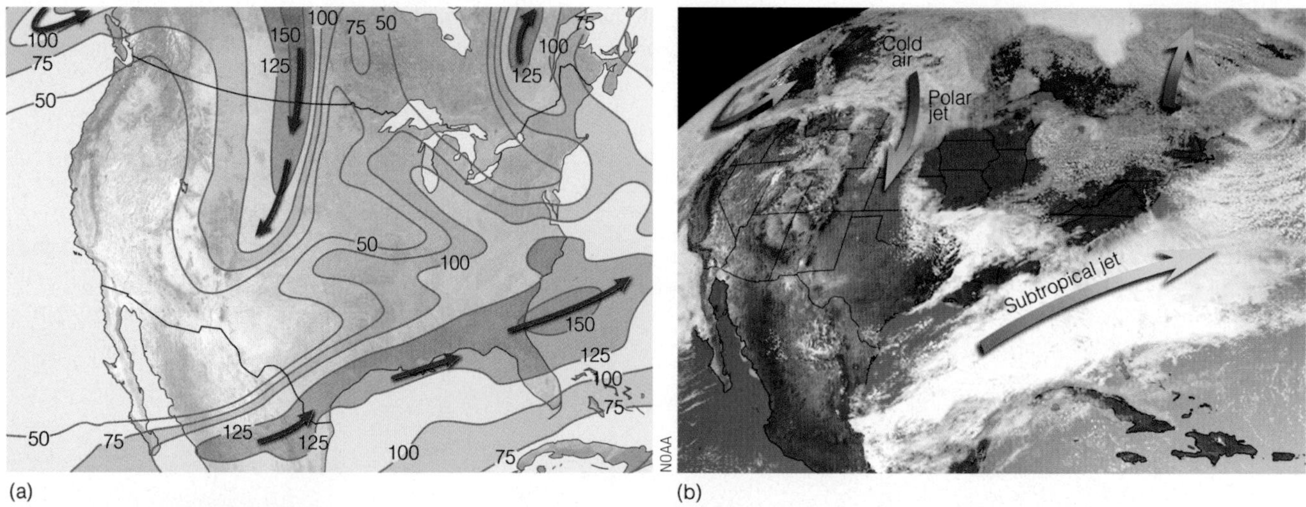

FIGURE 8.12 (a) Position of the polar jet stream (blue arrows) and the subtropical jet stream (orange arrows) at the 300-mb level (about 9 km or 30,000 ft above sea level) on March 9, 2005. Solid lines are lines of equal wind speed (isotachs) in knots. (b) Satellite image showing clouds and positions of the jet streams for the same day.

polar jet stream and the subtropical jet stream at the 300-mb level (near 9 km or 30,000 ft) on March 9, 2005. The fastest flowing air, or *jet core*, is represented by the heavy dark arrows. The map shows a strong polar jet stream sweeping south over the Great Plains with an equally strong subtropical jet over the Gulf states. Notice that the polar jet has a number of loops, with one off the west coast of North America and another over eastern Canada. Observe in the satellite image (Fig. 8.12b) that the polar jet stream (blue arrow) is directing cold, polar air into the Plains States, while the subtropical jet stream (orange arrow) is sweeping subtropical moisture, in the form of a dense cloud cover, over Florida.

The looping pattern of the polar jet stream has an important function. In the Northern Hemisphere, where the air flows southward, swiftly moving air directs cold air equatorward; where the air flows northward, warm air is carried toward the poles. Jet streams, therefore, play a major role in the global transfer of heat. Moreover, since jet streams tend to meander around the world, we can easily understand how pollutants or volcanic ash injected into the atmosphere in one part of the globe could eventually settle to the ground many thousands of miles downwind. And, as we will see in several chapters, the looping nature of the polar jet stream has an important role in the development of mid-latitude cyclonic storms, severe thunderstorms, and even tornadoes.

Although the polar and subtropical jets are the two most frequently in the news, there is another jet stream that plays a major role in the development of severe weather, the **low-level jet stream** that forms most frequently over the Great Plains of the United States, within a mile or so above the surface. During the summer, this jet (which usually has peak winds of less than 60 mi/hr) often contributes to the formation of thunderstorms and tornadoes by transporting humid air from the Gulf of Mexico rapidly northward.

In the vicinity of the polar front and subtropical jet streams there is usually a sharp change in wind speed and sometimes wind direction. This rapid change in wind speed or direction (or both) is called **wind shear**. Passengers inside an aircraft that flies in the vicinity of strong wind shear may be in for more than a bumpy ride as the plane may suddenly rise, then drop. In extreme events, passengers not strapped into their seats have been tossed about inside the cabin like beach balls. Some passengers have been seriously injured; some have died. If you are interested in knowing what causes this extreme bumpiness called **turbulence**, read the Focus section on pp. 220–221 before looking over the brief review.

BRIEF REVIEW

Before going on to the next section, here is a review of some of the important concepts presented so far:

- **The two major semipermanent subtropical highs that influence the weather of North America are the Pacific high situated off the west coast and the Bermuda high situated off the southeast coast.**
- **The polar front is a zone of low pressure where storms often form. It separates the mild westerlies of the middle latitudes from the cold, polar easterlies of the high latitudes.**

- In equatorial regions, the intertropical convergence zone (ITCZ) is a boundary where air rises in response to the convergence of the northeast trades and the southeast trades.
- In the Northern Hemisphere, the major global pressure systems and wind belts shift northward in summer and southward in winter.
- The northward movement of the Pacific high in summer tends to keep summer weather along the west coast of North America relatively dry.
- Jet streams exist where strong winds become concentrated in narrow bands. The polar-front jet stream forms along the polar front. The polar jet meanders in a wavy west-to-east pattern, becoming strongest in winter when the contrast in temperature along the front is greatest.
- The subtropical jet stream is found on the poleward side of the Hadley cell, between 20° and 30° latitude.

Up to this point we've looked at winds from a global perspective. We will now turn our attention to winds of a more regional nature. These smaller-scale circulations, sometimes called *local winds* or *mesoscale winds*, typically range in size from a few miles to more than 100 miles in diameter and often last from a few hours to several days.

Extreme (and Not So Extreme) Local Wind Systems

The first type of circulation we will explore is the monsoon. When most people hear the word *monsoon* they immediately think of heavy rain. In reality, monsoon winds are those that change direction on a seasonal basis, and one of those seasons may be extremely wet.

EXTREME WEATHER WATCH

Local wind systems exist throughout the world. For example, a local wind phenomenon, known as "la Lombarde," affects the Savoie region of France during the winter months. This southwesterly wind funnels moisture up the valleys of the area, resulting in intense local snowfalls. The town of Bessans once received a world-record snowfall of 67.8 in. in just 19 hours during a Lombarde event on April 5–6, 1959.

MONSOON WINDS The word *monsoon* derives from the Arabic *mausim,* which means seasons. As we just saw, a **monsoon wind system** is one that *changes direction seasonally,* blowing from one direction in summer and from the opposite direction in winter. This seasonal reversal of winds is especially well developed in eastern and southern Asia.

During the winter, the air over the Asian continent becomes much colder than the air over the ocean and a large, shallow high-pressure area develops over continental Siberia (see Fig. 8.13a). This large anticyclone produces a *clockwise* circulation of air that flows out over the Indian Ocean and South China Sea. Sinking air of the anticyclone and the downslope movement of northeasterly winds from the inland plateau provide eastern and southern Asia with generally fair weather. Hence, the *winter monsoon,* which lasts from about December through February, means clear skies (*dry season*), with surface winds that blow from land to sea.

In summer, the wind-flow pattern reverses itself as air over the continents becomes much warmer than air above the water. A shallow low-pressure area develops over the continental interior (see Fig. 8.13b). The

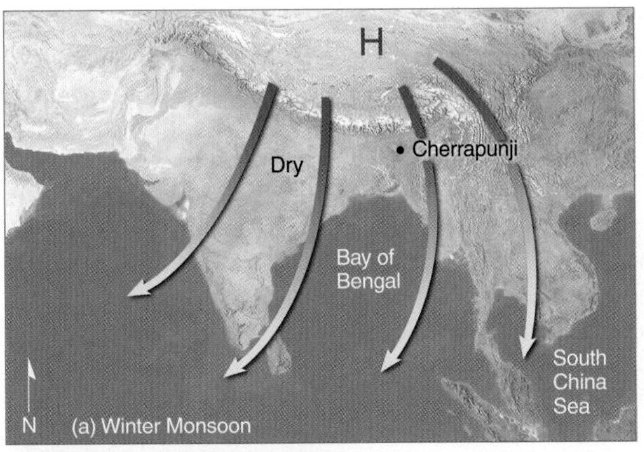

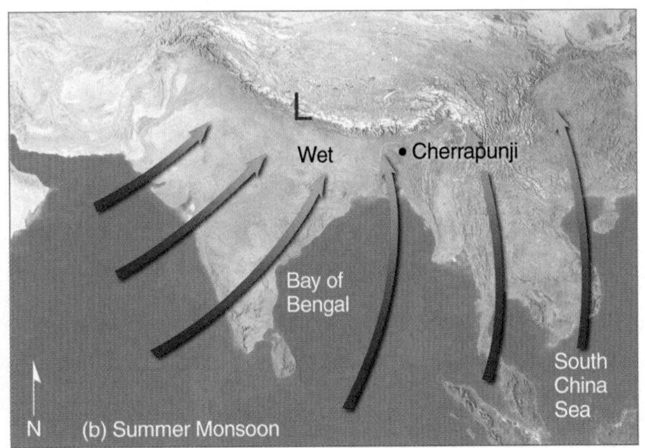

FIGURE 8.13 Changing annual wind-flow patterns associated with the winter and summer Asian monsoon.

heated air within the low rises, and the surrounding air responds by flowing *counterclockwise* into the low center. This condition results in moisture-bearing winds sweeping into the continent from the ocean. The humid air converges with a drier westerly flow, causing it to rise; further lifting is provided by hills and mountains. Lifting cools the air to its saturation point, resulting in heavy showers and thunderstorms. Thus, the *summer monsoon* of southeastern Asia, which lasts from about June through September, means wet, rainy weather (*wet season*) with winds that blow from sea to land. Although the majority of rain falls during the wet season, it does not rain all the time. In fact, rainy periods of between 15 to 40 days are often followed by several weeks of hot, sunny weather.

Summer monsoon rains over southern Asia can reach astonishing amounts. Located about 200 miles inland on the southern slopes of the Khasi Hills in northeastern India, Cherrapunji receives an average of 425 inches of rainfall each year, most of it during the summer monsoon between April and October (see Fig. 8.14). The summer monsoon rains are essential to the agriculture of that part of the world. With a population of over 900 million people, India depends heavily on the summer rains so that food crops will grow. The people also

FOCUS ON

EXTREME WEATHER

Aircraft Turbulence—Fasten Your Seat Belts

Anyone who has flown in an airplane has probably experienced some bumpiness in flight. Such bumpiness, called *turbulence,* can range from a small vibration to violent up and down motions that force passengers against their seats and toss objects throughout the cabin.

Turbulence can be created by a number of atmospheric conditions. For example, when an aircraft flies within or beneath a cumuliform cloud, the updrafts and downdrafts associated with the cloud can cause the plane to rise in the updraft and descend in the downdraft. This type of turbulence can occur in the absence of clouds when an aircraft flies through rising thermals (rising bubbles of air) that extend to considerable heights. Turbulence may also occur when an aircraft flies over a high mountain where swirling eddies of air have formed in the vicinity of wave clouds. But turbulence may suddenly and unexpectedly form in clear weather, especially near a jet stream in a zone of wind shear.

To better understand how turbulence forms along a zone of wind shear, imagine that, high in the atmosphere at an altitude of about 30,000 ft or so, there is a stable layer where the wind speed is strong but there is only small vertical wind speed shear (changing wind speed with height) as depicted in Fig. 1a. The top half of the layer slides over the bottom half, and the difference in wind speed between the two halves is relatively small. As long as the wind shear is small, turbulence is not likely to form. However, if the shear and the corresponding relative speed of these layers increase, wavelike undulations may form, as shown in Figs. 1b and 1c. When the shearing exceeds a certain value (Fig. 1d), the waves break into large swirls, with significant vertical movement. Swirling eddies such as these often form in the upper troposphere near jet streams, where large wind speed shears exist. When these huge eddies develop in clear air, this form of turbulence is referred to as *clear air turbulence,* or *CAT.* If wavelike clouds form in the region of wind shear, they are often called *billow clouds* (see Fig. 2).

The eddies associated with CAT may have diameters ranging from a couple of feet to several hundred feet or more. An unsuspecting aircraft entering such a region may be in for quite a ride. If the aircraft flies into a zone of descending air, it may drop suddenly, producing the sensation that there is no air to support the wings. Consequently, these regions have come to be known as *air pockets.*

Commercial aircraft entering an air pocket have dropped hundreds of feet, injuring passengers and flight attendants not strapped into their seats. For example, during December, 1997, a Boeing 747 at 33,000 ft over the Pacific Ocean, east of Tokyo, encountered a region of severe clear air turbulence and plunged about 1000 ft toward the earth before stabilizing. At least 110 people were injured, 12 seriously, and tragically, one person died of head injuries when hurled against the ceiling of the plane.

Clear air turbulence has occasionally caused structural damage to aircraft by breaking off vertical stabilizers and tail structures. Fortunately, the effects are usually not this dramatic.

depend on the rains for drinking water. Unfortunately, the monsoon can be unreliable in both duration and intensity. Since the monsoon is vital to the survival of so many people, it is no wonder that meteorologists have investigated it extensively. They have tried to develop methods of accurately forecasting the intensity and duration of the monsoon. With the aid of current research projects and the latest climate models, there is hope that monsoon forecasts will begin to improve in accuracy.

Monsoon wind systems exist in other regions of the world, such as Australia, Africa, and North and South America, where large contrasts in temperature develop between oceans and continents. (Usually, however, these systems are not as pronounced as in southeast Asia.) For example, a monsoonlike circulation exists in the southwestern United States, especially in Arizona, New Mexico, Nevada, and the southern part of California where spring and early summer are normally dry, as warm westerly winds sweep over the region. By mid-July, however, humid southerly or southeasterly winds are more common, and so are afternoon showers and thunderstorms (see ❯ Fig. 8.15 and ❯ Fig. 8.16).

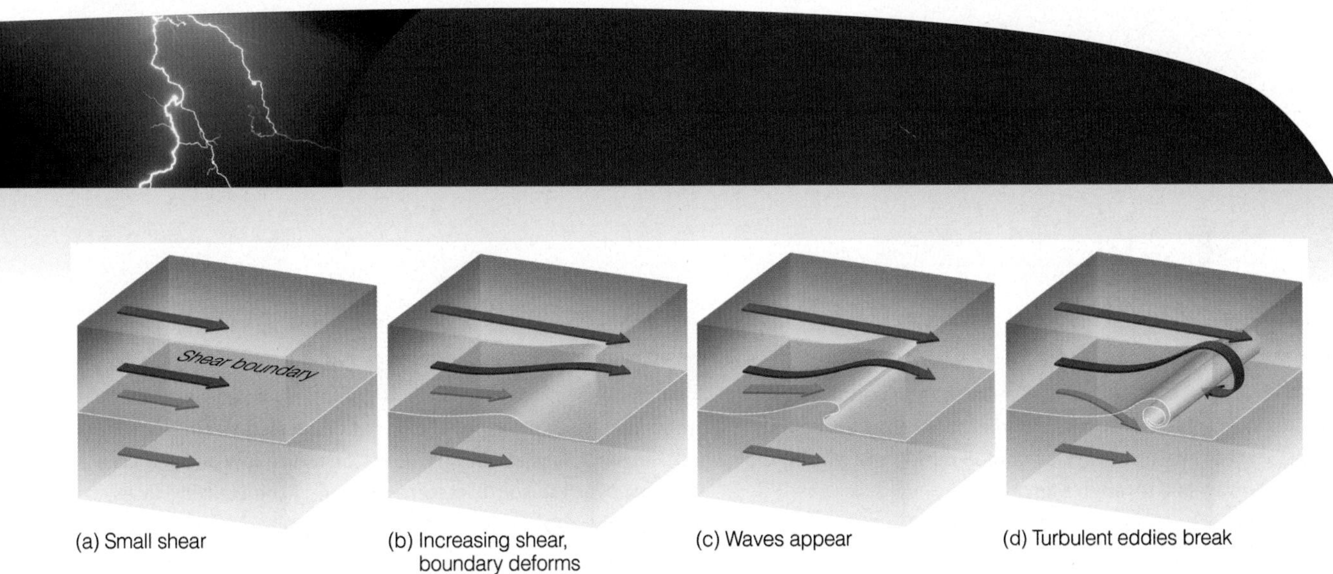

Figure 1 The formation of clear air turbulence (CAT) along a boundary of increasing wind speed shear in the vicinity of a jet stream. The wind in the top layer increases in speed from (a) through (d) as it flows over the bottom layer.

Figure 2 Billow clouds forming in a region of rapidly changing wind speed, called *wind shear*.

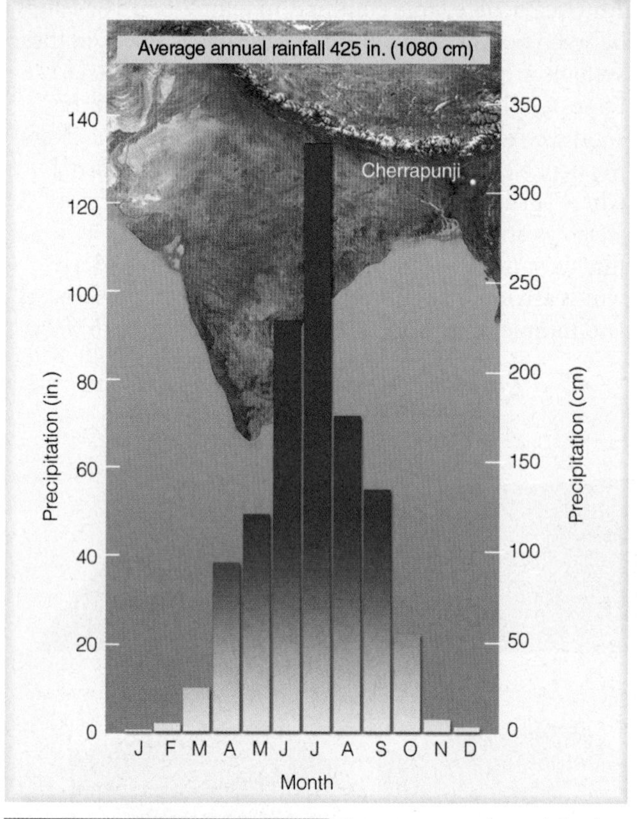

FIGURE 8.14 Average annual precipitation for Cherrapunji, India. Note the abundant rainfall during the summer monsoon (April through October) with the lack of rainfall during the winter monsoon (November through March).

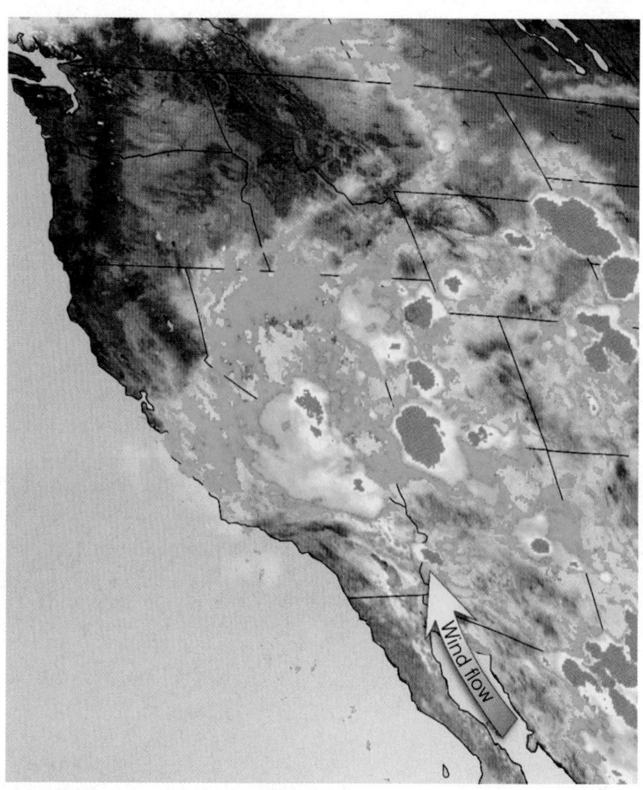

NOAA

FIGURE 8.15 Enhanced infrared satellite image with heavy arrow showing strong monsoonal circulation. Moist, southerly winds are causing showers and thunderstorms (yellow and red areas) to form over the southwestern section of the United States during July, 2001.

FIGURE 8.16

Clouds and thunderstorms forming over Arizona, as humid monsoonal air flows northward over the region during July, 2007.

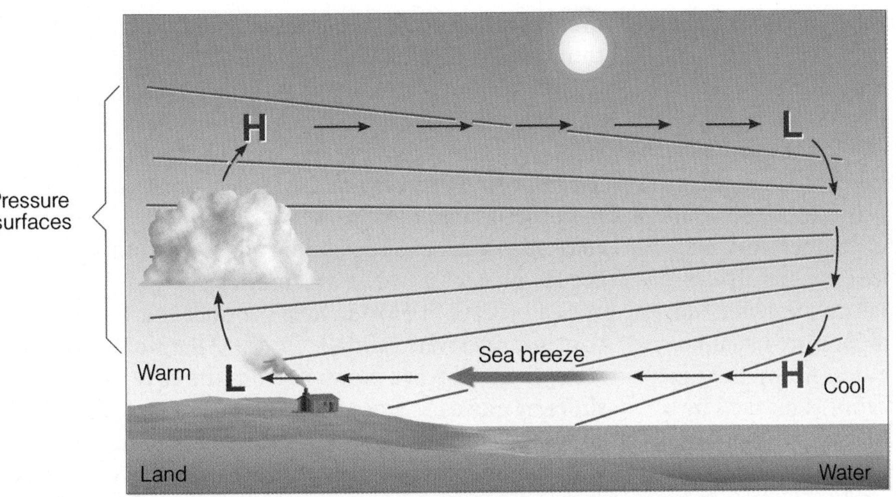

(a) Sea breeze

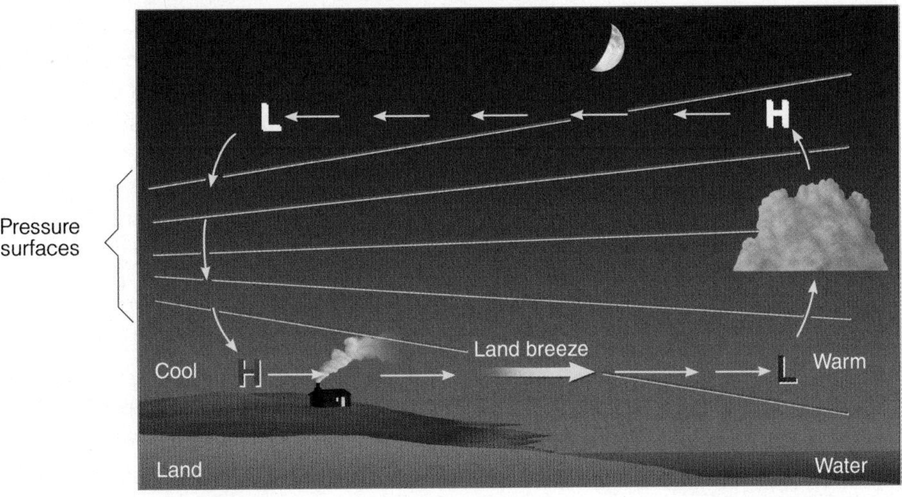

(b) Land breeze

FIGURE 8.17

Development of a sea breeze and a land breeze. (a) At the surface, a sea breeze blows from the water onto the land, whereas (b) the land breeze blows from the land out over the water. Notice that the pressure at the surface changes more rapidly with the sea breeze. This situation indicates a stronger pressure gradient force and higher winds with a sea breeze.

SEA AND LAND BREEZES The sea breeze is a type of circulation called a *thermal circulation.* The uneven heating rates of land and water cause these coastal winds. During the day, the land heats more quickly than the adjacent water, and the intensive heating of the air above produces a shallow low-pressure area (see Fig. 8.17a). The air over the water remains cooler than the air over the land and a shallow area of high pressure forms over the water. The overall effect of this pressure distribution is a **sea breeze** that blows at the surface from the sea toward the land. Since the strongest gradients of temperature and pressure occur near the land-water boundary, the strongest winds typically occur right near the beach and diminish inland. Further, since the greatest contrast in temperature between land and water usually occurs in the afternoon, sea breezes are strongest at this time. (The same type of breeze that develops along the shore of a large lake is called a **lake breeze.**)

At night, the land cools more quickly than the water. The air above the land becomes cooler than the air over the water, producing a distribution of pressure such as the one shown in Fig. 8.17b. With higher surface pressure now over the land, the surface wind reverses itself and becomes a *land breeze*—a breeze that flows from the land toward the water.* Temperature contrasts between land and water are generally much smaller at night; hence, land breezes are usually weaker than their daytime counterpart, the sea breeze. In regions where greater nighttime temperature contrasts exist, stronger land breezes occur over the water, off the coast. They are not usually noticed much on shore, but are frequently observed by ships in coastal waters.

Look at Fig. 8.17 again and observe that the rising air is over the land during the day and over the water during the night. Therefore, along the humid East Coast, daytime clouds tend to form over land and nighttime clouds over water. This explains why, at night, distant lightning flashes are sometimes seen over the ocean.

Sea breezes are best developed where large temperature differences exist between land and water. Such conditions prevail year-round in many tropical regions.

*Winds that blow from water to land, such as the sea breeze, are called *onshore winds.* Winds that blow from land to sea, such as the land breeze, are called *offshore winds.*

In middle latitudes, however, sea breezes are invariably spring and summer phenomena.

The leading edge of the sea breeze is called the **sea breeze front**. As the front moves inland, a rapid drop in temperature usually occurs just behind it. In some locations, this temperature change may be 10°F or more during the first hours—a refreshing experience on a hot, sultry day. In regions where the water temperature is warm, the cooling effect of the sea breeze is hardly evident. Since cities near the ocean usually experience the sea breeze by noon, their highest temperature usually occurs much earlier than in inland cities. Along the East Coast, the passage of the sea breeze front is marked by a wind shift, usually from west to east. In the cool ocean air, the relative humidity rises as the temperature drops. If the relative humidity increases to above 70 percent, water vapor begins to condense upon particles of sea salt or industrial smoke, producing haze. When the ocean air is highly concentrated with pollutants, the sea breeze front may meet relatively clear air and thus appear as a *smoke front,* or a *smog front.* If the ocean air becomes saturated, a mass of low clouds and fog will mark the leading edge of the marine air.

When there is a sharp contrast in air temperature across the frontal boundary, the warmer, lighter air will converge and rise. In many regions, this makes for good sea breeze glider soaring. If this rising air is sufficiently moist, a line of cumulus clouds will form along the sea breeze front, and, if the air is also conditionally unstable, thunderstorms may form.

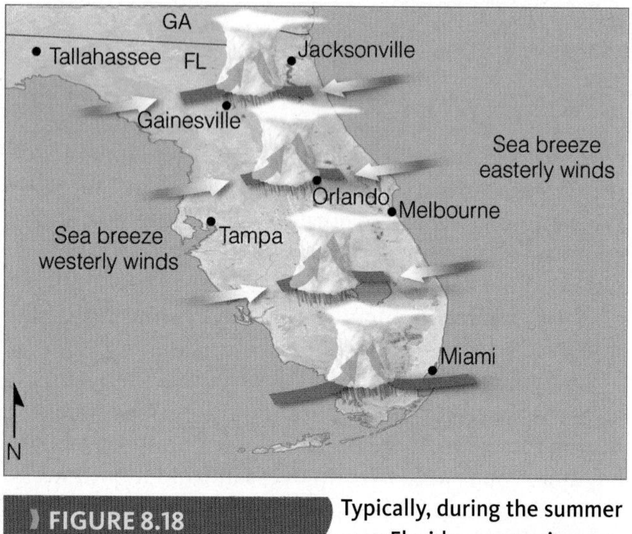

FIGURE 8.18 Typically, during the summer over Florida, converging sea breezes in the afternoon produce uplift that enhances thunderstorm development and rainfall. However, when westerly surface winds dominate and a ridge of high pressure forms over the area, thunderstorm activity diminishes, and dry conditions prevail.

When cool, dense, stable marine air encounters an obstacle, such as a row of hills, the heavy air tends to flow around them rather than over them. When the opposing breezes meet on the opposite side of the obstruction, they form what is called a *sea breeze convergence zone.* Such conditions are common along the Pacific Coast of North America.

Sea breezes in Florida help produce that state's abundant summertime rainfall. On the Atlantic side of the state, the sea breeze blows in from the east; on the Gulf shore, it moves in from the west (see Fig. 8.18). The convergence of these two moist wind systems, coupled with daytime convection, produces cloudy conditions and showery weather over the land. Over the water (where cooler, more stable air lies close to the surface), skies often remain cloud-free. On many days during June and July of 1998, however, Florida's converging wind system did not materialize. The lack of converging surface air and its accompanying showers left much of the state parched. Huge fires broke out over northern and central Florida, which left hundreds of people homeless and burned many thousands of acres of grass and woodlands. A weakened sea breeze and dry condition produced wild fires on numerous other occasions, including the spring of 2006.

In most areas, sea breezes tend to be best developed and therefore strongest in the afternoon when the greatest difference in air pressure exists between land and water. But, even in the absence of a sea breeze, winds in most regions tend to be strongest in the afternoon. If you are interested in knowing why this happens, read the Focus section on p. 225.

MOUNTAIN AND VALLEY BREEZES Mountain and valley breezes develop along mountain slopes. Observe in Fig. 8.19 that, during the day, sunlight warms the valley walls, which in turn warm the air in contact with them. The heated air, being less dense than the air of the same altitude above the valley, rises as a gentle upslope wind known as a **valley breeze**. At night, the flow reverses. The mountain slopes cool quickly, chilling the air in contact with them. The cooler, more-dense air glides downslope into the valley, providing a **mountain breeze**. (Because gravity is the force that directs these winds downhill, they are also referred to as *gravity winds,* or *nocturnal drainage winds.*) This daily cycle of wind flow is best developed in clear summer weather when prevailing winds are light.

In many areas, the upslope winds begin early in the morning, reach a peak speed of about 6 mi/hr by midday, and reverse direction by late evening. The downslope mountain breeze increases in intensity, reaching its peak in the early morning hours, usually just before sunrise. In the Northern Hemisphere, valley breezes are particularly well developed on south-facing slopes, where sunlight is most intense. On partially shaded north-facing slopes, the upslope breeze may be weak or absent. Since upslope winds begin soon after the sun's rays strike a hill, valley breezes typically begin first on the hill's east-facing side. In the late afternoon, this side of the mountain goes into shade first, producing the onset of downslope winds at an earlier time than experienced on west-facing slopes. Hence, it is possible for campfire smoke to drift downslope on one side of a mountain and upslope on the other side.

When the upslope winds are well developed and have sufficient moisture, they can reveal themselves as building cumulus clouds above mountain summits (see Fig. 8.20). Since valley breezes usually reach their maximum strength in the early afternoon, cloudiness, showers, and even thunderstorms are common over mountains during the warmest part of the day—a fact well known to climbers, hikers, and seasoned mountain picnickers.

FOCUS ON

A SPECIAL TOPIC

Windy Afternoons

On warm days when the weather is clear or partly cloudy, you may have noticed that the windiest time of the day is usually in the afternoon. The reason for this situation is due to several factors all working together: Surface heating, convection, and atmospheric stability.

We learned in Chapter 5 that in the early morning the atmosphere is most stable, meaning that the air resists up and down motions. As an example, consider the flow of air in the early morning as illustrated in Fig. 3a. Notice that weak winds exist near the surface with much stronger winds aloft. Because the atmosphere is stable, there is little vertical mixing between the surface winds and the winds higher up.

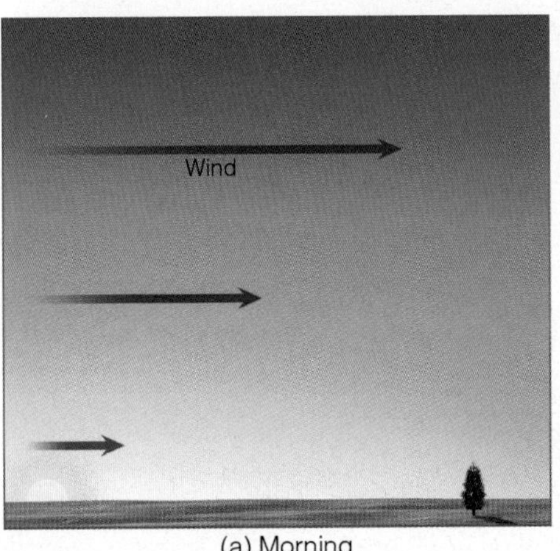

(a) Morning

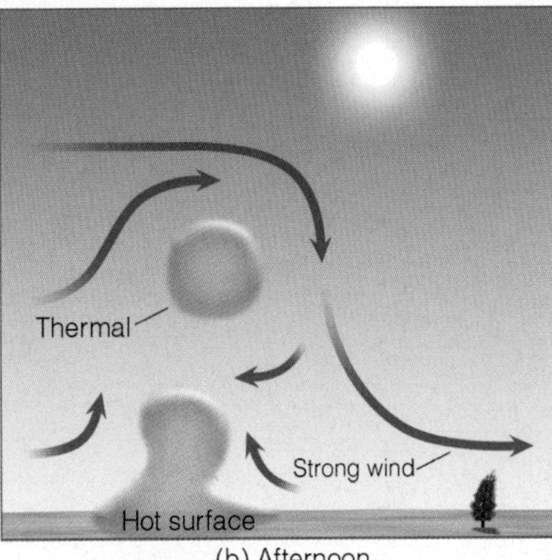

(b) Afternoon

Figure 3 (a) During the early morning, there is little exchange between the surface winds and the winds aloft. (b) In the afternoon, when the atmosphere is usually most unstable, convection in the form of rising thermals links surface air with the air aloft, causing strong winds from aloft to reach the ground, and produce strong, gusty surface winds.

As the day progresses, and the sun rises higher in the sky, the surface heats up and the lower atmosphere becomes more unstable. Over hot surfaces, the air begins to rise in the form of thermals that carry the slower-moving air with them (see Fig. 3b). At some level above the surface, the rising air links up with the faster-moving air aloft. If the air begins to sink as part of a convective circulation, it may pull some of the stronger winds aloft downward with it. If this sinking air should reach the surface, it produces a momentary gust of strong wind. In addition, this exchange of air increases the average wind speed at the surface. Because this type of air exchange is greatest on a clear day in the afternoon when the atmosphere is most unstable, we tend to experience the strongest, most gusty winds in the afternoon. At night, when the atmosphere stabilizes, the interchange between the surface air and the air aloft is at a minimum and the winds at the surface tend to die down.

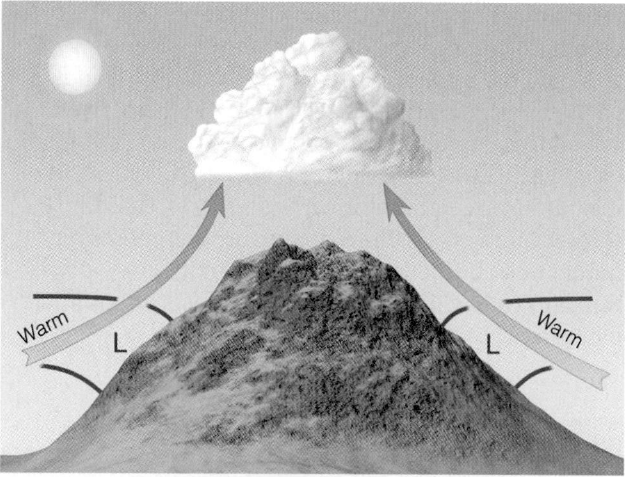

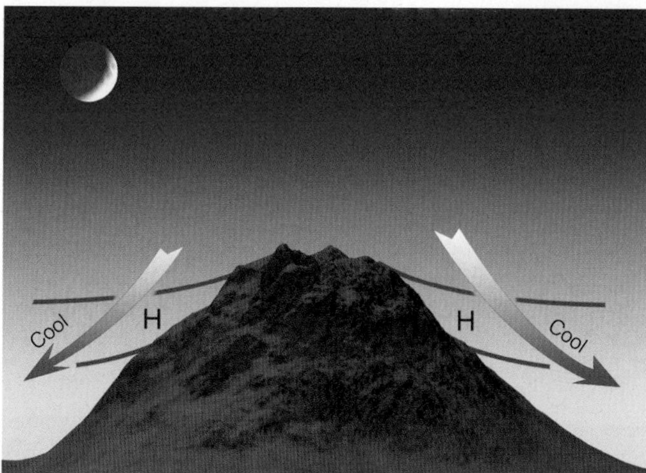

FIGURE 8.19 Valley breezes blow uphill during the day; mountain breezes blow downhill at night. (The L's and H's represent pressure, whereas the purple lines represent surfaces of constant pressure.)

FIGURE 8.20 As mountain slopes warm during the day, air rises and often condenses into cumuliform clouds, such as these forming over the Grand Tetons in Wyoming.

KATABATIC WINDS Although any downslope wind is technically a **katabatic wind**, the name is usually reserved for downslope winds that are much stronger than mountain breezes. Katabatic (or *fall*) winds can rush down elevated slopes at hurricane speeds, but most are not that intense and many are on the order of 10 mi/hr or less.

The ideal setting for a katabatic wind is an elevated plateau surrounded by mountains, with an opening that slopes rapidly downhill. When winter snows accumulate on the plateau, the overlying air grows extremely cold and a shallow dome of high pressure forms near the surface (see Fig. 8.21). Along the edge of the plateau, the horizontal pressure gradient force is usually strong enough to cause the cold air to flow across the isobars through gaps and saddles in the hills. Along the slopes of the plateau, the wind continues downhill as a gentle or moderate cold breeze. If the horizontal pressure gradient increases substantially, such as when a storm approaches, or if the wind is confined to a narrow canyon or channel, the flow of air can increase, often destructively, as cold air rushes downslope like water flowing over a fall.

Katabatic winds are observed in various regions of the world. For example, along the northern Adriatic coast in the former Yugoslavia, a polar invasion of cold air from Russia descends the slopes from a high plateau and reaches the lowlands as the *bora*—a cold, gusty, northeasterly wind with speeds sometimes in excess of 100 mi/hr. A similar, but often less violent, cold wind known as the *mistral* descends the western mountains into the Rhone Valley of France, and then out over the Mediterranean Sea. It frequently causes frost damage to exposed vineyards and makes people bundle up in the otherwise mild climate along the Riviera. Strong, cold katabatic winds also blow downslope off the icecaps in Greenland and Antarctica, occasionally with speeds greater than 100 mi/hr.

In North America, when cold air accumulates over the Columbia plateau,* it may flow westward through the Columbia River Gorge as a strong, gusty, and sometimes violent wind. Even though the sinking air warms by compression, it is so cold to begin with that it reaches the ocean side of the Cascade Mountains much colder than the marine air it replaces. The *Columbia Gorge wind* (called the *coho*) is often the harbinger of a prolonged cold spell.

Strong downslope katabatic-type winds funneled through a mountain canyon can do extensive damage. During January, 1984, a ferocious downslope wind blew through Yosemite National Park in California at speeds estimated at 105 mi/hr. The wind toppled many trees and, unfortunately, caused a fatality when a tree fell on a park employee sleeping in a tent.

CHINOOK WINDS The **chinook wind** is a warm, dry wind that descends the eastern slope of the Rocky Mountains. The region of the chinook is rather narrow (only several hundred miles wide) and extends from northeastern New Mexico northward into Canada. Similar winds occur along the leeward slopes of mountains in other regions of the world. In the European Alps, for example, such a wind is called a *foehn* and, in Argentina, a *zonda*. In South Africa, a hot, dry wind that blows downslope off the interior plateau, most frequently into west coastal areas, is the *berg*. When these winds move through an area, the temperature rises sharply, sometimes 40°F or more in one hour, and a corresponding sharp drop in the relative humidity occurs, occasionally to less than 5 percent.

Chinook-type winds occur when strong westerly winds aloft flow over a north-south-trending mountain range, such as the Rockies and Cascades. Such conditions can produce a trough of low pressure on the mountain's eastern side, a trough that tends to force the air downslope. As the air descends, it is compressed and warms. So the main source of warmth for a chinook is *compressional heating*, as potentially warmer (and drier) air is brought down from aloft.

*Information on geographic features and their location in North America is provided at the back of the book.

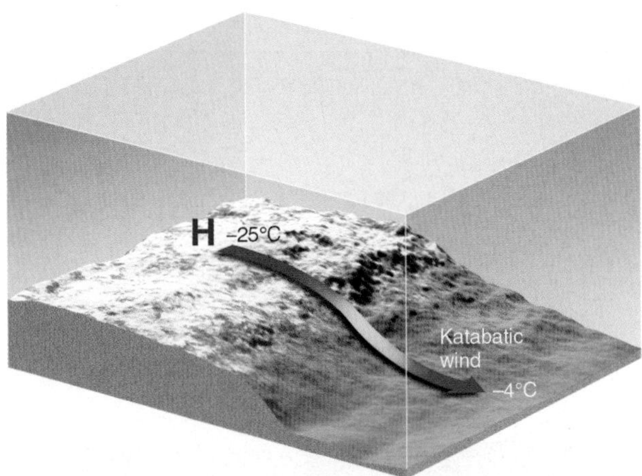

FIGURE 8.21 Strong katabatic winds can form where cold winds rush downhill from an elevated plateau covered with snow.

When clouds and precipitation occur on the mountain's windward side, they can enhance the chinook. For example, as the cloud forms on the upwind side of the mountain in Fig. 8.22, the release of latent heat inside the cloud supplements the compressional heating on the downwind side. This phenomenon makes the descending air at the base of the mountain on the downwind side warmer than it was before it started its upward journey on the windward side. The air is also drier, since much of its moisture was removed as precipitation on the windward side.

Along the front range of the Rockies, a bank of clouds forming over the mountains is a telltale sign of an impending chinook. This *chinook wall cloud* (which looks like a wall of clouds) usually remains stationary as air rises, condenses, and then rapidly descends the leeward slopes, often causing strong winds in foothill communities. Figure 8.23 shows how a chinook wall cloud appears as one looks west toward the Rockies from the Colorado plains. The photograph was taken on a winter afternoon with the air temperature about 20°F. That

EXTREME WEATHER WATCH

The most powerful and persistent katabatic winds in the world occur in Antarctica. The coastal research station at Cape Denison has an average annual wind speed of 50 mi/hr, with June, the windiest month, averaging 62 mi/hr. The local topography helps funnel and intensify the katabatic winds that affect this particular location.

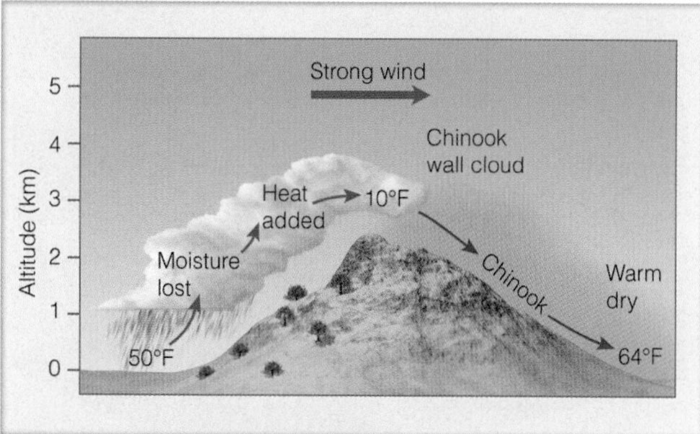

FIGURE 8.22 A chinook wind can be enhanced when clouds form on the mountain's windward side. Heat added and moisture lost on the upwind side produce warmer and drier air on the downwind side.

evening, the chinook moved downslope at high speeds through foothill valleys, picking up sand and pebbles (which dented cars and cracked windshields). The chinook spread out over the plains like a warm blanket, raising the air temperature the following day to a mild 60°F. The chinook and its wall of clouds remained for several days, bringing with it a welcomed break from the cold grasp of winter.

During chinook conditions, extremely strong winds will occasionally blow downslope along the eastern slope of the Rocky Mountains. Such winds are especially notorious in winter in Boulder, Colorado, where the average yearly windstorm damage is about $1 million. These *Boulder winds* have been recorded at over 100 mi/hr, damaging roofs, uprooting trees, overturning mobile homes, airplanes, and trucks, and sandblasting car windows (see Fig. 8.24). Although the causes of these high winds are not completely understood, some meteorologists believe that they may be associated with large vertically oriented spinning whirls of air that some scientists call *mountainadoes.* How these rapidly rotating vortices form is presently being investigated.

Strong downslope chinook winds have been associated with some of the most rapid changes in temperature ever measured. In fact, on January 11, 1980, due to a chinook wind, the air temperatures in Great Falls, Montana, rose from −32°F to 17°F (a 49°F rise in temperature) in just seven minutes. How such rapid changes in temperature can occur is illustrated in Fig. 8.25. Notice that a shallow layer of extremely cold air has moved out of Canada and is now resting against the Rocky Mountains. The cold air behaves just as any fluid, and, in some cases, atmospheric conditions may cause the air to move up and down much like water does when a bowl is rocked back and forth. This rocking motion can cause extreme temperature variations for cities located at the base of the hills along the periphery of the cold air-warm air boundary, as they are alternately in and then out of the cold air. Such a situation is held to be responsible for the extremely rapid two-minute temperature change of 49°F recorded at Spearfish, South Dakota, during the morning of January 22, 1943. On the same morning, in nearby Rapid City, the temperature fluctuated from −4°F at 5:30 A.M. to 54°F at 9:40 A.M., then down to 11°F at 10:30 A.M. and up to 55°F just 15 minutes later. At nearby cities, the undulating cold air produced similar temperature variations that lasted for several hours.

As a strong chinook wind moves over a heavy snow cover, it can melt and evaporate a foot of snow in less

FIGURE 8.23 A chinook wall cloud forming over the Colorado Rockies (viewed from the plains).

UCAR

FIGURE 8.24 Strong downslope winds in Boulder, Colorado, overturn this small aircraft during January, 1982.

than a day. This situation has led to some tall tales about these so-called "snow eaters." Canadian folklore has it that a sled-driving traveler once tried to outrun a chinook. During the entire ordeal his front runners were in snow while his back runners were on bare soil.

Actually, the chinook is important economically. It not only brings relief from the winter cold, but it uncovers prairie grass, so that livestock can graze on the open range. Also, these warm winds have kept railroad tracks clear of snow, so that trains can keep running. On the other hand, the drying effect of a chinook can create an extreme fire hazard. And when a chinook follows spring planting, the seeds may die in the parched soil. Along with the dry air comes a buildup of static electricity, making a simple handshake a shocking experience. These warm dry winds have sometimes adversely affected human behavior. During periods of chinook winds some people feel irritable and depressed and others become ill. The exact reason for this phenomenon is not clearly understood.

SANTA ANA WINDS A warm, dry wind that blows from the east or northeast into southern California is the **Santa Ana wind**. As the air descends from the elevated desert plateau, it funnels through mountain canyons in the San Gabriel and San Bernardino Mountains, finally spreading over the Los Angeles basin and San Fernando Valley and out over the Pacific Ocean (see Fig. 8.26). The wind often blows with exceptional speed—occasionally over 90 mi/hr—in the Santa Ana Canyon (the canyon from which it derives its name).

These warm, dry winds develop as a region of high pressure builds over the Great Basin. The clockwise circulation around the anticyclone forces air downslope from the high plateau. Thus, *compressional heating* provides the primary source of warming. The air is dry, since it originated in the desert, and it dries out even more as it is heated. Figure 8.27 shows a typical wintertime Santa Ana situation.

As the wind rushes through canyon passes, it lifts dust and sand and dries out vegetation, which sets the stage for serious brush fires, especially in autumn, when chaparral-covered hills are already parched from the dry summer.* One such fire in November of 1961—the infamous *Bel Air fire*—burned for three days, destroying 484 homes and causing over $25 million in damage. During October, 2003, massive wildfires driven by strong Santa Ana winds swept through Southern California. The fires charred more than 740,000 acres, destroyed over 2800 homes, took 20 lives, and caused over $1 billion in property damage. Only four years later (and after one of the driest years on record) in October, 2007, wildfires broke out again in Southern California. Pushed on by hellacious Santa Ana winds that gusted to over 80 mi/hr, the fires raced through dry vegetation, scorching everything in their paths. The fires, which extended

*Chaparral denotes a shrubby environment in which many of the plant species contain highly flammable oils.

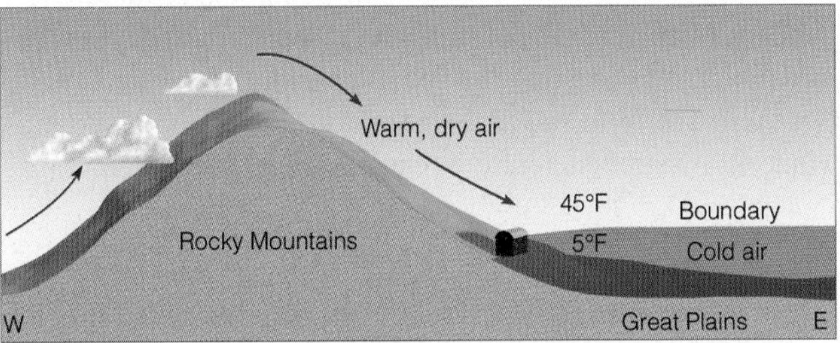

FIGURE 8.25 Extreme temperature changes brought on by a chinook wind. Cities near the warm air–cold air boundary can experience sharp temperature changes if cold air should rock up and down like water in a bowl.

FIGURE 8.26

Warm, dry Santa Ana winds sweep downhill through mountain canyons into Southern California. The large H represents higher air pressure over the elevated desert.

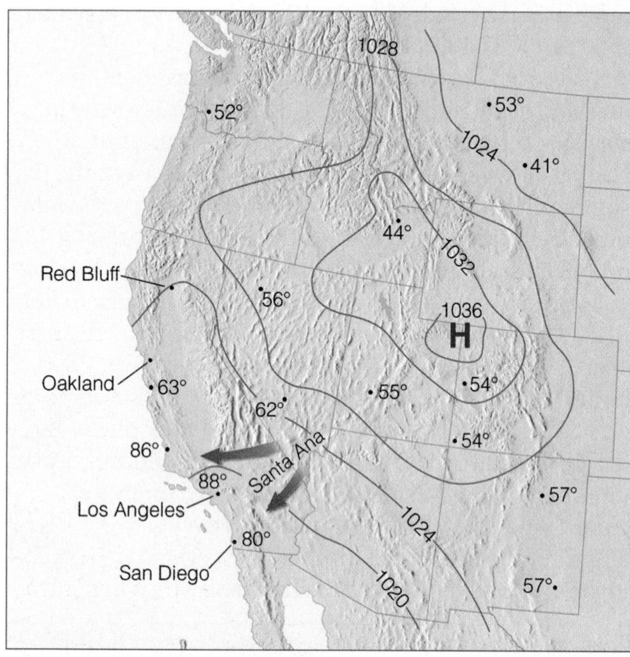

FIGURE 8.27 Surface weather map showing Santa Ana conditions in January. Maximum temperatures for this particular day are given in °F. Observe that the downslope winds blowing into southern California raised temperatures into the upper 80s, while elsewhere temperature readings were much lower.

FIGURE 8.28 Strong northeasterly Santa Ana winds on October 23, 2007, blew the smoke from massive wild fires (red dots) across southern California out over the Pacific Ocean.

from north of Los Angeles to the Mexican border (see Fig. 8.28), burned over 500,000 acres, destroyed more than 1800 homes, and took 8 lives. The total costs of the fires exceeded $1.5 billion.

Four hundred miles to the north of Los Angeles in Oakland, California, a ferocious Santa Ana-type wind (sometimes called the *Diablo wind*) was responsible for the disastrous Oakland hills fire during October, 1991.

The most catastrophic urban wildfire in the United States started in the parched Oakland hills, just east of San Francisco, where a firestorm driven by strong northeast winds blackened almost 2000 acres, damaged or destroyed over 3000 dwellings, caused almost $5 billion in damage, and took 25 lives (see Fig. 8.29). With the protective vegetation cover removed, the land is ripe for erosion, as winter rains may wash away topsoil and, in some areas, create serious mudslides. The adverse effects of a wind-driven Santa Ana fire may be felt long after the fire itself has been put out.

A similar downslope-type wind called a *California norther* can produce unbearably high temperatures in the northern half of California's Central Valley. On August 8, 1978, for example, a ridge of high pressure formed to the north of this region, while a thermal low was well entrenched to the south. This pressure pattern produced a north wind in the area. A summertime north wind in most parts of the country means cooler weather and a welcome relief from a hot spell, but not in Red Bluff, California, where the winds moved downslope off the mountains. Heated by compression, these winds increased the air temperature in Red Bluff to an unbelievable 119°F for two consecutive days—amazing when you realize that Red Bluff is located at about the same latitude as Philadelphia, Pennsylvania.

OTHER EXTREME WINDS OF INTEREST There are many other winds that can bring extreme weather conditions to a region. Some of these winds are extremely cold; others, excessively hot. In winter, for example, when an intense storm tracks east across the Great Plains of North America, cold northerly winds often plunge southward behind it. As the cold air moves through Texas, it may drop temperatures tens of degrees in a few hours. Such a cold wind is called a **Texas norther**, or *blue norther,* especially if accompanied by snow. If the cold air penetrates into Central America, it is known as a *norte.* Meanwhile, if the strong, cold winds over the plains states are accompanied by drifting, blowing, or falling snow, and the wind speed exceeds 35 mi/hr, the term *blizzard* is applied to this weather situation. A similar type of cold, snowy storm in Alaska is called the *burga.* Over the northern Siberian tundra of Russia, this cold, windy severe winter storm is called the *purga.*

In dry regions, where strong winds are able to lift and fill the air with particles of fine dust, huge **duststorms** may form. If the wind is capable of lifting larger sand-size particles, the storm is called a *sandstorm.* Such storms can grow to a very large size. As an example, an exceptionally large duststorm formed over the African Sahara during February, 2001. The storm—about the size of Spain—swept westward off the African coast, then northeastward (see Fig. 8.30). During the drought years of the 1930s, huge duststorms formed over the Great Plains of the United States. Some individual storms lasted for three days and spread dust for hundreds of miles over the Atlantic Ocean.

EXTREME WEATHER WATCH

Although Santa Ana winds are responsible for most of the destructive wildfires in southern California, there is a Santa Ana-type wind (called the *sundowner wind*) that moves into the Santa Barbara area. When weather conditions are right, strong offshore sundowner winds can form in the evening around sunset. These winds, which funnel into the area through gaps in the coastal mountains, originate in the warm inland valleys. Sundowner winds of up to 70 mi/hr fueled disastrous fires in Santa Barbara County during November, 2008, and again in May, 2009.

FIGURE 8.29 From atop his roof, a resident of Oakland's Rockridge district looks on in disbelief as his neighbors' homes are consumed in a raging firestorm on October 20, 1991.

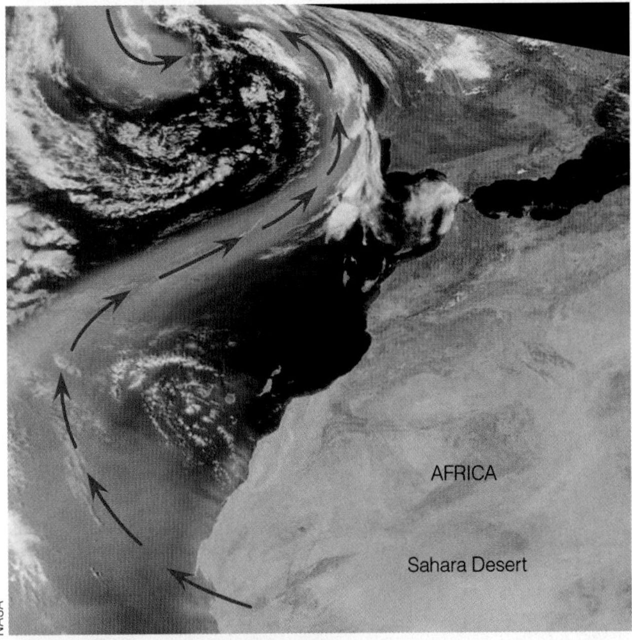

FIGURE 8.30 A large duststorm over the African Sahara Desert during February, 2001, sweeps westward off the coast, then northward into a mid-latitude cyclonic storm west of Spain, as indicated by the red arrows.

© John Coumerih

FIGURE 8.31 A haboob moving across northwest Kansas near the town of Bird City.

© Herbert Hoinkis

FIGURE 8.32 A dust devil forming above a field on a clear, hot summer day.

A spectacular example of a strong wind in combination with a duststorm or sandstorm is the **haboob** (from the Arabic *hebbe:* blown). The haboob usually forms as a cold downdraft inside an intense thunderstorm that sinks to the ground and spreads laterally outward. The leading edge of the cold air (called a gust front) lifts dust or sand into a huge tumbling dark cloud of wind-blown particles that can turn the sky dark (see Fig. 8.31). This darkening of the sky is why duststorms during the 1930s were sometimes referred to as **black blizzards**. The leading edge of the haboob—called the *dust wall*—may extend vertically to a height of more than 3000 ft with spinning whirls of dust forming along it. Surface wind speeds with a haboob are usually on the order of 30 mi/hr, although winds exceeding 60 mi/hr have been recorded. These storms are most common in the African Sudan (where about 24 occur each year) and in the desert southwest of the United States.

In dry areas, the wind may also produce rising, spinning columns of air that pick up dust or sand from the ground. Called **dust devils** or *whirlwinds*,* these rotating vortices generally form on clear, hot days over a dry surface where most of the sunlight goes into heating the surface, rather than evaporating water from vegetation (see Fig. 8.32).The atmosphere directly above the hot surface becomes unstable, convection sets in, and the heated air rises. Wind, often deflected by small topographic barriers, flows into this region, rotating the rising air (see Fig. 8.33). Depending on the nature of the topographic feature, the spin of a dust devil around its central eye may be cyclonic or anticyclonic, and both directions occur with about equal frequency.

Having diameters usually less than 10 feet and heights of less than 300 feet, most dust devils are small and last only a short time. There are, however, some dust devils of sizable dimension, extending upward from the surface for many hundreds of feet. Such whirlwinds are capable of considerable damage; winds exceeding 75 mi/hr may overturn mobile homes and tear the roofs off buildings. Fortunately, the majority of dust devils are small. Also keep in mind that dust devils *are not* tornadoes. The circulation of many tornadoes (as we will see in Chapter 12) usually descends downward from the base of a thunderstorm, whereas the circulation of a dust devil begins at the surface, normally in sunny weather, although some form beneath convective-type clouds.

Some of the hottest winds in the world blow across the Sahara Desert. The prevailing winds over North Africa are from the northeast. However, when a storm system is located west of Africa and southern Spain (position 1, Fig. 8.34), a hot, dry, and dusty easterly or southeasterly wind—the *leste*—blows over Morocco and out over the Atlantic. If the wind crosses the Mediterranean

*In Australia, the Aboriginal word *willy-willy* refers to a dust devil.

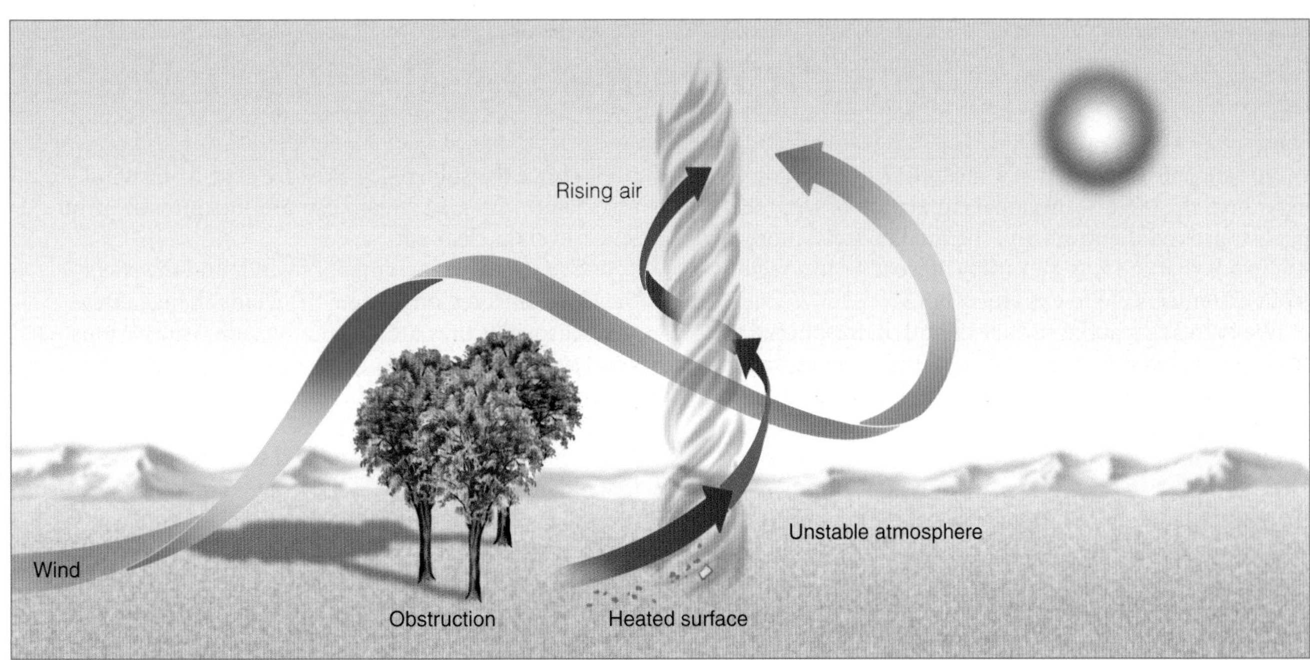

Active › FIGURE 8.33 The formation of a dust devil. On a hot, dry day, the atmosphere next to the ground becomes unstable. As the heated air rises, wind blowing past an obstruction twists the rising air, forming a rotating air column or *dust devil.* Air from the sides rushes into the rising column, lifting sand, dust, leaves, or any other loose material from the surface. Visit the Meteorology Resource Center to view this and other active figures at www.cengage.com/login

and enters southern Spain, it becomes the hot, dry *leveche.* When a low-pressure area is centered at position 2, a hot, dust-laden south or southeast wind—the *sirocco*—originates over the Sahara Desert and blows across North Africa.

A storm located still farther to the east (position 3) may produce a dry, hot southerly wind, called the *khamsin,* which blows over Egypt, the Red Sea, and Saudi Arabia. This wind can be very hot and raise air temperatures to 120°F, while lowering the relative humidity to less than 10 percent. If this wind moves into Israel, it is called the *Sharav.* As hot as these winds are, they don't hold a candle to one of the hottest winds on earth, the **simoom.** Called the "poison wind," this strong, dry, and dusty wind blows over Africa and the Arabian desert, often raising air temperatures in excess of 125°F—hot enough to cause heat stroke and death.

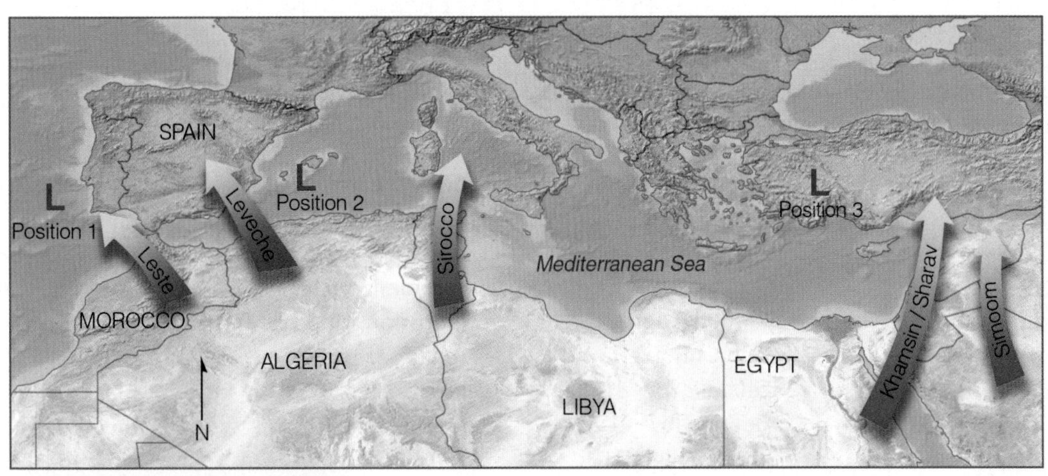

› FIGURE 8.34 Exceptionally hot, dry winds that form over North Africa.

Summary

In this chapter we looked at atmospheric circulations. We examined the large-scale global circulation of wind that persists around the world and the smaller-scale, more local, winds that can bring a variety of weather to a region—from extremely cold to extremely hot.

We found that at the surface in both hemispheres, the trade winds blow equatorward from the semipermanent high-pressure areas centered near 30° latitude. Near the equator, the trade winds converge along a boundary known as the intertropical convergence zone (ITCZ). Here, the air rises and condenses into huge thunderstorms that produce heavy rain shadows. Near 30° latitude, sinking air of the subtropical highs produces the major deserts of the world. On the poleward side of the subtropical highs are the prevailing westerly winds. The westerlies meet cold polar easterly winds along a boundary called the polar front, a zone of low pressure where middle-latitude cyclonic storms often form. The annual shifting of the major pressure areas and wind belts—northward in July and southward in January—strongly influences the annual precipitation of many regions.

Warm air aloft (high pressure) over low latitudes and cold air aloft (low pressure) over high latitudes produce westerly winds aloft in both hemispheres, especially at middle and high latitudes. The westerlies steer surface storms in a more or less west-to-east direction. Near the equator, easterly winds exist. The jet streams are located where strong winds concentrate into narrow bands. The polar jet stream, which is strongest in winter, forms in response to temperature contrasts along the polar front, whereas the subtropics jet stream forms at higher elevations above the subtropics. We saw that strong wind shear associated with a jet stream can produce aircraft turbulence, even in clear air.

We've examined a number of winds and the effect they have on a region. Where the winds change direction seasonally, they are called monsoon winds. Monsoon wind systems exist in many regions of the world, including North America and Asia. In India, where the monsoon winds become well developed, winters can be quite dry and summers extremely wet. Land and sea breezes are winds that blow in response to local pressure differences created by uneven heating and cooling rates of land and water.

Local winds that blow uphill during the day are called valley breezes, and those that blow downhill at night are called mountain breezes. A strong, cold downslope wind is the katabatic. A chinook wind is a warm, dry wind that blows downhill on the eastern side of the Rocky Mountains. This same type of wind in the Alps is the foehn. A warm, dry and often strong wind that blows into Southern California from the east or northeast is the Santa Ana wind, a wind that often produces extremely high fire danger. Over dry areas, duststorms, sandstorms, and the haboob can form. If rising air begins to spin, it may form into a dust devil.

Winds such as the burga in Alaska, the blizzard in the United States and Canada, and the purga in Russia are all often very cold, windy, and snowy. Other winds, such as the leste and sirocco, which originate over the Sahara Desert and blow across North Africa, are hot and dusty. Probably the hottest wind on earth is the simoom—the "poison wind" that can raise temperatures over the Arabian desert to dangerously high levels.

Key Terms

The following terms are listed (with page number) in the order they appear in the text. Define each. Doing so will aid you in reviewing the material covered in this chapter.

general circulation of the atmosphere, 208
Hadley cell, 208
doldrums, 209
subtropical highs, 209
horse latitudes, 210
trade winds, 210
intertropical convergence zone (ITCZ), 210
westerlies, 210
polar front, 210
subpolar low, 210
polar easterlies, 211
semipermanent highs and lows, 211
Bermuda high, 211
Pacific high, 211
Icelandic low, 211
Aleutian low, 211
Siberian high, 211
jet streams, 215
subtropical jet stream, 215
polar jet stream, 217
low level jet stream, 218
wind shear, 218
turbulence, 218
monsoon wind system, 219
sea breeze, 223
lake breeze, 223
land breeze, 223
sea breeze front, 224
valley breeze, 224
mountain breeze, 224
katabatic wind, 226
chinook wind, 227
Santa Ana wind, 229
Texas norther, 231
duststorms, 231
haboob, 232
black blizzards, 232
dust devils, 232
simoom, 233

Review Questions

1. If the earth did not rotate, how would surface winds tend to blow?
2. On a large circle, place the major surface pressure systems and major wind belts of the world at the approximate latitudes.
3. Why do large thunderstorms and heavy rain showers form along the ITCZ?
4. Why would you *not* expect a region of "horse latitudes" to form over the North Atlantic Ocean near 50°N?
5. Most of the United States is in what wind belt?
6. Explain how and why the average surface pressure features shift from summer to winter.
7. Describe how the general circulation helps to explain zones of abundant and sparse precipitation along a line that runs from the equator to the poles.
8. Why are summers dry in Southern California, and wet in Southern Georgia?
9. (a) Why do upper-level winds tend to blow from west to east?
   (b) Do these westerly winds explain why most severe weather phenomena, such as severe thunderstorms and tornadoes, tend to move toward the east?
10. (a) What are jet streams?
    (b) How are the polar jet stream and the subtropical jet stream similar?
    (c) How are they different?
11. Why is the polar jet stream stronger in winter than in summer?
12. Explain how wind shear in the vicinity of a jet stream can produce severe aircraft turbulence, even in clear air?
13. What is a monsoon wind system and how does it form over India?
14. Explain why cities in India, such as Cherrapunji, normally receive over 90 percent of their rainfall between April and October.
15. (a) How does a sea breeze differ from a land breeze?
    (b) Why are sea breezes usually stronger than land breezes?
16. Why are thunderstorms more likely to form over land with a sea breeze, and more likely to form over water with a strong land breeze?
17. You are fly-fishing in a mountain stream during the early morning: Would you expect the wind to be blowing upstream or downstream? Explain.
18. Which wind will most likely produce thunderstorms: A valley breeze or a mountain breeze? Why?
19. If katabatic winds blow downslope, why aren't they warm?
20. Explain why chinook winds are warm and dry.
21. What conditions must be in place for chinook winds to produce extreme temperature fluctuations?
22. Why are chinook winds called "snow eaters"?
23. What atmospheric conditions contribute to the development of a strong Santa Ana wind in Southern California? Why is a Santa Ana wind warm? Why is it also dry?
24. Why are haboobs more prevalent in Arizona than in Arkansas?
25. Describe how dust devils normally form.
26. In what part of the world would you most likely experience each of the following extreme winds, and what type of weather would each wind bring with it?
    (a) Texas norther
    (b) sirocco
    (c) burga
    (d) foehn
    (e) purga
    (f) California norther
    (g) Simoom
    (h) Columbia Gorge wind

Online Learning

STUDENT COMPANION WEBSITE: Visit this book's companion website at: www.cengage.com/ahrens/extreme1e and choose chapter 8 for many study aids and ideas for further reading and research. These include flashcards, practice quizzing, and web links.

METEOROLOGY RESOURCE CENTER: For students with access, log on at www.cengage.com/login for more assets, including animations, videos, and more. If your textbook did not come with access, visit www.CengageBrain.com to purchase.

Air Masses and Fronts

9

CONTENTS

AIR MASSES
FRONTS
SUMMARY
KEY TERMS
REVIEW QUESTIONS

An approaching front with showers and even a rainbow marches across western Iowa during June, 2009.

Extreme weather can form along fronts. A strong cold front can produce violent weather, such as severe thunderstorms, large hail, high surface winds, and even tornadoes. By the same token, warm fronts in winter often produce hazardous weather in the form of freezing rain, sleet, and fog. In this chapter, we will examine the typical weather that usually accompanies weather fronts. We will also look at some examples of extreme temperature changes that can take place along a cold front—where the temperature drops so quickly in such a short time that people on horseback have actually been frozen to their saddles. But first, we will examine air masses that bring with them the most brutal cold of winter and the oppressive scorching heat of summer.

Air Masses

An **air mass** is an extremely large body of air whose properties of temperature and humidity are fairly similar in any horizontal direction at any given altitude. Air masses may cover many thousands of square kilometers. In ❱ Fig. 9.1 a large winter air mass, associated with a high-pressure area, covers over half of the United States. Note that, although the surface air temperature and dew point vary somewhat, everywhere the air is cold and dry, with the exception of the zone of snow showers on the eastern shores of the Great Lakes. This cold, shallow anticyclone, steered by the winds aloft, will drift eastward, carrying with it the temperature and moisture characteristic of the region where the air mass formed; hence, in a day or two, cold air will be located over the central Atlantic Ocean. Part of weather forecasting is, then, a matter of determining air mass characteristics, predicting how and why they change, and in what direction the systems will move.

SOURCE REGIONS Regions where air masses originate are known as **source regions**. In order for a huge mass of air to develop uniform characteristics, its source region should be generally flat and of uniform composition with light surface winds. The longer the air remains stagnant over its source region, or the longer the path over which the air moves, the more likely it will acquire properties of the surface below. Consequently, ideal source regions are usually those areas dominated by surface high pressure. They include the ice- and snow-covered arctic plains in winter and subtropical oceans in summer. The middle latitudes, where surface temperatures and moisture characteristics vary considerably, are not good source regions. Instead, this region is a transition zone where air masses with different physical properties move in, clash, and produce an exciting array of weather activity.

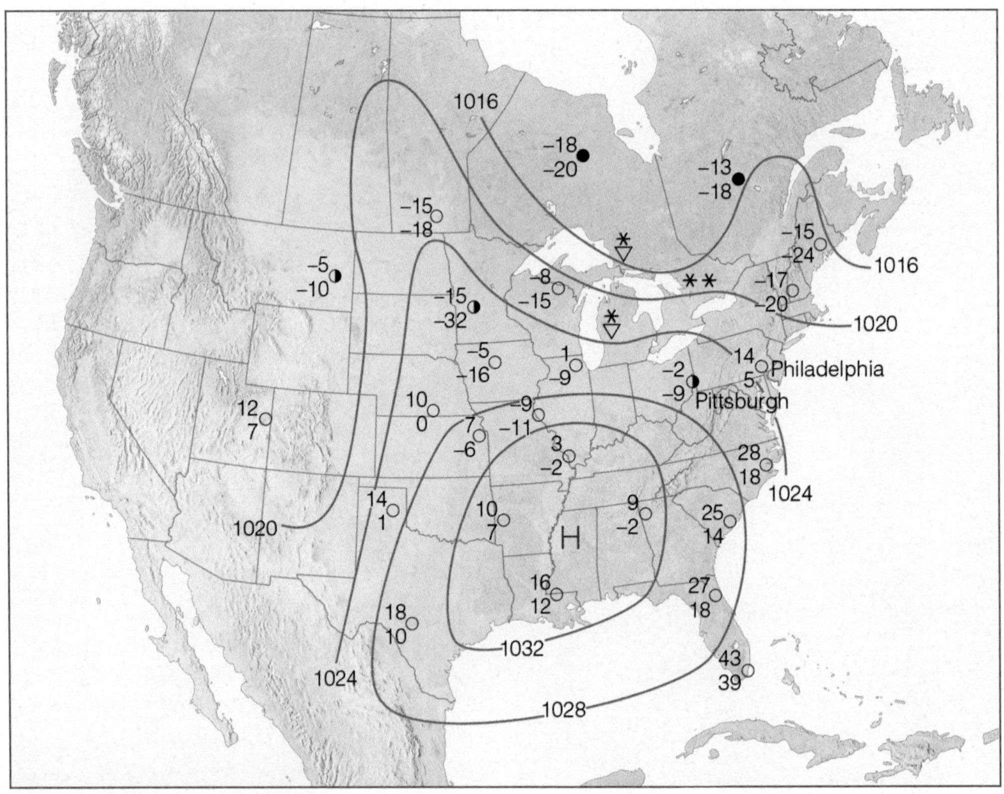

❱ **FIGURE 9.1** Here, a large, extremely cold winter air mass is dominating the weather over much of the United States. At almost all cities, the air is cold and dry. Upper number is air temperature (°F); bottom number is dew point (°F).

CLASSIFICATION Air masses are usually classified according to their temperature and humidity, both of which usually remain fairly uniform in any horizontal direction.* There are cold and warm air masses, humid and dry air masses. Air masses are grouped into five general categories according to their source region. Air masses that originate in polar latitudes are designated by the capital letter "P" (for *polar*); those that form in warm tropical regions are designated by the capital letter "T" (for *tropical*). If the source region is land, the air mass will be dry and the lowercase letter "c" (for *continental*) precedes the P or T. If the air mass originates over water, it will be moist—at least in the lower layers—and the lowercase letter "m" (for *maritime*) precedes the P or T. We can now see that polar air originating over land will be classified cP on a surface weather map, whereas tropical air originating over water will be marked as mT. In winter, an extremely cold air mass that forms over the arctic is designated as cA, *continental arctic*. Sometimes, however, it is difficult to distinguish between arctic and polar air masses, especially when the arctic air mass has traveled over warmer terrain. ❱ Table 9.1 lists the five basic air masses.

After the air mass spends some time over its source region, it usually begins to move in response to the winds aloft. As it moves away from its source region, it encounters surfaces that may be warmer or colder than itself. When the air mass is colder than the underlying surface, it is warmed from below, which produces instability at low levels. In this case, increased convection and turbulent mixing near the surface usually produce good visibility, cumuliform clouds, and showers of rain or snow. On the other hand, when the air mass is warmer than the surface below, the lower layers are chilled by contact with the cold earth. Warm air above cooler air produces stable air with little vertical mixing. This situation causes the accumulation of dust, smoke, and pollutants, which restricts surface visibilities. In moist air, stratiform clouds accompanied by drizzle or fog may form.

*In classifying air masses, it is common to use the *potential temperature* of the air. The potential temperature is the temperature that unsaturated (dry) air would have if moved from its original level to a pressure of 1000 millibars at the dry adiabatic rate (10°C/1000 m).

AIR MASSES OF NORTH AMERICA The principal air masses (with their source regions) that enter the United States are shown in ❱ Fig. 9.2. We are now in a position to study the formation and modification of each of these air masses and the type of weather that accompanies it.

CONTINENTAL POLAR (cP) AND CONTINENTAL ARCTIC (cA) AIR MASSES The bitterly cold weather that invades southern Canada and the United States in winter is associated with **continental polar** and **continental arctic air masses**. These air masses originate over the ice- and snow-covered regions of the arctic, northern Canada, and Alaska where long, clear nights allow for strong radiational cooling of the surface. Air in contact with the surface becomes quite cold and stable. Since little moisture is added to the air, it is also quite dry, and dew-point temperatures are often less than −22°F (−30°C). Eventually a portion of this cold air breaks away and, under the influence of the air flow aloft, moves southward as an enormous shallow high-pressure area, as illustrated in ❱ Fig. 9.3.

As the cold air moves into the interior plains, there are no topographic barriers to restrain it, so it continues southward, bringing with it cold wave warnings and frigid temperatures. The infamous Texas norther is associated with cold continental polar and arctic air. As the air mass moves over warmer land to the south, the air temperature moderates slightly as it is heated from below. However, even during the afternoon, when the surface air is most unstable, cumulus clouds are rare because of the extreme dryness of the air mass. At night, when the winds die down, rapid radiational surface cooling and clear skies combine to produce low minimum temperatures. If the cold air moves as far south as central or southern Florida, the winter vegetable crop may be severely damaged. When the cold, dry air mass

❱ Table 9.1

Air Mass Classification and Characteristics

SOURCE REGION	ARCTIC REGION (A)	POLAR (P)	TROPICAL (T)
Land	cA	cP	cT
Continental (c)	extremely cold, dry stable; ice- and snow-covered surface	cold, dry, stable	hot, dry, stable air aloft; unstable surface air
Water		mP	mT
Maritime (m)		cool, moist, unstable	warm, moist, usually unstable

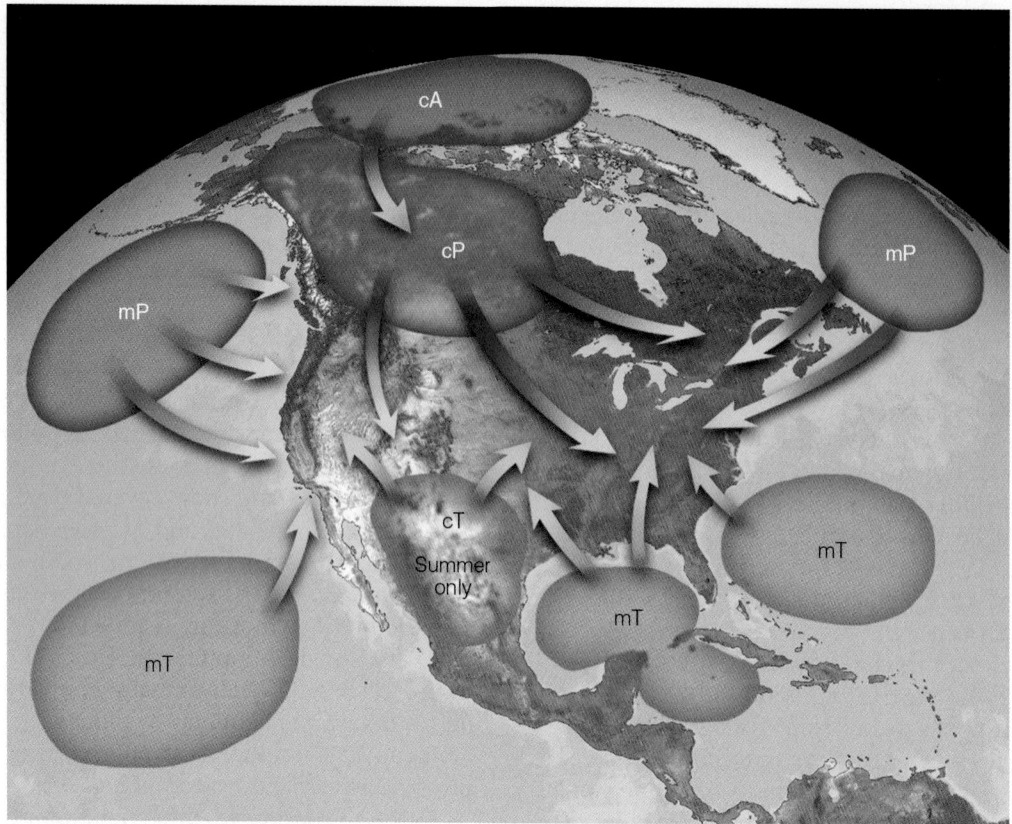

FIGURE 9.2 Air mass source regions and their paths.

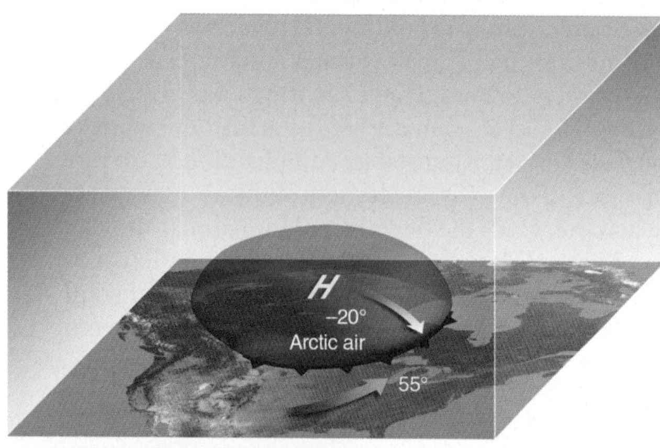

FIGURE 9.3 A shallow but large dome of extremely cold air—a continental arctic air mass—moves slowly southeastward across the upper plains. The leading edge of the air mass is marked by a cold front. (Numbers represent air temperature, °F.)

moves over a relatively warm body of water, such as the Great Lakes, heavy snow showers—called *lake-effect snows*—often form on the eastern shores.

Lake-effect snows, although often highly localized, can produce huge quantities of snow. On the downwind shores of the Great Lakes, snow depths after a lake-effect storm are often measured in feet. For example, Buffalo, New York, which lies to the east of Lake Erie, received nearly seven feet of snow in less than four days during December, 2001.* In winter, the generally fair weather accompanying polar and arctic air is due to the stable nature of the atmosphere aloft. Sinking air develops above the large dome of high pressure. The subsiding air warms by compression and creates warmer air, which lies above colder surface air. Therefore, a strong upper-level subsidence inversion often forms. Should the anticyclone stagnate over a region for several days, the visibility gradually drops as pollutants become trapped in the cold air near the ground. Usually, however, winds aloft move the cold air mass either eastward or southeastward.

The Rockies, Sierra Nevada, and Cascades normally protect the Pacific Northwest from the onslaught of polar and arctic air masses, but, occasionally, very cold air masses do invade these regions. When the upper-level winds over Washington and Oregon blow from the north or northeast on a trajectory beginning over northern Canada or Alaska, very cold air can slip over

*For more information on lake-effect snows, read the Focus sections on p. 136 in Chapter 5, and on p. 166 in Chapter 6.

the mountains and extend its icy fingers all the way to the Pacific Ocean. As the air moves off the high plateau, over the mountains, and on into the lower valleys, compressional heating of the sinking air causes its temperature to rise, so that by the time it reaches the lowlands, it is considerably warmer than it was originally. However, in no way would this air be considered warm. In some cases, the subfreezing temperatures slip over the Cascades and extend southward into the coastal areas of southern California.

A similar but less dramatic warming of cold air masses occurs along the eastern coast of the United States. Air rides up and over the lower Appalachian Mountains. Turbulent mixing and compressional heating increase the air temperatures on the downwind side. Consequently, cities located to the east of the Appalachian Mountains usually do not experience temperatures as low as those on the west side. In Fig. 9.1 notice that for the same time of day—in this case 7 A.M. EST—Philadelphia, with an air temperature of 14°F, is 16°F warmer than Pittsburgh, with an air temperature of −2°F.

Extremely Cold Outbreaks Figure 9.4 shows two upper-air wind patterns that led to extremely cold outbreaks of arctic air during December, 1989 and 1990. Upper-level winds typically blow from west to east, but, in both of these cases, the flow, as given by the heavy, dark arrows, had a strong north-south (meridional) trajectory. The H represents the positions of the cold surface anticyclones. Numbers on the map represent minimum temperatures (°F) recorded during the cold spells.

The arctic outbreak during 1989 lasted from about December 21 through December 28. East of the Rocky Mountains, over 350 record low temperatures were set between December 21 and 24. On December 22, the temperature plunged to −47°F at both Broadus and Hardin, Montana. In Kansas City, Missouri, the air temperature on December 22 dropped to −23°F, and only climbed to −8°F during the afternoon, making this one of the coldest days on record. The temperature remained below zero in Kansas City for about sixty hours. At one point, the frigid air stretched all the way from Northern Canada into the Deep South, where a hard freeze caused an estimated $480 million in damage to the fruit and vegetable crops in Texas and Florida.

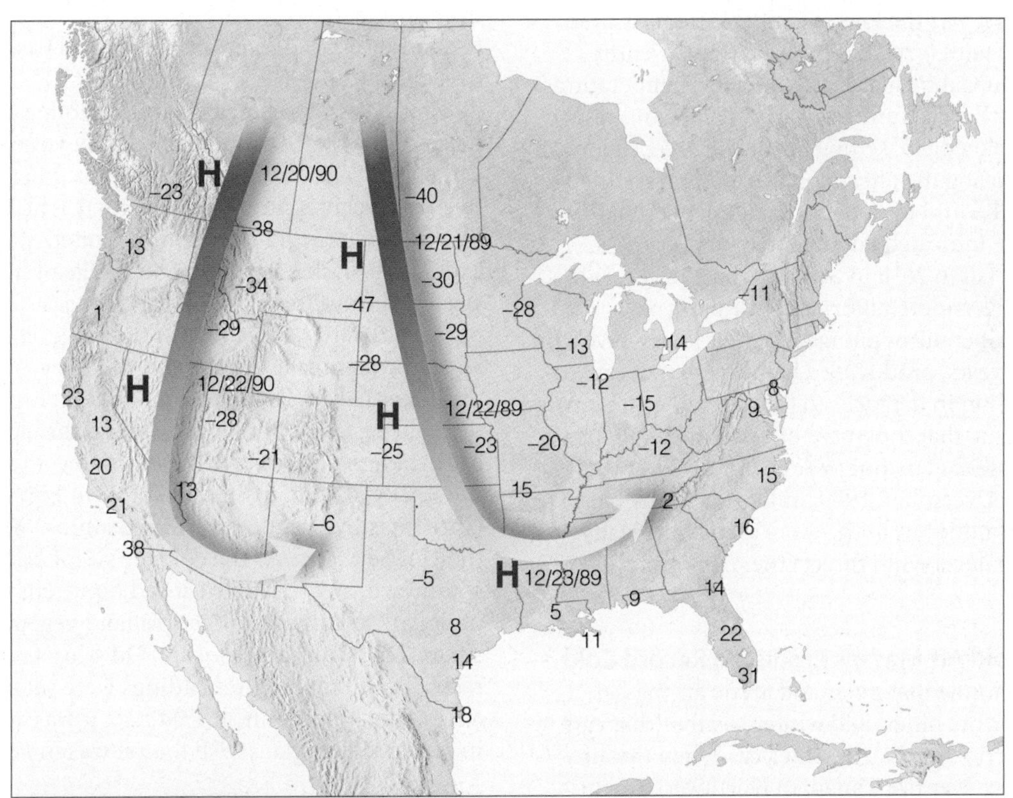

FIGURE 9.4 Average upper-level wind flow (heavy arrows) and surface position of anticyclones (H) associated with two extremely cold outbreaks of arctic air during December. Numbers on the map represent some minimum temperatures (°F) measured during each cold snap.

EXTREME WEATHER WATCH

The greatest cold wave since the establishment of the United States Weather Bureau (1870) occurred in early February, 1899, when an arctic air mass pushed all the way south to the Gulf of Mexico. For the only time on record, every state reported below zero temperatures, including Florida, where Tallahassee reached −2°F. Four inches of snow fell in New Orleans, where a record low temperature of 7°F was recorded. Other all-time cold records still standing include −15°F in Washington, D.C.; −9°F in Atlanta, Georgia; −23°F in Tulsa, Oklahoma; and −48°F in Billings, Montana.

During December, 1990, frigid arctic air moved south out of northern Canada into the United States west of the Rocky Mountains. The cold outbreak, which lasted from December 20 through December 23, was one of the coldest ever in the far west, with December 22 being one of the coldest days ever west of the Rocky Mountains. In Meadows Lodge, Idaho, the temperature plummeted to −55°F, which was the lowest temperature measured in the United States (excluding Alaska) for the entire year of 1990. In California's Central Valley, temperatures dropped to 19°F at Sacramento Airport, 18°F in Fresno, and 19°F in Bakersfield, setting an all-time record low for the city. During the afternoon, temperatures only climbed into the mid to upper 30s, and dew points registering 5°F indicated an extremely dry air mass. Temperatures fell to 20°F in Santa Barbara, California, and 22°F in Riverside, California. Over parts of California, temperatures plunged to their lowest levels in more than 40 years, producing a hard freeze that caused over $300 million in damages to the vegetable and citrus crops. It is ironic that the freeze of December, 1990, in California came exactly one year to the day after the hard freeze of December, 1989, in Florida. Look closely at Fig. 9.4 and notice in both severe arctic outbreaks how the upper-level wind directs the path of the air masses.

Extremely Cold Air Masses Produce a Record Cold Winter We know that polar and arctic air masses are responsible for the bitter cold winter weather that can cover wide sections of North America. When the air mass originates over the Canadian Northwest Territories, frigid air can bring record-breaking low temperatures. If one cold air mass after another should move into the United States, this pattern can make for an exceptionally cold winter.

The winter of 1983–1984 was one of the coldest on record across North America. Unseasonably cold weather arrived in December, which, for much of the United States, was one of the coldest Decembers since records have been kept. During the first part of the month, continental polar air covered most of the northern and central plains. As the cold air moderated slightly, far to the north a huge mass of bitter cold arctic air was forming over the frozen reaches of the Canadian Northwest Territories.

By midmonth, the frigid air, associated with a massive high-pressure area, covered all of northwest Canada. Meanwhile, an upper-level ridge was forming over Alaska. On the eastern side of the ridge, strong northerly winds associated with the polar jet stream directed the frigid air southward over the prairie provinces of Canada. A portion of the extraordinarily cold air broke away, and, like a large swirling bubble, moved as a cold, shallow anticyclone southward into the United States. Because the frigid air was accompanied in some regions by winds gusting to 50 mi/hr, one news reporter dubbed the onslaught of this arctic blast "the Siberian Express."

The Express dropped temperatures to some of the lowest readings ever recorded during the month of December. On December 22, Elk Park, Montana, recorded an unofficial low of −64°F, only 6°F higher than the all-time low of −70°F for the United States (excluding Alaska) recorded at Rogers Pass, Montana, on January 20, 1954.

The center of the massive anticyclone gradually pushed southward out of Canada. By December 24, its center was over eastern Montana (see ❯ Fig. 9.5), where the sea-level pressure at Miles City reached an incredible 1064 mb (31.42 in.)—a new United States record excluding Alaska. An enormous ridge of high pressure stretched from the Canadian arctic coast to the Gulf of Mexico. On the east side of the ridge, cold westerly winds brought lake-effect snows to the eastern shores of the Great Lakes. To the south of the high-pressure center, cold easterly winds, rising along the elevated plains, brought light amounts of *upslope snow*[*] to sections of the Rocky Mountain States. Notice in Fig. 9.5 that, on Christmas Eve, arctic air covered almost 90 percent of the United States. As the cold air swept eastward and southward, a hard freeze caused hundreds of millions of dollars in damage to the fruit and vegetable crops in Texas, Louisiana, and Florida. On Christmas Day, 125 record low temperature readings were set in twenty-four states. That afternoon, at 1:00 P.M., it was actually colder in Atlanta, Georgia, at 9°F than it was in Fairbanks,

[*]Upslope snow forms as cold air moving from east to west over the Great Plains gradually rises (and cools even more) as it approaches the Rocky Mountains.

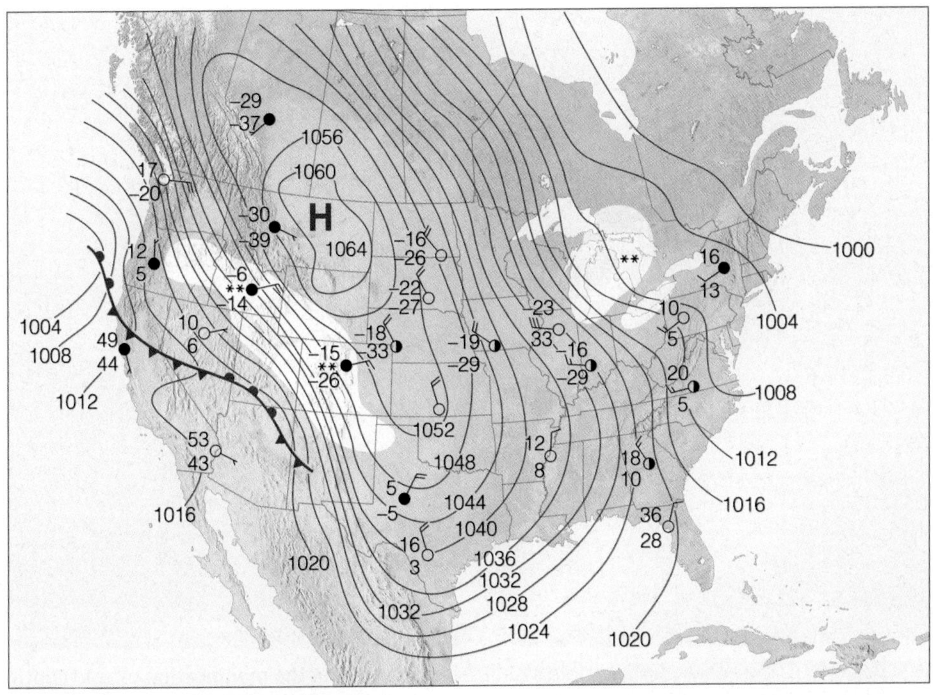

FIGURE 9.5 Surface weather map for 7 A.M., EST, December 24, 1983. Solid lines are isobars. Areas shaded white represent snow. An extremely cold arctic air mass covers nearly 90 percent of the United States. (Weather symbols for the surface map are given in Appendix C.)

Alaska (10°F). One of the worst cold waves to occur in December during the twentieth century continued through the week, as many new record lows were established in the Deep South from Texas to Louisiana.

By January 1, the extreme cold had moderated, as the upper-level winds became more westerly. These winds brought milder Pacific air eastward into the Great Plains. The warmer pattern continued until about January 10, when another cold blast decided to make a return visit. Driven by strong upper-level northerly winds, impulse after impulse of arctic air from Canada swept across the United States. On January 18, a record low of −65°F was recorded for the state of Utah at Middle Sinks.* On January 19, temperatures plummeted to a new low of −7°F for the airports in Philadelphia and Baltimore. Toward the end of the month, the upper-level winds once again became more westerly. Over much of the nation, the cold air moderated. But the extreme cold was to return at least one more time.

The beginning of February saw relatively warm air covering much of the nation from California to the Atlantic coast. On February 4, an arctic outbreak spread southward and eastward across the United States. Although freezing air extended southward into central Florida, the arctic blast ran out of steam, and a February heat wave soon engulfed most of the states east of the Rocky Mountains as warm, humid air from the Gulf of Mexico spread northward.

Even though February was a warm month over most of the United States, the winter of 1983–1984 (December, January, and February) will go down in the record books as one of the coldest winters for the United States as a whole since reliable record keeping began.

Modification of Cold Air Masses The continental polar air that moves into the United States in summer has properties much different from its winter counterpart. The source region remains the same but is now characterized by long summer days that melt snow and warm the land. The air is only moderately cool, and surface evaporation adds water vapor to the air. A summertime polar air mass usually brings relief from the oppressive heat in the central and eastern states, as cooler air lowers the air temperature to more comfortable levels. Daytime heating warms the lower layers, producing surface instability. The water vapor in the rising air may condense and create a sky dotted with fair weather cumulus clouds (cumulus humilis).

*Utah's current state record low is −69°F, recorded at Peter's Sink on February 1, 1985—the second coldest temperature recorded in the continental United States.

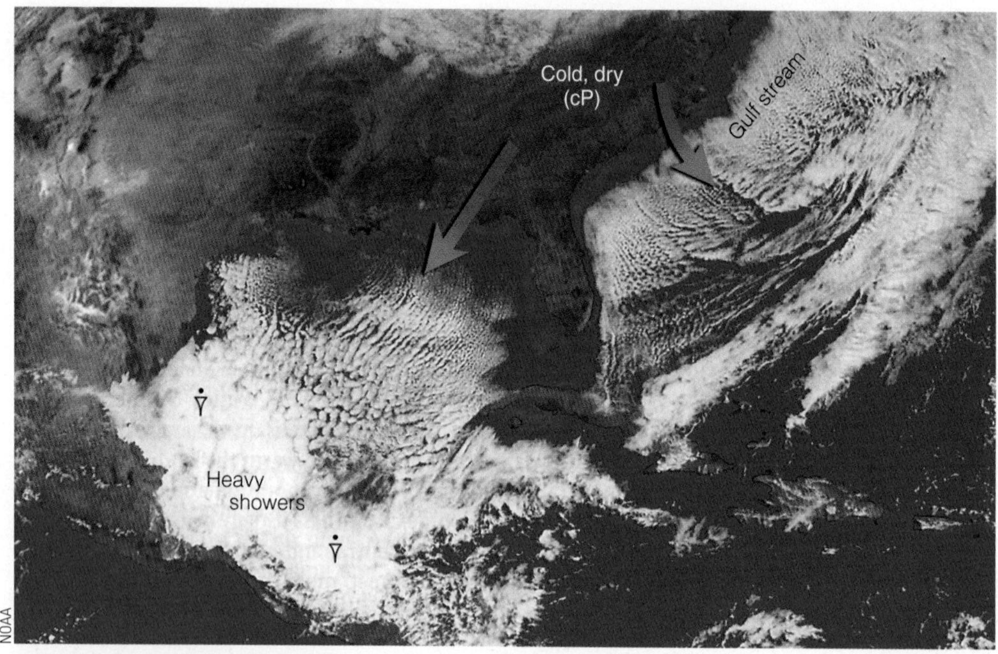

FIGURE 9.6 Visible satellite image showing the modification of cold continental polar air as it moves over the warmer Gulf of Mexico and the Atlantic Ocean.

When an air mass moves over a large body of water, its original properties may change considerably. For instance, cold, dry polar air moving over the Gulf of Mexico warms rapidly and gains moisture. The air quickly assumes the qualities of a maritime air mass. Notice in Fig. 9.6 that rows of cumulus clouds (*cloud streets*) are forming over the Gulf of Mexico parallel to northerly surface winds as the polar air is being warmed by the water beneath it, causing the air mass to destabilize. As the air continues its journey southward into Mexico and Central America, strong, moist northerly winds build into heavy clouds and showers along the northern coast. Hence, a once cold, dry, and stable air mass can be modified to such an extent that its original characteristics are no longer discernible. When this happens, the air mass often is given a new designation.

Notice also in Fig. 9.6 that a similar modification of cold, polar air is occurring along the Atlantic Coast, as northwesterly winds are blowing over the mild Atlantic. When this air encounters the much warmer Gulf Stream water, it warms rapidly and becomes conditionally unstable. Vertical mixing brings down faster-flowing cold air from aloft. This mixing creates strong, gusty surface winds and choppy seas, which can be hazardous to shipping.

MARITIME POLAR (mP) AIR MASSES During the winter, polar and arctic air originating over Asia and frozen polar regions is carried eastward and southward over the Pacific Ocean by the circulation around the Aleutian low. The ocean water modifies these cold air masses by adding warmth and moisture to them. Since this air travels across many hundreds or even thousands of miles of water, it gradually changes into *maritime polar air*.

By the time this **maritime polar air mass** reaches the Pacific Coast, it is cool, moist, and conditionally unstable. The ocean's effect is to keep air near the surface warmer than the air aloft. Temperature readings in the 40s and 50s (°F) are common near the surface, while air at an altitude of about 3000 feet or so above the surface may be at the freezing point. Within this colder air, characteristics of the original cold, dry air mass may still prevail. As the air moves inland, coastal mountains force it to rise, and much of its water vapor condenses into rain-producing clouds. In the colder air aloft, the rain changes to snow, with heavy amounts accumulating in mountain regions. Over the relatively warm open ocean, the cool moist air mass produces cumulus clouds that show up as tiny white splotches on a visible satellite image (see Fig. 9.7).

When the maritime polar air moves inland, it loses much of its moisture as it crosses a series of mountain ranges. Beyond these mountains, it travels over a cold, elevated plateau that chills the surface air and slowly transforms the lower level into dry, stable continental polar air. East of the Rockies this air mass is referred to as **Pacific air** (see Fig. 9.8). Here, it often brings fair

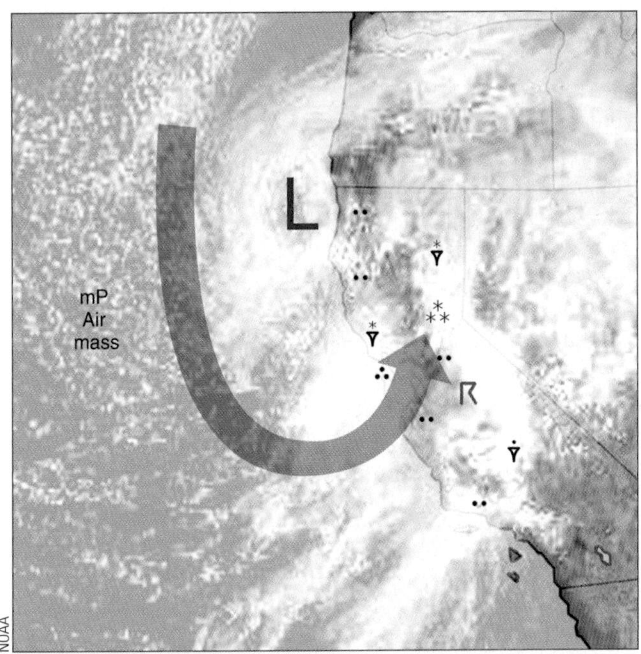

FIGURE 9.7 Clouds and air flow aloft (large blue arrow) associated with maritime polar air moving into California. The large L shows the position of an upper-level low. Regions experiencing precipitation are also shown. The small, white clouds over the open ocean are cumulus clouds forming in the conditionally unstable air mass. (Precipitation symbols are given in Appendix C.)

weather and temperatures that are cool but not nearly as cold as the continental polar and arctic air that invades this region from northern Canada. In fact, when Pacific air from the west replaces retreating cold air from the north, chinook winds often develop. Furthermore, when the modified maritime polar air replaces moist, tropical air, thunderstorms can form along the boundary separating the two air masses.

Along the East Coast, maritime polar air originates in the North Atlantic as continental polar air moves southward some distance off the Atlantic Coast. (Look back at Fig. 9.2, p. 240.) Steered by northeasterly winds, this cold, moist air then swings southwestward toward the northeastern states. Because the water of the North Atlantic is very cold and the air mass travels only a short distance, wintertime Atlantic maritime polar air masses are usually much colder than their Pacific counterparts. Because the prevailing winds aloft are westerly, Atlantic maritime polar air masses are also much less common.

Figure 9.9 illustrates a typical late winter or early spring surface weather pattern that carries maritime polar air from the Atlantic into the New England and middle Atlantic states. A slow-moving, cold anticyclone drifting to the east (north of New England) causes, to its south, a northeasterly flow of cold, moist air. The boundary separating this invading colder air from warmer air even farther south is marked by a stationary front. North of this front, northeasterly winds provide generally undesirable weather, consisting of damp air and low, thick clouds from which light precipitation falls in the form of rain, drizzle, or snow. When upper atmospheric conditions are right, a mid-latitude cyclone may develop along the stationary front, move eastward, and intensify near the shores of Cape Hatteras. Such storms, called *Hatteras lows*, sometimes swing northeastward along the coast, where they become *northeasters* (commonly called *nor'easters*) bringing with them strong northeasterly winds, heavy rain or snow, and coastal flooding. (Such developing storms will be treated in detail in Chapter 10.)

MARITIME TROPICAL (mT) AIR MASSES

The wintertime source region for Pacific **maritime tropical air masses** is the subtropical east Pacific Ocean. (Look back at Fig. 9.2, p. 240.) Air from this region must travel over 1000 miles of water before it reaches the southern

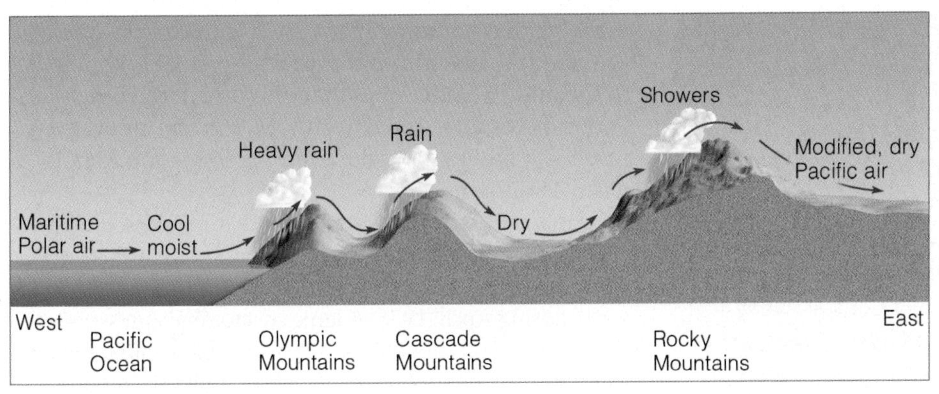

FIGURE 9.8 After crossing several mountain ranges, cool moist mP air from off the Pacific Ocean descends the eastern side of the Rockies as modified, relatively dry Pacific air.

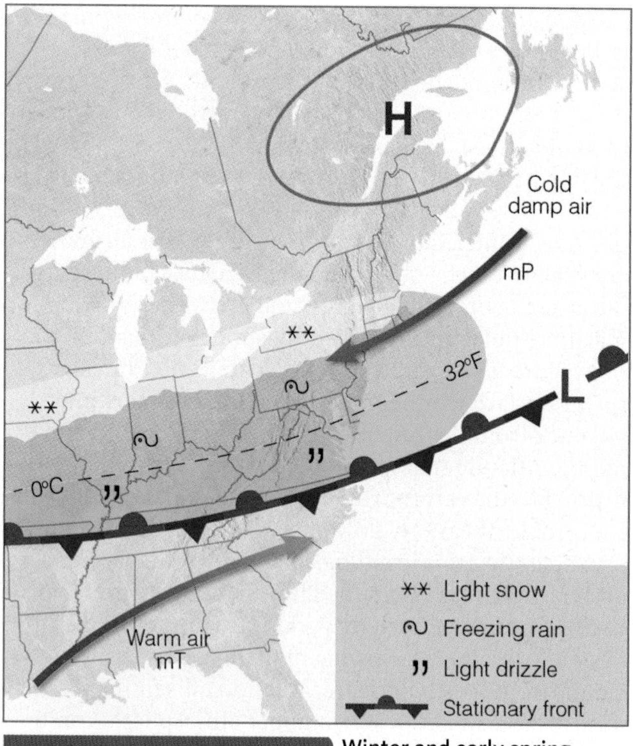

FIGURE 9.9 Winter and early spring surface weather pattern that usually prevails during the invasion of cold, moist maritime polar air into the mid-Atlantic and New England states. (Green-shaded area represents light rain and drizzle; pink-shaded region represents freezing rain and sleet; white-shaded area is experiencing snow.)

California coast. Consequently, these air masses are very warm and moist by the time they arrive along the West Coast. In winter, the warm air produces heavy precipitation usually in the form of rain, even at high elevations. Melting snow and rain quickly fill rivers, which overflow into the low-lying valleys, producing dangerous floods. The rapid snowmelt leaves local ski slopes barren, and the heavy rain can cause disastrous mud slides in the steep canyons.

Figure 9.10 shows maritime tropical air (usually referred to as *subtropical air*) streaming into northern California on January 1, 1997. The humid, subtropical air, which originated near the Hawaiian Islands, was termed by at least one forecaster as "*the pineapple express.*" After battering the Pacific Northwest with heavy rain, the pineapple express roared into northern and central California, causing catastrophic floods that sent over 100,000 people fleeing from their homes, mud slides that closed roads, property damage (including crop losses) that amounted to more than $1.5 billion, and eight fatalities. Yosemite National Park, which sustained over $170 million in damages due mainly to flooding, was forced to close for more than two months (see Fig. 9.11).

The warm, humid subtropical air that influences much of the weather east of the Rockies originates over the Gulf of Mexico and Caribbean Sea. In winter, cold polar air tends to dominate the continental weather scene, so maritime tropical air is usually confined to the Gulf and extreme southern states. Occasionally, a slow-moving cyclonic storm system over the Central Plains draws warm, humid air northward. Gentle south or southwesterly winds carry this air into the central and eastern parts of the United States in advance of the storm. Since the land is still extremely cold, air near the surface is chilled to its dew point. Fog and low clouds form in the early morning, dissipate by midday, and reform in the evening. This mild winter weather in the Mississippi and Ohio valleys lasts, at best, only a few days. Soon cold polar air will move down from the north behind the eastward-moving storm system. Along the boundary between the two air masses, the warm, humid air is lifted above the more-dense cold, polar air—a situation that often leads to heavy and widespread precipitation and storminess.

When a large mid-latitude cyclonic storm system stalls over the Central Plains, a constant supply of warm, humid air from the Gulf of Mexico can bring record-breaking maximum temperatures to the eastern half of the country. Sometimes the air temperatures are higher in the mid-Atlantic states than they are in the Deep South, as compressional heating warms the air even more as it moves downslope after crossing the Appalachian Mountains.

Figure 9.12 shows a surface weather map and the associated upper airflow (heavy arrow) that brought extremely warm maritime tropical air into the central and eastern states during April, 1976. A large surface high-pressure area centered off the southeast coast coupled with a strong southwesterly flow aloft carried warm moist air into the Midwest and East, causing a record-breaking April heat wave. The flow aloft prevented the surface low and the cold

EXTREME WEATHER WATCH

One of the most intense "pineapple expresses" on record occurred over the period of February 11–20,1986, when more than 41 inches of rain fell at a location in California's Napa Valley, and over 49 inches fell in the foothills of the Sierra Nevada. The heaviest 24-hour rainfall ever measured in California's Central Valley (17.60 inches) fell at Four Trees, and over 15 inches of rain fell on Atlas Mountain in Napa County on February 17 at the peak of the storm.

FIGURE 9.10 An infrared satellite image that shows maritime tropical air (heavy yellow arrow) moving into northern California on January 1, 1997. The warm, humid airflow (sometimes called "the pineapple express") produced heavy rain and extensive flooding in northern and central California.

polar air behind it from making much eastward progress, so that the warm spell lasted for five days. Providence, Rhode Island, experienced a record-breaking high temperature for April of 96°F. Note that, on the west side of the surface low, the winds aloft funneled cold air from the north into the western states, creating unseasonably cold weather from California to the Rockies. Hence, while people in the Southwest were huddled around heaters, others several thousand miles away in the Northeast were turning on air conditioners. We can see that it is the upper-level meridional flow, directing cold polar air southward and warm subtropical air northward, that makes these contrasts in temperature possible.

FIGURE 9.11 During January, 1997, a moist, subtropical air mass caused extensive flooding over parts of northern California, including Yosemite National Park, as shown here.

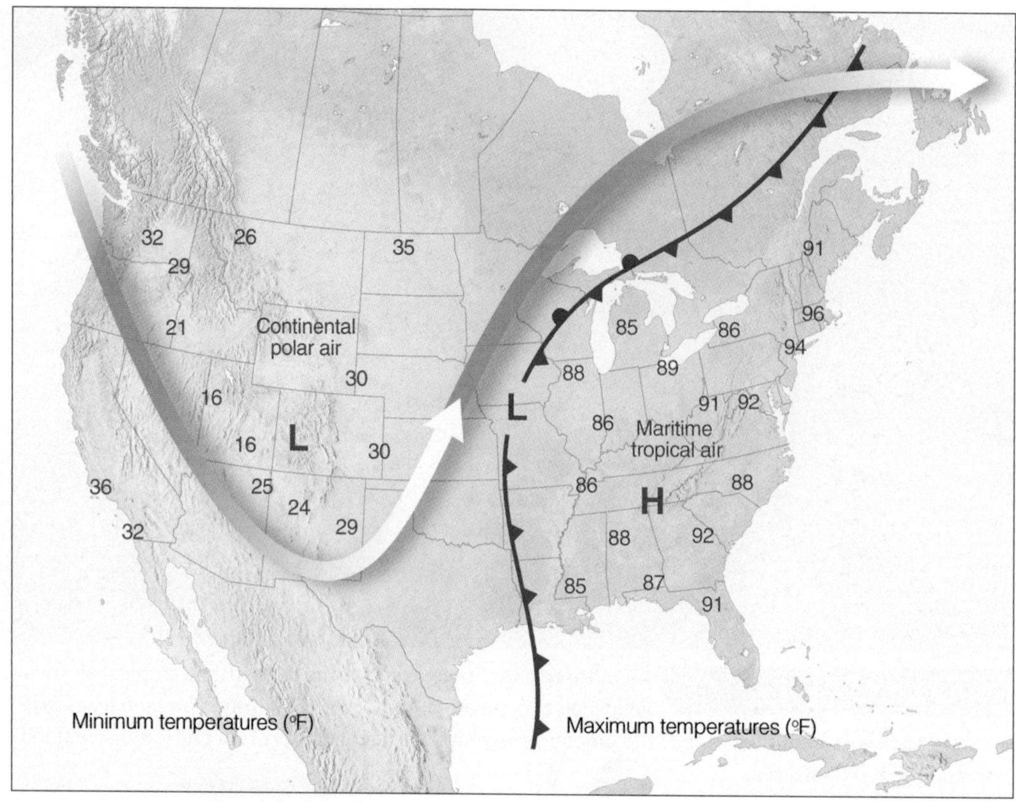

FIGURE 9.12 Weather conditions during an unseasonably hot spell in the eastern portion of the United States that occurred between the 15th and 20th of April, 1976. The surface low-pressure area and fronts are shown for April 17. Numbers to the east of the surface low (in red) are maximum temperatures recorded during the hot spell, while those to the west of the low (in blue) are minimum temperatures reached during the same time period. The heavy arrow is the average upper-level flow during the period. The purple L and H show average positions of the upper-level trough and ridge.

FIGURE 9.13 As moist, subtropical air is lifted along mountain ranges, such as the Sierra Nevada, towering clouds and thunderstorms often form in the afternoon.

In summer, the circulation of air around the Bermuda High (which sits off the southeast coast of North America; see Fig. 8.3b, p. 212) pumps warm, humid (mP) air northward from off the Gulf of Mexico and from off the Atlantic Ocean into the eastern half of the United States. As this humid air moves inland, it warms even more, rises, and frequently condenses into cumuliform clouds, which produce afternoon showers and thunderstorms. You can almost count on thunderstorms developing along the Gulf Coast every summer afternoon. As evening approaches, thunderstorm activity typically dies off. Nighttime cooling lowers the temperature of this hot, muggy air only slightly. Should the air become saturated, fog or low clouds usually form, and these normally dissipate by late morning as surface heating warms the air again.

A weak, but often persistent, flow around an upper-level anticyclone in summer will spread warm, humid air from the Gulf of Mexico and from the Gulf of California into the southern and central Rockies, where it causes afternoon thunderstorms. Occasionally, this easterly flow may work its way even farther west, producing shower activity in the otherwise dry southwestern desert.

During the summer, humid subtropical air originating over the southeastern Pacific and Gulf of California normally remains south of California. Occasionally, an upper-level southerly flow will spread this humid air northward into the southwestern United States, most often Arizona, Nevada, and the southern part of California. In many places, the moist, conditionally unstable air aloft only shows up as middle and high cloudiness, especially altocumulus and cirrocumulus castellanus that, remember from Chapter 5, indicates instability at cloud level. However, where the moist flow meets a mountain barrier, it usually rises and condenses into towering, shower-producing clouds (see ❱Fig 9.13).

CONTINENTAL TROPICAL (cT) AIR MASSES The only real source region for hot, dry **continental tropical air masses** in North America is found during the summer in northern Mexico and the adjacent arid southwestern United States. Here, the air mass is hot, dry, and conditionally unstable at low levels, with frequent dust devils forming during the day. Because of the low relative humidity (typically less than 10 percent during the afternoon), air must rise over 9000 feet before condensation begins. Furthermore, an upper-level ridge usually produces sinking air over the region, tending to make the air aloft rather stable and the surface air even warmer. Consequently, skies are generally clear, the weather is hot, and rainfall is practically nonexistent where continental tropical air masses prevail. If this air mass moves outside its source region and into the Great Plains and stagnates over that region for any length of time, a severe drought may result, accompanied by record-breaking high temperatures.

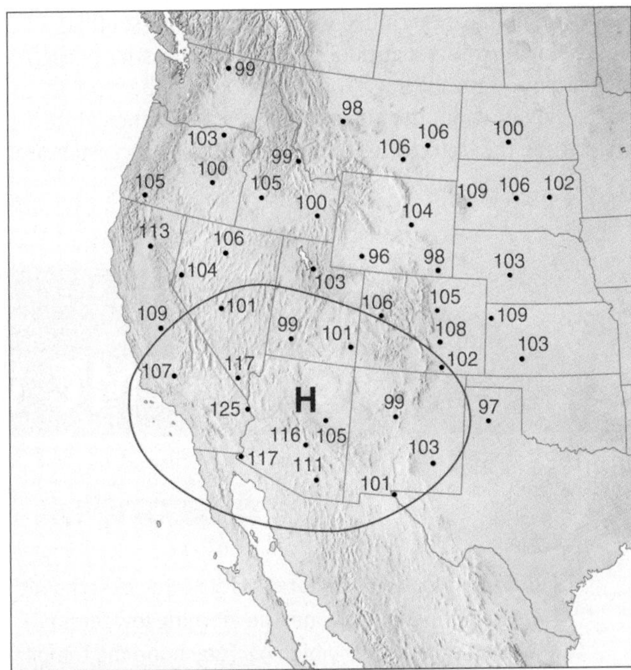

❱ **FIGURE 9.14** From July 14 through July 22, 2005, continental tropical air covered a large area of the southwestern United States. Numbers on the map represent maximum temperatures (°F) during this period. The large H with the isobar shows the upper-level position of the subtropical high. Sinking air associated with the high contributed to the hot weather. Winds aloft were weak, with the main flow over central Canada.

❱Figure 9.14 shows a weather map situation where continental tropical air produces hot, dry weather over a large portion of the southwestern United States during July, 2005. Notice that the heat is accompanied by a large upper-level area of high pressure. When the high becomes centered over a region, extremely high surface temperatures can occur. For example, during this July heat wave of 2005, Death Valley, California, reported a high temperature of 125°F or greater on seven consecutive days (from July 14 through July 20). Las Vegas, Nevada, set a record high temperature of 117°F on July 19 and experienced five days when the maximum temperatures exceeded 115°F. Farther east, Denver, Colorado, tied its all-time record high on July 20, when the temperature reached 105°F, and Goodland, Kansas, experienced its hottest day in 58 years when the temperature peaked at 109°F. All of these city temperatures are extremely high, but they do not come close to the 122°F reading in Phoenix, Arizona, on July 26, 1990. (More information on this scorching hot day is provided in the Focus section on p. 250.)

So far, we have examined the various air masses that enter North America annually. The characteristics of each depend upon the air mass source region and the type of surface over which the air mass moves. The winds aloft determine the trajectories of these air masses. Occasionally, an air mass will control the weather in a region for some time. These persistent weather conditions are sometimes referred to as *airmass weather.*

Airmass weather is especially common in the southeastern United States during summer as, day after day, humid subtropical air from the Gulf brings sultry

FOCUS ON EXTREME WEATHER

A Scorcher by Noon in June

June 26, 1990, was one hot day across the desert southwest. In Phoenix, Arizona, the morning low temperature was an exceptionally high 92°F; by noon the temperature climbed to an oven-like 114°F. And by late afternoon, the air temperature soared to an unbelievable 122°F, setting an all-time record high temperature for the city.

It was so hot at Phoenix's Sky Harbor airport, that fully loaded jet aircraft began to leave ruts in the softened airport pavement near the boarding gates. As a precautionary move, officials suspended aircraft operations because the extreme heat had also lowered the air density to a point where it could have dangerously reduced aircraft lift. Many flights were diverted to Tucson, Arizona, where the air temperature was only 117°F.

As we can see in Fig. 1, it was hot throughout the region. Notice that Yuma, Arizona, also hit a sweltering high temperature of 122°F. In Thermal, California, the temperature soared to 121°F, and in Palm Springs the temperature reached 119°F. The hot air mass even spread westward to downtown Los Angeles where the thermometer registered 112°F, setting an all-time record high. To make matters worse, smog levels in the city rose to unhealthful levels as a large upper-level high pressure area dominated the weather scene (see Fig. 2). Overall, June 26, 1990, was one of the hottest days ever in the southwest region of the United States.

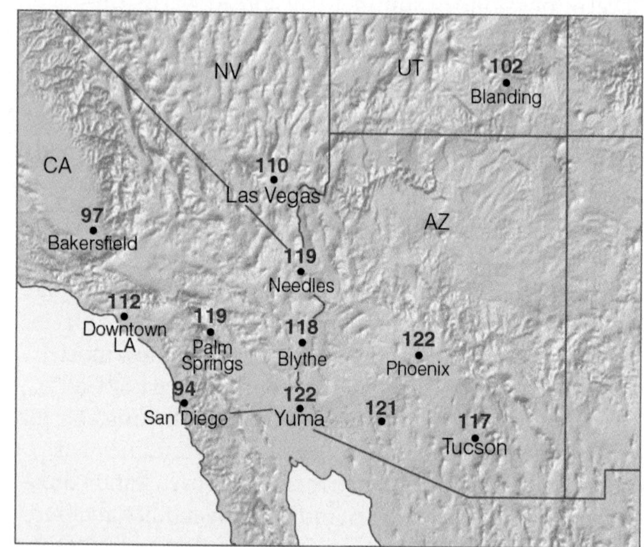

Figure 1 High temperatures throughout the southwestern United States on June 26, 1990.

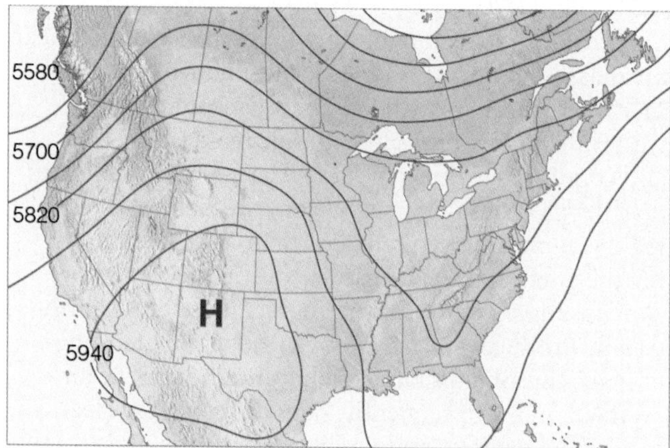

Figure 2 On June 26, 1990, the 500-mb chart shows that a large upper-level high covers most of the southwestern United States.

conditions and afternoon thunderstorms. It is also common in the Pacific Northwest in winter when conditionally unstable, cool moist air accompanied by widely scattered showers dominates the weather for several days or more. The real weather action, however, usually occurs not within air masses but at their margins, where air masses with sharply contrasting properties meet—in the zone marked by *weather fronts.*

BRIEF REVIEW

Before we examine fronts, here is a review of some of the important facts about air masses:

- **An air mass is a large body of air whose properties of temperature and humidity are fairly similar in any horizontal direction.**
- **Source regions for air masses tend to be generally flat, of uniform composition, and in an area of light winds, dominated by surface high pressure.**
- **Continental air masses form over land. Maritime air masses form over water. Polar air masses originate in cold, polar regions, and extremely cold arctic air masses form over arctic regions. Tropical air masses originate in warm, tropical regions.**
- **Continental polar (cP) air masses are cold and dry; continental arctic (cA) air masses are extremely cold and dry. It is the continental arctic air masses that produce the extreme cold of winter as they move across North America.**
- **Continental tropical (cT) air masses are hot and dry, and are responsible for the heatwaves of summer in the western half of the United States.**
- **Maritime polar (mP) air masses are cold and moist, and are responsible for the cold, damp and often wet weather along the northeast coast of North America, as well as for the cold, rainy winter weather along the west coast of North America.**
- **Maritime tropical (mT) air masses are warm and humid, and are responsible for the hot, muggy weather that frequently plagues the eastern half of the United States in summer.**

Fronts

Although we briefly looked at fronts in Chapter 1, we are now in a position to study them in depth, which will aid us in our understanding of the development of certain types of extreme weather. We will now learn about the general nature of fronts—how they move and what weather patterns are associated with them.

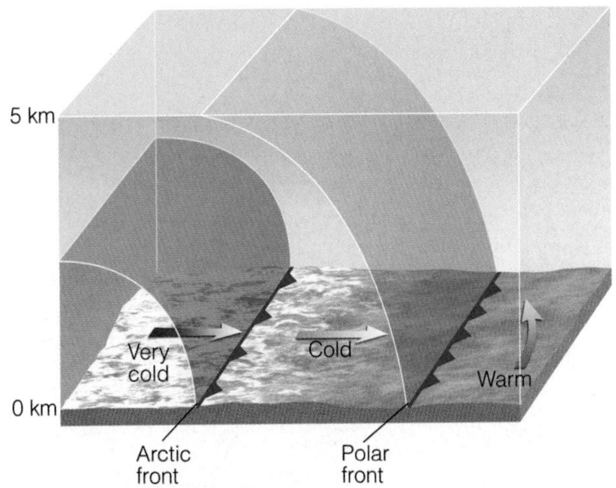

FIGURE 9.15 The polar front represents a cold frontal boundary that separates colder air from warmer air at the surface and aloft. The more shallow arctic front separates cold air from extremely cold air.

A **front** is the transition zone between two air masses of different densities.* Since density differences are most often caused by temperature differences, fronts usually separate air masses with contrasting temperatures. Often, they separate air masses with different humidities as well. Remember that air masses have both horizontal and vertical extent; consequently, the upward extension of a front is referred to as a *frontal surface*, or a *frontal zone.*

Figure 9.15 illustrates the vertical extent of two frontal zones—the *polar front* and the *arctic front.* The polar

*The word *front* is used to denote the clashing or meeting of two air masses, probably because it resembles the fighting in Western Europe during World War I, when the term originated.

EXTREME WEATHER WATCH

During the entire month of July, 1936, an immense dome of upper-level high pressure dominated most of the United States from the Rocky Mountains to the East Coast, resulting in the most intense heat wave ever recorded in the United States. Temperatures reached 121°F in North Dakota and Kansas. Manhattan's Central Park recorded 106°F, the hottest temperature ever recorded there. Thirteen state records set for heat during this period still stand today. Even nighttime temperatures remained at phenomenally high levels. Lincoln, Nebraska, had a low temperature of 91°F on July 25 before rising to 115°F later that day, the hottest night ever recorded in the United States outside of the desert Southwest.

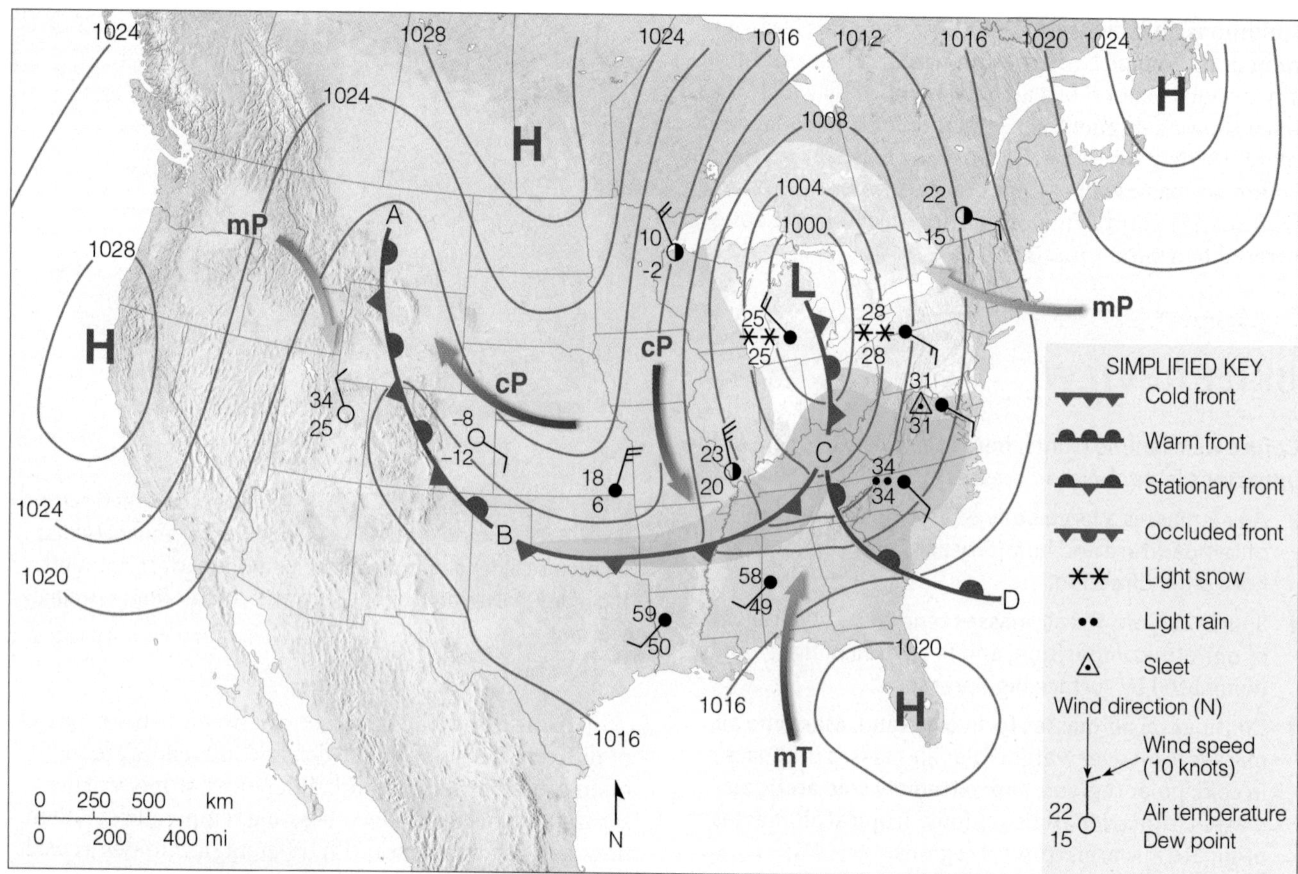

FIGURE 9.16 A surface winter weather map showing surface pressure systems, air masses, fronts, and isobars (in millibars) as solid gray lines. Large arrows in color show airflow. (Green-shaded area represents rain; pink-shaded area represents freezing rain and sleet; white-shaded area represents snow.)

front boundary, which extends upward to over 3 miles, separates warm humid air to the south from cold polar air to the north. The arctic front, which separates cold air from extremely cold arctic air, is much more shallow than the polar front and only extends upward to an altitude of about 1 mile or less. In the next several sections, as we examine fronts on a flat surface weather map, keep in mind that all fronts have horizontal and vertical extent.

Figure 9.16 shows a surface winter weather map illustrating four different fronts. Notice that the fronts are associated with lower pressure and that the fronts separate differing air masses. As we move from west to east across the map, the fronts appear in the following order: a stationary front between points A and B; a cold front between points B and C; a warm front between points C and D; and an occluded front between points C and L. Let's examine the properties of each of these fronts.

STATIONARY FRONTS A **stationary front** has essentially no movement.* On a colored weather map, it is drawn as an alternating red and blue line. Red semicircles face toward colder air on the red line and blue triangles point toward warmer air on the blue line. The stationary front between points A and B in Fig. 9.16 marks the boundary where cold, dense continental polar (cP) air from Canada butts up against the north-south trending Rocky Mountains. Unable to cross the barrier, the cold air shows little or no westward movement. The stationary front is drawn along a line separating the cP air from the milder, more humid maritime polar (mP) air to the west. Notice that the surface winds tend to blow parallel to the front, but in opposite directions on either side of it. Upper-level winds often blow parallel to a stationary front.

The weather along the front is clear to partly cloudy, with much colder air lying on its eastern side. Because both air masses are relatively dry, there is no precipitation. This is not, however, always the case. When warm moist air rides up and over the cold air, widespread cloudiness with light precipitation can cover a vast area. These are the conditions that prevail north of the east-west running stationary front depicted in Fig. 9.9, p. 246.

*They are usually called *quasi-stationary fronts* because they can show some movement.

If the warmer air to the west begins to move and replace the colder air to the east, the front in Fig. 9.16 will no longer remain stationary; it will become a warm front. If, on the other hand, the colder air slides up over the mountain and replaces the warmer air on the other side, the front will become a cold front. If either a cold front or a warm front stops moving, it becomes a stationary front.

COLD FRONTS The **cold front** between points B and C on the surface weather map (in Fig. 9.16) represents a zone where cold, dry, stable polar air is replacing warm, moist, conditionally unstable subtropical air. The front is drawn as a solid blue line with the triangles along the front showing its direction of movement. If you were standing on the surface ahead of the advancing cold front, it may appear similar to the approaching cold front shown in Fig. 9.17.

FIGURE 9.17 Clouds developing along an approaching cold front.

Typical Cold Fronts The weather in the immediate vicinity of the cold front in the southern United States in Fig. 9.16 is shown in Fig. 9.18. The data plotted on the map represent the current weather at selected cities. The station model used to represent the data at each reporting station is a simplified one that shows temperature, dew point, present weather, cloud cover, sea-level pressure, wind direction and speed. The little line in the lower right-hand corner of each station shows the *pressure tendency*—the pressure change, whether rising (/) or falling (\)—during the last three hours. (Appendix C explains the weather symbols and the station model more completely.)

The following criteria are used to locate a front on a surface weather map:

1. sharp temperature changes over a relatively short distance
2. changes in the air's moisture content (as shown by marked changes in the dew point)
3. shifts in wind direction
4. pressure and pressure changes
5. clouds and precipitation patterns

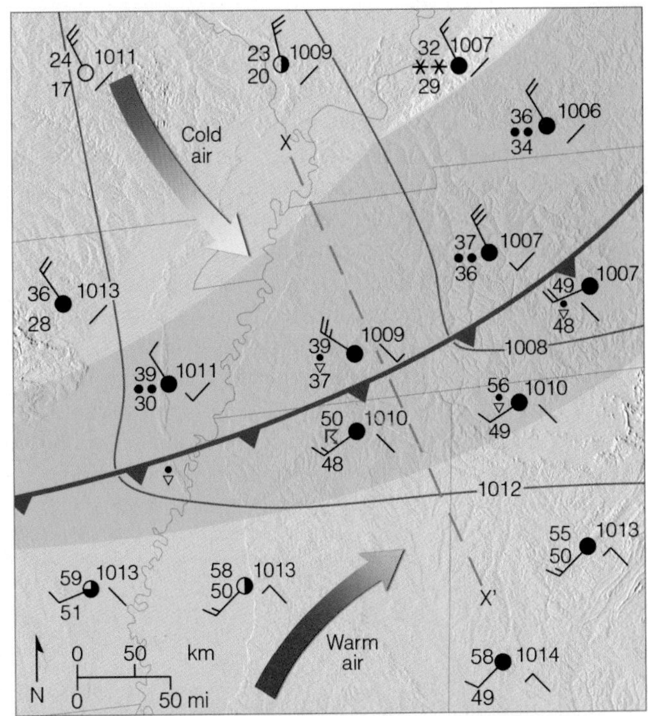

FIGURE 9.18 A closer look at the surface weather associated with the cold front situated in the southern United States in Fig. 9.16. (Gray lines are isobars. Green-shaded area represents rain; white-shaded area represents snow.)

In Fig. 9.18, we can see a large contrast in air temperature and dew point on either side of the front. There is also a wind shift from southwesterly ahead of the front, to northwesterly behind it. Notice that each isobar kinks as it crosses the front, forming an elongated area of low pressure—a *trough*—which accounts for the wind shift. Since surface winds normally blow across the isobars toward lower pressure, we find winds with a southerly component ahead of the front and winds with a northerly component behind it.

Since the cold front is a trough of low pressure, sharp changes in pressure can be significant in locating the

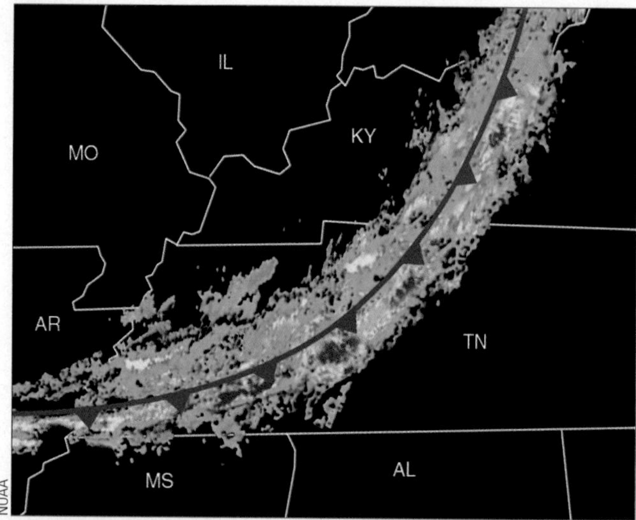

FIGURE 9.19 A Doppler radar image showing precipitation patterns along a cold front similar to the cold front in Fig. 9.18. Green represents light-to-moderate precipitation; yellow represents heavier precipitation; and red the most likely areas for thunderstorms. (The cold front is superimposed on the radar image.)

front's position. One important fact to remember is that the lowest pressure usually occurs just as the front passes a station. Notice that, as you move toward the front, the pressure drops, and, as you move away from it, the pressure rises. This is clearly shown by the pressure tendencies for each station on the map. Just before the front passes, the pressure tendency shows the atmospheric pressure is falling (\), while just behind the front, the pressure is now beginning to rise (✓), and farther behind the front, the pressure is rising steadily (/).

The precipitation pattern along the cold front in Fig. 9.18 might appear similar to the Doppler radar image shown in Fig. 9.19. The region in color extending from northeast to southwest represents precipitation along a cold front. Notice that light-to-moderate rain (color green) occurs over a wide area along the front, while the heavier precipitation (color yellow) tends to occur in a narrow band along the front itself. Thunderstorms (color red) do not occur everywhere, but only in certain areas along the front.

The cloud and precipitation patterns in Fig. 9.18 are shown in a side view of the front along the line X–X′ as illustrated in Fig. 9.20. We can see from Fig. 9.20 that, at the front, the cold, dense air wedges under the warm air, forcing the warm air upward, much like a snow shovel forces snow upward as the shovel glides through the snow. As the moist, conditionally unstable air rises, it condenses into a series of cumuliform clouds. Strong, upper-level westerly winds blow the delicate ice crystals (which form near the top of the cumulonimbus) into cirrostratus (Cs) and cirrus (Ci). These clouds usually appear far in advance of the approaching front. At the front itself, a relatively narrow band of thunderstorms (Cb) produces heavy showers with gusty winds. Behind the front, the air cools quickly. (Notice how the freezing level dips as it crosses the front.) The winds shift from southwesterly to northwesterly, pressure rises, and precipitation ends. As the air dries out, the skies clear, except for a few lingering fair weather cumulus clouds.

Observe that the leading edge of the front is steep. The steepness is due to friction, which slows the airflow near the ground. The air aloft pushes forward, blunting the frontal surface. If we could walk from where the front touches the surface back into the cold air, a distance of 50 km, the front would be about 1 km above us. Thus, the slope of the front—the ratio of vertical rise to horizontal distance—is 1:50. This is typical for a fast-moving cold front—those that move about 25 mi/hr. In a slower-moving cold front—one that moves about 10 mi/hr—the slope is much more gentle.

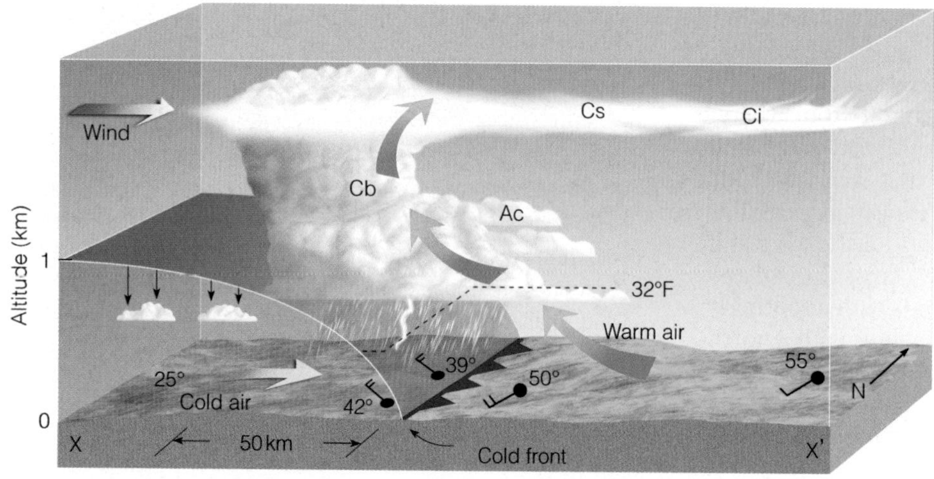

FIGURE 9.20 A vertical view of the weather across the cold front in Fig. 9.18 along the line X–X'.

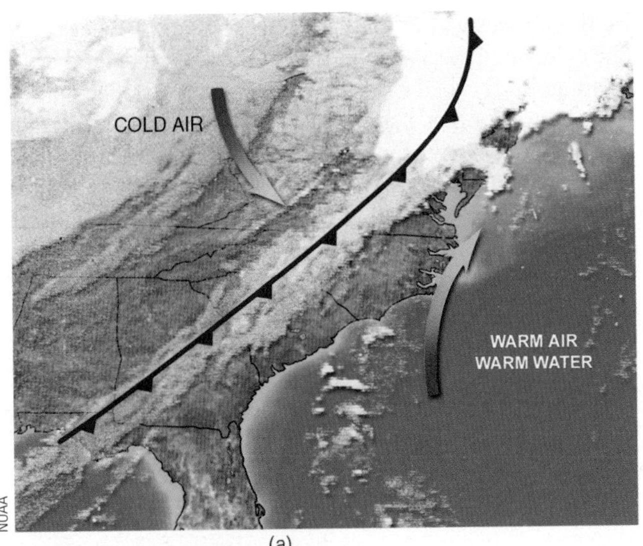

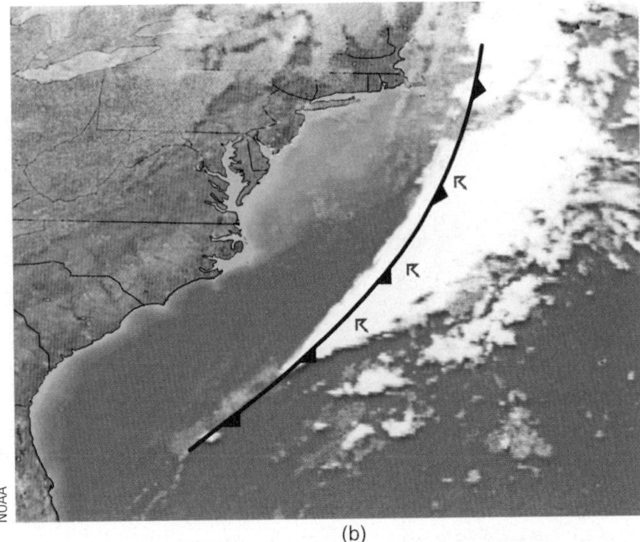

FIGURE 9.21 The infrared satellite image (a) shows a weakening cold front over land on Tuesday morning, November 21, intensifying into (b) a vigorous front over warm Gulf Stream water on Wednesday morning, November 22.

With slow-moving cold fronts, clouds and precipitation usually cover a broad area behind the front. When the ascending warm air is stable, stratiform clouds, such as nimbostratus, become the predominant cloud type and even fog may develop in the rainy area. Occasionally, out ahead of a fast-moving front, a line of active showers and thunderstorms, called a *squall line,* develops parallel to and often ahead of the advancing front, producing heavy precipitation and strong gusty winds.

As the temperature contrast across a front lessens, the front will often weaken and dissipate. Such a condition is known as **frontolysis**. On the other hand, an increase in the temperature contrast across a front can cause it to strengthen and regenerate into a more vigorous frontal system, a condition called **frontogenesis**.

An example of a regenerated front is shown in the infrared satellite images in Fig. 9.21. The cold front in Fig. 9.21a is weak, as indicated by the low clouds (gray tones) along the front. As the front moves offshore, over the warm Gulf Stream (Fig. 9.21b), it intensifies into a more vigorous frontal system as surface air becomes conditionally unstable and convective activity develops. Notice that the area of cloudiness is more extensive and thunderstorms are now forming along the frontal zone.

So far, we have considered the general weather patterns of "typical" cold fronts. There are, of course, many exceptions. In fact, no two fronts are exactly alike. In some, the cold air is very shallow; in others, it is much deeper. If the rising warm air is dry and stable, scattered clouds are all that form, and there is no precipitation. In extremely dry weather, a marked change in the dew point, accompanied by a slight wind shift, may be the only clue to a passing cold front.

During the winter, a series of cold polar or arctic outbreaks may travel across the United States so quickly that warm air is unable to develop ahead of the front. In this case, frigid arctic air associated with an arctic front usually replaces cold polar air, and a drop in temperature is the only indication that a front has moved through your area. Along the West Coast, the Pacific Ocean modifies the air so much that cold fronts, such as those described in the previous section, are never seen. In fact, as a cold front moves inland from the Pacific Ocean, the surface temperature contrast across the front may be quite small. Topographic features usually distort the wind pattern so much that locating the position of the front and the time of its passage is exceedingly difficult. In this case, the pressure tendency is the most reliable indication of a frontal passage.

In some instances along the West Coast, an approaching cold front (or upper-level trough) will cause cool marine air at the surface to *surge* into coastal and inland valleys. The cool air (which is often accompanied by a wind shift) may produce a sharp drop in air temperature. This may give the impression that a rather strong cold front has moved through, when in reality, the front may be many miles offshore.

"Back Door" Cold Fronts Cold fronts usually move toward the south, southeast, or east. But sometimes they will move southwestward. In New England, this movement occurs when northeasterly surface winds, blowing clockwise around an anticyclone centered to the north over Canada, push a cold front southwestward often as far south as Boston. Because the cold front moves in from

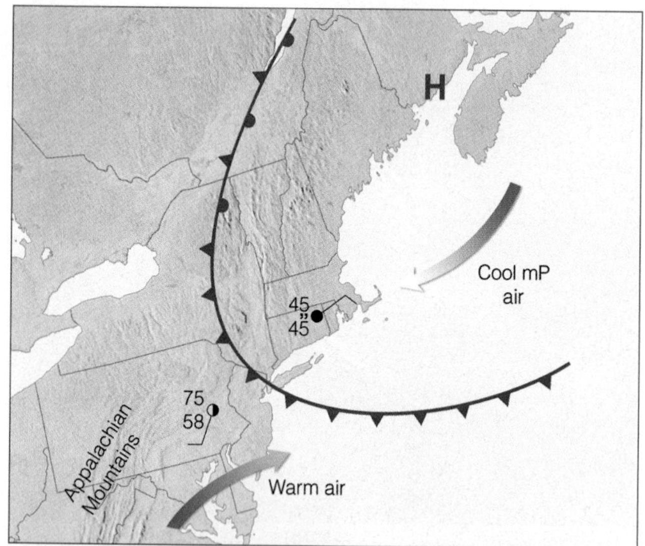

FIGURE 9.22 A "back door" cold front moving into New England during the spring. Notice that, behind the front, the weather is cold and damp with drizzle; while to the south, ahead of the front, the weather is partly cloudy and warm.

the east, or northeast, it is known as a **"back door" cold front**. As the front passes, westerly surface winds usually shift to easterly or northeasterly and temperatures drop as cool, moist air flows in off the Atlantic Ocean.

An example of a "back door" cold front is shown in Fig. 9.22. This is a springtime situation where, behind the front, the weather is cold and damp with drizzle, as northeasterly winds sweep into the region from off the chilly Atlantic. To the south of the front, the weather is much warmer. Should the front move through this area, the more summerlike weather would change, in a matter of hours, to more winterlike. The cold, dense air behind the front is rather shallow. Consequently, the Appalachian Mountains act as a dam to the front's forward progress, halting its westward movement. This situation, where the cold, damp air is confined to the eastern side of the mountains, is called **cold air damming**. The stalled cold front now becomes a stationary front. The cool air behind the front may linger for some time as warmer, less-dense air to the south rides up and over it. Forecasting how far south the "back door" cold front will move and when the entrenched cold air will leave can be a bit tricky.

Even though cold-front weather patterns have many exceptions, learning these patterns can be to your advantage if you live in an area that experiences well-defined cold fronts. Knowing them improves your own ability to forecast approaching thunderstorms and the threat of extreme weather. For your reference, Table 9.2 summarizes idealized cold-front weather in the Northern Hemisphere.

A Strong Cold Front During the passage of a strong cold front, the air temperature can drop dramatically. To illustrate this point look at Fig. 9.23, which represents a portion of a surface weather map for January 18, 1996, at 6 A.M. CST. Notice that a strong slow-moving cold front extends from Illinois southward into Arkansas. On the front's eastern side, warm air is lifted along the shallow front. Precipitation in the form of freezing rain falls directly behind the front, and snow is falling in a broad area farther to the west.

The dashed blue line on the map represents the freezing (32°F) isotherm, whereas the solid blue line represents the 0°F isotherm. Notice that the freezing isotherm lies directly along the front in eastern Missouri. About 75 miles to the east of the front in St. Louis, the air temperature is 63°F. This situation indicates that along the front there is a sharp temperature gradient where the air temperature drops by 31°F over a span of just 75 miles.*

*The actual temperature gradient directly along the cold front is probably even steeper than this value. It is impossible to tell the exact gradient along the front because there were no observations taken there.

Table 9.2

Typical Weather Conditions Associated with a Cold Front in the Northern Hemisphere

WEATHER ELEMENT	BEFORE PASSING	WHILE PASSING	AFTER PASSING
Winds	South or southwest	Gusty, shifting	West or northwest
Temperature	Warm	Sudden drop	Steadily dropping
Pressure	Falling steadily	Minimum, then sharp rise	Rising steadily
Clouds	Increasing Ci, Cs, then either Tcu* or Cb*	Tcu or Cb	Often Cu, Sc* when ground is warm
Precipitation	Short period of showers	Heavy showers of rain or snow, sometimes with hail, thunder, and lightning	Decreasing intensity of showers, then clearing
Visibility	Fair to poor in haze	Poor, followed by improving	Good, except in showers
Dew point	High; remains steady	Sharp drop	Lowering

*Tcu stands for towering cumulus, such as cumulus congestus; whereas Cb stands for cumulonimbus. Sc stands for stratocumulus.

EXTREME WEATHER WATCH

An extraordinarily sharp cold front rapidly swept through the Great Plains and Midwest on November 10–11,1911. The temperature on November 11 dropped from a record daily high of 76°F in Kansas City, Missouri, to a record daily low of 11°F later that same day. In Rapid City, South Dakota, the temperature dropped an incredible 75°F in just two hours—from 62°F to −13°F between 6 and 8 P.M. on November 10th. This event remains one of the most extreme cold frontal passages known on record in United States history.

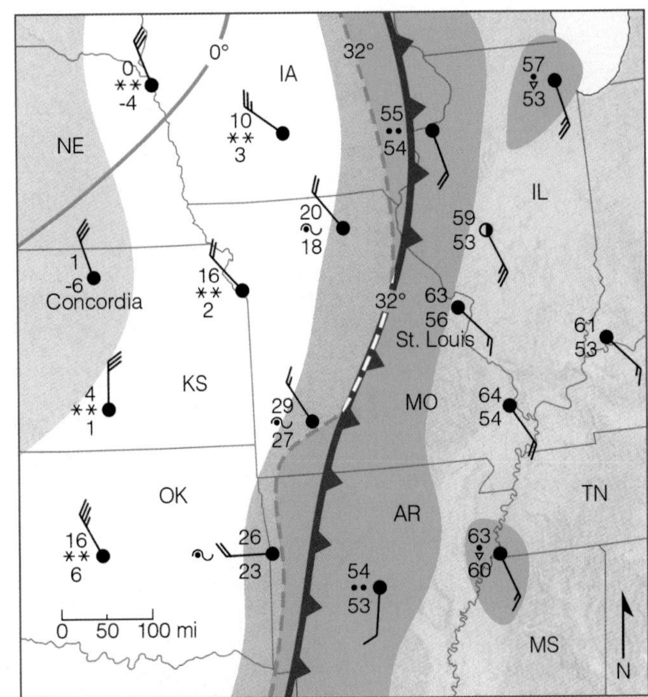

FIGURE 9.23 Surface weather map for the morning of January 18, 1996, shows a strong cold front stretching from Illinois to Arkansas. Ahead of the front, temperatures are in the 60s; behind the front, temperatures drop to below zero.

Behind the cold front, the air temperature continues to plummet as the temperature in Concordia, Kansas, is only 1°F. Hence, over a distance of about 400 miles (from St. Louis to Concordia), the air temperature drops by 62°F, indicating an average drop in temperature of about 1°F every 6 miles. (For information on an exceptionally strong cold front, read the Focus section below.)

WARM FRONTS In Fig. 9.16, p. 252, a **warm front** is drawn along the solid red line running from points C

FOCUS ON EXTREME WEATHER

The Extraordinary Cold Front of 1836

We know that the passage of a typical winter cold front is often accompanied by heavy showers of rain or snow, falling temperatures, and strong, gusty winds that usually change direction. But on December 21, 1836, a spectacular cold front (probably an arctic front) moved through Illinois. Although no reliable temperature records are available, estimates are that during the frontal passage air temperatures dropped almost instantly from a balmy 40°F to 0°F. An historical account of the frontal passage is provided in the following:

> About two o'clock in the afternoon it began to grow dark from a heavy, black cloud which was seen in the northwest. Almost instantly the strong wind, traveling at the rate of 70 miles an hour, accompanied by a deep bellowing sound, with its icy blast, swept over the land, and everything was frozen hard. The water in the little ponds in the roads froze in waves, sharp edged and pointed, as the gale had blown it. The chickens, pigs and other small animals were frozen in their tracks. Wagon wheels ceased to roll, froze to the ground. Men, going from their barns, or fields a short distance from their homes, in slush and water, returned a few minutes later walking on the ice. Those caught out on horseback were frozen to their saddles, and had to be lifted off and carried to the fire to be thawed apart. Two young men were frozen to death near Rushville. One of them was found with his back against a tree, with his horse's bridle over his arm and his horse frozen in front of him. The other was partly in a kneeling position, with a tinder box in one hand and a flint in the other, with both eyes wide open as if intent on trying to strike a light. Many other casualties were reported. As to the exact temperature, however, no instrument has left any record; but the ice was frozen in the stream, as variously reported, from six inches to a foot in thickness in a few hours.

John Moses, *Illinois: Historical and Statistical*

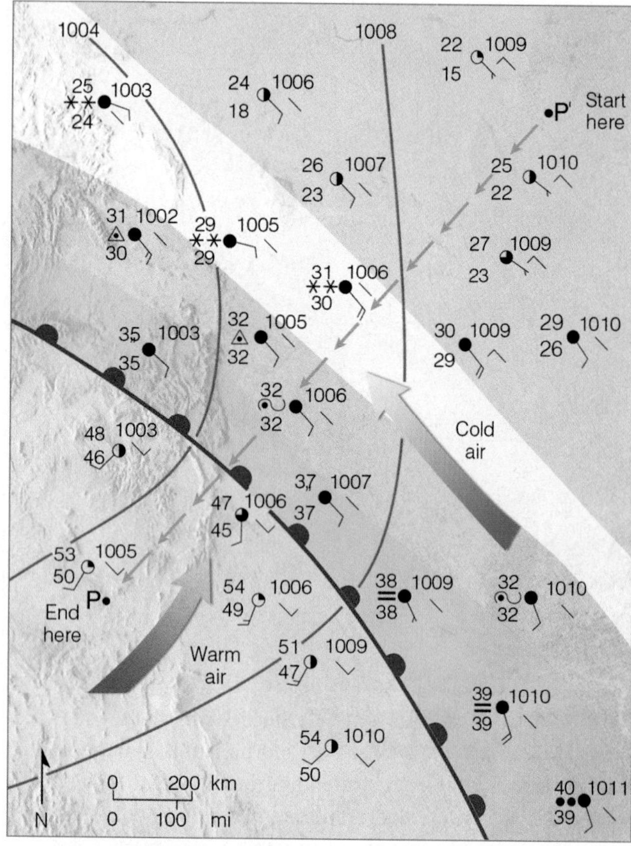

Active FIGURE 9.24 Surface weather associated with a typical warm front. A vertical view along the dashed line P–P′ is shown in Fig. 9.25. (Green-shaded area represents rain; pink-shaded area represents freezing rain and sleet; white-shaded area represents snow.) Visit the Meteorology Resource Center to view this and other active figures at academic.cengage.com/login

to D. Here, the leading edge of advancing warm, moist, subtropical (mT) air from the Gulf of Mexico replaces the retreating cold maritime polar air from the North Atlantic. The direction of frontal movement is given by the half circles, which point into the cold air; this front is heading toward the northeast. As the cold air recedes, the warm front slowly advances. The average speed of a warm front is about 10 mi/hr, or about half that of an average cold front. During the day, as mixing occurs on both sides of the front, its movement may be much faster. Warm fronts often move in a series of rapid jumps, which show up on successive weather maps. At night, however, radiational cooling creates cool dense surface air behind the front. This inhibits both lifting and the front's forward progress. When the forward surface edge of the warm front passes a station, the wind shifts, the temperature rises, and the overall weather conditions improve. To see why, we will examine the weather commonly associated with the warm front both at the surface and aloft.

Figure 9.24 is a surface weather map showing the position of a warm front and its associated weather. Figure 9.25 is a vertical view of the warm front in Fig. 9.24. Look at these two figures and observe that the warmer, less-dense air rides up and over the colder, more-dense surface air. This rising of warm air over cold, called **overrunning**, produces clouds and precipitation well in advance of the front's surface boundary. The warm front that separates the two air masses has an average slope of about 1:300*—a much more gentle or inclined slope than that of a typical cold front. Warm air overriding the cold air creates a stable atmosphere.

Suppose we are standing at the position marked P′ in Figs. 9.24 and 9.25. Note that we are over 1200 km or 750 mi ahead of where the warm front is touching the surface. Here, the surface winds are light and variable. The air is cold and about the only indication of an approaching warm front is the high cirrus clouds overhead. We know the front is moving slowly toward us and that within a day or so it will pass our area. Suppose that, instead of waiting for the front to pass us, we drive toward it, observing the weather as we go.

*The slope of 1:300 is a much more gentle slope than that of most warm fronts. Typically, the slope of a warm front is on the order of 1:150 to 1:200.

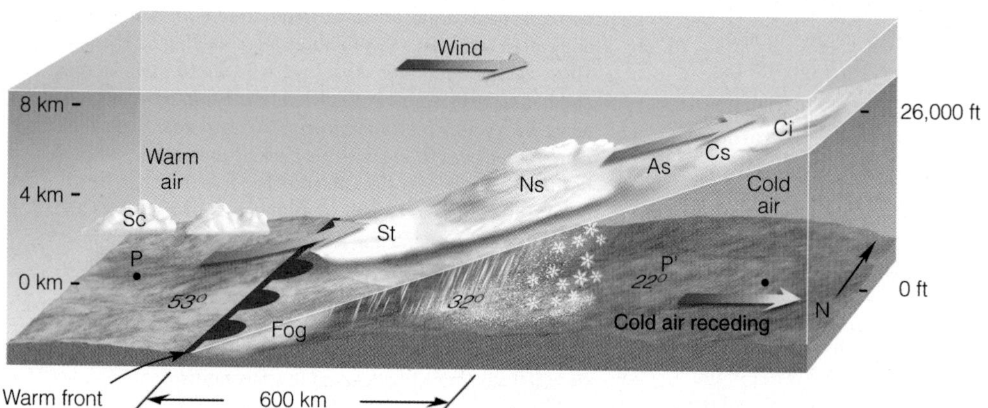

Active FIGURE 9.25 Vertical view of clouds, precipitation, and winds across the warm front in Fig. 9.24 along the line P–P′. Visit the Meteorology Resource Center to view this and other active figures at academic.cengage.com/login

Table 9.3

Typical Weather Conditions Associated with a Warm Front in the Northern Hemisphere

WEATHER ELEMENT	BEFORE PASSING	WHILE PASSING	AFTER PASSING
Winds	South or southeast	Variable	South or southwest
Temperature	Cool to cold, slow warming	Steady rise	Warmer, then steady
Pressure	Usually falling	Leveling off	Slight rise, followed by fall
Clouds	In this order: Ci, Cs, As, Ns, St, and fog; occasionally Cb in summer	Stratus-type	Clearing with scattered Sc, especially in summer, occasionally Cb in summer
Precipitation	Light-to-moderate rain, snow, sleet, or drizzle; showers in summer	Drizzle or none	Usually none; sometimes light rain or showers
Visibility	Poor	Poor, but improving	Fair in haze
Dew point	Steady rise	Steady	Rise, then steady

Heading toward the warm front, we notice that the cirrus (Ci) clouds gradually thicken into a thin, white veil of cirrostratus (Cs) whose ice crystals cast a halo around the sun.* Almost imperceptibly, the clouds thicken and lower, becoming altocumulus (Ac) and altostratus (As) through which the sun shows only as a faint spot against an overcast gray sky. Snowflakes begin to fall, and we are still over 600 km or 370 mi from the surface front. The snow increases, and the clouds thicken into a sheetlike covering of nimbostratus (Ns). The winds become brisk and out of the southeast, while the atmospheric pressure slowly falls. Within about 400 km or 250 mi of the front, the cold surface air mass is now quite shallow. The surface air temperature moderates and, as we approach the front, the light snow changes first into sleet. It then becomes freezing rain and finally rain and drizzle as the air temperature climbs above freezing. Overall, the precipitation remains light or moderate but covers a broad area. Moving still closer to the front, warm, moist air mixes with cold, moist air producing ragged wind-blown stratus (St) and fog. (Thus, flying in the vicinity of a warm front is quite hazardous.)

Finally, after a trip of over 1200 km or 750 mi, we reach the warm front's surface boundary. As we cross the front, the weather changes are noticeable, but much less pronounced than those experienced with the cold front; they show up more as a gradual transition rather than a sharp change. On the warm side of the front, the air temperature and dew point rise, the wind shifts from southeast to south or southwest, and the air pressure stops falling. The light rain ends and, except for a few stratocumulus (Sc), the fog and low clouds vanish.

This scenario of an approaching warm front represents average (if not idealized) warm-front weather in winter. In some instances, the weather can differ from this dramatically. For example, if the overrunning warm air is relatively dry and stable, only high and middle clouds will form, and no precipitation will occur. On the other hand, if the warm air is relatively moist and conditionally unstable (as is often the case during the summer), heavy showers can develop as thunderstorms become embedded in the cloud mass. If the cold surface air is rather shallow, even severe thunderstorms may form as humid air is lifted along the front.

Along the West Coast, the Pacific Ocean significantly modifies the surface air so that warm fronts are difficult to locate on a surface weather map. Also, not all warm fronts move northward or northeastward. On rare occasions, a front will move into the eastern seaboard from the Atlantic Ocean as the front spins all the way around a deep storm positioned off the coast. Cold northeasterly winds ahead of the front usually become warm northeasterly winds behind it. Even with these exceptions, knowing the normal sequence of warm-front weather will be useful, especially if you live where warm fronts become well developed. You can look for certain cloud and weather patterns and make reasonably accurate short-range forecasts of your own. Table 9.3 summarizes typical warm-front weather. (Before going on to the next section, you may wish to read the Focus section on p. 260, which illustrates how ice storms may form ahead of slow-advancing warm fronts.)

THE DRYLINE Drylines often play a key role in producing severe weather. **Drylines** are not warm fronts or cold fronts, but represent a narrow boundary where there is a steep horizontal change in moisture, so drylines separate moist air from dry air. Because dew-point temperatures may drop along this boundary by as much as 15°F per kilometer, drylines have been referred to as

*If the warm air is relatively unstable, ripples or waves of cirrocumulus clouds will appear as a "mackerel sky."

dew-point fronts.* Although drylines can occur in the United States as far north as the Dakotas, and as far east as the Texas-Louisiana border, they are most frequently observed in the western half of Texas, Oklahoma, and Kansas, especially during spring and early summer. In these locations, drylines tend to move eastward during the day, then westward toward evening. Drylines are observed in other regions of the world, too. They occur, for example, in Central West Africa and in India before the onset of the summer monsoon.

Figure 9.26 shows a dryline moving across Texas during May, 2001. Notice that the dryline is represented

*Recall from Chapter 3 that the dew-point temperature is a measure of the amount of water vapor in the air.

FOCUS ON EXTREME WEATHER

Warm Fronts and Ice Storms

Up to this point, we've examined idealized warm fronts on a surface weather map—like the one shown in Fig. 9.16, p. 252. Some warm fronts do look like this example; others, however, have an entirely different appearance. For instance, look at the warm front in Fig. 3. Notice that it has a wavelike shape as it approaches North Carolina from three different directions. So what causes the warm front to bend in this manner?

Look carefully at Fig. 3 and notice that at the surface cold air is flowing southwestward around a high-pressure area centered over southern Canada. As cold, dense, surface air pushes south into the southern states, it flows up against the Appalachian Mountains, which impede its westward progress. Since the shallow layer of cold air is unable to ride up and over the mountains, it becomes wedged along the mountains' eastern foothills. Recall from an earlier discussion that this trapping of cold air is called *cold air damming*.

The shallow layer of cold air usually becomes entrenched in low-lying areas and therefore retreats northward very slowly. Warm air pushing northward from the Gulf of Mexico rides up and over the cold, dry surface air as illustrated in Fig. 4. Clouds and precipitation often form in this rising warm air. When rain falls into the shallow, cold air, it may evaporate, chilling the air even more. Sometimes the rain freezes before reaching the ground, producing sleet; other times, the rain freezes on impact, producing freezing rain. If the frozen precipitation falls for many hours, severe ice storms may result, with heavy accumulations of ice causing treacherous driving conditions and downed power lines.

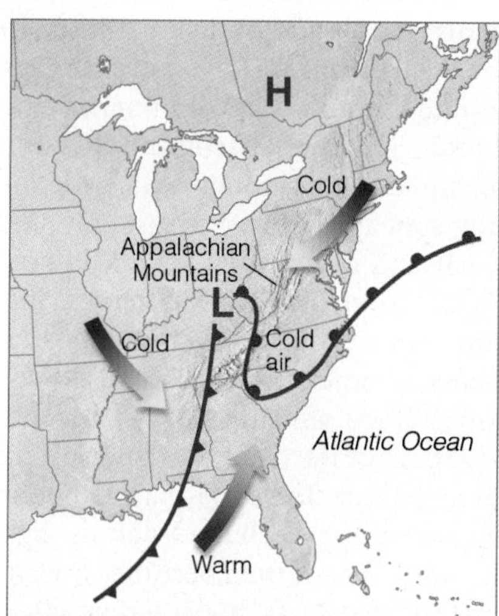

Figure 3 Surface weather map that shows a warm front moving very slowly. In advance of the front, cold air becomes entrenched on the eastern side of the Appalachian Mountains.

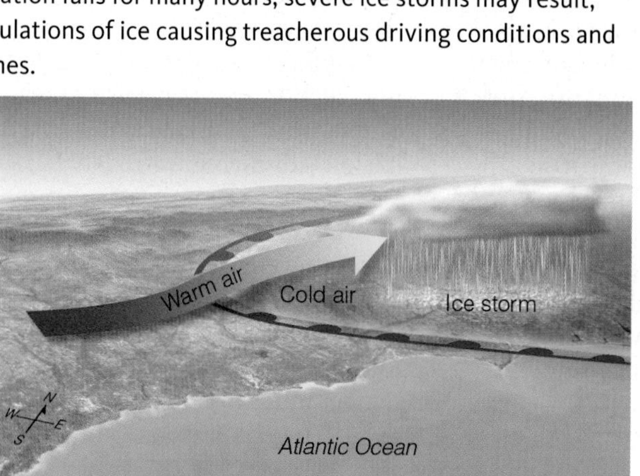

Figure 4 A three-dimensional view of Fig. 3 showing the warm front, overrunning warm air, and cold air trapped against the Appalachian Mountains. Notice that sleet and freezing rain are forming in the cold air, producing an ice storm.

as a line with brown half circles. Notice also that, to the west of the dryline, warm, dry continental tropical air is moving in from the southwest. Consequently, on this side, the weather is usually hot and dry with gusty southwesterly winds. To the east of the dryline, warm, very humid maritime tropical air is sweeping northward from the Gulf of Mexico. Here we typically find air temperatures to be slightly lower and the humidity (as indicated by the higher dew points) considerably higher than on the western side. The semicircles of the dryline point toward this humid air.

Even though the dryline represents a moisture boundary, its actual position on a weather map is plotted according to a shift in surface winds. When insects and insect-eating birds congregate along the dryline, Doppler radar may be able to locate it. On the radar screen, the echo from insects and birds shows up as a thin line, called a *fine line*.

Sometimes drylines are associated with mid-latitude cyclones, sometimes they are not. Cumulus clouds and thunderstorms often form along or to the east of the dryline. This cloud development is caused in part by daytime convection and a sloping terrain. The Central Plains area of North America is higher to the west and lower to the east. Convection over the elevated western Plains carries dry air high above the surface. Westerly winds sweep this dry air eastward over the lower plains where it overrides the slightly cooler but more humid air at the surface. This situation sets up a potentially unstable atmosphere that finds warm, dry air above warm, moist air. In regions where the air rises, cumulus clouds and organized bands of thunderstorms can form (see Fig. 9.27). We will examine in more detail the development of these storms in Chapter 11.

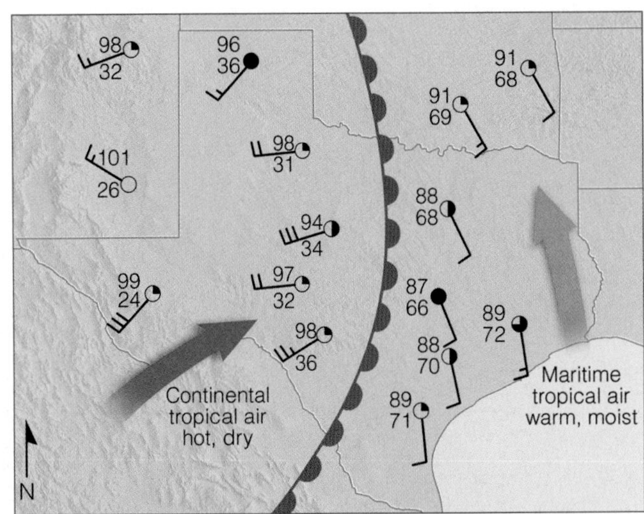

FIGURE 9.26 A dryline represents a narrow boundary where there is a steep horizontal change in moisture as indicated by a rapid change in dew-point temperature. Here, a dryline moving across Texas and Oklahoma separates warm, moist air from warm, dry air during an afternoon in May.

OCCLUDED FRONTS If a cold front catches up to and overtakes a warm front, the frontal boundary created between the two air masses is called an **occluded front**,

FIGURE 9.27 Cumulus clouds and thunderstorms developing along a dryline in Kansas during May, 2007.

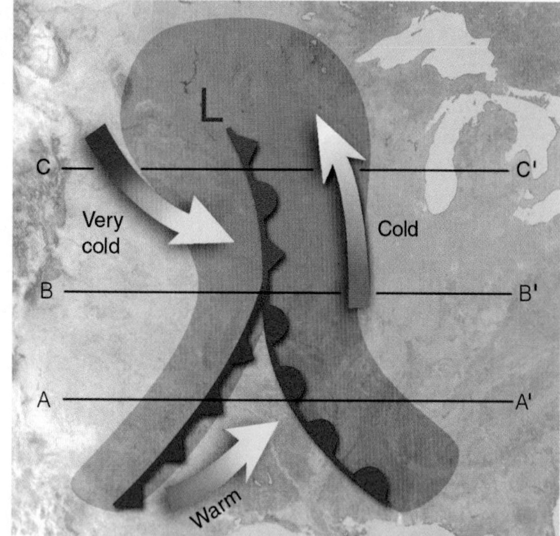

(d)

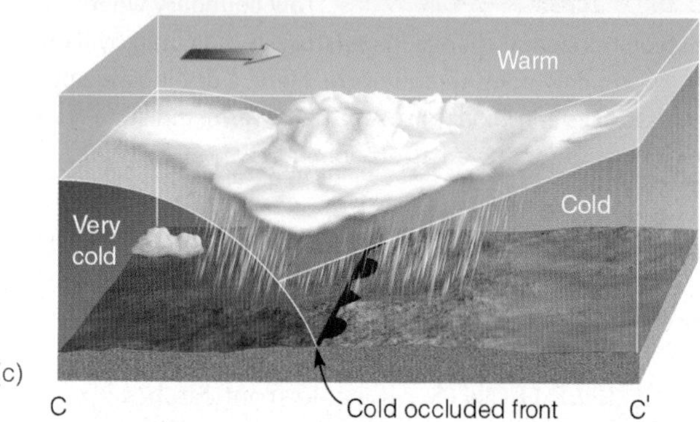

(c)

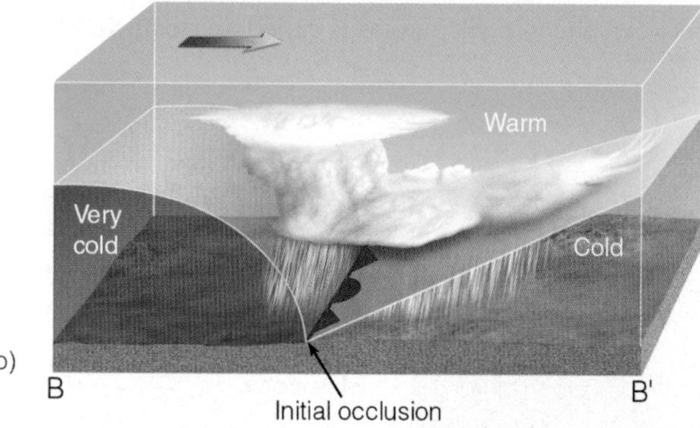

(b)

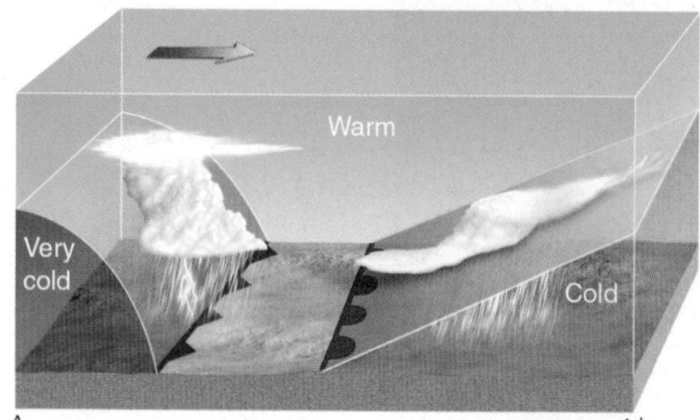

(a)

FIGURE 9.28 The formation of a cold-occluded front. The faster-moving cold front (a) catches up to the slower-moving warm front (b) and forces it to rise off the ground (c). [Green-shaded area in (d) represents precipitation.]

or, simply, an **occlusion** (meaning "closed off"). On the surface weather map, it is represented as a purple line with alternating cold-front triangles and warm-front half circles; both symbols point in the direction toward which the front is moving. Look back at Fig. 9.16, p. 252 and notice that the air behind the occluded front is colder than the air ahead of it. This is known as a *cold-type occluded front*, or **cold occlusion**. Let's see how this front develops.

The development of a cold occlusion is shown in Fig. 9.28. Along line A–A′, the cold front is rapidly approaching the slower-moving warm front. Along line B–B′, the cold front overtakes the warm front, and, as we can see in the vertical view across C–C′, underrides and lifts off the ground both the warm front and the warm air mass. As a cold-occluded front approaches, the weather sequence is similar to that of a warm front, with high clouds lowering and thickening into middle and low clouds, with precipitation forming well in advance of the surface front. Since the front represents a trough of low pressure, southeasterly winds and falling atmospheric pressure occur ahead of it.

The frontal passage, however, brings weather similar to that of a cold front: heavy, often showery precipitation with winds shifting to west or northwest. After a period of wet weather, the sky begins to clear, atmospheric pressure rises, and the air turns colder. The most violent weather usually occurs where the cold front is just overtaking the warm front, at the point of occlusion, where the greatest contrast in temperature occurs. Cold occlusions are the most prevalent type of front that moves into the Pacific coastal states and into interior North America. Occluded fronts frequently form over the North Pacific and North Atlantic, as well as in the vicinity of the Great Lakes.

Continental polar air over eastern Washington and Oregon may be much colder than milder maritime polar air moving inland from the Pacific Ocean. Figure 9.29 illustrates this situation. Observe that the air ahead of the warm front is colder than the air behind the cold front. Consequently, when the cold front catches up to and overtakes the warm front, the milder, lighter air behind the cold front is unable to lift the colder, heavier air off the ground. As a result, the cold front rides

*Due to the relatively mild winter air that moves into Europe from the North Atlantic, many of the occlusions that move into this region in winter are of the warm occlusion variety.

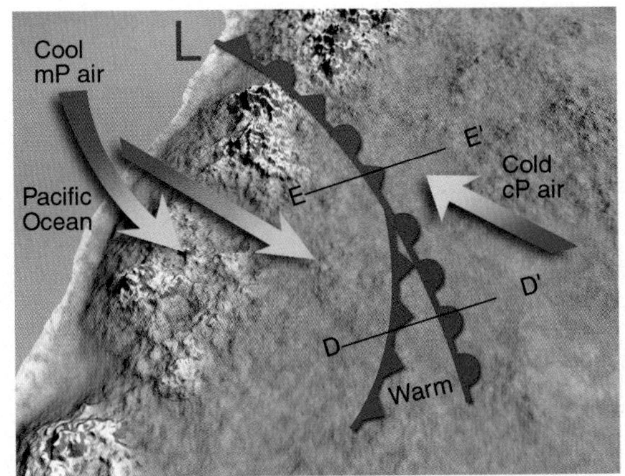

(c)

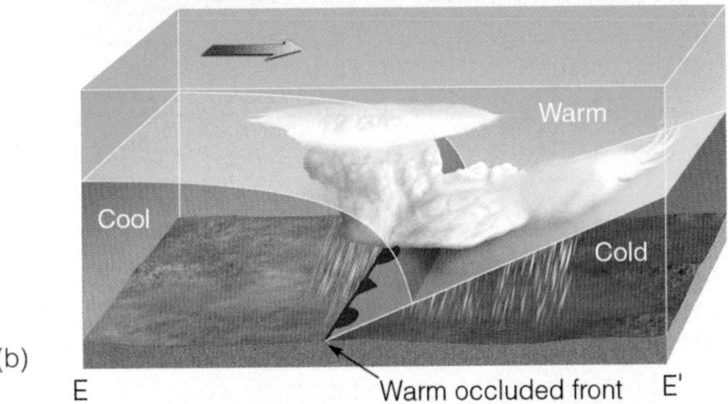

(b)

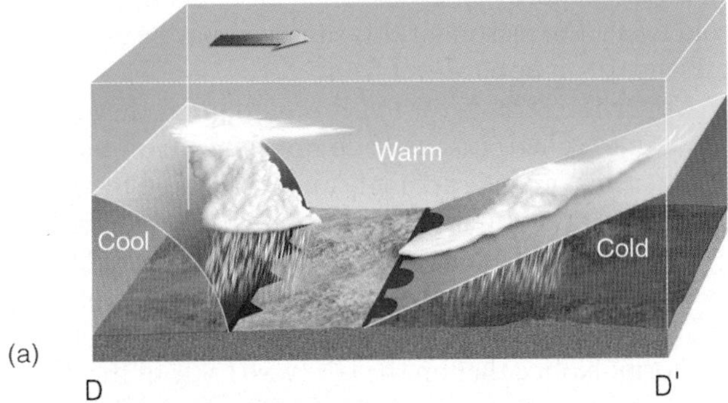

(a)

FIGURE 9.29 The formation of a warm-type occluded front. The faster-moving cold front in (a) overtakes the slower-moving warm front in (b). The lighter air behind the cold front rises up and over the denser air ahead of the warm front. Diagram (c) shows a surface map of the situation.

"piggyback" along the sloping warm front. This produces a *warm-type occluded front*, or a **warm occlusion**. The surface weather associated with a warm occlusion is similar to that of a warm front.*

Contrast Fig. 9.28 and Fig. 9.29. Note that the primary difference between the warm- and cold-type occluded front is the location of the upper-level front. In a warm occlusion, the upper-level cold front *precedes* the surface occluded front, whereas in a cold occlusion the upper warm front *follows* the surface occluded front.

In the world of weather fronts, occluded fronts are the mavericks. In our discussion, we treated occluded fronts as forming when a cold front overtakes a warm front. Some may form in this manner, but others apparently form as new fronts, which develop when a surface mid-latitude cyclonic storm intensifies in a region of cold air after its trailing cold and warm fronts have broken away and moved eastward. The new occluded front shows up on a surface chart as a trough of low pressure separating two cold air masses. Because of this, locating and defining occluded fronts at the surface is often difficult for the meteorologist. Similarly, you too may find it hard to recognize an occlusion. In spite of this, we will assume that the weather associated with occluded fronts behaves in a similar way to that shown in Table 9.4.

The frontal systems described in this chapter are actually part of a much larger storm system—the middle-latitude cyclone. Figure 9.30 shows the cold front, warm front, and occluded front in association

FIGURE 9.30 A visible satellite image showing a mid-latitude cyclonic storm with its weather fronts over the Atlantic Ocean during March, 2005. Superimposed on the photo is the position of the surface cold front, warm front, and occluded front. Precipitation symbols indicate where precipitation is reaching the surface.

Table 9.4

Typical Weather Most Often Associated with Occluded Fronts in North America

WEATHER ELEMENT	BEFORE PASSING	WHILE PASSING	AFTER PASSING
Winds	East, southeast, or south	Variable	West or northwest
Temperature			
(a) Cold-type occluded	Cold or cool	Dropping	Colder
(b) Warm-type occluded	Cold	Rising	Milder
Pressure	Usually falling	Low point	Usually rising
Clouds	In this order: Ci, Cs, As, Ns	Ns, sometimes Tcu and Cb	Ns, As, or scattered Cu
Precipitation	Light, moderate, or heavy precipitation	Light, moderate, or heavy continuous precipitation or showers	Light-to-moderate precipitation followed by general clearing
Visibility	Poor in precipitation	Poor in precipitation	Improving
Dew point	Steady	Usually slight drop, especially if cold-occluded	Slight drop, although may rise a bit if warm-occluded

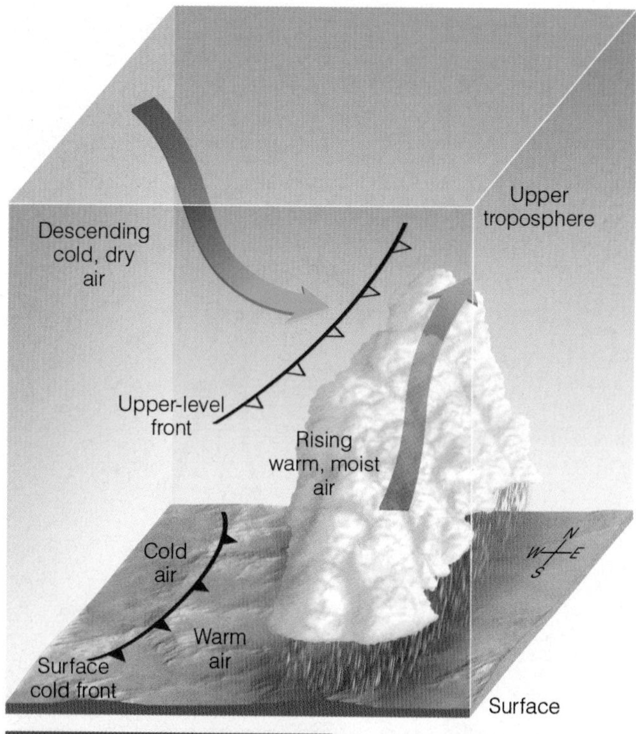

FIGURE 9.31 An upper-level front can form aloft where descending cold, dry air meets rising warm, moist air.

with a mid-latitude cyclonic storm. Notice that, as we would expect, clouds and precipitation form in a rather narrow band along the cold front, and in a much wider band with the warm and occluded fronts. In Chapter 10, we will look more closely at middle-latitude cyclonic storms, examining where, why, and how they form.

UPPER-LEVEL FRONTS Earlier we learned the precipitation associated with a cold front tends to form right along the front as warm humid air rises along the frontal boundary. But precipitation, as well as showers and thunderstorms, may form out ahead of an advancing cold front. The reason for this situation may be due to the presence of an **upper-level front**.

Upper-level fronts (or *upper cold fronts* as they are sometimes called) normally do not show up on a surface weather map and surprisingly are not marked by sharp changes in air temperature. Most often they are marked by a sharp change in humidity. We can obtain a better picture as to why this phenomenon happens by looking at Fig. 9.31, which illustrates how an upper-level front can form. Notice that on the left (west) side of the illustration very cold, dry air is descending from the upper troposphere. This descending air warms as it sinks, causing the relative humidity of the air to decrease. On the right (east) side of the illustration, warm, humid air rises from the surface.* As the air rises it cools and eventually becomes saturated. The boundary separating the moist air from the dry air is the upper-level front. Temperature contrasts along the front are small due to the fact that the cold, sinking air is warming and the warm, rising air is cooling. So by the time the sinking air and rising air are side by side there is little temperature difference between them. Because the dry air to the west is more dense than the moist air to the east, the drier air along the upper-level front often acts as a trigger, forcing the moist air to rapidly rise and condense into clouds that may develop into intense thunderstorms.

*The rising and sinking air described here is due to the air motions associated with an upper-level trough. This topic is covered in detail in Chapter 10.

Summary

In this chapter we looked at air masses and fronts, and examined the weather each brings to a region. We examined air masses and found that continental arctic air masses are responsible for the extremely cold (arctic) outbreaks of winter, whereas continental polar air masses are responsible for cold, dry weather in winter and cool, pleasant weather in summer. Maritime polar air, having traveled over an ocean for a considerable distance, brings cool, moist weather to an area. When maritime polar air moves inland from off the Pacific Ocean, it crosses several mountain ranges and reaches the Great Plains as dry, cool Pacific air. The extremely hot, dry weather of summer is associated with continental tropical air masses, whereas warm, muggy conditions are due to maritime tropical air masses. In winter, maritime tropical air flowing into the western United States can cause mud slides in canyons and flooding in low-lying valleys. Where air masses with sharply contrasting properties meet, we find weather fronts.

A front is a boundary between two air masses of different densities. Stationary fronts have essentially no movement, with cold air on one side and warm air on the other. Winds tend to blow parallel to the front, but in opposite directions on either side of it. Along the leading edge of a cold front, where colder air replaces warmer air, showers and thunderstorms are prevalent, especially if the warmer air is moist and conditionally unstable. If a cold front moves into New England from the east or northeast, it is called a "back door" cold front. Along a warm front, warmer air rides up and over colder surface air, producing widespread cloudiness and light-to-moderate precipitation that can cover thousands of square miles. When the rising air is conditionally unstable (such as it often is in summer), showers and thunderstorms may form ahead of and along the advancing warm front. Due to the overrunning of warm air over cold air, ice storms are more common with warm fronts. Cold fronts typically move faster and are more steeply sloped than warm fronts.

Drylines are not warm fronts but represent a boundary where hot, dry air replaces warm, humid air. The lifting of humid air in the vicinity of a dryline often produces afternoon thunderstorms. Occluded fronts, which are often difficult to locate and define on a surface weather map, may have characteristics of both cold and warm fronts. An upper-level front may form out ahead of an advancing cold front as cold, dry descending air meets warm, humid, rising air.

Key Terms

The following terms are listed (with page number) in the order they appear in the text. Define each. Doing so will aid you in reviewing the material covered in this chapter.

air mass, 238
source regions (for air masses), 238
continental polar air mass, 239
continental arctic air mass, 239
maritime polar air mass, 244
Pacific air, 244
maritime tropical air mass, 245
continental tropical air mass, 249
front, 251
stationary front, 252
cold front, 253
frontolysis, 255
frontogenesis, 255
"back door" cold front, 256
cold air damming, 256
warm front, 257
overrunning, 258
dryline, 259
occluded fronts (occlusion), 261
cold occlusion, 262
warm occlusion, 263
upper-level front, 264

Review Questions

1. Describe the characteristics of a good air mass source region.
2. What air mass brings the coldest winter weather to the United States ?
3. Why is continental polar air welcome to the Central Plains in summer but not welcome in the winter?
4. List the temperature and moisture characteristics of each of the major air masses.
5. Explain how arctic outbreaks can occur in the southern part of the United States. Where did the arctic air originate, and how is it able to move this far south?
6. In mountainous country, would afternoon thunderstorms be more likely with a maritime tropical air mass or with a continental tropical air mass? Explain.
7. Why do air temperatures tend to be a little higher on the eastern side of the Appalachian Mountains than on the western side, even though the same winter continental polar or continental arctic air mass dominates both areas?
8. Explain how the airflow aloft regulates the movement of air masses.
9. Why are maritime polar air masses along the East Coast of the United States usually colder than those along the nation's West Coast? Why are they also *less* prevalent?
10. The boundaries between neighboring air masses tend to be more distinct during the winter than during the summer. Explain why.
11. When a very cold air mass covers half of the United States, a very warm air mass frequently covers the other half. Explain how and why this situation often happens.
12. What type of air mass would be responsible for the weather conditions listed as follows?
    (a) heavy snow showers and low temperatures at Buffalo, New York
    (b) hot, muggy summer weather in the Midwest and the East
    (c) daily afternoon thunderstorms along the Gulf Coast
    (d) heavy snow showers along the western slope of the Rockies
    (e) refreshing, cool, dry breezes after a long summer hot spell on the Central Plains
    (f) heavy summer rainshowers in southern Arizona
    (g) drought with record-high temperatures over the desert southwest
    (h) persistent cold, damp weather with drizzle along the East Coast
    (i) summer afternoon thunderstorms forming along the eastern slopes of the Sierra Nevada
    (j) record low winter temperatures in South Dakota
    (k) heavy rain at high elevations along with flooding and mud slides along the west coast of North America.
13. Sketch side views of a typical cold front, warm front, and cold-occluded front. Include in each diagram cloud types and patterns, areas of precipitation, surface winds, and relative temperature on each side of the front.
14. In winter, the most extreme change in air temperature across a cold front will occur when the front separates what two air masses?
15. On a surface weather map, what do you know about a region where the word *frontogenesis* is marked?
16. Explain why barometric pressure usually falls with the approach of a cold front or occluded front.
17. How does the weather usually change along a dryline? Since hot, dry air is often located behind (to the west) of a dryline, why are drylines not classified as warm fronts?
18. Based on the following weather forecasts, what type of front will most likely pass the area?
    (a) Light rain and cold today, with temperatures just above freezing. Southeasterly winds shifting to westerly tonight. Turning colder with rain becoming heavy and possibly changing to snow.
    (b) Cool today with rain becoming heavy at times by this afternoon. Warmer tomorrow. Winds southeasterly becoming westerly by tomorrow morning.
    (c) Increasing cloudiness and warm today, with the possibility of thunderstorms by evening. Turning much colder tonight. Winds southwesterly, becoming gusty and shifting to northwesterly by tonight.

(d) Increasing high cloudiness and cold this morning. Clouds increasing and lowering this afternoon, with a chance of snow or rain tonight. Precipitation ending tomorrow morning. Turning much warmer. Winds light easterly today, becoming southeasterly tonight and southwesterly tomorrow.

19. During the spring, on a warm, sunny day in Boston, Massachusetts, the wind shifts from southwesterly to northeasterly and the weather turns cold, damp, and overcast. What type of front moved through the Boston area? From what direction did the front apparently approach Boston?

20. Why are ice storms more likely to form with a warm front than with a cold front?

21. In winter, cold-front weather is typically more violent than warm-front weather. Why? Explain why this situation is not necessarily true in summer.

Online Learning

STUDENT COMPANION WEBSITE: Visit this book's companion website at: www.cengage.com/ahrens/extreme1e and choose Chapter 9 for many study aids and ideas for further reading and research. These include flashcards, practice quizzing, and web links.

METEOROLOGY RESOURCE CENTER: For students with access, log on at www.cengage.com/login for more assets, including animations, videos, and more. If your textbook did not come with access, visit www.CengageBrain.com to purchase.

Mid-Latitude Cyclonic Storms

NASA

10

CONTENTS

POLAR FRONT THEORY

WHERE DO MID-LATITUDE CYCLONES TEND TO FORM?

DEVELOPING MID-LATITUDE CYCLONES

CONVEYOR BELT MODEL OF MID-LATITUDE CYCLONES

"STORM OF THE CENTURY"—THE MARCH STORM OF 1993

THE COLUMBUS DAY STORM—OCTOBER 12, 1962

MID-LATITUDE CYCLONES AND GREAT PLAINS BLIZZARDS

POLAR LOWS

SUMMARY

KEY TERMS

REVIEW QUESTIONS

A mid-latitude cyclonic storm south of Australia spins clockwise over the southern Pacific Ocean.

Early weather forecasters were aware that precipitation generally accompanied falling barometers and areas of low pressure. However, it was not until the early part of the twentieth century that scientists began to piece together the information that yielded the ideas of modern meteorology and cyclonic storm development.

Working largely from surface observations, a group of scientists in Bergen, Norway, developed a model explaining the life cycle of an *extratropical*, or *middle-latitude cyclonic storm*: that is, a storm that forms at middle and high latitudes outside of the tropics. This extraordinary group of meteorologists included Vilhelm Bjerknes, his son Jakob, Halvor Solberg, and Tor Bergeron. They published their *Norwegian cyclone model* shortly after World War I. It was widely acclaimed and became known as the "polar front theory of a developing wave cyclone" or, simply, the **polar front theory**. What these meteorologists gave to the world was a working model of how a mid-latitude cyclone progresses through the stages of birth, growth, and decay. An important part of the model involved the development of weather along the polar front. As new information became available, the original work was modified, so that, today, it serves as a convenient way to describe the structure and weather associated with a migratory middle-latitude cyclonic storm system.

In the following sections we will first examine, from a surface perspective, how a mid-latitude cyclone develops along the polar front. Then we will examine how the winds aloft influence the developing surface storm. Later on, we will obtain a three-dimensional view of a mid-latitude cyclone by observing how ribbons of air glide through the storm system. Toward the end of the chapter, we will look at some individual storms whose extreme weather made the record books.

Polar Front Theory

The development of a mid-latitude cyclone, according to the Norwegian model, begins along the polar front. Remember from the discussion of the general circulation in Chapter 8 that the polar front is a semicontinuous global boundary separating cold polar air from warm subtropical air. Because the mid-latitude cyclonic storm forms and moves along the polar front in a wavelike manner, the developing storm is called a **wave cyclone**.

STAGES OF A DEVELOPING STORM The stages of a developing wave cyclone are illustrated in the sequence of surface weather maps shown in ⟫Fig. 10.1. Figure 10.1a shows a segment of the polar front as a stationary front. It represents a trough of lower pressure with higher pressure on both sides. Cold air to the north and warm air to the south flow parallel to the front, but in opposite directions. This type of flow sets up a cyclonic wind shear. You can conceptualize the shear more clearly if you place a pen between the palms of your hands and move your left hand toward your body; the pen turns counterclockwise, cyclonically.

Under the right conditions (described later in this chapter), a wavelike kink forms on the front, as shown in Fig. 10.1b. The wave that forms is known as a *frontal wave* or an *incipient cyclone*. Watching the formation of a frontal wave on a weather map is like watching a water wave from its side as it approaches a beach: It first builds, then breaks, and finally dissipates, which is why a mid-latitude cyclonic storm system is known as a *wave cyclone*. Figure 10.1b shows the newly formed wave with a cold front pushing southward and a warm front moving northward.

The region of lowest pressure (called the *central pressure*) is at the junction of the two fronts. As the cold air displaces the warm air upward along the cold front, and as *overrunning* occurs ahead of the warm front, a narrow band of precipitation forms (shaded green area). Steered by the winds aloft, the system typically moves east or northeastward and gradually becomes a fully developed *open wave* in 12 to 24 hours (see Fig. 10.1c). The central pressure is now much lower, and several isobars encircle the wave's apex. These more tightly packed isobars create a stronger cyclonic flow, as the winds swirl counterclockwise and inward toward the low's center. Precipitation forms in a wide band *ahead* of the warm front and *along a narrow band* of the cold front. The region of warm air between the cold and warm fronts is known as the **warm sector**. Here, the weather tends to be partly cloudy, although scattered showers may develop if the air is conditionally unstable.

Energy for the storm is derived from several sources. As the air masses try to attain equilibrium, warm air rises and cold air sinks, transforming potential energy into kinetic energy—energy of motion. Condensation supplies energy to the system in the form of latent heat. And, as the surface air converges toward the low's center, wind speeds may increase, producing an increase in kinetic energy.

As the open wave moves eastward, its central pressure continues to decrease, and the winds blow more vigorously as the wave quickly develops into a **mature cyclone**. The faster-moving cold front constantly inches closer to the warm front, squeezing the warm sector into a smaller area, as shown in Fig. 10.1d. In this model, the cold front eventually overtakes the warm front and the system becomes occluded. At this point, the storm is usually most intense, with clouds and precipitation covering a large area.

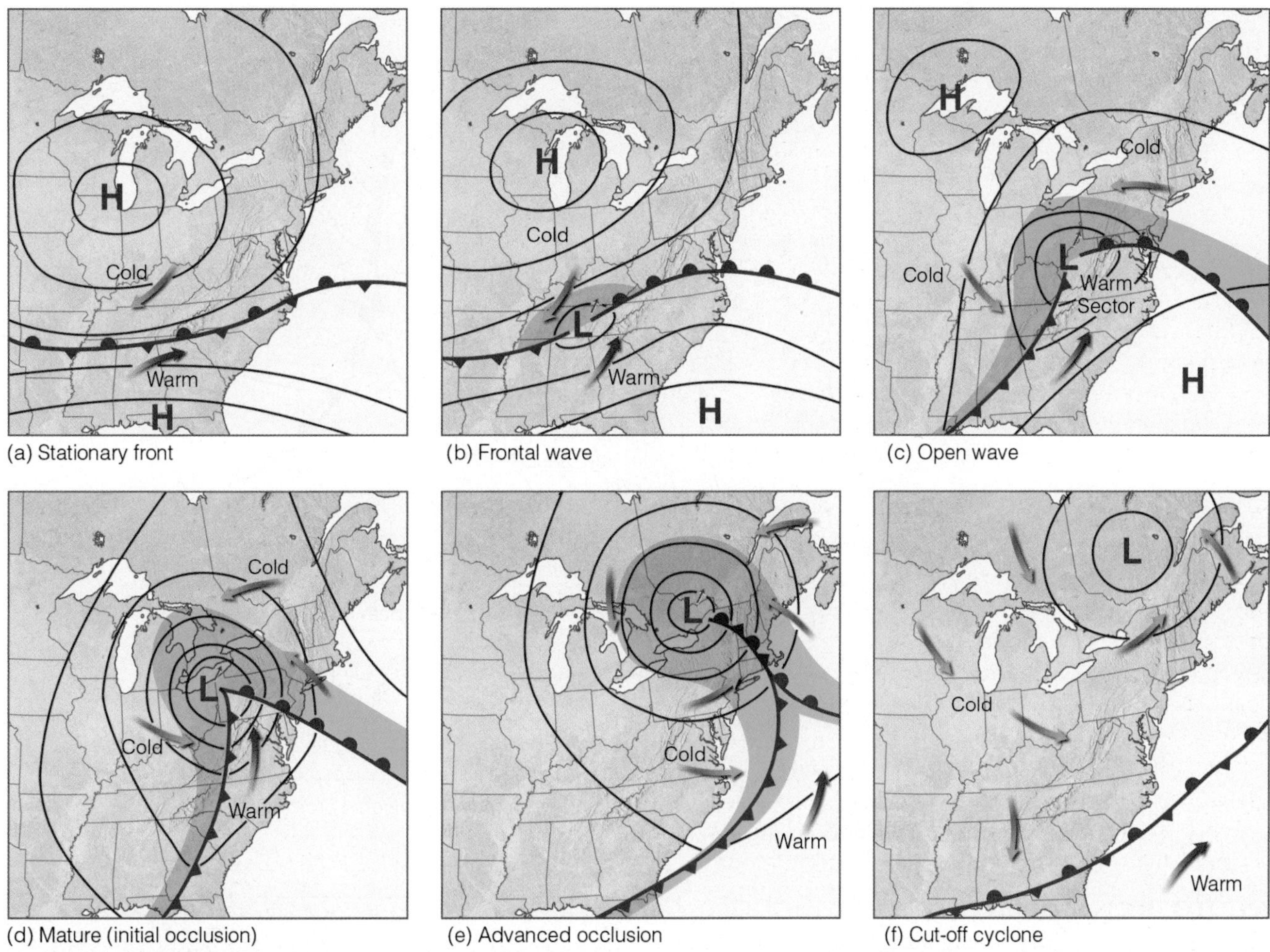

FIGURE 10.1 The idealized life cycle of a mid-latitude cyclone (a through f) in the Northern Hemisphere based on the polar front theory. As the life cycle progresses, the system moves northeastward in a dynamic fashion. The small arrow next to each L shows the direction of storm movement.

The point of occlusion where the cold front, warm front, and occluded front all come together in Fig. 10.1e is referred to as the *triple point*. Notice that in this region the cold and warm fronts appear similar to the open-wave cyclone in Fig. 10.1c. It is here where a new wave (called a **secondary low**) will occasionally form, move eastward, and intensify into a cyclonic storm. The center of the intense storm system shown in Fig. 10.1e gradually dissipates, because cold air now lies on both sides of the occluded front. The warm sector is still present, but is far removed from the center of the storm. Without the supply of energy provided by the rising warm, moist air, the old storm system dies out and gradually disappears (see Fig. 10.1f). We can think of the sequence of a developing wave cyclone as a whirling eddy in a stream of water that forms behind an obstacle, moves with the flow, and gradually vanishes downstream. The entire life cycle of a wave cyclone can last from a few days to over a week.

Figure 10.2 shows a series of wave cyclones in various stages of development along the polar front in winter. Such a succession of storms is known as a *"family" of cyclones*. Observe that to the north of the front are cold anticyclones; to the south over the Atlantic Ocean is the warm, semipermanent Bermuda high. The polar front itself has developed into a series of loops, and at the apex of each loop is a cyclonic storm system. The cyclone over the northern plains (Low 1) is just forming; the one along the East Coast (Low 2) is an open wave; and the occluded system near Iceland (Low 3) is dying out. If the average rate of movement of a wave cyclone from birth to decay is 25 mi/hr, then it is entirely possible for a storm to develop over the central part of the United States, intensify into a large storm over New England, become occluded over the ocean, and reach the coast of England in its dissipating stage less than a week after it formed.

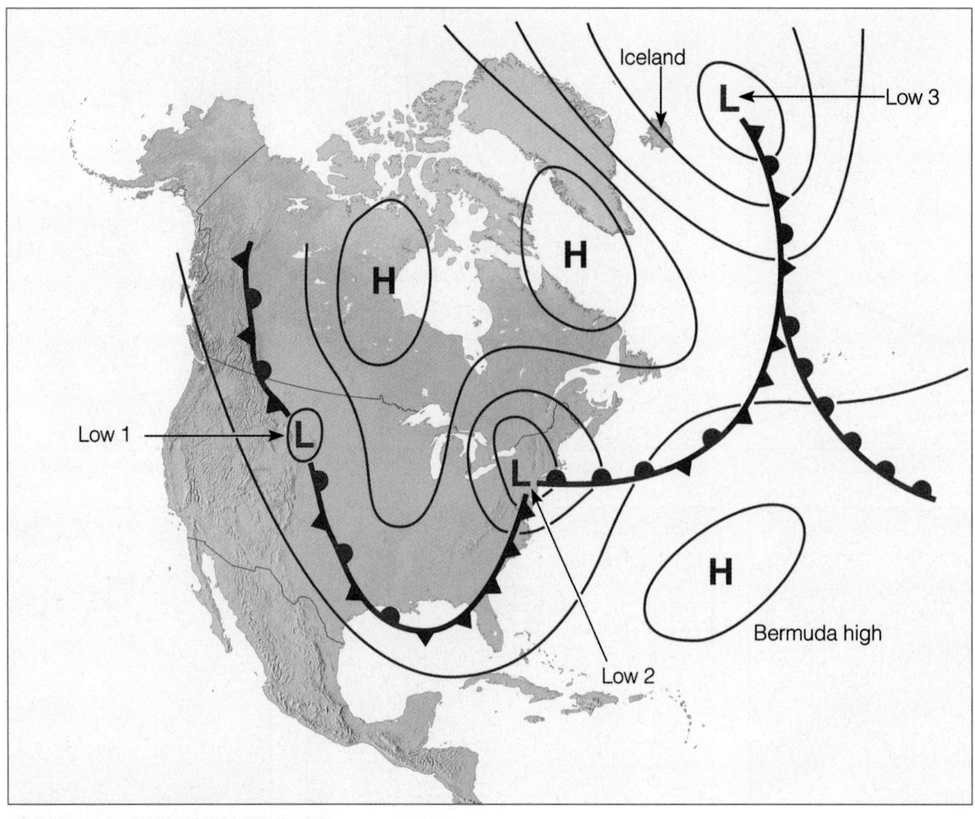

FIGURE 10.2 A series of wave cyclones (a "family" of cyclones) forming along the polar front.

A NOTORIOUS STORM FOLLOWS THE MODEL

Figure 10.3 shows a developing wave cyclone over Kansas on the morning of November 9, 1975. The storm is a classic open wave cyclone that, in less than 24 hours, will develop into a history-making storm. While over Kansas, the storm's central pressure is only about 1000 mb (29.53 in.). Notice that the storm takes a northeasterly path, and by the morning of November 10, 1975, the storm has deepened into a monster of a mature cyclone with a central pressure of 982 mb (29.00 in.). Now over Lake Superior, the storm (with its tightly packed isobars) generates high winds and huge waves that proved fatal to a ship named the *Edmund Fitzgerald.*

The *Edmund Fitzgerald*, a huge 729-foot-long freighter, set out from Superior, Wisconsin, with a crew of 29 on November 9, 1975. Loaded with 26,000 tons of iron ore pellets, the ship's destination was a steel plant in Detroit, Michigan. As the vessel moved toward eastern Lake Superior on November 10, the occluded front associated with the deep low-pressure area (shown in Fig. 10.3) swept by the ship, and strong northeasterly winds shifted to strong northwesterly winds. Storm warnings had been issued by the National Weather Service, and in the afternoon, as the *Fitzgerald* reached the eastern part of the lake, northwest winds of 85 mi/hr were reported. The *Arthur M. Anderson*, a sister ship,

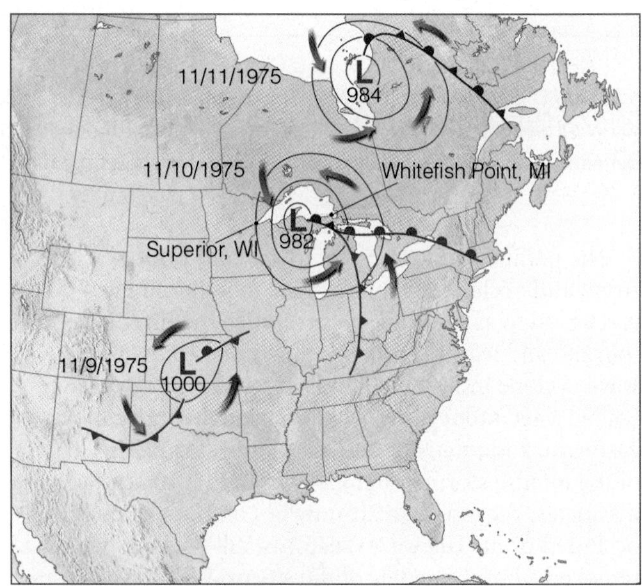

FIGURE 10.3 The development of a wave cyclone that formed into an extremely strong and deadly mid-latitude cyclonic storm. Notice how closely the cyclone adheres to the polar front model described in Fig. 10.1.

kept in constant contact with the *Fitzgerald* as it followed about 10 miles behind. By 7:00 P.M. that evening, high winds and waves, reported to be over 25 feet high, were pounding both vessels. The *Fitzgerald* was in trouble; it was listing badly and taking on water. With his ship in peril, the *Fitzgerald's* captain sent a final radio message stating that "It's one of the worst seas I've ever been in." About 10 minutes later, the crippled ship and its crew of 29 went to the bottom of Lake Superior, about 20 miles northwest of Whitefish Point, Michigan.*

The deadly, intense mid-latitude cyclone moved into northern Canada, and by the morning of November 11, the storm was in an advanced occluded stage (Fig. 10.3). Cut off from its energy source, it gradually dissipated and eventually died out, becoming a swirl of air embedded in the circulation of the atmosphere.

Up to now, we have considered the polar front model of a developing wave cyclone, which represents a rather simplified version of the stages that an extratropical cyclonic storm system must go through. In fact, few (if any) storms adhere to the model exactly. Nevertheless, it serves as a good foundation for understanding the structure of storms. So keep the model in mind as you read the following sections.

Where Do Mid-Latitude Cyclones Tend to Form?

Any development or strengthening of a mid-latitude cyclone is called **cyclogenesis**. There are regions of North America that show a propensity for cyclogenesis, including the Gulf of Mexico, the Atlantic Ocean east of the Carolinas, and the eastern slope of high mountain ranges, such as the Rockies and the Sierra Nevada. For example, when a westerly flow of air crosses a north-to-south trending mountain range, the air on the downwind (leeward) side tends to curve cyclonically, as shown in ❯ Fig. 10.4. This curving of air adds to the developing or strengthening of a cyclonic storm. Such storms that form on the leeward side of a mountain are called **lee-side lows**.

Another region of cyclogenesis lies near Cape Hatteras, North Carolina, where warm Gulf Stream water can supply moisture and warmth to the region south of a stationary front, thus increasing the contrast between air masses to a point where storms may suddenly spring up along the front. These cyclones, called **northeasters** or *nor'easters,* normally move northeastward along the Atlantic Coast, bringing high winds and heavy snow or rain to coastal areas. Before the age of modern satellite imagery and weather prediction, such coastal storms would often go undetected during their formative stages, and sometimes an evening weather forecast of "fair and colder" along the eastern seaboard would have to be changed to "heavy snowfall" by morning. Fortunately, with today's weather information-gathering and forecasting techniques, these storms rarely strike by surprise. (Additional information on northeasters is given in the Focus section on p. 274.)

*The *Edmund Fitzgerald* was the largest of more than 1000 ships of all sizes to be wrecked or sunk by violent weather on the Great Lakes. Ferocious late autumn storms on the Great Lakes such as this one are often referred to as "the Witches of November," whose tempers are responsible for the sinking of so many ships and the taking of so many lives.

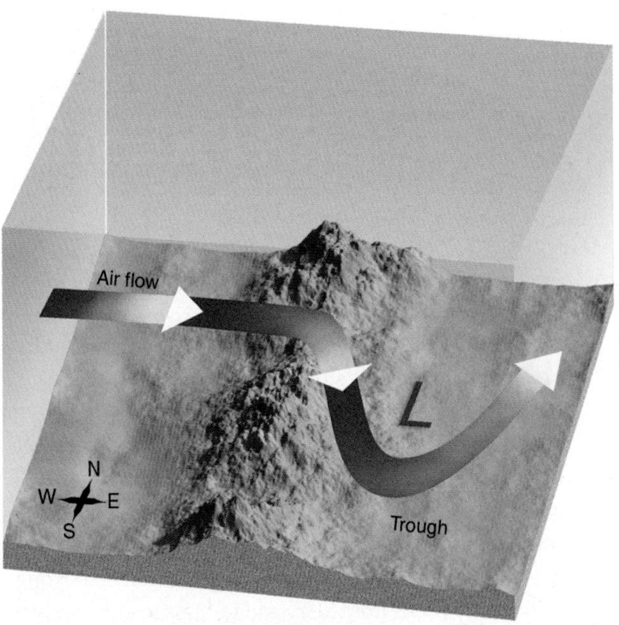

❯ **FIGURE 10.4** As westerly winds blow over a mountain range, the airflow is deflected in such a way that a trough forms on the downwind (leeward) side of the mountain. Troughs and developing cyclonic storms that form in this manner are called *lee-side lows.*

EXTREME WEATHER WATCH

Mariners of the Great Lakes look forward to November 10th with dread. Aside from this being the date of the sinking of the *Edmund Fitzgerald* in 1975, it is also the date of the deadliest storm in Great Lakes history. In 1913, the infamous storm, known as the "White Hurricane," lashed Lake Erie and Lake Huron with blizzard conditions and 80 to 100 mi/hr winds. Cleveland, Ohio, recorded its greatest one-day snowfall ever of 22.2 inches. On the lakes, ten large, 300-foot-long ore ships were lost and another seven were totally wrecked. As many as 300 sailors lost their lives, with the exact total never known.

FOCUS ON

EXTREME WEATHER

Northeasters

Northeasters (commonly called *nor'easters*) are mid-latitude cyclonic storms that develop or intensify off the eastern seaboard of North America then move northeastward along the coast. They often bring gale force northeasterly winds to coastal areas, along with heavy rain, snow, or sleet. They usually deepen and become most intense off the coast of New England. The ferocious northeaster of December, 1992 (shown in Fig. 1), produced strong northeasterly winds from Maryland to Massachusetts. Huge waves accompanied by hurricane-force winds that reached 90 miles per hour in Wildwood, New Jersey, pounded the shoreline, causing extensive damage to beaches, beachfront homes, sea walls, and boardwalks. Heavy snow and rain, which lasted for several days, coupled with high winds and high tides, put many coastal areas and highways under water, including parts of the New York City subway. Another strong northeaster dumped between one and three feet of snow over portions of the northeast during late March, 1997.

During March, 1962, an extremely powerful northeaster off the mid-Atlantic coast became the most destructive storm ever along the mid-Atlantic seaboard. High winds and high waves coupled with an extremely high tide produced a surge of water over 10 feet high, which caused extensive beach erosion and damaged or destroyed over 5000 buildings and every pier along coastal Delaware and New Jersey (see Fig. 2).

Studies suggest that some of the northeasters, which batter the coastline in winter, may actually possess some of the characteristics of a tropical hurricane. For example, the northeaster shown in Fig. 1 actually developed something like a hurricane's "eye" as the winds at its center went calm when it moved over Atlantic City, New Jersey. (We will examine hurricanes, and their characteristics in more detail in Chapter 13).

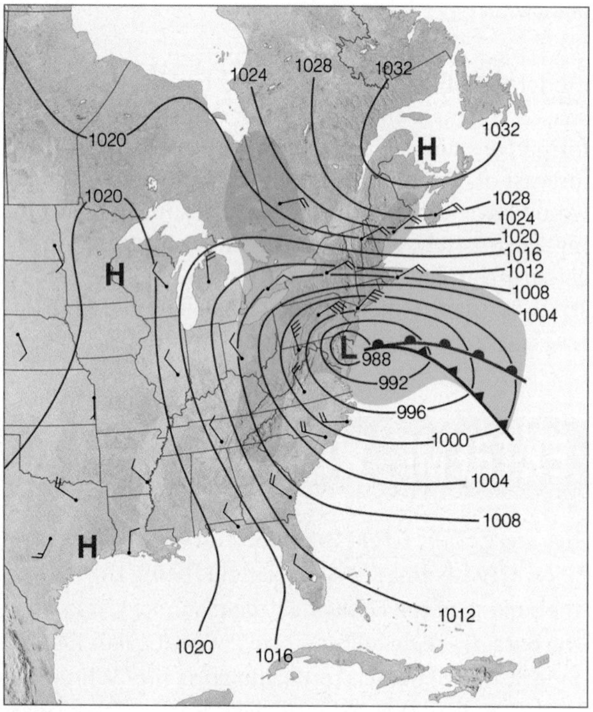

Figure 1 The surface weather map for 7:00 A.M. (EST) December 11, 1992, shows an intense low-pressure area (central pressure 988 mb, or 29.18 in.), which is generating strong northeasterly winds and heavy precipitation (area shaded green) from the mid-Atlantic states into New England. This northeaster devastated a wide area of the eastern seaboard causing damage in the hundreds of millions of dollars.

Figure 2 A couple inspects the remains of their beach house in Harvey Cedars, New Jersey, that was destroyed by the March Storm of 1962.

Figure 10.5 shows the typical paths taken in winter by mid-latitude cyclones and anticyclones. Notice in Fig. 10.5a that some of the lows are named after the region where they form, such as the *Hatteras low* which develops off the coast near Cape Hatteras, North Carolina. The *Alberta Clipper* forms (or redevelops) on the eastern side of the Rockies in Alberta, Canada, then rapidly skirts across the northern tier states. The *Colorado low*, in contrast, forms (or redevelops) on the eastern side of the Rockies. Notice that the lows generally move eastward or northeastward, whereas the highs typically move southeastward, then eastward. Later in this chapter we will see why lows and highs tend to take these paths.

When mid-latitude cyclones deepen rapidly (in excess of 24 mb in 24 hours), the term *explosive cyclogenesis*, or "*bomb*," is sometimes used to describe them. As an example, explosive cyclogenesis occurred in a storm that developed over the warm Atlantic just east of New Jersey on September 10, 1978. As the central pressure of the storm dropped nearly 60 mb (1.8 in.) in 24 hours, hurricane force winds battered the ocean liner *Queen Elizabeth II* and sank the fishing vessel *Captain Cosmo*.

Some frontal waves form suddenly, grow in size, and develop into huge cyclonic storms, such as the storm shown in Fig. 10.3. They slowly dissipate with the entire process taking several days to a week to complete. Other frontal waves remain small and never grow into a giant weather-producer. Why is it that some frontal waves develop into huge cyclonic storms, whereas others simply dissipate in a day or so?

This question poses one of the real challenges in weather forecasting. The answer is complex. Indeed, there are many surface conditions that do influence the formation of a mid-latitude cyclone, including mountain ranges and land-ocean temperature contrasts. However, the real key to the development of a wave cyclone is found in the *upper-wind flow*, in the region of the high-level westerlies. Therefore, before we can arrive at a reasonable answer to our question, we need to see how the winds aloft influence surface pressure systems.

Developing Mid-Latitude Cyclones

Developing mid-latitude cyclonic storms are deep low-pressure areas that usually intensify with height and, therefore, appear on surface weather maps and on upper-level charts. On an upper-level chart, the storm usually appears as either a closed low or a trough.

Suppose the upper-level low is directly above the surface low as illustrated in Fig. 10.6. Notice that only at the surface (because of friction) do the winds blow inward toward the low's center. As these winds converge (flow together), the air "piles up." This piling up of air, called **convergence**, causes air density to increase directly above the surface low. This increase in mass causes surface pressures to rise; gradually, the low fills and the surface low dissipates. The same reasoning can be applied to surface anticyclones. Winds blow outward,

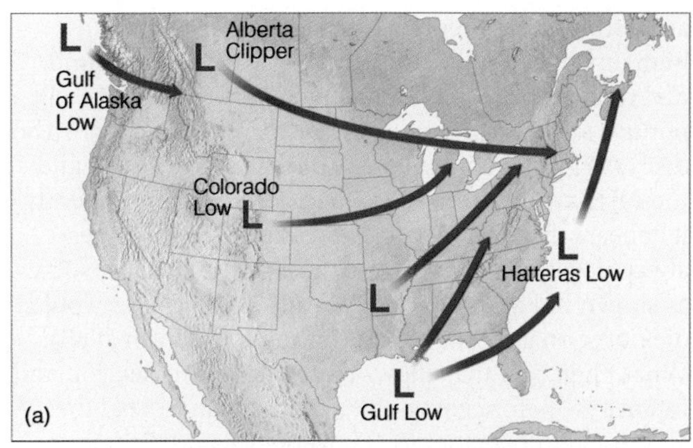

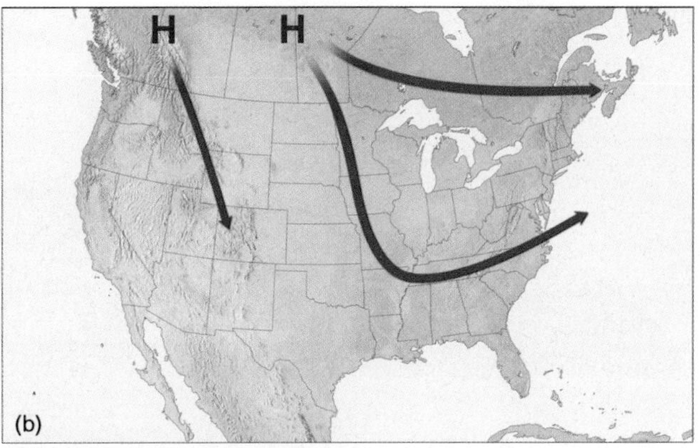

FIGURE 10.5 (a) Typical paths of winter mid-latitude cyclones. The lows are named after the region where they form. (b) Typical paths of winter anticyclones.

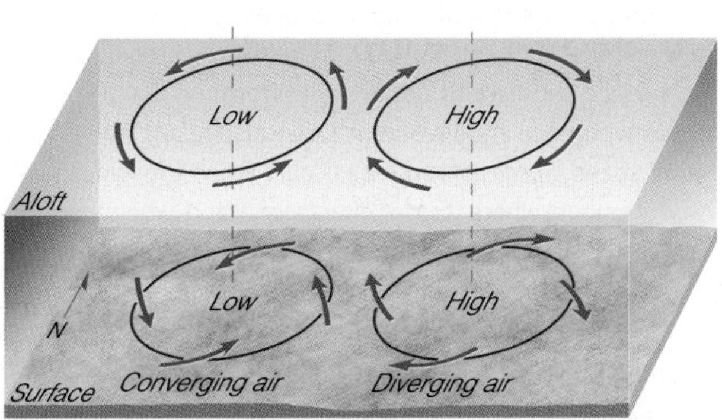

FIGURE 10.6 If lows and highs aloft were always directly above lows and highs at the surface, the surface systems would quickly dissipate.

away from the center of a surface high. If a closed high or ridge lies directly over the surface anticyclone, **divergence** (the spreading out of air) at the surface will remove air from the column directly above the high. The decrease in mass causes the surface pressure to fall and the surface high-pressure area to weaken. Consequently, it appears that, if upper-level pressure systems were always located directly above those at the surface (such as shown in Fig. 10.6), cyclones and anticyclones would die out soon after they form (if they could form at all). What, then, is it that allows these systems to develop and intensify? (Before going on, you may wish to read the additional information on convergence and divergence given in the Focus section below.)

THE ROLE OF CONVERGENCE AND DIVERGENCE

For mid-latitude cyclones and anticyclones to maintain themselves or intensify, the winds aloft must blow in such a way that zones of converging and diverging air form. For example, notice in Fig. 10.7 that the surface winds are converging about the center of the low; while aloft, directly above the low, the winds are diverging. For the surface low to develop into a major storm system, *upper-level divergence of air must be greater than surface convergence of air;* that is, more air must be removed above the storm than is brought in at the surface. When this event happens, surface air pressure decreases, and we say that the storm system is *intensifying* or *deepening*. If the reverse should occur (more air flows in at the surface than is removed at the top), surface pressure will rise, and the storm system will weaken and gradually dissipate in a process called *filling*.

FOCUS ON
A SPECIAL TOPIC

A Closer Look at Convergence and Divergence

We know that *convergence* is the piling up of air above a region, while *divergence* is the spreading out of air above some region. Convergence and divergence of air may result from changes in wind direction and wind speed. For example, convergence occurs when moving air is tunneled into an area, much in the way cars converge when they enter a crowded freeway. Divergence occurs when moving air spreads apart, much as cars spread out when a congested two-lane freeway becomes three lanes. On an upper-level chart, this type of convergence (also called *confluence*) occurs when contour lines move closer together, as a steady wind flows parallel to them (see the upper-level chart in Fig. 3). On the same chart, this type of divergence (also called *diffluence*) occurs when the contour lines move apart as a steady wind flows parallel to them. Notice that below the area of divergence lies the surface middle-latitude cyclonic storm.

Convergence and divergence may also result from changes in wind speed. *Speed convergence* occurs when the wind slows down as it moves along, whereas *speed divergence* occurs when the wind speeds up. We can grasp these relationships more clearly if we imagine air molecules to be marching in a band. When the marchers in front slow down, the rest of the band members squeeze together, causing convergence; when the marchers in front start to run, the band members spread apart, or diverge.

In summary, *speed convergence* takes place when the wind speed decreases downwind, and *speed divergence* takes place when the wind speed increases downwind.

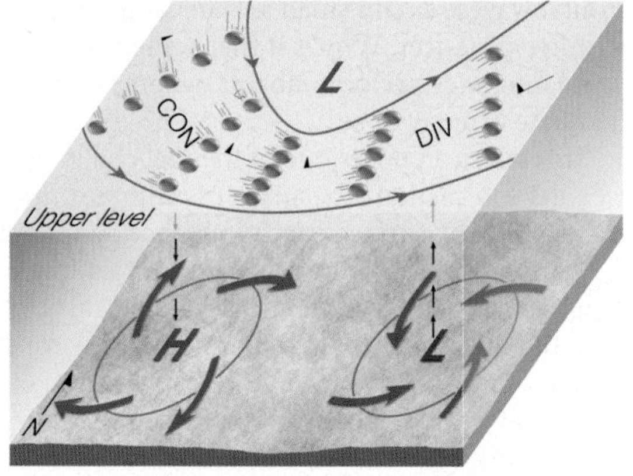

Figure 3 The formation of convergence (CON) and divergence (DIV) of air with a constant wind speed (indicated by flags) in the upper troposphere. Circles represent air parcels that are moving parallel to the contour lines on a constant pressure chart. Below the area of convergence the air is sinking, and we find the surface high (H). Below the area of divergence the air is rising, and we find the surface low (L).

Notice also in Fig. 10.7 that surface winds are diverging about the center of the high, while aloft, directly above the anticyclone, they are converging. In order for the surface high to strengthen, *upper-level convergence of air must exceed low-level divergence of air* (more air must be brought in above the anticyclone than is removed at the surface). When this occurs, surface air pressure increases, and we say that the high-pressure area is *building*.

In Fig. 10.7, the convergence of air aloft causes an accumulation of air above the surface high, which allows the air to sink slowly and replace the diverging surface air. Above the surface low, divergence allows the converging surface air to rise and flow out the top of the column.

We can see from Fig. 10.7 that, when an upper-level trough is as sufficiently deep as is illustrated here, a region of converging air usually forms on the west side of the trough and a region of diverging air forms on the east side. (For reference, compare Fig. 10.7 with Fig. 3 on p. 276.) Aloft, the area of diverging air is directly above the surface low, and the area of convergence is directly above the surface high. This configuration means that, for a surface mid-latitude cyclone to intensify, the upper-level trough of low pressure must be located *behind* (or to the *west* of) the surface low. When the upper-level trough is in this position, the atmosphere is able to redistribute its mass, as regions of low-level convergence are compensated for by regions of upper-level divergence, and vice versa.

Winds aloft steer the movement of the surface pressure systems. Since the winds above the surface low in Fig. 10.7 are blowing from the southwest, the surface low should move northeastward. The northwesterly winds above the surface high should direct it toward the southeast. These paths are typical of the average movement of surface pressure systems in the eastern two-thirds of the United States, as shown in Fig. 10.5 on p. 275.

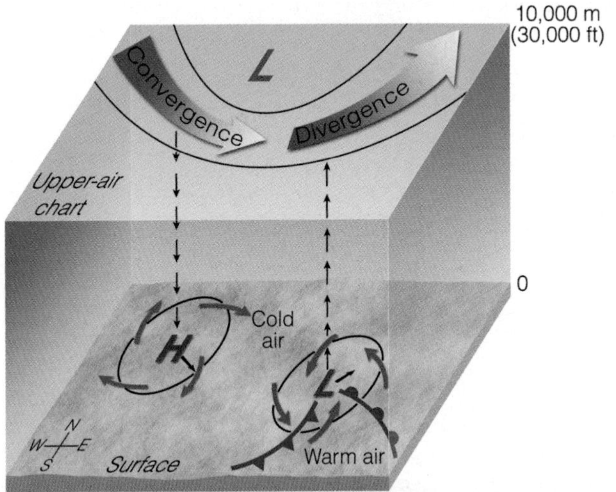

Active › FIGURE 10.7 Convergence, divergence, and vertical motions associated with surface pressure systems. Notice that for the surface storm to intensify, the upper trough of low pressure must be located to the left (or west) of the surface low. Visit the Meteorology Resource Center to view this and other active figures at academic.cengage.com/login

WAVES IN THE WESTERLIES Regions of strong upper-level divergence and convergence typically occur when well-developed waves exist in the flow aloft. The flow above the middle latitudes usually consists of a series of waves in the form of troughs and ridges. The distance from trough to trough (or ridge to ridge) is known as the *wavelength*. When the wavelength is on the order of many thousands of miles, the wave is called a **longwave**. Observe in › Fig. 10.8a that the length of the longwave is greater than the width of North America. Typically, at any given time, there are between three and six longwaves looping around the earth. These longwaves are also known as *Rossby waves*, after C. G. Rossby, a famous meteorologist who carefully studied their motion. In Fig. 10.8a, we can see that embedded in longwaves are **shortwaves**, which are small disturbances or ripples that move with the wind flow.

By comparing Fig. 10.8a with Fig 10.8b we can see that while the longwaves move eastward very slowly, the shortwaves move fairly quickly around the longwaves. Generally, shortwaves deepen (that is, increase in size) when they approach a longwave trough and weaken (become smaller) when they approach a ridge. Also, notice in Fig. 10.8b that, when a shortwave moves into a longwave trough, the trough tends to deepen. Look at shortwave 3 in Fig. 10.8b.

Notice in Fig. 10.8b that there are regions where the wind (blue and red arrows) cross the isotherms (dashed red lines). In these regions, there is an important process taking place called *temperature advection*. Where the wind crosses the isotherm in such a way that colder air is replacing warmer air, the transport of colder air into a region is called **cold advection**. In Fig. 10.8b, cold advection is represented by blue arrows. Where the wind crosses the isotherms in such a way that warmer air is replacing colder air, the transport of warmer air into a region is called **warm advection**. On the map, warm advection is represented by red arrows. Because temperature advection plays a major part in the development of a mid-latitude cyclonic storm, we will examine its important role in the following section.

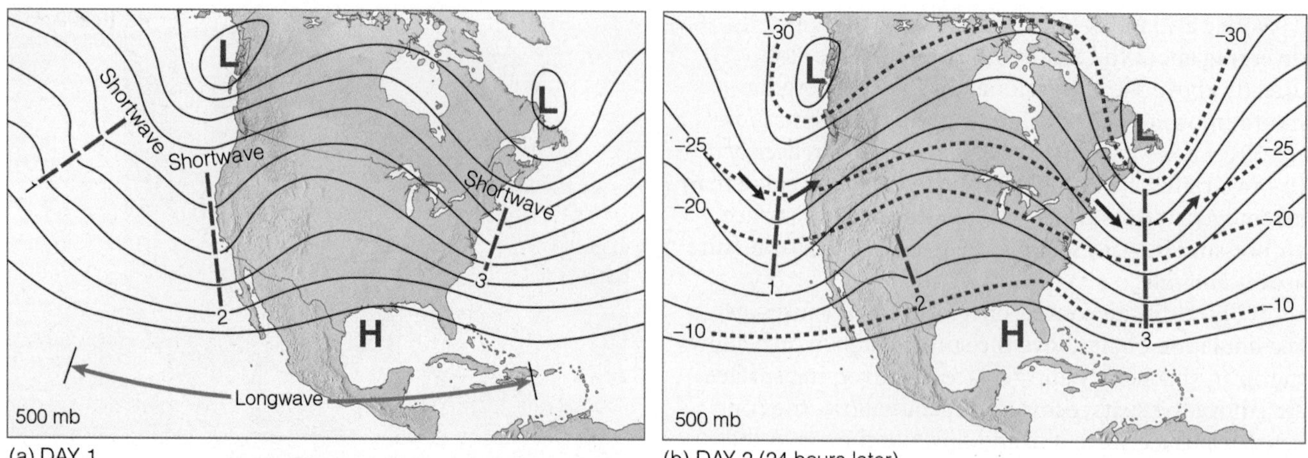

Active FIGURE 10.8 (a) Upper-air chart showing a longwave with three shortwaves (heavy dashed lines) embedded in the flow. (b) Twenty-four hours later the shortwaves have moved rapidly around the longwave. Notice that the shortwaves labeled 1 and 3 tend to deepen the longwave trough, while shortwave 2 has weakened as it moves into a ridge. Notice also that as the longwave deepens in diagram (b), its length actually shortens. Dashed lines are isotherms in °C. Solid lines are contours. Blue arrows indicate cold advection and red arrows, warm advection. Visit the Meteorology Resource Center to view this and other active figures at academic. Cengage.com/login

UPPER-AIR SUPPORT FOR THE DEVELOPING STORM To better understand how a wave cyclone may develop and intensify into a huge mid-latitude cyclonic storm, we need to examine atmospheric conditions at the surface and aloft. Suppose that a portion of a longwave trough at the 500-mb level lies directly above a surface stationary front, as illustrated in ❯ Fig. 10.9a. On the 500-mb chart, contour lines (solid lines) and isotherms (dashed lines) parallel each other and are crowded close together. Colder air is located in the northern half of the map, while warmer air is located to the south. Winds are blowing at fairly high velocities, which produce a sharp change in wind speed—a strong *wind speed shear*—from the surface up to this level. Suppose a shortwave moves through this region, disturbing the flow as shown in Fig. 10.9b.

As the flow aloft becomes disturbed, it begins to lend support for the intensification of surface pressure systems, as a region of converging air forms above position 1 in Fig. 10.9b and a region of diverging air forms above

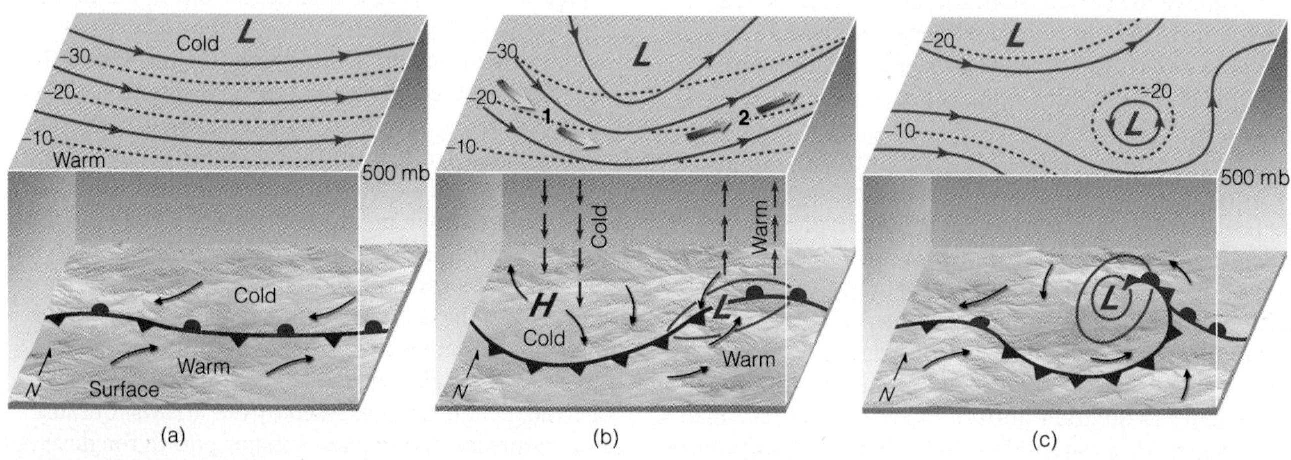

❯ FIGURE 10.9 An idealized 3-D view of the formation of a mid-latitude cyclone. (a) A longwave trough at 500 mb lies parallel to and directly above the surface stationary front. (b) A shortwave (not shown) disturbs the flow aloft, initiating temperature advection (blue arrow, cold advection; red arrow, warm advection). The upper trough intensifies and provides the necessary vertical motions (as shown by vertical arrows) for the development of the surface wave cyclone. (c) The surface storm occludes, and without upper-level divergence to compensate for surface converging air, the storm system dissipates.

position 2.* The converging air aloft causes the surface air pressure to rise in the region marked *H* in Fig. 10.9b. Surface winds begin to blow out away from the region of higher pressure, and the air aloft gradually sinks to replace it. Meanwhile, diverging air aloft causes the surface air pressure to decrease beneath position 2, in the region marked *L* on the surface map. This initiates rising air, as the surface winds blow in toward the region of lower pressure. As the converging surface air develops cyclonic spin, cold air flows southward and warm air northward. We can see in Fig. 10.9b that the western half of the stationary front is now a cold front and the eastern half a warm front. Cold air moves in behind the cold front, while warm air slides up along the warm front. These regions of cold and warm advection occur all the way up to the 500-mb level, about 18,000 feet above the surface.

On the 500-mb chart in Fig. 10.9b, cold advection is occurring at position 1 (blue arrows) as the wind crosses the isotherms, bringing cold air into the trough. The cold advection makes the air more dense and lowers the height of the air column from the surface up to the 500-mb level. (Recall that, on a 500-mb chart, lower heights mean the same as lower pressures.) Consequently, the pressure in the trough lowers and the trough deepens. The deepening of the upper trough causes the contour lines to crowd closer together and the winds aloft to increase. Meanwhile, at position 2 warm advection is taking place (red arrows), which has the effect of raising the height of a column of air; here, the 500-mb heights increase and a ridge builds (strengthens). Therefore, *the overall effect of differential temperature advection is to amplify the upper-level wave.* As the trough aloft deepens, its curvature increases, which in turn increases the region of divergence above the developing surface storm. At this point, the surface mid-latitude cyclone rapidly develops as surface pressures fall.

Regions of cold and warm advection are associated with vertical motions. Where there is cold advection, some of the cold, heavy air sinks; where there is warm advection, some of the warm, light air rises. Hence, due to advection, air must be sinking in the vicinity of position 1 and rising in the vicinity of position 2.

The sinking of cold air and the rising of warm air provide energy for a developing cyclone, as potential energy is transformed into kinetic energy. Further, if clouds form, condensation in the ascending air releases latent heat, which warms the air. The warmer air lowers the surface pressure, which strengthens the surface low even more. So, we now have a full-fledged middle-latitude cyclonic storm with all of the necessary ingredients for its development.

Eventually, the warm air curls around the north side of the low, and the storm system occludes (see Fig. 10.9c). Some storms may continue to deepen, but most do not as they move out from under the region of upper-level divergence. Additionally, at the surface the storm weakens as the supply of warm air is cut off and cold, dry air behind the cold front (called a *dry slot*) is drawn in toward the surface low.

Sometimes, an upper-level pool of cold air (which has broken away from the main flow) lies almost directly above the surface low. Occasionally the upper low will break away entirely from the main flow, producing a **cut-off low**, which often appears as a single contour line on an upper-level chart. When the upper low lies directly above the surface low (as in the Fig. 10.9c), the storm system is said to be *vertically stacked.* Usually the isotherms around the upper low parallel the contour lines, which indicates that no significant temperature advection is occurring. Without the necessary energy transformations, the surface system gradually dissipates. As its winds slacken and its central pressure gradually rises, the low slowly fills. The upper-level low, however, may remain stationary for many days. If air is forced to ascend into this cold pocket, widespread clouds and precipitation may persist for some time, even though the surface storm system itself has moved east out of the picture.

THE ROLE OF THE JET STREAM Jet streams play an additional part in the formation of surface mid-latitude cyclones. When the polar jet stream flows in a wavy west-to-east pattern (as illustrated in ❯ Fig. 10.10a), deep troughs and ridges exist in the flow aloft. Notice that, in the trough, the area shaded orange represents a strong core of winds called the *jet stream core*, or **jet streak**. The curving of the jet stream coupled with the changing wind speeds around the jet streak produce regions of strong convergence and divergence along the flanks of the jet. (More on this topic is provided in the Focus section on p. 280.)

The region of diverging air above the surface low (marked D in Fig. 10.10a) draws warm surface air upward to the jet stream, which quickly sweeps the air downstream. Since the air above the mid-latitude cyclone is being removed more quickly than converging surface winds can supply air to the storm's center, the central pressure of the storm drops rapidly. As surface pressure gradients increase, the wind speed increases. Above the high-pressure area, a region of converging air (marked C in Fig. 10.10a) feeds cold air downward into the anticyclone to replace the diverging surface air. Hence, we find the jet stream *removing air above the surface cyclone and supplying air to the surface*

*Look back at Fig. 10.7, p. **277**, and the upper-air chart in Fig. 3 on p. 276, and note the regions of converging air and diverging air on these maps.

FOCUS ON

A SPECIAL TOPIC

Jet Streaks and Storms

Figure 4 shows an area of maximum winds, a *jet streak*, on a 300-mb chart. Jet streaks, which have winds of at least 50 miles per hour, are important in the development of surface mid-latitude cyclones because areas of convergence and divergence form at specific regions around them. To understand why, consider air moving through a straight jet streak (shaded area) in Fig. 5. As the air enters the front of the streak (known as the *entrance region*), it increases in speed; as it leaves the rear of the streak (known as the *exit region*), it decreases in speed. At this elevation in the atmosphere (about 33,000 feet above the surface), the wind flow is nearly in geostrophic balance with the pressure gradient force (directed north) and the Coriolis force (directed south). As the air enters the jet streak, it increases in speed because the contour lines are closer together, causing an increase in the pressure gradient force. The pressure gradient force temporarily exceeds the Coriolis force, and the air swings slightly to the north across the contour lines, which causes a piling up of air and *strong convergence at point 1*. Weak divergence occurs at point 2.

Toward the middle of the jet streak, the increase in wind speed causes the Coriolis force to increase and the wind to become nearly geostrophic again. However, as the air exits the jet streak, the pressure gradient force is reduced as the contour lines spread farther apart. Hence, the Coriolis force temporarily exceeds the pressure gradient force, causing the air to cross the contour lines and swing slightly to the south. This process produces *strong divergence at point 3* and weak convergence at point 4.

The conditions described so far exist in a straight jet streak that shows no curvature. When the jet stream becomes wavy, and the jet streak exhibits cyclonic curvature (as it does in Fig. 6), the areas of weak divergence (at point 2) and weak convergence (at point 4) all but disappear. What we are left with is a curving jet streak that exhibits strong divergence at point 3 (in the left exit region) and strong convergence at point 1 (in the left entrance region). Notice in Fig. 6 that below the area of strong divergence the air rises, cools and, if sufficiently moist, condenses into clouds. Moreover, the removal of air in the region of strong divergence causes surface pressures to fall, which results in the development of an area of surface low pressure.

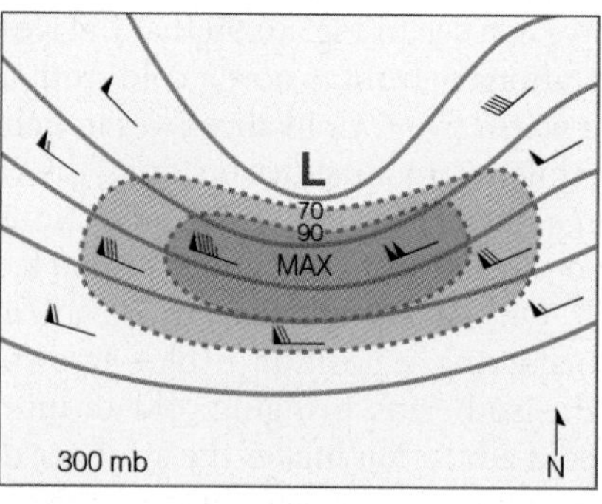

Figure 4 A portion of a 300-mb chart (about 33,000 ft above sea level) that shows the core of the jet—the region of maximum winds (MAX)—called a *jet streak*. Dashed lines are equal lines of wind speed (isotachs) in miles per hour.

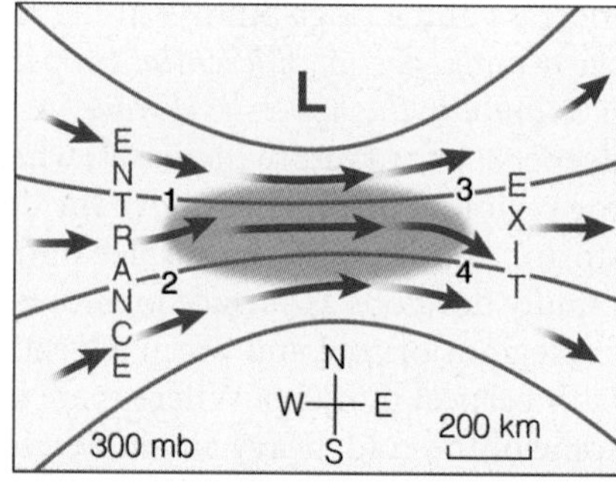

Figure 5 Changing air motions within a straight jet streak (shaded area) cause strong convergence of air at point 1 (left entrance region) and strong divergence at point 3 (left exit region).

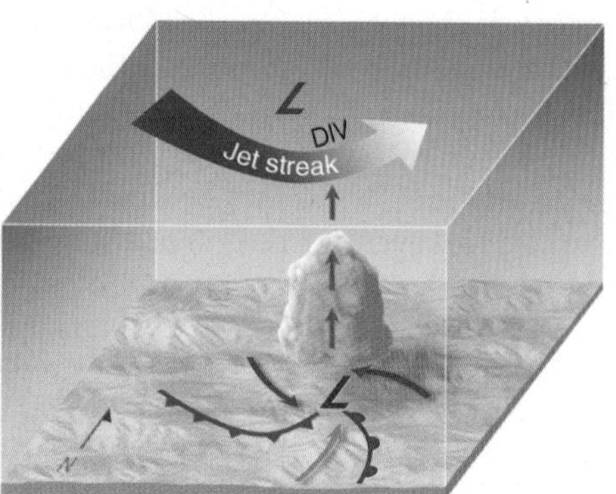

Figure 6 An area of strong divergence (DIV) can form with a curving jet streak. Below the area of divergence are rising air, clouds, and the developing mid-latitude cyclonic storm.

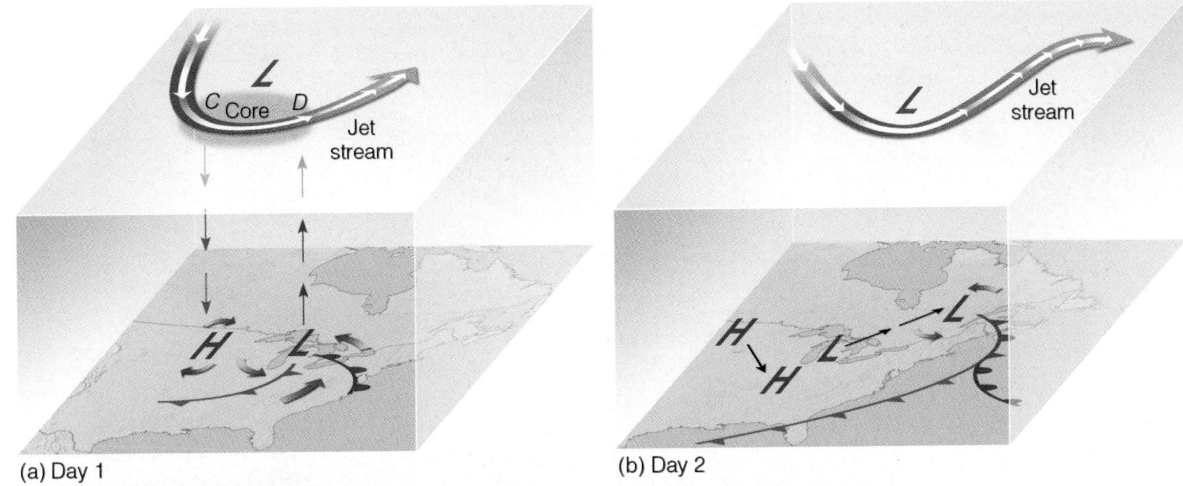

FIGURE 10.10 (a) As the polar jet stream and its area of maximum winds (the jet streak, or core) swings over a developing mid-latitude cyclone, an area of divergence (*D*) draws warm surface air upward, and an area of convergence (*C*) allows cold air to sink. The jet stream removes air above the surface storm, which causes surface pressures to drop and the storm to intensify. (b) When the surface storm moves northeastward and occludes, it no longer has the upper-level support of diverging air, and the surface storm gradually dies out.

anticyclone. Additionally, the sinking of cold air and the rising of warm air provide energy for the developing cyclone as potential energy is transformed into energy of motion (kinetic energy).

As the jet stream steers the storm along (toward the northeast, in this case), the surface storm occludes, and cold air surrounds the surface low (see Fig. 10.10b). Since the surface low has moved out from under the pocket of diverging air aloft, the occluded storm gradually fills as the surface air flows into the system.

Since the polar jet stream is strongest and moves farther south in winter, we can see why mid-latitude cyclonic storms are better developed and move more quickly during the coldest months. During the summer when the polar jet shifts northward, developing mid-latitude cyclonic storm activity shifts northward and occurs principally over the Canadian provinces of Alberta and the Northwest Territories.

In general, we now have a fairly good picture as to why some surface lows intensify into huge mid-latitude cyclones while others do not. For a surface cyclonic storm to intensify, there must be an upper-level counterpart—a trough of low pressure—that lies to the *west* of the surface low. As shortwaves disturb the flow aloft, they cause regions of differential temperature advection to appear, leading to an intensification of the upper-level trough. At the same time, the polar jet forms into waves and swings slightly south of the developing storm. When these conditions exist, zones of converging and diverging air, along with rising and sinking air, provide energy conversions for the storm's growth. With this atmospheric situation, storms may form even where there are no pre-existing fronts.* In regions where the upper-level flow is not disturbed by shortwaves or where no upper trough or jet stream exists, the necessary vertical and horizontal motions are insufficient to enhance cyclonic storm development and we say that the surface storm does not have the proper *upper-air support*. The horizontal and vertical motions, cloud patterns, and weather that typically occur with a developing open-wave cyclone are summarized in Fig. 10.11.

*It is interesting to note that the beginning stage of a wave cyclone almost always takes place when an area of upper-level divergence passes over a surface front. However, even if initially there are no fronts on the surface map, they may begin to form where air masses having contrasting properties are brought together in the region where the surface air rises and the surrounding air flows inward.

EXTREME WEATHER WATCH

A powerful mid-latitude storm can at times result in a variety of extreme weather over a relatively small area. During March, 1932, a strong storm system traversed the state of Illinois. In the north-central part of the state, heavy snow with drifts of up to ten feet high accompanied the storm. At the same time over the central and southern sections of the state, deadly tornadoes and severe thunderstorms pounded the area with large hail over 3.5 inches in diameter.

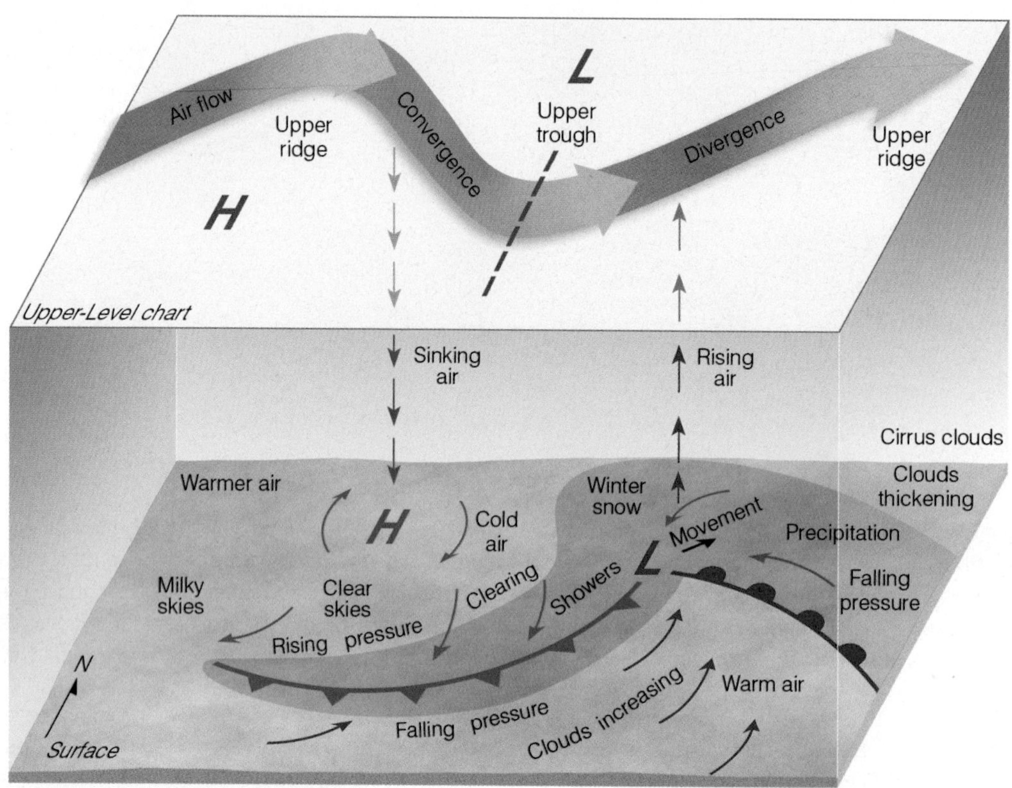

FIGURE 10.11 Summary of clouds, weather, vertical motions, and upper-air support associated with a developing mid-latitude cyclone.

Conveyor Belt Model of Mid-Latitude Cyclones

A three-dimensional model of a developing mid-latitude cyclone that typically forms along the east coast of North America is illustrated in Fig. 10.12. The model describes rising and sinking air as traveling along three main "conveyor belts." Just as people ride escalators to higher levels in a department store, so air glides along through a constantly evolving mid-latitude cyclone. According to the **conveyor belt model**, a warm air stream (known as the *warm conveyor belt*—orange arrow in Fig. 10.12) originates at the surface in the warm sector, ahead of the cold front. As the warm air stream moves northward, it slowly rises along the sloping warm front, up and over the cold air below. As the rising air cools, water vapor condenses, and clouds form well out ahead of the surface low and its surface warm front. From these clouds, steady precipitation usually falls in the form of rain or snow. Aloft, the warm air flow gradually turns toward the northeast, parallel to the upper-level winds.

Directly below the warm conveyor belt, a cold, relatively dry airstream—the *cold conveyor belt*—moves slowly westward (see Fig. 10.12). As the air moves west

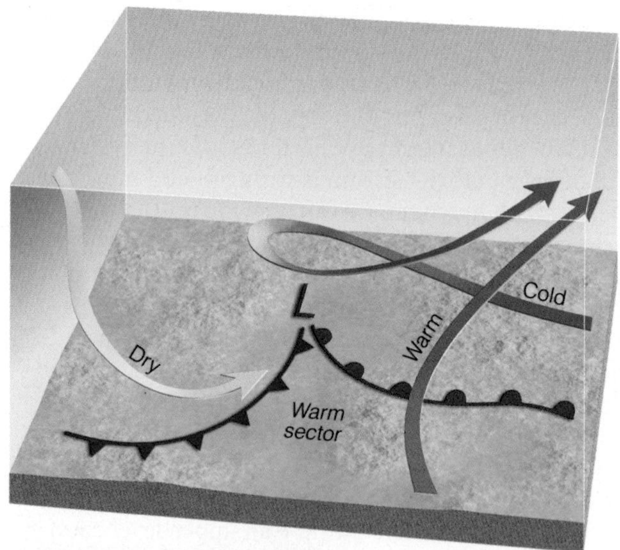

FIGURE 10.12 The conveyor belt model of a developing mid-latitude cyclone. The warm conveyor belt (in orange) rises along the warm front, causing clouds and precipitation to cover a vast area. The cold conveyor belt (in blue) slowly rises as it carries cold, moist air westward ahead of the warm front but under the rising warm air. The cold conveyor belt lifts rapidly and wraps counterclockwise around the center of the surface low. The dry conveyor belt (in yellow) brings very dry, cold air downward from the upper troposphere.

FIGURE 10.13 Visible satellite image of a mature mid-latitude cyclone with the three conveyor belts superimposed on the storm. As in Fig. 10.12, the warm conveyor belt is in orange, the cold conveyor belt is in blue, and the dry conveyor belt (forming the *dry slot*) is in yellow.

ahead of the warm front, precipitation and surface moisture evaporates into the cold air, making it moist. As the cold, moist airstream moves into the vicinity of the surface low, rising air gradually forces the cold conveyor belt upward. As the cold, moist air sweeps northwest of the surface low, it often brings heavy winter snowfalls to this region of the cyclone. The rising air stream usually turns counterclockwise, around the surface low, first heading south, then northeastward, when it gets caught in the upper air flow. It is the counterclockwise turning of the cold conveyor belt that produces the comma-shaped cloud similar to the one shown in Fig. 10.13.

The last conveyor belt is a dry one that forms in the cold, very dry region of the upper troposphere. Called the *dry conveyor belt,* and shaded yellow in Fig. 10.12, this airstream slowly descends from the northwest behind the surface cold front, where it brings generally clear, dry weather and, occasionally, blustery winds. If a branch of the dry air sweeps into the storm, it produces a clear area called a **dry slot**, which appears to pinch off the comma cloud's head from its tail. This phenomenon tends to show up on satellite images as the mid-latitude storm becomes more fully developed (see Fig. 10.13).

We are now in a position to tie together many of the concepts we have learned about developing mid-latitude cyclones by examining a monstrous storm—one for the record books—that formed during March, 1993. (Before going on to the next section, however, you may wish to read how storms that form along the east coast are classified. This information is provided in the Focus section on p. 284.)

BRIEF REVIEW

Up to this point, we have looked at the structure and development of mid-latitude cyclones. Here is a summary of a few of the important ideas presented so far:

- **The polar front (or Norwegian) model of a developing mid-latitude cyclonic storm represents a simplified but useful model of how an ideal storm progresses through the stages of birth, maturity, and dissipation.**
- **For a surface mid-latitude cyclone to develop or intensify (deepen), the upper-level low must be located to the west of (behind) the surface low.**
- **For a surface mid-latitude cyclonic storm to form, there must be an area of upper-level diverging air above the surface low. For the surface storm to intensify, the region of upper-level diverging air must be greater than surface converging air (that is, more air must be removed above the storm than is brought in at the surface).**
- **When the upper-air flow develops into waves, winds often cross the isotherms, producing regions of cold advection and warm advection, which tend to amplify the wave. At the same time, vertical air motions begin to enhance the formation of the surface storm as the rising of warm air and the sinking of cold air provide the proper energy conversion for the storm's growth.**
- **When the polar-front jet stream develops into a looping wave, it provides an area of upper-level diverging air for the development of surface mid-latitude cyclonic storms.**
- **The curving nature of the polar-front jet stream tends to direct surface mid-latitude cyclonic storms northeastward and surface anticyclones southeastward.**

"Storm of the Century"— The March Storm of 1993

In mid-March, 1993, the most massive and disruptive middle-latitude cyclonic storm in United States' history swept across the eastern third of the country. Variously named the "Storm of the Century" or "Super Storm," this cyclonic storm claimed some 318 lives (including 48 lost at sea) and caused hundreds of millions of dollars in damage. This whopper of a storm affected the lives of

FOCUS ON
A SPECIAL TOPIC

Ranking East Coast Storms

Nor'easters are mid-latitude cyclonic storms that form along the eastern seaboard of North America, gain in intensity and move northeastward. These storms often produce high winds and copious precipitation. In winter, heavy snow can fall over a wide area, sometimes crippling major East Coast cities, such as Washington, D.C., Philadelphia, New York, and Boston. The impact these snowstorms have along the East Coast is described in the *Northeast Snowfall Impact Scale (NESIS)*.

The NESIS, developed by Paul Kocin (formerly of NOAA and *The Weather Channel*) and Louis Uccellini (director of the National Weather Service, National Center for Environmental Prediction), takes into account two main storm factors: (1) the distribution and amount (depth) of snowfall over a region extending from southern Virginia to New England and (2) the population density impacted by the storm. The calculated NESIS value for each storm ranges from 1 (smaller storms) to greater than 10 (huge storms). The storm's NESIS value is then placed into one of five categories, ranging from category 1 (a notable storm) to category 5 (an extreme storm). Extreme storms are those that produce very heavy snowfall over a vast region that includes major metropolitan areas (see Table 1). Table 2 shows the top ten snowstorms that adversely affected the northeastern United States from 1955 to 2009. Notice that the March, 1993, storm described on p. 283 is ranked number one.

Table 1

***Northeast Snowfall Impact Scale* (NESIS) Categories, Values, and Storm Impact**

STORM CATEGORY	NESIS VALUE	STORM DESCRIPTION
1	1.0–2.5	Notable
2	2.5–4.0	Significant
3	4.0–6.0	Major
4	6.0–10.0	Crippling
5	10.0+	Extreme

Table 2

The Top Ten Snowstorms That Adversely Affected the Northeast United States (from Southern Virginia to New England) from 1955 to 2009, according to NESIS

STORM RANK	DATE	NESIS VALUE	NESIS CATEGORY	STORM DESCRIPTION
1	March 12–14, 1993	13.20	5	Extreme
2	January 6–8, 1996	11.78	5	Extreme
3	February 15–18, 2003	8.91	4	Crippling
4	March 2–5, 1960	8.77	4	Crippling
5	February 2–5, 1961	7.06	4	Crippling
6	January 11–14, 1964	6.91	4	Crippling
7	January 21–24, 2005	6.80	4	Crippling
8	January 19–21, 1978	6.53	4	Crippling
9	December 25–28, 1969	6.29	4	Crippling
10	February 10–12, 1983	6.25	4	Crippling

more Americans (over 100 million) than any storm of any kind previously or since including, for the first time, causing the closure of every major airport on the East Coast, outside of Florida.

The beginning stage of this super storm is shown in ▶Fig. 10.14, as a small frontal wave forms just off the Texas coast on the morning of March 12, 1993. With a central pressure near 1000 mb (29.53 in.), there is nothing special about this frontal wave. However, in just twenty-one hours this storm will move over northern Florida and deepen into an open-wave cyclone with a central pressure near 975 mb (28.79 in.). This extremely low pressure is extraordinary when compared to the central pressure of 996 mb (29.41 in.) found in a typical open-wave cyclone.

We can obtain a more detailed picture of this open-wave cyclone by looking at ▶Fig. 10.15, which shows the surface weather map for the morning of March 13, 1993. Notice that a strong cold front stretches from the storm's center through western Florida. Behind the front, cold arctic air pours into the Deep South. Ahead of the advancing front, a band of heavy thunderstorms (along a squall line) is pounding Florida with heavy rain, high winds, and tornadoes. Warm humid air in the warm sector is streaming northward, overrunning cold surface air ahead of the warm front, which is causing precipitation in the form of rain, snow, and sleet to fall over a vast area extending from Florida to New York.

An enhanced satellite image of the open wave cyclone over Florida is shown in ▶Fig. 10.16. Notice that the

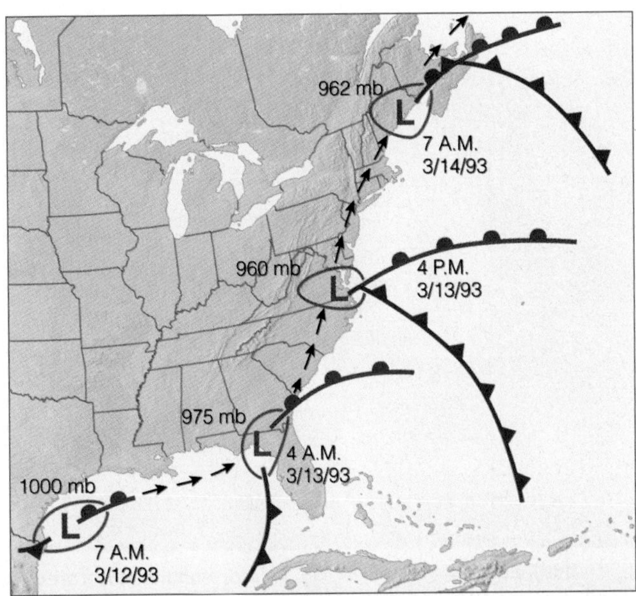

▶ **FIGURE 10.14** The development of the ferocious mid-latitude cyclonic storm of March, 1993. A small wave in the western Gulf of Mexico intensifies into a deep open-wave cyclone over Florida. It moves northeastward and becomes occluded over Virginia where its central pressure drops to 960 mb (28.35 in.). As the occluded storm continues its northeastward movement, it gradually fills and dissipates. The number next to the storm is its central pressure in millibars. Arrows show direction of movement. Time is Eastern Standard Time.

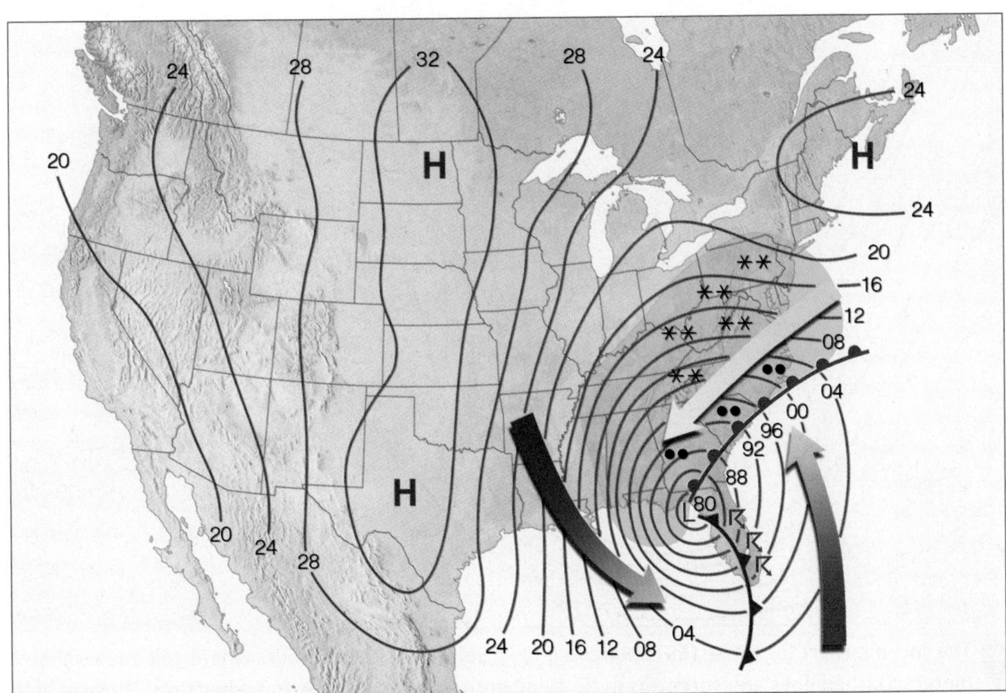

▶ **FIGURE 10.15** Surface weather map for 4 A.M (EST) on March 13, 1993. Lines on the map are isobars. A reading of 96 is 996 mb and a reading of 00 is 1000 mb. (To obtain the proper pressure in millibars, place a 9 before those readings between 80 and 96, and place a 10 before those readings of 00 or higher.) Green shaded areas are receiving precipitation. Heavy arrows represent surface winds. The orange arrow represents warm, humid air; the light blue arrow, cold, moist air; and the dark blue arrow, cold, arctic air.

FIGURE 10.16 A color-enhanced infrared satellite image that shows a developing mid-latitude cyclone at 2 A.M. (EST) on March 13, 1993. The darkest shades represent clouds with the coldest and highest tops. The dark cloud band moving through Florida represents a line of severe thunderstorms. Notice that the cloud pattern is in the shape of a comma.

storm's cloud band is in the shape of a comma that covers the entire eastern seaboard. Such **comma clouds** indicate that the storm is still developing and intensifying. Notice also that the center of the storm shown in Fig. 10.15 is located near the head of the comma cloud and the cold front is positioned along the comma cloud's tail. The area of clear weather near the comma's head represents a *dry slot*, where dry air is sweeping into the storm from aloft.

The 500-mb chart for the morning of March 13 (see Fig. 10.17) shows that a deep trough extending southward out of Canada lies to the west of the surface low. Around the trough where strong winds cross the isotherms, there is temperature advection. Notice that warm advection (indicated by red barbs) is occurring on the trough's eastern side, ahead of the surface warm front shown in Fig. 10.15. Cold advection (indicated by blue barbs) is occurring on the trough's western side, behind the surface cold front. As temperature advection deepens the trough, rising and sinking air provide energy for the developing surface storm.

Higher up, in the upper troposphere, a strong jet stream and jet streak swing southward over northern Florida (see Fig. 10.18). On the eastern side of the jet

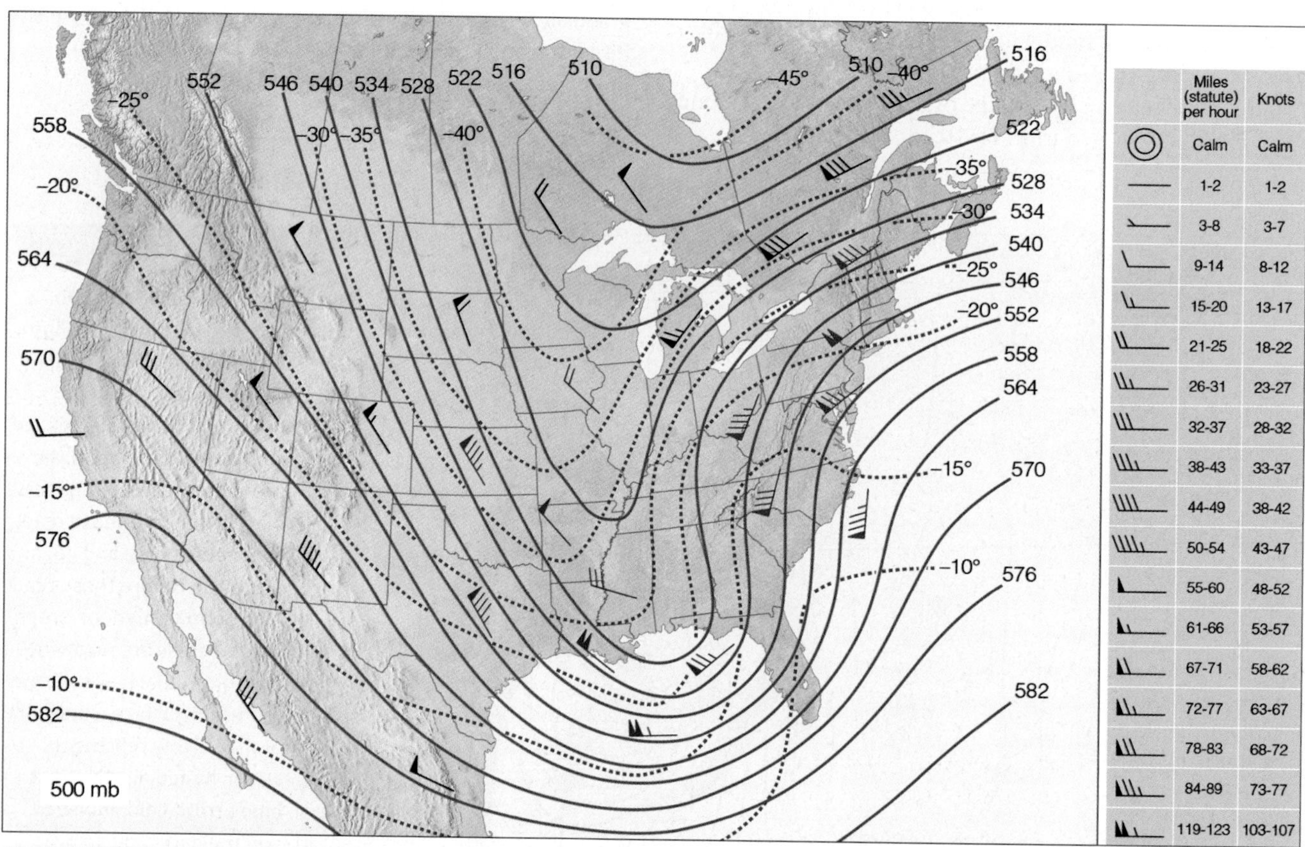

FIGURE 10.17 The 500-mb chart for 7 A.M. (EST) March 13, 1993. Solid lines are contours where 564 equals 5640 meters. Dashed lines are isotherms in °C. Wind entries in red show warm advection. Those in blue show cold advection. Those in black indicate no appreciable temperature advection is occurring.

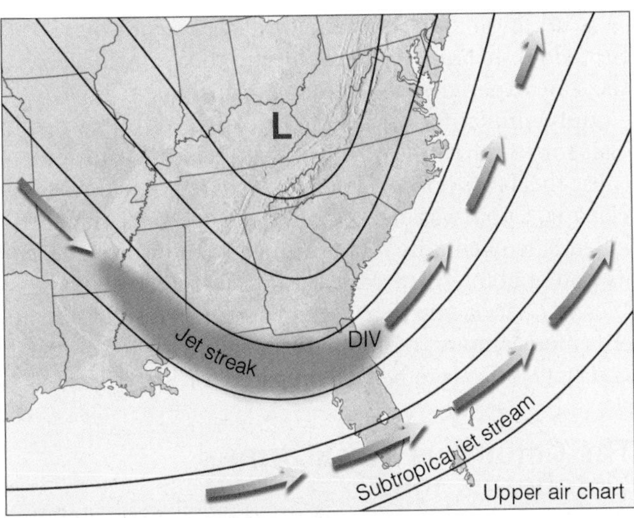

FIGURE 10.18 Air flow aloft at an altitude above 30,000 ft (10,000 m) on March 13, 1993. Notice that a jet streak (orange shade) swings over northern Florida. The letters DIV represent an area of strong divergence that formed above the surface low.

Table 10.1

Snowfall Totals for the Storm of March 12–14, 1993

CITY	AMOUNT/RECORD
Mt. LeConte, TN	60 in. State record for single storm
Mt. Mitchell, NC	50 in. Includes 36" in 24 hours, state record
Syracuse, NY	43 in. Single storm record
Seven Springs, PA	40 in. 24-hour state record
Beckley, WV	30 in. 24-hour record
Hazard, KY	25 in. 24-hour state record
Mountain City, GA	24 in. 24-hour state record
Pittsburg, PA	24 in. 24-hour record
Asheville, NC	17 in. 24-hour record
Birmingham, AL	13 in. 24-hour and single storm record
Century, FL	6 in. Unofficial 24-hour state record

streak, divergence forms over the surface low, causing surface air to rapidly rise. The rising air is then swept northeastward by the jet stream and the surface low deepens into an intense area of low pressure.

We now have a pretty good picture as to why this storm developed into such a deep low-pressure area. The storm began on March 12 as a frontal wave off the Texas coast. In the upper air, a shortwave, moving rapidly around a longwave, disturbed the flow, setting up the necessary ingredients for the surface storm's development. By the morning of March 13, the storm had intensified into a deep open wave cyclone centered over Florida. In the upper air, a region of diverging air associated with a jet streak (see Fig. 10.18) and positioned above the storm caused the storm's surface pressures to drop rapidly. Upper-level southwesterly winds, as shown in Fig. 10.18, directed the surface low northeastward, where it became occluded over Virginia during the afternoon of March 13 (look back at Fig. 10.14, p. 285). At this point, the storm's central pressure dropped to an incredibly low 960 mb (28.35 in.), a pressure comparable to a Category 3 hurricane.*

Although the surface winds were quite strong and gusty, they were not as strong as those in a Category 3 hurricane because the isobars around the storm were spread farther apart than in a hurricane and because surface friction slowed the winds. Higher up, however, away from the influence of the surface, the winds were much stronger as indicated by a wind gust of 144 mi/hr reported at the top of 6,088-foot Mount Washington, New Hampshire.

The upper trough remained to the west of the surface low, and the storm continued its northeastward movement. Look back at Fig. 10.14 and observe that by the morning of March 14 the storm (which was now a deep, occluded system) had weakened slightly and was centered along the coast of Maine. Moving out from under its area of upper-level divergence, the storm weakened even more as it continued its northeastward journey out over the North Atlantic.

This storm produced the greatest distribution of significant snow of any storm in modern times. It blanketed 50 billion tons of snow from Alabama to Canada, setting many local and state snowfall records (see Table 10.1). Fierce winds piled the snow into huge drifts that closed roads, leaving hundreds of motorists stranded (see Fig. 10.19). In addition to the snow, very cold

*As we will see in Chapter 13, a pressure of 960 mb is equivalent to the pressure in a Category 3 hurricane on the Saffir-Simpson scale, which ranges from 1 to 5, with 5 being the strongest.

EXTREME WEATHER WATCH

A devastating mid-latitude cyclonic storm with winds of 135 mi/hr swept across France and Germany during December, 1999. Dubbed Europe's "Storm of the Century," it almost completely wiped out the historic gardens of Versailles, France, where it destroyed almost 10,000 trees, some of which had stood since the French Revolution.

Table 10.2

Record Low Temperatures during the Storm of March, 1993

CITY	RECORD LOW TEMPERATURE (°F)
Caribou, ME	−12
Burlington, VT	−12
Elkins, WV	−5
Pittsburgh, PA	1
Asheville, NC	2
Birmingham, AL	2
Greensboro, NC	8
Washington, D.C.	15
New York City (JFK), NY	15
Atlanta, GA	18
Mobile, AL	21
Daytona Beach, FL	21

arctic air poured into the eastern half of the country, setting many record low temperatures (see Table 10.2). The storm produced wind gusts exceeding 100 mi/hr from Florida to Nova Scotia, Canada, cut off electricity to more than 3 million people, damaged or destroyed hundreds of homes, and set record low barometric pressure readings in a dozen states, including a record-low reading of 961 mb (28.38 in.) in White Plains, New York.

© Stephen S. Chang

FIGURE 10.19 Over 17 inches of snow covers the ground in Asheville, North Carolina, during the March, 1993, superstorm.

In addition to the above, Florida was struck by 15 tornadoes that killed 44 people, and the Gulf Coast of the state experienced a 12-foot storm surge in Taylor County, similar to what might be expected from a strong Category 3 hurricane. A tornado in Reynosa, Mexico (near the Texas border) left 5000 people homeless. Havana, Cuba, was blacked out by high winds that also killed 3 people in the city. Fortunately, the storm and its potential intensity was well forecast by the National Weather Service up to a week in advance, resulting in an enormous amount of publicity and allowing for public and state officials to be well prepared for its arrival.

The Columbus Day Storm—October 12, 1962

One of the strongest and most destructive mid-latitude cyclonic storms ever to hit Washington and Oregon roared into the Pacific Northwest on Columbus Day, October 12, 1962. The storm originated west of California, as the moist tropical remains of Typhoon Freda drifted into a region favorable for cyclonic storm development.

With upper-level support, the storm quickly deepened and moved northward along the coast. Figure 10.20 shows the storm on the morning of October 12, with the numbers on the map representing peak wind gusts at various locations. As the storm headed northward, it deepened even more, and its central pressure dropped to 960 mb (28.35 in.). The storm's extreme pressure gradient produced fierce winds, especially along the coast. Cape Blanco, Oregon, for example, reported a wind gust of 145 mi/hr and in Astoria the winds demolished a cannery as they peaked at 96 mi/hr.

Farther inland, the National Weather Service office in Portland reported a wind gust of 80 mi/hr just before a power outage left the city in darkness. In the downtown area, many windows were blown out by high winds that were unofficially measured at 116 mi/hr on the Morrison Bridge. Salem, Oregon, had a wind gust of 90 mi/hr, Eugene, 86 mi/hr, and Corvallis reported an incredible gust of 127 mi/hr.

In Seattle, Washington, the storm with its high winds hit the city after dark and temporarily closed the World's Fair. The highest wind speed reported at the National Weather Service office was 65 mi/hr, although stronger winds were reported in cities north and south of Seattle (see Fig. 10.20).

The damage from this storm covered a vast area, as the extreme winds blew down millions of trees—an estimated 10 to 15 billon board feet of timber. Hence, this storm is often referred to as the "big blow down." Downed trees closed side streets and major highways, as Highway 101 between Eureka, California, and Crescent City, California, was closed due to felled redwood trees. Farther south the storm caused millions of dollars in damage to California's wine industry, when heavy rain fell on grapes just before they were harvested. And in San Francisco, California, the storm annoyed baseball fans as the World Series game between the Giants and the Yankees was cancelled. Tragically, the storm claimed 46 lives.

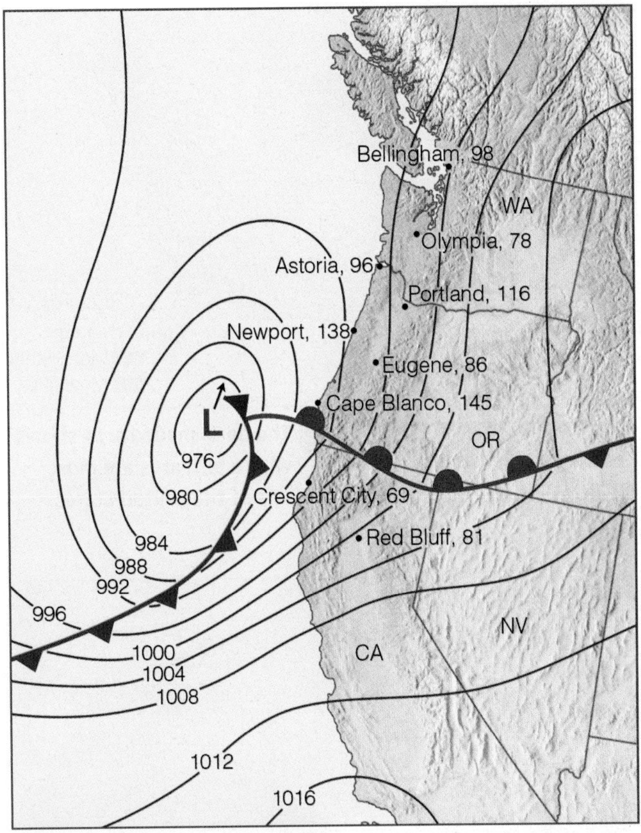

FIGURE 10.20 Surface map on the morning of October 12, 1962, shows the center of the destructive mid-latitude cyclonic storm that pounded the Pacific Northwest with extremely high winds and heavy rain, just off the coast of California and Oregon. Numbers next to the cities represent peak winds (mi/hr) during the storm.

Mid-Latitude Cyclones and Great Plains Blizzards

In Chapter 6 we learned that a *blizzard* is a storm characterized by low temperatures and strong winds bearing large amounts of snow that greatly reduces visibility. Although the word "blizzard" popularly refers to any cold snowstorm with exceptionally high winds, the United States National Weather Service only issues a blizzard warning when the following conditions are expected to exist for three hours or longer: winds of at least 35 mi/hr with falling or blowing snow that reduces visibility to less than a quarter of a mile.

In a severe blizzard, temperatures can drop well below 0°F, winds can exceed 50 mi/hr, and large quantities of falling and blowing snow can reduce visibility to near zero. Blizzards are extremely dangerous storms. Snow can pile up against a house, trapping its occupants inside (see Fig. 10.21). Cattle can freeze in their tracks. Frostbite on exposed human skin can occur in minutes. Streets and major highways can become impassable, leaving motorists stranded in their vehicles. The cost of

FIGURE 10.21

The Great Blizzard of January 26, 1978, across the upper Ohio valley and Great Lakes was a ferocious storm with winds in Ohio that gusted to over 80 mi/hr and snow drifts in some regions that covered the roofs of houses.

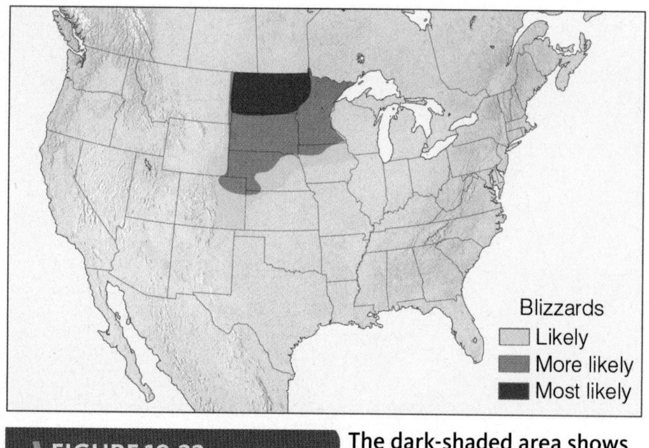

FIGURE 10.22 The dark-shaded area shows where blizzards are most likely to develop over the Great Plains of the United States.

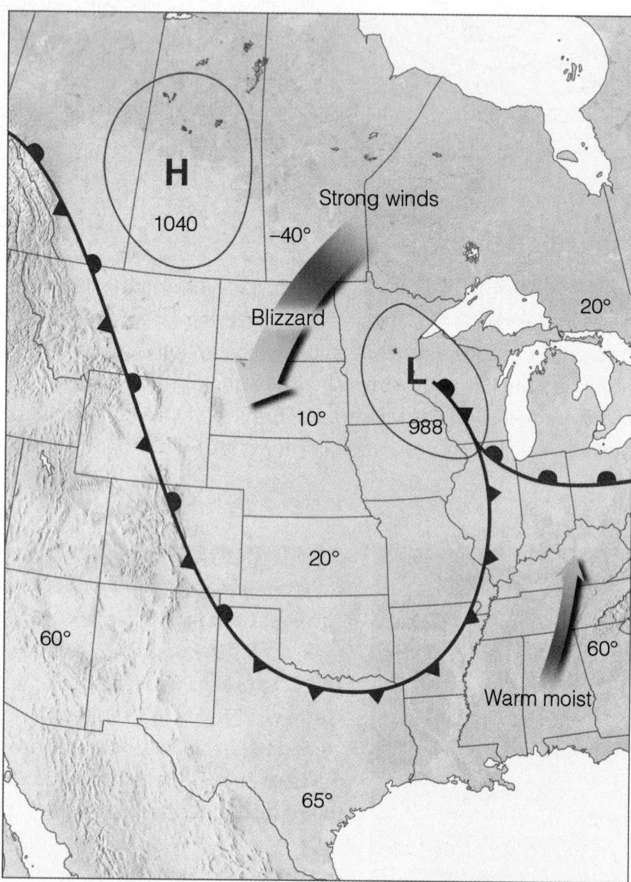

FIGURE 10.23 Blizzards often form on the northwest side of mature mid-latitude cyclonic storms, where tight pressure gradients produce strong winds and where warm, moist air, having lifted along the warm front, wraps around the northwest side of the low, where it produces heavy snow.

a severe blizzard can run into the millions of dollars, due mainly to lost livestock and damaged property.

Figure 10.22 shows the region over the Great Plains most likely to experience a blizzard. Although these storms can occur in all northern states, notice that blizzards are most common in a belt that stretches from eastern Colorado northward into South Dakota, North Dakota, and Minnesota. Blizzards are also common in southcentral Canada, in the provinces of Alberta, Saskatchewan, Manitoba, and Ontario.

All the necessary ingredients for a blizzard—wind, cold air, and snow—are wrapped up in a mature mid-latitude cyclonic storm. Figure 10.23 shows the typical surface weather conditions that produce blizzards over the Great Plains. Notice that a deep area of low pressure (a mature mid-latitude cyclone) is centered over the northern Plains with a cold front (or arctic front) extending southward. Northwest of the low exists a strong area of high pressure associated with an extremely cold arctic air mass. Blizzards tend to form to the northwest of the low's center, where steep pressure gradients between the low- and high-pressure areas produce extremely strong winds that push southward, providing the winds and cold air of the blizzard.

Moisture for the blizzard actually sweeps northward ahead of the advancing cold front. Here, warm, moist air from the Gulf of Mexico pushes northward and eventually rides up along the warm front. As the moist air rises, it wraps around the counterclockwise spinning low, producing a wrap-around band of clouds and precipitation on the low's northwest side. In winter, heavy snow can fall from this **wrap-around band**. As the snow falls into the bitter cold surface air it remains light and fluffy, and the

EXTREME WEATHER WATCH

The two deadliest blizzards in the history of the United States ironically occurred during the same year: 1888. In mid-January a ferocious storm swept across the Great Plains from Texas to the Dakotas and into Wisconsin. It became known as the "Children's Blizzard" because of the many dozens of school children frozen to death on their way to school. The blizzard took 237 lives and wiped out the Plains' free-range livestock industry. Another, and more infamous, blizzard during mid-March, raked the northeast United States, paralyzing New York City and dropping up to 50 inches of snow near Albany, New York. An estimated 400 lives were lost, some people literally frozen in their tracks in downtown Manhattan.

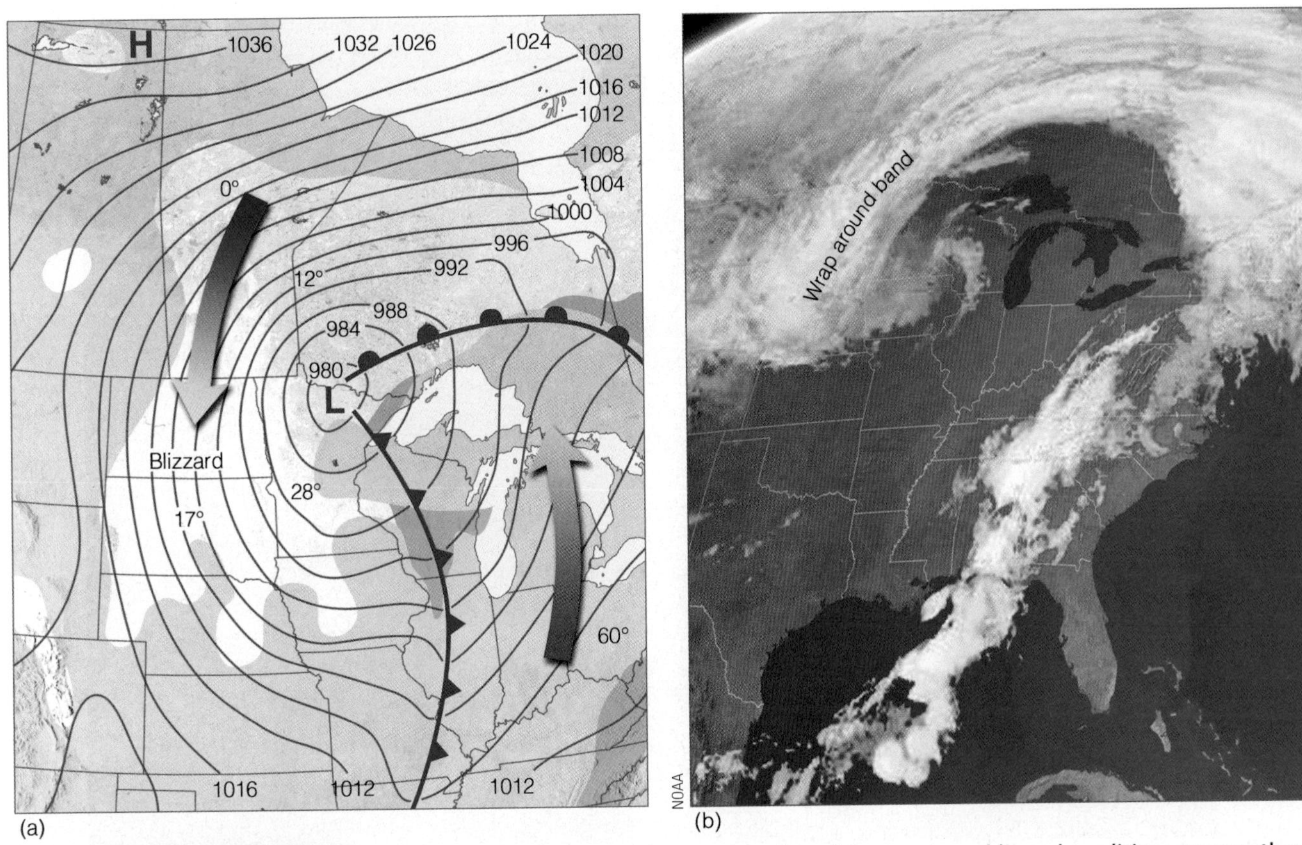

FIGURE 10.24 (a) Surface weather map for the morning of April 6, 1997, shows blizzard conditions over southern Canada, North Dakota, and South Dakota. Areas shaded white are receiving snow, while areas shaded green are receiving rain. (b) Infrared satellite image for the morning of April 6, 1997. Notice the wrap-around cloud band is over the region experiencing snow.

strong surface winds blow it about, which greatly reduces visibility. Many of the mature cyclones that produce blizzards over the Great Plains form on the eastern side of the Rockies, especially in eastern Colorado, and follow a path similar to the one shown in Fig. 10.5a on p. 275.

Figure 10.24a shows the surface weather conditions during a blizzard on the morning of April 6, 1997. The wrap-around band that brought the heavy snowfall to the region is shown in the infrared satellite image (see Fig. 10.24b). The storm dumped up to 3 feet of snow over portions of North and South Dakota. High winds, gusting to 60 mi/hr, knocked out power and piled snow into huge drifts over 20-feet high. In the Black Hills of South Dakota, Buskala Ranch reported a snowfall total of 53.1 inches while Bismarck, North Dakota, reported 17.3 inches, which brought Bismarck's seasonal snowfall total to a record 106.6 inches. In North Dakota, about 10 percent of the state's cattle were lost, and total damages from the storm exceeded $21 million.

The melting of heavy snow from this storm and from earlier storms produced historic flooding in North Dakota. In mid-to-late April, the Red River, which runs northward along the North Dakota-Minnesota border, rose to record high levels. The city of Grand Forks (population 41,000) was particularly hard hit by flooding as approximately 90 percent of the city was under water at one point and over 1000 homes were damaged or completely destroyed by flood waters (see Fig. 10.25). (Information on how to survive a blizzard is given in the Focus section on p. 293.)

Polar Lows

Up to now, we have concentrated on middle-latitude cyclones, especially those storms that form along the polar front. There are storms, however, that develop over polar water behind (or poleward of) the main polar front. Such storms are called **polar lows**. Although polar lows develop in both hemispheres, our discussion will center on those storms that form in the Northern Hemisphere in the cold polar air of the North Pacific, North Sea, and North Atlantic, especially in the region south of Iceland.

With diameters typically less than 600 miles, polar lows are generally smaller in size than their mid-latitude cousin, the wave cyclone that tends to form along the polar front. Some polar lows have a comma-shaped

AP Photo

FIGURE 10.25 The melting of snow from several blizzards during the winter of 1997 caused the Red River (which runs between North Dakota and Minnesota) to overflow its banks, flooding much of Grand Forks, North Dakota, during April, 1997.

STEVEN BUSINGER/University of Hawaii

FIGURE 10.26 An enhanced infrared satellite image of an intense polar low situated over the Norwegian Sea, north of the Arctic Circle. Notice that convective clouds swirl counterclockwise about a clear area, or eye. Surprising similarities exist between polar lows and tropical hurricanes described in Chapter 13.

cloud band. Others have a tight spiral of convective clouds that swirls counterclockwise about a clear area, or "eye," which resembles the eye of a tropical hurricane (see Fig. 10.26). In fact, like hurricanes, these smaller intense storms normally have a warmer central core, strong winds (often gale force or higher), and heavy showery precipitation that, unlike a hurricane, is in the form of snow.*

Polar lows typically form during the winter, from November through March. During this time, the sun is low on the horizon and absent for extended periods. This situation allows the air next to snow-and-ice-covered surfaces to cool rapidly and become incredibly cold. As this frigid air sweeps off the winter ice that covers much of the Arctic Ocean, it may come in contact with warmer (but still cold) air that is resting above a

*Tropical storms such as hurricanes are covered in Chapter 13. As we will see, the input of heat from the ocean surface into the hurricane increases as the wind speed increases, causing a positive feedback. Moreover, in hurricane environments, the ocean and air temperatures are about the same, whereas in the Arctic, the transfer of sensible heat from the surface is large because the air-sea temperature difference is large.

relatively warm ocean current. Where these two masses of contrasting air meet, the boundary separating them is the arctic front described in Chapter 9.

Along the arctic front, the warmer (less-dense) air rises, while the much colder (more-dense) air slowly sinks beneath it. As the warm air rises, some of its water vapor condenses, resulting in the formation of clouds and the release of latent heat, which warms the atmosphere. The warmer air has the effect of lowering the surface air pressure. Meanwhile, at the ocean surface there is a transfer of sensible heat from the relatively warm water to the cold air above. This transfer drives convective updrafts directly from the surface. In addition, it tends to destabilize the atmosphere, as heat is gained at the surface and lost to space at the top of the clouds as they radiate infrared energy upward.

FOCUS ON

EXTREME WEATHER

Blizzard Safety Tips

If you are caught in a blizzard, what do you do? If inside a stranded vehicle, *do not go outside and start walking*. The odds of surviving the ordeal are much greater if you stay inside your vehicle, away from the strong winds and blowing snow. Not only will the vehicle protect you from the extremely low wind-chill temperature, but it will also trap some of your body heat, thus keeping the inside air warmer than the outside air. Moreover, people outside in a blizzard often become disoriented by the blinding snow, and are unable to find their way back to their vehicle.

While inside your vehicle, try to keep warm by wrapping yourself in blankets or cloth seat coverings. Make sure you have several large candles. Just one lit candle can keep the inside of your car much warmer than you might think. Run your car heater for only about ten minutes each hour and make sure you roll down the windows, just a little, so that carbon monoxide—a poisonous, odorless gas—will not build up inside the vehicle. Periodically go outside and check the vehicle's tailpipe to make sure snow has not blocked it, forcing exhaust back into your vehicle. In a severe blizzard, stranded cars can be buried for hours or even days, so make sure your vehicle is visible to rescuers by: (1) turning on the vehicle's dome light at night while the engine is running, and (2) tying a piece of colored cloth (not white) to the vehicle's antenna or door.

If you are going to be traveling and severe winter weather is likely to occur, make sure you take along a number of items, including a *winter weather survival kit* that includes:

- blankets and sleeping bags
- extra clothing
- a flashlight and batteries
- a knife
- high-calorie nonperishable foods
- drinking water
- several large candles
- matches or a lighter
- a can for melting snow for extra drinking water
- a bag of sand or cat litter
- a shovel
- first aid kit
- a tow rope
- jumper cables
- a compass
- several flares
- a tool kit
- a fully charged cell phone
- a large container with a cover for sanitary purposes

If you are caught outside in a blizzard, look for any dry shelter. Cover all exposed parts of your body to help trap heat and prevent frostbite. If no dry shelter is available, look for a windbreak, such as large shrubbery or a stranded locked vehicle. Build a snowcave for protection from the wind. If possible, build a small fire, or light a large candle. The fire not only provides heat but also helps rescuers locate you. To prevent dehydration, melt snow, then drink it. Don't eat snow—it lowers your body temperature. Try to keep warm by moving about. If hypothermia starts to set in, don't fall asleep. The best advice, of course, is to make sure you obtain up-to-the-minute weather information on your NOAA weather radio so that you won't put yourself in a situation where you might be caught outside in a blizzard.

The storm's development is enhanced if an upper trough lies to the west of the surface system and a shortwave disturbs the flow aloft. Similarly, the storm may intensify if a band of maximum winds—a jet streak—moves over the surface storm and a region of upper-level divergence draws the surface air upward. The developing cyclonic storm may attain a central pressure of 980 mb (28.94 in.) or lower and produce high winds that create huge waves and hazardous seas for shipping. Generally, polar lows dissipate rapidly when they move over land.

Summary

In this chapter, we discussed where, why, and how a mid-latitude cyclone forms. We began by examining the early polar front theory proposed by Norwegian scientists after World War I. We saw that the wave cyclone goes through a series of stages from birth to maturity to finally decay as an occluded storm.

We looked at the important effect that the upper air flow has on the intensification and movement of surface mid-latitude cyclones. We saw that when an upper-level trough lies to the west of a surface low-pressure area and when a shortwave disturbs the flow aloft, horizontal and vertical air motions begin to enhance the formation of the surface storm. Aloft, a region of divergence removes air from above the surface mid-latitude cyclone. The rising of warm air and the sinking of cold air provide the proper energy conversions for the storm's growth, as potential energy is transformed into kinetic energy. A region of maximum winds—a jet streak—associated with the polar jet stream provides additional support as an area of divergence removes air above the surface mid-latitude cyclone, allowing it to develop into a deep low-pressure area. The curving nature of the jet stream tends to direct mid-latitude cyclones northeastward and anticyclones southeastward.

We looked at a number of infamous mid-latitude cyclones that produced extreme weather over a vast region. We examined storms, such as the nor'easter, that move northeastward along the coast of North America, often crippling major cities with heavy snowfall, and storms that slam into the Pacific Northwest with high winds and heavy precipitation. We looked at how mid-latitude cyclonic storms provide the necessary elements for the formation of blizzards in the upper Great Plains. Finally, we examined polar lows—intense storms that form over water in the cold air of the polar regions.

Key Terms

The following terms are listed (with page number) in the order they appear in the text. Define each. Doing so will aid you in reviewing the material covered in this chapter.

polar front theory, 270
wave cyclone, 270
warm sector, 270
mature cyclone, 270
secondary low, 271
cyclogenesis, 273
lee-side low, 273
northeasters, 273
convergence, 275
divergence, 276
longwave, 277
shortwave, 277
cold advection, 277
warm advection, 277
cut-off low, 279
jet streak, 279
conveyor belt model, 282
dry slot, 283
comma cloud, 286
wrap-around band, 290
polar lows, 291

Review Questions

1. Why are mid-latitude cyclonic storms called "extratropical cyclones"?
2. On a piece of paper, draw the different stages of a mid-latitude cyclonic storm as it goes through birth to decay according to the polar front (Norwegian) model.
3. Why do some mid-latitude cyclones quickly "die out" after they become occluded?
4. List four regions in North America where mid-latitude cyclones tend to develop or redevelop.
5. What are northeasters (nor'easters)? Why are they given that name? What type of extreme weather do nor'easters bring to northeastern North America? In Fig. 10.5a, p. 275, which of the low-pressure areas on the map could develop into nor'easters?
6. What is an Alberta Clipper? Where does it form, and how does it move?
7. Explain how converging air and diverging air forms (a) at the surface and (b) aloft.
8. Explain this fact: Without upper-level divergence, a surface open wave cyclone would probably persist for less than a day.
9. If upper-level diverging air above a surface area of low pressure exceeds converging air around the surface low, will the surface low-pressure area weaken or intensify? Explain.
10. How are longwaves in the upper-level westerlies different from shortwaves?
11. What are the necessary ingredients for a mid-latitude cyclonic storm to develop into a huge storm system?
12. Why do surface storms tend to dissipate ("fill") when the upper-level low and the surface low become vertically stacked?
13. How does cold advection differ from warm advection?
14. How does the polar jet stream influence the formation of a mid-latitude cyclone?
15. Explain why, even though the polar jet stream coincides with the polar front, some surface regions are more favorable for the development of mid-latitude cyclones than others.
16. Using a diagram, explain why a surface high-pressure area over North Dakota will typically move southeastward while, at the same time, a deep mid-latitude cyclone over the Great Lakes will generally move northeastward.
17. What are the sources of energy for a developing mid-latitude cyclone?
18. What are the roles of warm, cold, and dry conveyor belts in the development of a mid-latitude cyclonic storm system?
19. Why was the March Storm of 1993 given a Category 5 ranking on the Northeast Snowfall Impact Scale (NESIS)?
20. Describe the weather conditions necessary for a blizzard to form over the northern Great Plains.
21. On a satellite image of a mid-latitude cyclonic storm, where would you look for a wrap-around band of clouds and precipitation?
22. If you are stuck in a stranded car in a blizzard, what items would you want to have with you? Would it be wise to leave your car and seek help? Explain.
23. What are polar lows? How and where do they form? How do they produce hazardous weather for shipping?

Online Learning

STUDENT COMPANION WEBSITE: Visit this book's companion website at: www.cengage.com/ahrens/extreme1e and choose Chapter 10 for many study aids and ideas for further reading and research. These include flashcards, practice quizzing, and web links.

METEOROLOGY RESOURCE CENTER: For students with access, log on at www.cengage.com/login for more assets, including animations, videos, and more. If your textbook did not come with access, visit www.CengageBrain.com to purchase.

Thunderstorms

© Mike Hollingshead

11

CONTENTS

THUNDERSTORM DEVELOPMENT
THUNDERSTORMS AND FLOODING
DISTRIBUTION OF THUNDERSTORMS
LIGHTNING AND THUNDER
SUMMARY
KEY TERMS
REVIEW QUESTIONS

This supercell thunderstorm near York, Nebraska, on June 17, 2009, had recently produced a Tornado in Aurora, Nebraska.

It probably comes as no surprise that *thunderstorms* are merely storms containing lightning and thunder. Sometimes, thunderstorms bring cooling breezes on a hot, muggy day; other times, they bring unwelcome high winds, heavy rain, and large hail. A thunderstorm may be a single cumulonimbus cloud, or several thunderstorms may form into a cluster. In some cases, a line of thunderstorms will form and extend for hundreds of miles. In this chapter, we will examine the many aspects of thunderstorms and the various types of extreme weather they produce.

Thunderstorm Development

Thunderstorms are *convective storms* that form with rising air. So the birth of a thunderstorm often begins when warm, moist air rises in a conditionally unstable environment.* The rising air may be a parcel of air ranging in size from a large balloon to a city block, or an entire layer, or slab of air, may be lifted. As long as a rising air parcel is warmer (less dense) than the air surrounding it, there is an upward-directed *buoyant force* acting on it. The warmer the parcel is compared to the air surrounding it, the greater the buoyant force and the stronger the convection. The trigger (or "forcing mechanism") needed to start air moving upward may be:

1. unequal heating at the surface
2. the effect of terrain, or the lifting of air along shallow boundaries of converging surface winds
3. diverging upper-level winds, coupled with converging surface winds and rising air
4. warm air rising along a frontal zone

Usually, several of these mechanisms work together with vertical wind shear to generate severe thunderstorms.

Although we often see thunderstorms forming where the surface air is quite buoyant (that is, warm and humid), they may also form when the surface air temperature is no more than 50°F (10°C). This latter situation often occurs in winter along the west coast of North America, when cold air aloft moves over the region. The cold air aloft destabilizes the atmosphere to the point where air parcels, given an initial push upward, are able to continue their upward journey because they remain warmer (less dense) than the colder air surrounding them. The cold air aloft may even produce sufficient instability to generate thunderstorms in wintertime snowstorms.

*A conditionally unstable atmosphere exists when cold, dry air aloft overlies warm, moist surface air. Additional information on atmospheric instability is given in Chapter 5 beginning on p. 128.

Most thunderstorms that form over North America are short-lived, produce rain showers, gusty surface winds, thunder and lightning, and sometimes small hail. Many have an appearance similar to the mature thunderstorm shown in ❯Fig. 11.1. The majority of these storms do not reach severe status. *Severe thunderstorms* are defined by the National Weather Service as having at least one of the following: large hail with a diameter of at least three-quarters of an inch and/or surface wind gusts of 58 mi/hr (50 knots) or greater, or produces a tornado.

Scattered thunderstorms (sometimes called "pop-up" storms) that typically form on warm, humid days are often referred to as *ordinary cell thunderstorms** or *air-mass thunderstorms* because they tend to form in warm, humid air masses away from significant weather fronts. Ordinary cell (air mass) thunderstorms can be considered "simple storms" because they rarely become severe, typically are less than one-half mile wide, and they go through a rather predictable life cycle from birth to maturity to decay that usually takes less than an hour to complete. However, under the right atmospheric conditions (described later in this chapter), more intense "complex thunderstorms" may form, such as the *multicell thunderstorm* and the *supercell thunderstorm*—a huge rotating storm that can last for hours and produce severe weather such as strong surface winds, large damaging hail, flash floods, and violent tornadoes.

We will examine the development of ordinary cell (air mass) thunderstorms first, before we turn our attention to the more complex multicell and supercell storms.

ORDINARY CELL THUNDERSTORMS

Ordinary cell (air mass) thunderstorms or, simply, *ordinary thunderstorms*, tend to form in a region where there is limited wind shear—that is, where the wind speed and wind direction *do not* abruptly change with increasing height above the surface. Many ordinary thunderstorms appear to form as parcels of air are lifted from the surface by turbulent overturning in the presence of wind. Moreover, ordinary storms often form along shallow zones where surface winds converge. Such zones may be due to any number of things, such as topographic irregularities, sea-breeze fronts, or the cold outflow of air from inside a thunderstorm that reaches the ground and spreads horizontally. These converging wind boundaries are normally zones of contrasting air temperature and humidity and, hence, air density.

Extensive studies indicate that ordinary thunderstorms go through a cycle of development from birth to maturity to decay. The first stage is known as the **cumulus stage**, or *growth stage*. As a parcel of warm, humid air rises, it cools and condenses into a single

*In convection, the cell may be a single updraft or a single downdraft, or a combination of the two.

cumulus cloud or a cluster of clouds (see Fig. 11.2a). If you have ever watched a thunderstorm develop, you may have noticed that at first the cumulus cloud grows upward only a short distance, then it dissipates. The top of the cloud dissipates because the cloud droplets evaporate as the drier air surrounding the cloud mixes with it. However, after the water drops evaporate, the air is more moist than before. So, the rising air is now able to condense at successively higher levels, and the cumulus cloud grows taller, often appearing as a rising dome or tower.

As the cloud builds, the transformation of water vapor into liquid or solid cloud particles releases large quantities of latent heat, a process that keeps the rising air inside the cloud warmer (less dense) than the air surrounding it. The cloud continues to grow in the

FIGURE 11.1

An ordinary thunderstorm in its mature stage. Note the distinctive anvil top.

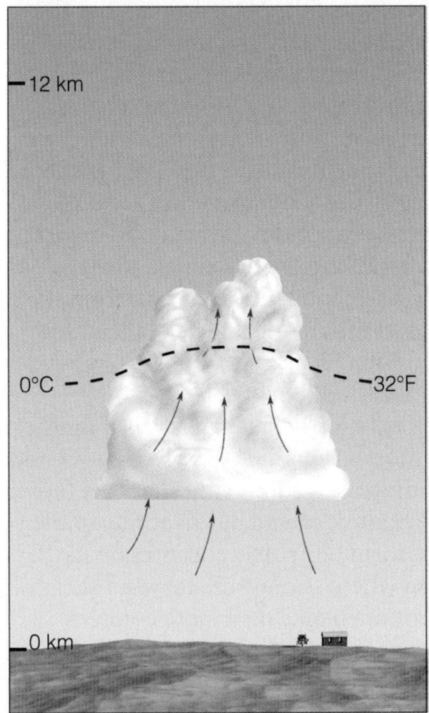

(a) Cumulus

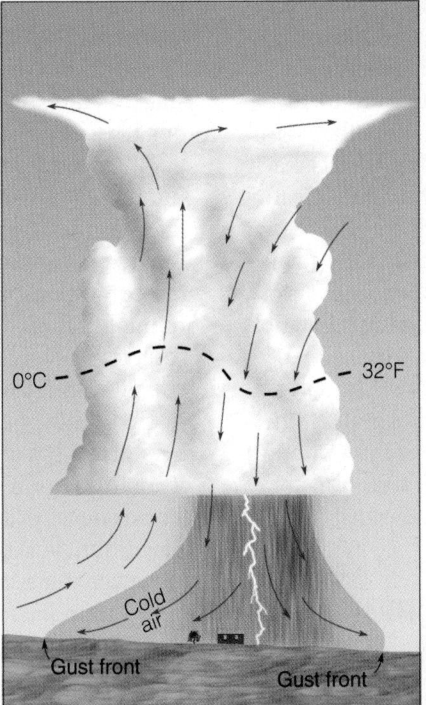

(b) Mature

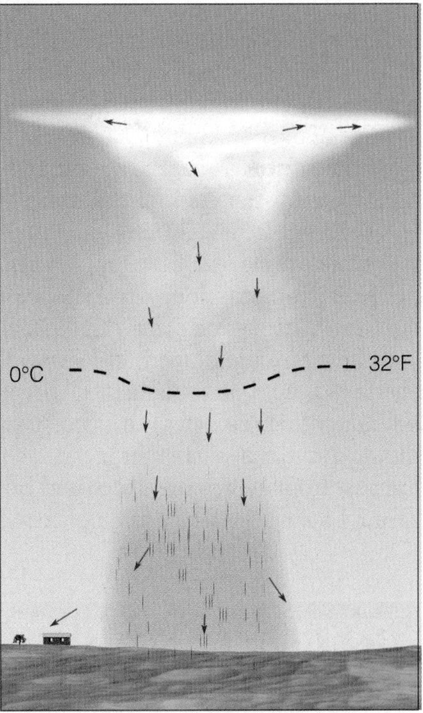

(c) Dissipating

FIGURE 11.2

Simplified model depicting the life cycle of an ordinary cell thunderstorm that is nearly stationary as it forms in a region of low wind shear. (Arrows show vertical air currents. Dashed line represents freezing level, 32°F isotherm.)

unstable atmosphere as long as it is constantly fed by rising air from below. In this manner, a cumulus cloud may show extensive vertical development and grow into a towering cumulus cloud (cumulus congestus) in just a few minutes. During the cumulus stage, there normally is insufficient time for precipitation to form, and the updrafts keep water droplets and ice crystals suspended within the cloud. Also, there is no lightning or thunder during this stage.

As the cloud builds well above the freezing level, the cloud particles grow larger and heavier as they collide and join with one another. Eventually, the rising air is no longer able to keep them suspended, and they begin to fall. While this phenomenon is taking place, drier air from around the cloud is being drawn into it in a process called *entrainment*. The entrainment of drier air causes some of the raindrops to evaporate, which chills the air. The air, now colder and heavier than the air around it, begins to descend as a **downdraft**. The downdraft may be enhanced as falling precipitation drags some of the air along with it.

The appearance of the downdraft marks the beginning of the **mature stage**. The downdraft and updraft within the mature thunderstorm now constitute the cell. In some storms, there are several cells, each of which may last for less than 30 minutes.

During its mature stage, the thunderstorm is most intense. The top of the cloud, having reached a stable region of the atmosphere (which may be the stratosphere), begins to take on the familiar anvil shape, as upper-level winds spread the cloud's ice crystals horizontally (see Fig. 11.2b). The cloud itself may extend upward to an altitude of over 12 km (40,000 ft) and be more than a mile in diameter near its base. Updrafts and downdrafts reach their greatest strength in the middle of the cloud, creating severe turbulence. Lightning and thunder are also present in the mature stage. Heavy rain (and occasionally small hail) falls from the cloud. And, at the surface, there is often a downrush of cold air with the onset of precipitation.

Where the cold downdraft reaches the surface, the air spreads out horizontally in all directions. The surface boundary that separates the advancing cooler air from the surrounding warmer air is called a *gust front*. Along the gust front, winds rapidly change both direction and speed. Look at Fig. 11.2b and notice that the gust front forces warm, humid air up into the storm, which enhances the cloud's updraft. In the region of the downdraft, rainfall may or may not reach the surface, depending on the relative humidity beneath the storm. In the dry air of the desert Southwest, for example, a mature thunderstorm may look ominous and contain all of the ingredients of any other storm, except that the raindrops evaporate before reaching the ground. However, intense downdrafts from the storm may reach the surface, producing strong, gusty winds and a gust front.

After the storm enters the mature stage, it begins to dissipate in about 15 to 30 minutes. The **dissipating stage** occurs when the updrafts weaken as the gust front moves away from the storm and no longer enhances the updrafts. At this stage, as illustrated in Fig. 11.2c, downdrafts tend to dominate throughout much of the cloud. The reason the storm does not normally last very long is that the downdrafts inside the cloud tend to cut off the storm's fuel supply by destroying the humid updrafts. Deprived of the rich supply of warm, humid air, cloud droplets no longer form. Light precipitation now falls from the cloud, accompanied by only weak downdrafts. As the storm dies, the lower-level cloud particles evaporate rapidly, sometimes leaving only the cirrus anvil as the reminder of the once mighty presence (see ❱ Fig. 11.3). A single ordinary cell thunderstorm may go through its three stages in one hour or less.

Not only do these thunderstorms produce summer rainfall for a large portion of the United States, but they also bring with them momentary cooling after an oppressively hot day. The cooling comes during the mature stage, as the downdraft reaches the surface in the form of a blast of welcome relief. Sometimes, the air temperature may lower as much as 18°F (10°C) in just a few minutes. Unfortunately, the cooling effect often is short-lived, as the downdraft diminishes or the thunderstorm moves on. In fact, after the storm has ended, the air temperature usually rises; and as the moisture from the rainfall evaporates into the air, the humidity increases, sometimes to a level where it actually feels more oppressive after the storm than it did before.

Up to this point, we've looked at ordinary cell thunderstorms that are short-lived, rarely become severe, and form in a region with weak vertical wind shear. As these storms develop, the updraft eventually gives way to the downdraft, and the storm ultimately collapses on itself. However, in a region where strong vertical wind shear exists, thunderstorms often take on a more complex structure. Strong, vertical wind shear can cause the storm to tilt in such a way that it becomes a multicell thunderstorm—a thunderstorm with more than one cell.

EXTREME WEATHER WATCH

On July 13, 1999, in Sattley, California, a strong downdraft from a mature thunderstorm dropped the air temperature from 97°F at 4:00 P.M. to a chilly 57°F one hour later.

FIGURE 11.3 A dissipating thunderstorm near Naples, Florida. Most of the cloud particles in the lower half of the storm have evaporated.

MULTICELL THUNDERSTORMS Thunderstorms that contain a number of cells, each in a different stage of development, are called **multicell thunderstorms** (see Fig. 11.4). Such storms tend to form in a region of moderate-to-strong vertical wind speed shear. Look at Fig. 11.5 and notice that on the left side of the illustration the wind speed increases rapidly with height, producing strong wind speed shear. This type of shearing causes the cell inside the storm to tilt in such a way that the updraft actually rides up and over the downdraft. Note that the rising updraft is capable of generating new cells that go on to become mature thunderstorms. Notice also that precipitation inside the storm does not fall into the updraft (as it does in the ordinary cell thunderstorm), so the storm's fuel supply is not cut off and the storm complex can survive for a long time. Because the likelihood that a thunderstorm will become severe increases with the length of time the storm exists, long-lasting multicell storms can become intense and produce severe weather.

When convection is strong and the updraft intense (as it is in Fig. 11.5), the rising air may actually intrude well into the stable stratosphere, producing an **overshooting top**. As the air spreads laterally into the anvil, sinking air in this region of the storm can produce beautiful mammatus clouds. At the surface, below the thunderstorm's cold downdraft, the cold, dense air (called a *cold pool*) may cause the surface air pressure to rise—sometimes several millibars. The relatively small, shallow area of high pressure is called a *mesohigh* (meaning "mesoscale high").

The Gust Front When the cold downdraft reaches the earth's surface, it pushes outward in all directions, producing a strong **gust front** that represents the leading edge of the cold outflowing air (see Fig. 11.6). Gust fronts can be over 1000 feet deep and move with speeds of between 10 and 30 mi/hr. To an observer on the ground, the passage of the gust front resembles that of a cold front. During its passage, the temperature drops sharply and the wind shifts and becomes strong

FIGURE 11.4

This multicell storm complex is composed of a series of cells in successive stages of growth. The thunderstorm in the middle is in its mature stage, with a well-defined anvil. Heavy rain is falling from its base. To the right of this cell, a thunderstorm is in its cumulus stage. To the left, a well-developed cumulus congestus cloud is about ready to become a mature thunderstorm. With new cells constantly forming, the multicell storm complex can exist for hours.

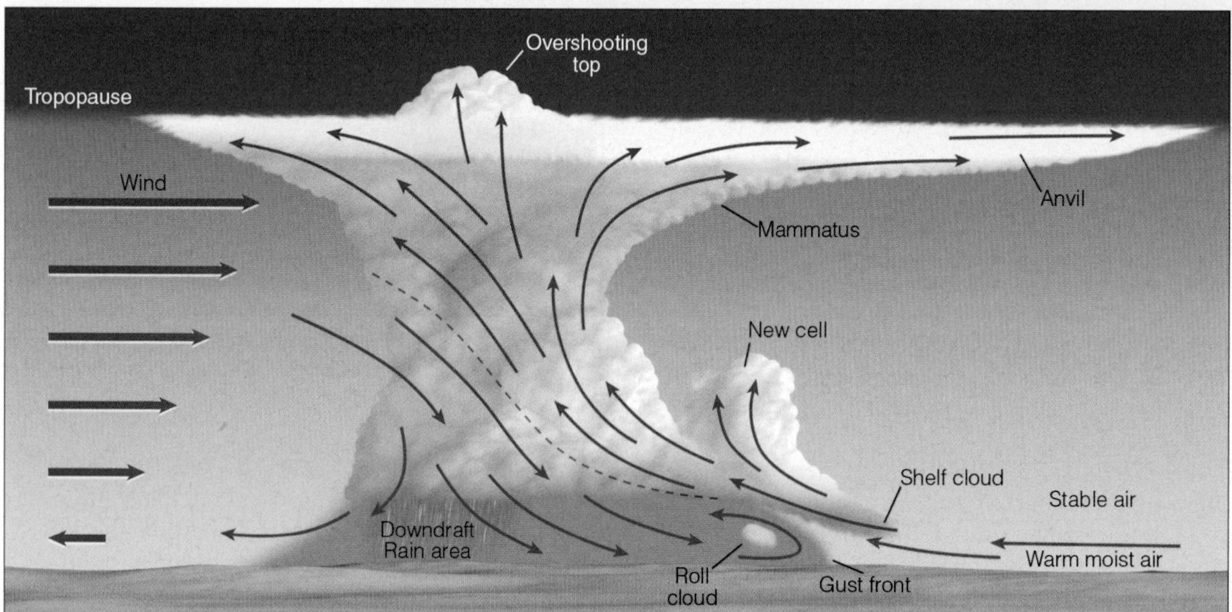

Active FIGURE 11.5 A simplified model describing air motions and other features associated with an intense multicell thunderstorm that has a tilted updraft. The severity depends on the intensity of the storm's circulation pattern. Visit the Meteorology Resource Center to view this and other active figures at www. cengage.com/login

and gusty, with speeds occasionally exceeding 60 mi/hr. These high winds behind a strong gust front are called **straight-line winds** to distinguish them from the rotating winds of a tornado. As we will see later in this chapter, straight-line winds are capable of inflicting a great deal of damage such as blowing down trees and overturning mobile homes.

Along the leading edge of the gust front, the air is quite turbulent. Here, strong winds can pick up loose dust and soil and lift them into a huge tumbling cloud (see Fig. 11.7).* The cold surface air behind the gust front may even linger close to the ground for hours, well after thunderstorm activity has ceased.

As warm, moist air rises along the forward edge of the gust front, a **shelf cloud** (also called an *arcus cloud*) may form, such as the one shown in Fig. 11.8. These clouds are especially prevalent when the atmosphere is very stable near the base of the thunderstorm. Look

*In dry, dusty areas or desert regions, the leading edge of the gust front is the haboob described in Chapter 8, p. 232.

again at Fig. 11.5 and notice that the shelf cloud is attached to the base of the thunderstorm. Occasionally, an elongated ominous-looking cloud forms just behind the gust front. These clouds, which appear to slowly spin about a horizontal axis, are called **roll clouds** (see ❱ Fig. 11.9).

When the atmosphere is conditionally unstable, the leading edge of the gust front may force the warm, moist air upward, producing a complex of multicell storms, each with new gust fronts. These gust fronts may then merge into a huge gust front called an **outflow boundary**. Along the outflow boundary, air is forced upward, often generating new thunderstorms (see ❱ Fig. 11.10). Thus, a single multicell thunderstorm can spawn a series of storms that move across the landscape.

Microbursts Beneath an intense thunderstorm, the downdraft may become localized so that it hits the ground and spreads horizontally in a radial burst of wind, much like water pouring from a tap and striking the sink below. (Look at the downdraft in Fig. 11.6.) Such downdrafts are called **downbursts**. A relatively small downburst (where winds extend for no more than 2.5 miles) is called a **microburst**. In spite of its relatively small size, an intense microburst can induce damaging straight-line winds as high as 150 mi/hr. ❱ Figure 11.11 shows the dust clouds generated from a microburst north of Denver, Colorado. Since a microburst is an intense downdraft, its leading edge can evolve into a gust front.

Microbursts are capable of blowing down trees and inflicting heavy damage upon poorly built structures as well as upon sailing vessels that encounter microbursts over open water. In fact, microbursts may be responsible for some damage once attributed to tornadoes. Moreover, microbursts and their accompanying *wind shear* (that is, rapid changes in wind speed and wind direction) appear

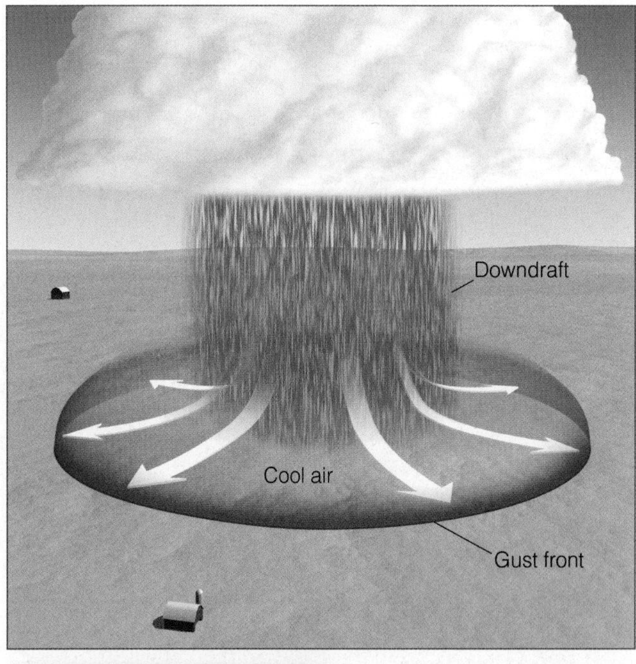

❱ **FIGURE 11.6** **When a thunderstorm's downdraft reaches the ground, the air spreads out, forming a gust front.**

❱ **FIGURE 11.7**

A swirling mass of dust forms along the leading edge of a gust front as it moves across western Nebraska.

FIGURE 11.8 A dramatic example of a shelf cloud (or arcus cloud) associated with an intense thunderstorm. The photograph was taken in the Philippines as the thunderstorm approached from the northwest.

FIGURE 11.9 A roll cloud forming behind a gust front.

to be responsible for several airline crashes. When an aircraft flies through a microburst at a relatively low altitude, say 1000 ft above the ground, it first encounters a headwind that generates extra lift. This is position (a) in ❯ Fig. 11.12. At this point, the aircraft tends to climb (it gains lift), and if the pilot noses the aircraft downward there could be grave consequences, for in a matter of seconds the aircraft encounters the powerful downdraft (position b), and the headwind is replaced by a tail wind (position c). This situation causes a sudden loss of lift and a subsequent decrease in the performance of the aircraft, which is now accelerating toward the ground.

One accident attributed to a microburst occurred north of Dallas–Fort Worth Regional Airport during August, 1985. Just as an aircraft was making its final approach, it encountered severe wind shear beneath a small but intense thunderstorm. The aircraft then dropped to the ground and crashed, killing over 100 passengers. To detect the hazardous wind shear associated with microbursts, many major airports use a high resolution Doppler radar. The radar uses algorithms that are computer programmed to detect microbursts and low-level wind shear.

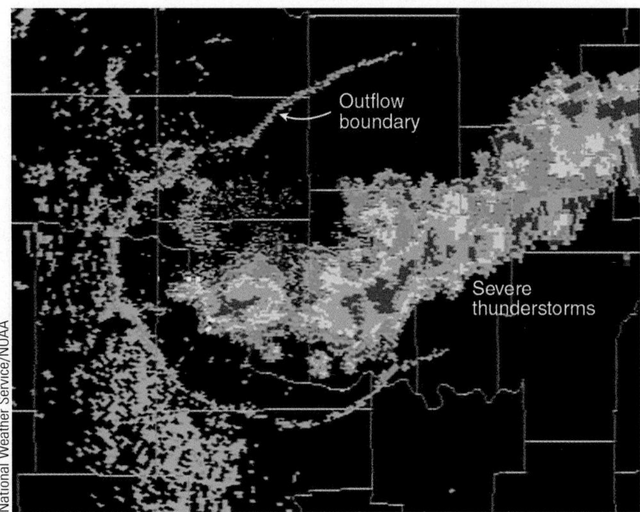

National Weather Service/NOAA

❯ FIGURE 11.10 Radar image of an outflow boundary. As cool (more-dense) air from inside the severe thunderstorms (red and orange colors) spreads outward, away from the storms, it comes in contact with the surrounding warm, humid (less-dense) air, forming a density boundary (blue line) called an *outflow boundary* between cool air and warm air. Along the outflow boundary, new thunderstorms often form.

© C. Donald Ahrens

❯ FIGURE 11.11 Dust clouds rising in response to the outburst winds of a microburst north of Denver, Colorado.

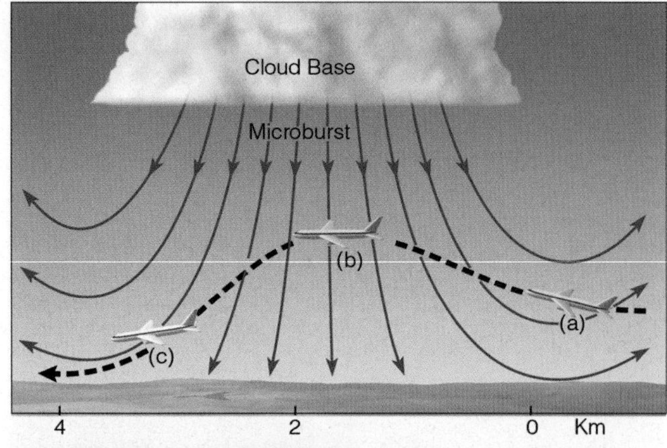

FIGURE 11.12 Flying into a microburst. At position (a), the pilot encounters a headwind; at position (b), a strong downdraft; and at position (c), a tailwind that reduces lift and causes the aircraft to lose altitude.

The leading edge of a microburst can contain an intense horizontally rotating vortex that is often filled with dust in a relatively dry region. In eastern Colorado, many microbursts emanate from virga—rain falling from a cloud but evaporating before reaching the ground. Apparently, in these "dry" microbursts (Fig. 11.11), evaporating rain cools the air. The cooler heavy air then plunges downward through the warmer lighter air below. In humid regions, many microbursts are "wet" in that they are accompanied by blinding rain.

Microbursts can be associated with severe thunderstorms, producing strong, damaging winds. But studies show that they can also occur with ordinary cell thunderstorms and with clouds that produce only isolated showers—clouds that may or may not contain thunder and lightning.

Heat Bursts Up to this point, you might think that thunderstorm downdrafts are always cool. Most are cool, but occasionally they can be extremely hot. For example, during the evening of May 22, 1996, in the town of Chickasha, Oklahoma, a blast of hot, dry air from a dissipating thunderstorm raised the surface air temperature from 88°F to 102°F in just 25 minutes. Such sudden warm downbursts are called **heat bursts**.

Apparently, the heat burst originates high up in the thunderstorm and warms by compressional heating as it plunges toward the surface. In cold downbursts, the descending air is colder than its surroundings as a result of the evaporation of the falling precipitation. Heat bursts, however, most likely contain little precipitation during most of their descent and probably form during the thunderstorm's dissipating stage, after most of the precipitation has fallen from the cloud.

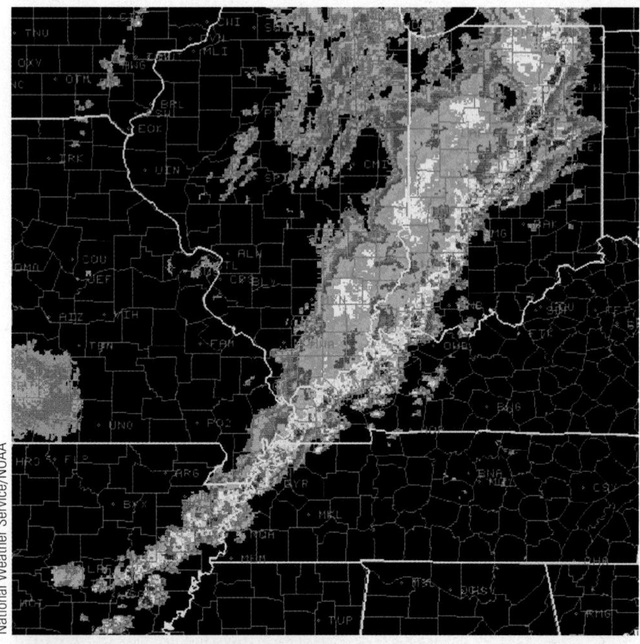

National Weather Service/NOAA

FIGURE 11.13 A Doppler radar composite showing a pre-frontal squall line extending from Indiana southwestward into Arkansas. Severe thunderstorms (red and orange colors) associated with the squall line produced large hail and high winds during October, 2001.

SQUALL-LINE THUNDERSTORMS A *squall* is a "violent burst of wind" and a **squall line** is an active line of thunderstorms that often produce strong gusty winds. A squall line may form as a line of thunderstorms extending for hundreds of miles *directly along* a cold front; or it may form as a line of thunderstorms in the warm air 50 to 200 miles out *ahead* of the advancing front. These *pre-frontal squall-line* thunderstorms of the middle latitudes represent some of the largest and most severe types of squall lines, with huge thunderstorms* often causing severe weather over much of its entire length (see Fig. 11.13). Squall lines may also form when thunderstorms organize into a cluster of storms that move in tandem across the landscape.

As warm, humid air rises along a cold front, it is easy to see how a line of thunderstorms could form along the advancing front. But what causes the air to rise so that thunderstorms are able to form many miles ahead of the advancing cold front? Models that simulate the formation of pre-frontal squall line thunderstorms suggest that initially, convection begins along the cold front, then reforms farther away. Moreover, the surging nature of the main cold front itself, or developing cumulus

*Within a squall line there may be multicell thunderstorms, as well as supercell storms—violent thunderstorms that contain a single rapidly rotating updraft. We will look more closely at supercells in the next section.

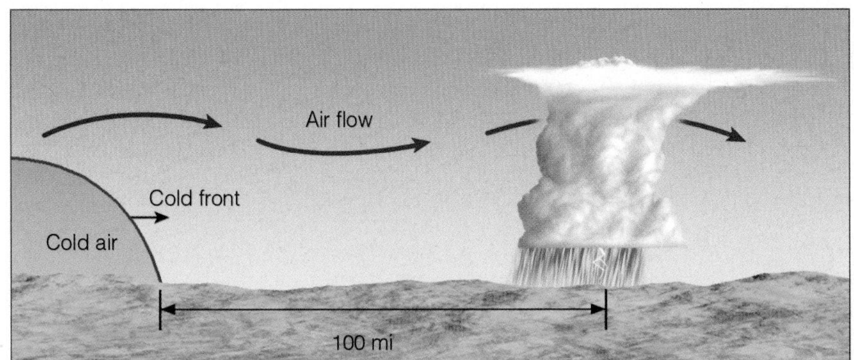

FIGURE 11.14

Pre-frontal squall-line thunderstorms may form ahead of an advancing cold front as the upper-air flow develops waves downwind from the cold front.

FIGURE 11.15

A model describing air motions and precipitation associated with a squall line that has a trailing stratiform cloud layer.

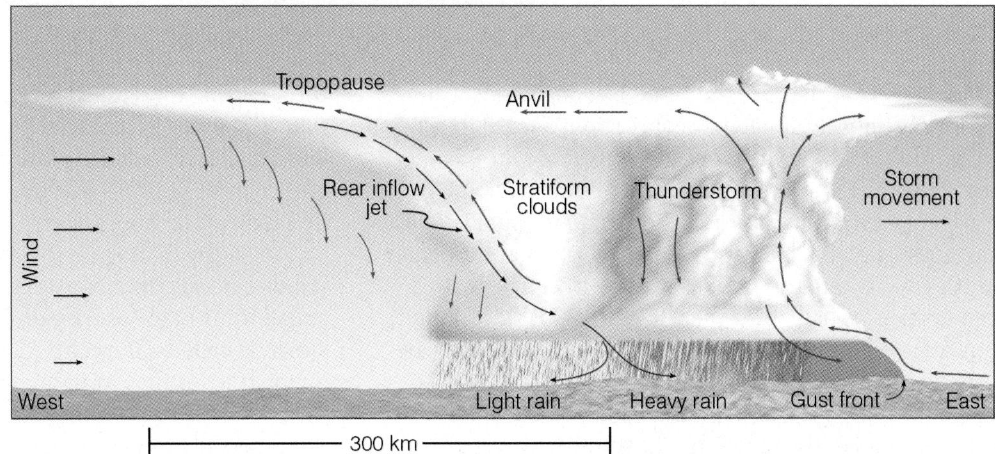

clouds along the front, may cause the air aloft to develop into waves called *gravity waves* (see Fig. 11.14), much like the waves that form downwind of a mountain chain. Out ahead of the cold front, the rising motion of the wave may be the trigger that initiates the development of cumulus clouds and a pre-frontal squall line. In some instances, low-level converging air is better established out ahead of the advancing cold front. In still other instances, an upper-level front may precede a surface front and initiate uplift, similar to the example shown in Fig. 9.31 on p. 264 in Chapter 9.

Rising air along the frontal boundary (and along the gust front), coupled with the tilted nature of the updraft, promotes the development of new cells as the storm moves along. Hence, as old cells decay and die out, new ones constantly form, and the squall line can maintain itself for hours on end. Occasionally, a new squall line will actually form out ahead of the front as the gust front pushes forward, beyond the main line of storms.

Squall lines that exhibit weaker updrafts and downdrafts tend to be more shallow than pre-frontal squall lines, and usually they have shorter life spans. These storms are referred to as *ordinary squall lines*. Severe weather may occur with them, but more typically they form as a line of thunderstorms that exhibit characteristics of ordinary cell thunderstorms. Ordinary squall lines may form along a gust front, with a stationary front, with a weak wave cyclone, or where no large-scale cyclonic storms are present. Many of the ordinary squall lines that form in the middle latitudes exhibit a structure similar to squall lines that form in the tropics.

In some squall lines, the leading area of thunderstorms and heavy precipitation is followed by a region of extensive stratified clouds and light precipitation (see Fig. 11.15). The stratiform clouds represent a region where the anvil cloud trails behind the thunderstorm.

The downdraft to the rear of the storm in Fig. 11.15 forms as some of the falling precipitation evaporates and chills the air. The heavy cooler air then descends, dragging some of the surrounding air with it. If the cool air rapidly descends, it may concentrate into a rather narrow band of fast-flowing air called the *rear inflow jet*. Sometimes the rear inflow jet will bring with it the stong upper-level winds from aloft. Should these winds reach the surface, they rush outward producing damaging *straight-line winds* that may exceed 100 mi/hr (see Fig. 11.16).

As the strong winds rush forward along the ground, they sometimes push the squall line outward so that it appears as a *bow* (or a series of bows) on the radar screen. Such a bow-shaped squall line is called a **bow echo** (see

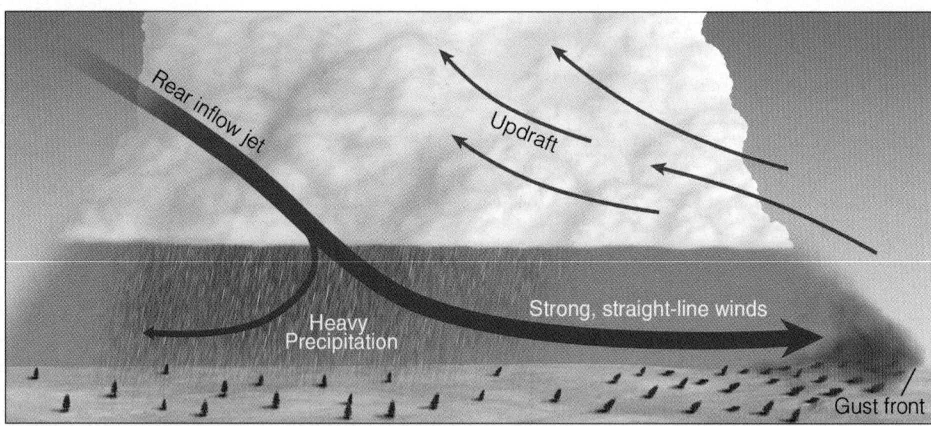

FIGURE 11.16

A side view of the lower half of a squall-line thunderstorm with the rear inflow jet carrying strong winds from high altitudes down to the surface. These strong winds push forward along the surface, causing damaging straight-line winds that may exceed 100 mi/hr. If the high winds extend horizontally for a considerable distance, the wind storm is called a *derecho*.

Fig. 11.17). Sometimes the rush of strong winds will produce relatively small bows only about 5 to 10 miles long (*mini-bows*). If the wind shear ahead of the advancing squall line is strong, much larger bows (over 100 miles long) may form, similar to the bow echo in Fig. 11.17. The strongest straight-line winds tend to form near the center of the bow, where the sharpest bending occurs. Tornadoes can form, especially near the left (northern) side of the bow where cyclonic rotation often develops, but they are usually small and short-lived.

Look at Fig. 11.17 closely and observe that at the northern end of the bow echo the thunderstorms are organizing into a region of cyclonic rotation. In this region, the release of large quantities of latent heat during cloud formation leads to the development of a small area of low pressure that spins counterclockwise. Because the size of the spinning low (called a *vortex*) is relatively small—less than 150 miles in diameter—the spinning low and its band of clouds is called a **mesoscale convective vortex**. Notice in Fig. 11.18 that the vortex becomes more well defined several hours later.

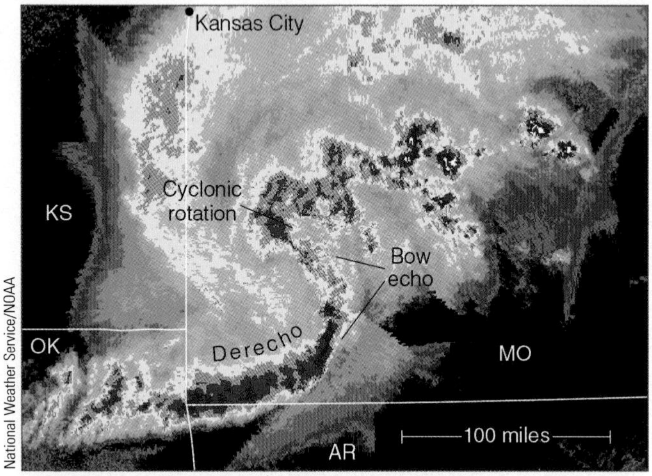

National Weather Service/NOAA

FIGURE 11.17 **A Doppler radar image showing an intense squall line in the shape of a bow—called a *bow echo*—moving eastward across Missouri on the morning of May 8, 2009. The strong thunderstorms (red and orange in the image) are producing damaging straight-line winds over a wide area. Damaging straight-line wind that extends for a good distance along a squall line is called a *derecho*.**

When the damage associated with the straight-line winds extends for a considerable distance along the squall line's path (as they do in Fig. 11.17), the windstorm is called a **derecho** (day-ray-sho), after the Spanish word for "straight ahead." Typically, derechoes form in the early evening and last throughout the night. An especially powerful derecho roared through New York State during the early morning of July 15, 1995, where it blew down millions of trees in Adirondack State Park. In an average year about 20 derechoes occur in the United States. During July, 2005, two derechoes within three days moved through the St. Louis, Missouri, metro area. With winds gusting to over 80 mi/hr, they downed trees and power lines all across the region, leaving half a million residents without power.

It is common for the damaging effects of a derecho to be attributed to a tornado. The degree of damage from a derecho can be extensive, just as with a tornado (see Fig. 11.19). However, with a derecho, debris is blown in one direction and generally over a wide area, whereas debris with a tornado is usually thrown in many directions. Moreover, tornado winds typically cut a circular swath through an area.

Squall lines are one type of convective phenomenon called a *Mesoscale Convective System* (*MCS*). Squall lines come under this heading because they are driven by convective processes and because they are mesoscale (middle scale) in size.[*] Mesoscale convective systems are organized thunderstorms that can take on a variety of configurations, from the elongated squall line, to the circular mesoscale convective vortex, to the much larger *Mesoscale Convective Complex* described in the next section.

[*]You may recall from Chapter 1, p. 19, that a mesoscale (middle scale) system typically ranges in size from a few miles up to about 100 miles.

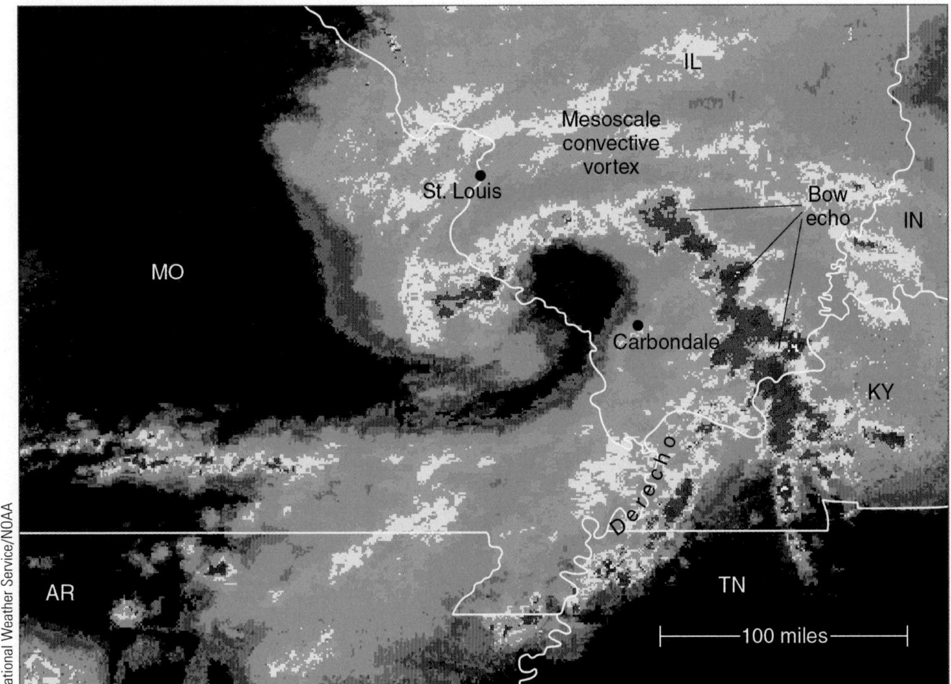

FIGURE 11.18

Doppler radar image showing that the bow echo in Fig. 11.17 has developed into a *mesoscale convective vortex* by noon on May 8, 2009. Strong, straight-line winds are still occurring with severe thunderstorms, as Carbondale, Illinois, reported an unofficial wind gust of 106 mi/hr.

FIGURE 11.19 A building near Hudson Oaks, Texas, experiences the damaging effects of a derecho on June 1, 2004. The windstorm produced extensive damage over a wide swath across north Texas and Louisiana.

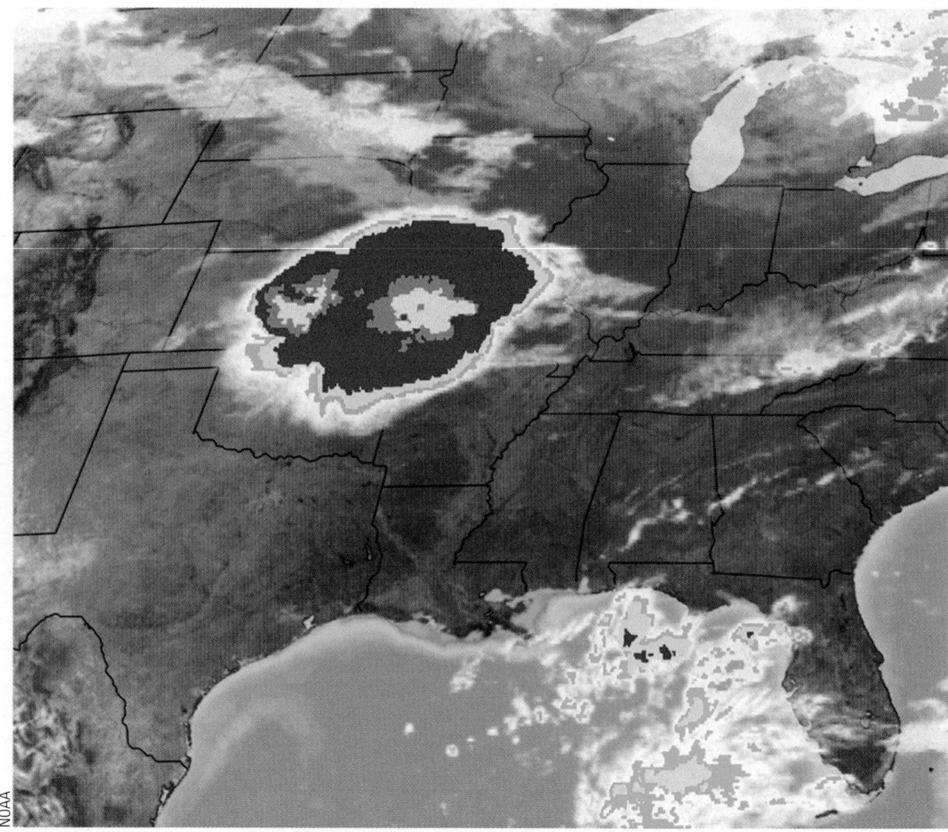

FIGURE 11.20

An enhanced infrared satellite image showing the cold cloud tops (dark red and orange colors) of a Mesoscale Convective Complex extending from central Kansas across western Missouri. This organized mass of multicell thunderstorms brought hail, heavy rain, and flooding to this area.

MESOSCALE CONVECTIVE COMPLEXES Where conditions are favorable for convection, a number of individual multicell thunderstorms may occasionally grow in size and organize into a large circular convective weather system. These convectively driven systems, called **Mesoscale Convective Complexes (MCCs)**, are quite large—they can be as much as 1000 times larger than an individual ordinary cell thunderstorm. In fact, they are often large enough to cover an entire state, an area in excess of 100,000 square kilometers (see Fig. 11.20).

Within the MCCs, the individual thunderstorms apparently work together to generate a long-lasting (more than 6 hours) weather system that moves slowly (normally less than 20 mi/hr) and often exists for periods exceeding 12 hours. Thunderstorms that comprise MCCs support the growth of new thunderstorms as well as a region of widespread precipitation. These systems are beneficial, as they provide a significant portion of the growing season rainfall over much of the corn and wheat belts of the United States. However, MCCs can also produce a wide variety of severe weather, including hail, high winds, destructive flash floods, and tornadoes.

Mesoscale Convective Complexes tend to form during the summer in regions where the upper-level winds are weak, which is often beneath a ridge of high pressure. If a weak cold front should stall beneath the ridge, surface heating and moisture may be sufficient to generate thunderstorms on the cool side of the front. Often moisture from the south is brought into the system by a low-level jet stream often found within 5000 ft of the surface. In addition, the low-level jet can provide shearing so that multicell storms can form. Most MCCs reach their maximum intensity in the early morning hours, which is partly due to the fact that the low-level jet reaches its maximum strength late at night or in the early morning. Moreover, at night, the cloud tops cool rapidly by emitting infrared energy to space. Gradually, the atmosphere destabilizes as a vast amount of latent heat is released in the lower and middle part of the clouds. Within the multicell storm complex new thunderstorms form as older ones dissipate. With only weak upper-level winds, most MCCs move southeastward very slowly.

Figure 11.21 shows an MCC moving slowly south-southeastward into Kentucky. Notice that the leading edge of the complex is the region of most active convection and strongest thunderstorms. Here, the thunderstorms have actually organized into a squall line. Behind the squall line is a region of stratified-type clouds (called a *stratified region*) where mostly less-intense rainfall is falling over a wide area.

BRIEF REVIEW

In the last several sections, we examined different types of thunderstorms. Before we look at supercell storms, here is a list of some of the important concepts we covered so far:

- All thunderstorms need three basic ingredients: (1) moist surface air; (2) a conditionally unstable atmosphere; and (3) a mechanism "trigger" that forces the air to rise.
- Ordinary cell (air mass) thunderstorms tend to form where warm, humid air rises in a conditionally unstable atmosphere and where vertical wind shear is weak. They are usually short-lived and go through their life cycle of growth (cumulus stage), maturity (mature stage), and decay (dissipating stage) in less than an hour. They rarely produce severe weather.
- An ordinary cell thunderstorm dies because its downdraft falls into the updraft, which cuts off the storm's fuel supply.
- As wind shear increases (and the winds aloft become stronger), multicell thunderstorms are more likely to form as the storm's updraft rides up and over the downdraft. The tilted nature of the storm allows new cells to form as old ones die out.
- Multicell storms often form as a complex of storms, such as the squall line (a long line of thunderstorms that form along or out ahead of a frontal boundary) and the Mesoscale Convective Complex (a large circular cluster of thunderstorms).
- The stronger the convection and the longer a multistorm system exists, the greater the chances of the thunderstorm becoming severe.
- A gust front, or outflow boundary, represents the leading edge of cool air that originates inside a thunderstorm, reaches the surface as a downdraft, and moves outward away from the thunderstorm.
- Strong downdrafts of a thunderstorm—called downbursts (or microbursts if the downdrafts are smaller than 2.5 mi [4 km])—have been responsible for several airline crashes, because upon striking the surface, these winds produce extreme wind shear—rapid changes in wind speed and wind direction.
- A derecho is a strong straight-line wind produced by strong downdrafts from intense thunderstorms that often appear as a bow (bow echo) on a radar screen.

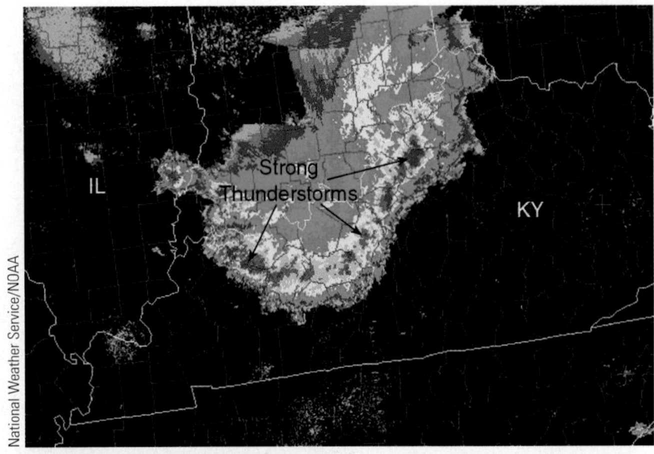

National Weather Service/NOAA

FIGURE 11.21 A Doppler radar image of a Mesoscale Convective Complex moving south-southeastward into Kentucky. The areas of red and orange represent active convection and strong thunderstorm activity, whereas the green and blue represent light precipitation.

SUPERCELL THUNDERSTORMS In a region where there is strong vertical wind shear (both speed and direction shear), the thunderstorm may form in such a way that the outflow of cold air from the downdraft never undercuts the updraft. In such a storm, the wind shear may be so strong as to create horizontal spin, which, when tilted into the updraft, causes it to rotate. A large, long-lasting thunderstorm with a single violently rotating updraft is called a **supercell**.* It is the rotating aspect of the supercell that can lead to the formation of tornadoes.

Figure 11.22 shows a supercell near Sioux City, Iowa. The internal structure of a supercell is organized in such a way that the storm may maintain itself as a single entity for hours. Storms of this type are capable of producing an updraft that may exceed 100 mi/hr, damaging surface winds, and large tornadoes. Violent updrafts keep hailstones suspended in the cloud long enough for them to grow to considerable size—sometimes to the size of grapefruits. Once they are large enough, they may fall out the bottom of the cloud with the downdraft, or the violent spinning updraft may whirl them out the side of the cloud or even from the base of the anvil. Aircraft have actually encountered hail in clear air several miles from a storm. In some cases, the top of the storm may extend to as high as 60,000 ft (18 km) above the surface, and the width of the storm may exceed 25 mi (40 km).

Although no two supercells are exactly alike, for convenience they are often divided into three types.

*Smaller thunderstorms that occur with rotating updrafts are referred to as *mini supercells.*

© Mike Hollingshead

FIGURE 11.22 A supercell thunderstorm near Sioux City, Iowa, on May 28, 2004.

Classic (CL) supercells, for example, are well balanced storms that produce heavy rain, large hail, high surface winds, and the majority of tornadoes. The classic supercell serves as an excellent model for all supercells, and is the one normally shown in diagrams. Supercells that produce heavy precipitation and large hail, which appears to fall in the center of the storm, are called *HP supercells*, (for *H*igh *P*recipitation). Such storms often produce extreme downdrafts (downbursts) and flash flooding. If tornadoes are present, it is often difficult to see them, as they tend to form in the area of heavy precipitation. A supercell characterized by little precipitation is referred to as an *LP supercell* (for *L*ow *P*recipitation). These storms, which are capable of producing tornadoes and large hail, often have a vertical tower that, due to the storm's rotation, resembles a corkscrew.

A model of a classic supercell with many of its features is given in Fig. 11.23. In the diagram, we are viewing the storm from the southeast, and the storm is moving from southwest to northeast. The rotating air column on the south side of the storm, usually 3 to 6 mi (about 5 to 10 km) across, is called a **mesocyclone** (meaning "mesoscale cyclone"). The rotating updraft associated with the mesocyclone is so strong that precipitation cannot fall through it. This situation produces a rain-free area (called a *rain-free base*) beneath the updraft. Observe that the increasing wind speed with increasing height above the surface produces strong wind speed shear.

Strong southwesterly winds aloft usually blow the precipitation northeastward. Notice that large hail, having remained in the cloud for some time, usually falls just north of the updraft, and the heaviest rain occurs just north of the falling hail, with the lighter rain falling in the northeast quadrant of the storm. If low-level humid air is drawn into the updraft, a rotating cloud, called

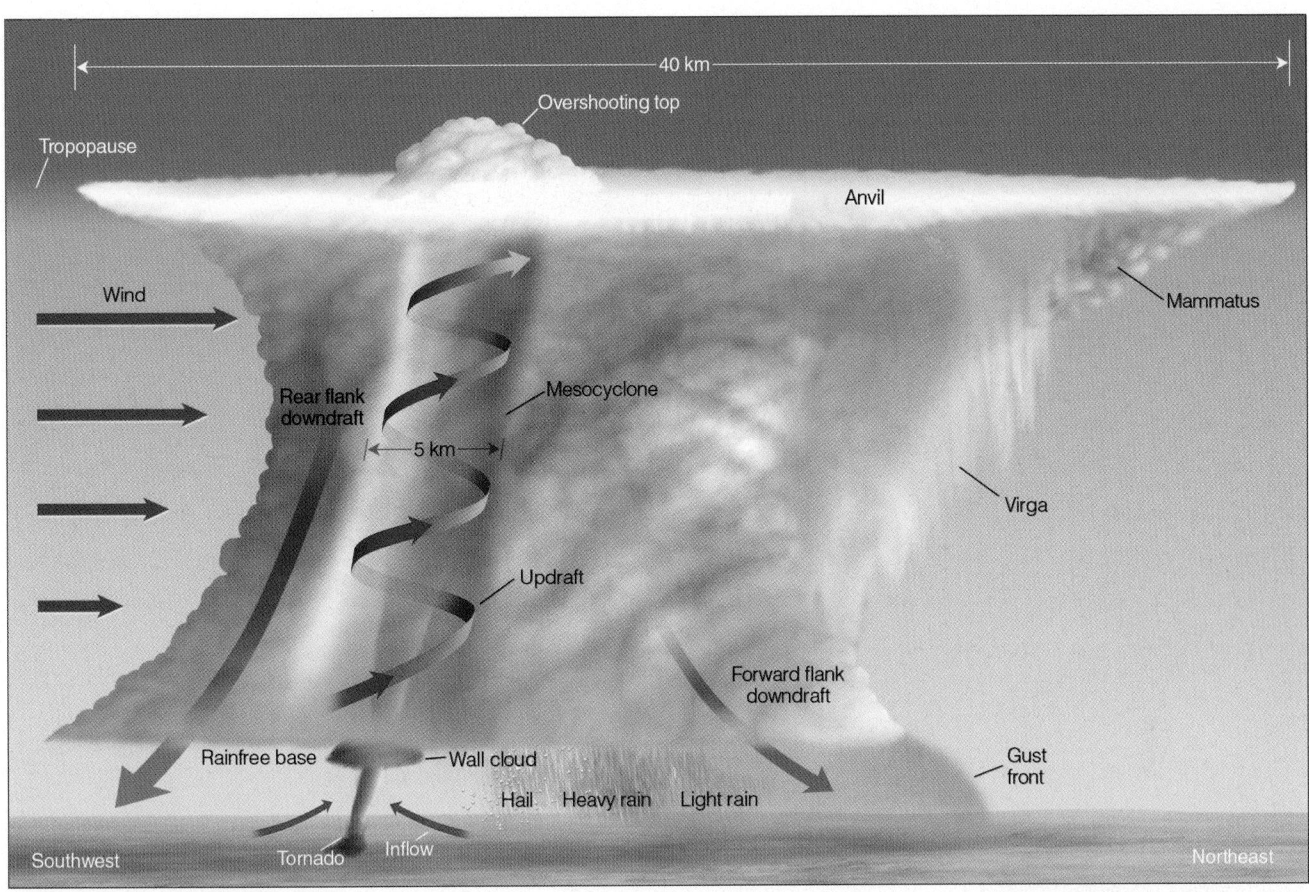

FIGURE 11.23 Some of the features associated with a classic tornado-breeding supercell thunderstorm as viewed from the southeast. The storm is moving to the northeast.

a **wall cloud**, may descend from the base of the storm (see Fig. 11.24). Notice also in Fig. 11.23 that mammatus clouds often form in the anvil and that the forward flank downdraft produces a strong gust front.

We can obtain a better picture of how wind shear plays a role in the development of supercell thunderstorms by observing Fig. 11.25. The illustration represents atmospheric conditions during the spring over the Central Plains. At the surface, we find an open-wave middle-latitude cyclone with cold, dry air moving in behind a cold front, and warm humid air pushing northward from the Gulf of Mexico behind a warm front. Above the warm surface air, a wedge or "tongue" of warm, moist air is streaming northward. It is in this region we find a relatively narrow band of strong winds called the *low-level jet*.* Winds in the low-level jet may exceed 55 mi/hr. Directly above the moist layer is a wedge of cooler, drier air moving in from the southwest.

*The low-level jet often forms at night, but it can form at any time along the front due to strong pressure gradients found there.

Higher up, at the 500-mb level (about 18,000 ft above the surface), a trough of low pressure exists to the west of the surface low. At the 300-mb level (about 30,000 ft above the ground), the polar front jet stream swings over the region, often with an area of maximum wind (a *jet streak*) above the surface low. At this level, the jet stream provides an area of divergence that enhances surface convergence and rising air. The stage is now set for the development of supercell thunderstorms.

The light yellow area on the surface map (Fig. 11.25) shows where supercells are likely to form. They tend to form in this region because: (1) the position of cold air above warm air produces a conditionally unstable atmosphere; (2) this is a region where warm humid air is being lifted from the surface; and (3) strong vertical wind shear induces rotation.

Rapidly increasing wind speed just above the surface provides strong wind speed shear. Within this region, wind shear causes the air to spin about a horizontal axis. You can obtain a better idea of this spinning by placing a pen (or pencil) in your left hand, parallel to the

FIGURE 11.24 A wall cloud photographed southwest of Norman, Oklahoma.

table. Now take your right hand and push it over the pen away from you. The pen rotates much like the air rotates. If you tilt the spinning pen into the vertical, the pen rotates counterclockwise from the perspective of looking down on it. A similar situation occurs with the rotating air. As the spinning air rotates counterclockwise about a horizontal axis, an updraft from a developing thunderstorm can draw the spinning air into the cloud, causing the updraft to rotate. It is this rotating updraft that is characteristic of all supercells. The increasing wind speed with height up to the 300-mb level, coupled with the changing wind direction with height from more southerly at low levels to more westerly at high levels, further induces storm rotation.*

Ahead of the advancing cold front, we might expect to observe many supercells forming as warm, conditionally unstable air rises from the surface. Often, however, numerous supercells do not form as the atmospheric conditions that promote the formation of large supercell thunderstorms tend to prevent many smaller ones from forming. To see why, we need to examine the vertical profile of temperature and moisture—a sounding—in the warm air ahead of the advancing cold front.

Figure 11.26 shows a typical sounding of temperature and dew point in the warm air before supercells form. From the surface up to 800 mb—in a layer of air perhaps 6000 ft (2000 m) thick—the air is warm, very humid, and conditionally unstable. At 800 mb, a shallow inversion (or simply a very stable layer), has formed due to descending air from aloft. This stable layer acts like a *cap* (or *lid*) on the humid air below. Above the stable layer, the air is cold and much drier, as indicated by the low dew point temperatures. The cooling of this upper layer is due, mainly, to cold air moving in from the west, as illustrated in Fig. 11.25. This situation where cold, dry air overlies warm, humid air produces a type of atmospheric instability called *convective instability,* which means that the atmosphere will destabilize even more if a layer of air is somehow forced to rise. In addition, the warm, humid air beneath the capping stable layer represents potential energy for the thunderstorm. This potential energy is converted into kinetic energy when parcels of air are lifted and form into clouds.

*As we will see later in Chapter 12, it is this rotation that sets the stage for tornado development.

The lifting of warm surface air can occur at the frontal zones, but the air may also begin to rise anywhere in the region of warm air when the surface air heats up during the day. However, in the morning, the stable layer acts as a lid on rising thermals and only small cumulus clouds form. As the day progresses (and the surface air heats even more), rising air breaks through the stable layer at isolated places and clouds build rapidly, sometimes explosively, as the moist air is vented upward through the opening. Thus, we can see that the stable layer prevents many small thunderstorms from forming. When the surface air is finally able to puncture the stable layer, a jet streak associated with the upper-level jet stream (at the 300-mb level) rapidly draws the moist air up into the cold unstable air, and a large supercell quickly develops to great height.

Violent thunderstorms have very strong updrafts. A measure of how much energy is available to produce these updrafts is the *C*onvective *A*vailable *P*otential *E*nergy (*CAPE*). We know that thunderstorms begin as parcels of air rise from the surface, eventually become saturated and condense into clouds. CAPE is a measure of how rapidly an air parcel will rise inside a cloud (its positive buoyancy) when the parcel becomes warmer than the air surrounding it. The higher the value of CAPE, the more likely a supercell will form. For example, a value of CAPE ranging from 0 to 500 joules*/kg means that there is only a marginal chance for the development of strong thunderstorms, whereas values greater than 3500 indicate that the environment is ripe for strong convection (strong updrafts) and the formation of supercell thunderstorms.

Most thunderstorms move roughly in the direction of the winds in the middle troposphere. However, most supercell storms are *right-movers*; that is, they move to the right of the steering winds aloft. These right-movers tend to move about 30 degrees to the right of the mean wind in the middle troposphere. Apparently, the rapidly rising air of the storm's updraft interacts with increasing horizontal winds that change direction with height (from more southerly to more westerly) in such a way that vertical pressure gradients are able to generate new updrafts on the right side of the storm. Hence, as the storm moves along, new cells form to the right of the winds aloft.**

THUNDERSTORMS AND THE DRYLINE Thunderstorms may form along or just east of a boundary called a *dryline*. Recall from Chapter 9 that the dryline

*Recall from Chapter 2 that a joule is a unit of energy where one joule equals 0.239 calories.

**Some thunderstorms move to the left of the steering winds aloft. This movement may happen as a thunderstorm splits into two storms, with the northern half of the storm often being a left-mover, and the southern half a right-mover.

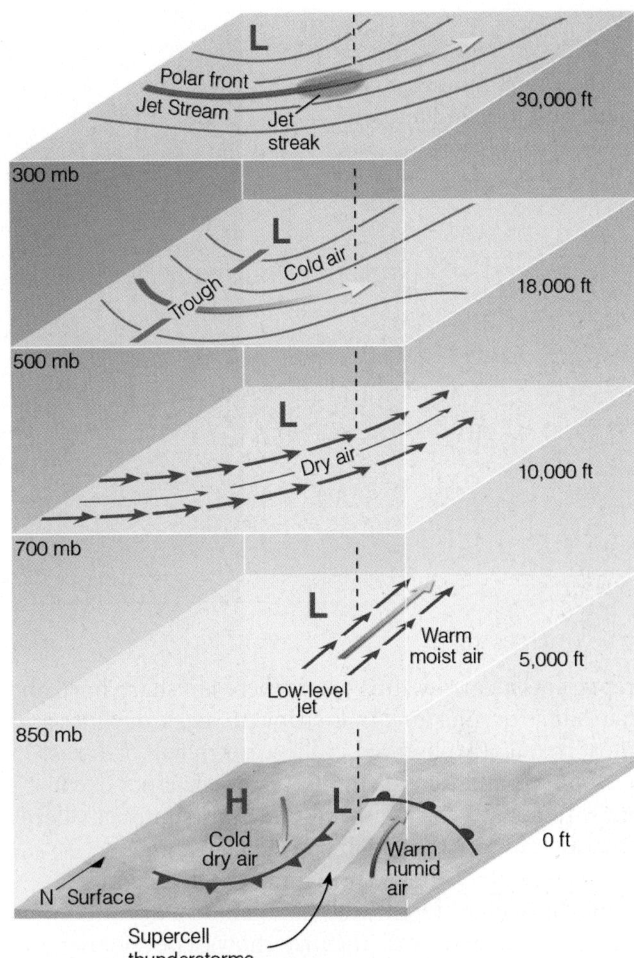

FIGURE 11.25 Conditions leading to the formation of severe thunderstorms, especially supercells. The area in yellow is where supercell thunderstorms are likely to form.

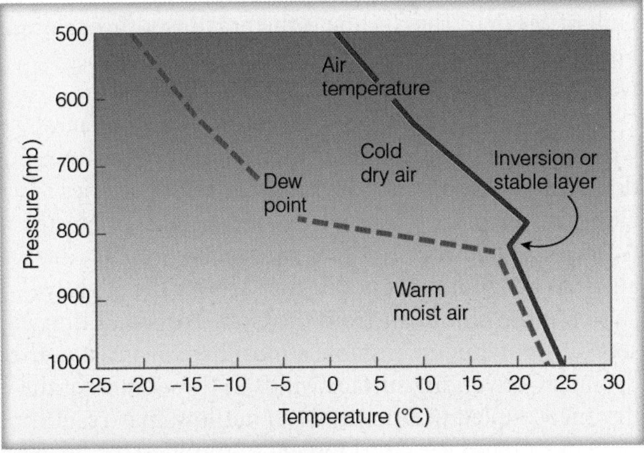

FIGURE 11.26 A typical sounding of air temperature and dew point that frequently precedes the development of supercell thunderstorms.

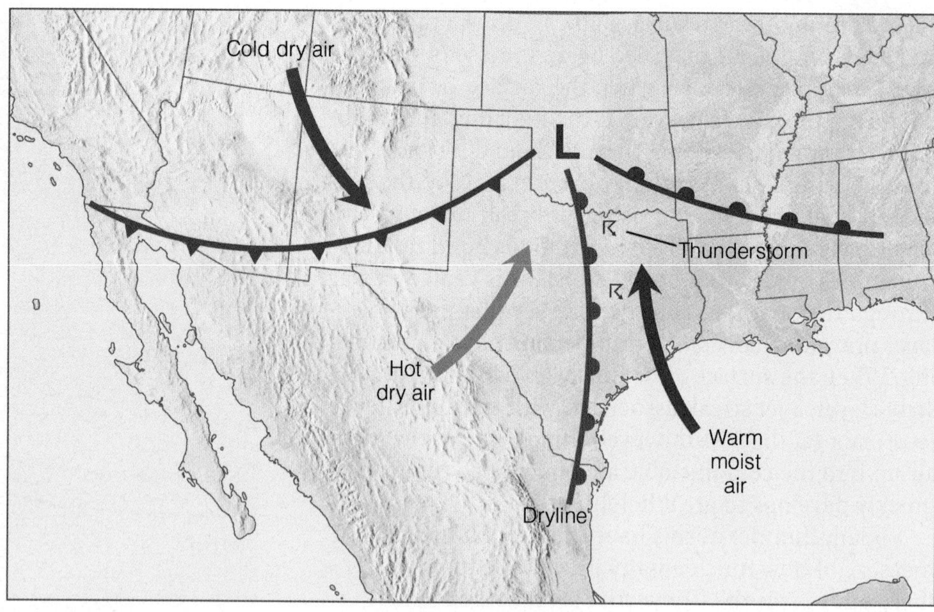

FIGURE 11.27

Surface conditions that can produce a dryline with intense thunderstorms.

represents a narrow zone where there is a sharp horizontal change in moisture. In the United States, drylines are most frequently observed in the western half of Texas, Oklahoma, and Kansas. In this region, drylines occur most frequently during spring and early summer, where they are observed about 40 percent of the time.

Figure 11.27 shows springtime weather conditions that can lead to the development of a dryline and intense thunderstorms. The map shows a developing mid-latitude cyclone with a cold front, a warm front, and three distinct air masses. Behind the cold front, cold dry continental polar air or modified cool dry Pacific air pushes in from the northwest. In the warm air, ahead of the cold front, warm dry continental tropical air moves in from the southwest. Farther east, warm but very humid maritime tropical air sweeps northward from the Gulf of Mexico. The dryline is the north–south-oriented boundary that separates the warm, dry air and the warm, humid air.

Along the cold front—where cold, dry air replaces warm, dry air—there is insufficient moisture for thunderstorm development. The moisture boundary lies along the dryline. Because the Central Plains of North America are elevated to the west, some of the hot, dry air from the southwest is able to ride over the slightly cooler, more humid air from the Gulf. This condition sets up a potentially unstable atmosphere just east of the dryline. Converging surface winds in the vicinity of the dryline, coupled with upper-level outflow, may result in rising air and the development of thunderstorms. As thunderstorms form, the cold downdraft from inside the storm may produce a blast of cool air that moves along the ground as a gust front and initiates the uplift necessary for generating new (possibly more severe) thunderstorms.

Thunderstorms and Flooding

Although organized thunderstorms, such as squall lines and Mesoscale Convective Complexes, produce much of the welcome summer rain that falls over the Great Plains of North America, these same storms are quite capable of producing floods and flash floods. Such flooding often results where thunderstorms either stall or move very slowly over an area. Flooding may also result when the storm moves quickly, but keeps passing over the same region—a situation called *training.*

Flooding can also occur in mountain valleys when storms overflow their banks after heavy downpours. This situation happened in Colorado's Big Thompson Canyon on July 31, 1976, and is illustrated in the Focus section on p. 317.

Distribution of Thunderstorms

It is estimated that more than 50,000 thunderstorms occur each day throughout the world. Hence, over 18 million occur annually. The combination of warmth and moisture make equatorial landmasses especially conducive to thunderstorm formation. Here, thunderstorms occur on about one out of every three days (see Fig. 11.28). Thunderstorms are also prevalent over water along the intertropical convergence zone, where the low-level convergence of air helps to initiate uplift. The heat energy liberated in these storms helps the earth maintain its heat balance by distributing heat poleward.

FOCUS ON
EXTREME WEATHER

A Terrifying Flash Flood

July 31, 1976, was like any other summer day in the Colorado Rockies, as small cumulus clouds with flat bases and dome-shaped tops began to develop over the eastern slopes near the Big Thompson and Cache La Poudre rivers. At first glance, there was nothing unusual about these clouds, as almost every summer afternoon they form along the warm mountain slopes. Normally, strong upper-level winds push them over the plains, causing rainshowers of short duration. But the cumulus clouds on this day were different. For one thing, they were much lower than usual, indicating that the southeasterly surface winds were bringing in a great deal of moisture. Also, their tops were somewhat flattened, suggesting that an inversion aloft was stunting their growth. But these harmless-looking clouds gave no clue that later that evening in the Big Thompson Canyon more than 135 people would lose their lives in a terrible flash flood.

By late afternoon, a few of the cumulus clouds were able to puncture the inversion. Fed by moist southeasterly winds, these clouds soon developed into gigantic multicell thunderstorms with tops exceeding 60,000 ft (18 km). By early evening, these same clouds were producing incredible downpours in the mountains.

In the narrow canyon of the Big Thompson River, some places received as much as 12 inches of rain in the four hours between 6:30 P.M. and 10:30 P.M. local time. This is an incredible amount of precipitation, considering that the area normally receives about 16 inches for an entire year. The heavy downpours turned small creeks into raging torrents, and the Big Thompson River was quickly filled to capacity. Where the canyon narrowed, the river overflowed its banks and water covered the road. The relentless pounding of water caused the road to give way.

Figure 1 This car is one of more than 400 destroyed by flood waters in the Big Thompson Canyon on July 31, 1976.

Soon cars, tents, mobile homes, resort homes, and campgrounds were being claimed by the river (see Fig. 1). Where the debris entered a narrow constriction, it became a dam. Water backed up behind it, then broke through, causing a wall of water to rush downstream.

Figure 2 shows the weather conditions during the evening of July 31, 1976. A cool front moved through earlier in the day and is now south of Denver. The weak inversion layer associated with the front kept the cumulus clouds from building to great heights earlier in the afternoon. However, the strong southeasterly surface flow behind the cool front pushed unusually moist air upslope along the mountain range. Heated from below, the conditionally unstable air eventually punctured the inversion and developed into a huge multicell thunderstorm complex that remained nearly stationary for several hours due to the weak southerly winds aloft.

The deluge may have deposited 7.5 inches of rain on the main fork of the Big Thompson River in about one hour. Of the approximately 2000 people in the canyon that evening, over 135 lost their lives, and property damage exceeded $35.5 million.

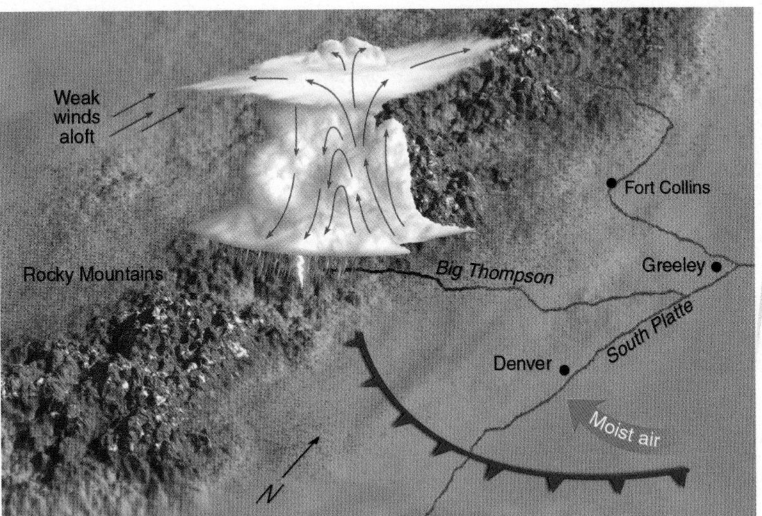

Figure 2 Weather conditions that led to the development of intense multicell thunderstorms, which remained nearly stationary over the Big Thompson Canyon in the Colorado Rockies. The arrows within the thunderstorm represent air motions.

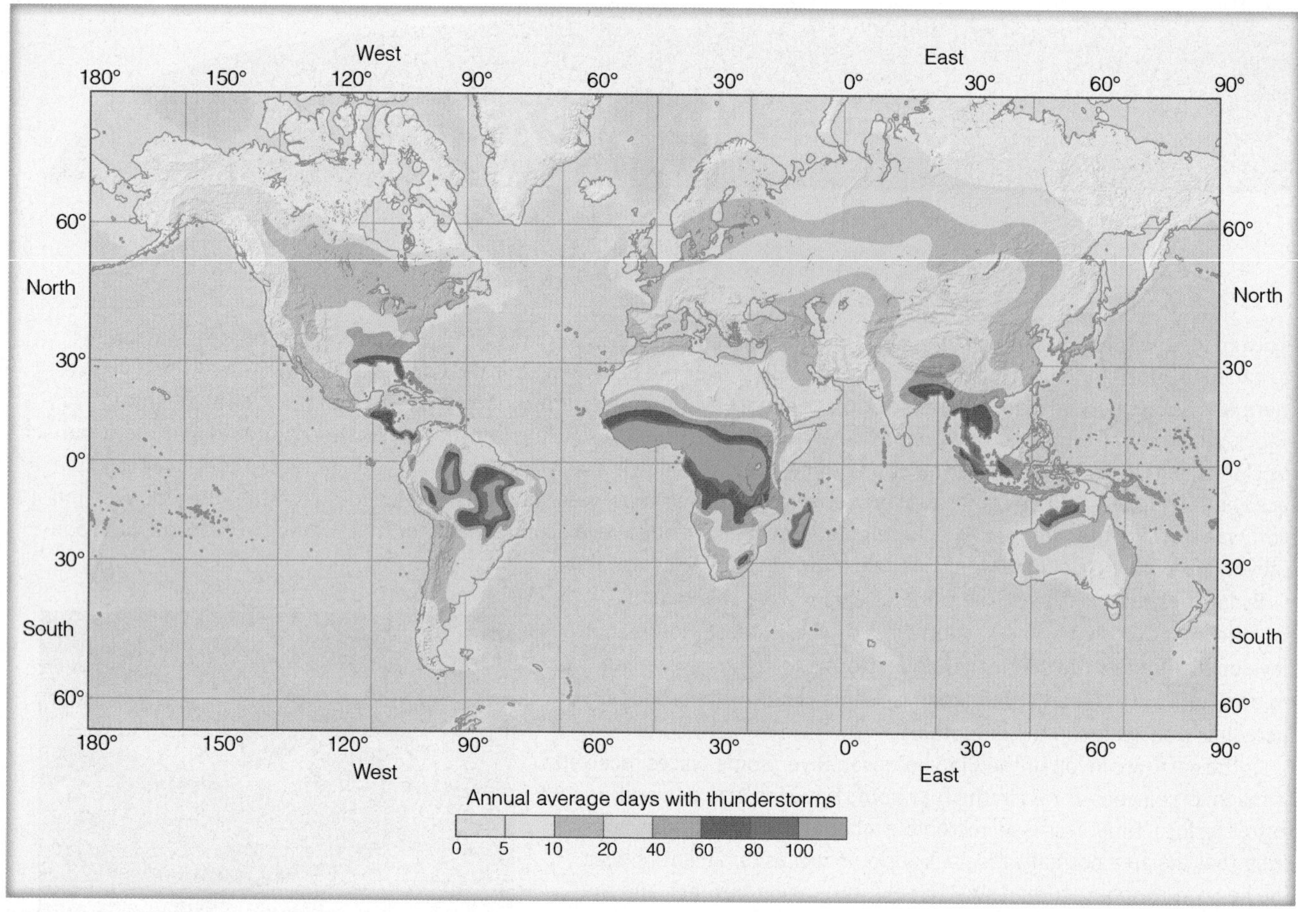

FIGURE 11.28 Average number of days each year on which thunderstorms occur throughout the world.

Thunderstorms are much less prevalent in dry climates, such as the polar regions and the desert areas dominated by subtropical highs.

Figure 11.29 shows the average annual number of days having thunderstorms in various parts of the United States. Notice that they occur most frequently in the southeastern states along the Gulf Coast with a maximum in Florida. A secondary maximum exists over the central Rockies. The region with the fewest thunderstorms is the Pacific coastal and interior valleys.

EXTREME WEATHER WATCH

Possibly the region with the most thunderstorms each year is Africa. Lake Victoria near Kampala, Uganda, experiences thunderstorms on average 242 days a year. Many of these storms form at night as breezes from off the land converge, forcing moist surface air to rise and condense into huge cumulonimbus clouds.

In many areas, thunderstorms form primarily in summer during the warmest part of the day when the surface air is most unstable. There are some exceptions, however. During the summer in the valleys of central and southern California, dry, sinking air produces an inversion that inhibits the development of towering cumulus clouds. In these regions, thunderstorms are most frequent in winter and spring, particularly when cold, moist, conditionally unstable air aloft moves over moist, mild surface air. The surface air remains relatively warm because of its proximity to the ocean. Over the Central Plains, thunderstorms tend to form more frequently at night. These storms may be caused by a low-level southerly jet stream that forms at night, and not only carries humid air northward but also initiates areas of converging surface air, which helps to trigger uplift. As the thunderstorms build, their tops cool by radiating infrared energy to space. This cooling process tends to destabilize the atmosphere, making it more suitable for nighttime thunderstorm development.

At this point, it is interesting to compare Fig. 11.29 with Fig. 11.30. Notice that, even though the greatest frequency of thunderstorms is near the Gulf Coast, the

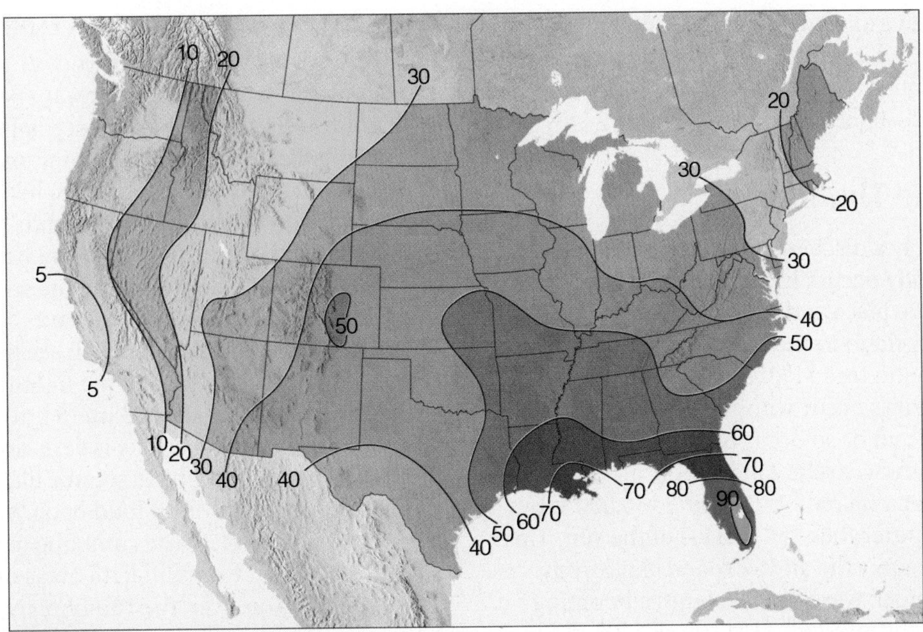

FIGURE 11.29 The average number of days each year on which thunderstorms are observed throughout the United States. (Due to the scarcity of data, the number of thunderstorms is underestimated in the mountainous far west.)

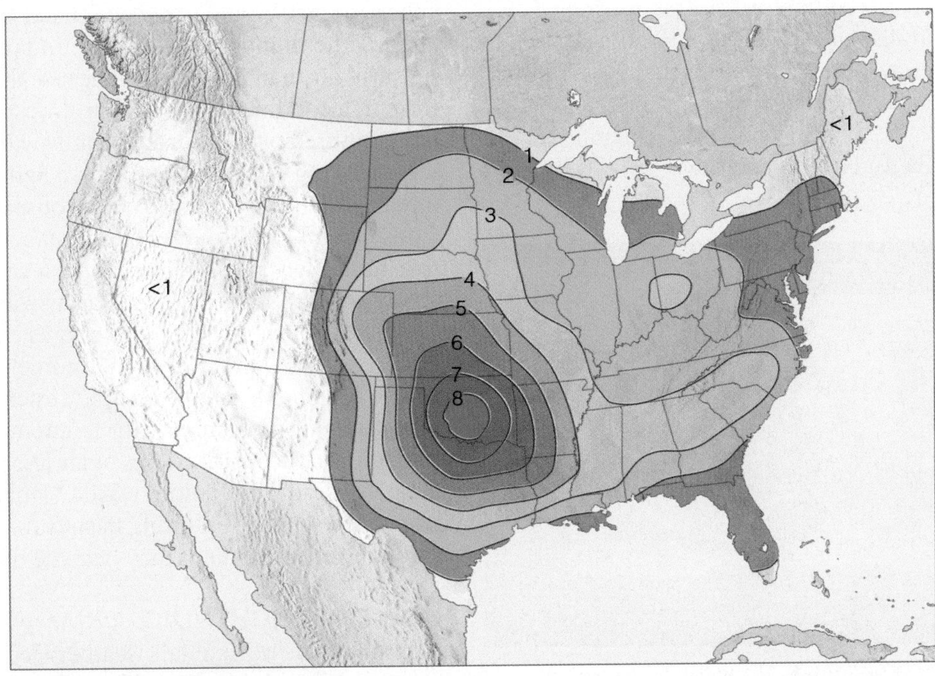

FIGURE 11.30 The average number of days each year on which large hail (with a diameter of 3/4 inch or greater) is observed throughout the United States from 1980 to 1999. (NASA)

greatest frequency of large hailstones is over the Great Plains, and especially over Oklahoma. The reason for this situation is that conditions over the Great Plains are more favorable for the development of severe thunderstorms, and especially supercells that have strong updrafts capable of keeping hailstones suspended within the cloud for a long time so that they can grow to an appreciable size before plunging to the ground.*

*The formation of hail is detailed in Chapter 6 on p. 159.

Now that we have looked at the development and distribution of thunderstorms, we are ready to examine an interesting, though yet not fully understood, aspect of all thunderstorms—lightning.

Lightning and Thunder

Lightning is simply a discharge of electricity, a giant spark, which usually occurs in mature thunderstorms. Lightning may take place within a cloud, from one cloud to another, from a cloud to the surrounding air, or from a cloud to the ground (see Fig. 11.31). (The majority of lightning strikes occur within the cloud, while only about 20 percent or so occur between cloud and ground.) The lightning stroke can heat the air through which it travels to an incredible 54,000°F (30,000°C), which is 5 times hotter than the surface of the sun. This extreme heating causes the air to expand explosively, thus initiating a shock wave that becomes a booming sound wave—called **thunder**—that travels outward in all directions from the flash.

A sound occasionally mistaken for thunder is the **sonic boom**. Sonic booms are produced when an aircraft exceeds the speed of sound at the altitude at which it is flying. The aircraft compresses the air, forming a shock wave that trails out as a cone behind the aircraft. Along the shock wave, the air pressure changes rapidly over a short distance. The rapid pressure change causes the distinct boom. (Exploding fireworks generate a similar shock wave and a loud bang.)

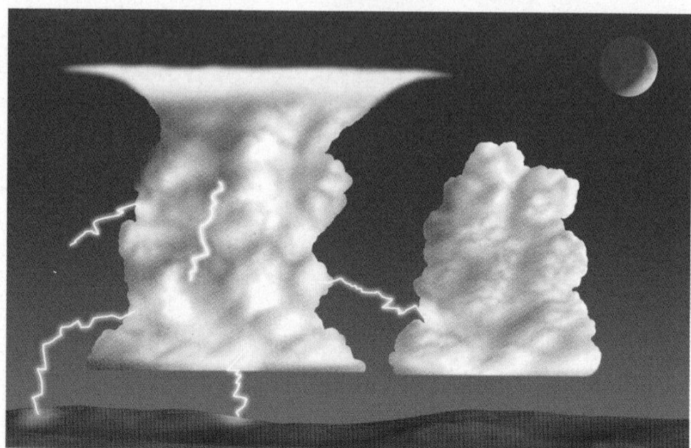

FIGURE 11.31 The lightning stroke can travel in a number of directions. It can occur within a cloud, from one cloud to another cloud, from a cloud to the air, or from a cloud to the ground. Notice that the cloud-to-ground lightning can travel out away from the cloud, then turn downward, striking the ground many miles from the thunderstorm. When lightning behaves in this manner, it is often described as a *"bolt from the blue."*

HOW FAR AWAY IS THE LIGHTNING?—START COUNTING When you see a flash of lightning, how can you tell how far away (or how close) it is? Light travels so fast that you see light instantly after a lightning flash. But the sound of thunder, traveling at only about 1100 ft/sec, takes much longer to reach your ear. If you start counting seconds from the moment you see the lightning until you hear the thunder, you can determine how far away the stroke is. Because it takes sound about 5 seconds to travel one mile (about 3 seconds for one kilometer), if you see lightning and hear the thunder 15 seconds later, the lightning stroke (and the thunderstorm) is about 3 mi (5 km) away.

When the lightning stroke is very close—on the order of 300 feet or less—thunder sounds like a clap or a crack followed immediately by a loud bang. When it is farther away, it often rumbles. The rumbling can be due to the sound emanating from different areas of the stroke (see Fig. 11.32). Moreover, the rumbling is accentuated when the sound wave reaches an observer after having bounced off obstructions, such as hills and buildings.

In some instances, lightning is seen but no thunder is heard. Does this mean that thunder was not produced by the lightning? Actually, there is thunder, but the atmosphere refracts (bends) and attenuates the sound waves, making the thunder inaudible. Sound travels faster in warm air than in cold air.* Because thunderstorms form in a conditionally unstable atmosphere, where the temperature normally drops rapidly with height, a sound wave moving outward away from a lightning stroke will often bend upward, away from an observer at the surface. Consequently, an observer closer than about 3 mi (5 km) to a lightning stroke will usually hear thunder, whereas an observer about 5 mi (8 km) away will not.

However, even when a viewer is as close as several kilometers to a lightning flash, thunder may not be heard. For one thing, the complex interaction of sound waves and air molecules tends to attenuate the thunder. In addition, turbulent eddies of air less than 150 feet in diameter scatter the sound waves. Hence, when thunder from a low-energy lightning flash travels several miles through turbulent air, it may become inaudible.

HOW DOES LIGHTNING FORM? What causes lightning? The normal fair weather electric field of the atmosphere is characterized by a negatively charged surface and a positively charged upper atmosphere. For lightning to occur, separate regions containing opposite electrical charges must exist within a cumulonimbus cloud. Exactly how this charge separation comes about is not totally comprehended; however, there are several theories to account for it.

*The speed of sound in calm air is equal to $20\sqrt{T}$, where T is the air temperature in Kelvins.

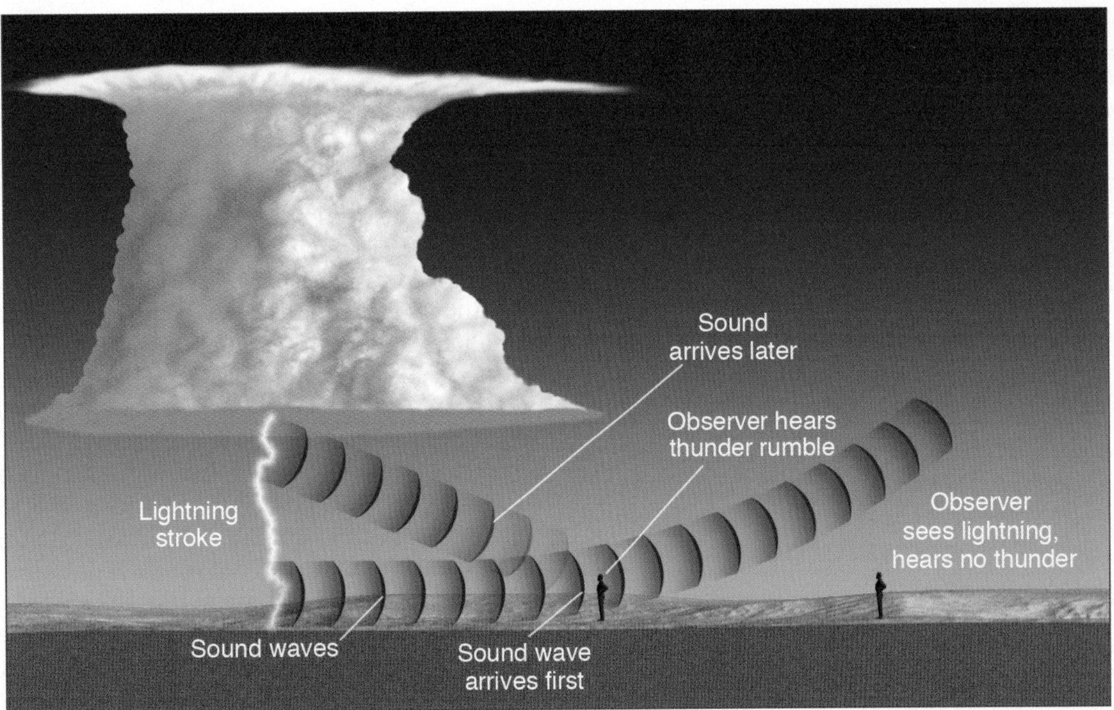

FIGURE 11.32 Thunder travels outward from the lightning stroke in the form of waves. If the sound waves from the lower part of the stroke reach an observer before the waves from the upper part of the stroke, the thunder appears to rumble. If the sound waves bend upward away from an observer, the lightning stroke may be seen, but the thunder will not be heard.

Electrification of Clouds One theory proposes that clouds become electrified when graupel (small ice particles called *soft hail*) and hailstones fall through a region of supercooled liquid droplets and ice crystals. As liquid droplets collide with a hailstone, they freeze on contact and release latent heat. This process keeps the surface of the hailstone warmer than that of the surrounding ice crystals. When the warmer hailstone comes in contact with a colder ice crystal, an important phenomenon occurs: *There is a net transfer of positive ions (charged molecules) from the warmer object to the colder object.* Hence, the hailstone (larger, warmer particle) becomes negatively charged and the ice crystal (smaller, cooler particle) positively charged, as the positive ions are incorporated into the ice crystal (see Fig. 11.33). The same effect occurs when colder, supercooled liquid droplets freeze on contact with a warmer hailstone and tiny splinters of positively charged ice break off. These lighter, positively charged particles are then carried to the upper part of the cloud by updrafts. The larger hailstones (or graupel), left with a negative charge, either remain suspended in an updraft or fall toward the bottom of the cloud. By this mechanism, the cold upper part of the cloud becomes positively charged, while the middle of the cloud becomes negatively charged. The lower part of the cloud is generally of negative and mixed charge except for an

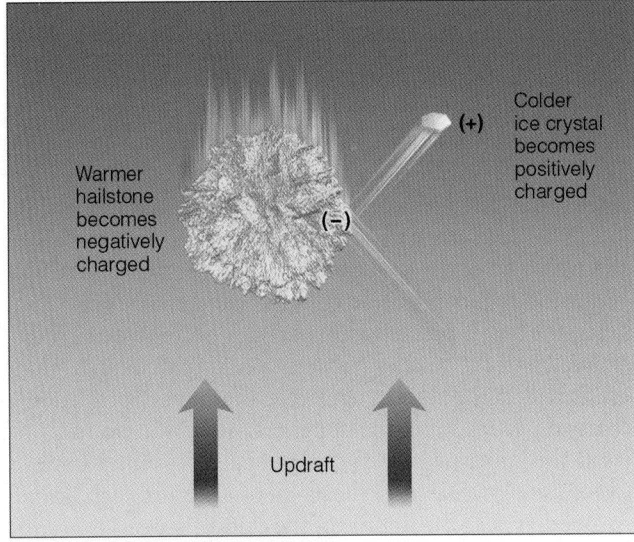

FIGURE 11.33 When the tiny colder ice crystals come in contact with the much larger and warmer hailstone (or graupel), the ice crystal becomes positively charged and the hailstone negatively charged. Updrafts carry the tiny positively charged ice crystal into the upper reaches of the cloud, while the heavier hailstone falls through the updraft toward the lower region of the cloud.

FIGURE 11.34 The generalized charge distribution in a mature thunderstorm.

occasional positive region located in the falling precipitation near the melting level (see Fig. 11.34).

Another school of thought proposes that during the formation of precipitation, regions of separate charge exist within tiny cloud droplets and larger precipitation particles. In the upper part of these particles we find negative charge, while in the lower part we find positive charge. When falling precipitation collides with smaller particles, the larger precipitation particles become negatively charged and the smaller particles, positively charged. Updrafts within the cloud then sweep the smaller positively charged particles into the upper reaches of the cloud, while the larger negatively charged particles either settle toward the lower part of the cloud or updrafts keep them suspended near the middle of the cloud.

The Lightning Stroke Because unlike charges attract one another, the negative charge at the bottom of the cloud causes a region of the ground beneath it to become positively charged. As the thunderstorm moves along, this region of positive charge follows the cloud like a shadow. The positive charge is most dense on protruding objects, such as trees, poles, and buildings. The difference in charges causes an electric potential between the cloud and ground. In dry air, however, a flow of current does not occur because the air is a good electrical insulator. Gradually, the electrical potential gradient builds, and when it becomes sufficiently large (on the order of one million volts per meter), the insulating properties of the air break down, a current flows, and lightning occurs.

Cloud-to-ground lightning begins within the cloud when the localized electric potential gradient exceeds 3 million volts per meter along a path perhaps 150 ft (about 50 m) long. This situation causes a discharge of electrons to rush toward the cloud base and then toward the ground in a series of steps (see Fig. 11.35a). Each discharge covers about 150 to 300 ft (about 50 to 100 m), then stops for about 50-millionths of a second, then occurs again over another 150 feet or so. This **stepped leader** is very faint and is usually invisible to the human eye. As the tip of the stepped leader approaches the ground, the potential gradient (the voltage per meter) increases, and a current of positive charge starts upward from the ground (usually along elevated objects) to meet it (see Fig. 11.35b). After they meet, large numbers of electrons flow to the ground and a much larger, more luminous **return stroke** about an inch or so in diameter surges upward to the cloud along the path followed by the stepped leader (Fig. 11.35c). Hence, the downward flow of electrons establishes the bright channel of upward propagating current. Even though the bright return stroke travels from the ground up to the cloud, it happens so quickly—in one ten-thousandth of a second—that our eyes cannot resolve the motion, and we see what appears to be a continuous bright flash of light (see Fig. 11.36).

Sometimes there is only one lightning stroke, but more often the leader-and-stroke process is repeated in the same ionized channel at intervals of about four-hundredths of a second. The subsequent leader, called a **dart leader**, proceeds from the cloud along the same channel as the original stepped leader; however, it proceeds downward more quickly because the electrical resistance of the

EXTREME WEATHER WATCH

Some lightning bolts are up to 100 times more powerful than others. The most powerful lightning bolts, sometimes called "superbolts," often originate from the upper region of a thunderstorm. An example of such a superbolt occurred on April 2, 1959, when lightning struck a cornfield near Leland, Illinois, and left a hole a foot deep and 12 feet wide. The lightning bolt was so powerful that it shattered windows in homes a mile away from the cornfield.

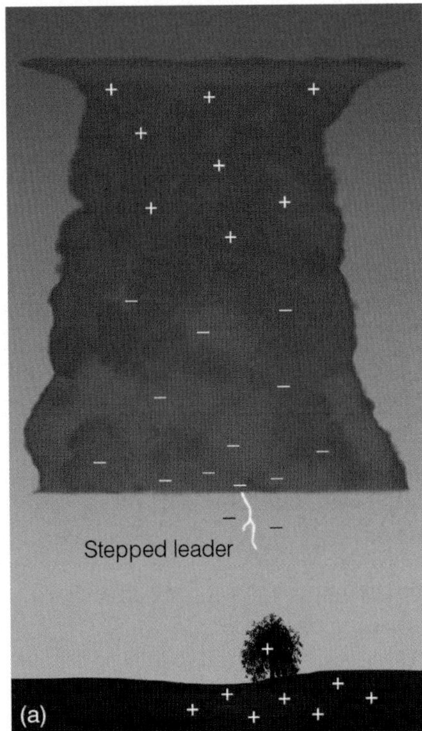

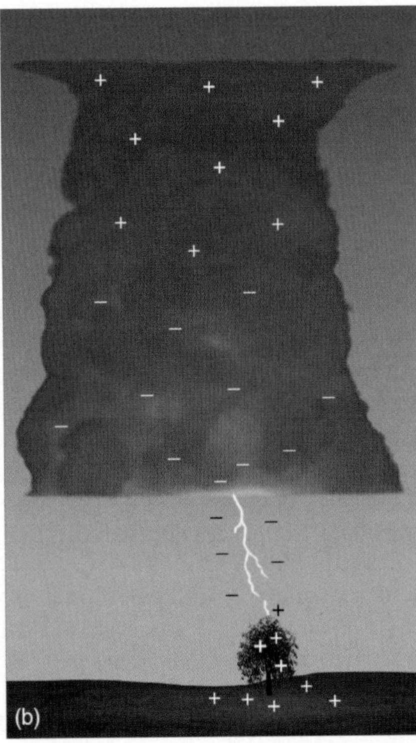

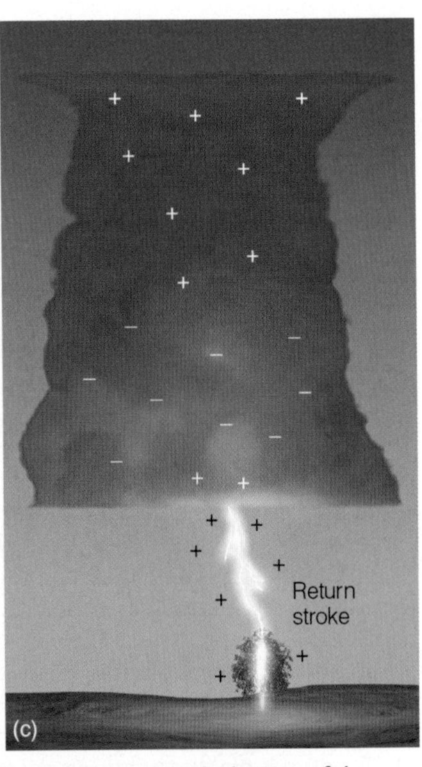

Active FIGURE 11.35 The development of a lightning stroke. (a) When the negative charge near the bottom of the cloud becomes large enough to overcome the air's resistance, a flow of electrons—the stepped leader—rushes toward the earth. (b) As the electrons approach the ground, a region of positive charge moves up into the air through any conducting object, such as trees, buildings, and even humans. (c) When the downward flow of electrons meets the upward surge of positive charge, a strong electric current—a bright return stroke—carries positive charge upward into the cloud. Visit the Meteorology Resource Center to view this and other active figures at www.cengage.com/login

FIGURE 11.36 Time exposure of an evening thunderstorm with an intense lightning display near Denver, Colorado. The bright flashes are return strokes. The lighter forked flashes are probably stepped leaders that did not make it to the ground.

path is now lower. As the leader approaches the ground, normally a less energetic return stroke than the first one travels from the ground to the cloud. Typically, a lightning flash will have three or four leaders, each followed by a return stroke. A lightning flash consisting of many strokes (one photographed flash had 26 strokes) usually lasts less than a second. During this short period of time, our eyes may barely be able to perceive the individual strokes, and the flash appears to flicker.

The lightning described so far (where the base of the cloud is negatively charged and the ground positively charged) is called *negative cloud-to-ground lightning*, because the stroke carries negative charges from the cloud to the ground. About 90 percent of all cloud-to-ground lightning is negative. However, when the base of the cloud is positively charged and the ground negatively charged, a *positive cloud-to-ground lightning* flash may result. Positive lightning, most common with supercell thunderstorms, has the potential to cause more damage because it generates a much higher current level and its flash lasts for a longer duration than negative lightning.

THE DIFFERENT FORMS OF LIGHTNING In ❯ Fig. 11.37, you can see that lightning can take on a variety of shapes and forms, some of which are described below.

Forked Lightning The most common form of lightning is **forked lightning**, which normally has a crooked or forked appearance with branches extending outward from the main flash (see Fig. 11.36 and Fig. 11.37a). Forked lightning often forms where a dart leader moving toward the ground deviates from the original path taken by the stepped leader.

Heat Lightning Distant lightning from thunderstorms that is seen but not heard is commonly called **heat lightning** because it frequently occurs on hot summer nights when the overhead sky is clear. As the light from distant electrical storms is refracted through the atmosphere, air molecules and fine dust scatter the shorter wavelengths of visible light, often causing heat lightning to appear orange to a distant observer (see Fig. 11.37b).

Sheet Lightning When either the lightning flash occurs inside a cloud or intervening clouds obscure the flash, such that a portion of the cloud (or clouds) appears as a luminous white sheet, the bright part of the cloud (or clouds) is called **sheet lightning** (see Fig. 11.37c).

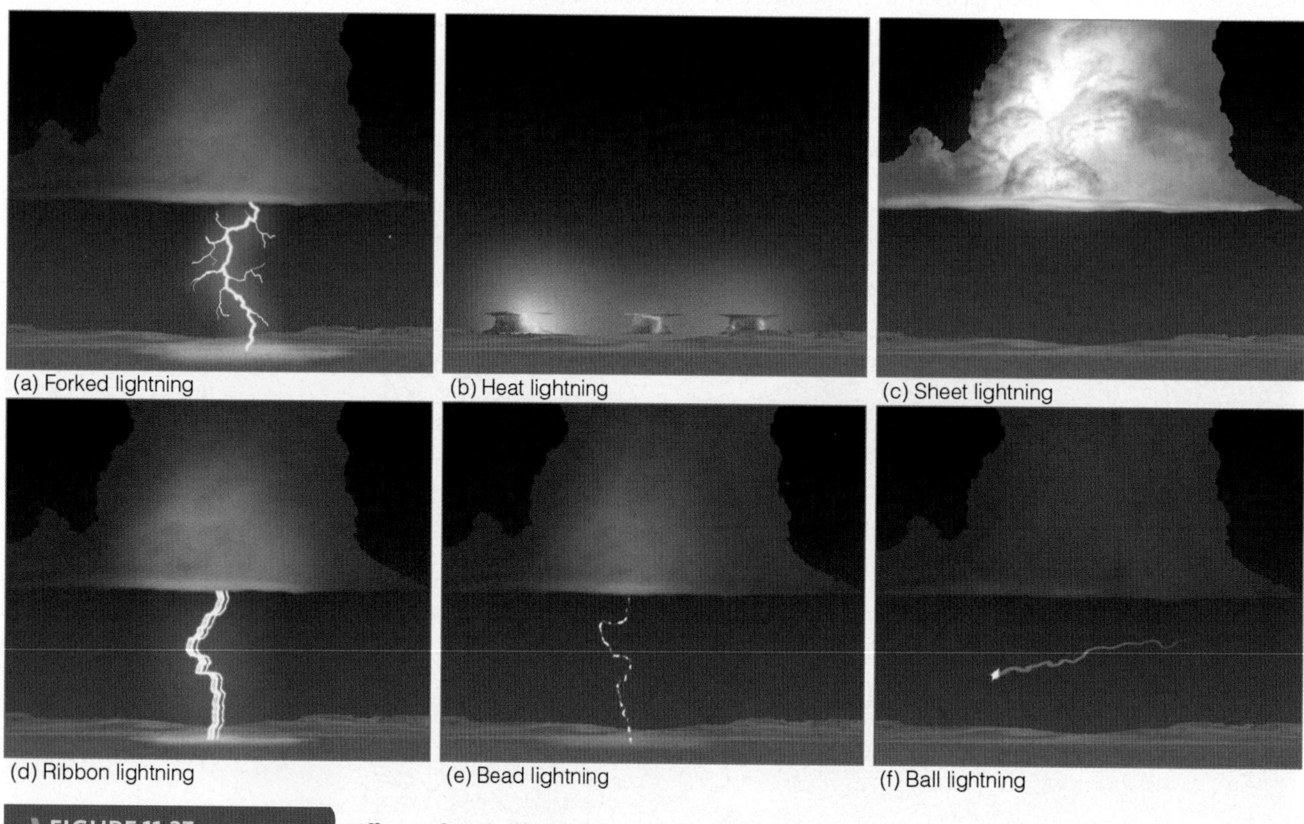

❯ **FIGURE 11.37** Different forms of lightning.

Ribbon Lightning An interesting form of lightning is **ribbon lightning**, which forms when a strong wind moves the ionized channel between each return strike, causing the lightning to appear as a ribbon hanging from the cloud (see Fig. 11.37d).

Bead Lightning If the lightning channel breaks up or appears to break up, the lightning can appear as a series of beads tied to a string (see Fig. 11.37e). Called **bead lightning**, it is almost impossible to see with the naked eye, although it has been photographed with high-speed cameras.

Ball Lightning One of the most fascinating areas of lightning research deals with **ball lightning**. This form of lightning, which looks like a luminous sphere (about the size of a basketball), may appear to float in the air or slowly dart about for several seconds (see Fig. 11.37f). Although many theories have been proposed, the actual cause of ball lightning remains an enigma, and for this reason there are scientists who dispute the existence of ball lightning as a distinct form of lightning.

Dry Lightning When cloud-to-ground lightning occurs with thunderstorms that do not produce rain, the lightning is often called **dry lightning**. Such lightning often starts forest fires in regions of dry timber.

St. Elmo's Fire As the electric potential near the ground increases, a current of positive charge moves up pointed objects, such as antennas and masts of ships. However, instead of a lightning stroke, a luminous greenish or bluish halo may appear above them, as a continuous supply of sparks—a *corona discharge*—is sent into the air. This electric discharge, which can cause the top of a ship's mast to glow, is known as **St. Elmo's Fire**, named after the patron saint of sailors (see ❯ Fig. 11.38). St. Elmo's Fire is also seen around power lines and the wings of aircraft. When St. Elmo's Fire is visible and a thunderstorm is nearby, a lightning flash may occur in the near future, especially if the electric field of the atmosphere is increasing.

Earlier, we learned that lightning occurs with mature thunderstorms. But lightning may also occur in snowstorms, in dust storms, in the gas cloud of an erupting volcano, and on very rare occasions in nimbostratus clouds. Lightning may also shoot from the top of thunderstorms into the upper atmosphere, producing an eerie-looking blue lightning. More on this topic is given in the Focus section on p. 326.

WHEN LIGHTNING STRIKES To protect a building and its occupants inside from the damaging effects of lightning, many structures have lightning rods. The rod is made of metal and has a pointed tip, which extends well above the structure (see ❯ Fig. 11.39). The positive charge concentration will be maximum on the tip of the rod, thus increasing the probability that the lightning will strike the tip and follow the metal rod harmlessly down into the ground, where the other end is deeply buried.

When lightning enters sandy soil, the extremely high temperature of the stroke may fuse sand particles together, producing a rootlike system of tubes called a **fulgurite**, after the Latin word for "lightning" (see ❯ Fig. 11.40). When lightning strikes an object such as a car, it normally leaves the passengers unharmed because

❯ FIGURE 11.38 St. Elmo's fire tends to form above objects such as aircraft wings, ship's masts, and flag poles.

EXTREME WEATHER WATCH

The most lightning-prone area in the United States is located in Pasco County, Florida, just north of Tampa Bay. Here, in a typical year, about 40 lightning flashes occur over an area of one square mile.

it usually takes the quickest path to the ground along the outside metal casing of the vehicle. The lightning then jumps to the road through the air, or it enters the roadway through the tires (see Fig. 11.41). The same type of protection is provided by the metal skin of a jet airliner, as hundreds of aircraft are struck by lightning each year.

Because a single lightning stroke may involve a current as great as 100,000 amperes, animals and humans can be electrocuted when struck by lightning. The average yearly death toll in the United States attributed to lightning is nearly 100, with Florida accounting for the most fatalities. Many victims are struck in open places, riding on farm equipment, playing golf, attending sports events, or sailing in a small boat. Some live to tell about it, as did the retired champion golfer Lee Trevino. Others are less fortunate.

About 10 percent of people struck by lightning are killed. Most die from cardiac arrest. Consequently, when you see someone struck by lightning, immediately give

FOCUS ON

A SPECIAL TOPIC

Strange Lightning in the Upper Atmosphere

For many years airline pilots reported seeing strange bolts of light shooting upward, high above the tops of intense thunderstorms. These faint, mysterious flashes did not receive much attention, however, until they were first photographed in 1989. Photographs from sensitive, low-light-level cameras onboard jet aircraft revealed that the mysterious flashes were actually a colorful display called *red sprites* and *blue jets*, which seemed to dance above the clouds.

Sprites are massive, but dim, light flashes that appear directly above an intense thunderstorm system (see Fig. 3). Usually red, and lasting but a few thousandths of a second, sprites tend to form almost simultaneously with lightning in the cloud below and with severe thunderstorms that have positive cloud-to-ground lightning strokes. (Most cloud-to-ground lightning is negative.) Although it is not entirely clear how they form, the thinking now is that sprites form when positive lightning disrupts the atmosphere's electrical field in such a way that charged particles in the upper atmosphere are accelerated downward toward the thunderstorm and upward to higher levels in the atmosphere.

Blue jets usually dart upward in a conical shape from the tops of thunderstorms that are experiencing vigorous lightning activity (Fig. 3). Although faint, blue jets can be seen with the naked eye. They are not well understood, but appear to transfer large amounts of electrical energy into the upper atmosphere.

*ELVES** as illustrated in Fig. 3 appear as a faint red halo—too faint to be seen with the naked eye, only with sensitive cameras. They occur in the ionized region of the upper atmosphere. ELVES occur at night and are extremely short-lived. They appear to form when a lightning bolt from an intense thunderstorm gives off a strong electromagnetic pulse (EMP) that causes electrons in the ionosphere to collide with molecules that become excited and give off light.

The roles that red sprites, blue jets, and ELVES play in the earth's global electrical system have yet to be determined.

*The acronym ELVES is from *E*missions of *L*ight and *V*ery low frequency from lightning-induced *E*lectro-Magnetic Pulsation *S*ources.

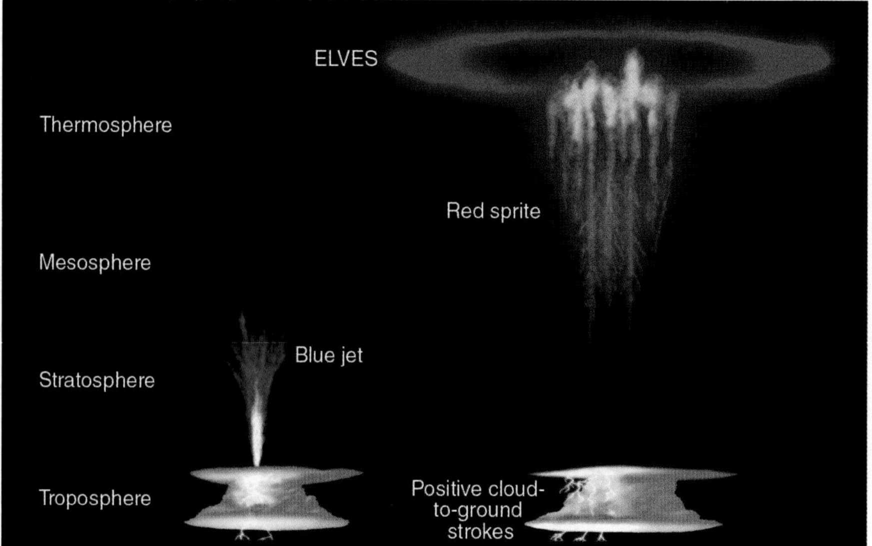

Figure 3 Various electrical phenomena observed in the upper atmosphere.

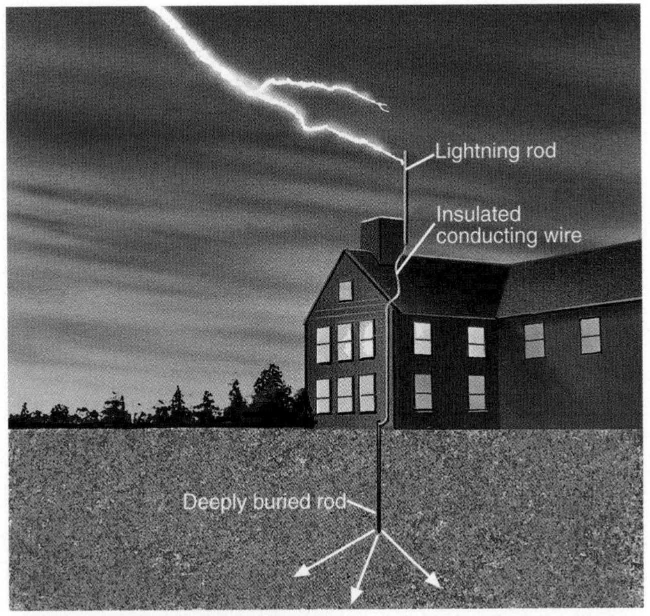

FIGURE 11.39 The lightning rod extends above the building, increasing the likelihood that lightning will strike the rod rather than some other part of the structure. After lightning strikes the metal rod, it follows an insulated conducting wire harmlessly into the ground.

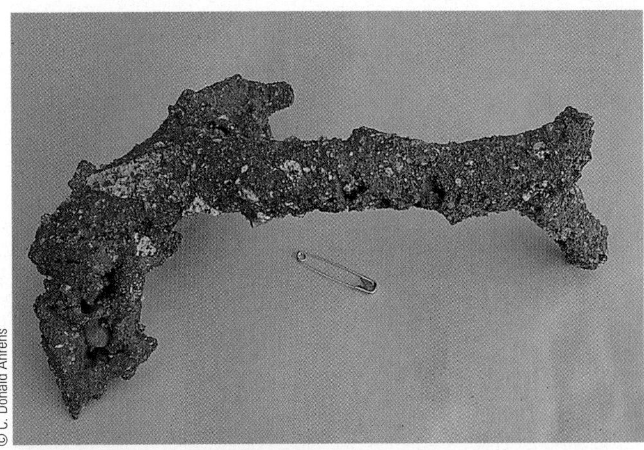
© C. Donald Ahrens

FIGURE 11.40 A fulgurite that formed by lightning fusing sand particles.

© C. Donald Ahrens

FIGURE 11.41

The four marks on the road surface represent areas where lightning, after striking a car traveling along south Florida's Sunshine State Parkway, entered the roadway through the tires. Lightning flattened three of the car's tires and slightly damaged the radio antenna. The driver and a six-year-old passenger were taken to a nearby hospital, treated for shock, and released.

CPR (cardiopulmonary resuscitation), as lightning normally leaves its victims unconscious without heartbeat and without respiration. Those who do survive often suffer from long-term psychological disorders, such as personality changes, depression, and chronic fatigue.

Many lightning fatalities occur in the vicinity of relatively isolated trees (see Fig. 11.42). As a tragic example, during June, 2004, three people were killed near Atlanta, Georgia, seeking shelter under a tree. Because a positive charge tends to concentrate in upward projecting objects, the upward return stroke that meets the stepped leader is most likely to originate from such objects. Clearly, sitting under a tree during an electrical storm is not wise. So what *should* you do?

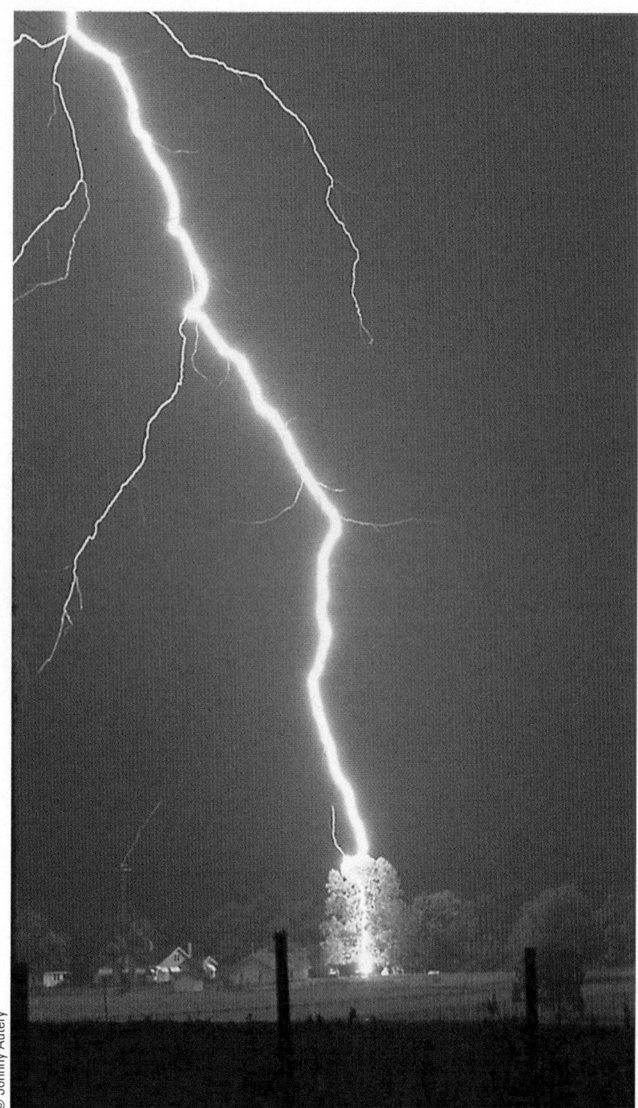

© Johnny Autery

FIGURE 11.42 A cloud-to-ground lightning flash hitting a 65-foot sycamore tree. It should be apparent why one should *not* seek shelter under a tree during a thunderstorm.

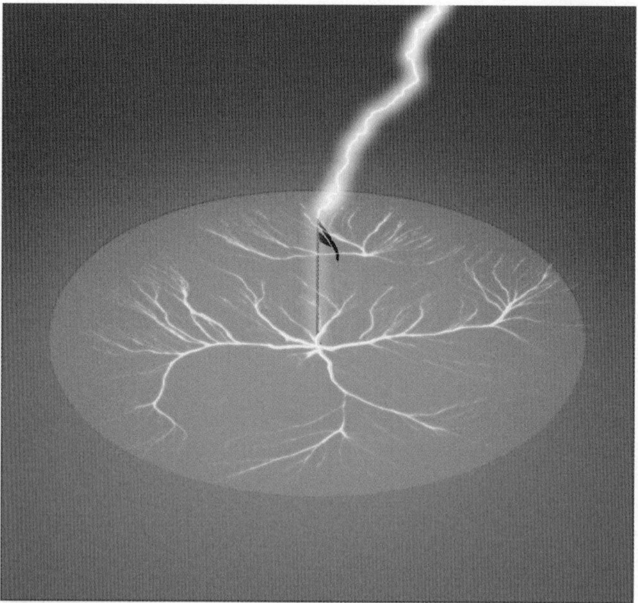

FIGURE 11.43 When lightning strikes an elevated object, such as a golf pin, the lightning moves downward, then outward through the ground in all directions.

© Mary McQuilken

FIGURE 11.44 Lightning can be both hair-raising and deadly. This photograph, taken by Mary McQuilken, shows her younger brother, Sean (on the left), and older brother, Michael (on the right), standing beneath a thunderstorm atop Moro Rock in California's Sequoia National Park. Shortly after this photo was taken, Sean was struck by lightning and seriously injured, and a nearby hiker was killed by the same lightning strike.

When caught outside in a thunderstorm, the best protection, of course, is to get inside a building. But stay away from electrical appliances and corded phones, and avoid taking a shower. Automobiles with metal frames and trucks (but not golf carts) may also provide protection. If no such shelter exists, be sure to avoid elevated places and isolated trees. If you are on level ground, try to keep your head as low as possible, but do not lie down. Because lightning channels usually emanate outward through the ground at the point of a lightning strike, a surface current may travel through your body and injure or kill you (see Fig. 11.43). Therefore, crouch down as low as possible and minimize the contact area you have with the ground by touching it with only your toes or your heels.

There are some warning signs to alert you to a strike. If your hair begins to stand on end or your skin begins to tingle and you hear clicking sounds, beware—lightning may be about to strike. And if you are standing upright, you may be acting as a lightning rod (see Fig. 11.44).

LIGHTNING DETECTION AND SUPPRESSION For many years, lightning strokes were detected primarily by visual observation. Today, cloud-to-ground lightning is located by means of an instrument called a *lightning direction-finder*, which works by detecting the radio waves produced by lightning. Such waves are called *sferics*, a contraction from their earlier designation, *atmospherics*. A web of these magnetic devices is a valuable tool in pinpointing lightning strokes throughout the contiguous United States, Canada, and Alaska. Lightning detection devices allow scientists to examine in detail the lightning activity inside a storm as it intensifies and moves (see ❱ Fig. 11.45). This gives forecasters a better idea where intense lightning strokes might be expected.* Moreover, satellites now have the capability of providing more lightning information than ground-based sensors, because satellites can continuously detect all forms of lightning over land and over water (see ❱ Fig. 11.46). Lightning information correlated with satellite images provides a more complete and precise structure of a thunderstorm.

Each year, approximately 10,000 fires are started by lightning in the United States alone and over $50 million worth of timber is destroyed. For this reason, tests have been conducted to see whether the number of cloud-to-ground lightning discharges can be reduced. One technique that has shown some success in suppressing lightning involves seeding a cumulonimbus cloud with hair-thin pieces of aluminum about 4 inches long. The idea is that these pieces of metal will produce many tiny sparks, or *corona discharges*, and prevent the electrical potential in the cloud from building to a point where lightning occurs. While the results of this experiment are inconclusive, many forestry specialists point out that nature itself may use a similar mechanism to prevent excessive lightning damage. The long, pointed needles of pine trees may act as tiny lightning rods, diffusing the concentration of electric charges and preventing massive lightning strokes.

*In fact, with the aid of these instruments and computer models of the atmosphere, the National Weather Service currently issues lightning probability forecasts for the western United States.

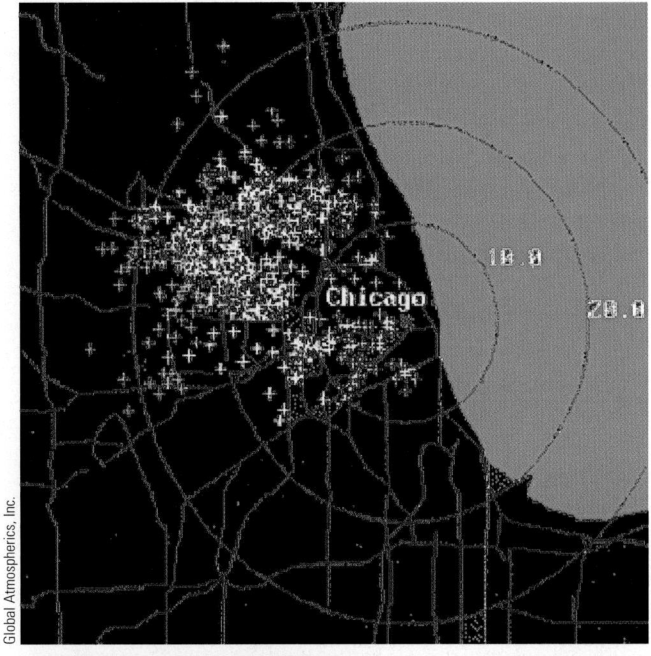

Global Atmospherics, Inc.

❱ FIGURE 11.45 Cloud-to-ground lightning strikes in the vicinity of Chicago, Illinois, as detected by the National Lightning Detection Network.

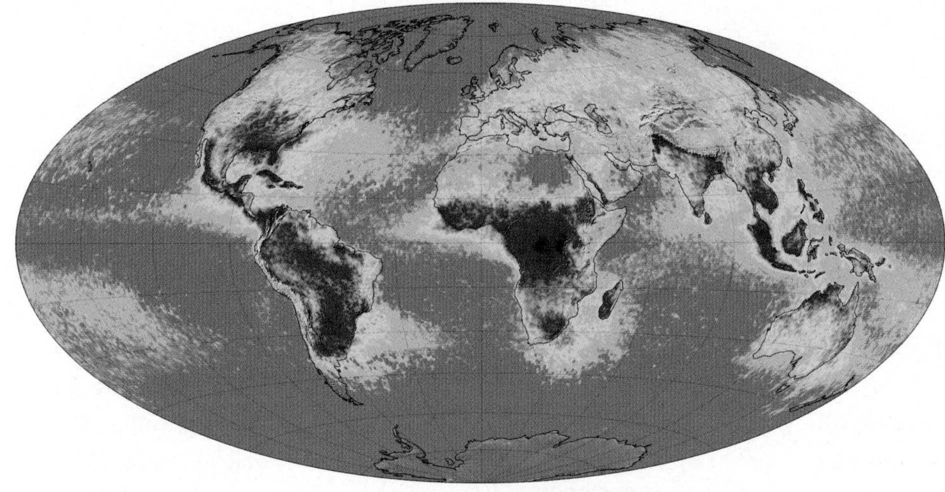

❱ FIGURE 11.46 The average yearly number of lightning flashes per square kilometer based on data collected by NASA satellites between 1995 and 2002.

Summary

In this chapter, we examined the formation of thunderstorms—convective storms containing thunder and lightning—and the extreme weather they produce. The ingredients for the isolated ordinary cell thunderstorm are humid surface air, plenty of sunlight to heat the ground, a conditionally unstable atmosphere, a "trigger" to start the air rising, and weak vertical wind shear. When these conditions prevail, and the air begins to rise, small cumulus clouds may grow into towering clouds and thunderstorms within 30 minutes.

When conditions are ripe for thunderstorm development, and moderate or strong vertical wind shear exists, the updraft in the thunderstorm may tilt and ride up and over the downdraft. As the forward edge of the downdraft (the gust front) pushes outward along the ground, the air is lifted and new cells form, producing a multicell thunderstorm—a storm with cells in various stages of development. Some multicell storms form as a complex of thunderstorms, such as the squall line (which forms as a line of thunderstorms either along or out ahead of an advancing cold front) and the Mesoscale Convective Complex (which forms as a cluster of storms). When convection in the multicell storm is strong, it may produce severe weather, such as strong damaging surface winds, hail, and flooding.

Supercell thunderstorms are large, intense thunderstorms with a single rotating updraft. The updraft and the downdraft in a supercell are nearly in balance, so that the storm may exist for many hours. Supercells are capable of producing severe weather, including strong damaging tornadoes.

When a thunderstorm's downdraft is intense, it is called a *downburst*. A relatively small downburst is called a *microburst*. When the downdraft is warm and produces a sudden rise in surface air temperature, it is called a *heat burst*. A cold downdraft upon reaching the earth's surface can push outward in all directions, creating a gust front. Strong straight-line winds behind a gust front can exceed 100 mi/hr.

When a squall line on a radar screen takes on the shape of a bow, it is called a *bow echo*. If the northern end of the bow begins to rotate counterclockwise, it may form into a mesoscale convective vortex. When damaging straight-line winds extend for a considerable distance along a squall line, the windstorm is called a *derecho*.

Lightning is a discharge of electricity that occurs in mature thunderstorms. Lightning occurs inside a cloud, or it can travel in a variety of ways: from cloud to cloud, from cloud to air, or from cloud to ground. The lightning stroke momentarily heats the air to an incredibly high temperature. The rapidly expanding air produces a sound called *thunder*.

When caught outside in a thunderstorm, never seek shelter under an isolated tree, as the lightning frequently strikes the tops of tall trees because the positive charge near the ground tends to concentrate on elevated objects. To protect yourself from lightning, try to seek shelter inside a building immediately.

Key Terms

The following terms are listed (with page number) in the order they appear in the text. Define each. Doing so will aid you in reviewing the material covered in this chapter.

ordinary cell (air mass) thunderstorms, 298
cumulus stage, 298
downdraft, 300
mature stage, 300
dissipating stage, 300
multicell thunderstorm, 301
overshooting top, 301
gust front, 301
straight-line winds, 302
shelf cloud, 302
roll cloud, 303
outflow boundary, 303
downburst, 303
microburst, 303
heat burst, 306
squall line, 306
bow echo, 307
mesoscale convective vortex, 308
derecho, 308
Mesoscale Convective Complex, 310
supercell, 311
mesocyclone, 312
wall cloud, 313
lightning, 320
thunder, 320
sonic boom, 320
stepped leader, 322
return stroke, 322
dart leader, 322
forked lightning, 324
heat lightning, 324
sheet lightning, 324
ribbon lightning, 325
bead lightning, 325
ball lightning, 325
dry lightning, 325
St. Elmo's Fire, 325
fulgurite, 325

Review Questions

1. What are thunderstorms?
2. What atmospheric conditions are necessary for the development of ordinary cell thunderstorms?
3. Describe the stages of development of an ordinary cell (air mass) thunderstorm.
4. How do downdrafts form in thunderstorms?
5. Why do ordinary cell thunderstorms most frequently form in the afternoon?
6. Explain why ordinary cell thunderstorms tend to dissipate much sooner than multicell storms.
7. How does the National Weather Service define a severe thunderstorm?
8. What is necessary for a multicell thunderstorm to form?
9. (a) How do gust fronts form?
   (b) What type of weather does a gust front bring when it passes?
10. (a) Describe how a microburst forms.
    (b) Why is the term *wind shear* often used in conjunction with a microburst?
11. What is a derecho, and how does it form?
12. How does a squall line differ from a Mesoscale Convective Complex (MCC)?
13. Give a possible explanation for the generation of a pre-frontal squall-line thunderstorm.
14. Where are the strongest straight-line winds normally observed with a squall line?
15. What is a bow echo, and how does it appear on a radar screen?
16. How does the cell in an ordinary cell thunderstorm differ from the cell in a supercell thunderstorm?
17. Describe the atmospheric conditions at the surface and aloft that are necessary for the development of most supercell thunderstorms. (Include in your answer the role that the low-level jet plays in the rotating updraft.)
18. What is the difference between an HP supercell and an LP supercell?
19. When thunderstorms are *training*, what are they doing?
20. In what region in the United States do dryline thunderstorms most frequently form? Why there?
21. (a) Where does the highest frequency of thunderstorms occur in the United States? Why there?
    (b) Where does the highest frequency of thunderstorms occur in the world? Why there?
22. Describe one process by which thunderstorms become electrified.
23. Explain how a cloud-to-ground lightning stroke develops.
24. How does negative cloud-to-ground lightning differ from positive cloud-to-ground lightning?
25. How is thunder produced?
26. Why is it unwise to seek shelter under an isolated tree during a thunderstorm? If caught out in the open, what should you do?

Online Learning

STUDENT COMPANION WEBSITE: Visit this book's companion website at: www.cengage.com/ahrens/extreme1e and choose Chapter 11 for many study aids and ideas for further reading and research. These include flashcards, practice quizzing, and web links.

METEOROLOGY RESOURCE CENTER: For students with access, log on at www.cengage.com/login for more assets, including animations, videos, and more. If your textbook did not come with access, visit www.CengageBrain.com to purchase.

Tornadoes

© Eric Nguyen

12

CONTENTS

WHAT IS A TORNADO?
TORNADO LIFE CYCLE
TORNADO OCCURRENCE AND DISTRIBUTION
TORNADO WINDS
TORNADO OUTBREAKS
TORNADO FORMATION
TORNADIC WINDS AND DOPPLER RADAR
WATERSPOUTS
SUMMARY
KEY TERMS
REVIEW QUESTIONS

A supercell thunderstorm produces a tornado near Attica, Kansas, on May 24, 2004.

On the afternoon of March 18, 1925, the sky turned dark green over Murphysboro, Illinois. The wind began to blow when suddenly a huge rolling black cloud appeared on the horizon. With no discernable funnel, the mile-wide tornado roared through the town, leveling 40 percent of it and killing 234 inhabitants. The tornado continued its destructive path northeastward through the states of Illinois and Indiana. All tolled, the tornado (or a series of tornadoes spawned from the same thunderstorm) obliterated 4 towns, killed an estimated 695 people, and injured more than 2000.

Large tornadoes and even small ones are capable of inflicting great damage upon a community. Every year, tornadoes kill people and livestock, and destroy property worth many millions of dollars. In this chapter, we will examine the atmospheric conditions that lead to the formation of tornadoes. We will also look at some of the more infamous, destructive, and deadly tornadoes to hit the United States. Hopefully, by the end of this chapter you will have insight as to what you should do if ever confronted by a tornado—the most intense of all atmospheric circulations.

What Is a Tornado?

A **tornado** is a rapidly rotating column of air that blows around a small area of intense low pressure with a circulation that reaches the ground. A tornado's circulation is present on the ground either as a funnel-shaped cloud or as a swirling cloud of dust and debris. Sometimes called *twisters* or *cyclones*, tornadoes can assume a variety of shapes and forms that range from twisting rope-like funnels, to cylindrical-shaped funnels, to massive black wedge-shaped funnels, to funnels that resemble an elephant's trunk hanging from a large cumulonimbus cloud (see ❱ Fig. 12.1). Most tornadoes have wind speeds of less than 135 mi/hr, although violent tornadoes may have winds exceeding 250 mi/hr. A **funnel cloud** is a tornado whose circulation has not reached the ground. When viewed from above, the majority of North American tornadoes rotate counterclockwise about their central core of low pressure. A few have been seen rotating clockwise, but those are rare.

The diameter of most tornadoes is between 300 and 2000 ft (about 100 to 600 m), although some are less than 20 feet wide and others have diameters exceeding one mile. In fact, on May 22, 2004, one of the largest tornadoes on record touched down near Hallam, Nebraska,

❱ FIGURE 12.1 A tornado and a rainbow form over south-central Kansas during June, 2004. White spots in the sky are descending hailstones

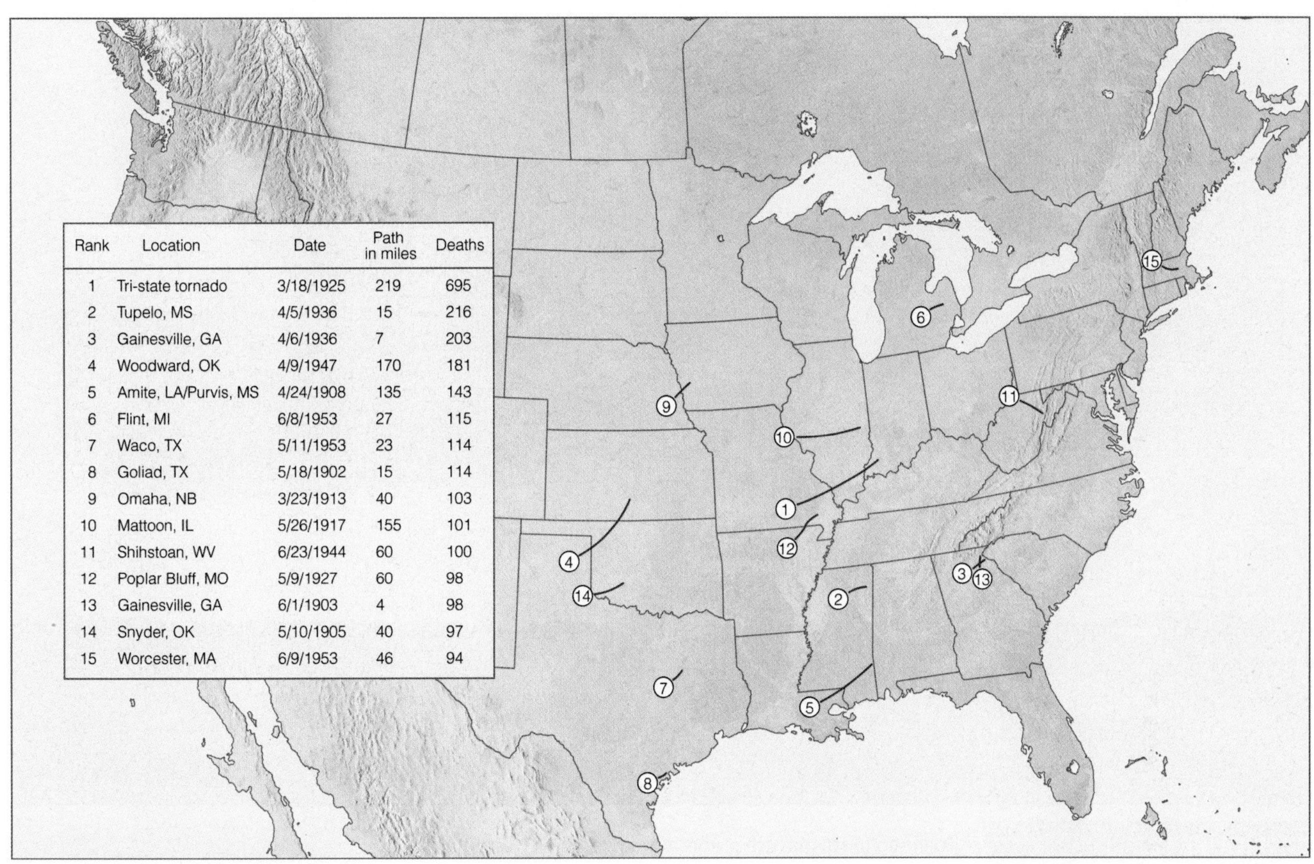

Rank	Location	Date	Path in miles	Deaths
1	Tri-state tornado	3/18/1925	219	695
2	Tupelo, MS	4/5/1936	15	216
3	Gainesville, GA	4/6/1936	7	203
4	Woodward, OK	4/9/1947	170	181
5	Amite, LA/Purvis, MS	4/24/1908	135	143
6	Flint, MI	6/8/1953	27	115
7	Waco, TX	5/11/1953	23	114
8	Goliad, TX	5/18/1902	15	114
9	Omaha, NB	3/23/1913	40	103
10	Mattoon, IL	5/26/1917	155	101
11	Shihstoan, WV	6/23/1944	60	100
12	Poplar Bluff, MO	5/9/1927	60	98
13	Gainesville, GA	6/1/1903	4	98
14	Snyder, OK	5/10/1905	40	97
15	Worcester, MA	6/9/1953	46	94

FIGURE 12.2 The fifteen deadliest tornadoes in the United States from 1900 to present. Red lines on the map represent the total distance (path length) over which the tornado moved.

with a diameter of about 2.5 mi (4 km). Tornadoes that form ahead of an advancing cold front are often steered by southwesterly winds and, therefore, tend to move from the southwest toward the northeast at speeds usually between 25 and 50 mi/hr. However, some have been clocked at speeds greater than 80 mi/hr. Most tornadoes last only a few minutes and have an average path length of about 4 mi (7 km). There are cases where they have reportedly traveled for hundreds of miles and have existed for many hours, such as the one that reportedly lasted over 7 hours and cut a path 290 mi (470 km) long through portions of Illinois and Indiana on May 26, 1917.

We can obtain a clearer picture of how tornadoes tend to move by examining Fig. 12.2, which lists the fifteen deadliest tornadoes in the United States from 1900 to the present and shows the paths they took. Notice that, in general, most move toward the northeast, although there are a number of exceptions. For example, look at the direction of movement of the tornadoes in West Virginia and Massachusetts. Also, notice that the deadliest tornado of all stayed on the ground for the longest distance—291 miles—and lasted for 3 hours and 30 minutes, which gives this tornado an average speed of nearly 63 mi/hr. We now know that it is extremely unlikely that a tornado will remain on the ground for such a long distance, so it is likely that the damage path measured for this tornado (and possibly others) represents the paths of several tornadoes, each spawned by the same supercell thunderstorm as it moved along.

EXTREME WEATHER WATCH

The deadliest tornadoes in the world have occurred in Bangladesh. In fact, on April 26, 1989, the deadliest tornado in recorded history killed 1300 people in a village north of the capital city of Dhaka. The high death toll was a result of population density and poor infrastructure. Tornadoes in Bangladesh occur most frequently at the beginning of the monsoon season, typically in April and May, when cool, dry air spills off the Tibetan Plateau and collides with warm, moist air streaming up from the Bay of Bengal.

FIGURE 12.3 A tornado in its mature stage roars over the Great Plains.

Tornado Life Cycle

Major tornadoes usually evolve through a series of stages. The first stage is the *dust-whirl stage*, where dust swirling upward from the surface marks the tornado's circulation on the ground and a short funnel often extends downward from the thunderstorm's base. Damage during this stage is normally light. As the tornado increases in strength, it enters its *mature stage*. During this stage, damage normally is most severe as the funnel reaches its greatest width and is almost vertical (see Fig. 12.3). As the tornado moves out of its mature stage, the width of the funnel shrinks and becomes more tilted. At the surface, the width of the damage swath narrows, although the tornado may still be capable of inflicting intense and sometimes violent damage. The final stage, called the *decay stage*, usually finds the tornado stretched into the shape of a rope. Normally, the tornado becomes greatly contorted before it finally dissipates. Although these are the main stages of a major tornado, minor tornadoes may actually skip the mature stage, and go directly into the decay stage. However, when a tornado reaches its mature stage, its circulation usually stays in contact with the ground until it dissipates.

Tornado Occurrence and Distribution

Tornadoes occur in many parts of the world, but no country experiences more tornadoes than the United States, which, in recent years, averages more than 1000 annually and experienced a record 1722 tornadoes during 2004. Although tornadoes have occurred in every state, including Alaska and Hawaii, the greatest number occur in the tornado belt, or **tornado alley**, of the Central Plains, which stretches from central Texas to Nebraska (see Fig. 12.4). The tornado maximum that stretches across Alabama and Mississippi is sometimes called "Dixie Alley."*

The Central Plains region is most susceptible to tornadoes because it often provides the proper atmospheric setting for the development of the severe thunderstorms** that spawn tornadoes. You may recall from Chapter 11 on p. 313 that over the Central Plains (especially in spring) warm, humid surface air is overlain by cooler,

*Many of the tornadoes that form along the Gulf Coast are generated by thunderstorms embedded within the circulation of hurricanes.

**Another reason that the Central Plains of the United States experiences so many tornadoes—more than anywhere else in the world—is the fact that there are no large west-to-east trending mountain ranges in the heart of North America to impede the clashing of warm, humid, subtropical air with cold, dry polar air.

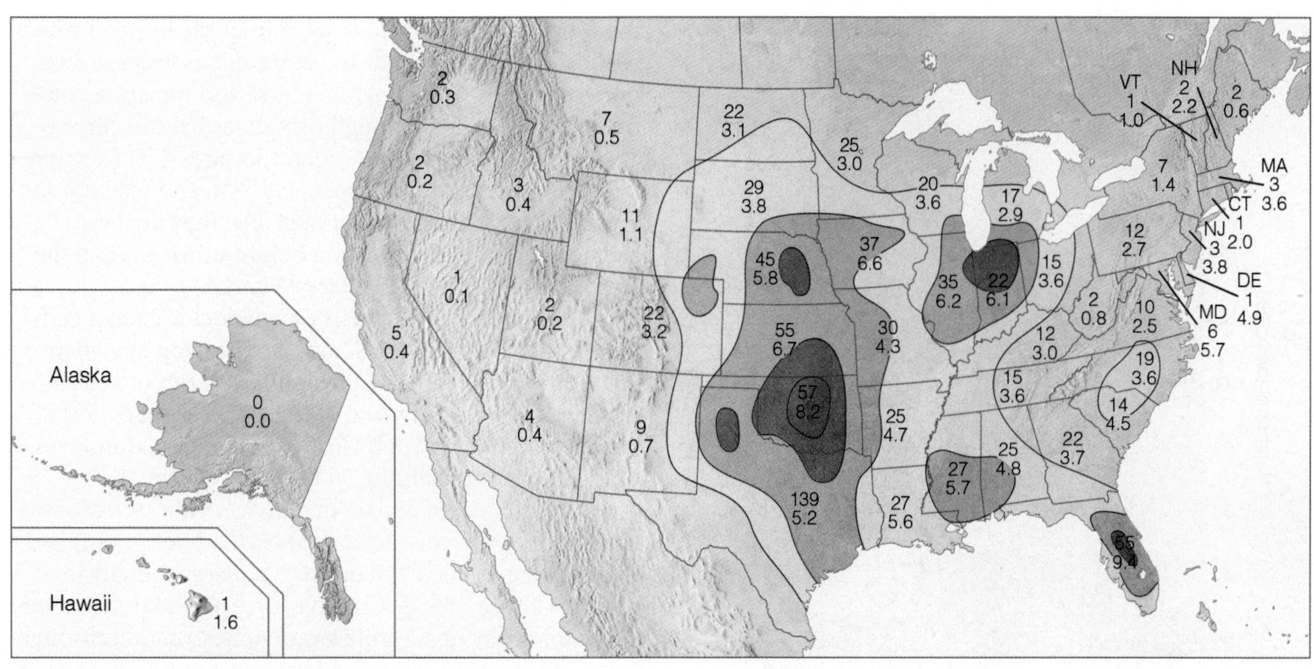

FIGURE 12.4 Tornado incidence by state. The upper figure shows the average annual number of tornadoes observed in each state from 1953–2004. The lower figure is the average annual number of tornadoes per 10,000 square miles in each state during the same period. The darker the shading, the greater the frequency of tornadoes. (NOAA)

drier air aloft, producing a conditionally unstable atmosphere. When a strong vertical wind shear exists (usually provided by a low-level jet and by the polar jet stream) and the surface air is forced upward, large supercell thunderstorms capable of spawning tornadoes may form (see Fig. 12.5). Therefore, tornado frequency is highest during the spring and lowest during the winter when the warm surface air is normally absent.

The frequency of tornadic activity shows a seasonal shift. For example, during the winter, tornadoes are most likely to form over the southern Gulf states when the polar-front jet is above this region, and the contrast between warm and cold air masses is greatest. In spring, humid Gulf air surges northward; contrasting air masses and the jet stream also move northward and tornadoes become more prevalent from the southern Atlantic states westward into the southern Great Plains. In summer, the contrast between air masses lessens, and the jet stream is normally near the Canadian border; hence, tornado activity tends to be concentrated from the northern plains eastward to New York State.

In Fig. 12.6 we can see that about 70 percent of all tornadoes in the United States develop from March to July. The month of May normally has the greatest number of tornadoes* (the average is about 6 per day) while

*During May, 2003, a record 516 tornadoes touched down in the United States (an average of over 16 per day)—the most in any month ever.

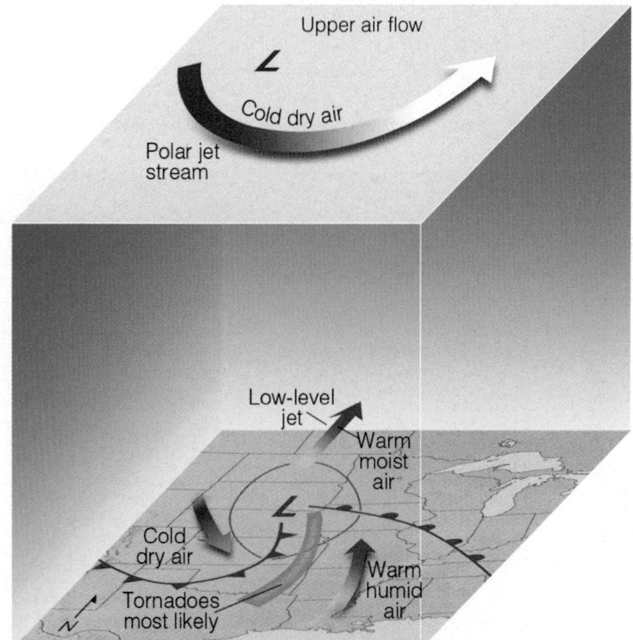

FIGURE 12.5 Tornadoes are most likely to form in a conditionally unstable atmosphere where there is strong vertical wind shear for supercell thunderstorm development. Here, wind speeds increase with height and winds change direction with height, usually from more southerly at the surface to more westerly at higher levels. The region favorable for tornado development is shown in yellow.

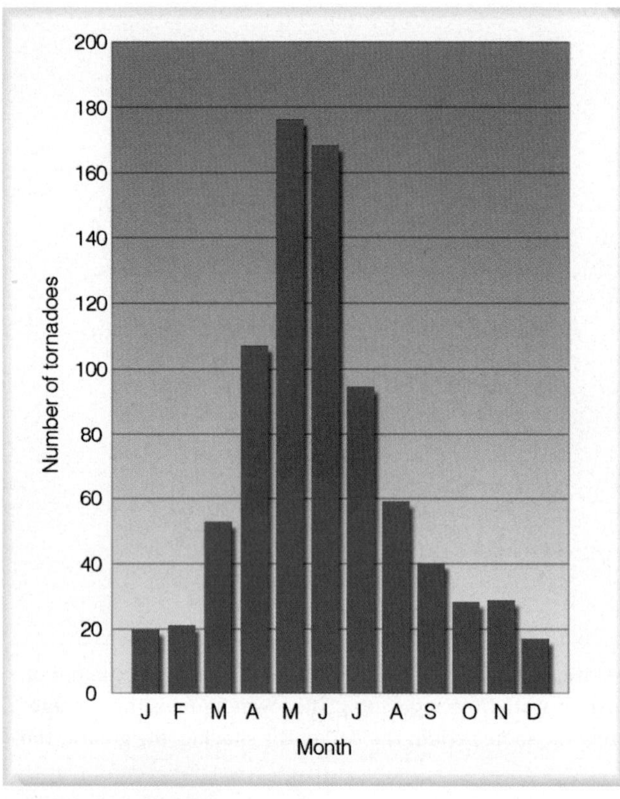

FIGURE 12.6 Average number of tornadoes during each month in the United States. (NOAA)

the most violent tornadoes seem to occur in April when vertical wind shear tends to be present as well as when horizontal and vertical temperature and moisture contrasts are greatest. Although tornadoes have occurred at all times of the day and night, they are most frequent in the late afternoon (between 4:00 P.M. and 6:00 P.M.), when the surface air is most unstable; they are least frequent in the early morning before sunrise, when the atmosphere is most stable (see Fig. 12.7).

Although large, destructive tornadoes are most common in the Central Plains, they can develop anywhere if conditions are right. For example, a series of at least 36 tornadoes, more typical of those that form over the plains, marched through North and South Carolina on March 28, 1984, claiming 59 lives and causing hundreds of millions of dollars in damage. One tornado was enormous, with a diameter of at least 2.5 mi (4 km) and winds that exceeded 230 mi/hr. No place is totally immune to a tornado's destructive force. On March 1, 1983, a rare tornado cut a 3-mile swath of destruction through downtown Los Angeles, California, damaging more than 100 homes and businesses, and injuring 33 people.

Tornadoes are also rare in Utah, with only about two per year being reported. However, during August, 1999, a tornado rampaged through downtown Salt Lake City. While only on the ground for about 5 miles, the tornado damaged more than 120 homes, injured a dozen people, produced a fatality, and caused over $50 million in damage.

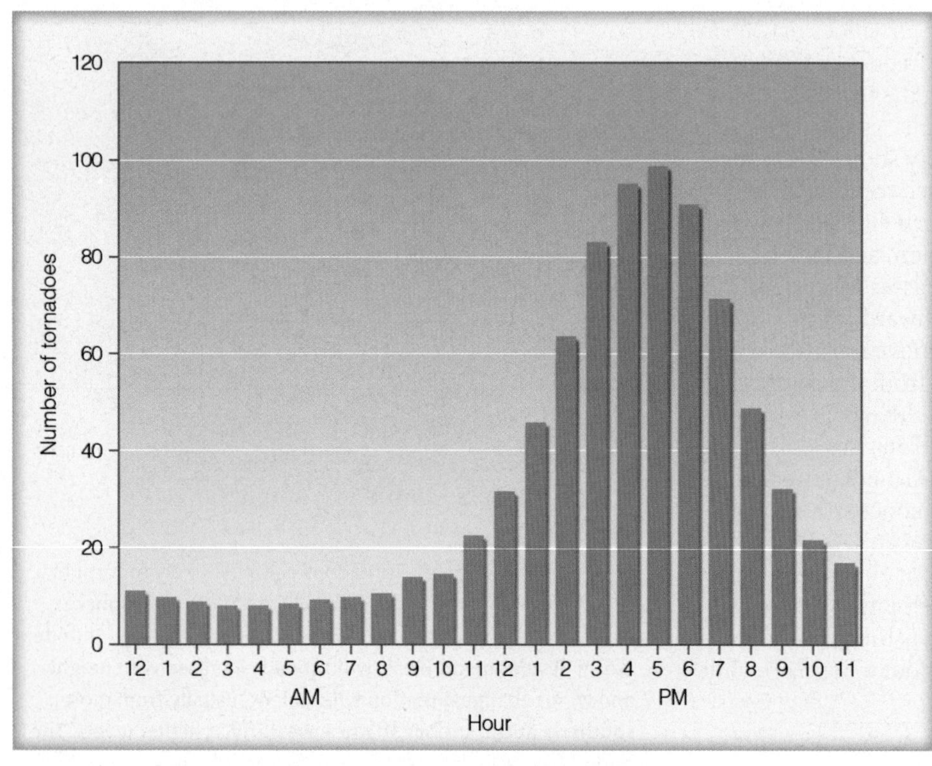

FIGURE 12.7 Average number of tornadoes reported in the United States for every hour of the day from 1950 to 2008. (This figure is an average and does not represent the hourly distribution for any one state, as tornadoes in Florida tend to be more frequent in the early afternoon.)

FIGURE 12.8 A large wedge-shaped tornado moves northwestward directly for Windsor, Colorado, on May 22, 2008. The photo (taken by a webcam) shows hail the size of golf balls falling from the thunderstorm and covering the ground.

During May, 2008, an unusually large and powerful tornado, packing winds of at least 150 mi/hr, ripped a 35-mile path of destruction through northern Colorado and straight through the town of Windsor. This large wedge-shaped storm (see Fig. 12.8) measured nearly a mile wide at times and stayed on the ground for 45 minutes. It caused millions of dollars in damage and took one life. Although tornadoes of this size are rare along the Front Range of the Rockies, what is even more unusual is that this tornado took a path from southeast to northwest, which is quite different from the southwest to northeast direction taken by most twisters.

In the central part of the United States, where tornadoes are most prevalent, the statistical chance that a tornado will strike a particular place in any given year is quite small. However, tornadoes can provide many exceptions to statistics. Oklahoma City, for example, has been struck by tornadoes at least 34 times in the past 100 years. And the little town of Codell, Kansas, was hit by tornadoes in 3 consecutive years—1916, 1917, and 1918—and each time on the same date: May 20! Considering the many millions of tornadoes that must have formed during the geological past, it is likely that at least one actually moved across the land where your home is located, especially if it is in the Central Plains.

Tornado Winds

The strong winds of a tornado can destroy buildings, uproot trees, and hurl all sorts of lethal missiles into the air. People, animals, and home appliances all have been picked up, carried several miles, then deposited. Tornadoes have accomplished some astonishing feats, such as lifting a railroad coach with its 117 passengers and dumping it in a ditch 80 feet away. Showers of toads and frogs have poured out of a cloud after tornadic winds sucked them up from a nearby pond. Other oddities include chickens losing all of their feathers, pieces of straw being driven into metal pipes, and frozen hot dogs being driven into concrete walls. Miraculous events have occurred, too. In one instance, a schoolhouse was demolished and the 85 students inside were carried over 100 yards without one of them being killed. How fast must the wind be blowing for a tornado to pick up a large animal, such as a cow, and whirl it through the air? If you are not sure, read the Focus section on p. 342.

Our earlier knowledge of the furious winds of a tornado came mainly from observations of the damage done and the analysis of motion pictures. Today more accurate wind measurements are made with Doppler radar. Because of the destructive nature of the tornado, it was once thought that it packed winds greater than

FIGURE 12.9 The total wind speed of a tornado is greater on one side than on the other. When facing an onrushing tornado, the strongest winds will be on your left side.

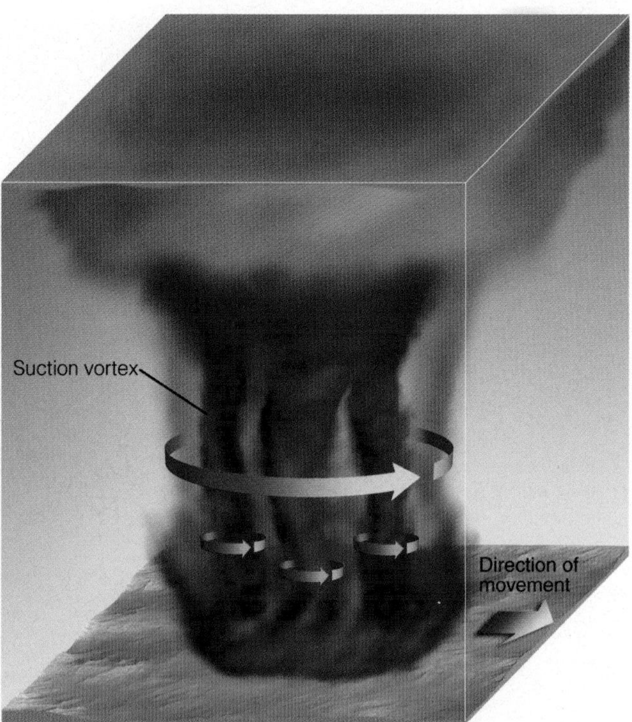

FIGURE 12.10 A powerful multi-vortex tornado with three suction vortices.

500 mi/hr. However, studies conducted after 1973 reveal that even the most powerful twisters seldom have winds exceeding 250 mi/hr, and most tornadoes have winds of less than 135 mi/hr. Nevertheless, being confronted with even a small tornado can be terrifying.

When a tornado is approaching from the southwest, its strongest winds are on its southeast side. We can see why in Fig. 12.9. The tornado is heading northeast at 50 mi/hr. If its rotational speed is 100 mi/hr, then its forward speed will add 50 mi/hr to its southeastern side (position D) and subtract 50 mi/hr from its northwestern side (position B). Hence, the most destructive and extreme winds will be on the tornado's southeastern side.

Many violent tornadoes contain smaller whirls that rotate within them. Such tornadoes are called **multi-vortex tornadoes** and the smaller whirls are called **suction vortices** (see Fig. 12.10). Suction vortices are only about 30 feet or so in diameter, but they rotate very fast and can do a great deal of damage.

The damage caused by a multi-vortex tornado may not be evenly distributed along the path of the twister. The reason for this effect is that the rotating suction vortices add rotating winds to the spinning tornado, as shown in Fig. 12.11. For example, if the tornado in Fig. 12.11 is moving northeast at a speed of 50 mi/hr, and its rotating winds are 100 mi/hr, then the wind speed on the tornado's southeast side would be 150 mi/hr. If a suction vortex is spinning at 50 mi/hr and located in the large tornado's southeastern side, then the total wind speed for the tornado on this side would be 50 + 100 + 50, or 200 mi/hr. This situation, where stronger winds exist at different regions within a large violent multi-vortex tornado, explains why some homes in the tornado's path are completely obliterated while others sustain considerable damage, but are left standing.

SEEKING SHELTER The high winds of the tornado cause the most damage as walls of buildings buckle and collapse when blasted by the extreme wind force and by debris carried by the wind. Also, as high winds blow over a roof, lower air pressure forms above the roof. The greater air pressure inside the building then lifts the roof just high enough for the strong winds to carry it away. A similar effect occurs when the tornado's intense low-pressure center passes overhead. Because the pressure in the center of a tornado may be more than 100 mb (3 in.) lower than that of its surroundings, there is a momentary drop in outside pressure when the tornado is above the structure. It was once thought that opening windows and allowing inside and outside pressures to equalize would minimize the chances of the building exploding.

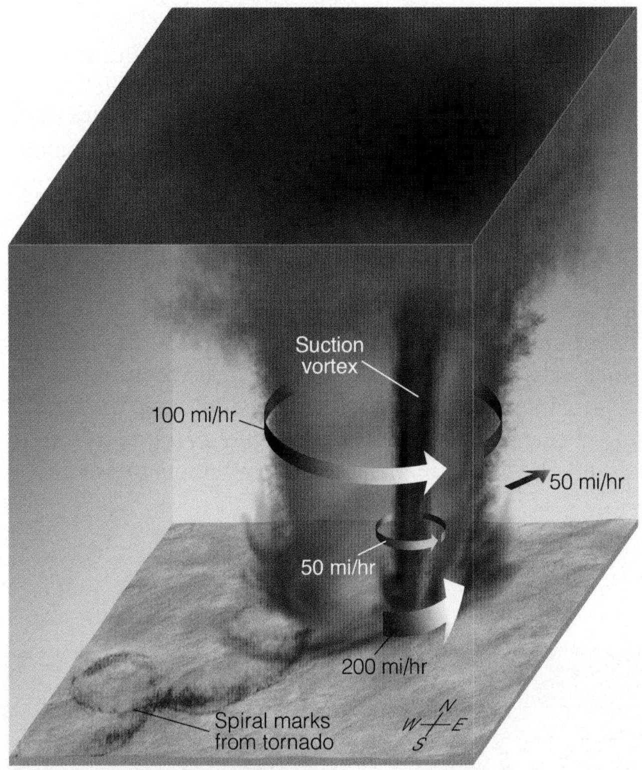

FIGURE 12.11 A suction vortex (or multiple suction vortices) within a large tornado can add substantial wind speed to the tornado. Tornadoes containing suction vortices often produce spiral ground markings.

However, it is now known that opening windows during a tornado actually increases the pressure on the opposite wall and *increases* the chances that the building will collapse. (The windows are usually shattered by flying debris anyway.) So stay away from windows. Damage from tornadoes may also be inflicted on people and structures by flying debris. Hence, the wisest course to take when confronted with an approaching tornado is to *seek shelter immediately*.

At home, take shelter in a basement. In a large building without a basement, the safest place is usually in a small room, such as a bathroom, closet, or interior hallway, preferably on the lowest floor and near the middle of the edifice. Pull a mattress around you as the handles on the side make it easy to hang onto. Wear a bike or football helmet to protect your head from flying debris. At school, move to the hallway and lie flat with your head covered. In a mobile home, leave immediately and seek substantial shelter. If none exists, lie flat on the ground in a depression or ravine. Don't try to outrun an oncoming tornado in a car or truck, as tornadoes often cover erratic paths with speeds sometimes exceeding 80 mi/hr. Stop your car and let the tornado go by or turn around on the road's shoulder and drive in the opposite direction. And do not take shelter under a freeway overpass, as the tornado's winds are actually funneled (strengthened) by the overpass structure. If caught outdoors in an open field, look for a ditch, streambed, or ravine, and lie flat with your head covered.

When tornadoes are likely to form during the next few hours, a **tornado watch** is issued by the Storm Prediction Center in Norman, Oklahoma, to alert the public that tornadoes may develop within a specific area during a certain time period. Many communities have trained volunteer spotters, who look for tornadoes after the watch is issued. (If a tornado is spotted in the watch area, keep abreast of its movement by listening to the NOAA Weather Radio.) Once a tornado is spotted—either visually or on a radar screen—a **tornado warning** is issued by the local National Weather Service Office.* In some communities, sirens are sounded to alert people of the approaching storm. Radio and television stations interrupt regular programming to broadcast the warning. Although not completely effective, this warning system is apparently saving many lives. Despite the large increase in population in the tornado belt during the past 30 years, tornado-related deaths have actually shown a decrease (see Table 12.1).

In recent years, however, an alarming statistic shows that over 45 percent of all tornado fatalities occurred in mobile homes. For example, in 2007, tornadoes claimed 81 lives, 52 in mobile homes (64 percent), 16 in permanent structures, 10 in businesses, 2 in vehicles, and only 1 outside in an open area.

*In October, 2007, the National Weather Service launched a new, more specific tornado warning system called *Storm Based Warnings*. The new system provides more precise information on where a tornado is located and where it is heading.

Table 12.1

Average Annual Number of Tornadoes and Tornado Deaths by Decade*

DECADE	TORNADOES/YEAR	DEATHS/YEAR
1950–59	480	148
1960–69	681	94
1970–79	858	100
1980–89	819	52
1990–99	1,220	56
2000–07	1,319†	52

*Slightly less than a ten-year period.
†More tornadoes are being reported as populations increase and tornado-spotting technology improves.

Look at Table 12.1 and observe that the average yearly number of tornadoes reported in the United States over the past six decades appears to be increasing. This increase may be due to the fact that more people now live in areas where tornadoes are prevalent; hence, many tornadoes that would have gone undetected in years past are now reported. Another reason for the tornado increase may be due to the improvements in radar technology that makes identifying a tornado easier today than in the past. Although the overall yearly average number of tornadoes reported in the United States appears to be increasing, this increase has not been observed everywhere, as the state of Oklahoma—situated in the heart of tornado alley—has not witnessed an increase.

THE FUJITA SCALE In the 1960s, the late Dr. T. Theodore Fujita, a noted authority on tornadoes at the University of Chicago, proposed a scale (called the **Fujita Scale**) for classifying tornadoes according to their rotational wind speed. The tornado winds are estimated based on the damage caused by the storm. However, classifying a tornado based solely on the damage it causes is rather subjective. But the scale became widely used and is presented in ❯ Table 12.2.

The original Fujita scale, implemented in 1971, was based mainly on tornado damage incurred by a frame house. Because there are many types of structures susceptible to tornado damage, a new scale came into effect in February, 2007. Called the *Enhanced Fujita Scale*, or simply the **EF Scale,** the new scale attempts to provide

FOCUS ON
A SPECIAL TOPIC

Flying Cows or "the Herd Shot Around the World"

It seems as if every movie that includes tornadoes has an obligatory flying cow scene. But can a cow be lifted airborne by a tornado? With wind speeds exceeding 150 mi/hr, it is likely that cows (and many unanchored objects) could be blown sideways. There is, in fact, ample evidence of cows being hurt and even killed by violent winds of a tornado blowing them off their feet. But airborne?

In the infamous 1925 Tri-State Tornado, there is documentation of a cow being picked up and dropped on the roof of a destroyed building. In 1915, near Great Bend, Nebraska, a tornado "flew" five horses a quarter of a mile and landed them unhurt and apparently still hitched to the same rail, giving new meaning to the term "quarter horse." Given these and other remarkable stories of airborne animals, it seems clear that such occurrences do happen. So, at what wind speed can this happen?

Figure 1 Can cows be lifted by a tornado and whirled through the air?

Not surprisingly, there have been no studies of cow flight versus wind speed. Identifying a specific speed would be difficult because it is less likely that the horizontal speed is as important on animal levitation as the vertical velocity. For an animal to be airborne, the upward motion needs to overcome gravity. The vertical wind speed in tornadoes has not been measured, though estimates in excess of 100 mi/hr straight up have been suggested. But these fast upward motions probably don't occur until the air is a few hundred feet above the surface. So for a large animal like a cow to become airborne it needs to be lifted first to a point where it can be levitated by the updraft, and the mechanisms to explain how this phenomenon can happen remain a mystery.

Table 12.2

Original Fujita Scale for Damaging Winds

SCALE	CATEGORY	MI/HR	KNOTS	EXPECTED DAMAGE
F0	Weak	40–72	35–62	Light: tree branches broken, sign boards damaged
F1		73–112	63–97	Moderate: trees snapped, windows broken
F2	Strong	113–157	98–136	Considerable: large trees uprooted, weak structures destroyed
F3		158–206	137–179	Severe: trees leveled, cars overturned, walls removed from buildings
F4	Violent	207–260	180–226	Devastating: frame houses destroyed
F5*		261–318	227–276	Incredible: structures the size of autos moved over 100 meters, steel-reinforced structures highly damaged

*The scale continues up to a theoretical F12. Very few (if any) tornadoes have wind speeds in excess of 318 mi/hr.

a wide range of criteria in estimating a tornado's winds by using a set of 28 damage indicators. These indicators include items such as small barns, mobile homes, schools, and trees. Each item is then examined for the degree of damage it sustained. The combination of the damage indicators along with the degree of damage provides a range of probable wind speeds and an EF rating for the tornado. The wind estimates for the EF scale are given in Table 12.3.

Figure 12.12 shows a house situated somewhere on the Great Plains of the United States or Canada, and Fig. 12.13 shows the damaging effect that tornadoes ranging in intensity from EF0 to EF5 can have on this house and its surroundings. Notice that an EF0 tornado causes only minimal damage, whereas an EF5 completely demolishes the house and sweeps it off its foundation.

Statistics reveal that the majority of tornadoes are relatively weak, with wind speeds less than about 110 mi/hr. Only a few percent each year are classified as violent, with perhaps one or two EF5 tornadoes reported annually (although several years may pass without the United States experiencing an EF5). However, it is the violent tornadoes that account for the majority of tornado-related deaths. As an example, a powerful F5 tornado marched through the southern part of Andover, Kansas, on the evening of April 26, 1991. The tornado, which stayed on the ground for nearly 70 miles, destroyed more than 100 homes and businesses, injured several hundred people, and out of the 39 tornado fatalities in 1991, this violent tornado alone took the lives of 17.

The first tornado with an EF5 (or F5) rating since May 3, 1999, roared through the town of Greensburg, Kansas, on the evening of May 4, 2007. The tornado with winds estimated at 205 mi/hr and a width approaching 2 miles, completely destroyed over 95 percent of the town, and left about 5 percent severely damaged (see Fig. 12.14). The tornado took 11 lives and probably

Table 12.3

Modified (EF) Fujita Scale for Damaging Winds

EF SCALE	MI/HR*	KNOTS
EF0	65–85	56–74
EF1	86–110	75–95
EF2	111–135	96–117
EF3	136–165	118–143
EF4	166–200	144–174
EF5	>200	>174

*The wind speed is a 3-second gust estimated at the point of damage, based on a judgment of damage indicators.

EXTREME WEATHER WATCH

The strongest tornado ever recorded west of the Great Plains occurred on July 21, 1987, in the unlikely location of northwest Wyoming's Teton Wilderness Area, just south of Yellowstone National Park. The F4 twister carved a mile-wide path for 24 miles across mountainous terrain, over 8000 feet above sea level. It even crossed the Continental Divide, where the elevation was 10,170 feet above sea level. Fortunately, the area affected was uninhabited, so no injuries or property damage were reported.

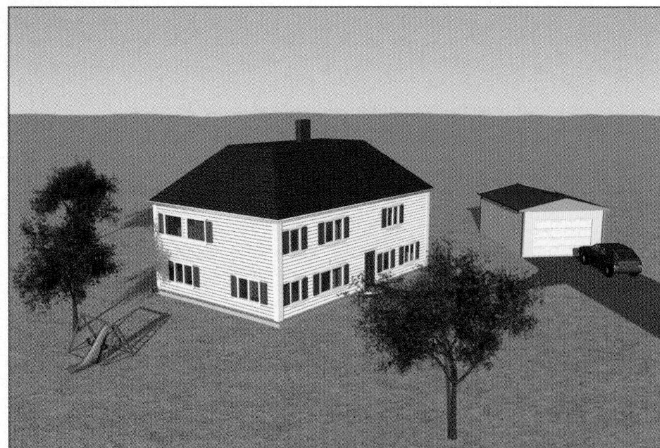

FIGURE 12.12 A house situated on the Great Plains. Observe in Fig. 12.13 how tornadoes of varying EF intensity can damage this house and its surroundings.

FIGURE 12.13 Damage to the house in Fig. 12.12 and its surroundings caused by tornadoes of varying EF intensity.

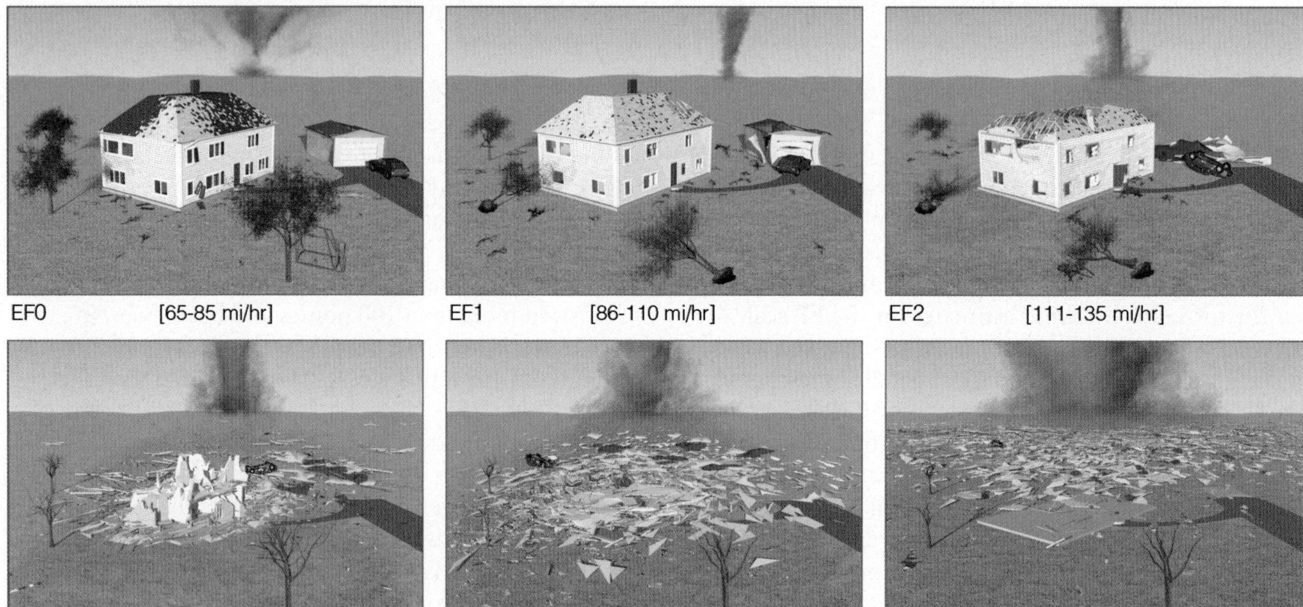

FIGURE 12.14 Aerial view of Greensburg, Kansas, (a) before the tornado of May 4, 2007, and (b) several days after the EF5 tornado destroyed the town.

more would have perished had it not been for the tornado warning issued by the National Weather Service and the sirens in the town signaling "take cover" about 20 minutes before the tornado struck.

One important reason for the number of deaths and extensive damage being caused by violent tornadoes is that, as the wind speed doubles, the force of the wind exerted on an object increases by a factor of four. Hence, the 200 mi/hr winds of an EF4 tornado exert four times as much force on a building as do the 100 mi/hr winds of an EF1.

Tornado Outbreaks

Each year, tornadoes take the lives of many people. The yearly average is less than 100, although over 100 may die in a single day. The deadliest tornadoes are those that occur in *families*; that is, different tornadoes spawned by the same thunderstorm. (Some thunderstorms produce a sequence of several tornadoes over 2 or more hours and over distances of 50 miles or more.) Tornado families often are the result of a single, long-lived supercell thunderstorm. When a large number of tornadoes (typically 6 or more) form over a particular region, this constitutes what is termed a **tornado outbreak**.

Table 12.4 shows the 10 deadliest tornado outbreaks in the United States since 1900. The deadly outbreak of March 18, 1925, contained the infamous tri-state tornado that ripped through Murphysboro, Illinois, killing 234 people (see Fig. 12.2 on p. 335). A more recent violent outbreak occurred on April 3 and 4, 1974. During a 16-hour period, 148 tornadoes cut through parts of 13 states from Illinois to Alabama and eastward to Virginia, killing 315 people, injuring more than 6100, and causing an estimated $600 million in damages. This outbreak is the largest in the United States ever (see Table 12.5). Some of these tornadoes in this *super outbreak* were among the most powerful ever witnessed, as at least 6 tornadoes reached F5 intensity and 24 were classified as F4. A town especially hard hit by an F5 tornado was Xenia, Ohio. The tornado tore through the town during the afternoon of April 3, destroying 300 homes, taking 34 lives, and injuring 1600 others. The combined path of all the tornadoes during this outbreak amounted to 2598 mi (4182 km), well over half of the total path for an average year.

Figure 12.15a shows the surface weather map for the morning of April 3, 1974. Notice that warm, humid air is streaming northward ahead of a cold front and a low-pressure area centered over Kansas. The low moved rapidly northeastward, and 24 hours later was centered over Michigan (see Fig. 12.15b). Warm, humid surface air overlain by cold, dry air aloft set up a conditionally unstable atmosphere. Strong upper-level southwesterly winds helped produce strong vertical wind shear for the development of a series of squall lines that contained supercell thunderstorms that produced an incredible 148 tornadoes, their paths illustrated by solid red lines in Fig. 12.15b.

A particularly devastating outbreak occurred on May 3, 1999, when 78 tornadoes marched across parts of Texas, Kansas, and Oklahoma. One tornado (an F5), whose width at times reached one mile and whose wind speed was measured by Doppler radar at 318 mi/hr

Table 12.4

The 10 Deadliest Tornado Outbreaks in the United States Since 1900

RANKING	DATE	LOCATION	STRONG AND VIOLENT TORNADOES*	DEATHS
1	March 18, 1925	MO, IL, IN, KY	9	747
2	April 5-6, 1936	AR, TN, AL, MS, GA, SC	12	454
3	March 21-22, 1932	AL, TN, KY, GA, SC	36	330
4	April 23-24, 1908	AR, NE, TX, AL, LA, MS	30	324
5	April 3-4, 1974	United States and Canada	95	315
6	April 11, 1965	IA, WI, IL, IN, MI, OH	38	256
7	June 8-9, 1953	MI, MA	14	236
8	April 20, 1920	MS, AL, TN	31	224
9	May 8-9, 1927	TX, KS, AR, MO, IL	32	217
10	March 23-24, 1913	NE, IA, AR, IL, IN, MO, LA	14	199

*This number only represents an estimate of the strong and violent tornadoes associated with each outbreak.

Table 12.5

The 10 Largest Tornado Outbreaks (Greatest Number of Tornadoes) in the United States

RANKING	DATE	DEATHS	TOTAL NUMBER OF TORNADOES
1	April 3-4, 1974	315	148
2	May 3-5, 2007	14	123
3	September 19-23, 1967	5	111
4	May 26-27, 1973	22	99
5	November 21-23, 1992	26	95
6	May 4-5, 2003	37	94
7	February 5-6, 2008	57	87
8	May 22-23, 2004	1	85
9	May 18-19, 1995	4	80
10	March 28-31, 2007	5	80

(276 knots), moved through the southwestern section of Oklahoma City and cut through the suburb of Moore, Del City and Midwest City (see Fig. 12.16). Within its 40-mile path, it damaged or destroyed thousands of homes, injured nearly 600 people, claimed 38 lives, and caused over $1 billion in property damage, making this tornado the costliest on record.

More lives would have undoubtedly been lost during this devastating tornado outbreak had it not been for advanced warning provided by state, local, and private meteorologists and the National Weather Service. Many people living in inadequate structures, such as mobile homes, during the outbreak survived the powerful tornado winds by abandoning their homes in favor of

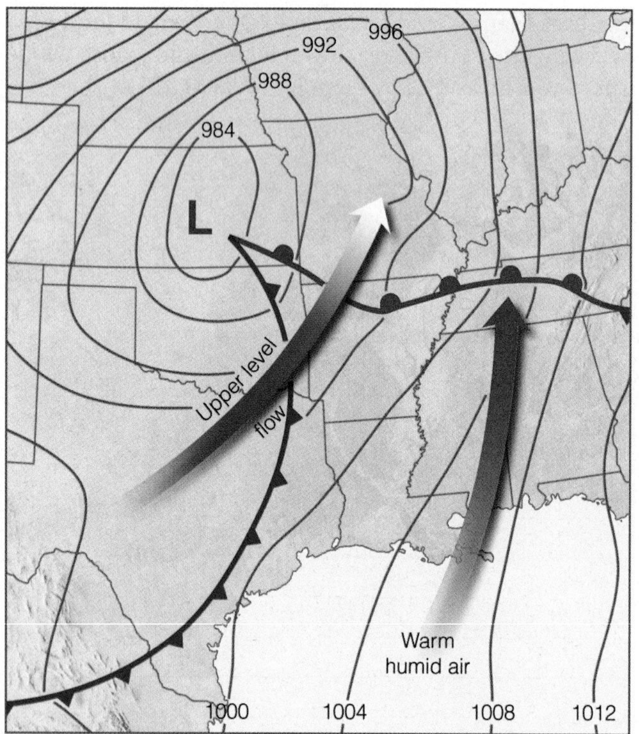

(a) April 3,1974

(b) April 4, 1974

FIGURE 12.15 (a) Surface weather map for the morning (8:00 A.M. CST) of April 3, 1974. (b) Surface weather map 24 hours later on April 4, 1974. The solid red lines on the map show the direction and total path length of the tornadoes during the superoutbreak of 1974.

© AP Photo/J. Pat Carter

FIGURE 12.16 The deadly tornado that destroyed scores of homes in Oklahoma City on May 3, 1999, smashes into a house near Newcastle, Oklahoma. The flash at the bottom of the tornado is a power line exploding.

© 1999 C. Doswell

FIGURE 12.17 Total destruction caused by an EF5 tornado that devastated parts of Oklahoma on May 3, 1999.

muddy ditches that were below ground level and out of the path of windblown debris. Many who stayed in the confines of their inadequate homes perished when the tornado literally blew their homes away, leaving only the foundation (see Fig. 12.17).

Glance back at Table 12.5 and notice that even though tornado outbreaks are more common during the spring, they can occur in all seasons of the year. For example, in late November, 1992, a widespread tornado outbreak extending from the Midwest to the Deep South, produced 95 twisters. And a late winter outbreak during February 5 and 6, 2008 (often referred to as "*the super Tuesday outbreak*"), produced 87 tornadoes from Illinois to Alabama, causing $500 million in damages and taking 57 lives.

Tornado Formation

Although not everything is known about the formation of a tornado, we do know that tornadoes tend to form with intense thunderstorms and that a conditionally unstable atmosphere is essential for their development. Most often they form with supercell thunderstorms in an environment with strong vertical wind shear. The rotating air of the tornado may begin within a thunderstorm and work its way downward, or it may begin at the surface and work its way upward. First, we will examine tornadoes that form with supercells; then we will examine nonsupercell tornadoes.

SUPERCELL TORNADOES Tornadoes that form with supercell thunderstorms are called **supercell tornadoes**. In Chapter 11, we learned that a supercell is a thunderstorm that has a single rotating updraft that can exist for hours. Recall also that supercells form in a region of strong, vertical wind shear that causes the updraft inside the storm to rotate (see Fig. 12.18).

How does the supercell updraft develop rotation? In Fig. 12.19a notice that there is wind direction shear, as the surface winds are southerly, and several thousand feet above the surface, they are northerly. There is also wind speed shear as the wind speed increases rapidly with height. This wind shear causes the air near the surface to rotate about a horizontal axis much like a pencil rotates around its long axis. Such horizontal tubes of spinning

EXTREME WEATHER WATCH

Can tornadoes "glow" at night? Although very rare, tornadoes have in fact been observed to glow at night. The best documented luminous tornado occurred in Toledo, Ohio, during the historic Palm Sunday tornado outbreak of April 11, 1965, when dozens of people reported seeing it, and someone actually photographed it. The cause of the glow remains a mystery, although some scientists believe the glow is due to static charges produced by the friction of flying debris within the funnel, whereas others propose that the glow comes from the reflection of light from the ground.

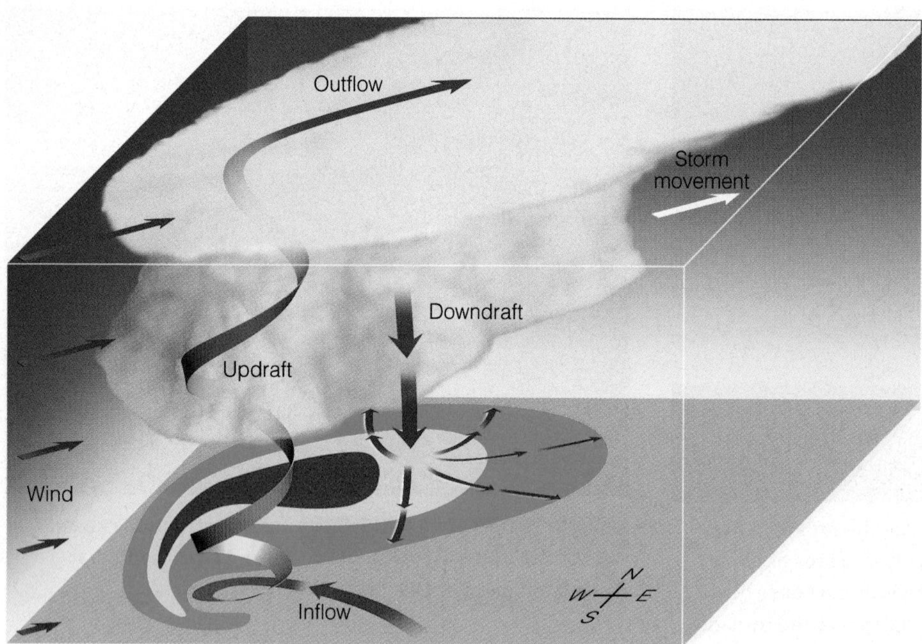

FIGURE 12.18

As warm, humid air is drawn into the supercell thunderstorm, it spins counterclockwise as it rises. Near the top of the storm, strong winds aloft push the rising air to the northeast. Heavy precipitation falling northeast of the updraft produces a strong downdraft. The separation of the updraft from the downdraft helps the storm maintain itself as a single entity, capable of existing for hours. (Regions beneath the supercell receiving precipitation are shown with colors: light precipitation is shown in green, heavier precipitation in yellow and regions receiving very heavy rain and hail are shown in red.)

air are called *vortex tubes.* (These spirally vortex tubes also form when a southerly low-level jet exists just above southerly surface winds.) If the strong updraft of a developing thunderstorm should tilt the rotating tube upward and draw it into the storm, as illustrated in Fig. 12.19b, the tilted rotating tube then becomes a rotating air column inside the storm.* The rising, spinning air is now part of the storm's structure called the *mesocyclone*—an area of lower pressure (a small cyclone) perhaps 3 to 6 mi (5 to 10 km) across. The rotation of the updraft lowers the pressure in the mid-levels of the thunderstorm, which acts to increase the strength of the updraft.*

As we learned in Chapter 11 the updraft is so strong in a supercell (sometimes 100 mi/hr) that precipitation

*The rising tube produces a core of both counterclockwise and clockwise rotation. The clockwise rotating core is not supported by the Coriolis force, and generally weakens rapidly.

*You can obtain an idea of what might be taking place in the supercell by stirring a cup of coffee or tea with a spoon and watching the low pressure form in the middle of the beverage.

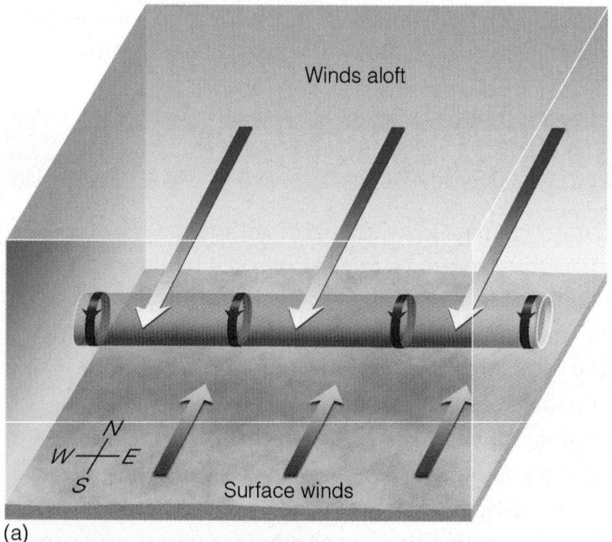

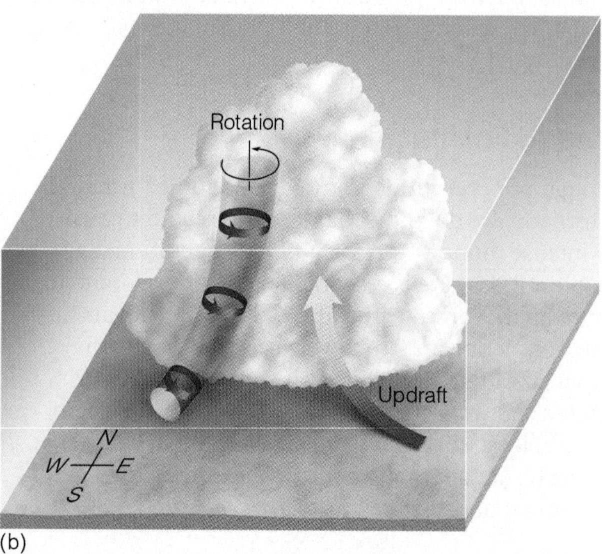

FIGURE 12.19 (a) A spinning vortex tube created by wind shear. (b) The strong updraft in the developing thunderstorm carries the vortex tube into the thunderstorm producing a rotating air column that is oriented in the vertical plane.

cannot fall through it. Southwesterly winds aloft usually blow the precipitation northeastward. If the mesocyclone persists, it can circulate some of the precipitation counterclockwise around the updraft. This swirling precipitation shows up on the radar screen, whereas the area inside the mesocyclone (nearly void of precipitation at lower levels) does not. The region inside the supercell where radar is unable to detect precipitation is known as the *bounded weak echo region (BWER)*. Meanwhile, as the precipitation is drawn into a cyclonic spiral around the mesocyclone, the rotating precipitation may, on the Doppler radar screen, unveil itself in the shape of a hook, called a **hook echo**, as shown in ⟫ Fig. 12.20.

At this point in the storm's development, the updraft, the counterclockwise swirling precipitation, and the surrounding air may all interact to produce the *rear-flank downdraft* (to the south of the updraft), as shown in ⟫ Fig. 12.21. The strength of the downdraft is driven by the amount of precipitation-induced cooling in the upper levels of the storm. The rear-flank downdraft appears to play an important role in producing tornadoes in classic supercells.

When the rear-flank downdraft strikes the ground, it may (under favorable shear conditions) interact with the *forward-flank downdraft* beneath the mesocyclone to initiate the formation of a tornado. At the surface, the

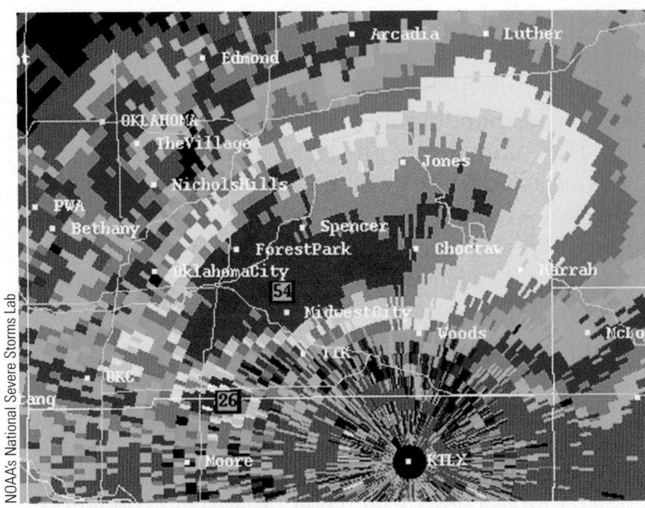

⟫ **FIGURE 12.20** A tornado-spawning supercell thunderstorm over Oklahoma City on May 3, 1999, shows a hook echo in its rainfall pattern on a Doppler radar screen. The colors red and orange represent the heaviest precipitation. Compare this precipitation pattern with the precipitation pattern illustrated in Fig. 12.18.

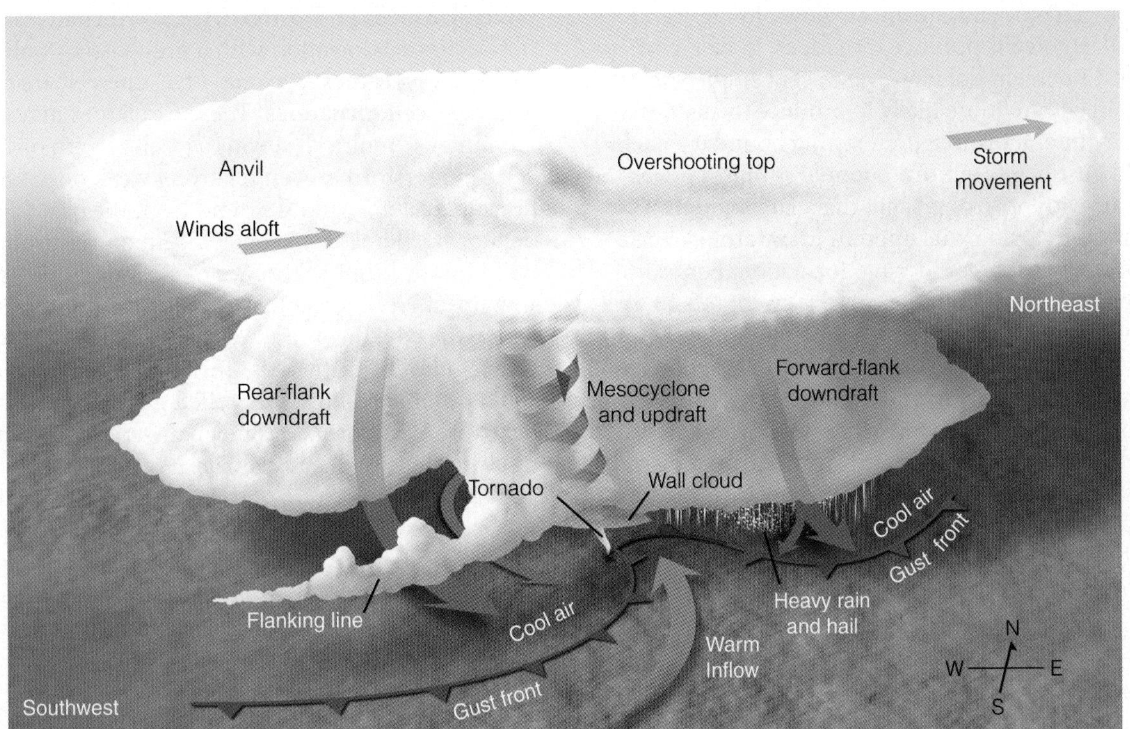

⟫ **FIGURE 12.21** A classic tornadic supercell thunderstorm showing updrafts and downdrafts, along with surface air flowing counterclockwise and in toward the tornado. The flanking line is a line of cumulus clouds that form as surface air is lifted into the storm along the gust front.

cool air of the rear-flank downdraft (and the forward-flank downdraft) sweeps around the center of the mesocyclone, effectively cutting off the rising air from the warmer surrounding air. The lower half of the updraft now rises more slowly. The rising updraft, which we can imagine as a column of air, now shrinks horizontally and stretches vertically. This *vertical stretching* of the spinning column of air causes the rising, spinning air to spin faster.* If this stretching process continues, the rapidly rotating air column may shrink into a narrow column of rapidly rotating air—a *tornado vortex*.

As air rushes upward and spins around the low-pressure core of the vortex, the air expands, cools, and, if sufficiently moist, condenses into a visible cloud—the *funnel cloud*. As the air beneath the funnel cloud is drawn into its core, the air cools rapidly and condenses, and the funnel cloud descends toward the surface. Upon reaching the ground, the tornado's circulation usually picks up dirt and debris, making it appear both dark and ominous. While the air along the outside of the funnel is spiraling upward, Doppler radar reveals that, within the core of violent tornadoes, the air is descending toward the extreme low pressure at the ground (which may be 100 mb lower than that of the surrounding air). As the air descends, it warms, causing the cloud droplets to evaporate. This process leaves the core free of clouds. Supercell tornadoes usually develop near the right rear sector of the storm, on the southwestern side of a northeastward-moving storm, as shown in Fig. 12.21.

Not all supercells produce tornadoes; in fact, perhaps less than 15 percent do. However, recent studies reveal that supercells are more likely to produce tornadoes when they interact with a pre-existing boundary, such as an old gust front (outflow boundary) that supplies the surface air with horizontal spin that can be tilted and lifted into the storm by its updraft. Many atmospheric situations may suppress tornado formation. For example, if the precipitation in the cloud is swept too far away from the updraft, or if too much precipitation wraps around the mesocyclone, the necessary interactions that produce the rear flank downdraft are disrupted, and a tornado is not likely to form. Moreover, tornadoes are not likely to form if the supercell is fed warm, moist air that is elevated above a deep layer of cooler surface air.

As we have seen, the first sign that a supercell is about to give birth to a tornado is the sight of *rotating clouds* at the base of the storm.** If the area of rotating clouds lowers, it becomes the *wall cloud*. Notice in Fig. 12.21 that the tornado extends from within the wall cloud to the earth's surface. Sometimes the air is so dry that the swirling, rotating wind remains invisible until it reaches the ground and begins to pick up dust. Unfortunately, people have mistaken these "invisible tornadoes" for dust devils, only to find out (often too late) that they were not. Occasionally, the funnel cannot be seen due to falling rain, clouds of dust, or darkness. Even when not clearly visible, many tornadoes have a distinctive roar that can be heard as the tornado approaches. This sound, which has been described as "a roar like a thousand freight trains," appears to be loudest when the tornado is touching the surface. However, not all tornadoes make this sound and, when these storms strike, they become silent killers.

Certainly, the likelihood of a thunderstorm producing a tornado increases when the storm becomes a supercell, but not all supercells produce tornadoes. And not all tornadoes come from rotating thunderstorms (supercells).

Watching a tornado (from a safe distance) can be an exhilarating experience. Today, many people (professionals and amateurs alike) spend days, weeks, and sometimes months hunting down these elusive violent storms. But tornado-chasing can be very dangerous, as illustrated in the Focus section on p. 351.

NONSUPERCELL TORNADOES Tornadoes that do not occur in association with a pre-existing wall cloud (or a mid-level mesocyclone) of a supercell are called **nonsupercell tornadoes**. These tornadoes may occur with intense multicell storms as well as with ordinary cell thunderstorms, even relatively weak ones.* Some nonsupercell tornadoes extend from the base of a thunderstorm whereas others may begin on the ground and build upwards in the absence of a condensation funnel.

Nonsupercell tornadoes may form along a gust front where the cool downdraft of the thunderstorm forces warm, humid air upwards. Tornadoes that form along a gust front are commonly called **gustnadoes**. These relatively weak tornadoes normally are short-lived and rarely inflict significant damage. Gustnadoes are often seen as a rotating cloud of dust or debris rising above the surface (see ❱ Fig. 12.22).

Occasionally, rather weak, short-lived tornadoes will occur with rapidly building cumulus congestus clouds. Tornadoes such as these commonly form over east-central Colorado. Because they look similar to

*As the rotating air column stretches vertically into a narrow column, its rotational speed increases. This situation is called the *conservation of angular momentum*.

**Occasionally, people will call a sky dotted with mammatus clouds "a tornado sky." Mammatus clouds may appear with both severe and nonsevere thunderstorms as well as with a variety of other cloud types (see Chapter 4). Mammatus clouds are not funnel clouds, do not rotate, and their appearance has no relationship to tornadoes.

*Ordinary cell thunderstorms and multicell thunderstorms are described in Chapter 11, beginning on p. 298.

FOCUS ON
EXTREME WEATHER

Takin' a Chase on the Wild Side

On May 22, 2008, a team of undergraduate students from the University of Michigan and their professor positioned themselves for an encounter with a supercell thunderstorm near Oberlin, Kansas. The goal on this day was to film the genesis of the thunderstorm and hope that it might ultimately produce a tornado.

As the team sat pondering their next move, a dark cloud formed to their south. The cloud seemed to extend to the ground, making it impossible to distinguish any features. Radar indicated that the cloud contained rotation, but many clouds on this day had rotation, and not one had spawned a tornado. Suddenly, the cloud lifted, and a suspicious formation became visible beneath the cloud base. The sky at this point was so dark that blinking flasher lights from passing cars reflected off the dry road surface. Strong turbulent motions were visible inside the cloud as it moved northward toward the team. The approaching storm prompted the team to quickly move to the east.

Suddenly a large v-shaped tornado descended from the approaching thunderstorm, which made staying in their new location very dangerous. Two of the three vehicles in the team sped to the east to avoid the tornado, while the third vehicle stayed behind with a plan to head west if necessary. Despite staying safely to the west of the tornado, the car that stayed behind experienced a wind gust of 135 mi/hr and a 15-mb drop in pressure in less than two minutes as the twister passed nearby (see Fig. 2).

Meanwhile, the team in one of the cars that sped east of the storm was able to film the formation of the tornado, which quickly grew into a massive tornado with multiple vortices (see Fig. 3). The other car that sped east was behind the first car and unfortunately was overtaken by blinding rain and debris. In the nervous moments that followed, the driver of the vehicle had to stop in the middle of the highway and turn the vehicle into the wind to take advantage of the car's aerodynamic design. Straw, branches and all sorts of other debris pelted the vehicle. The winds increased dramatically and the car was rocked and pushed backward across the highway by winds estimated at over 150 mi/hr. Fortunately, the car remained upright.

After the tornado passed, the wheel wells of the car (which had been blasted by the twister) were totally stuffed with debris, and straw was sticking out of every exposed crack in the vehicle's doors and windows. It was a close call, but everyone was safe. The lesson here is that chasing tornadoes requires an understanding of storms and their behavior. Even with such knowledge, chasing storms can be dangerous.

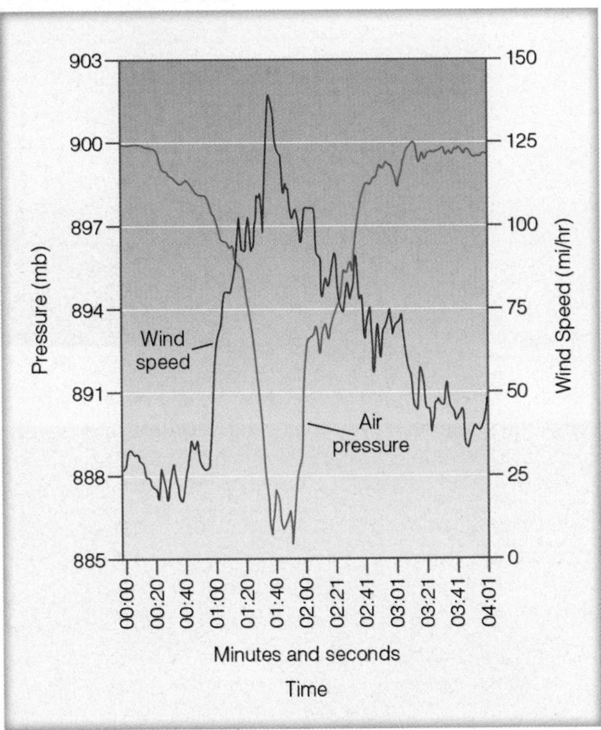

Figure 2 Data collected by undergraduate students near Oberlin, Kansas, as a tornado passed less than a mile from their location. Notice the air pressure drop of nearly 15 millibars in less than 2 minutes, and the wind gust of 135 mi/hr.

Figure 3 The multivortex tornado near Oberlin, Kansas, on May 22, 2008, just about the time it overtook a team of undergraduate students and their professor in a car.

© Perry Samson

FIGURE 12.22

A gustnado that formed along a gust front swirls across the plains of eastern Nebraska.

NCAR/UCAR/NSF

FIGURE 12.23

A well-developed landspout moves over eastern Colorado.

waterspouts that form over water, they are sometimes called **landspouts*** (see Fig. 12.23).

Figure 12.24 illustrates how a landspout can form. Suppose, for example, that the winds at the surface converge along a boundary, as illustrated in Fig. 12.24a. (The wind may converge due to topographic irregularities or any number of other factors, including temperature and moisture variations.) Notice that along the boundary, the air is rising, condensing, and forming into a cumulus congestus cloud. Notice also that along the surface at the boundary there is horizontal rotation (spin) created by the wind blowing in opposite directions along the boundary. If the developing cloud should move over the region of rotating air, the spinning air may be drawn up into the cloud by the storm's updraft. As the spinning, rising air shrinks in diameter, it produces a tornado-like structure, a *landspout* similar to the one shown in Fig. 12.23. Landspouts usually dissipate when rain falls through the cloud and destroys the updraft. Tornadoes may form in this manner along many types of converging wind boundaries, including sea breezes and gust fronts. Nonsupercell tornadoes and funnel clouds may also form with thunderstorms when cold air aloft (associated with an upper-level trough) moves over a region. Common along the west coast of North America, these short-lived tornadoes are sometimes called **cold-air funnels** (see Fig. 12.25).

*Landspouts occasionally form on the backside of a squall line where southerly winds ahead of a cold front and northwesterly winds behind it create swirling eddies that can be drawn into thunderstorms by their strong updrafts.

EXTREME WEATHER WATCH

Tornadoes have been reported in almost all countries of the world. In fact, on September 7, 2009, a particularly powerful tornado ripped through towns along the Argentina-Brazil border, where it killed at least 10 people in the town of San Pedro, Argentina.

Tornadic Winds and Doppler Radar

Most of our knowledge about what goes on inside a tornado-generating thunderstorm has been gathered through the use of **Doppler radar**. Remember from Chapter 6 on p. 168 that a radar transmitter sends out microwave pulses and that, when this energy strikes an object, a small fraction is scattered back to the antenna. Precipitation particles are large enough to bounce microwaves back to the antenna. Consequently, as we saw earlier, the colorful area on the radar screen in Fig. 12.20, p. 349, represents precipitation intensity inside a supercell thunderstorm.

Doppler radar can do more than measure rainfall intensity, it can actually measure the speed at which precipitation is moving horizontally toward or away from the radar antenna. Because precipitation particles are carried by the wind, Doppler radar can peer into a severe storm and reveal its winds.

Doppler radar works on the principle that, as precipitation moves toward or away from the antenna, the returning radar pulse will change in frequency. A similar change occurs when the high-pitched sound (high frequency) of an approaching noise source, such as a siren or train whistle, becomes lower in pitch (lower frequency) after it passes by the person hearing it. This change in frequency in sound waves or microwaves is called the *Doppler shift* and this, of course, is where the Doppler radar gets its name.

FIGURE 12.25 A funnel cloud—called a *cold air funnel*—descends from a thunderstorm in California's Central Valley near Lodi.

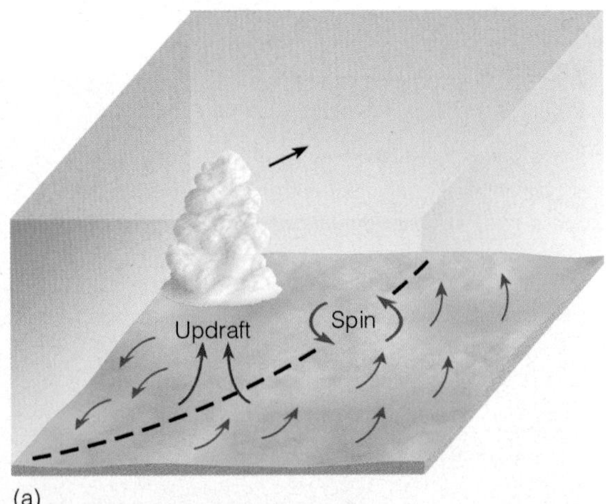

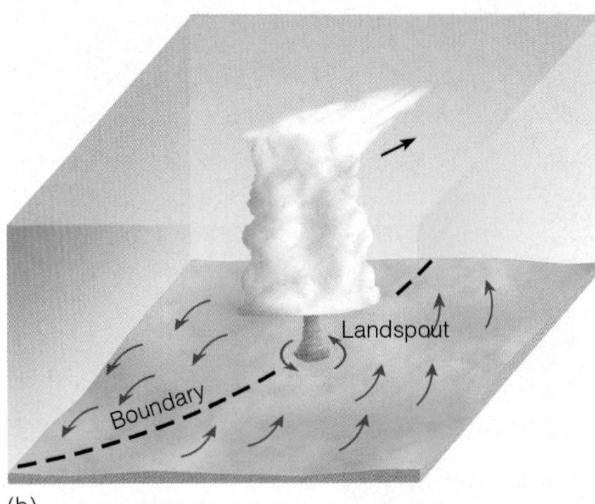

FIGURE 12.24 (a) Along the boundary of converging winds, the air rises and condenses into a cumulus congestus cloud. At the surface the converging winds along the boundary create a region of counterclockwise spin. (b) As the cloud moves over the area of rotation, the updraft draws the spinning air up into the cloud producing a nonsupercell tornado, or landspout. (Modified after Wakimoto and Wilson)

A single Doppler radar cannot detect winds that blow parallel to the antenna. Consequently, two or more units probing the same thunderstorm are needed to give a complete three-dimensional picture of the winds within the storm. To help distinguish the storm's air motions, wind velocities can be displayed in color. Color contouring the wind field gives a good picture of the storm (see ❯ Fig. 12.26).

Even a single Doppler radar can uncover many of the features of a severe thunderstorm. For example, studies conducted in the 1970s revealed, for the first time, the existence of the swirling winds of the mesocyclone inside a supercell storm. Mesocyclones have a distinct image (signature) on the radar display. Tornadoes also have a distinct signature on the radar screen, known as the *tornado vortex signature (TVS)*, which shows up as a region of rapidly (or abruptly) changing wind directions within the mesocyclone. (Look at Fig. 12.26.)

Unfortunately, the resolution of the Doppler radar is not high enough to measure actual wind speeds of most small tornadoes. However, a new and experimental Doppler system—called *Doppler lidar*—uses a light beam (instead of microwaves) to measure the change in frequency of falling precipitation, cloud particles, and dust. Because it uses a shorter wavelength of radiation, it has a narrower beam and a higher resolution than does Doppler radar. In an attempt to obtain tornado wind information at fairly close range (less than 6 miles), smaller portable Doppler radar units (Doppler on wheels) are peering into tornado-generating storms (see ❯ Fig. 12.27).

The network of more than 150 Doppler radar units deployed at selected weather stations within the continental United States is referred to as **NEXRAD** (an acronym for *NEX*t Generation Weather *RAD*ar). The NEXRAD system consists of the WSR-88D* Doppler radar and a set of computers that perform a variety of functions.

The computers take in data, display them on a monitor, and run computer programs called *algorithms*, which, in conjunction with other meteorological data, detect severe weather phenomena, such as storm cells, hail, mesocyclones, and tornadoes. Algorithms provide a great deal of information to the forecasters that allows them to make better decisions as to which thunderstorms are most likely to produce severe weather and possible flash flooding. In addition, the algorithms give advanced and improved

*The name WSR-88D stands for *W*eather *S*urveillance *R*adar, *1988 D*oppler.

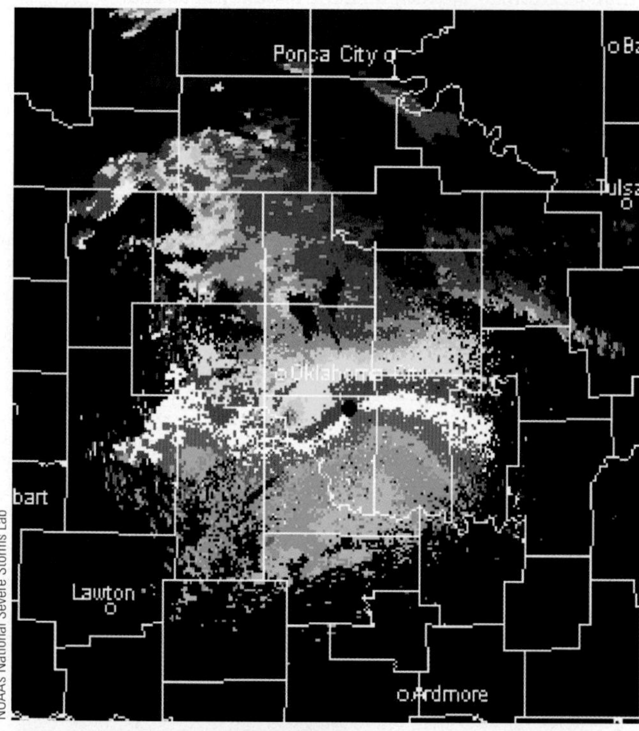

NOAA's National Severe Storms Lab

❯ **FIGURE 12.26** Doppler radar display of winds associated with the supercell storm that moved through parts of Oklahoma City during the afternoon of May 3, 1999. The close packing of the horizontal winds blowing toward the radar (green and blue shades), and those blowing away from the radar (yellow and red shades), indicate strong cyclonic rotation and the presence of a tornado.

© Howard B. Bluestein

❯ **FIGURE 12.27** Graduate students from the University of Oklahoma use a portable Doppler radar to probe a tornado near Hodges, Oklahoma.

warning of an approaching tornado. More reliable warnings, of course, will cut down on the number of false alarms.

Because the Doppler radar shows horizontal air motion within a storm, it can help to identify the magnitude of other severe weather phenomena, such as gust fronts, derechoes, microbursts, and wind shears that are dangerous to aircraft. Certainly, as more and more information from Doppler radar becomes available, our understanding of the processes that generate severe thunderstorms and tornadoes will be enhanced, and hopefully there will be an even better tornado and severe storm warning system, resulting in fewer deaths and injuries.

In an attempt to unravel some of the mysteries of the tornado, several studies are under way. In one study, called *VORTEX 2*, scientists using an armada of observational vehicles, and state-of-the-art equipment, including mobile Doppler radar, pursued tornado-generating thunderstorms over portions of the Central Plains during the spring of 2009. (Additional information on VORTEX 2 is provided in the Focus section below.)

FOCUS ON

A SPECIAL TOPIC

A Field Study to Explore Tornadoes

During the spring and early summer of 2009, a field research project in the Central Plains of the United States got underway to try and determine the origin, evolution, and structure of tornadoes. Called *VORTEX 2* (Verification of the Origins of Rotation in Tornadoes Experiment 2), the study consisted of more than 100 scientists, students, and technicians, all hoping to get an up close look at a supercell thunderstorm with a tornado.

Armed with a host of equipment, including more than 40 vehicles, researchers set out to look for supercell thunderstorms. Some of their tools included mobile Doppler radar units mounted on trucks, instruments placed on tripods (called *StickNets*) that were to be placed in a line perpendicular to an approaching supercell, and instruments attached to the tops of cars (mobile *MesoNets*) that were to measure the atmospheric conditions surrounding a supercell. Other instruments included lasers for measuring rain droplet size, high-tech balloons for measuring winds aloft, and unmanned small planes designed to measure temperature, pressure, humidity, and winds surrounding the storm.

Figure 4 **A tornado in southeastern Wyoming encountered by *VORTEX 2* on June 5, 2009.**

A typical day in the field actually begins a day or so earlier when researchers look over models and forecast charts to see where atmospheric conditions will be favorable for supercell thunderstorm development. Once it is determined where supercells will most likely form, a decision is made to move the field operation to that spot. The idea, of course, is to try and surround the developing storm with as much field equipment as possible; StickNets are set up ahead of the storm; mobile MesoNets take weather readings ahead of and behind the storm to see how atmospheric conditions are changing; and mobile radars are placed away from the advancing storm, as it takes time to set up and move a radar platform. While all of this activity is going on, balloons are being released, and filming of the storm is taking place from many vantage points. Meanwhile, back at the center for operations, scientists poured over models and weather maps to see where the entire team should be positioned on upcoming days.

In June, 2009, during the last week of *VORTEX 2*, the team encountered a significant tornado in southeastern Wyoming (see Fig. 4). Researchers arrived about 20 minutes before the tornado formed, giving them ample time to set up their equipment and move into position to observe the twister. As the tornado passed through the array of measuring devices, Doppler radar was able to measure its winds. The information gathered from this storm and others will help researchers understand how tornadoes form and why some supercells produce them while others do not.

The observations from *VORTEX 2* and other field studies are providing valuable information on the inner workings of supercell tornado-generating thunderstorms. In addition, laboratory models of tornadoes in chambers (called *vortex chambers*), along with mathematical computer models, are offering new insights into the formation and development of these fascinating storms.

Waterspouts

A **waterspout** is a rotating column of air that is connected to a cumuliform cloud over a large body of water. (See ❯ Fig. 12.28.) The waterspout may be a tornado that formed over land and then traveled over water. In such a case, the waterspout is sometimes referred to as a *tornadic waterspout*. Such tornadoes can inflict major damage to ocean-going vessels, especially when the tornadoes are of the supercell variety. Strong waterspouts that form over water then move over land can

© G. Kaufman

Active ❯ FIGURE 12.28 A powerful waterspout moves across Lake Tahoe, California. Visit the Meteorology Resource Center to view this and other active figures at academic.cengage.com/login

cause considerable damage. For example, on August 30, 2009, an intense waterspout formed over the warm Gulf of Mexico, then moved onshore into Galveston, Texas, where it caused EF1 damage over several blocks and injured three people. Waterspouts not associated with supercells that form over water, especially above warm, tropical coastal waters (such as in the vicinity of the Florida Keys, where almost 100 occur each month during the summer), are often referred to as *"fair weather" waterspouts*.* These waterspouts are generally much smaller than an average tornado, as they have diameters usually between 10 and 300 ft (about 3 and 100 m). Fair weather waterspouts are also less intense, as their rotating winds are typically less than 50 mi/hr. In addition, they tend to move more slowly than tornadoes and they only last for about 10 to 15 minutes, although some have existed for up to one hour.

Fair weather waterspouts tend to form in much the same way that landspouts do—when the air is conditionally unstable and cumulus clouds are developing. Some form with small thunderstorms, but most form with developing cumulus congestus clouds whose tops are frequently no higher than 12,000 ft (3,600 m) and do not extend to the freezing level. Apparently, the warm, humid air near the water helps to create atmospheric instability, and the updraft beneath the resulting cloud helps initiate uplift of the surface air. Studies even suggest that gust fronts and coverging sea breezes may play a role in the formation of some of the waterspouts that form over the Florida Keys.

The waterspout funnel is similar to the tornado funnel in that both are clouds of condensed water vapor with converging winds that rise about a central core. Contrary to popular belief, the waterspout does not draw water up into its core; however, swirling spray may be lifted several meters when the waterspout funnel touches the water. Apparently, the most destructive waterspouts are those that begin as tornadoes over land, then move over water.

*"Fair weather" waterspouts may form over any large body of warm water. Hence, they occur frequently over the Great Lakes in summer.

Summary

Tornadoes are rapidly rotating columns of air with a circulation that reaches the ground. The rotating air of the tornado may begin within the thunderstorm or it may begin at the surface and extend upward. The majority of tornadoes have wind speeds of less than 135 mi/hr and are usually less than one half mile wide. Violent tornadoes can have wind speeds in excess of 175 mi/hr and widths greater than 2 miles. Many large, violent tornadoes contain smaller whirls (suction vortices) embedded within them. Large, violent tornadoes are capable of inflicting great damage and are responsible for most deaths.

Although tornadoes can form anywhere in the world, they are most common in the Central Plains of North America, where a conditionally unstable atmosphere and strong wind shear provide favorable atmospheric conditions for their development. Tornadoes can occur in any month, but they are most common in April, May, and June. And they most frequently form in the afternoon between 4:00 and 6:00 P.M. local time. The Fujita scale, developed in the late 1960s and modified in 2007, uses damage indicators to estimate the winds of a tornado. The new enhanced Fujita (EF) scale has a range from EF1 to EF5, with the winds of an EF5 exceeding 200 mi/hr.

Tornadoes can form with supercells, as well as with less intense nonsupercell thunderstorms. Landspouts and gustnadoes are examples of tornadoes that do not form from supercells. With the aid of Doppler radar, scientists are probing tornado-spawning thunderstorms, hoping to better predict tornadoes and to better understand where, when, and how they form.

A normally small and less destructive cousin of the tornado is the "fair weather" waterspout that commonly forms above the warm waters of the Florida Keys and the Great Lakes in summer.

Key Terms

The following terms are listed (with page numbers) in the order they appear in the text. Define each. Doing so will aid you in reviewing the material covered in this chapter.

tornado, 334
funnel cloud, 334
tornado alley, 336
multi-vortex tornado, 340
suction vortices, 340
tornado watch, 341
tornado warning, 341
Fujita scale, 342
EF scale, 342
tornado outbreak, 345
supercell tornadoes, 347
hook echo, 349
nonsupercell tornadoes, 350
gustnadoes, 350
landspouts, 352
cold-air funnels, 352
Doppler radar, 353
NEXRAD, 354
waterspout, 356

Review Questions

1. What is a tornado? Give some average statistics about tornado size, winds, and direction of movement.
2. What is the primary difference between a tornado and a funnel cloud?
3. Why do the majority of tornadoes move from southwest to northeast?
4. Is it more likely for a tornado to have a rotating wind speed of 125 mi/hr or 200 mi/hr?
5. If you see a tornado and it appears to be stretched into the shape of a rope, what stage of development is the tornado in?
6. Why is the Central Plains of the United States most susceptible to tornado development than any other region of the world?
7. What is "tornado alley" and where is it located?
8. During what months are tornadoes most frequent in the United States?
9. Explain both how and why there is a shift in tornado activity from winter to summer.
10. Tornadoes most frequently occur during what hours of the day?
11. Why should you *not* open windows when a tornado is approaching?
12. If a tornado is moving toward the northeast, on which side of the tornado (northwest or southeast) would the winds be strongest?
13. Explain how suction vortices can add wind speed to a large multi-vortex tornado.
14. Explain how a violent multi-vortex tornado will completely demolish some homes in its path, and yet leave others standing.
15. How does a tornado *watch* differ from a tornado *warning*?
16. If you are in a single story house (without a basement) during a tornado warning, what should you do?
17. If you are in a mobile home during a tornado warning, what should you do?
18. Give two reasons for the possible increase in the average number of tornadoes reported in the United States during the past six decades.
19. How does the new enhanced Fujita scale differ from the old Fujita scale? Why was the old Fujita scale modified?
20. Is it unusual or common to have three EF5 tornadoes in a year?
21. Explain why "tornado families" are often so deadly.
22. What atmospheric conditions are necessary for the formation of supercells that spawn tornadoes.
23. Give an explanation as to how a supercell thunderstorm can develop a rotating updraft.

24. Supercell thunderstorms that produce tornadoes form in a region of strong wind shear. Explain how the wind usually changes in speed and direction to produce this shear.
25. If you are looking up and see rotating clouds associated with a supercell above you, would the clouds (as you view them) most likely be rotating clockwise or counterclockwise? (This question is a bit tricky, so it may be helpful to make a sketch on a piece of paper.)
26. Explain how a nonsupercell tornado, such as a landspout, might form.
27. Are gustnadoes classified as supercell or nonsupercell tornadoes? Why?
28. Describe how Doppler radar measures the winds inside a severe thunderstorm.
29. How has Doppler radar helped in the prediction of severe weather and tornadoes?
30. What atmospheric conditions lead to the formation of "fair weather" waterspouts?

Online Learning

STUDENT COMPANION WEBSITE: Visit this book's companion website at: www.cengage.com/ahrens/extreme1e and choose Chapter 12 for many study aids and ideas for further reading and research. These include flashcards, practice quizzing, and web links.

METEOROLOGY RESOURCE CENTER: For students with access, log on at www.cengage.com/login for more assets, including animations, videos, and more. If your textbook did not come with access, visit www.CengageBrain.com to purchase.

Hurricanes

13

CONTENTS

TROPICAL WEATHER
ANATOMY OF A HURRICANE
HURRICANE FORMATION AND DISSIPATION
NAMING HURRICANES AND TROPICAL STORMS
DEVASTATING WINDS, FLOODING, AND THE STORM SURGE
EXTREME FLOODING WITH A TROPICAL STORM
SOME NOTABLE HURRICANES
OTHER DEVASTATING HURRICANES
HURRICANE WATCHES, WARNINGS, AND FORECASTS
MODIFYING HURRICANES
HURRICANES IN A WARMER WORLD
SUMMARY
KEY TERMS
REVIEW QUESTIONS

The high winds of Hurricane Katrina blow the roof off a building in Kenner, Louisiana, on the morning of August 29, 2005.

During August, 2005, Hurricane Katrina unleashed its devastating fury on the Gulf Coast of the United States, with sustained winds of 127 mi/hr and a surge of water over 20 feet high. The storm took more than 1500 lives, caused over $200 billion in property damage, and displaced over one million people from their homes—a humanitarian crisis on a scale unseen in the United States since the Great Depression.

Hurricanes, such as Katrina, are born over tropical waters. Here, the warm water and a rich supply of water vapor provide sufficient energy for the hurricane to grow into a ferocious storm, capable of generating enormous waves, heavy rain, and high winds. To better understand how hurricanes and other tropical systems are able to produce such extreme weather, we need to turn our attention to the tropics, the birthplace of the hurricane.

Tropical Weather

In the broad belt around the earth known as the tropics—the region between 23½° north and south of the equator—the weather is much different from that of the middle latitudes. In the tropics, the noon sun is always high in the sky, and so diurnal and seasonal changes in temperature are small. The daily heating of the surface and high humidity favor the development of cumulus clouds and afternoon thunderstorms. Most of these are individual thunderstorms that are not severe. Sometimes, however, they group together into loosely organized systems called *non-squall clusters*. On other occasions, the thunderstorms will align into a row of vigorous convective cells known as a *tropical squall cluster*, or *squall line*. The passage of a squall line is usually noted by a sudden wind gust followed immediately by a heavy downpour that may produce more than one inch of rainfall in about every 30 minutes. This deluge is then followed by several hours of relatively steady rainfall. Many of these tropical squall lines are similar to the middle-latitude ordinary squall lines described in Chapter 11 on p. 397.

As it is warm all year long in the tropics, the weather is not characterized by four seasons which, for the most part, are determined by temperature variations. Rather, most of the tropics are marked by seasonal differences in precipitation. The greatest cloudiness and precipitation occur during the high-sun period, when the intertropical convergence zone moves into the region. Even during the dry season, precipitation can be irregular, as periods of heavy rain, lasting for several days, may follow an extreme dry spell.

The winds in the tropics generally blow from the east, northeast, or southeast. Because the variation of sea-level pressure is normally quite small, drawing isobars on a weather map provides little useful information. Instead of isobars, **streamlines** that depict wind flow are drawn. Streamlines are useful because they show where surface air converges and diverges. Occasionally, the streamlines will be disturbed by a weak trough of low pressure called a **tropical wave**, or **easterly wave**, because it tends to move from east to west (see Fig. 13.1).

Tropical waves have wavelengths on the order of 1500 mi (2400 km) and travel from east to west at speeds between 10 and 20 mi/hr. Look at Fig. 13.1 and observe that, on the western side of the trough (heavy dashed green line), where easterly and northeasterly surface winds diverge, sinking air produces generally fair weather. On its eastern side, southeasterly surface winds converge. The converging air rises, cools, and often condenses into showers and thunderstorms. Consequently, the main area of showers forms *behind* the trough. Occasionally, an easterly wave will intensify and grow into a hurricane.

Anatomy of a Hurricane

A **hurricane** is an intense storm of tropical origin, with sustained winds exceeding 74 mi/hr (64 knots), which forms over the warm northern Atlantic and eastern

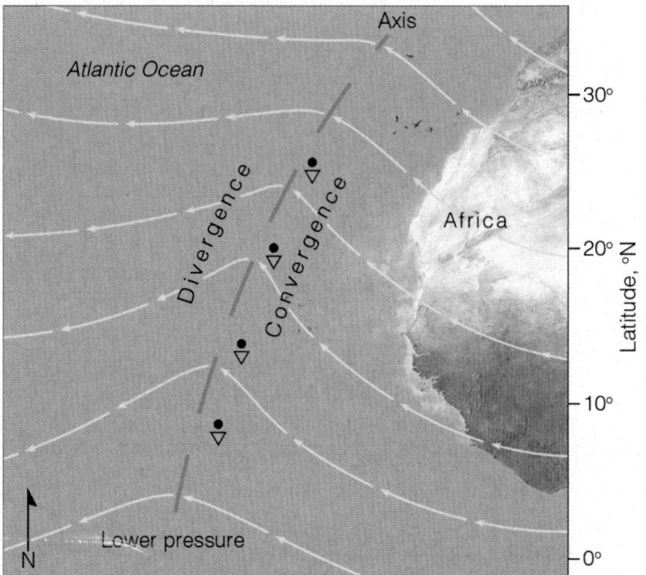

FIGURE 13.1 A tropical wave (also called an easterly wave) moving off the coast of Africa over the Atlantic. The wave is shown by the bending of streamlines—lines that show wind-flow patterns. (The heavy dashed green line is the axis of the trough.) The wave moves slowly westward, bringing fair weather on its western side and rain showers on its eastern side.

FIGURE 13.2

Hurricane Elena over the Gulf of Mexico about 80 miles southwest of Apalachicola, Florida, as photographed from the space shuttle *Discovery* during September, 1985. Because this storm is situated north of the equator, surface winds are blowing counterclockwise about its center (eye). The central pressure of the storm is 955 mb, with sustained winds of 120 mi/hr (105 knots) near the center.

North Pacific oceans. This same type of storm is given different names in different regions of the world. In the western North Pacific, it is called a **typhoon**, in India a *cyclone,* and in Australia a *tropical cyclone.* By international agreement, **tropical cyclone** is the general term for all hurricane-type storms that originate over tropical waters. For simplicity, we will refer to all of these storms as hurricanes.*

Figure 13.2 is a photo of Hurricane Elena situated over the Gulf of Mexico. The storm is approximately 300 mi (500 km) in diameter, which is about average for hurricanes. The area of broken clouds at the center is its **eye**. Elena's eye is almost 25 mi (40 km) wide. Within the eye, winds are light and clouds are mainly broken. The surface air pressure is very low, nearly 955 mb (28.20 in.).** Notice that the clouds align themselves into spiraling bands (called *spiral rain bands)* that swirl in toward the storm's center, where they wrap themselves around the eye. Surface winds increase in speed as they blow counterclockwise and inward toward this center. (In the Southern Hemisphere, the winds blow clockwise around the center.) Adjacent to the eye is the **eyewall**, a ring of intense thunderstorms that whirl around the storm's center and may extend upward to almost 59,000 ft (18 km) above sea level. Within the eyewall, we find the heaviest precipitation and the strongest winds, which, in this storm, are 120 mi/hr, with peak gusts of 135 mi/hr.

If we were to venture from west to east (left to right) at the surface through the storm in Fig. 13.2, what might we experience? As we approach the hurricane, the sky becomes overcast with cirrostratus clouds; barometric pressure drops slowly at first, then more rapidly as we move closer to the center. Winds blow from the north and northwest with ever-increasing speed as we near the eye. The high winds, which generate huge waves over 30 feet high, are accompanied by heavy rainshowers. As we move into the eye, the winds slacken, rainfall ceases, and the sky brightens, as middle and high clouds appear overhead. The atmospheric pressure is now at its lowest point (965 mb), some 50 mb lower than the pressure measured on the outskirts of the storm. The brief respite ends as we enter the eastern region of the eyewall. Here, we are greeted by heavy rain and strong southerly winds. As we move away from the eyewall, the pressure rises, the winds diminish, the heavy rain lets up, and eventually the sky begins to clear.

This brief, imaginary venture raises many unanswered questions. Why, for example, is the surface

*The word *hurricane* derives from the Taino language of Central America. The literal translation of the Taino word *hurucan* is "god of evil." The word *typhoon* comes from the Chinese word *taifung,* meaning "big wind."

**An extreme low pressure of 870 mb (25.70 in.) was recorded in Typhoon Tip while over the tropical Pacific Ocean during October, 1979, and Hurricane Wilma had a pressure reading of 882 mb (26.04 in.) while over the Gulf of Mexico during October, 2005.

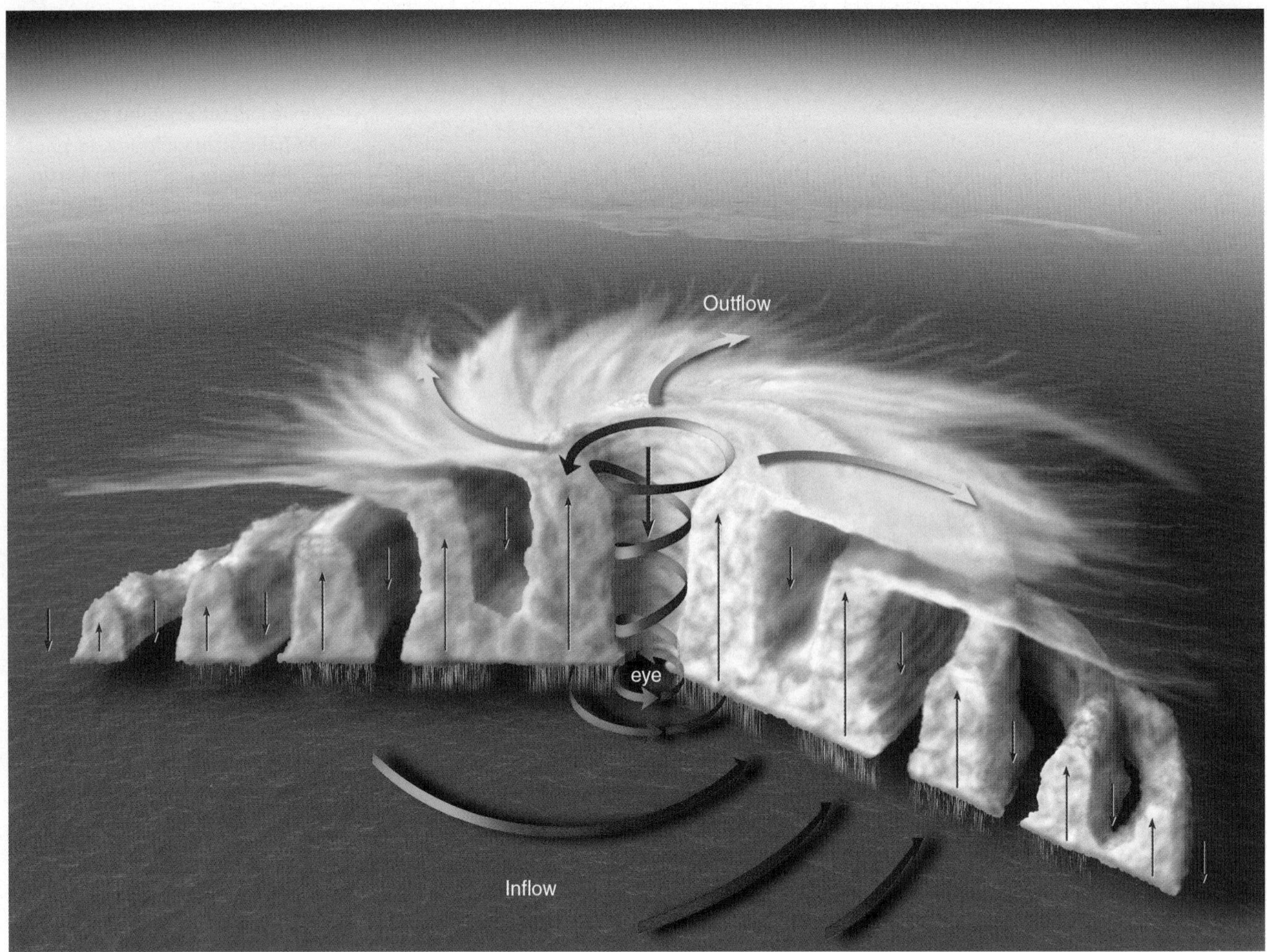

Active ❯ FIGURE 13.3 A model that shows a vertical view, a profile, of air motions and clouds in a typical hurricane in the Northern Hemisphere. The diagram is exaggerated in the vertical. Visit the Meteorology Resource Center to view this and other active figures at academic.cengage.com/login

pressure lowest at the center of the storm? And why is the weather clear almost immediately outside the storm area? To help us answer such questions, we need to look at a vertical view, a profile, of the hurricane along a slice that runs through its center. A model that describes such a profile is given in ❯ Fig 13.3.

The model shows that the hurricane is composed of an organized mass of thunderstorms* that are an integral part of the storm's circulation. Near the surface, moist tropical air flows in toward the hurricane's center. Adjacent to the eye, this air rises and condenses into huge cumulonimbus clouds that produce heavy rainfall, as much as 10 inches per hour. Near the top of the clouds, the relatively dry air, having lost much of its moisture, begins to flow outward away from the center. This diverging air aloft actually produces a clockwise (*anticyclonic* in the Northern Hemisphere) flow of air about 100 miles from the eye. As this outflow reaches the storm's periphery, it begins to sink and warm, inducing clear skies. In the vigorous convective clouds of the eyewall, the air warms due to the release of large quantities of latent heat. This produces slightly higher pressures aloft, which initiate downward air motion within the eye. As the air descends, it warms by compression. This process helps to account for the warm air and the absence of convective clouds in the eye of the storm (see ❯ Fig. 13.4).

As surface air rushes in toward the region of much lower surface pressure, it should expand and cool, and we might expect to observe cooler air around the eye, with warmer air farther away. But, apparently, so much heat is added to the air from the warm ocean surface that the surface air temperature remains fairly uniform throughout the hurricane.

*These huge convective cumulonimbus clouds have surprisingly little lightning (and, hence, thunder) associated with them. Even so, for simplicity we will refer to these clouds as thunderstorms throughout this chapter.

FIGURE 13.4

The cloud mass is Hurricane Katrina's eyewall, and the clear area is Katrina's eye photographed inside the eye on August 28, 2005, from a NOAA reconnaissance (hurricane hunter) aircraft.

Figure 13.5 is a three-dimensional radar composite of Hurricane Katrina as it passes over the central area of the Gulf of Mexico. Compare Katrina's features with those of typical hurricanes illustrated in Fig. 13.2 and Fig. 13.3. Notice that the strongest radar echoes (heaviest rain) near the surface are located in the eyewall, adjacent to the eye.

Hurricane Formation and Dissipation

We are now left with an important question: Where and how do hurricanes form? While there is no widespread agreement on how hurricanes actually form, it is known that certain necessary ingredients are required before a weak tropical disturbance will develop into a full-fledged hurricane.

THE RIGHT ENVIRONMENT Hurricanes form over tropical waters where the winds are light, the humidity is high in a deep layer extending up through the troposphere, and the surface water temperature is warm, typically 80°F (26.5°C) or greater, over a vast area* (see Fig. 13.6). These conditions usually prevail over the tropical and subtropical North Atlantic and North Pacific oceans during the summer and early fall; hence, the hurricane season normally runs from June through November. Fig. 13.7 shows the number of tropical

*It was once thought that for hurricane formation, the ocean must be sufficiently warm through a depth of about 650 ft (200 m). It is now known that hurricanes can form in the eastern North Pacific when the warm layer of ocean water is only about 65 ft (20 m) deep.

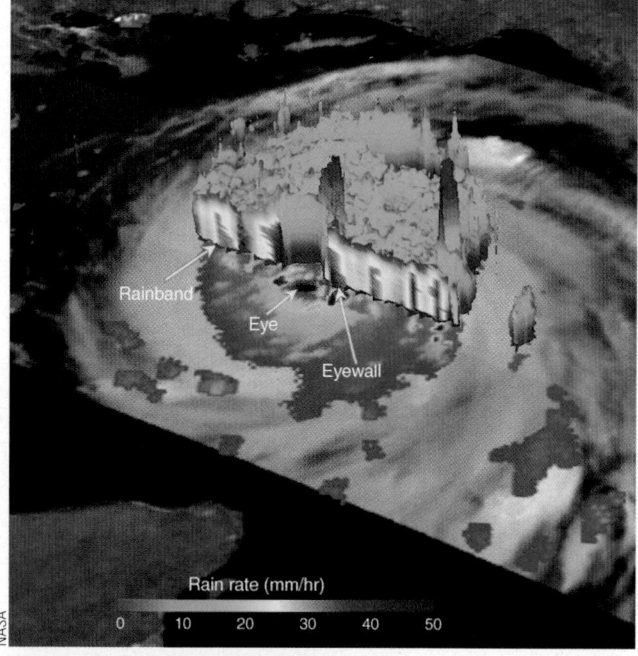

FIGURE 13.5 A three-dimensional TRMM satellite view of Hurricane Katrina passing over the central Gulf of Mexico on August 28, 2005. The cutaway view shows concentric bands of heavy rain (red areas inside the clouds) encircling the eye. Notice that the heaviest rain (large red area) occurs in the eyewall. The isolated tall cloud tower (in red) in the northern section of the eyewall indicates a cloud top of 52,000 ft (16 km) above the ocean surface. Such tall clouds in the eyewall often indicate that the storm is intensifying.

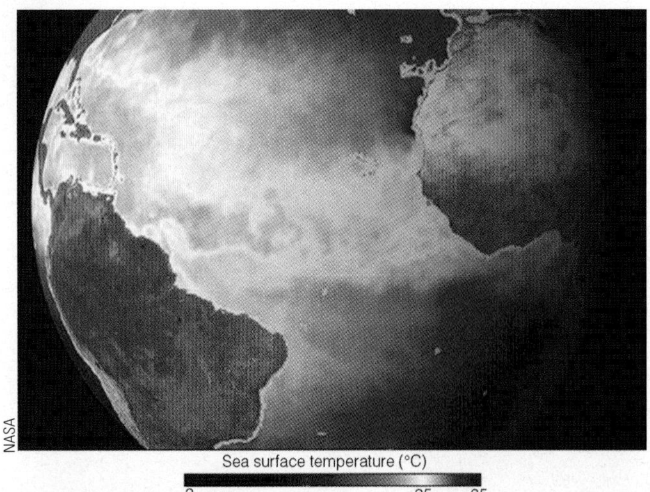

FIGURE 13.6 Hurricanes form over warm, tropical waters. This image shows where sea surface temperatures in the tropical Atlantic exceed 28°C (82°F)—warm enough for tropical storm development—during May, 2002.

storms and hurricanes that formed over the tropical Atlantic during the past 100 years. Notice that hurricane activity picks up in August, peaks in September, then drops off rapidly.

For a mass of unorganized thunderstorms to develop into a hurricane, the surface winds must converge. In the Northern Hemisphere, converging air spins counterclockwise about an area of surface low pressure. Because this type of rotation will not develop on the equator where the Coriolis force is zero (see Chapter 7), hurricanes form in tropical regions, usually between 5° and 20° latitude. (In fact, about two-thirds of all tropical cyclones form between 10° and 20° of the equator.)

Hurricanes do not form spontaneously, but require some kind of "trigger" to start the air converging. We know, for example, from Chapter 8 that surface winds converge along the intertropical convergence zone (ITCZ). Occasionally, when a wave forms along the ITCZ, an area of low pressure develops, convection becomes organized, and the system grows into a hurricane. Weak convergence also occurs on the eastern side of a tropical wave, where hurricanes have been known to form. In fact, many if not most Atlantic hurricanes can be traced to tropical waves that form over Africa. However, only a small fraction of all of the tropical disturbances that form over the course of a year ever grow into hurricanes. Studies suggest that major Atlantic hurricanes are more numerous when the western part of Africa is relatively wet. Apparently, during the wet years, tropical waves are stronger, better organized, and more likely to develop into strong Atlantic hurricanes.

Convergence of surface winds may also occur along a pre-existing atmospheric disturbance, such as a front that has moved into the tropics from middle latitudes. Although the temperature contrast between the air on both sides of the front is gone, converging winds may still be present so that thunderstorms are able to organize.

Even when all of the surface conditions appear near perfect for the formation of a hurricane (for example, warm water, humid air, converging winds, and so forth), the storm may not develop if the weather conditions aloft are not just right. For instance, in the region of the trade winds, and especially near latitude 20°, the air is

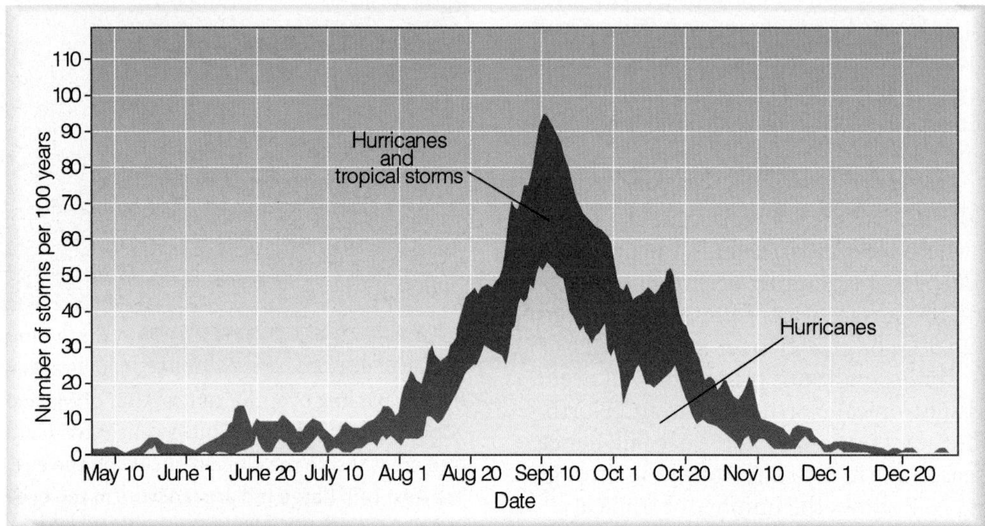

FIGURE 13.7 The total number of hurricanes and tropical storms (red shade) and hurricanes only (yellow shade) that have formed during the past 100 years in the Atlantic Basin—the Atlantic Ocean, the Caribbean Sea, and the Gulf of Mexico. (NOAA)

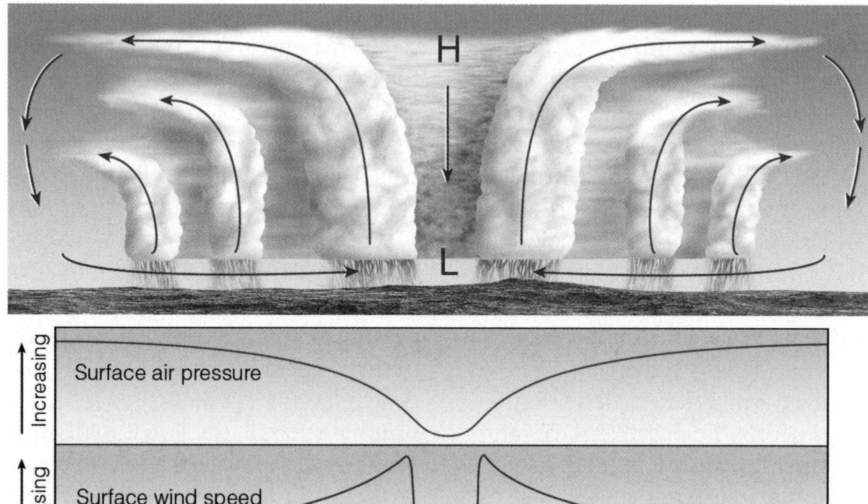

FIGURE 13.8 The top diagram shows an intensifying tropical cyclone. As latent heat is released inside the clouds, the warming of the air aloft creates an area of high pressure, which induces air to move outward, away from the high. The warming of the air lowers the air density, which in turn lowers the surface air pressure. As surface winds rush in toward the surface low, they extract sensible heat, latent heat, and moisture from the warm ocean. As the warm, moist air flows in toward the center of the storm, it is swept upward into the clouds of the eyewall. As warming continues, surface pressure lowers even more, the storm intensifies, and the winds blow even faster. This situation increases the transfer of heat and moisture from the ocean surface. The middle diagram illustrates how the air pressure drops rapidly as you approach the eye of the storm. The lower diagram shows how surface winds normally reach maximum strength in the region of the eyewall.

often sinking in association with the subtropical high. The sinking air warms and creates an inversion above the surface, known as the **trade wind inversion**.* When the inversion is strong, it can inhibit the formation of intense thunderstorms and hurricanes. Also, hurricanes do not form where the upper-level winds are strong, creating strong wind shear. Strong wind shear tends to disrupt the organized pattern of convection and disperses heat and moisture, which are necessary for the growth of the storm.

The situation of strong winds aloft typically occurs over the tropical Atlantic during a major El Niño event, a situation where extensive ocean warming occurs over the eastern tropical Pacific Ocean. As a consequence, during El Niño there are usually fewer Atlantic hurricanes than normal. However, the warmer water of El Niño in the northern tropical Pacific favors the development of hurricanes in that region. During the cold water episode in the eastern tropical Pacific (known as La Niña), winds aloft over the tropical Atlantic usually weaken and become easterly—a condition that favors hurricane development.*

*The height of the base of the trade wind inversion varies from about 1600 ft (500 m) over the eastern part of the ocean to about 6500 ft (2000 m) over the extreme western part.

THE DEVELOPING STORM The energy for a hurricane comes from the direct transfer of sensible heat and latent heat from the warm ocean surface. For a hurricane to form, a cluster of thunderstorms must become organized around a central area of surface low pressure. But it is not totally clear how this process occurs. One theory proposed that a hurricane forms in the following manner: Suppose, for example, that the trade wind inversion is weak and that thunderstorms start to organize along the ITCZ, or along a tropical wave. In the deep, moist conditionally unstable environment, a huge amount of latent heat is released inside the clouds during condensation. The process warms the air aloft, causing the temperature near the cluster of thunderstorms to be much higher than the air temperature at the same level farther away. This warming of the air aloft causes a region of higher pressure to form in the upper troposphere (see Fig. 13.8).

*El Niño and La Niña are covered more thoroughly in Chapter 14, beginning on p. 418.

This situation causes a horizontal pressure gradient aloft that induces the air aloft to move outward, away from the region of higher pressure in the anvils of the cumulonimbus clouds. This diverging air aloft, coupled with warming of the vertical air column, causes the surface pressure to drop and a small area of surface low pressure to form. The air now begins to spin counterclockwise (Northern Hemisphere) and in toward the region of surface low pressure. As the air moves inward, its speed increases, just as ice skaters spin faster as their arms are brought in close to their bodies (a phenomenon called the *conservation of angular momentum*).

As the air moves over the warm water, small swirling eddies transfer heat energy from the ocean surface into the overlying air. The warmer the water and the greater the wind speed, the greater the transfer of sensible and latent heat into the air above. As the air sweeps in toward the center of lower pressure, the rate of heat transfer increases because the wind speed increases. Similarly, the higher wind speed causes greater evaporation rates, and the overlying air becomes nearly saturated. The turbulent eddies then transfer the warm, moist air upward, where the water vapor condenses to fuel new thunderstorms. As the surface air pressure lowers, wind speeds increase, more evaporation occurs at the ocean surface, and thunderstorms become more organized. At the top of the thunderstorms, heat is lost by the clouds radiating infrared energy to space.

The driving force behind a hurricane is similar to that of a heat engine. In a heat engine, heat is taken in at a high temperature, converted into work, then ejected at a low temperature. In a hurricane, heat is taken in near the warm ocean surface, converted to kinetic energy (energy of motion or wind), and lost at its top through radiational cooling.

In a heat engine, the amount of work done is proportional to the difference in temperature between its input and output region. The maximum strength a hurricane can achieve is proportional to the difference in air temperature between the tropopause and the surface, and to the potential for evaporation from the sea surface. As a consequence, the warmer the ocean surface, the lower the minimum pressure of the storm, and the higher its winds. Because there is a limit to how intense the storm can become, peak wind gusts seldom exceed 200 mi/hr.

After a hurricane forms, it may go through an internal cycle of intensification. In strong hurricanes, for example, the eyewall may become encircled by a second eyewall, as another band of strong thunderstorms forms perhaps 3 to 15 miles out from the original eyewall. The growing outer eyewall cuts off the moisture supply to the original eyewall, causing it to dissipate. The dissipation of the original eyewall and the formation of a new one farther out from the eye is called **eyewall replacement**. As the replacement of the eyewall is taking place, the central pressure of the storm may rise, and its maximum winds may lessen. Eventually, however, the newly formed eyewall will usually contract toward the center of the storm as the hurricane re-intensifies.

THE STORM DIES OUT If the hurricane remains over warm water, it may survive for a long time. For example, Hurricane Tina (1992) traveled for thousands of miles over deep, warm, tropical waters and maintained hurricane force winds for 24 days, making it one of the longest-lasting North Pacific hurricanes on record. However, most hurricanes last for less than a week.

Hurricanes weaken rapidly when they travel over colder water and lose their heat source. Studies show that if the water beneath the eyewall of the storm (the region of thunderstorms adjacent to the eye) cools by 4.5°F (2.5°C), the storm's energy source is cut off, and the storm will dissipate. Even a small drop in water temperature beneath the eyewall will noticeably weaken the storm. A hurricane can also weaken if the layer of warm water beneath the storm is shallow. In this situation, the strong winds of the storm generate powerful waves that produce turbulence in the ocean water under the storm. Such turbulence creates currents that bring to the surface cooler water from below. If the storm is moving slowly, it is more likely to lose intensity, as the eyewall will remain over the cooler water for a longer period.

Hurricanes also dissipate rapidly when they move over a large landmass. Here, they not only lose their energy source, but friction with the land surface also causes surface, winds to decrease and blow more directly into the storm, an effect that causes the hurricane's central pressure to rise. And a hurricane, or any tropical system for that matter, will rapidly dissipate should it move into a region of strong vertical wind shear.

Our understanding of hurricane behavior is far from complete. However, with the aid of computer model

EXTREME WEATHER WATCH

When conditions are favorable, tropical cyclones can intensify remarkably fast in a very short period of time. Super Typhoon Forrest, which formed in the Western Pacific in September, 1983, recorded a central pressure drop of 99 mb (3 in.) in just one 24-hour period, while its winds simultaneously increased from 75 mi/hr to 175 mi/hr. Hurricane Wilma intensified so rapidly while approaching the Yucatan Peninsula on October 18, 2005, that in just two hours its central pressure fell 43 mb (1.32 in.).

simulations and research projects such as *RAINEX** (*Rain*band and Intensity Change *Ex*periment), scientists are gaining new insight into how tropical cyclones form, intensify, and ultimately die.

INVESTIGATING THE STORM There are a variety of ways to obtain information about a developing hurricane and its environment. Visible, infrared, and enhanced infrared satellite images all provide a bird's eye view of the storm, while sophisticated onboard radar instruments can actually peer into the storm and unveil its clouds as a three-dimensional image (see Fig. 13.5, p. 365). There are even satellites equipped with onboard instruments capable of obtaining surface wind information in and around the storm (see Fig. 13.9). A visible satellite image can be important in determining whether a developing hurricane will continue to strengthen. For example, the huge thunderstorms in the eyewall of the storm often produce a dense cirrus cloud shield that extends outward away from the eye, as illustrated in Fig. 13.3 on p. 364. If the storm in a visible satellite image has a well-defined eye and a dense cirrus cloud shield when it reaches hurricane strength, the storm will most likely continue to strengthen, as there appears to be insufficient wind shear to tear it apart.

Detailed information about a hurricane can also come from aircraft that fly directly into the storm. These so-called *hurricane hunters* carry instruments directly on the aircraft as well as instruments, such as the *dropsonde,* that are dropped from the aircraft into the storm. On its way down to the ocean surface, the dropsonde measures air temperature, humidity, and atmospheric pressure, which are transmitted back to the aircraft. Because the dropsonde is equipped with a Global Positioning System (GPS) that constantly monitors its changing position, it has the capability of providing wind information as well. Another temperature-measuring device dropped from the aircraft is the *bathythermograph,* which falls into the ocean where it measures water temperature as it slowly descends beneath the surface. Other probes dropped into the sea measure the speed of ocean currents and the salinity (saltiness) of the water, an important factor in determining water density.

HURRICANE STAGES OF DEVELOPMENT

Hurricanes go through a set of stages from birth to death. Initially, a *tropical disturbance* shows up as a mass of thunderstorms with only slight wind circulation. The tropical disturbance becomes a **tropical depression** when the winds increase to between 23 and 39 mi/hr and several closed isobars appear about its center on a surface weather map. When the isobars are packed together and the winds are between 40 and 74 mi/hr, the tropical depression becomes a **tropical storm**. (At this point, the storm gets a name.) The tropical storm is classified as a *hurricane* only when its winds exceed 74 mi/hr (64 knots).

Figure 13.10 shows four tropical systems in various stages of development. Moving from east to west, we see a weak tropical disturbance (a tropical wave) crossing over Panama. Farther west, a tropical depression is organizing around a developing center with winds less than 30 mi/hr. In a few days, this system will develop into a hurricane. Farther west is a full-fledged hurricane with peak winds in excess of 125 mi/hr. The swirling band of clouds to the northwest is Emilia; once a hurricane (but now with winds less than 45 mi/hr), it is rapidly weakening over colder water.

*The *RAINEX* project consisted of reconnaissance aircraft flying into several hurricanes during the hurricane season of 2005. Equipped with sophisticated scientific instruments, including advanced Doppler radar, the mission obtained high resolution data on each storm's structure, cloud configuration, and winds.

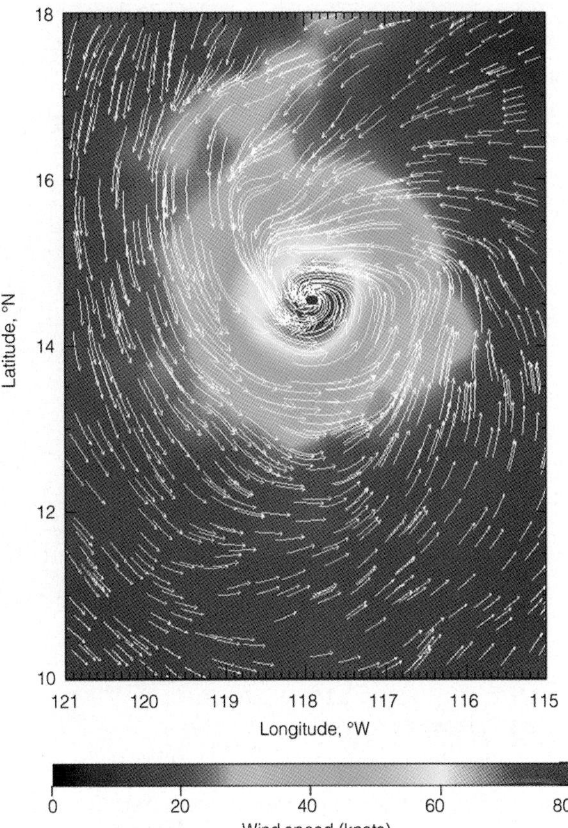

FIGURE 13.9 Arrows show surface winds spinning counterclockwise around Hurricane Dora situated over the eastern tropical Pacific during August, 1999. Colors indicate surface wind speeds. Notice that winds of 80 knots (92 mi/hr) are encircling the eye (the dark dot in the center). Wind speed and direction obtained from Quik-SCAT satellite. (NASA/JPL)

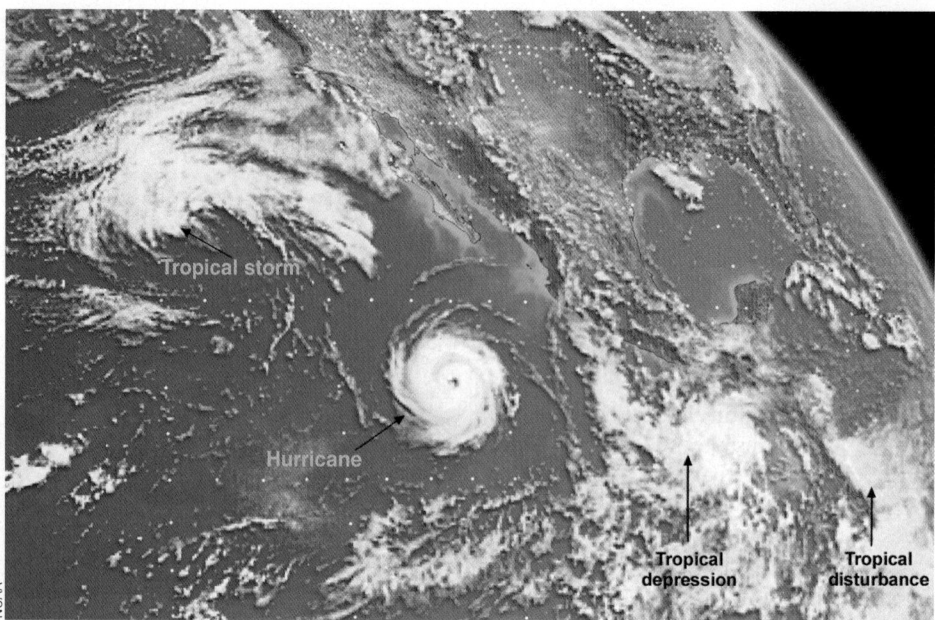

FIGURE 13.10

Visible satellite image showing four tropical systems, each in a different stage of its life cycle.

BRIEF REVIEW

Before reading the next several sections, here is a review of some of the important points about hurricanes.

- Hurricanes are tropical cyclones, comprised of an organized mass of thunderstorms.
- Hurricanes have peak winds about a central core (eye) that exceed 74 mi/hr (64 knots).
- The strongest winds and the heaviest rainfall normally occur in the eyewall—a ring of intense thunderstorms that surround the eye.
- Hurricanes form over warm tropical waters, where light surface winds converge, the humidity is high in a deep layer, and the winds aloft are weak.
- For a mass of thunderstorms to organize into a hurricane there must be some mechanism that triggers the formation, such as converging surface winds along the ITCZ, a pre-existing atmospheric disturbance, such as a weak front from the middle latitudes, or a tropical wave.
- Hurricanes derive their energy from the warm, tropical oceans and by evaporating water from the ocean's surface. Heat energy is converted to wind energy when the water vapor condenses and latent heat is released inside deep convective clouds.
- When hurricanes lose their source of warm water (either by moving over colder water or over a large landmass), they dissipate rapidly.
- The three primary stages in a developing hurricane are: tropical depression, tropical storm, and hurricane.

HURRICANE MOVEMENT Figure 13.11 shows where most hurricanes are born and the general direction in which they move, whereas Fig. 13.12 shows the actual paths taken by all hurricanes from 1985 to 2005. Notice that hurricanes that form over the warm, tropical North Pacific and North Atlantic are steered by easterly winds and move west or northwestward at about 10 mi/hr for a week or so. Gradually, they swing poleward around the subtropical high, and when they move far enough north, they become caught in the westerly flow, which curves them to the north or northeast. In the middle latitudes, the hurricane's forward speed normally increases, sometimes to more than 55 mi/hr. The actual path of a hurricane (which appears to be determined by the structure of the storm and the storm's interaction with the environment) may vary considerably. Some take erratic paths and make odd turns that occasionally catch weather forecasters by surprise (see Fig. 13.13). There have been many instances where a storm heading directly for land suddenly veered away and spared the region from almost certain disaster. As a case in point, Hurricane Elena, with peak winds of 100 mi/hr, moved northwestward into the Gulf of Mexico on August 29, 1985. It then veered eastward toward the west coast of Florida. After stalling offshore, it headed northwest. After weakening, it then moved onshore near Biloxi, Mississippi, on the morning of September 2.

Look at Fig. 13.11 and notice there that it appears as if hurricanes do not form over the South Atlantic and the eastern South Pacific—directly east and west of South America. Cooler water, vertical wind shear, and the

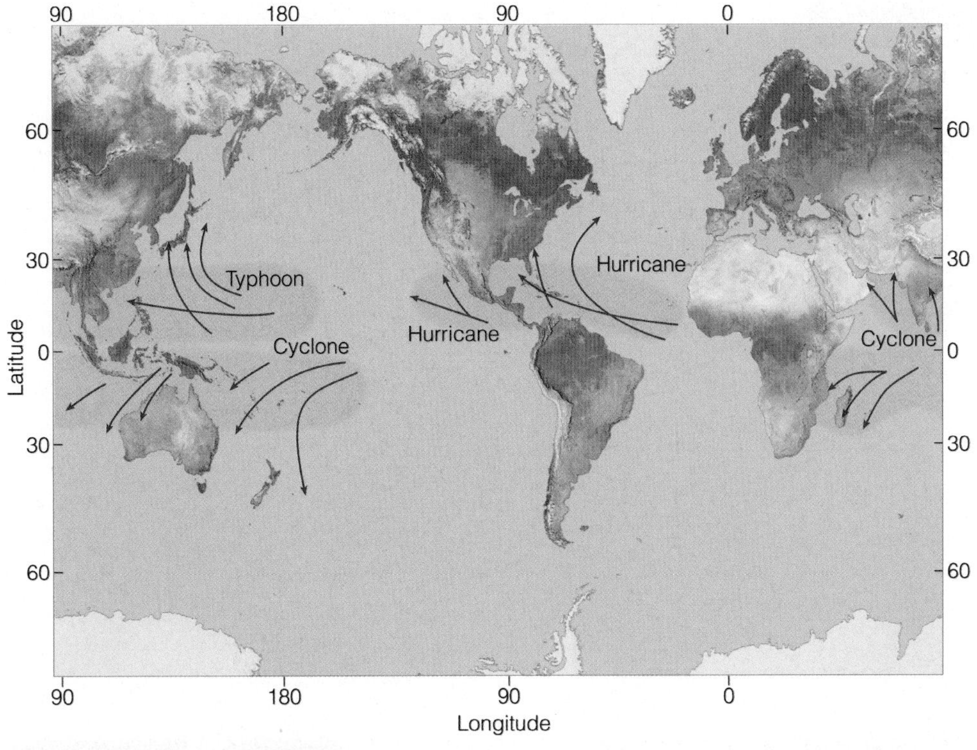

FIGURE 13.11 Regions where tropical storms form (orange shading), the names given to storms, and the typical paths they take (red arrows).

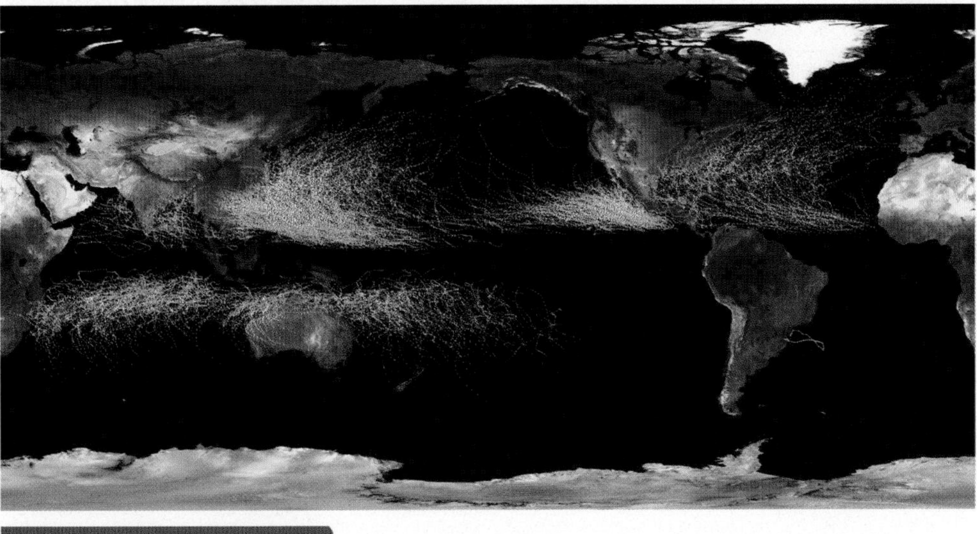

FIGURE 13.12 Paths taken by tropical cyclones worldwide from 1985 to 2005.

unfavorable position of the ITCZ discourages hurricanes from developing in these regions. Then, guess what? For the first time since satellites began observing the south Atlantic, a tropical cyclone formed off the coast of Brazil during March, 2004. The path of the storm shows up in Fig. 13.12 as a single line off the east coast of Brazil. A visible satellite image of the storm is given in Fig. 13.14. So rare are tropical cyclones in this region that no government agency has an effective warning system for them, which is why the tropical cyclone was not given a name.

Eastern Pacific Hurricanes As we saw in an earlier section, many hurricanes form off the coast of Mexico

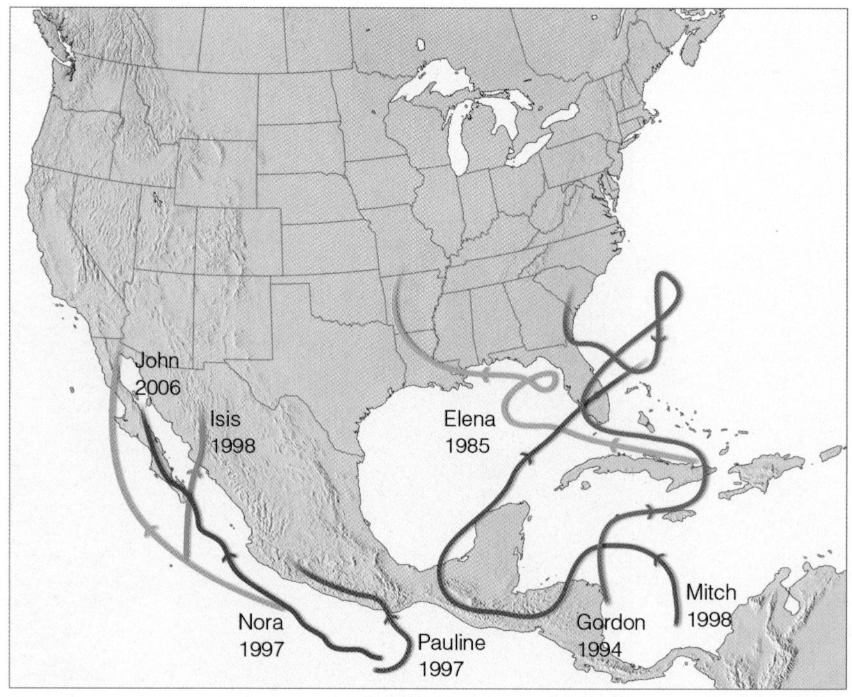

FIGURE 13.13

Some erratic paths taken by hurricanes.

FIGURE 13.14

An extremely rare tropical cyclone (with no name) near 28°S latitude spins clockwise over the south Atlantic off the coast of Brazil during March, 2004. Due to cool water and vertical wind shear, storms rarely form in this region of the Atlantic Ocean. In fact, this is the only tropical storm ever officially reported there.

over the North Pacific. In fact, this area usually spawns about nine hurricanes each year, which is slightly more than the yearly average of six storms born over the tropical North Atlantic. We can see in Fig. 13.11 that eastern North Pacific hurricanes normally move westward, away from the coast, and so little is heard about them. When one does move northwestward, it normally weakens rapidly over the cool water of the North Pacific. Occasionally, however, one will curve northward or even northeastward and slam into Mexico, causing destructive flooding. Hurricane Tico left 25,000 people homeless and caused an estimated $66 million in property damage after passing over Mazatlán, Mexico, in October, 1983. The remains of Tico even produced record rains and flooding in Texas and Oklahoma. Even less frequently, a hurricane will stray far enough north to bring summer rains to southern California and Arizona, as did

FEMA News Photo

FIGURE 13.15 Major damage to buildings caused by Hurricane Iniki on the Island of Kauai during September, 1992.

the remains of Hurricane Nora during September, 1997. (Nora's path is shown in Fig. 13.13.) The only hurricane on record to reach the west coast of the United States with sustained hurricane-force winds did so in October, 1858, when a hurricane slammed into the extreme southern part of California near San Diego.

The Hawaiian Islands, which are situated in the central North Pacific between about 20° and 23°N, appear to be in the direct path of many eastern Pacific hurricanes and tropical storms. By the time most of these storms reach the islands, however, they have weakened considerably, and pass harmlessly to the south or northeast. The exceptions were Hurricane Iwa during November, 1982, and Hurricane Iniki during September, 1992. Iwa lashed part of Hawaii with 115 mi/hr winds and huge surf, causing an estimated $312 million in damages. Iniki, the worst hurricane to hit Hawaii in the twentieth century, battered the island of Kauai with torrential rain, sustained winds of 131 mi/hr that gusted to 161 mi/hr and 20-foot waves that crashed over coastal highways. Major damage was sustained by most of the hotels and about 50 percent of the homes on the island. Iniki (the costliest hurricane in Hawaiian history with damage estimates of $1.8 billion) flattened sugarcane fields, destroyed the macadamia nut crop, injured about 100 people, and caused at least 7 deaths (see Fig. 13.15).

EXTREME WEATHER WATCH

The western North Pacific is the most active tropical cyclone region in the world, with an average 32 storms developing every year (more than twice the Atlantic Basin average). Also, many of the world's most intense tropical cyclones on record have formed there. The most powerful of all was Super Typhoon Tip that developed on October 5, 1979, and grew into a monster with a circulation and cloud formation 1350 miles in diameter (the distance between Key West, Florida, and Amarillo, Texas!). Tip's barometric pressure bottomed out at a world-record 870 mb (25.69 in.) with 190 mi/hr sustained winds near its eye. Fortunately, the storm never made landfall before dying out east of Japan on October 21, 1979.

North Atlantic Hurricanes Hurricanes that form over the tropical North Atlantic also move westward or northwestward on a collision course with Central or North America. Most hurricanes, however, swing away from land and move northward, parallel to the coastline of the United States.* A few storms, perhaps three per year, move inland, bringing with them high winds, huge waves, and torrential rain that may last for days.

Figure 13.16 shows the regions where Atlantic Basin hurricanes tend to form and the typical paths they take during the active hurricane months of August, September, and October. Observe that, during August, hurricanes are most likely to form over the western tropical Atlantic, where they then either track westward into the Gulf of Mexico toward Texas, or they move northwestward into Florida, or they follow a path parallel to the coast of the United States. In September notice that the region where hurricanes are most likely to form stretches westward into the Gulf of Mexico and northward along the Atlantic seaboard. Typical hurricane paths take them into the central Gulf of Mexico or northeastward out over the Atlantic. Should an Atlantic hurricane track close to the coastline, it could make landfall anywhere from Florida to the mid-Atlantic states. In October, hurricanes are most likely to form in the western Caribbean and adjacent to the coast of North America, where they tend to take a more northerly trajectory.

A hurricane moving northward over the Atlantic will normally survive as a hurricane for a much longer time than will its counterpart at the same latitude over the eastern Pacific. The reason for this situation is that an Atlantic hurricane moving northward will usually stay over warmer water, whereas an eastern Pacific hurricane heading north will quickly move over much cooler water and, with its energy source cut off, will rapidly weaken.

Up to this point, it is probably apparent that tropical cyclones called hurricanes are similar to middle-latitude cyclones in that, at the surface, both have central cores of low pressure and winds that spiral counterclockwise (in the Northern Hemisphere) about their respective centers. However, there are many differences between the two systems, which are described in the Focus section on p. 375.

*Sometimes hurricanes that remain over water and pose no threat to land are called "fish hurricanes" because their greatest impact is on the fish in the open ocean.

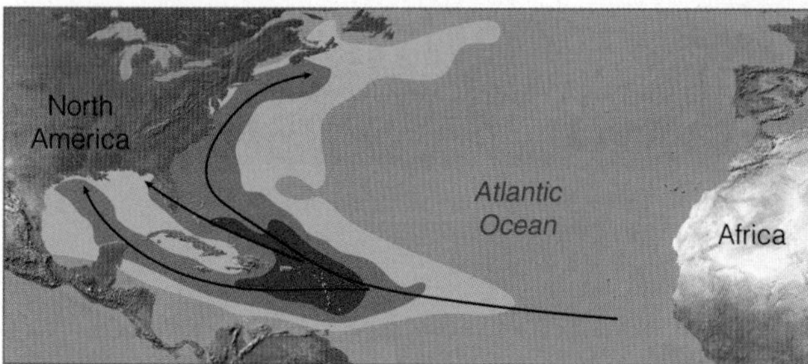

(a) August

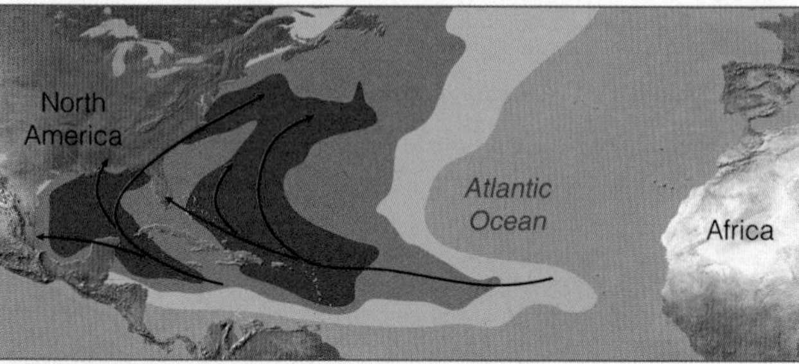

(b) September

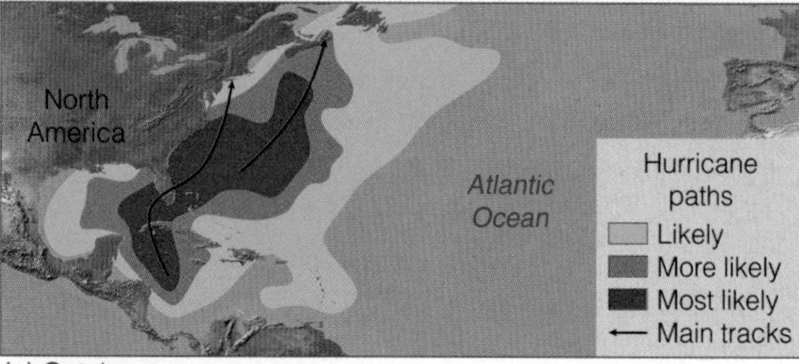

(c) October

FIGURE 13.16 Regions where Atlantic Basin hurricanes tend to form and the paths they are most likely to take during the months of (a) August, (b) September, and (c) October. (Data from NOAA)

Naming Hurricanes and Tropical Storms

In an earlier section, we learned that hurricanes are given a name when they reach tropical storm strength. Before hurricanes and tropical storms were assigned names, they were identified according to their latitude and longitude. This method was confusing, especially when two or more storms were present over the same ocean. To reduce the confusion, hurricanes were identified by letters of the alphabet. During World War II, names like Able and Baker were used. (These names correspond to the radio code words associated with each letter of the alphabet.) This method also seemed cumbersome so, beginning in 1953, the National Weather Service began using female names to identify hurricanes. The list of names for each year was in alphabetical order, so that the names of the season's first storm began with the letter *A*, the second with *B*, and so on.

From 1953 to 1977, only female names were used. However, beginning in 1978, tropical storms in the

FOCUS ON
A SPECIAL TOPIC

How Do Hurricanes Compare with Middle-Latitude Cyclones?

By now, it should be apparent that a hurricane is much different from the mid-latitude cyclone that we discussed in Chapter 10. A hurricane derives its energy from the warm water and the latent heat of condensation, whereas the mid-latitude storm derives its energy from horizontal temperature contrasts. The vertical structure of a hurricane is such that its central column of air is warm from the surface upward; consequently, hurricanes are called *warm-core lows*. A hurricane weakens with height, and the area of low pressure at the surface may actually become an area of high pressure above 40,000 ft (12 km). Mid-latitude cyclones, on the other hand, are *cold-core lows* that usually intensify with increasing height, with a cold upper-level low or trough often existing above, or to the west of the surface low.

A hurricane usually contains an eye where the air is sinking, while mid-latitude cyclones are characterized by centers of rising air. Hurricane winds are strongest near the surface, whereas the strongest winds of the mid-latitude cyclone are found aloft in the jet stream.

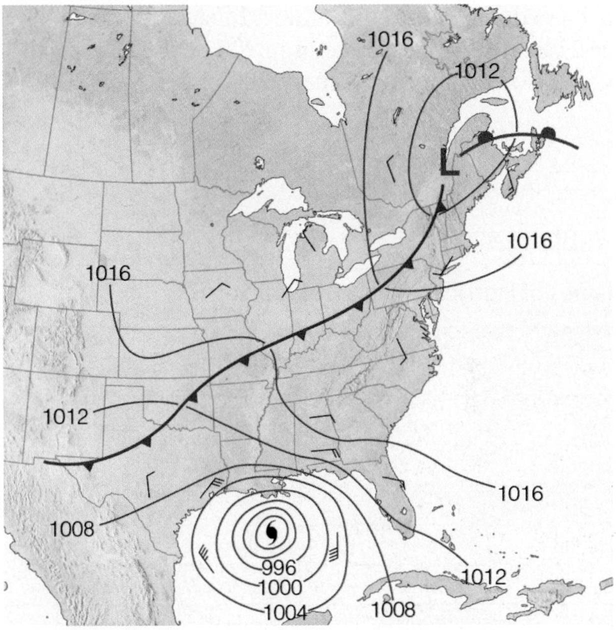

Figure 1 Surface weather map for the morning of September 23, 2005, showing Hurricane Rita over the Gulf of Mexico and a middle-latitude storm system north of New England.

Further contrasts can be seen on a surface weather map. Figure 1 shows Hurricane Rita over the Gulf of Mexico and a mid-latitude storm north of New England. Around the hurricane, the isobars are more circular, the pressure gradient is much steeper, and the winds are stronger. The hurricane has no fronts and is smaller (although Rita is a large Category 5 hurricane). There are similarities between the two systems: Both are areas of surface low pressure, with winds moving counterclockwise about their respective centers.

It is interesting to note that some northeasters (winter storms that move northeastward along the coastline of North America, bringing with them heavy precipitation, high surf, and strong winds) may actually possess some of the characteristics of a hurricane. For example, a particularly powerful northeaster during January, 1989, was observed to have a cloud-free eye, with surface winds in excess of 90 mi/hr spinning about a warm inner core. Moreover, some *polar lows*—lows that develop over polar waters during winter—may exhibit many of the observed characteristics of a hurricane, such as a symmetric band of thunderstorms spiraling inward around a cloud-free eye, a warm-core area of low pressure, and strong winds near the storm's center. In fact, when surface winds within these polar storms reach 60 mi/hr, they are sometimes referred to as *Arctic hurricanes*.

Even though hurricanes weaken rapidly as they move inland, their circulation may draw in air with contrasting properties. If the hurricane links with an upper-level trough, it may actually become a mid-latitude cyclone. Swept eastward by upper-level winds, the remnants of an Atlantic hurricane can become an intense mid-latitude autumn storm in Europe.

eastern Pacific were alternately assigned female and male names, but not just English names, as Spanish and French ones were used too. This practice began for North Atlantic hurricanes in 1979. If a storm causes great damage and becomes infamous as a Category 3 or higher, its name is retired for at least ten years.

Table 13.1 gives the proposed list of names for both North Atlantic and eastern Pacific hurricanes. If the number of named storms in any year should exceed the names on the list, as occurred in 2005, then tropical storms are assigned names from the Greek alphabet, such as Alpha, Beta, and Gamma. In fact, the last of the 27 named tropical systems in 2005 was Zeta, which actually formed during January, 2006.

Devastating Winds, Flooding, and the Storm Surge

When a hurricane is approaching from the south, its highest winds are usually on its eastern (right) side. The reason for this phenomenon is that the winds that push the storm along add to the winds on the east side and subtract from the winds on the west (left) side. The hurricane illustrated in Fig. 13.17 is moving northward along the east coast of the United States with winds of 100 mi/hr swirling counterclockwise about its center. Because the storm is moving northward at about 25 mi/hr, sustained winds on its eastern side are about 125 mi/hr, while on its western side, winds are only 75 mi/hr.

Even though the hurricane in Fig. 13.17 is moving northward, there is a net transport of water directed

Table 13.1

Names of Hurricanes and Tropical Storms

NORTH ATLANTIC HURRICANE NAMES				EASTERN NORTH PACIFIC HURRICANE NAMES			
2010	2011	2012	2013	2010	2011	2012	2013
Alex	Arlene	Alberto	Andrea	Agatha	Adrian	Aletta	Alvin
Bonnie	Bret	Beryl	Barry	Blas	Beatriz	Bud	Barbara
Colin	Cindy	Chris	Chantal	Celia	Calvin	Carlotta	Cosme
Danielle	Don	Debby	Dorian	Darby	Dora	Daniel	Dalila
Earl	Emily	Ernesto	Erin	Estelle	Eugene	Emilia	Erick
Fiona	Franklin	Florence	Fernand	Frank	Fernanda	Fabio	Flossie
Gaston	Gert	Gordon	Gabrielle	Georgette	Greg	Gilma	Gil
Hermine	Harvey	Helene	Humberto	Howard	Hilary	Hector	Henriette
Igor	Irene	Isaac	Ingrid	Isis	Irwin	Ileana	Ivo
Julia	Jose	Joyce	Jerry	Javier	Jova	John	Juliette
Karl	Katia	Kirk	Karen	Kay	Kenneth	Kristy	Kiko
Lisa	Lee	Leslie	Lorenzo	Lester	Lidia	Lane	Lorena
Matthew	Maria	Michael	Melissa	Madeline	Max	Miriam	Manuel
Nicole	Nate	Nadine	Nestor	Newton	Norma	Norman	Narda
Otto	Ophelia	Oscar	Olga	Orlene	Otis	Olivia	Octave
Paula	Philippe	Patty	Pablo	Paine	Pilar	Paul	Priscilla
Richard	Rina	Rafael	Rebekah	Roslyn	Ramon	Rosa	Raymond
Shary	Sean	Sandy	Sebastien	Seymour	Selma	Sergio	Sonia
Tomas	Tammy	Tony	Tanya	Tina	Todd	Tara	Tico
Virginie	Vince	Valerie	Van	Virgil	Veronica	Vicente	Velma
Walter	Whitney	William	Wendy	Winifred	Wiley	Willa	Wallis
				Xavier	Xina	Xavier	Xina
				Yolanda	York	Yolanda	York
				Zeke	Zelda	Zeke	Zelda

eastward toward the coast. To understand this behavior, recall from Chapter 7 that as the wind blows over open water, the water beneath is set in motion. If we imagine the top layer of water to be broken into a series of layers, then we find each layer moving to the *right* of the layer above (Northern Hemisphere). This type of movement (bending) of water with depth (called the *Ekman Spiral)* causes a net transport of water (known as *Ekman transport)* to the right of the surface wind in the Northern Hemisphere. Hence, the north wind on the hurricane's left (western) side causes a net transport of water toward the shore. Here, the water piles up and rapidly inundates the region.

The high winds of a hurricane also generate large waves, sometimes 30 to 50 feet high. These waves move outward, away from the storm, in the form of *swells* that carry the storm's energy to distant beaches. Consequently, the effects of the storm may be felt days before the hurricane arrives.

Although the hurricane's high winds inflict a great deal of damage, it is the huge waves, high seas, and *flooding* that normally cause most of the destruction. The flooding is also responsible for the loss of many lives. In fact, the majority of hurricane-related deaths during the past century has been due to flooding. The flooding is due, in part, to winds pushing water onto the shore and to the heavy rains, which may exceed 25 inches in 24 hours.* Flooding is also aided by the low pressure of the storm. The region of low pressure allows the ocean level to rise, much like a soft drink rises up a straw as air is withdrawn. The combined effect of high water (which is usually well above the high-tide level), high winds, and the net Ekman transport toward the coast, produces the **storm surge**—an abnormal rise in the ocean level—which inundates low-lying areas and turns beachfront homes into piles of splinters (see ❱Fig.13.18). The storm surge is particularly damaging when it coincides with normal high tides.

*Hurricanes may sometimes have a beneficial aspect, in the sense that they can provide much needed rainfall in drought-stricken areas.

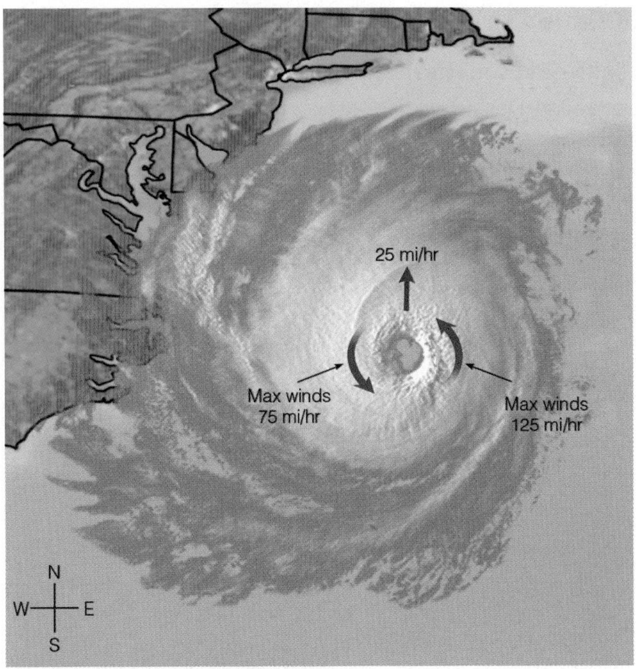

❱ FIGURE 13.17 A hurricane moving northward will have higher sustained winds on its eastern side than on its western side. If the hurricane moves from east to west, highest sustained winds will be on its northern side.

❱ FIGURE 13.18 When a storm surge moves in at high tide it can inundate and destroy a wide swath of coastal lowlands.

Table 13.2

Saffir-Simpson Hurricane Damage-Potential Scale

SCALE NUMBER	CENTRAL PRESSURE		WINDS		STORM SURGE		
(CATEGORY)	mb	in.	mi/hr	knots	ft	m	DAMAGE
1	≥980*	≥28.94	74–95	64–82	4–5	~1.5	Damage mainly to trees, shrubbery, and unanchored mobile homes
2	965–979	28.50–28.91	96–110	83–95	6–8	~2.0–2.5	Some trees blown down; major damage to exposed mobile homes; some damage to roofs of buildings
3	945–964	27.91–28.47	111–130	96–113	9–12	~2.5–4.0	Foliage removed from trees; large trees blown down; mobile homes destroyed; some structural damage to small buildings
4	920–944	27.17–27.88	131–155	114–135	13–18	~4.0–5.5	All signs blown down; extensive damage to roofs, windows, and doors; complete destruction of mobile homes; flooding inland as far as 10 km (6 mi); major damage to lower floors of structures near shore
5	<920	<27.17	>155	>135	>18	>5.5	Severe damage to windows and doors; extensive damage to roofs of homes and industrial buildings; small buildings overturned and blown away; major damage to lower floors of all structures less than 4.5 m (15 ft) above sea level within 500 m of shore

*Symbol > means "greater than"; < means "less than"; ≥ means "equal to or greater than"; ~ means "approximately equal to."

In an effort to estimate the possible damage a hurricane's sustained winds and storm surge could do to a coastal area, the **Saffir-Simpson scale** was developed (see Table 13.2). The scale numbers (which range from 1 to 5) are based on actual conditions at some time during the life of the storm. As the hurricane intensifies or weakens, the category, or scale number, is reassessed accordingly. Major hurricanes are classified as Category 3 and above. In the western Pacific, a typhoon with sustained winds of at least 150 mi/hr (130 knots)—at the upper end of the wind speed range in Category 4 on the Saffir-Simpson scale—is called a **super typhoon.** Figure 13.19 illustrates how the storm surge changes along the coast as hurricanes with increasing intensity make landfall.*

Figure 13.20 shows the number of hurricanes that have made landfall along the coastline of the United States from 1900 through 2007. Out of a total of 181

*Landfall is the position along the coast where the center of a hurricane passes from ocean to land.

Normal high tide

Category 1 [4-foot rise]

Category 3 [12-foot rise]

Category 5 [20-foot rise]

FIGURE 13.19 The changing of the ocean level as different category hurricanes make landfall along the coast. The water typically rises about 4 feet with a Category 1 hurricane, but may rise to 22 feet (or more) with a Category 5 storm.

hurricanes striking the American coastline, 73 (40 percent) were major hurricanes—Category 3 or higher. Hence, along the Gulf and Atlantic coasts, on the average, about five hurricanes make landfall every three years, two of which are major hurricanes with winds in excess of 110 mi/hr (95 knots) and a storm surge exceeding 8 ft (2.5 m).

Although the high winds of a hurricane can devastate a region, considerable damage may also occur from hurricane-spawned tornadoes. About one-fourth of the hurricanes that strike the United States produce tornadoes. In fact, in 2004 six tropical systems produced just over 300 tornadoes in the southern and eastern United States. The exact mechanism by which these tornadoes form is not totally clear; however, studies suggest that surface topography may play a role by initiating the convergence (and hence, rising) of surface air. Moreover, tornadoes tend to form in the right front quadrant of an advancing hurricane,* where vertical wind speed shear is greatest. Studies also suggest that swathlike areas of extreme damage once attributed to tornadoes may actually be due to strong downdrafts (microbursts) associated with the large, intense thunderstorms around the eyewall.

In examining the extensive damage wrought by Hurricane Andrew during August, 1992, researchers theorized that the areas of most severe damage might have been caused by small whirling eddies perhaps 100 to 300 feet in diameter that occur in narrow bands. Many scientists today believe those rapidly rotating eddies were, in fact, small tornadoes. Lasting for about 10 seconds, the vortices appeared to have formed in a region of strong wind speed shear in the hurricane's eyewall, where the air was rapidly rising. As intense updrafts stretched the vortices vertically, they shrank horizontally, which induced them to spin faster, perhaps as fast as 80 mi/hr. When the rotational winds of a vortex are added to the hurricane's steady wind, the total wind speed over a relatively small area may increase substantially. In the case of Hurricane Andrew, isolated wind speeds may have reached 200 mi/hr over narrow stretches of south Florida.

Up until 2005, the annual death toll from hurricanes in the United States, over a span of about 30 years, averaged less than 50 persons.** Most of these fatalities were due to flooding. This relatively low total was due in part to the advanced warning provided by the National Weather Service and the fact that only a few really intense storms had made landfall during this time. But the hurricane death toll in the United States rose dramatically in 2005 when Hurricane Katrina slammed into Mississippi and Louisiana.

As Hurricane Katrina moved toward the coast, evacuation orders were given to residents living in low-lying areas, including the city of New Orleans.

*In the northeast quadrant of the hurricane shown in Fig. 13.17, on p. 377.

**In other countries, the annual death toll was considerably higher. Estimates are that more than 3000 people died in Haiti from flooding and mud slides when Hurricane Jeanne moved through the Caribbean during September, 2004.

EXTREME WEATHER WATCH

The storm surge that drowned at least 300,000 people (some estimates say 500,000) in Bangladesh during the Great Bhola Cyclone of 1970 was estimated to have been as high as 40 feet along portions of the heavily populated shoreline of the Bay of Bengal. A storm surge of 42 feet, the highest ever recorded in the world, was observed during Australia's famous Bathurst Bay cyclone of March 5, 1899. The greatest measured storm surge on record in the United States was 27.8 feet at Pass Christian, Mississippi, during the passage of Hurricane Katrina on August 29, 2005, although estimates indicate that a higher surge occurred just east of Pass Christian, where flood gauge instruments failed.

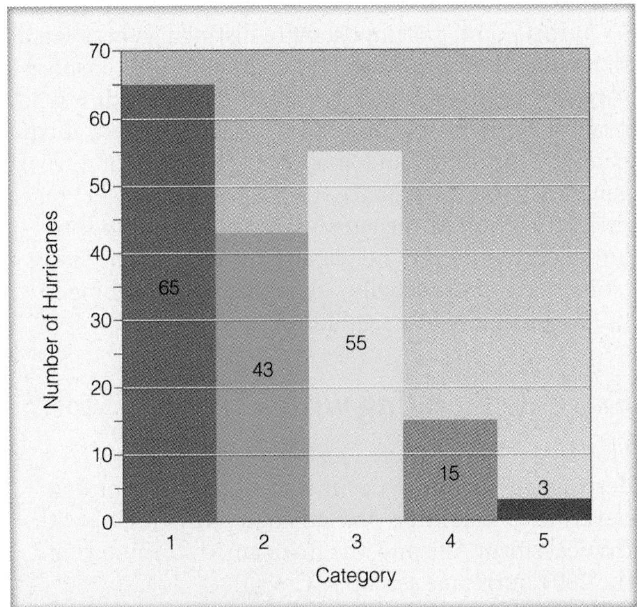

FIGURE 13.20 The number of hurricanes (by each category) that made landfall along the coastline of the United States from 1900 through 2007. All of the hurricanes struck the Gulf or Atlantic coasts. Categories 3, 4, and 5 are considered major hurricanes.

Many thousands of people moved to higher ground but, unfortunately, many people either refused to leave their homes or had no means of leaving, and were forced to ride out the storm. Tragically, more than 1500 people died either from Katrina's huge storm surge and high winds that demolished countless buildings, or from the flooding in New Orleans, when several levees broke and parts of the city were inundated with water over 20 feet deep. As the population density continues to increase in vulnerable coastal areas, the potential for another hurricane-caused disaster increases also.

Even when a hurricane does not make landfall, it can be deadly. For example, during August, 2009, as Hurricane Bill moved northeast more than 150 miles off the coast of Maine, thousands of people flocked to Maine's rocky shoreline to observe the huge waves generated by Bill. Tragically, an unusually large wave washed several people from a rocky cliff into the churning ocean below, including a 7-year-old girl who drowned in the surf.

The aftermath of an intense hurricane can be devastating. The supply of fresh drinking water may be contaminated and food may become scarce, as grocery stores and markets are forced to close, and stay closed for days or even weeks. Roads may be blocked by fallen trees and debris, or by sand that was deposited during the storm surge. Electrical and telephone service may be disrupted or completely lost. And many people may be displaced from their damaged or destroyed homes. Even the cleanup efforts can prove deadly, as poisonous snakes often find their way into various nooks and crannies of the debris.

At this point, it is important to distinguish between a *storm surge* and a *tsunami.* A storm surge is an onshore surge of ocean water usually caused by the winds of a tropical cyclone pushing sea water onto the coast. A tsunami, on the other hand, is a wave (or a series of waves) generated by a disturbance (usually an earthquake) on the ocean floor. As the tsunami wave moves into shallow water, it builds in height and rushes onto the land (sometimes unexpectedly), sweeping up everything in its path, including cars, buildings, and people.

Extreme Flooding with a Tropical Storm

Flooding is not restricted just to hurricanes, as destructive floods can occur with tropical storms that never reach hurricane strength. Such was the case with tropical storm Allison—the first named storm during the 2001 hurricane season.

In late May, 2001, Allison began as a tropical wave that moved westward across the Atlantic. The wave continued its westward journey, and by the first of June it had moved across Central America and out over the Pacific Ocean. Here, it organized into a band of thunderstorms and a tropical depression. Upper-level winds guided the depression northward over Central America, then out over the Gulf of Mexico, where the warm water fueled the circulation; and just east of Galveston, Texas, the depression became tropical storm Allison. Packing winds of 61 mi/hr, Allison made landfall over the east end of Galveston Island on June 5. It drifted inland and weakened (see Fig. 13.21).

On the eastern side of the storm, heavy rain fell over parts of Texas and Louisiana. Some areas of southeast Texas received as much as 10 inches of rain in less than five hours. Homes, streets, and highways flooded as heavy rain continued to pound the area. But the worst was yet to come.

On June 7, as the upper-level winds began to change, the remnants of Allison drifted southwestward toward Houston. Heavy rain fell over southeast Texas and Louisiana, where several tornadoes touched down. Over the Houston area, more than 20 inches of rain fell within a 12-hour period, submerging a vast part of the city. In six days the Port of Houston received a staggering 35 inches of rain.

The center of circulation drifted southward, moving off the Texas coast and out over the Gulf of Mexico on the evening of June 9. The flow aloft then guided the storm northeastward, where the storm made landfall

FIGURE 13.21 Visible satellite image showing the remains of tropical storm Allison centered over Texas on the morning of June 6, 2001. Heavy rain is falling from the thick clouds over Louisiana and eastern Texas.

again, but this time in southeastern Louisiana. Heavy rain continued to pound Louisiana, creating one of the worst floods on record—a station in southern Louisiana reported a rainfall total of 30 inches.

On June 11, a zone of maximum winds aloft (a jet streak) associated with the subtropical jet stream enhanced the outflow above the surface storm, and the remains of tropical storm Allison actually began to intensify over land. As the storm entered Mississippi, its central pressure lowered, wind gusts reached 60 mi/hr, and the center of circulation developed a weak-looking eye (see Fig. 13.22). As the system trekked eastward, it weakened and lost its eye, but continued to dump heavy rain over the southern Gulf States. Eventually, on June 14, the storm reached the Carolina coast.

Unfortunately, the storm slowed, then turned northward over North Carolina. Flooding became a major problem—Doppler radar estimated that 21 inches of rain had fallen over parts of the state. Severe weather broke out in Georgia and in the Carolinas, where some areas reported hail and downed trees due to gusty winds. The storm moved northeastward, parallel to the coast. A cold front moving in from the west eventually hooked up with the moisture from Allison. This situation caused heavy rain to fall over the mid-Atlantic states and southern New England. The storm finally accelerated to the northeast, away from the coast on June 18.

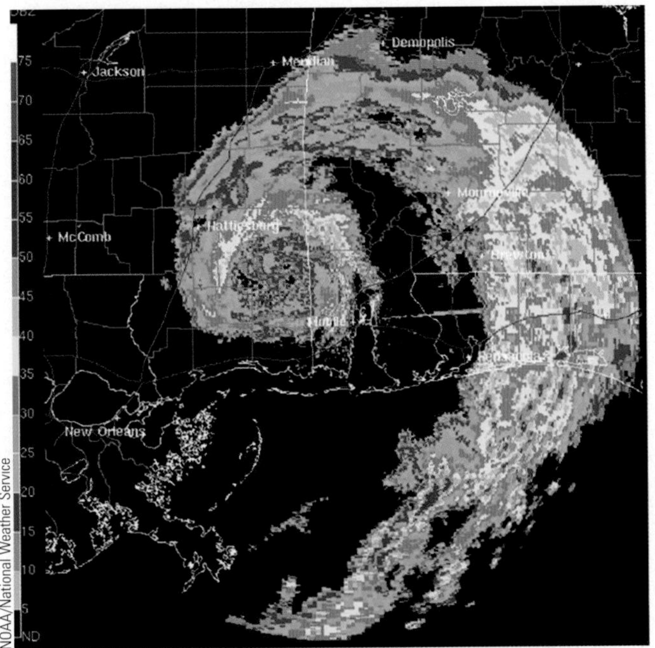

NOAA/National Weather Service

FIGURE 13.22 Doppler radar display on June 11, 2001, showing bands of heavy rain swirling counterclockwise into the center of once tropical storm Allison. The center of the storm, which is over Mississippi, has actually deepened and formed somewhat of an eye.

EXTREME WEATHER WATCH

Amazing rainfall totals have occurred during the passage of tropical cyclones over land areas. In early August, 2009, Typhoon Morakot slowly traversed Taiwan, dumping an incredible 114 inches of rain in 72 hours on Mount Alishan at the southern end of the island. Hurricane Wilma, the most intense Atlantic hurricane on record, deposited more than 65 inches of rain in 24 hours on Isla Mujeras in the Yucatan of Mexico during its passage on October 21 and 22, 2005. This total remains a Western Hemisphere record for 24-hour rainfall.

Allison, which never developed hurricane strength winds, claimed the lives of 43 people, whose deaths were mainly due to flooding. The total damage from the storm totaled in the billions of dollars, with the Houston area alone sustaining over $2 billion in damage. If all the rain that fell from Allison could be placed in Texas, it would cover two-thirds of the state with water a foot deep. (To date, Allison is the only tropical storm to have its name retired.)

Some Notable Hurricanes

CAMILLE, 1969 Hurricane Camille (1969) stands out as one of the most intense hurricanes to reach the coastline of the United States during the twentieth century (see Table 13.3). With a central pressure of 909 mb, tempestuous winds reaching 184 mi/hr (160 knots) and a storm surge more than 23 feet above the normal high-tide level, Camille, as a Category 5 storm, unleashed its fury on Mississippi, destroying thousands of buildings. During its rampage, it caused an estimated $1.5 billion in property damage and took more than 200 lives.

HUGO, 1989 During September, 1989, Hurricane Hugo, born as a cluster of thunderstorms, became a tropical depression off the coast of Africa, southeast of the Cape Verde Islands. The storm grew in intensity, tracked westward for several days, then turned northwestward, striking the island of St. Croix with sustained winds of 144 mi/hr. After passing over the eastern tip of Puerto Rico, this large, powerful hurricane took aim at the coastline of South Carolina. With maximum winds estimated at about 138 mi/hr and a central pressure near 934 mb, Hugo made landfall as a Category 4 hurricane near Charleston, South Carolina, about midnight on

Table 13.3

The Thirteen Most Intense Hurricanes (at Landfall) to Strike the United States from 1900 through 2007

RANK	HURRICANE (MADE LANDFALL)	YEAR	CENTRAL PRESSURE (MILLIBARS/INCHES)	CATEGORY	DEATH TOLL
1	Florida (Keys)	1935	892/26.35	5	408
2	Camille (Mississippi)	1969	909/26.85	5	256
3	Andrew (South Florida)	1992	922/27.23	5	53
4	Katrina (Louisiana)	2005	920/27.17	3*	>1500
5	Florida (Keys)/South Texas	1919	927/27.37	4	>600†
6	Florida (Lake Okeechobee)	1928	929/27.43	4	>2000
7	Donna (Long Island, New York)	1960	930/27.46	4	50
8	Texas (Galveston)	1900	931/27.49	4	>8000
9	Louisiana (Grand Isle)	1909	931/27.49	4	350
10	Louisiana (New Orleans)	1915	931/27.49	4	275
11	Carla (South Texas)	1961	931/27.49	4	46
12	Hugo (South Carolina)	1989	934/27.58	4	49
13	Florida (Miami)	1926	935/27.61	4	243

*Although the central pressure in Katrina's eye was quite low, Katrina's maximum sustained winds of 110 knots at landfall made it a Category 3 storm.
†More than 500 of this total were lost at sea on ships. (The > symbol means "greater than.")

September 21 (see Fig. 13.23). The high winds and storm surge, which ranged between 8 and 20 feet, hurled a thundering wall of water against the shore. This knocked out power, flooded streets, and caused widespread destruction to coastal communities. The total damage in the United States attributed to Hugo was over $7 billion, with a death toll of 21 in the United States and 49 overall. (Since 1955, no hurricane hunter reconnaissance aircraft has been lost over the tropical Atlantic, although one close call occurred when a reconnaissance plane flew into Hurricane Hugo. What happened to the aircraft and the crew is chronicled in the Focus section on pp. 384–385.)

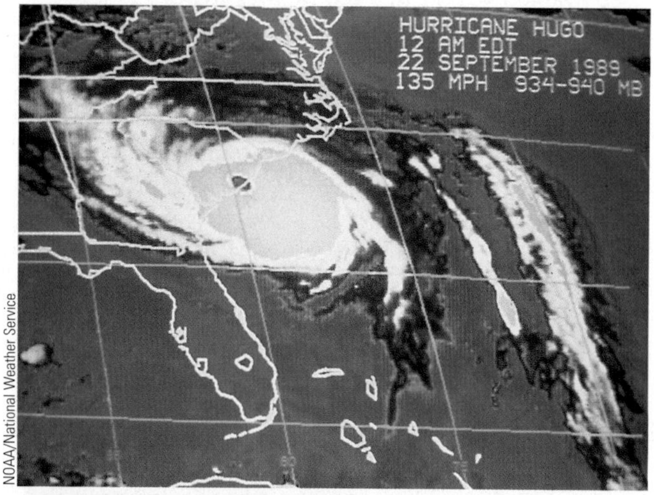

NOAA/National Weather Service

FIGURE 13.23 A color-enhanced infrared satellite image of Hurricane Hugo with its eye over the coast near Charleston, South Carolina.

ANDREW, 1992 Another devastating hurricane during the twentieth century was Hurricane Andrew. On August 21, 1992, as tropical storm Andrew churned westward across the Atlantic it began to weaken, prompting some forecasters to surmise that this tropical storm would never grow to hurricane strength. But Andrew moved into a region favorable for hurricane development. Even though it was outside the tropics near latitude 25°N, warm surface water and weak winds aloft allowed Andrew to intensify rapidly. And in just two days Andrew's winds increased from 52 mi/hr to 140 mi/hr, turning an average tropical storm into one of the most intense hurricanes to strike Florida in the past 105 years (see Table 13.3).

With winds of at least 155 mi/hr and a powerful storm surge, Andrew made landfall south of Miami on the morning of August 24 (see Fig.13.24). The eye of the storm moved over Homestead, Florida. Andrew's fierce winds completely devastated the area (see Fig. 13.25) as 50,000 homes were destroyed, trees were leveled (see Fig. 13.26), and steel-reinforced tie

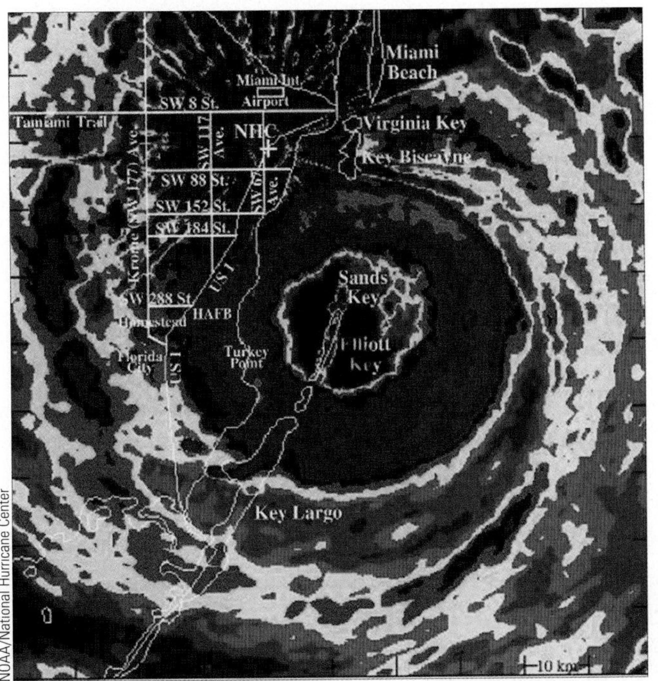

FIGURE 13.24 Color radar image of Hurricane Andrew as its moves onshore over south Florida on the morning of August 24, 1992. The dark red and purple show where the heaviest rain is falling. Miami Beach is just to then north of the eye, and the National Hurricane Center (NHC) is about 20 miles to the northwest of the eye.

FIGURE 13.26 The remains of a two-story apartment complex in Homestead, Florida, after Hurricane Andrew slammed the area with winds exceeding 150 mi/hr.

FIGURE 13.25 A community in Homestead, Florida, devasted by Hurricane Andrew during August, 1992.

beams weighing tons were torn free of townhouses and hurled as far as several blocks. Swaths of severe damage led scientists to postulate that peak winds may have approached 200 mi/hr. Such winds may have occurred with small tornadoes, which added substantially to the storm's wind speed. In an instant, a wind gust of 164 mi/hr blew down a radar dome and inactivated several satellite dishes on the roof of the National Hurricane Center (NHC) in Coral Gables. Observations reveal that some of Andrew's destruction may have been caused by microbursts in the severe thunderstorms of the eyewall. The hurricane roared westward across southern Florida, weakened slightly, then regained strength over the warm Gulf of Mexico. Surging northwestward, Andrew slammed into Louisiana as a Category 3 with 130 mi/hr winds on the evening of August 25.

All told, Hurricane Andrew was one of the costliest natural disasters ever to hit the United States. It destroyed or damaged over 200,000 homes and businesses, left more than 160,000 people homeless, caused over $30 billion in damages, and took 53 lives, including 41 in Florida.

IVAN, 2004 Hurricane Ivan was an interesting but costly hurricane. It moved onshore just west of Gulf Shores, Alabama, on September 15, 2004 (see ❯ Fig 13.27) as a strong Category 3 hurricane with winds 120 mi/hr and a storm surge of about 16 feet. The strongest winds and greatest damage occurred over an area near the border between Alabama and Florida (see Fig. ❯ 13.28). As Ivan moved inland, it weakened and

FOCUS ON

EXTREME WEATHER

Hunting Hugo

In September, 1989, the crew of the NOAA *Flight 42* Hurricane Hunter aircraft unwittingly became the first ever to fly into a powerful Category 5 hurricane at a low altitude, and live to tell about it.

The entire crew of *Flight 42* were experienced hurricane hunters, with many flights under their belts. The plan on this day was to fly into what appeared to be a rather weak Category 1 hurricane named Hugo, located out over the tropical Atlantic Ocean.

Because Hugo was a weak hurricane, the decision was made to push the limits of safe hurricane flying by going into the eyewall at 1500 feet—the altitude where the hurricane's winds and turbulence are the strongest. It was a decision that almost proved fatal.

Fifteen minutes from Hugo's first spiral band the flight director received his first radar view of the storm. It had an impressive symmetry, with two major spiral bands and a relatively small 12-mile diameter eye (see Fig. 2). Unnoticed on the radar screen, however, was that Hugo's strongest winds were actually higher than the highest wind speed for which the radar system had been calibrated.

As the aircraft entered the eyewall, it grew dark inside the cabin. Powerful wind gusts tore at the aircraft, slamming it from side to side, while torrential rains pelted the plane. Wingtips flexed up and down between three

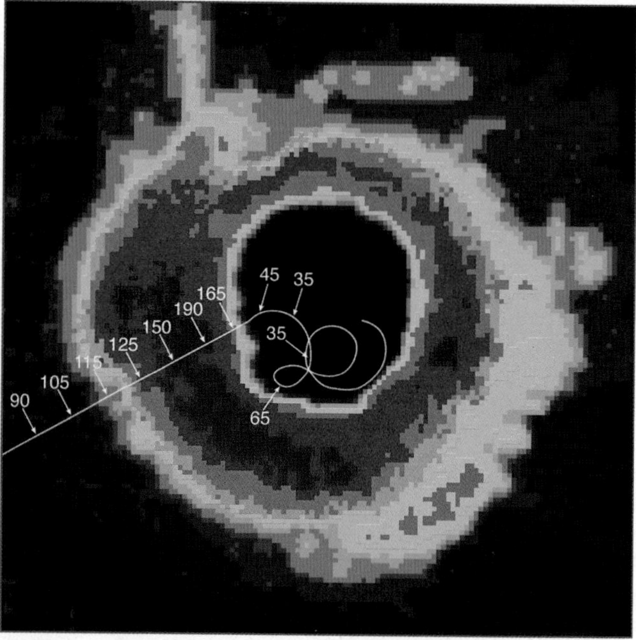

Figure 2 This radar image shows the track of NOAA *Flight 42* into the eye of Hurricane Hugo. Numbers printed on the radar screen show the wind speeds encountered by the flight. Dark red and purple represent the heaviest precipitation in the eyewall.

FIGURE 13.27 Visible satellite image of Hurricane Ivan as it makes landfall near Gulf Shores, Alabama, on September 15, 2004. Ivan is a major hurricane with winds of 121 mi/hr and a surface air pressure of 945 mb (27.91 in.).

eventually linked up with a mid-latitude low. The remains of Ivan then split from the low and drifted southward, eventually ending up in the Gulf of Mexico, where it regained tropical storm strength. It made landfall for the second time along the Gulf Coast, but this time as a tropical depression. All told, Ivan took 25 lives in the United States, produced a record 117 tornadoes over the southern and eastern states, and caused an estimated $14 billion in damages. (Ivan was one of five

to six feet. The pilots tried to climb to 5000 feet where it would be safer, but the turbulence was so violent that they could not keep the plane under control. Wind speeds indicated that Hugo was not a weak Category 1 hurricane, but a rare Category 5 storm with winds exceeding 155 mi/hr.

A fierce updraft wrenched the aircraft, first slamming everyone into their seats with twice the force of gravity, then, seconds later, a strong downdraft left everyone dangling weightless. A huge updraft then lifted the plane violently, creating a shower of flying gear throughout the cabin.

Suddenly, the sky lightened, the clouds thinned, and the rain abated. The plane was at the very edge of the eyewall. But just as the calm set in, disaster struck. Thick, dark clouds suddenly enveloped the aircraft. An enormous gust of wind hit the plane, causing gear, loosened by the previous extreme turbulence, to fly about the cabin and bounce off walls, ceilings, and even crew members. Then another huge gust of wind staggered the plane. At this point, the number three engine failed, and debris hung from the number four engine as flames shot from it.

With two engines damaged, both on the same wing, *Flight 42* was in trouble. The plane began a dive toward the ocean below. For several terrifying seconds, the crew watched the massive waves grow larger as the plane rapidly approached them. Everyone on board knew that the odds of recovering from the dive were not good. But miraculously the crew, through skillful maneuvering, was able to bring the aircraft out of the dive, just as a monstrous wave nearly swamped the aircraft. The plane and its crew had survived, but now they were trapped inside Hugo's eye with only two working engines.

As they circled inside the eye, the aircraft had fallen so low that the crew could see under the bottom edge of the eyewall clouds. A decision was made to jettison anything nonessential in an attempt to lighten the plane, so they could climb to a higher altitude. Slowly the plane rose. An Air Force reconnaissance plane, moving through Hugo's eye, advised *Flight 42* to follow them through what appeared to be a weaker portion of the eyewall.

As they entered the eyewall, darkness fell and intense blasts of turbulent wind rocked the airplane. Wind speeds shot up to 170 mi/hr and gusted to 190 mi/hr (see Fig. 2). After several long minutes, the aircraft was out of the eyewall. They had made it! The excellent piloting that got them this far brought the aircraft and crew home safely.

Analysis of this flight revealed that the plane had hit a tornado-like vortex embedded in the eyewall. Research now suggests that similar vortices may be responsible for some of the incredible damage hurricanes inflict when they strike land. *Flight 42* encountered this vortex first hand, and thankfully survived to provide us with this harrowing tale.

USGS

(a)

(b)

FIGURE 13.28 Beach homes along the Gulf Coast at Orange Beach, Alabama (a) before, and (b) after Hurricane Ivan made landfall during September, 2004. (Red arrows are for reference.)

hurricanes to make landfall in the United States during 2004. Out of the five hurricanes that hit the United States, four impacted the state of Florida. More information on the record-setting Atlantic hurricane seasons of 2004 and 2005 is given in the Focus section on p. 389.

KATRINA, 2005 Hurricane Katrina was the most costly hurricane to ever hit the United States. Forming over warm tropical water south of Nassau in the Bahamas, Katrina became a tropical storm on August 24, 2005, and a Category 1 hurricane just before making landfall in south Florida on August 25. (Katrina's path is given in Fig. 3 on p. 389.) It moved southwestward across Florida and out over the eastern Gulf of Mexico. As Katrina moved westward, it passed over a deep band of warm water called the *Loop current* that allowed Katrina to rapidly intensify. Within 12 hours, the hurricane increased from a Category 3 to a Category 5 storm with winds of 175 mi/hr and a central pressure near 902 mb (see Fig. 13.29).

Over the Gulf of Mexico, Katrina gradually turned northward toward Mississippi and Louisiana. As the powerful Category 5 hurricane moved slowly toward the coast, its rainbands near the center of the storm began to converge toward the storm's eye. This process cut off moisture to the eyewall. As the old eyewall dissipated, a new one formed farther away in a phenomenon called *eyewall replacement*. The replacement of the eyewall weakened the storm such that Katrina made landfall near Buras, Louisiana, on August 29 (see Fig. 13.30) as a strong Category 3 hurricane with sustained winds of 127 mi/hr, a central pressure of 920 mb, and a storm surge between 20 and 30 feet.

Katrina's strong winds and high storm surge on its eastern side devastated southern Mississippi, with Biloxi, Gulfport, and Pass Christian being particularly hard hit (see Fig. 13.31). The winds demolished all but the strongest structures, and the huge storm surge scoured areas up to 10 miles inland.

New Orleans and the surrounding parishes actually escaped the brunt of Katrina's winds, as the eye passed just to the east of the city (Fig. 13.30). However, the combination of high winds, large waves, and a huge storm surge caused disastrous breeches in the levee system that protects New Orleans from the Mississippi River, the Gulf of Mexico, and Lake Pontchartrain. When the levees gave way, water up to 20 feet deep invaded a large part of the city, tragically before thousands of people could escape (see Fig.13.32). Less than a month later, powerful Hurricane Rita with sustained winds of 175 mi/hr moved over the Gulf of Mexico, south of New Orleans. Strong, tropical storm–force easterly winds, along with another storm surge, caused some of the repaired levees to break again, flooding parts of the city that just days earlier had been pumped dry. The death toll due to Hurricane Katrina climbed to more than 1500 and the devastation wrought by the storm totaled more than $75 billion. Although Katrina may well be the most expensive hurricane on record, tragically it is not the deadliest.

NOAA

FIGURE 13.29 Visible satellite image of Hurricane Katrina over the Gulf of Mexico. With sustained winds of 175 mi/hr and a central pressure near 902 mb (26.64 in.), this large and powerful Category 5 hurricane takes aim on Louisiana and Mississippi.

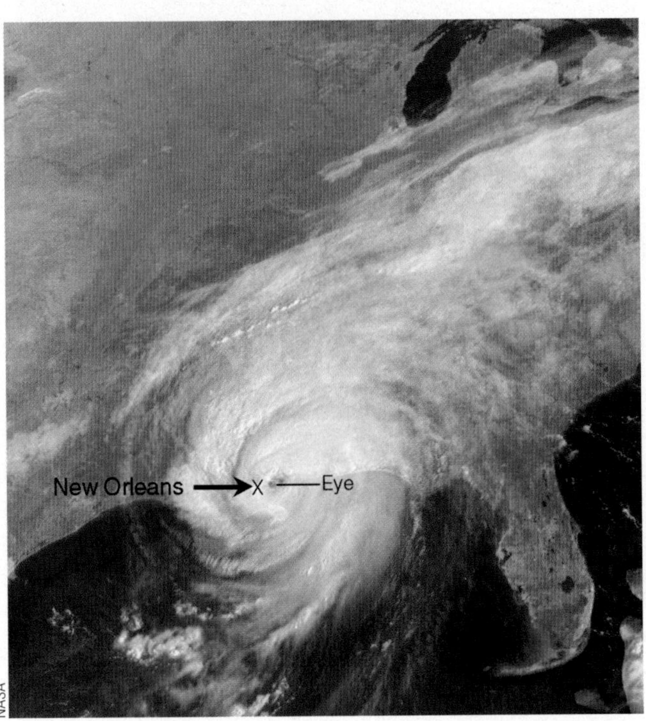

FIGURE 13.30 Hurricane Katrina just after making landfall along the Mississippi/Louisiana coast on the morning of August 29, 2005. Shown here, the storm is moving north with its eye due east of New Orleans. At landfall, Katrina had sustained winds of 127 mi/hr, a central pressure of 920 mb (27.17 in.), and a storm surge over 20 feet.

Other Devastating Hurricanes

Before the era of satellites and radar, there were many instances in the United States where hurricanes had caused catastrophic loss of life (see Table 13.4). In 1900, more than 8000 people (and perhaps as many as 12,000) lost their lives when a hurricane slammed into Galveston, Texas, with a huge storm surge over 15 feet high. At one point during the storm, the water rose in town 4 feet in less than one minute. As flood waters washed over the offshore island on which the city was built, the entire city became submerged, and every residence was destroyed (see Fig. 13.33).

FIGURE 13.31

High winds and huge waves crash against a boat washed onto Highway 90 in Gulfport, Mississippi, as Hurricane Katrina makes landfall on the morning of August 29, 2005.

AP Photo/John Bazemore

Vincent Laforet/The New York Times/Redux Pictures

FIGURE 13.32

Flood waters inundate New Orleans, Louisiana, during August, 2005, after the winds and storm surge from Hurricane Katrina caused several levee breaks.

In October, 1893, nearly 1800 people perished on the Gulf Coast of Louisiana as a giant storm surge swept that region. Spectacular losses were not confined to the Gulf Coast, as nearly 1000 people lost their lives in Charleston, South Carolina, during August of the same year.

Notice in Table 13.4 that hurricanes have taken many lives as far north as New England. For example, in 1938 a powerful September hurricane roared northward along the east coast of North America (see Fig. 13.34). The hurricane slammed into the south shore of Long Island as a strong Category 3 storm with a central pressure of 946 mb (27.94 in.) and a storm surge exceeding 15 feet. After blasting Long Island, the hurricane moved northward, making a second landfall in Connecticut. Winds on the eastern side of the storm peaked at 125 mi/hr in Providence, Rhode Island, and the Blue Hill Observatory near Boston, Massachusetts, reported a wind gust of 186 mi/hr. The storm then headed north northwest into Vermont. The hurricane damaged or destroyed more than 25,000 homes, took over 600 lives, and caused an estimated $400 million in damages (adjusted to over $5.5 billion in today's costs).

FOCUS ON

A SPECIAL TOPIC

The Record-Setting Atlantic Hurricane Seasons of 2004 and 2005

Both 2004 and 2005 were active years for hurricane development over the tropical North Atlantic. During 2004, nine storms became full-fledged hurricanes. Out of the five hurricanes that made landfall in the United States, three (Charley, Frances, and Jeanne) plowed through Florida, and one (Ivan) came onshore just west of the Florida panhandle (see Fig. 3), making this the first time since record-keeping began in 1861 that four hurricanes have impacted the state of Florida in one year. Total damage in the United States from the four hurricanes exceeded $40 billion.

Then, in 2005, a record twenty-seven named storms developed (the most in a single season), of which fifteen (another record) reached hurricane strength. The 2005 Atlantic hurricane season also had four hurricanes (Emily, Katrina, Rita, and Wilma) reach Category 5 intensity for the first time since reliable record-keeping began. And hurricane Wilma had the lowest central pressure ever measured in an Atlantic hurricane—882 mb (26.04 in.). Out of five hurricanes that made landfall in the United States, three (Dennis, Katrina, and Wilma) made landfall in hurricane-wary Florida and one (Ophelia) skirted northward along Florida's east coast, giving Florida the dubious distinction of being the only state on record to experience eight hurricanes during the span of sixteen months (Fig. 3). Total damage in the United States from the five hurricanes that made landfall exceeded $100 billion.

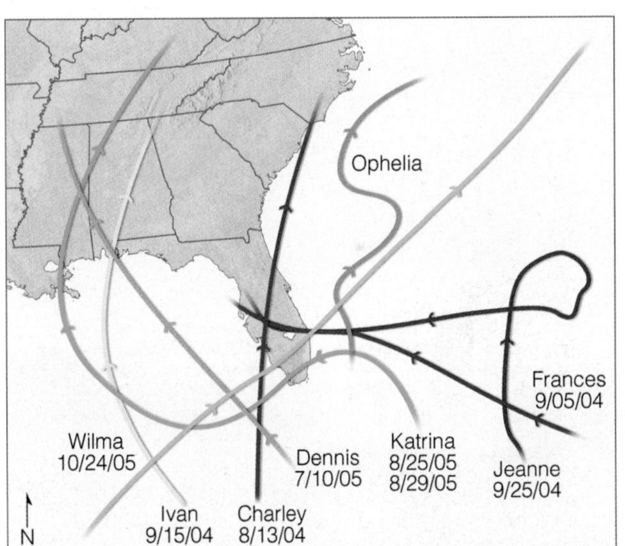

Figure 3 The paths of eight hurricanes that impacted Florida during 2004 and 2005. Notice that in 2004 hurricanes Frances and Jeanne made landfall at just about the same spot along Florida's southeast coast. The date under the hurricane's name indicates the date the hurricane made landfall.

Apparently, in 2004 and in 2005, very warm ocean water and weak vertical wind shear provided favorable conditions for hurricane development. In previous years, winds associated with a persistent upper-level trough over the eastern United States steered many tropical systems away from the coast before they could make landfall. However, in 2004 and in 2005, an area of high pressure replaced the trough, and winds tended to steer tropical cyclones on a more westerly track, toward the coastline of North America.

Table 13.4

The Fifteen Deadliest Hurricanes in the United States

RANK	HURRICANE (MADE LANDFALL)	MONTH/YEAR	DEATH TOLL
1	Galveston, Texas	September, 1900	>8000
2	Florida (Lake Okeechobee)	September, 1928	>2000
3	Louisiana/Mississippi	October, 1893	>1800
4	Katrina (Louisiana)	August, 2005	>1500
5	South Carolina/Georgia	August, 1893	>1000
6	Georgia/South Carolina	August, 1881	>700
7	New England	September, 1938	638
8	Florida Keys/South Texas	September, 1919	>600
9	Georgia/South Carolina	September, 1804	>500
10	Corpus Christy, Texas	September, 1919	>450
11	North Carolina (Capes)	September, 1857	424
12	Florida (Keys)	September, 1935	408
13	Louisiana	August, 1856	400
14	New England	September, 1944	390
15	Audrey (Louisiana)	June, 1957	390

FIGURE 13.33 Total destruction to the harbor in Galveston, Texas, inflicted by the Great Galveston Hurricane of September 8, 1900.

The statistics so far are small compared to the more than 300,000 lives taken as a killer tropical cyclone and storm surge ravaged the coast of Bangladesh with flood waters in 1970. In April, 1991, a similar cyclone devastated the area with reported winds of 145 mi/hr and a storm surge of 23 feet. In all, the storm destroyed 1.4 million houses and killed 140,000 people and 1 million cattle. Again in November, 2007, Tropical Cyclone Sidr, a Category 4 storm with winds of 150 mi/hr, moved into the region, killing thousands of people, damaging or destroying over one million houses, and flooding more than two million acres. Estimates are Cyclone Sidr adversely affected more than 8.5 million people. Unfortunately, the potential for a repeat of this type of disaster remains high in Bangladesh, as many people live along the relatively low, wide floodplain that slopes outward to the bay and, historically, this region is in a path frequently taken by tropical cyclones.

In May, 2008, Tropical Cyclone Nargis took aim on Bangladesh (see Fig. 13.35), but instead moved east striking Myanmar (Burma). Although the cyclone was accompanied by strong winds, it was the 16-foot storm surge and 8-foot waves that caused much of the damage. Nargis pushed flood waters inland for at least 30 miles. In this region, millions of people live in flood-prone homes less than 10 feet above sea level. The cyclone killed at least 140,000 people as flooding washed away entire villages, in some places without leaving a single structure.

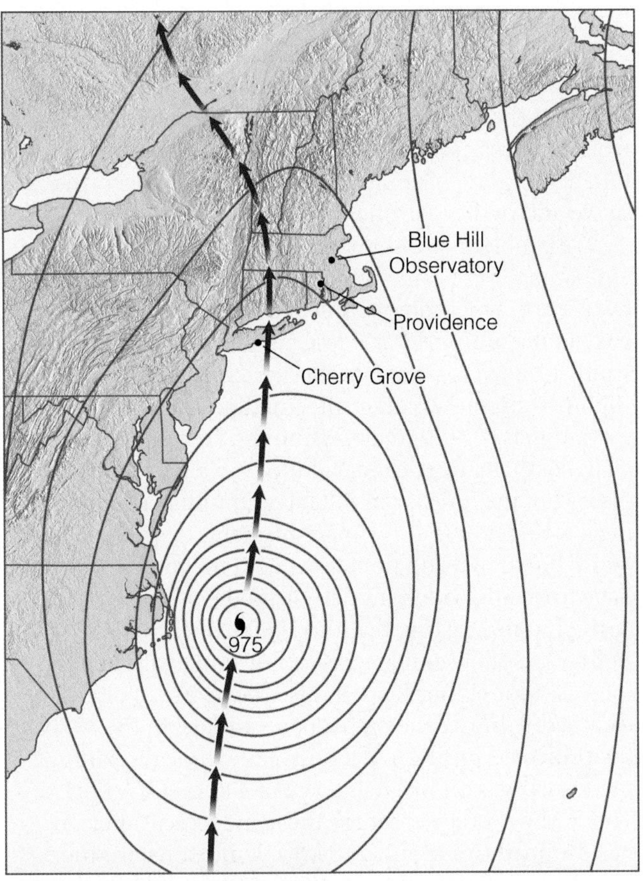

FIGURE 13.34 A major hurricane off the coast of North Carolina on the morning of September 21, 1938, is rapidly moving northward toward Long Island and New England. The path of this destructive storm is indicated by the heavy arrows.

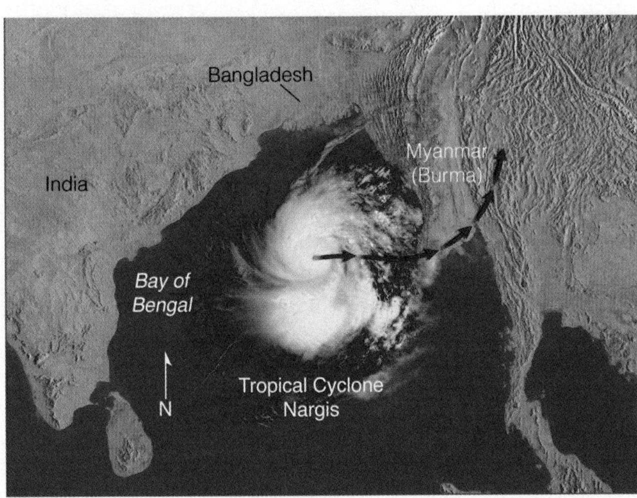

FIGURE 13.35 Visible satellite image of Tropical Cyclone Nargis on May 2, 2008, as it begins to move eastward over the Bay of Bengal toward Myanmar (Burma), where its storm surge and flood waters killed more than 140,000 people. (The red dashed lines show the path of Nargis.)

During late October, 1998, Hurricane Mitch became the most deadly hurricane to strike the Western Hemisphere since the Great Hurricane of 1780, which claimed approximately 22,000 lives in the eastern Caribbean. Mitch's high winds, huge waves (estimated maximum height 44 ft), and torrential rains destroyed vast regions of coastal Central America (for Mitch's path, see Fig.13.13, p. 372). In the mountainous regions of Honduras and Nicaragua, rainfall totals from the storm may have reached 75 inches. The heavy rains produced floods and deep mud slides that swept away entire villages, including the inhabitants. Mitch caused over $5 billion in damages, destroyed hundreds of thousands of homes, and killed over 11,000 people. More than 3 million people were left homeless or were otherwise severely affected by this deadly storm.

Hurricane Watches, Warnings, and Forecasts

With the aid of ship reports, satellites, radar, buoys, and reconnaissance aircraft, the location and intensity of hurricanes are pinpointed and their movements carefully monitored. When a hurricane poses a direct threat to an area, a **hurricane watch** is issued, typically 24 to 48 hours before the storm arrives, by the National Hurricane Center in Miami, Florida, or by the Pacific Hurricane Center in Honolulu, Hawaii. When it appears that the storm will strike an area, a **hurricane warning** is issued (see Fig.13.36). Along the east coast of North America, the warning is accompanied by a probability. The probability gives the percent chance of the hurricane's center passing within 65 mi (105 km) of a particular community. The warning is designed to give residents ample time to secure property and, if necessary, to evacuate the area.

Because hurricane-force winds can extend a considerable distance on either side of where the storm is expected to make landfall, a hurricane warning is issued for a rather large coastal area, usually about 340 mi (550 km) in length. Since the average swath of hurricane damage is normally about one-third this length, much of the area is "over-warned." As a consequence, many people in a warning area feel that they are needlessly forced to evacuate. The evacuation order is given by local authorities* and typically only for those low-lying coastal areas directly affected by the storm surge. People at higher elevations or farther from the coast are not usually requested to leave, in part because

*In the state of New Jersey, the Board of Casinos and the Governor must be consulted before an evacuation can be ordered.

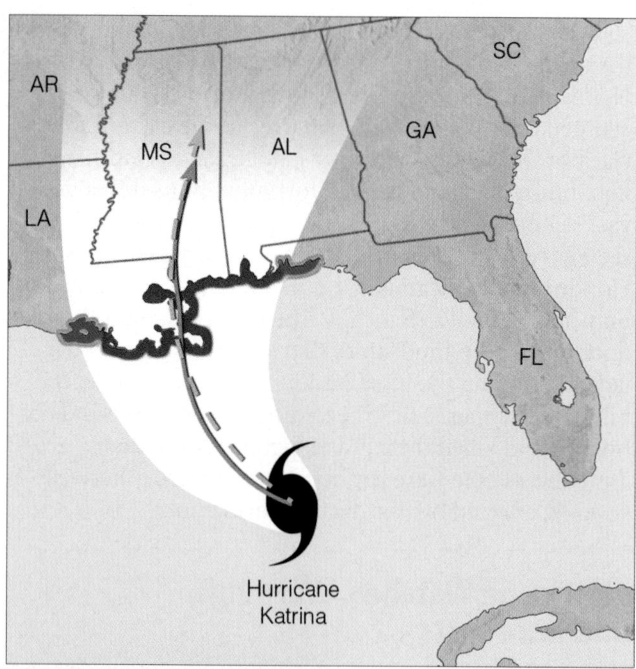

FIGURE 13.36 Hurricane Katrina over the Gulf of Mexico with sustained winds of 145 mi/hr on August 28, 2005, at 1 A.M. CDT. The current movement of the storm is west-northwest at 8 mi/hr. The dashed orange line shows the hurricane's projected path; the solid purple line, the hurricane's actual path. Areas under a hurricane warning are in red. Those areas under a hurricane watch are in pink, while those areas under a tropical storm warning are in blue.

of the added traffic problems this would create. This issue has engendered some controversy in the wake of Hurricane Andrew, since its winds were so devastating over inland south Florida during August, 1992. The time it takes to complete an evacuation puts a special emphasis on the timing and accuracy of the warning.

As Hurricane Katrina approaches land (see Fig. 13.36), will it intensify, maintain its strength, or weaken? Also, will it continue to move in the same direction and at the same speed? Such questions have challenged forecasters for some time. To forecast the intensity and movement of a hurricane, meteorologists use numerical weather prediction models, which are computer models that represent the hurricane and its environment in a greatly simplified manner.

Information from satellites, buoys, and reconnaissance aircraft (that deploy dropsondes* into the eye of the storm) are fed into the models. The models then forecast the intensity and movement of the storm. There are a variety of forecast models, each one treating some aspect of the atmosphere (such as evaporation of water from the ocean's surface) in a slightly different manner. Often, the models do not agree on where the storm will move and on how strong it will be.

The problem of different models forecasting different paths for the same hurricane has been addressed by using the method of *ensemble forecasting,* which is based on running several forecast models (or different simulations of the same model), each beginning with slightly different weather information. If the forecast models (or different versions of the same model) all agree that a hurricane will move in a particular direction, the forecaster will have confidence in making a forecast of the storm's movement. If, on the other hand, the models do not agree, then the forecaster will have to decide which model (or models) is most likely correct in forecasting the hurricane's track.

The use of ensemble forecasting along with better forecast models has helped raise the level of skill in forecasting hurricane paths. For example, in the 1970s, the projected position of a hurricane three days into the future was off by an average of 440 mi (708 km). Today, the average error for the same forecast period has dropped to 173 mi (278 km). Unfortunately, the forecasting of hurricane intensity has shown little improvement since the early 1990s.

To help predict hurricane intensity, forecasters have been using statistical models that compare the behavior of the present storm with that of similar tropical storms in the past. The results using these models have not been encouraging. Another more recent model uses the depth of warm ocean water in front of the storm's path to predict the storm's intensity. Recall from an earlier discussion (p. 368) that if the reservoir of warm water ahead of the storm is relatively shallow, ocean waves generated by the hurricane's wind will turbulently bring deeper, cooler water to the surface. The cooler water will cut off the storm's energy source, and the hurricane will weaken. On the other hand, should a deep layer of warm water exist ahead of the hurricane, cooler water will not be brought to the surface, and the storm will either maintain its strength or intensify, as long as other factors remain the same. So, knowing the depth of warm surface water ahead of the storm is important in predicting whether a hurricane will intensify or weaken.*

*Recall from p. 369 that dropsondes are instruments (radiosondes) that are dropped from reconnaissance aircraft into a storm. As the instrument descends, it measures and relays data on temperature, pressure, and humidity back to the aircraft. Also obtained are data concerning wind speed and wind direction.

*Sophisticated satellite instruments carefully measure ocean height, which is translated into ocean temperature beneath the sea surface. This information is then fed into the forecast models.

As new hurricane-prediction models with greater resolution are implemented, and as our understanding of the nature of hurricanes increases, forecasting hurricane intensification and movement should improve.

Modifying Hurricanes

Because of the potential destruction and loss of lives that hurricanes can inflict, attempts have been made to reduce their winds by seeding them with silver iodide. The idea is to seed the clouds just outside the eyewall with just enough artificial ice nuclei so that the latent heat given off will stimulate cloud growth in this area of the storm. These clouds, which grow at the expense of the eyewall thunderstorms, actually form a new eyewall farther away from the hurricane's center. (The process of eyewall replacement is described on p. 368.) As the storm center widens, its pressure gradient should weaken, which may cause its spiraling winds to decrease in speed.

During project STORMFURY, a joint effort of the National Oceanic and Atmospheric Administration (NOAA) and the U.S. Navy, several hurricanes were seeded by aircraft. In 1963, shortly after Hurricane Beulah was seeded with silver iodide, surface pressure in the eye began to rise and the region of maximum winds moved away from the storm's center. Even more encouraging results were obtained from the multiple seeding of Hurricane Debbie in 1969. After one day of seeding, Debbie showed a 30 percent reduction in maximum winds. But many hurricanes that are not seeded show this type of behavior. So, the question remains: Would the winds have lowered naturally had the storm not been seeded? Several studies even cast doubt upon the theoretical basis for this kind of hurricane modification because hurricanes appear to contain too little supercooled water and too much natural ice. Consequently, there are many uncertainties about the effectiveness of seeding hurricanes in an attempt to reduce their winds, and all endeavors to modify hurricanes have been discontinued since the 1970s.

Other ideas have been proposed to weaken the winds of a hurricane. One idea is to place some form of oil (monomolecular film) on the water to retard the rate of evaporation and hence cut down on the release of latent heat inside the clouds. Some sailors, even in ancient times, would dump oil into the sea during stormy weather, claiming it reduced the winds around the ship. At this point, it's interesting to note that a recent mathematical study suggests that ocean spray has an effect on the winds of a hurricane. Apparently, the tiny spray reduces the friction between the wind and the sea surface. Consequently, with the same pressure gradient, the more ocean spray, the higher the winds. If this idea proves correct, limiting ocean spray from entering the air above may reduce the storm's winds. Perhaps the ancient sailors knew what they were doing after all.

Hurricanes in a Warmer World

In the Focus section on p. 389, we saw that 2005 was a record year for Atlantic hurricanes, with 27 named storms, 15 hurricanes, and 5 storms reaching Category 5 status on the Saffir-Simpson scale. Was the record hurricane year 2005 related to global warming?

We know that hurricanes are fueled by warm tropical water — the warmer the water, the more fuel available to drive the storm. A mere 1°F (0.6°C) increase in sea-surface temperature will increase the maximum winds of a hurricane by about 5 mi/hr, everything else being equal. During May, 2005, just before the hurricane season got underway, the surface water temperature over the tropical North Atlantic was considerably warmer than normal (see ❯ Fig. 13.37). Moreover, studies conducted by the National Center for Atmospheric Research (NCAR) in Boulder, Colorado, found that between June and October, 2005, sea-surface temperatures in the tropical Atlantic were about 1.6°F (0.9°C) warmer than the long-time (1901–1970) average for that region. The study concluded that about half of the warming (about 0.7°F) was due to global warming caused by increasing concentrations of greenhouse gases in the atmosphere. These findings suggest that global warming (described more completely in Chapter 15) may have had an effect on the intensity of some storms and on the number of tropical storms that formed, since the ocean surface remained warmer than normal well past October. Interestingly, during the hurricane season of 2007, only one weak hurricane (Humberto) made landfall in the United States, but two storms (Hurricanes Dean and Felix) reached Category 5 status over the warm Gulf of Mexico before weakening and making landfall south of the United States.

Climate models predict that, as the world warms, sea-surface temperatures in the tropics will rise by about 3.6°F (2°C) by the end of this century. Should these projections prove correct, a hurricane forming in today's atmosphere with maximum sustained winds of 145 mi/hr (a Category 4 storm) could, in the warmer world, have maximum sustained winds of 160 mi/hr (a Category 5 storm). As sea-surface temperatures rise, will hurricanes become more frequent? Presently, there is no clear answer to this question as some climate models predict more hurricanes, whereas others predict fewer storms.

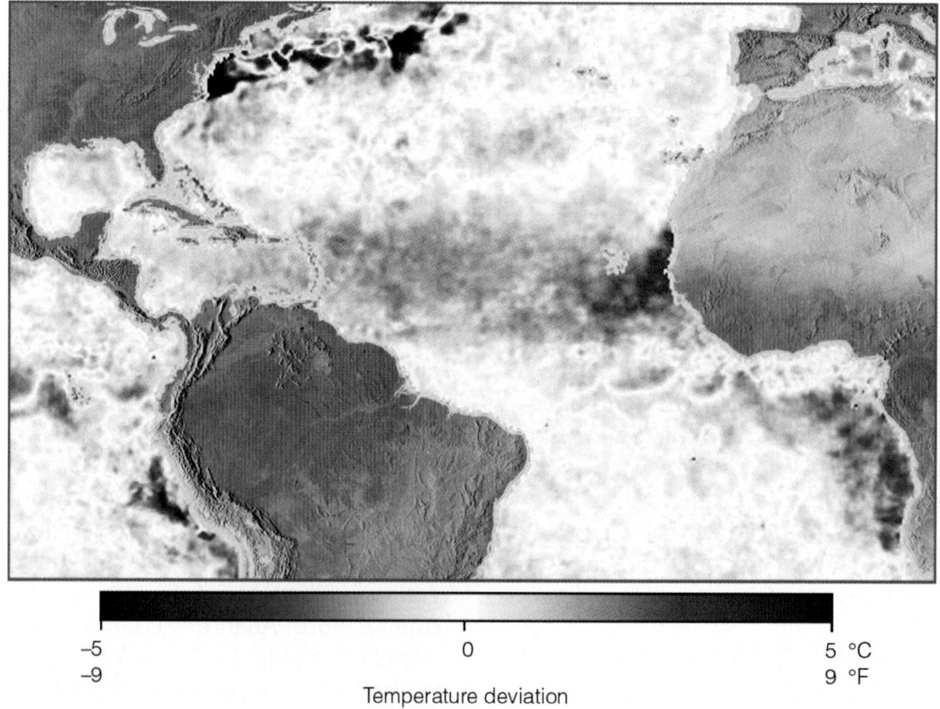

FIGURE 13.37 Sea-surface temperature departures from the twelve-year average (1985–1997) on May 30, 2005. Notice that the darker the red, the warmer the surface water. (Data from NOAA)

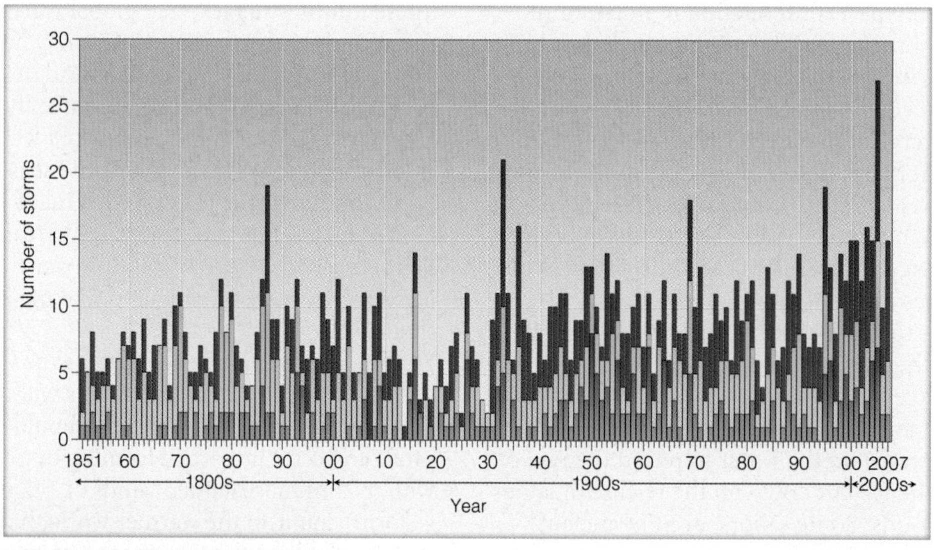

FIGURE 13.38 The total number of tropical storms and hurricanes (red bars), hurricanes only (yellow bars), and Category 3 hurricanes or greater (green bars) in the Atlantic Basin for the period 1851 through 2007. (NOAA)

Since the earth's surface is gradually warming, are today's hurricanes more intense than those of the past? Several studies suggest that the frequency of major hurricanes (Category 3 and above) has been increasing (see Fig. 13.38). The problem with these studies is that reliable records of tropical cyclones have only been available since the 1970s, when observations from satellites became more extensive. Sophisticated instruments today allow scientists to peer into hurricanes and examine their structure and winds with much greater clarity than in the past. Any trend in hurricane frequency or intensity will likely become clearer when more reliable information on past tropical cyclone activity becomes available, especially from the ongoing investigation of sea sediment cores, which hold clues to past tropical cyclone occurrences.

Summary

In this chapter, we examined hurricanes and the extreme weather they produce. Hurricanes are tropical cyclones with winds that exceed 74 mi/hr (64 knots) and blow counterclockwise about their centers in the Northern Hemisphere. A hurricane consists of a mass of organized thunderstorms that spiral in toward the extreme low pressure of the storm's eye. The most intense thunderstorms, the heaviest rain, and the highest winds occur outside the eye, in the region known as the eyewall. In the eye itself, the air is warm, winds are light, and skies may be broken or overcast.

Hurricanes (and all tropical cyclones) are born over warm tropical waters where the air is humid, surface winds converge, and thunderstorms become organized in a region of weak upper-level winds. Surface convergence may occur along the ITCZ, on the eastern side of a tropical wave, or along a front that has moved into the tropics from higher latitudes. If the disturbance becomes more organized, it becomes a tropical depression. If central pressures drop and surface winds increase, the depression becomes a tropical storm. At this point, the storm is given a name. Some tropical storms continue to intensify into full-fledged hurricanes, as long as they remain over warm water and are not disrupted by strong vertical wind shear.

The energy source that drives the hurricane comes primarily from the warm tropical oceans and from the release of latent heat. A hurricane is like a heat engine in that energy for the storm's growth is taken in at the surface in the form of sensible and latent heat, converted to kinetic energy in the form of winds, then lost at the cloud tops through radiational cooling.

The easterly winds in the tropics usually steer hurricanes westward. In the Northern Hemisphere, most storms then gradually swing northwestward around the subtropical high. If the storm moves into middle latitudes, the prevailing westerlies steer it northeastward. Because hurricanes derive their energy from the warm surface water and from the latent heat of condensation, they tend to dissipate rapidly when they move over cold water or over a large mass of land, where surface friction causes their winds to decrease and flow into their centers.

Although the high winds of a hurricane can inflict a great deal of damage, it is usually the huge waves and the flooding associated with the storm surge that cause the most destruction and loss of life. The Saffir-Simpson hurricane scale was developed to estimate the potential destruction that a hurricane can cause. The scale ranges from 1 to 5 with major hurricanes classified as Category 3 and above. Of all the hurricanes to strike the United States in the past 100 years or so, about 40 percent were major hurricanes with winds in excess of 111 mi/hr. Although hurricanes typically cause the most destruction with their high winds, heavy rain, and huge storm surges (which may exceed 20 feet), tropical storms that never reach hurricane strength can cause massive flooding over a vast area.

Hurricane watches are issued when a hurricane poses a direct threat to an area. Hurricane warnings are issued when it appears that a hurricane will strike an area. In predicting hurricane intensity, it is important to know the depth of warm water directly in the path of the storm, and the air temperature in the storm's eye and eyewall. In recent years, ensemble forecasting, along with better forecasting models, has helped raise the level of skill in forecasting hurricane paths. Although attempts have been made in the past to modify hurricanes through cloud seeding, uncertainties in the effectiveness of seeding hurricanes have caused all modification efforts to be discontinued.

Key Terms

The following terms are listed (with page number) in the order they appear in the text. Define each. Doing so will aid you in reviewing the material covered in this chapter.

streamlines, 362
tropical (easterly) wave, 362
hurricane, 362
typhoon, 363
tropical cyclone, 363
eye, 363
eyewall, 363
trade wind inversion, 367
eyewall replacement, 368
tropical depression, 369
tropical storm, 369
storm surge, 377
Saffir-Simpson scale, 378
super typhoon, 378
hurricane watch, 391
hurricane warning, 391

Review Questions

1. What is a tropical (easterly) wave? How do these waves generally move in the Northern Hemisphere? Are showers found on the eastern or western side of the wave?
2. Why are streamlines, rather than isobars, used on surface weather maps in the tropics?
3. What is the name given to a hurricane-like storm that forms over the western North Pacific Ocean?
4. Describe the horizontal and vertical structure of a hurricane.
5. Why are skies often clear or partly cloudy in a hurricane's eye?
6. What conditions at the surface and aloft are necessary for hurricane development?
7. List three "triggers" that help in the initial stage of hurricane development.
8. (a) Hurricanes are sometimes described as a heat engine. What is the "fuel" that drives the hurricane?

   (b) What determines the maximum strength (the highest winds) that the storm can achieve?
9. Would it be possible for a hurricane to form over land? Explain.
10. If a hurricane with sustained winds of 100 mi/hr is moving westward at 20 mi/hr, will the strongest winds and greatest damage potential be on its northern or southern side? Explain. If the same hurricane turns northward, will the strongest winds and greatest damage potential be on its eastern or western side?
11. What factors tend to weaken hurricanes?
12. Distinguish among a tropical depression, a tropical storm, and a hurricane.
13. In what ways is a hurricane different from a mid-latitude cyclone? In what ways are these two systems similar?
14. Why do most hurricanes move westward over tropical waters?
15. (a) In the Atlantic Basin, when does hurricane season peak?

    (b) Would you expect more Atlantic hurricanes to form in October or July? Explain.
16. Describe how the area where hurricanes tend to form and the paths they tend to take along the east coast of the United States changes from August to October.
17. If the high winds of a hurricane are not responsible for inflicting the most damage, what is?
18. Most hurricane-related deaths are due to what?
19. Explain how a storm surge forms. How does it inflict damage in hurricane-prone areas?
20. Hurricanes are given names when the storm is in what stage of development?
21. When Hurricane Andrew moved over south Florida during August, 1992, what was it that apparently caused the relatively small areas of extreme damage?

22. As Hurricane Katrina moved toward the Louisiana coast, it underwent eyewall replacement. What actually happened to the eyewall during this process?
23. How do meteorologists forecast the intensity and paths of hurricanes?
24. How does a hurricane watch differ from a hurricane warning?
25. Why have hurricanes been seeded with silver iodide?
26. Give two reasons why hurricanes are more likely to strike New Jersey than Oregon.

Online Learning

STUDENT COMPANION WEBSITE: Visit this book's companion website at: www.cengage.com/ahrens/extreme1e and choose Chapter 13 for many study aids and ideas for further reading and research. These include flashcards, practice quizzing, and web links.

METEOROLOGY RESOURCE CENTER: For students with access, log on at www.cengage.com/login for more assets, including animations, videos, and more. If your textbook did not come with access, visit www.CengageBrain.com to purchase.

Weather Forecasting

UCAR/Carlye Calvin

14

CONTENTS

ACQUISITION OF WEATHER INFORMATION

WEATHER FORECASTING TOOLS

WEATHER FORECASTING METHODS

PREDICTING SHORT-TERM HAZARDOUS AND SEVERE WEATHER EVENTS

PREDICTING LONG-TERM WEATHER AND CLIMATE PATTERNS

SUMMARY

KEY TERMS

REVIEW QUESTIONS

Clouds dissipating at sunset over Saw Hill Ponds near Boulder, Colorado, suggest a forecast of clearing weather by morning.

Weather forecasts are issued to save lives, to save property and crops, and to report what to expect in our atmospheric environment. In the case of extreme weather events, forecasts are designed to warn people of impending weather hazards, and, in some cases, give them ample time to evacuate the risk area. As we saw in Chapter 12, even though the population of the United States has increased, the annual average number of deaths caused by tornadoes over the past 50 years has actually decreased due mainly to improvements in severe storm forecasting. However, far too many people are complacent, and tend to ignore watches and warnings of impending severe weather.

One reason for this complacency is that presently, when a tornado watch is issued, it is impossible to forecast exactly where a tornado will touch down. Consequently, tornado watches cover a rather large geographic region to alert as many people as possible to the potential of severe weather. Some, however, do not understand the difference between a tornado watch and a tornado warning, so they feel that if a tornado did not develop within the watch area, the forecast must be in error. Because of this situation, some people lose confidence in severe weather forecasts and develop a false sense of security when a tornado warning is issued.

The forecasting of where and when a hurricane will make landfall when the storm is many miles off shore can be especially challenging. An inaccurate prediction could lead to expensive and unnecessary evacuations. On the other hand, not providing sufficient time for evacuation leaves the population at risk. In this case, making an inaccurate forecast can impact the lives of millions of people.

How accurate are forecasts provided by the National Weather Service? Will forecasters ever be able to predict the weather accurately more than 10 days into the future? How are forecasts made and why do they sometimes go awry? What steps are being taken to improve the forecasting art? These are just a few of the questions we will address in this chapter.

Acquisition of Weather Information

Weather forecasting basically entails predicting how the present state of the atmosphere will change. Consequently, if we wish to make a weather forecast, present weather conditions over a large area must be known. To obtain this information, a network of observing stations is located throughout the world. Over 10,000 land-based stations and hundreds of ships and buoys provide surface weather information four times a day. Most airports observe conditions hourly. Additional information, especially upper-air data, is supplied by radiosondes, aircraft, and satellites.

A United Nations agency—the World Meteorological Organization (WMO)—consists of over 175 nations. The WMO is responsible for the international exchange of weather data and certifies that the observation procedures do not vary among nations, an extremely important task, since the observations must be comparable.

Weather information from all over the world is transmitted electronically to a branch of the National Weather Service (NWS), the National Center for Environmental Prediction (NCEP), which is located in Camp Springs, Maryland, just outside Washington, D.C. Here, the massive job of analyzing the data, preparing weather maps and charts, and predicting the weather on a global and national scale begins. From NCEP, weather information is transmitted to private and public agencies, such as weather forecast offices that use the information to issue local and regional weather forecasts.

The public hears weather forecasts over radio or television. Many stations hire private meteorological companies or professional meteorologists to make their own forecasts aided by NCEP material or to modify a weather service forecast. Other stations hire meteorologically untrained announcers who paraphrase or read the forecasts of the National Weather Service word for word.

Today, the forecaster has access to many hundreds of maps and charts, as well as vertical profiles (called *soundings*) of temperature, dew point, and winds. Also available are visible and infrared satellite images, as well as Doppler radar information that can detect and monitor the severity of precipitation and thunderstorms.

When severe or hazardous weather is likely, the National Weather Service issues advisories in the form of weather watches and warnings. A **watch** indicates that atmospheric conditions favor hazardous weather occurring over a particular region during a specified time period, but the actual location and time of the occurrence is uncertain. A **warning**, on the other hand, indicates that hazardous weather is either imminent or actually occurring within the specified forecast area. *Advisories* are issued to inform the public of less hazardous conditions caused by wind, dust, fog, snow, sleet, or freezing rain. (Additional information on watches, warnings, and advisories is given in the Focus section on p. 401.)

FOCUS ON

EXTREME WEATHER

Watches, Warnings, and Advisories

As we have seen, where severe or hazardous weather is either occurring or possible, the National Weather Service issues a forecast in the form of a watch or warning. The public, however, is not always certain as to what this forecast actually means. For example, a *high wind warning* indicates that there will be high winds—but how high and for how long? The following describes a few of the various watches, warnings, and advisories issued by the National Weather Service and the necessary precautions that should be taken during the event.

Wind advisory Issued when sustained winds reach 25 to 39 mi/hr or wind gusts are up to 57 mi/hr.

High wind warning Issued when sustained winds are at least 40 mi/hr, or wind gusts exceed 57 mi/hr. Caution should be taken when driving high-profile vehicles, such as trucks, trailers, and motor homes.

Wind-chill advisory Issued for wind-chill temperatures of –30° to –35°F or below.*

Heat advisory/warning Advisory issued when the daytime Heat Index is expected to reach 105°F for 3 hours or more and nighttime lows do not drop below 80°F. Warning issued when Heat Index reaches 115°F or above.

Flash-flood watch Heavy rains may result in flash flooding in the specified area. Be alert and prepared for the possibility of a flood emergency that will require immediate action.

Flash-flood warning Flash flooding is occurring or is imminent in the specified area. Move to safe ground immediately.

Figure 1 **Flags indicating advisories and warnings in maritime areas.**

Urban and small stream advisory Issued when flooding is occurring in small streams, streets, or in low-lying areas, such as railroad underpasses and urban storm drains.

Severe thunderstorm watch Thunderstorms (with winds exceeding 57 mi/hr and/or hail three-fourths of an inch or more in diameter) are possible.

Severe thunderstorm warning Severe thunderstorms have been visually sighted or indicated by Doppler radar. Be prepared for lightning, heavy rains, strong winds, and large hail. (Tornadoes can form with severe thunderstorms.)

Tornado watch Issued to alert people that tornadoes may develop within a specified area during a certain time period.

Tornado warning Issued to alert people that a tornado has been spotted either visually or by Doppler radar. Take shelter immediately.

Snow advisory In nonmountainous areas, expect a snowfall of 2 in. or more in 12 hours, or 3 in. or more in 24 hours.*

Winter storm warning (formerly heavy snow warning) In nonmountainous areas, expect a snowfall of 4 in. or more in 12 hours or 6 in. or more in 24 hours. (Where heavy snow is infrequent, a snowfall of several inches may justify a warning.)*

Blizzard warning Issued when falling or blowing snow and winds of at least 35 mi/hr frequently restrict visibility to less than 1/4 mile for several hours.

Dense fog advisory Issued when fog limits visibility to less than 1/4 mile, or in some parts of the country to less than 1/8 mile.

WARNINGS OVER THE WATER

Small craft advisories Issued to alert mariners that weather or sea conditions might be hazardous to small boats. Expect winds of 18 to 34 knots (21 to 39 mi/hr). Figure 1 displays the posted advisory and warning flags.

Gale warning Winds will range between 34 and 47 knots (39 to 54 mi/hr) in the forecast area.

Storm warning Winds in excess of 47 knots (54 mi/hr) are to be expected in the forecast area.

Hurricane watch Issued when a tropical storm or hurricane becomes a threat to a coastal area. Be prepared to take precautionary action in case hurricane warnings are issued.

Hurricane warning Issued when it appears that the storm will strike an area within 24 hours. Expect wind speeds in excess of 64 knots (74 mi/hr).

* It should be noted that watches, warnings, or advisories for wind chill or for snowfall-related events (e.g., winter storms, etc.) may use different criteria in different regions. For example, mountainous areas that experience frequent heavy snow may have higher snowfall criteria, whereas areas with infrequent snow may have lower snowfall criteria. Similarly, in areas that experience frequent extreme cold, wind chills may have lower (colder) criteria for advisories.

Weather Forecasting Tools

To help forecasters handle all the available charts and maps, high-speed data modeling systems using computers are employed by the National Weather Service. The communication system in use today is known as **AWIPS** (*A*dvanced *W*eather *I*nteractive *P*rocessing *S*ystem). The AWIPS system is shown in ⟫ Fig. 14.1.

The AWIPS system has data communications, storage, processing, and display capabilities (including graphical overlays) to better help the individual forecaster extract and assimilate information from the mass of available data. In addition, AWIPS is able to process information received from the Doppler radar system (the WSR-88D), satellite imagery, and the *A*utomated *S*urface *O*bserving *S*ystems (ASOS) that are operational at selected airports and other sites throughout the United States. The ASOS system is designed to provide nearly continuous information about wind, temperature, pressure, cloud-base height, and runway visibility at various airports. Meteorologists are hopeful that information from all of these sources will improve the accuracy of weather forecasts by providing previously unobtainable data for integration into numerical models. Moreover, much of the information from ASOS and Doppler radar is processed by software according to predetermined formulas, or *algorithms,* before it goes to the forecaster. Certain criteria or combinations of measurements can alert the forecaster to an impending weather situation, such as the severe weather illustrated in ⟫ Fig. 14.2.

A software component of AWIPS (called the Interactive Forecast Preparation System) allows forecasters to look at the daily prediction of weather elements, such as temperature and dew point, in a grid format with spacing as small as 1.2 mi (2 km). Presenting the data in this format allows the forecaster to predict the weather more precisely over a relatively small area.

© Jan Null

⟫ FIGURE 14.1 The AWIPS computer work station provides various weather maps and overlays on different screens.

Composite Algorithm Output

STM ID	AZ	RAN	MESO	HAIL	SVRH	SIZE	MAXZ
4	303	111	NONE	100%	100%	>300	74
14	294	128	NONE	100%	100%	>300	75
6	299	115	NONE	100%	100%	>300	74
2	179	161	NONE	70%	50%	<100	62
5	336	166	NONE	70%	0%	<100	50

ORLANDO
KISSIMMEE
<0 3 9 14 20 25 31 36 42 47 53 58 64 >65 RF
Ctr az, ran: 298,107 k Mag: 6X Sel az, ran: 0, 0 k Hgt: 0.0 k Val: 0029.0
Vol scan: 22 Sweep: 0 Mode PPI Angle: 0.5 Nyquist: 31
Radar: KMLB Site: KMLB MMDDYY: 03/26/92 HHMMSS: 00:01:24

J.T. Johnson, National Severe Storms Lab

⟫ FIGURE 14.2 Doppler radar data from Melbourne, Florida, on March 25, 1992, during the time of a severe hailstorm that caused $60 million in damages in the Orlando area. In the table near the top of the display, the hail algorithm determined that there was 100 percent probability that the storm was producing hail and severe hail. The algorithm also estimated the maximum size of the hailstones to be greater than 3 inches. A forecaster can project the movement of the storm and adequately warn those areas in the immediate path of severe weather.

With so much information at the forecaster's disposal, it is essential that the data be easily accessible and in a format that allows several weather variables to be viewed at one time. The **meteogram** is a chart that shows how one or more weather variables has changed at a station over a given period of time. As an example, the chart may represent how air temperature, dew point, and sea-level pressure have changed over the past five days, or it may illustrate how these same variables are projected to change over the next five days (see ⟫ Fig. 14.3).

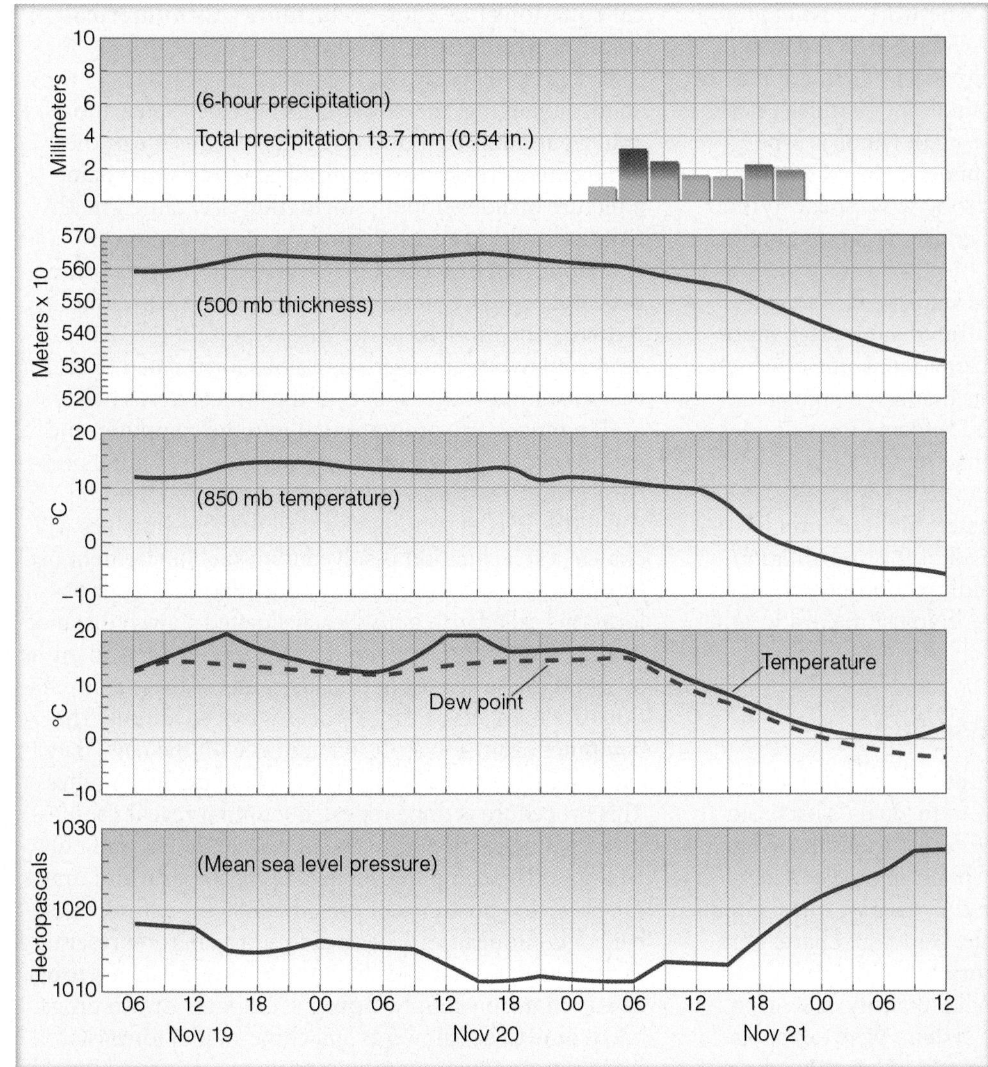

FIGURE 14.3

Meteogram illustrating predicted weather at the surface and aloft at St. Louis, Missouri, from 6 a.m., November 19, 2007, to noon on November 21, 2007. The forecast is derived from the Global Forecast System (GFS) model. (NOAA)

FIGURE 14.4 (below)

A sounding of air temperature, dew point, and winds at Pittsburgh, PA, on January 14, 1999. Looking at this sounding, a forecaster would see that saturated air extends up to a pressure near 820 millibars (about 4000 feet) above the surface. The forecaster would also observe that below-freezing temperatures only exist in a shallow layer near the surface and that the freezing rain presently falling over the Pittsburgh area would continue or possibly change to rain, as cold easterly surface winds are swinging around to warmer southwesterly winds aloft.

Another aid in weather forecasting is the use of **soundings**—a two-dimensional vertical profile of temperature, dew point, and winds (see Fig. 14.4).* The analysis of a sounding can be especially helpful when making a short-range forecast that covers a relatively small area, such as the mesoscale. The forecaster examines the sounding of the immediate area (or closest proximity), as well as the soundings of those sites upwind, to see how the atmosphere might be changing. Computer programs then automatically calculate from the sounding a number of meteorological *indexes* that can aid the forecaster in determining the likelihood of smaller-scale weather phenomena, such as thunderstorms, tornadoes, and hail. Soundings also provide information that can aid in the prediction of fog, air pollution alerts, and the downwind mixing of strong winds.

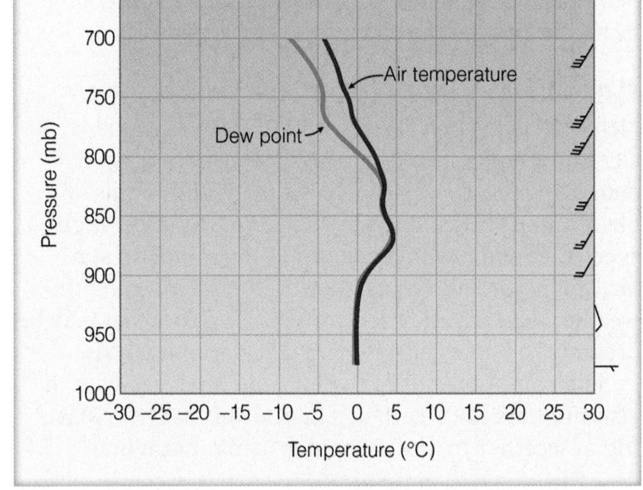

*A sounding is obtained from a radiosonde. For additional information on the radiosonde see the Focus section in Chapter 3 on p. 75.

In the central United States, a network of *wind profilers* (see Chapter 7, p. 192) is providing forecasters with hourly wind speed and wind direction information at 72 different levels in a column of air about 10 miles thick. The almost continuous monitoring of winds is especially beneficial when briefing pilots on areas of strong headwinds and on regions of strong wind shear. Wind information from the profilers is also integrated into computer forecasting models.

Satellite information is also a valuable tool for the forecaster. Visible, enhanced infrared, and water vapor images provide a wealth of information, some of which comes from inaccessible regions that can be plugged into forecast models. This added information provides a clearer representation of the atmosphere.*

Up to this point, we have examined some of the weather data and tools a forecaster might use in making a weather prediction. With all of this information available to the forecaster, including hundreds of charts and maps, just *how* does a meteorologist make a weather forecast?

Weather Forecasting Methods

As late as the mid-1950s, all weather maps and charts were plotted by hand and analyzed by individuals. Meteorologists predicted the weather using certain rules that related to the particular weather system in question. For short-range forecasts of six hours or less, surface weather systems were moved along at a steady rate. Upper-air charts were used to predict where surface storms would develop and where pressure systems aloft would intensify or weaken. The predicted positions of these systems were extrapolated into the future using linear graphical techniques and current maps. Experience played a major role in making the forecast. In many cases, these forecasts turned out to be amazingly accurate. They were good but, with the advent of modern computers, along with our present observing techniques, today's forecasts are even better.

THE COMPUTER AND WEATHER FORECASTING: NUMERICAL WEATHER PREDICTION Modern electronic computers can analyze large quantities of data extremely fast. Each day the many thousands of observations transmitted to NCEP are fed into a high-speed computer, which plots and draws lines on surface and upper-air charts. Meteorologists interpret the weather patterns and then correct any errors that may be present. The final chart is referred to as an **analysis**.

The computer not only plots and analyzes data, it also predicts the weather. The routine daily forecasting of weather by the computer using mathematical equations has come to be known as **numerical weather prediction**.

Because the many weather variables are constantly changing, meteorologists have devised **atmospheric models** that describe the present state of the atmosphere. These are not physical models that paint a picture of a developing storm; they are, rather, mathematical models consisting of many mathematical equations that describe how atmospheric temperature, pressure, winds, and moisture will change with time. Actually, the models do not fully represent the real atmosphere but are approximations formulated to retain the most important aspects of the atmosphere's behavior.

The models are programmed into the computer, and surface and upper-air observations of temperature, pressure, moisture, winds, and air density are fed into the equations. To determine how each of these variables will change, each equation is solved for a small increment of future time—say, five minutes—for a large number of locations called *grid points,* each situated a given distance apart.* In addition, each equation is solved for as many as 50 levels in the atmosphere. The results of these computations are then fed back into the original equations. The computer again solves the equations with the new "data," thus predicting weather over the following five minutes. This procedure is done repeatedly until it reaches some desired time in the future, usually 6, 12, 24, 36, and out to 84 hours. The computer then analyzes the data and draws the projected positions of pressure systems with their isobars or contour lines. The final forecast chart representing the atmosphere at a specified future time is called a **prognostic chart**, or, simply, a **prog**. Computer-drawn progs have come to be known as "machine-made" forecasts.

The computer solves the equations more quickly and efficiently than could be done by hand. For example, just to produce a 24-hour forecast chart for the Northern Hemisphere requires many hundreds of millions of mathematical calculations. It would, therefore, take a group of meteorologists working full time with hand calculators years to produce a single chart; by the time the forecast was available, the weather for that day would already be ancient history.

The forecaster uses the progs as a guide to predicting the weather. At present, there are a variety of models (and, hence, progs) from which to choose, each producing a slightly different interpretation of the weather for the same projected time and atmospheric level (see Fig. 14.5). The differences between progs may result from the way the models use the equations, or the distance between grid points, called *resolution.*

* Information provided by satellites is located in various sections of this book. For example, see Chapter 4, p. 114.

* Some models have a grid spacing as small as 0.2 mi, whereas the spacing in others exceeds 60 mi. There are models that actually describe the atmosphere using a set of mathematical equations with wavelike characteristics rather than a set of discrete numbers associated with grid points.

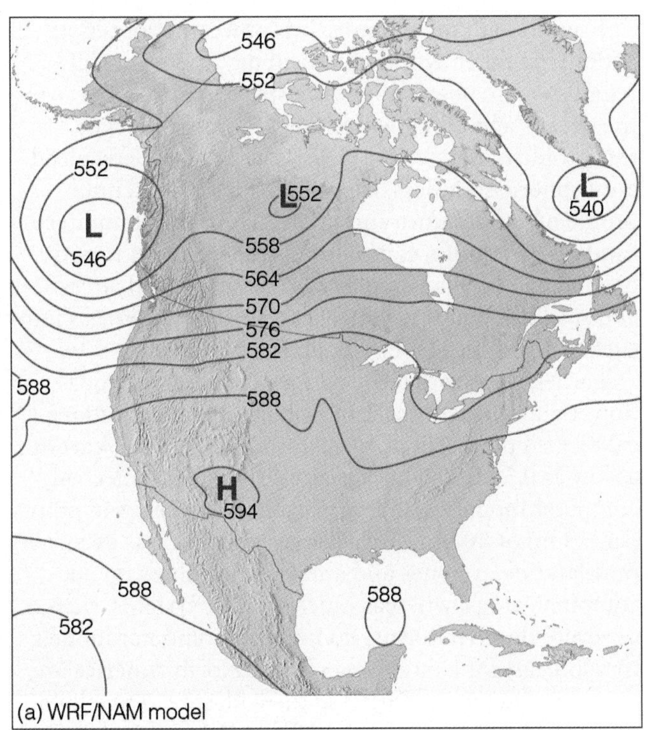

(a) WRF/NAM model

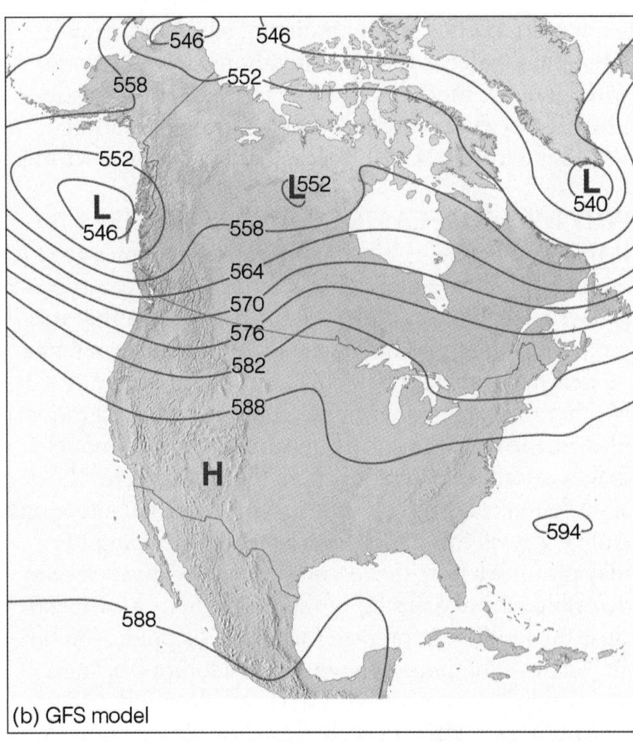

(b) GFS model

FIGURE 14.5 Two 500-mb progs for 7 P.M. EST, July 12, 2006—48 hours into the future. Prog (a) is the WRF/NAM model, with a resolution (grid spacing) of 7 mi (about 12 km), whereas prog (b) is the GFS model with a resolution of about 37 mi (60 km). Solid lines on each map are height contours, where 570 equals 5700 meters. Notice how the two progs (models) agree on the atmosphere's large-scale circulation. The main difference between the progs is in the way the models handle the low off the west coast of North America. Model (a) predicts that the low will dig deeper along the coast, while model (b) predicts a more elongated west-to-east (zonal) low. (The abbreviation WRF/NAM stands for Weather Research Forecast/North American Mesoscale Model, and GFS stands for Global Forecast System.)

Some models predict some features better than others: One model may work best in predicting the position of troughs on upper-level charts, whereas another forecasts the position of surface lows quite well. Some models even forecast the state of the atmosphere 384 hours (16 days) into the future. Look at Fig. 14.6 and notice that model (b) in Fig. 14.5 with a resolution of about 40 miles actually did a better job of forecasting the structure of the low off the west coast of North America than did model (a) with a resolution of only 7 miles.

A good forecaster knows the idiosyncrasies of each model [such as model (a) and model (b) in Fig.14.5] and carefully scrutinizes all the progs. The forecaster then makes a prediction based on the *guidance* from the computer, a personalized practical interpretation of the weather situation and any local geographic features that influence the weather within the specific forecast area.

Currently, forecast models predict the weather reasonably well 4 to 6 days into the future. The models

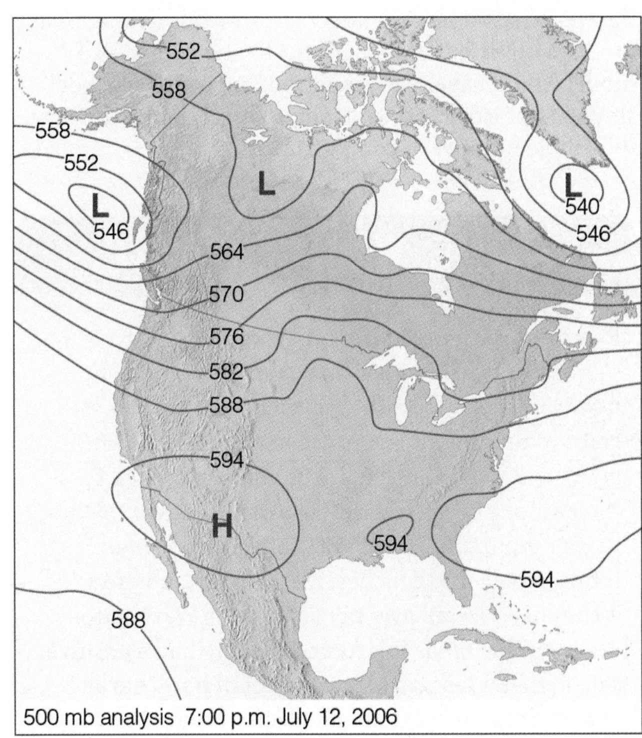

500 mb analysis 7:00 p.m. July 12, 2006

FIGURE 14.6 The 500-mb analysis for 7 P.M. EST, July 12, 2006.

tend to do a better job at predicting temperature and jet-stream patterns than precipitation. However, even with all of the modern advances in weather forecasting provided by ever more powerful computers, National Weather Service (NWS) forecasts are sometimes wrong.

WHY NWS FORECASTS GO AWRY AND STEPS TO IMPROVE THEM Why do forecasts sometimes go wrong? There are a number of reasons why forecasts can go awry. Why, for example, will the heavy snow that was predicted sometimes not materialize? For one, computer models have inherent flaws that limit the accuracy of weather forecasts. For example, computer-forecast models idealize the real atmosphere, meaning that each model makes certain assumptions about the atmosphere. These assumptions may be on target for some weather situations and be way off for others. Consequently, the computer may produce a prog that on one day comes quite close to describing the actual state of the atmosphere, and not so close on another. A forecaster who bases a prediction on an "off day" computer prog may find a forecast of "rain and windy" turning out to be a day of "clear and colder."

Another forecasting problem arises because the majority of models are not global in their coverage, and errors are able to creep in along the model's boundaries. For example, a model that predicts the weather for North America may not accurately treat weather systems that move in along its boundary from the western Pacific. This kind of inaccuracy is probably why model (b) in Fig. 14.5—a global model with a lower resolution—actually did a better job in predicting the low off the west coast than did model (a), which is a nonglobal model with a higher resolution. Obviously, a global model would usually be preferred. But a global model of similar sophistication with a high resolution requires an incredible number of computations.

EXTREME WEATHER WATCH

In recent years, numerical prediction models have performed well in forecasting storm development intensification, as well as movement along the east coast of the United States. However, back on February 18 and 19, 1979, the prediction models failed to forecast one of the heaviest snowstorms ever to hit Washington, D.C. The models called for snowfall totals of only 4 to 6 inches, when, in fact, up to 27 inches of snow fell over portions of the Washington metropolitan area. This forecast failure had a positive side in that it led to a reassessment of how cold air and warm air interact at mid-levels of the atmosphere during the period of rapid storm development.

Even though many thousands of weather observations are taken worldwide each day, there are still regions where observations are sparse, particularly over the oceans and at higher latitudes. To help alleviate this problem, the newest *GOES* satellite, with advanced atmospheric sounders, is providing a more accurate profile of temperature and humidity for the computer models. Wind information now comes from a variety of sources, such as Doppler radar, commercial aircraft, and satellites that translate ocean surface roughness into surface wind speed (see Chapter 7, p. 192).

Earlier, we saw that the computer solves the equations that represent the atmosphere at many locations called grid points, each spaced from 62 mi (100 km) to as low as 0.3 mi (0.5 km) apart. As a consequence, on computer models with large spacing between grid points (say 25 mi or 40 km), weather systems, such as extensive mid-latitude cyclones and anticyclones, show up on computer progs, whereas much smaller systems, such as severe thunderstorms, do not. The computer models that forecast for a large area such as North America are, therefore, better at predicting the widespread precipitation associated with a large cyclonic storm than local showers and thunderstorms. In summer, when much of the precipitation falls as local showers, a computer prog may have indicated fair weather, while outside it is pouring rain and hailing.

To capture the smaller-scale weather features as well as the terrain of the region, the distance between grid points on some models is being reduced. For example, the forecast model known as MM5 has a grid spacing as low as 0.3 mi (0.5 km). This model predicts mesoscale atmospheric conditions over a limited region, such as a coastal area where terrain might greatly impact the local weather. The problem with models that have a small grid spacing (high resolution) is that, as the horizontal spacing between grid points decreases, the number of computations increases. When the distance is halved, there are 8 times as many computations to perform, and the time required to run the model goes up by a factor of 16.

Another forecasting problem is that many computer models cannot adequately interpret many of the factors that influence surface weather, such as the interactions of water, ice, surface friction, and local terrain on weather systems. Many large-scale models now take mountain regions and oceans into account. Some models (such as the MM5) take even smaller factors into account—features that large-scale computers miss due to their longer grid spacing. Given the effect of local terrain, as well as the impact of some of the other problems previously mentioned, computer models that forecast the weather over a vast area do an inadequate job of predicting local weather conditions, such as surface temperatures, winds, and precipitation.

Even with better observing techniques and near perfect computer models, there are countless small, unpredictable atmospheric fluctuations that fall under the heading of **chaos**. For example, tiny eddies are much smaller than the grid spacing on the computer model and, therefore, go unmeasured. These small disturbances, as well as small errors (uncertainties) in the data, generally amplify with time as the computer tries to project the weather further and further into the future. After a number of days, these initial imperfections tend to dominate, and the forecast shows little or no accuracy in predicting the behavior of the real atmosphere. In essence, what happens is that the small uncertainty in the initial atmospheric conditions eventually leads to a huge uncertainty in the model's forecast.

Because of the atmosphere's chaotic nature, meteorologists are turning to a technique called **ensemble forecasting** to improve short- and medium-range forecasts. The ensemble approach is based on running several forecast models—or different versions (simulations) of a single model—each beginning with slightly different weather information to reflect the errors inherent in the measurements.

Suppose, for example, a forecast model predicts the state of the atmosphere 24 hours into the future. For the ensemble forecast, the entire model simulation is repeated, but only after the initial conditions are "tweaked" just a little. The "tweaking," of course, represents the degree of uncertainty in the observations. Repeating this process several times creates an ensemble of forecasts for a range of small initial changes.

▶ Figure 14.7 shows an ensemble 500-mb forecast chart for July 21, 2005 (48 hours into the future), using the global atmospheric circulation model. The chart is constructed by running the model 15 different times, each time starting with slightly different initial conditions. Notice that the red contour line (which represents

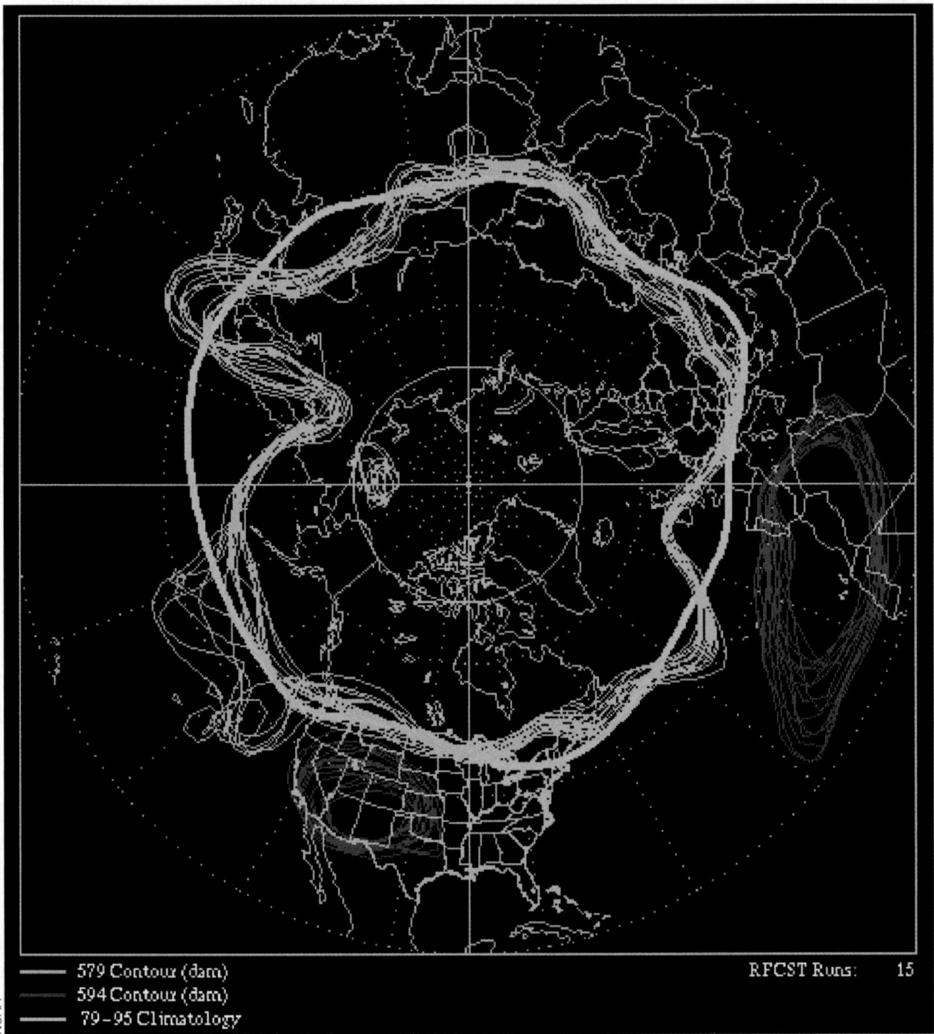

▶ **FIGURE 14.7**

Ensemble 500-mb forecast chart for July 21, 2005 (48 hours into the future). The chart is constructed by running the model 15 different times, each time beginning with a slightly different initial condition. The blue lines represent the 5790-meter contour line; the red lines, the 5940-meter contour line; and the green line, the 500-mb 25-year average, called *climatology*.

a height of 5940 meters) circles the southwestern United States, indicating a high degree of confidence in the model for that region. Here, a large upper-level high pressure area covers the region, and so a forecast for the southwestern United States would be "very hot and dry." The blue scrambled contour lines (representing a height of 5790 meters) off the west coast of North America indicate a great deal of uncertainty in the forecast model. As the forecast goes further and further into the future, the lines look more and more like scrambled spaghetti, which is why an ensemble forecast chart such as this one is often referred to as a *spaghetti plot.*

If, at the end of a specific time, the progs, or model runs, match each other fairly well, as they do over the southwestern United States in Fig. 14.7, the forecast is considered *robust.* This situation allows the forecaster to issue a prediction with a high degree of confidence. If the progs disagree, as they do off the west coast of North America in Fig. 14.7, the forecaster with little faith in the computer model prediction issues a forecast with limited confidence. In essence, *the less agreement among the progs, or model runs, the less predictable the weather.* Consequently, it would not be wise to make outdoor plans for Saturday when on Monday the weekend forecast calls for "sunny and warm" with a low degree of confidence.

In summary, imperfect numerical weather predictions may result from flaws in the computer models, from errors that creep in along the models' boundaries, from the sparseness of data, and/or from inadequate representation of many pertinent processes, interactions, and inherently chaotic behavior that occurs within the atmosphere.

Up to this point, we have looked primarily at weather forecasts made by high-speed computers using atmospheric models. There are, however, other forecasting methods, many of which have stood the test of time and are based mainly on the experience of the forecaster. Many of these techniques are of value, but often they give more of a general overview of what the weather should be like, rather than a specific forecast.

OTHER FORECASTING METHODS Probably the easiest weather forecast to make is a **persistence forecast,** which is simply a prediction that future weather will be the same as present weather. If it is snowing today, a persistence forecast would call for snow through tomorrow. Such forecasts are most accurate for time periods of several hours and become less and less accurate after that.

Another method of forecasting is the **steady-state,** or **trend forecast**. The principle involved here is that surface weather systems tend to move in the same direction and at approximately the same speed as they have been moving, providing no evidence exists to indicate otherwise. Suppose, for example, that a cold front is moving eastward at an average speed of 25 mi/hr and it is 75 miles west of your home. Using the steady-state method, we might extrapolate and predict that the front should pass through your area in three hours.

The **analogue method** is yet another form of weather forecasting. Basically, this method relies on the fact that existing features on a weather chart (or a series of charts) may strongly resemble features that produced certain weather conditions sometime in the past. To the forecaster, the weather map "looks familiar," and for this reason the analogue method is often referred to as **pattern recognition.** A forecaster might look at a prog and say "I've seen this weather situation before, and this happened." Prior weather events can then be utilized as a guide to the future. The problem here is that, even though weather situations may appear similar, they are never *exactly* the same. There are always sufficient differences in the variables to make applying this method a challenge.

Presently, **statistical forecasts** are made routinely of weather elements based on the past performance of computer models. Known as *Model Output Statistics,* or MOS, these predictions, in effect, are statistically weighted analogue forecast corrections incorporated into the computer model output. For example, a forecast of tomorrow's maximum temperature for a city might be derived from a statistical equation that uses a numerical model's forecast of relative humidity, cloud cover, wind direction, and air temperature.

When the Weather Service issues a forecast calling for rain, it is usually followed by a probability. For example: "The chance of rain is 60 percent." Does this mean (a) that it will rain on 60 percent of the forecast area or (b) that there is a 60 percent chance that it will rain within the forecast area? Neither one! The expression means that there is a 60 percent chance that any random place in the forecast area, such as your home, will receive measurable rainfall.* Looking at the forecast in another way, if the forecast for 10 days calls for a 60 percent chance of rain, it should rain where you live on 6 of those days. The verification of the forecast (as to whether it actually rained or not) is usually made at the Weather Service office, but remember that the computer models forecast for a given region, not for an individual location.

An example of a **probability forecast** using climatological data is given in Fig. 14.8. The map shows the probability of a "White Christmas"—one inch or more of snow on the ground—across the United States. The map is based on the average of 30 years of data and gives the likelihood of snow in terms of a probability. For instance, the chances are greater than 90 percent (9 Christmases out of 10) that portions of northern Minnesota, Michigan, and Maine will experience a White

* The 60 percent chance of rain does not apply to a situation that involves rain showers. In the case of showers, the percentage refers to the expected area over which the showers will fall.

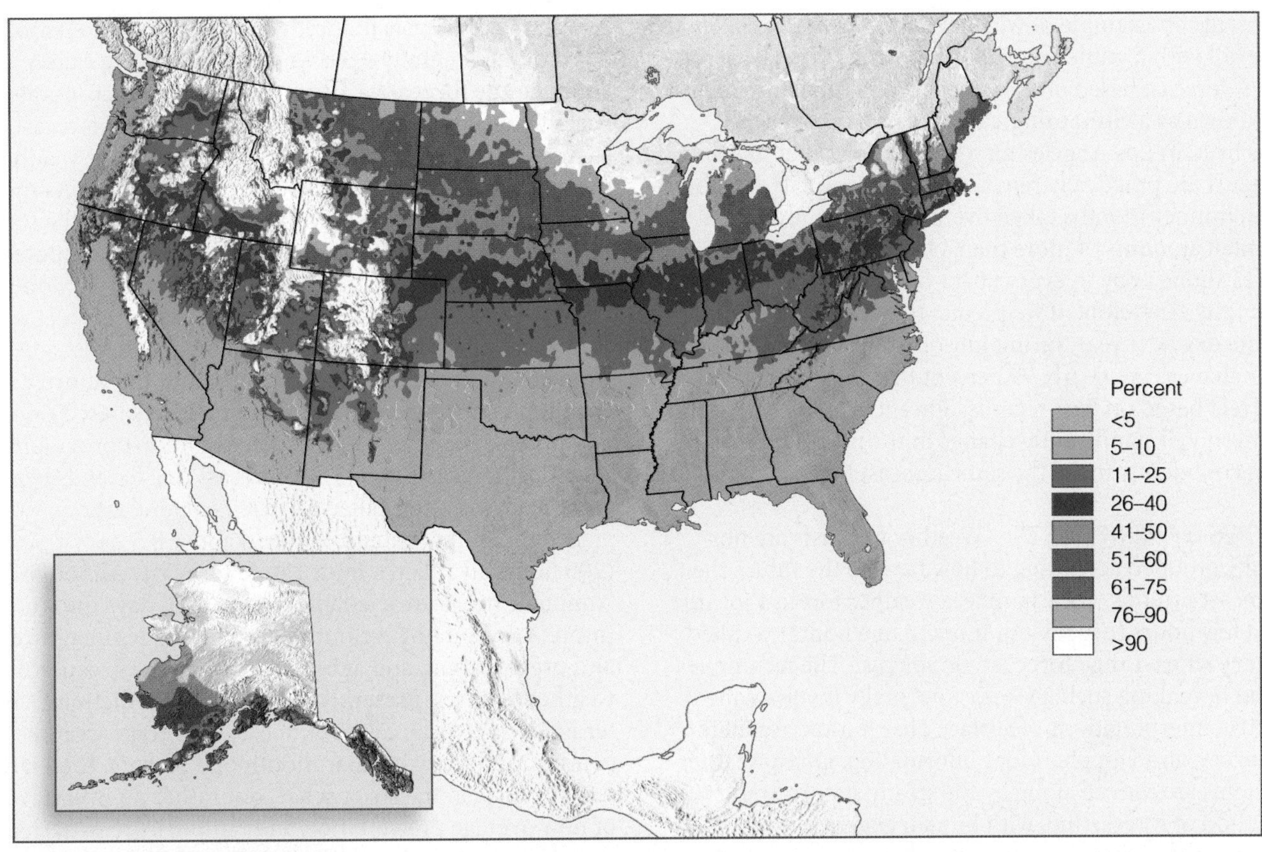

FIGURE 14.8 Probability of a "White Christmas"—one inch or more of snow on the ground—based on a 30-year average. The probabilities do not include the mountainous areas in the western United States. (NOAA)

Christmas. In Chicago, it is close to 50 percent; and in Washington, D.C., about 20 percent. Many places in the far west and south have probabilities less than 5 percent, but nowhere is the probability exactly 0, for there is always some chance (no matter how small) that a mantle of white will cover the ground on Christmas Day.

Predicting the weather by **weather types** employs the analogue method. In general, weather patterns are categorized into similar groups or "types," using such criteria as the position of the subtropical highs, the upper-level flow, and the prevailing storm track. As an example, when the Pacific high is weak or depressed southward and the flow aloft is zonal (west-to-east), surface storms tend to travel rapidly eastward across the Pacific Ocean and into the United States without developing into deep systems. But when the Pacific high is to the north of its normal position and the upper airflow is meridional (north-south), looping waves form in the flow with surface lows usually developing into huge storms. As we saw in Chapter 10, these upper-level longwaves move slowly, usually remaining almost stationary for perhaps a few days to a week or more. Consequently, the particular surface weather at different positions around the wave is likely to persist for some time. Figure 14.9

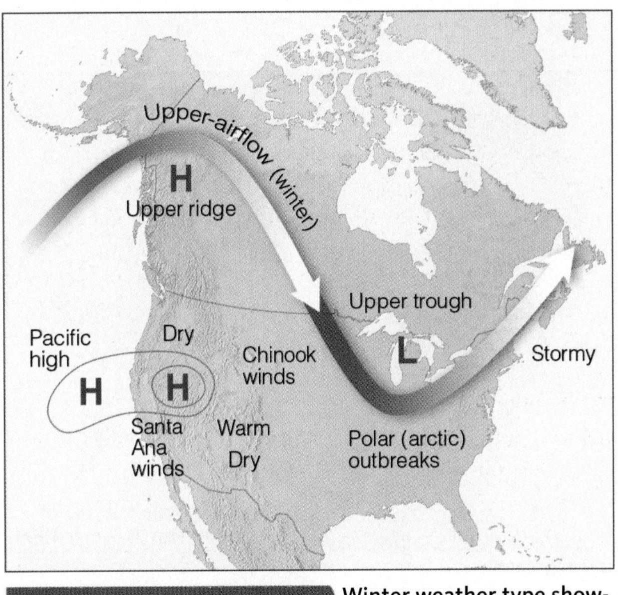

FIGURE 14.9 Winter weather type showing upper-airflow (heavy arrow), surface position of Pacific high, and general weather conditions that should prevail.

presents an example of weather conditions most likely to prevail with a winter meridional weather type.

A forecast based on the climate* of a particular region is known as a **climatological forecast**. Anyone who has lived in Los Angeles for a while knows that July and August are practically rain-free. In fact, rainfall data for the summer months taken over many years reveal that rainfall amounts of more than a trace occur in Los Angeles about 1 day in every 90, or only about 1 percent of the time. Therefore, if we predict that it will not rain on some day next year during July or August in Los Angeles, our chances are nearly 99 percent that the forecast will be correct based on past records. Since it is unlikely that this pattern will significantly change in the near future, we can confidently make the same forecast for the year 2020.

TYPES OF FORECASTS Weather forecasts are normally grouped according to how far into the future the forecast extends. For example, a weather forecast for up to a few hours (usually not more than 6 hours) is called a **very short-range forecast,** or *nowcast.* The techniques used in making such a forecast normally involve subjective interpretations of surface observations, satellite imagery, and Doppler radar information. Often weather systems are moved along by the steady state or trend method of forecasting, with human experience and pattern recognition coming into play.

* The climate of a region represents the total accumulation of daily and seasonal weather events for a specific interval of time, most often 30 years.

Weather forecasts that range from about 6 hours to a few days (generally 2.5 days or 60 hours) are called **short-range forecasts.** The forecaster may incorporate a variety of techniques in making a short-range forecast, such as satellite imagery, Doppler radar, surface weather maps, upper-air winds, and pattern recognition. As the forecast period extends beyond about 12 hours, the forecaster tends to weight the forecast heavily on computer-drawn progs and statistical information, such as Model Output Statistics (MOS).

A **medium-range forecast** is one that extends from about 3 to 8.5 days (200 hours) into the future. Medium-range forecasts are almost entirely based on computer-derived products, such as forecast progs and statistical forecasts (MOS). A forecast that extends beyond 3 days is often called an *extended forecast.*

A forecast that extends beyond about 8.5 days (200 hours) is called a **long-range forecast.** Although computer progs are available for up to 16 days into the future, they are not accurate in predicting temperature and precipitation, and at best only show the broad-scale weather features. Presently, the Climate Prediction Center issues forecasts, called *outlooks*, of average weather conditions for a particular month or a season. These are not forecasts in the strict sense, but rather an overview of how average precipitation and temperature patterns may compare with normal conditions. ▶Figure 14.10 gives a typical 90-day outlook.

Initially, outlooks were based mainly on the relationship between the projected average upper-air flow and

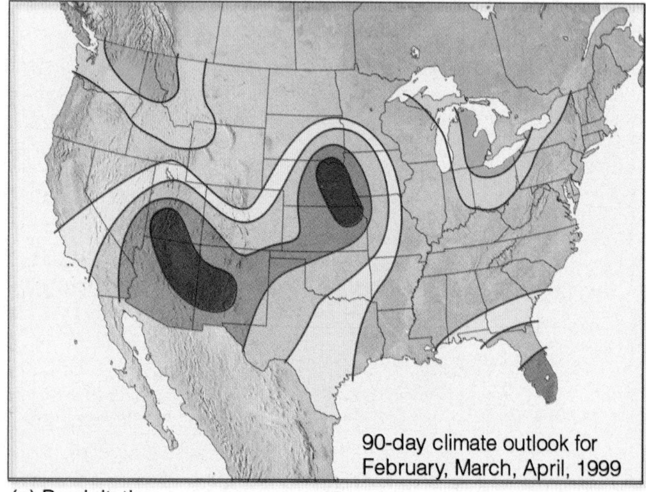

(a) Precipitation

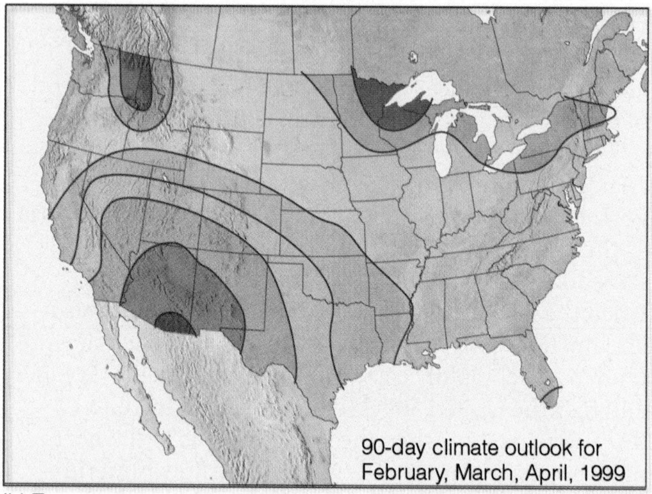

(b) Temperature

▶ FIGURE 14.10 The 90-day outlook for (a) precipitation and (b) temperature for February, March, and April, 1999. For precipitation (a), the darker the green color the greater the probability of precipitation being above normal, whereas the deeper the red color the greater the probability of precipitation being below normal. For temperature (b), the darker the orange/red colors the greater the probability of temperatures being above normal, whereas the darker the blue color, the greater the probability of temperatures being below normal. (National Weather Service/NOAA)

the surface weather conditions that the type of flow will create. Today, many of the outlooks are based on persistence statistics that carry over the general weather pattern from immediately preceding months, seasons, and years. In addition, long-range forecasts are made from models that link the atmosphere with the ocean surface temperature. (We will examine this link between the ocean and the atmosphere and the role it plays in predicting long-term weather and climate patterns later in this chapter.)

In most locations throughout North America, the weather is fair more often than rainy. Consequently, there is a forecasting bias toward fair weather, which means that, if you made a forecast of "no rain" where you live for each day of the year, your forecast would be correct more than 50 percent of the time. But did you show any *skill* in making your correct forecast? What constitutes skill, anyway? And how accurate are the forecasts issued by the National Weather Service?

EXTREME WEATHER WATCH

What is Groundhog Day? Groundhog Day (February 2) is a day that is supposed to represent the midpoint of winter—halfway between the winter solstice and the vernal equinox. Years ago, in an attempt to forecast what the remaining half of winter will be like, people placed the burden of weather prognostication on various animals, such as the groundhog (which is actually a woodchuck). Folklore says that if the groundhog emerges from his burrow and sees (or casts) his shadow on the ground and then returns to his burrow, there will be six more weeks of winter weather. One can only wonder whether it is really the groundhog's shadow that drives him back into his burrow or the people standing around gawking at him.

ACCURACY AND SKILL IN FORECASTING In spite of the complexity and ever-changing nature of the atmosphere, forecasts made for between 12 and 24 hours are usually quite accurate. Those made for between 2 and 5 days are fairly good. Beyond about 7 days, due to the chaotic nature of the atmosphere, computer prog forecast accuracy falls off rapidly. Although weather predictions made for up to 3 days are by no means perfect, they are far better than simply flipping a coin. But how accurate are they?

One problem with determining forecast accuracy is deciding what constitutes a right or wrong forecast. Suppose tomorrow's forecast calls for a minimum temperature of 35°F. If the official minimum turns out to be 37°F, is the forecast incorrect? Is it as incorrect as one 10 degrees off? By the same token, what about a forecast for snow over a large city, and the snow line cuts the city in half with the southern portion receiving heavy amounts and the northern portion none? Is the forecast right or wrong? At present, there is no clear-cut answer to the question of determining forecast accuracy.

How does forecast accuracy compare with forecast skill? Suppose you are forecasting the daily summertime weather in Los Angeles. It is not raining today and your forecast for tomorrow calls for "no rain." Suppose that tomorrow it doesn't rain. You made an accurate forecast, but did you show any skill in so doing? Earlier, we saw that the chance of measurable rain in Los Angeles on any summer day is very small indeed; chances are good that day after day it will not rain. For a forecast to show skill, it should be better than one based solely on the current weather (*persistence*) or on the "normal" weather (*climatology*) for a given region. Therefore, during the summer in Los Angeles, a forecaster will have many accurate forecasts calling for "no measurable rain," but will need skill to predict correctly on which summer days it will rain. So, if on a sunny July day in Los Angeles you forecast rain for tomorrow and it rains, you showed skill in making your forecast because your forecast was better than both persistence and climatology.

Meteorological forecasts, then, show skill when they are more accurate than a forecast utilizing only persistence or climatology. Persistence forecasts are usually difficult to improve upon for a period of time of several hours or less. Weather forecasts ranging from 12 hours to a few days generally show much more skill than those of persistence. However, as the range of the forecast period increases, because of chaos the skill drops quickly. The 6- to 14-day mean outlooks both show some skill (which has been increasing over the last several decades) in predicting temperature and precipitation, although the accuracy of precipitation forecasts is less than that for temperature. Presently, 7-day forecasts now show about as much skill as 3-day forecasts did a decade ago. Beyond 15 days, specific forecasts are only slightly better than climatology.

Forecasting large-scale weather events several days in advance (such as the blizzard of 1996 along the eastern seaboard of the United States) is far more accurate than forecasting the precise evolution and movement of small-scale, short-lived weather systems, such as tornadoes and severe thunderstorms. In fact, 3-day forecasts of the development and movement of a major low-pressure system show more skill today than 36-hour forecasts did 15 years ago.

Even though the *precise* location where a tornado will form is presently beyond modern forecasting techniques, the general area where the storm is *likely* to form can often be predicted up to 3 days in advance. With improved observing systems, such as Doppler radar and advanced satellite imagery, the lead time of watches and warnings for severe storms has increased. In fact, the lead time* for tornado warnings has more than doubled over the last decade with the average lead time today being close to 15 minutes.

Although scientists may never be able to skillfully predict the weather beyond about 15 days using available observations, the prediction of *climatic trends* appears to be more promising. Whereas individual weather systems vary greatly and are difficult to forecast very far in advance, global-scale patterns of winds and pressure frequently show a high degree of persistence and predictable change over periods of a few weeks to a month or more. With the latest generation of high-speed super computers, general circulation models (GCMs)** are doing a far better job at predicting large-scale atmospheric behavior than did the earlier models.

As new knowledge and methods of modeling are fed into the GCMs, it is hoped that they will become a reliable tool in the forecasting of weather and climate. (In Chapter 15 we will examine in more detail the climatic predictions based on numerical models.)

BRIEF REVIEW

Up to this point, we have looked at the various methods of weather forecasting. Before going on, here is a review of some of the important ideas presented so far:

- **Available to the forecaster are a number of tools that can be used when making a forecast, including surface and upper-air maps, computer progs, meteograms, soundings, Doppler radar, and satellite information.**
- **The forecasting of weather by high-speed computers is known as *numerical weather prediction*. Mathematical models that describe how atmospheric temperature, pressure, winds, and moisture will change with time are programmed into the computer. The computer then draws surface and upper-air charts, and produces a variety of forecast charts called *progs*.**
- **After a number of days, atmospheric chaos and small errors in the data greatly limit the accuracy of weather forecasts.**
- **Ensemble forecasting is a technique based on running several forecast models (or different versions of a single model), each beginning with slightly different weather information to reflect errors in the measurements.**
- **A *persistence forecast* is a prediction that future weather will be the same as the present weather, whereas a *climatological forecast* is based on the climatology of a particular region.**
- **For a forecast to show skill, it must be better than a persistence forecast or a climatological forecast.**
- **Weather forecasts for up to a few hours are called *very short-range forecasts;* those that range from about 6 hours to a few days are called *short-range forecasts; medium-range forecasts* extend from about 3 to 8.5 days into the future, whereas *long-range forecasts* extend beyond, about 8.5 days.**
- **Forecasting large-scale weather events several days into the future is far more accurate than predicting precisely where a supercell thunderstorm or tornado will form.**
- **Seasonal outlooks provide an overview of how temperature and precipitation patterns may compare with normal conditions.**

Predicting Short-Term Hazardous and Severe Weather Events

Figure 14.11 represents a simplified surface weather map during a morning in late winter. A single isobar is drawn around the pressure centers to show their positions without cluttering the map. Notice that warm winds are drawn with red arrows, and cold winds with blue arrows. Our goal here is to predict where hazardous and severe weather might develop. We will begin our forecasting by examining the weather in the western United States.

FORECASTING WEATHER IN THE WEST Notice in Fig. 14.11 that no significant weather is taking place in the northwestern part of the United States. High pressure dominates the region, and the upper-air flow is directing storms to the north into Canada. In fact, observe in Fig. 14.12 that a strong upper-level ridge covers the region. This area of high pressure produces sinking air and generally fair weather. With this weather pattern in place, many locations might set record-high temperatures for the day. So a short-range forecast for a city such as Seattle, Washington, might be:

> "Generally fair weather, with light winds and near record-high temperatures."

A city situated in a mountain valley might have a much different weather forecast than that of Seattle. For example, Medford, Oregon, sits in a valley surrounded by mountains. Cold air often settles into the valley, and with no strong

* *Lead time* is the interval of time between the issue of the warning and actual observance of the tornado.

** The GCMs are numerical computer models that simulate global patterns of wind, pressure, and temperature, and how these variables change over time.

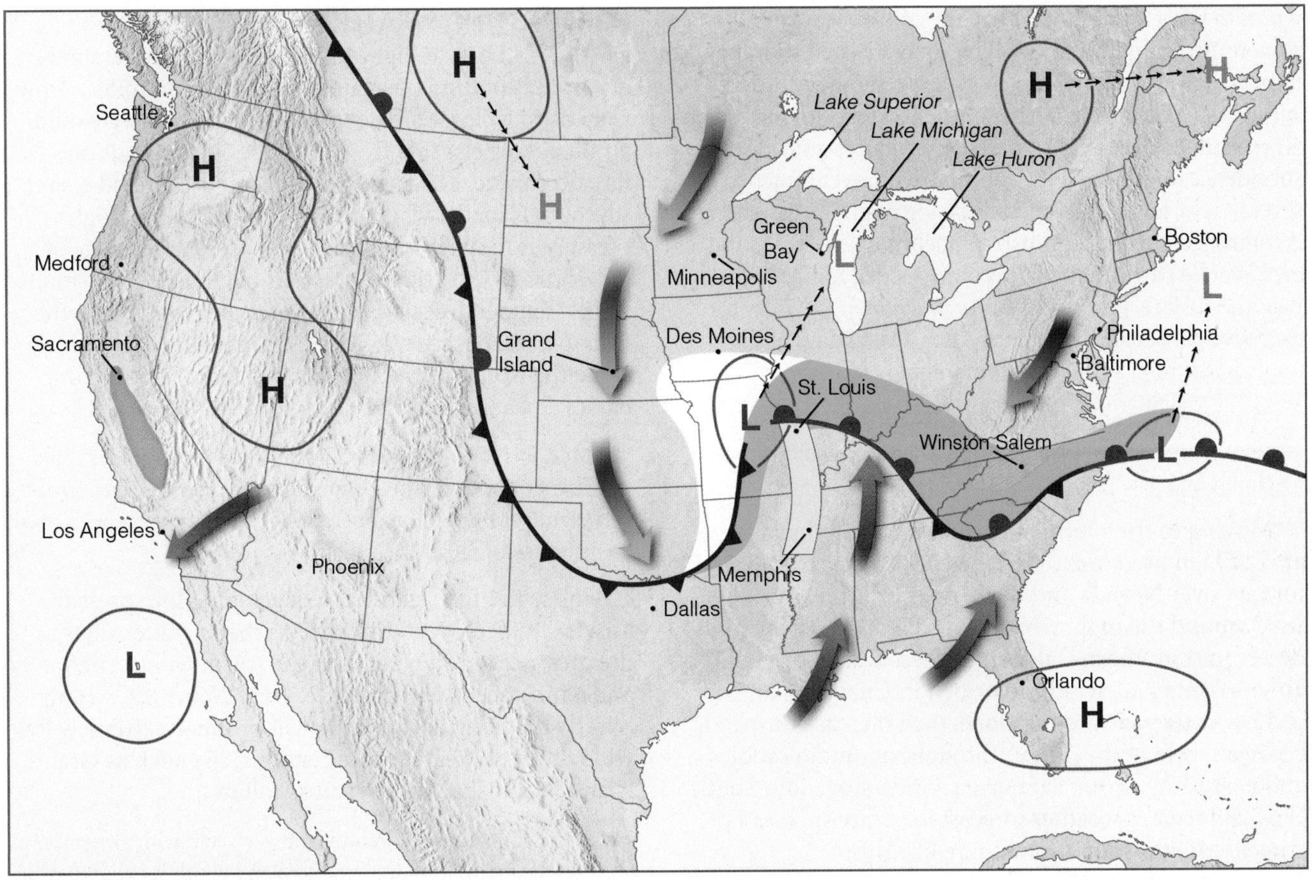

FIGURE 14.11 Surface weather map during the morning in late winter. Areas of high pressure are shown with a large H; areas of low pressure with a large L. Red arrows represent warm winds; blue arrows represent cold winds. Dashed lines show predicted movement of pressure systems. Areas shaded green are receiving rain, whereas areas shaded white are receiving snow, and those shaded pink are receiving freezing rain.

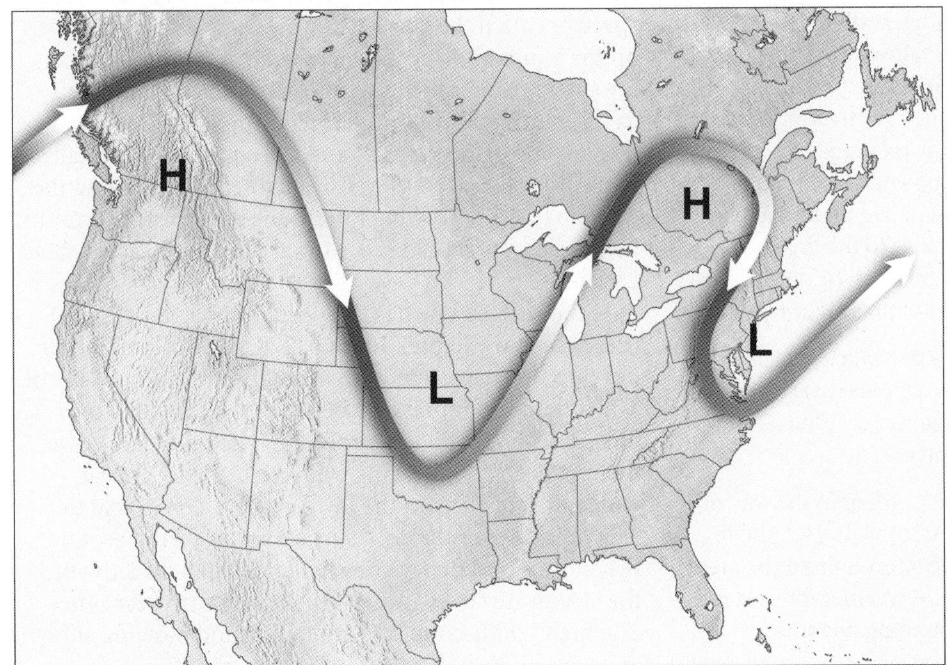

FIGURE 14.12 Upper-air flow for the same day and time as the surface map in Fig. 14.11.

winds to stir it up or move it out, the cold air stays there, sometimes for many days. Sinking air associated with the upper ridge produces warm air, which rests above the cold surface air. Recall from Chapter 5, p. 127, that sinking, warm air positioned above cold surface air produces a strong subsidence inversion. If the cold air becomes entrenched for several days, low clouds and patchy fog might form. And if controlled burning of brush and debris has begun in the area, smoke particles might become trapped in the cold air near the surface, producing an *Air Stagnation Advisory* for excessive particulate pollutants. After several days of stagnant air, the forecast for Medford might read:

"Air stagnation advisory. Light winds and high levels of smoke, especially near burn areas. Continued cold with low clouds and patchy fog."

Moving to the south, notice in Fig. 14.11 that a large area of high pressure, called the *Great Basin High* has formed over Nevada and Utah. As the air flows clockwise around the high, winds blow downhill off the high desert into southern California. These warm, dry winds are the Santa Ana winds described in Chapter 8 on p. 229. As they blow downslope, they increase in speed, especially where they move through mountain saddles and passes. As strong Santa Ana winds move into Southern California, a weather forecast for a city such as Los Angeles, might sound something like this:

"Sunny and very warm and dry today with strong and gusty northeasterly Santa Ana winds. Extreme fire danger. High wind warning posted over mountain passes for winds gusting to over 60 mi/hr."

To the north of Los Angeles in the Central Valley of California, the strong upper-level ridge and the Great Basin High are producing generally fair weather throughout the region. Weak pressure gradients are producing weak surface winds. Because the valley is surrounded (for the most part) by mountains, cool air is being trapped at the surface. Clear skies at night allow for rapid radiational cooling and for the formation of late night and early morning radiation fog that lingers close to the ground and produces hazardous driving conditions. A weather forecast for a city in the valley, such as Sacramento, might read:

"Dense fog advisory. Fair during the day with late night and early morning dense fog which can reduce visibility to less than a quarter of a mile and produce hazardous driving conditions, especially in low-lying areas."

Look at Fig. 14.11 and notice that nothing in the way of extreme weather is occurring over Arizona. In fact, the region is influenced by high pressure. So throughout the area, the weather is clear and quite warm. A weather forecast for Phoenix, Arizona, might sound something like this:

"Clear and quite warm with near record-high temperatures."

FORECASTING WEATHER FOR THE GREAT PLAINS Look at Fig. 14.11 and notice that a stationary front is butting up against the Rocky Mountains from Canada to Colorado. The front separates relatively mild air on its western side from bitterly cold arctic air on its eastern side. The extremely cold air is centered over southern Canada, where a large shallow area of high pressure is moving southeastward. The leading edge of this arctic air is a cold front that is beginning to push into Texas. Temperatures are expected to drop rapidly as the front passes. Because there is little moisture in the air along the front, no precipitation will occur as the front passes. So, a weather forecast for Dallas, Texas, might be:

"Mild this morning, turning windy and much colder this afternoon with rapidly dropping temperatures. Hard freeze warning. Minimum temperatures by tomorrow morning will be in the low 20s."

Notice in Fig. 14.11 that a developing low-pressure area is situated over Missouri. As the low intensifies, the pressure gradient around the storm will steepen and wind speeds will increase. Over Nebraska, strong, bitterly cold northerly winds will produce extremely low wind chills. A weather forecast for a city such as Grand Island, Nebraska, might read as follows:

"Wind chill advisory. Windy and very cold with air temperatures dropping to well below zero and winds gusting to over 30 mi/hr."

Upper-level winds tend to steer surface pressure systems. As a consequence, notice in Fig. 14.11 that the low over Missouri is projected to move northeasterly to a position over Lake Michigan in about 24 hours. The speed at which the surface low moves will depend in part on the winds aloft. As a general rule of thumb, surface low-pressure areas tend to move in the same direction as the winds at the 500-mb level (roughly 18,000 ft or 5600 m above the surface) and at a speed of about half the wind speed at this level. Therefore, if the winds at the 500-mb level are blowing toward the northeast at 50 mi/hr, the surface low should move toward the northeast at about 25 mi/hr.

If the surface low has the proper upper-air support (described in Chapter 10 on p. 278), the low will likely develop into a deep mature mid-latitude cyclonic storm. The difference in pressure between the deep low over Lake Michigan and the strong high-pressure area over the northern Great Plains will produce very strong winds. If moisture on the low's eastern side is able to ride up and over the cold surface air and wrap around the surface low, heavy snow will likely fall directly into the strong surface winds on the low's northwest side. The high winds coupled with falling and blowing snow will likely produce blizzard conditions over portions of Minnesota and Wisconsin. Therefore, a weather forecast for Minneapolis, Minnesota, might be:

> "Blizzard warning. Very cold with strong northeasterly winds, gusting to over 40 mi/hr. Falling and blowing snow will reduce visibility at times to less than a quarter of a mile."

As the low over Missouri moves toward the northeast, snow will likely fall over eastern Nebraska, eastern South Dakota, Iowa, and Wisconsin. In fact, due to heavy snow and high winds, a winter storm warning might be posted in a number of locations including Des Moines, Iowa, where the weather forecast might be:

> "Winter storm warning. Snow, heavy at times, cold and windy with total snow accumulation between 6 and 10 inches."

As the low tracks northeastward, cities situated on the east side of the storm's path will likely receive rain, whereas cities on the west side of the low will likely receive snow, or perhaps, rain changing to snow. What about a city such as Green Bay, Wisconsin, which is just to the west of the storm's projected path? Green Bay appears to be right on the border between rain and snow. A forecasting tool weather forecasters often use to predict whether rain or snow will fall over a region is the *thickness chart.* More information on the use of this chart and whether rain or snow will fall at Green Bay is given in the Focus section on p. 416.

As the low-pressure area moves over Lake Michigan, it will more than likely be a deep low-pressure area. Strong winds over the Lakes will produce huge waves and hazardous seas. Over Lake Huron and Lake Michigan, very strong southwesterly winds (possibly exceeding 40 mi/hr) will require gale warnings to be posted. And over Lake Superior, strong northeasterly winds may exceed 55 mi/hr, prompting the forecasters to issue storm warnings.

FORECASTING SOUTHERN STORMS Again, look back at Fig. 14.11, p. 413, and notice that over the southern states, ahead of the advancing cold front, warm, humid air is streaming northward from the Gulf of Mexico. The combination of warm, humid surface air overlain by cold air aloft sets up a conditionally unstable atmosphere. This atmospheric situation coupled with strong vertical wind shear (southerly surface winds and southwesterly winds aloft), produces an environment ripe for severe thunderstorm development, especially where the surface air is being lifted by the advancing cold front. On the map, the region shaded yellow represents the area where severe thunderstorms are most likely to form, and where a severe thunderstorm watch has been issued. The watch area will likely extend farther into Illinois as the warm front advances northward.

Where, exactly, in the severe thunderstorm watch area will severe thunderstorms form, and which thunderstorms will most likely produce a tornado? These questions are impossible to answer at this time. However, forecasters using a variety of indices determine which *areas* are *most likely* to produce severe thunderstorms and tornadoes.

One very useful index for predicting the intensity of thunderstorms and the likelihood of tornado development is the *C*onvective *A*vailable *P*otential *E*nergy or *CAPE*, which we briefly looked at in Chapter 11, on p. 315. To determine CAPE you need a temperature sounding (vertical profile of air temperature and dew point). In the sounding, you lift a parcel of air from the surface and compare the rising parcel's temperature with that of the surrounding air.* (This concept is illustrated in Chapter 5, Fig. 5.9.) The warmer the rising air parcel compared to its surrounding environment, the greater the value of CAPE.

The value of CAPE represents the rising parcel's positive buoyancy, which is a measure of the amount of energy available to create updrafts in a thunderstorm. The value of CAPE is also proportional to the rate at which the air rises inside the storm. Thus, CAPE is an indication of the storm's intensity or strength. Table 14.1 shows how the

* To compute CAPE, you calculate on a chart the area between the temperature of the rising parcel and that of the surrounding air. This value is then converted to energy, or CAPE.

Table 14.1

Values of CAPE Related to Atmospheric Stability, Convection, and the Likelihood of Thunderstorms and Tornadoes

CAPE VALUE (J/Kg)*	ATMOSPHERIC STABILITY	CONVECTION (UPDRAFTS)	FORECAST
0–1000	Marginally unstable	Weak	Isolated (ordinary cell/airmass) thunderstorms possible
1000–2500	Moderately unstable	Moderate	Intense thunderstorms possible
2500–3500	Very unstable	Strong	Severe thunderstorms possible
> 3500	Extremely unstable	Very strong	Severe thunderstorms possibly with tornadoes

* The units of CAPE, J/Kg, represent the amount of energy in joules (J) in one kilogram of air where 1 joule (J) equals 0.24 calories.

FOCUS ON

A SPECIAL TOPIC

Forecasting Rain or Snow with the Thickness Chart

The thickness chart can be a valuable forecasting tool for the meteorologist. It can help identify air masses and locate fronts, and a prognostic thickness chart (*thickness prog*) can help predict the daily max and min temperature. It can also help predict whether falling precipitation will be in the form of rain or snow. What, then, is a thickness chart?

A thickness chart shows the difference in height between two constant pressure surfaces. The vertical depth or thickness between any two pressure surfaces is related to the average air temperature between the two surfaces. Recall from Chapter 7, p. 176, that air pressure decreases more rapidly with height in cold air than in warm air. This fact is illustrated in Fig. 2 which shows a 1000-mb pressure surface and a 500-mb pressure surface. The difference in height between these two pressure surfaces is called the *1000-mb to 500-mb thickness*.* Notice that the vertical distance (thickness) between these two pressure surfaces is greater in warm air than in cold air. Consequently, warm air produces high thickness, and cold air low thickness.

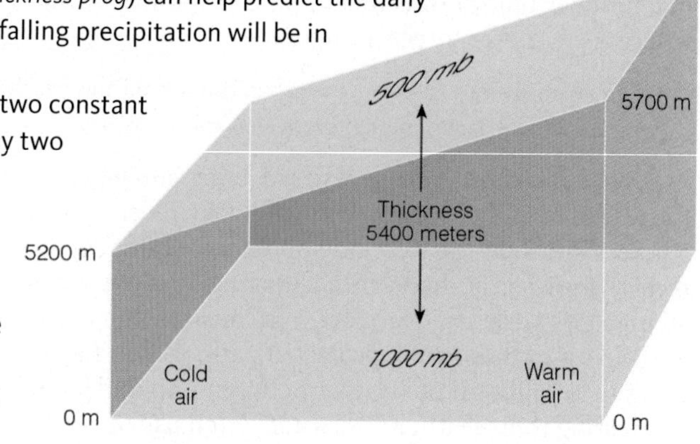

Figure 2 The vertical separation (thickness) between the 1000-mb pressure surface and the 500-mb pressure surface is greater in warm air than in cold air.

When 1000-mb to 500-mb thickness lines are drawn on a chart, they may appear similar to those shown in Fig. 3. On the chart, regions of low thickness correspond to cold air, and regions of high thickness to warm air. There are a number of forecasting rules for predicting air temperature and precipitation using this chart. One forecasting rule is that the 5400-meter thickness line often represents the dividing line between rain and snow, especially for cities receiving precipitation east of the Rocky Mountains. If precipitation is falling, cities with a thickness greater than 5400 m should be receiving rain, whereas cities with a thickness less than 5400 m should be receiving snow.

Suppose Fig. 3 represents a thickness prog for a time when the low-pressure area moves over Lake Michigan in Fig. 14.11. Notice that the 5400-m thickness line is to the south of Green Bay, Wisconsin, indicating that Green Bay will most likely receive snow from this storm.

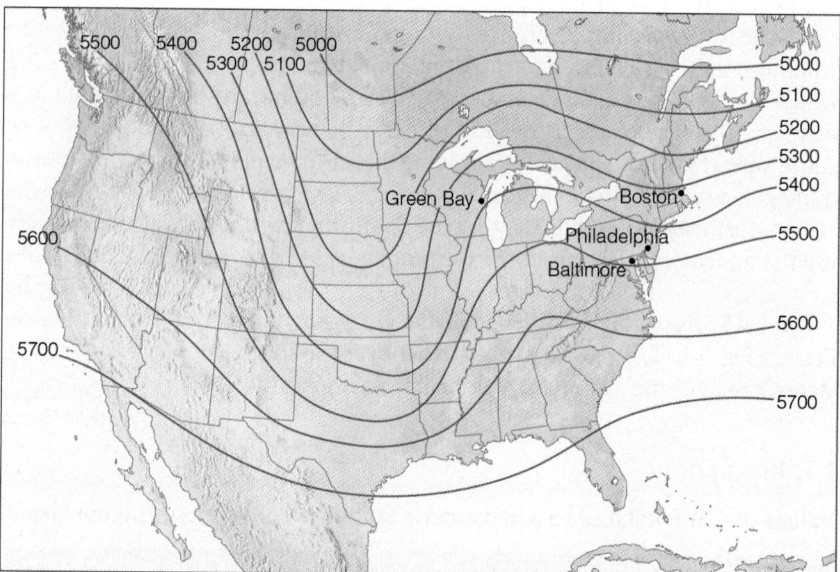

Figure 3 A 1000-mb to 500-mb thickness chart during late winter. The lines on the chart represent the vertical depth in meters between the 1000-mb and 500-mb pressure surfaces. Low thickness lines correspond to cold air, and high thickness lines to warm air. For reference, the 5400-meter thickness line represents a vertical layer of air 5400 meters thick with an average temperature of 19°F (−7°C). The 5200 thickness line is roughly the boundary for arctic air.

* Because the 1000-mb pressure surface is often close to the surface of the earth, the 1000-mb to 500-mb thickness is sometimes referred to as *the surface to 500-mb thickness.*

different values of CAPE relate to atmospheric stability and the likelihood of severe thunderstorm development.

Observe in Fig. 14.11, p. 413, that both St. Louis, Missouri, and Memphis, Tennessee, are in the severe thunderstorms watch box area. Suppose we had soundings available to use for both cities. Further suppose that from these soundings we calculated the value of CAPE to be 2800 J/Kg for St. Louis and 3600 J/Kg for Memphis. From Table 14.1 we can see that severe thunderstorm development is possible for St. Louis and even more likely for Memphis. Moreover, in Memphis, there is an even greater likelihood that a severe thunderstorm will spawn a tornado. So a weather forecast for St. Louis might read:

"Severe thunderstorm watch. Thunderstorms, some possibly severe, with strong winds, large hail, and heavy rain. Possible flooding in low-lying areas."

A forecast for Memphis might be similar to that in St. Louis, but with a stronger threat of severe weather:

"Severe thunderstorm watch. A likelihood of severe thunderstorms capable of producing damaging high winds, large hail, and heavy rain. Severe thunderstorms can produce tornadoes."

A FORECAST OF SOUTHEASTERN ICE Notice on the weather map (Fig. 14.11, p. 413) that in the southern states the warm front becomes a stationary front over Georgia and South Carolina. On the north side of the stationary front, a shallow layer of below-freezing cold air has pushed into the region. On the south side of the front, warm, humid air is flowing northward. This warm less-dense air rides up and over the shallow layer of cold air, producing a dense cloud layer from which rain is falling. As the rain descends into the cold air, it freezes on contact with objects on the surface. If air temperatures remain below freezing for a number of hours, the freezing rain will accumulate into a layer of ice capable of downing power lines and producing treacherous driving conditions. Should the freezing rain continue for many hours, the region could experience an *ice storm.* A short-range weather forecast for a city within the region of freezing rain such as Winston-Salem, North Carolina, might read:

"Winter storm warning. Freezing rain will produce hazardous driving conditions. Possible ice storm. Temperatures remaining just below freezing with light winds."

A FORECAST OF HOT AND DRY FOR FLORIDA Notice in Fig. 14.11, p. 413, that an area of high pressure exists at the surface over Florida. This weather situation limits convection over the state, as sinking air dominates the region. Should this weather pattern exist for an extended period of time (several weeks), hot and dry weather will persist, and even an extensive drought may set in. This hot, dry weather pattern may even make the state ripe for wildfires that could burn thousands of acres of brush and timber. With this weather pattern, a weather forecast for Orlando, Florida, might sound like this:

"Sunny and continued hot and dry with near record-high temperatures."

WEATHER FORECAST FOR THE NORTHEAST
Again, look back at Fig. 14.11, p. 413, and notice that a small area of low pressure is situated just off the coast of North Carolina. If the low has the proper upper-level support (described in Chapter 10 on p. 278), the low might develop into a deep mid-latitude cyclonic storm that moves northeast along the coast as a nor'easter. If the low intensifies and moves northeastward as indicated in Fig. 14.11, its counterclockwise circulation should spread moisture from off the Atlantic Ocean into the mid-Atlantic states, which will likely fall as heavy precipitation. But will the precipitation fall as rain or snow? If the storm moves up the coast slowly, cities such as Baltimore, Maryland, and Philadelphia, Pennsylvania, will likely experience rain, as warm air spreads into these areas from off the ocean. But what about cities farther to the north, will they experience snow?

As we saw earlier, one tool for predicting rain or snow is the thickness chart. Suppose we want to predict whether Boston, Massachusetts, will receive rain or snow. If we look at the thickness prog in the Focus section on p. 416, we will see that about the time the low-pressure area is projected to be off the New Jersey coast, the thickness value over Boston is less than 5400 m, indicating snow, at least initially. But, as the storm moves up the coast, the snow may change to rain as warm air is brought into the Boston area from off the Atlantic. This influx of warm air makes predicting whether rain or snow will fall along the coast very tricky, even with a thickness prog. Because Philadelphia and Baltimore both have thickness values greater than 5400 m, a forecast for these cities would be for rain. In fact, if the progs show that heavy rain will fall over a particular region (such as over southeastern Pennsylvania), swollen streams will likely overflow their banks and a flood advisory may have to be issued by the forecaster. Therefore, a short-range weather forecast for Philadelphia might sound like this:

"Urban and small stream advisory. Heavy rain today with strong northeasterly winds. Rain amounts may exceed 2 inches in some locations. Small streams may overflow their banks, and water may pond in low-lying areas."

The short-term forecasting of hazardous and extreme weather is more involved than this brief introduction might lead you to believe. In fact, this section is intended to give you a few forecasting tips as well as an overview of many of the extreme weather topics covered so far in this book. The next section takes a look at how long-range patterns of weather and climate can be predicted based on the interactions that take place between the ocean and the atmosphere.

Predicting Long-Term Weather and Climate Patterns

Earlier in this chapter we learned that the ocean surface temperature in one part of the world can influence the weather and climate in another. These types of widely spaced interactions between ocean surface temperature and weather conditions are called **teleconnections**. In the following sections we will investigate a number of these types of interactions, the first of which is *El Niño* and *La Niña*.

EL NIÑO, LA NIÑA, AND THE SOUTHERN OSCILLATION What is El Niño? The answer to this question begins along the west coast of South America, where the cool Peru current sweeps northward, bringing cool water to coastal areas (see Fig. 14.13). Here, southerly winds promote the rising of cold, nutrient-rich water (upwelling) that gives rise to large fish populations, especially anchovies. Near the end of the calendar year, a warm current of nutrient-poor tropical water often moves southward, replacing the cold, nutrient-rich surface water. Because this condition frequently occurs around Christmas, local residents call it *El Niño* (Spanish for boy child, referring to the Christ child).

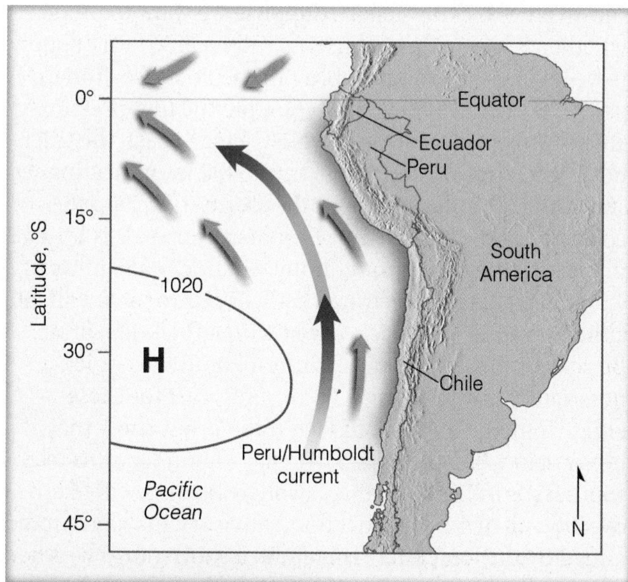

FIGURE 14.13 A large area of high pressure positioned off the west coast of South America produces southerly winds that cause cool surface water to drift northward as the Peru/Humboldt Current (dark blue arrows). As southerly winds blow parallel to coastal margins, the Coriolis effect bends moving surface water to the left (Southern Hemisphere), which induces upwelling—the rising of cold water from below.

In most years, the warming lasts for only a few weeks to a month or more, after which weather patterns usually return to normal and fishing improves. However, when El Niño conditions last for many months, and a more extensive ocean warming occurs, the economic results can be catastrophic. This extremely warm episode, which occurs at irregular intervals of two to seven years and covers a large area of the tropical Pacific Ocean, is now referred to as a *major El Niño event*, or simply **El Niño**.*

During a major El Niño event, large numbers of fish and marine plants may die. Dead fish and birds may litter the water and beaches of Peru; their decomposing carcasses deplete the water's oxygen supply, which leads to the bacterial production of huge amounts of smelly hydrogen sulfide. The El Niño of 1972–1973 reduced the annual Peruvian anchovy catch from 10.3 million metric tons in 1971 to 4.6 million metric tons in 1972. Since much of the harvest of this fish is converted into fishmeal and exported for use in feeding livestock and poultry, the world's fishmeal production in 1972 was greatly reduced. Countries such as the United States that rely on fishmeal for animal feed had to use soybeans as an alternative. This raised poultry prices in the United States by more than 40 percent.

Why does the ocean become so warm over the eastern tropical Pacific? Normally, in the tropical Pacific Ocean, the trades are persistent winds that blow westward from a region of higher pressure over the eastern Pacific toward a region of lower pressure centered near Indonesia (see Fig. 14.14a). The trades create upwelling that brings cold water to the surface. As this water moves westward, it is heated by sunlight and the atmosphere. Consequently, in the Pacific Ocean, surface water along the equator usually is cool in the east and warm in the west. In addition, the dragging of surface water by the trades raises sea level in the western Pacific and lowers it in the eastern Pacific, which produces a thick layer of warm water over the tropical western Pacific Ocean and a weak ocean current (called the *countercurrent*) that flows slowly eastward toward South America.

* It was thought that El Niño was a local event that occurs along the west coast of Peru and Ecuador. It is now known that the ocean-warming associated with a major El Niño can cover an area of the tropical Pacific much larger than the continental United States.

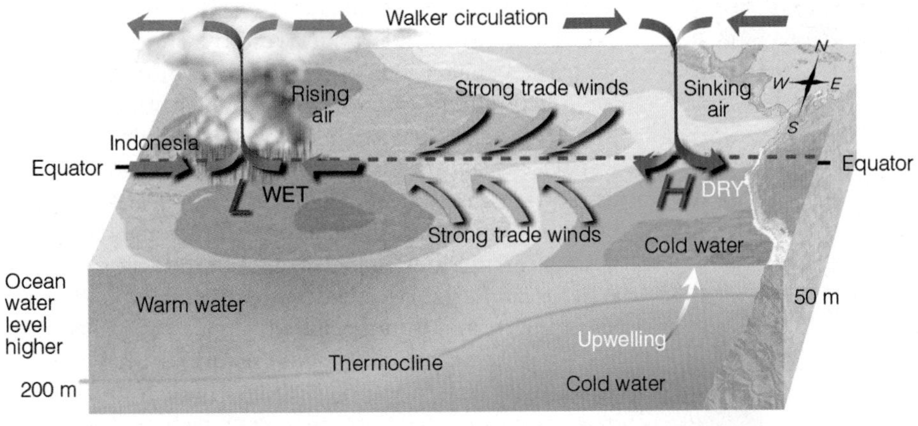

(a) Non-El Niño conditions

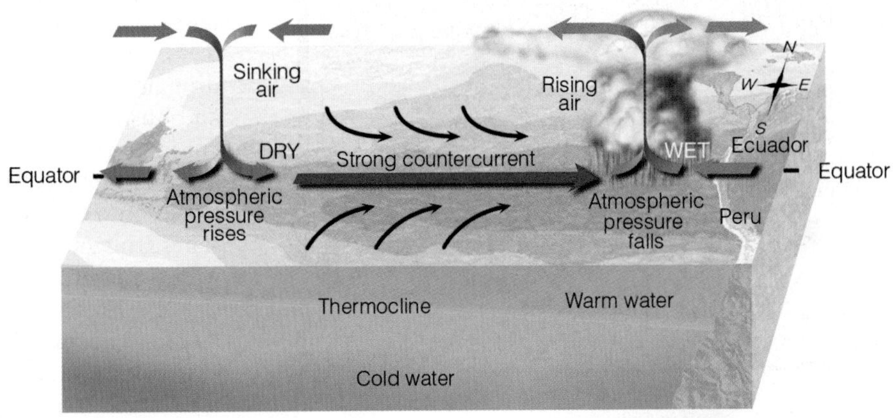

(b) El Niño Conditions

FIGURE 14.14 In diagram (a), under ordinary conditions higher pressure over the southeastern Pacific and lower pressure near Indonesia produce easterly trade winds along the equator. These winds promote upwelling and cooler ocean water in the eastern Pacific, while warmer water prevails in the western Pacific. The trades are part of a circulation (called the *Walker circulation*) that typically finds rising air and heavy rain over the western Pacific and sinking air and generally dry weather over the eastern Pacific. When the trades are exceptionally strong, water along the equator in the eastern Pacific becomes quite cool. This cool event is called La Niña. During El Niño conditions—diagram (b)—atmospheric pressure decreases over the eastern Pacific and rises over the western Pacific. This change in pressure causes the trades to weaken or reverse direction. This situation enhances the countercurrent that carries warm water from the west over a vast region of the eastern tropical Pacific. The thermocline, which separates the warm water of the upper ocean from the cold water below, changes as the ocean conditions change from non-El Niño to El Niño.

Every few years, the surface atmospheric pressure patterns break down, as air pressure rises over the region of the western Pacific and falls over the eastern Pacific (see Fig. 14.14b). This change in pressure weakens the trades, and, during strong pressure reversals, east winds are replaced by west winds. The west winds strengthen the countercurrent, causing warm water to head eastward toward South America over broad areas of the tropical Pacific. Toward the end of the warming period, which may last between one and two years, atmospheric pressure over the eastern Pacific reverses and begins to rise, whereas, over the western Pacific, it falls. This seesaw pattern of reversing surface air pressure at opposite ends of the Pacific Ocean is called the **Southern Oscillation**. Because the pressure reversals and ocean warming are more or less simultaneous, scientists call this phenomenon the *El Niño/Southern Oscillation* or **ENSO** for short. Although most ENSO episodes follow a similar evolution, each event has its own personality, differing in both strength and behavior.

During especially strong ENSO events (such as in 1982–1983 and 1997–1998) the easterly trades may actually become westerly winds, as illustrated in Fig. 14.14b. As these winds push eastward, they drag surface water

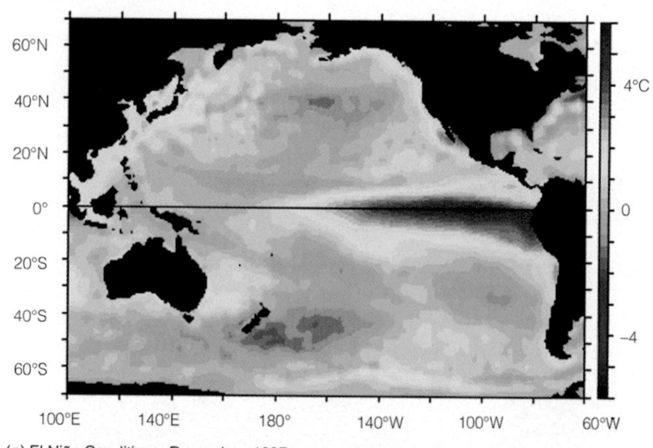

(a) El Niño Conditions, December, 1997

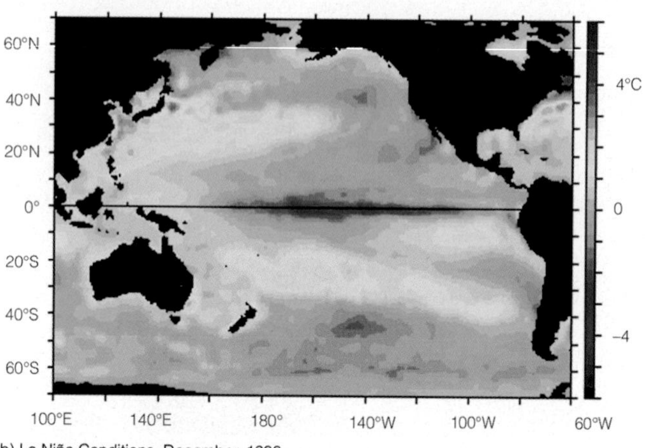

(b) La Niña Conditions, December, 1998

FIGURE 14.15 **(a) Average sea surface temperature departures from normal as measured by satellite. During El Niño conditions, upwelling is greatly diminished and warmer than normal water (deep red color) extends from the coast of South America westward, across the Pacific. (b) During La Niña conditions, strong trade winds promote upwelling, and cooler than normal water (dark blue color) extends over the eastern and central Pacific. (NOAA/PHEL/TAO)**

with them. This dragging raises sea level in the eastern Pacific and lowers sea level in the western Pacific. The eastward-moving water gradually warms under the tropical sun, becoming as much as 11°F (6°C) warmer than normal in the eastern equatorial Pacific. Gradually, a thick layer of warm water pushes into coastal areas of Ecuador and Peru, choking off the upwelling that supplies cold, nutrient-rich water to South America's coastal region. The unusually warm water may extend from South America's coastal region for many thousands of miles westward along the equator (see Fig. 14.15a).

Such a large area of abnormally warm water can have an effect on global wind patterns. The warm tropical water fuels the atmosphere with additional warmth and moisture, which the atmosphere turns into additional storminess and rainfall. The added warmth from the oceans and the release of latent heat during condensation apparently influence the westerly winds aloft in such a way that certain regions of the world experience too much rainfall and flooding, whereas others have too little and experience drought. Meanwhile, over the warm tropical central Pacific, the frequency of typhoons usually increases. However, over the tropical Atlantic, between Africa and Central America, strong winds aloft tend to disrupt the organization of thunderstorms that is necessary for hurricane development; hence, there are fewer hurricanes in this region during strong El Niño events. And, during strong El Niño events there is a tendency for monsoon conditions over India to weaken.

Although the actual mechanism by which changes in surface ocean temperatures influence global wind patterns is not fully understood, the by-products are plain to see. For example, during exceptionally warm El Niños, drought is normally felt in Indonesia, southern Africa, and Australia, while heavy rains and flooding often occur in Ecuador and Peru. In the Northern Hemisphere, a strong subtropical westerly jet stream normally directs storms into California and heavy rain into the Gulf Coast states. The total damage worldwide due to flooding, winds, and drought may exceed many billions of dollars.

Following an ENSO event, the trade winds usually return to normal. However, if the trades are exceptionally strong, unusually cold surface water moves over the central and eastern Pacific, and the warm water and rainy weather is confined mainly to the western tropical Pacific (see Fig. 14.15b). This cold-water episode, which is the opposite of El Niño conditions, has been termed **La Niña** (the girl child).

EXTREME WEATHER WATCH

The strong El Niño of 1982 and 1983 disturbed the weather worldwide. For example, in California many areas experienced their rainiest season ever. Musolitt, in northern California, recorded 185.5 inches of precipitation in 1983, a record for a calendar year in California. In South America, Guayaquil, Ecuador, recorded 165.6 inches of rain between November, 1982, and June, 1983, nearly four times the average for this period. The worst drought on record hit southern Australia, which culminated during February, 1983, with the *Ash Wednesday Fire* that killed 72 people and burned over 2000 homes in southeast Australia. During the drought, a dust storm plowed through Melbourne, dumping more than 11,000 tons of dry soil onto the city in 40 minutes.

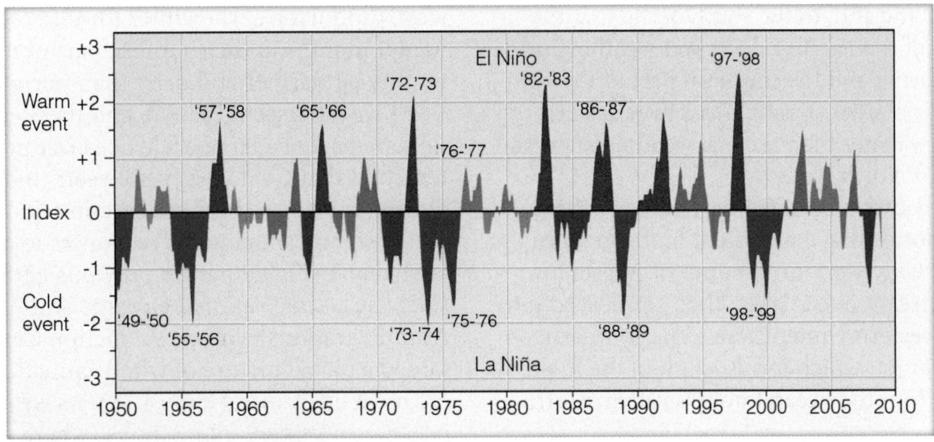

FIGURE 14.16 The Ocean Niño Index (ONI). The numbers on the left side of the diagram represent a running 3-month mean for sea surface temperature variations (from normal) over the tropical Pacific Ocean from latitude 5°N to 5°S and from longitude 120°W to 170°W. Warm El Niño episodes are in red; cold La Niña episodes are in blue. Warm and cold events occur when the deviation from the normal is 0.5 or greater. An index value between 0.5 and 0.9 is considered weak; an index value between 1.0 and 1.4 is considered moderate, and an index value of 1.5 or greater is considered strong. (Courtesy of NOAA and Jan Null.)

Figure 14.16 shows warm events, El Niño years, in red and cold events, or La Niña years, in blue. Notice that the two strongest El Niños were 1982–83 and 1997–98. The weaker El Niño events also have an effect on the Northern Hemisphere's weather patterns. For example, during the El Niño of 1986–87, the subtropical jet stream (being fueled by the warm tropical waters and huge thunderstorms) curved its way over the southeastern United States, where it brought abundant rainfall to a region that, during the previous summer, had suffered through a devastating drought. During the El Niño of 1991–92, the subtropical jet stream once again swung over North America. Water evaporating from the warm tropical oceans fueled huge thunderstorms. The subtropical jet stream initially swept this moisture into Texas, where it caused extensive flooding.

Figure 14.17a illustrates typical winter weather patterns over North America during El Niño conditions. Notice that a persistent trough of low pressure forms

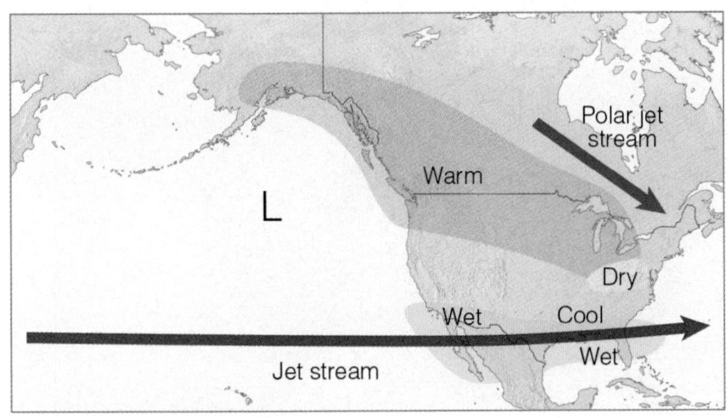

(a) El Niño winter conditions

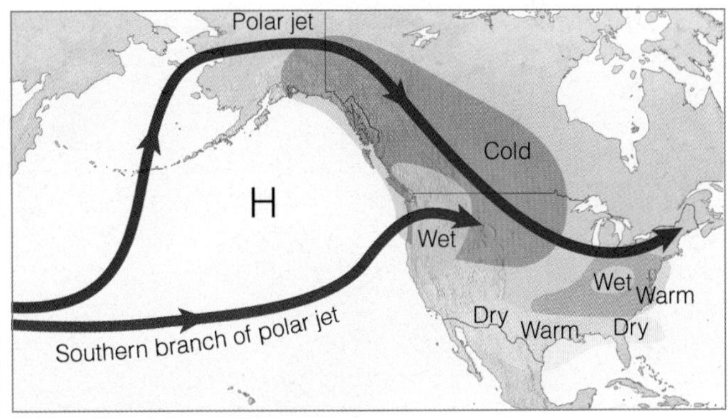

(b) La Niña winter conditions

FIGURE 14.17 Typical winter weather patterns across North America during an El Niño warm event (a) and during a La Niña cold event (b). During El Niño conditions, a persistent trough of low pressure forms over the north Pacific and, to the south of the low, the jet stream (from off the Pacific) steers wet weather and storms into California and the southern part of the United States. During La Niña conditions, a persistent high-pressure area forms south of Alaska forcing the polar jet stream and accompanying cold air over much of western North America. The southern branch of the polar jet stream directs moist air from the ocean into the Pacific Northwest, producing a wet winter for that region. These associations aid forecasters in their ability to predict long-term weather and climate patterns.

over the north Pacific and, to the south of the low, the jet stream (from off the Pacific) steers wet weather and storms into California and the southern part of the United States. A weak polar jet stream forms over eastern Canada allowing warmer than normal weather to prevail over a large part of North America.

Figure 14.17b shows typical winter weather patterns with a La Niña. Notice that a persistent high-pressure area (called a *blocking high*) forms south of Alaska forcing the polar jet stream into Alaska, then southward into Canada and the western United States. The southern branch of the polar jet, which forms south of the high, directs moist air from the ocean into the Pacific northwest, producing a wet winter for that region. Meanwhile, winter months in the southern part of the United States tend to be warmer and drier than normal.

As we have seen, El Niño and the Southern Oscillation are part of a large-scale ocean-atmosphere interaction that can take several years to run its course. During this time, there are certain regions in the world where significant climatic responses to an ENSO event are likely. Using data from previous ENSO episodes, scientists at the National Oceanic and Atmospheric Administration's Climate Prediction Center have obtained a global picture of where climatic abnormalities are most likely (see Fig. 14.18). As we saw earlier, such ocean-atmosphere interactions (where warm or cold

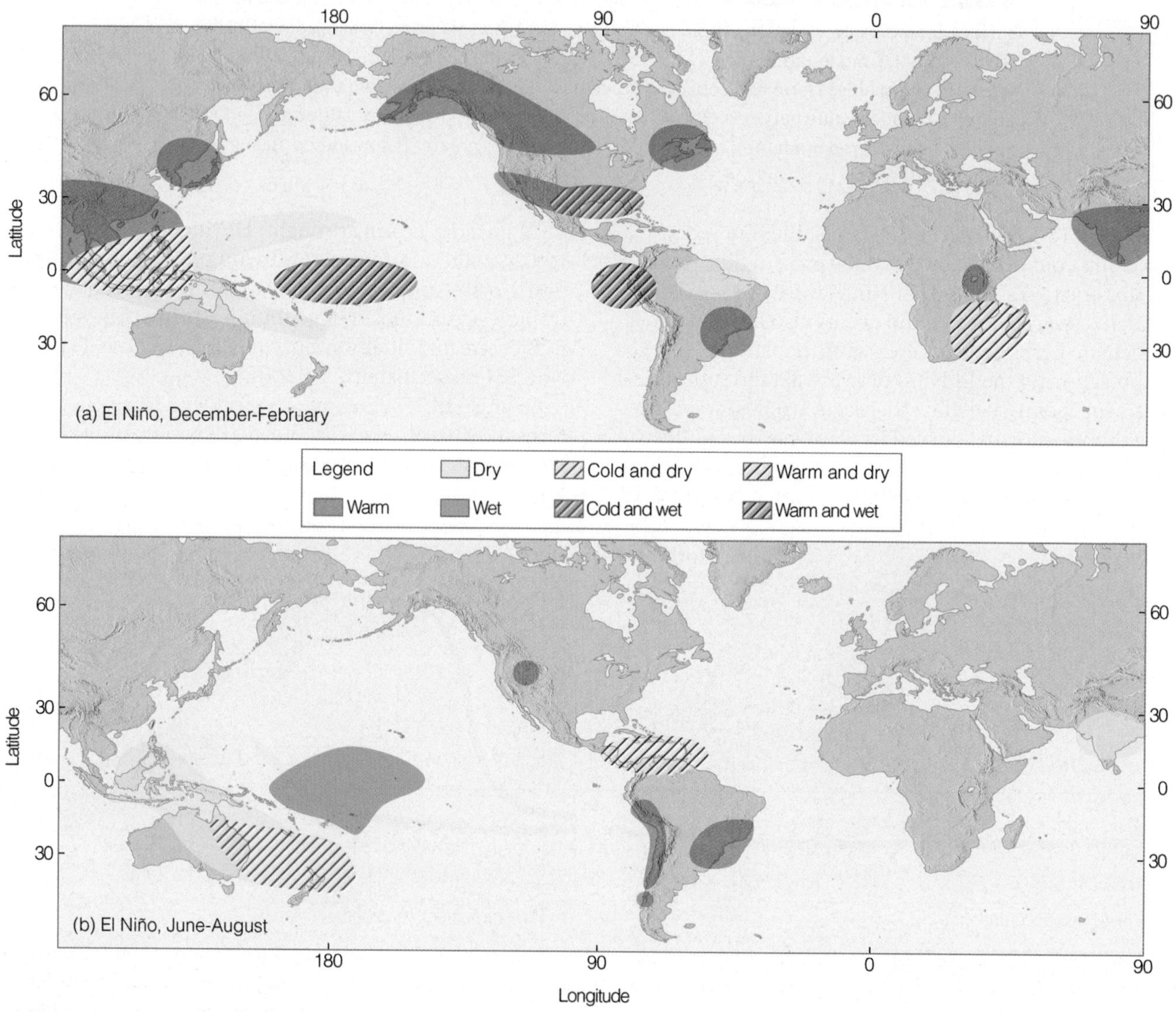

FIGURE 14.18 Regions of climatic abnormalities associated with El Niño–Southern Oscillation conditions (a) during December through February and (b) during June through August. A strong ENSO event may trigger a response in nearly all indicated areas, whereas a weak event will likely play a role in only some areas. (After NOAA Climate Prediction Center.)

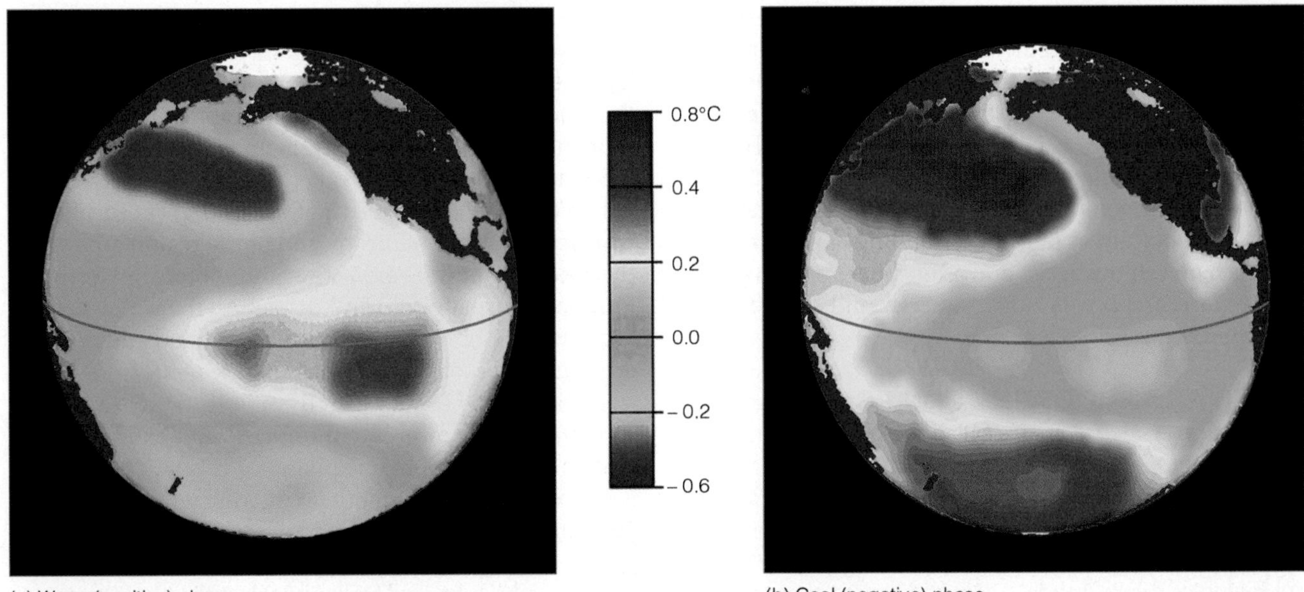

FIGURE 14.19 Typical winter sea surface temperature departure from normal in °C during the Pacific Decadal Oscillation's warm phase (a) and cool phase (b). (*Source*: JISAO, University of Washington, obtained via http://www.tao.atmos.washington.edu/pdo. Used with permission of N. Mantua.)

surface ocean temperatures can influence weather and climate patterns in a distant part of the world) are called *teleconnections.*

Based on teleconnections, weather and climate seasonal forecasts have shown promise. For example, as the tropical equatorial Pacific became much warmer than normal during the spring and early summer of 1997, forecasters predicted a wet rainfall season over central and southern California. Although the heavy rain did not begin until late November, the weather during the winter of 1997–1998 was wet and wild. Storm after storm pounded the region, causing heavy rain, mud slides, road closures, and millions of dollars in damage.

Up to this point, we have looked at El Niño, La Niña, and the Southern Oscillation and how the reversal of surface ocean temperatures and atmospheric pressure combine to influence regional and global weather and climate patterns. There are other interactions between ocean surface temperatures and changes in atmospheric pressure and winds that can have an effect on the forecasting of large-scale weather and climate patterns. Some of these are described in the following sections.

PACIFIC DECADAL OSCILLATION Over the Pacific Ocean, changes in surface water temperatures appear to influence winter weather along the west coast of North America. In the mid 1990s, scientists at the University of Washington, while researching connections between Alaskan salmon production and Pacific climate, identified a long-term Pacific Ocean temperature fluctuation, which they called the **Pacific Decadal Oscillation (PDO)** because the ocean surface temperature reverses every 20 to 30 years. The Pacific Decadal Oscillation is like ENSO in that it has a warm phase and a cool phase, but its temperature behavior is much different from that of El Niño.

During the warm (or positive) phase, unusually warm surface water exists along the west coast of North America, whereas over the central North Pacific, cooler than normal surface water prevails (see Fig. 14.19a). At the same time, the Aleutian low in the Gulf of Alaska strengthens, which causes more Pacific storms to move into Alaska and California. This situation causes winters, as a whole, to be warmer and drier over northwestern North America. Elsewhere, winters tend to be drier over the Great Lakes, and cooler and wetter in the southern United States. Meanwhile, during this warm phase, salmon populations increase in Alaska and diminish along the Pacific Northwest coast. The latest warm phase began in 1977 and ended in the late 1990s.

The present cool (or negative) phase finds cooler-than-average surface water along the west coast of North America and an area of warmer-than-normal surface water extending from Japan into the central North Pacific (see Fig. 14.19b). Winters in the cool phase tend to be cooler and wetter than average over northwestern North America, wetter over the Great Lakes, and warmer and drier in the southern United States. Salmon fishing diminishes in Alaska and picks up along the Pacific Northwest Coast.

The climate patterns described so far only represent average conditions, as individual years within either phase may vary considerably. In fact, the Pacific Ocean temperature pattern in a particular phase may even reverse for a few years, as it did from 1958 to 1960. These small reversals can make it difficult to decipher exactly when the ocean temperature changes from one phase to another. Hopefully, as our understanding of the interactions between the ocean and atmosphere improves, climate forecasts across North America and elsewhere will improve as well.

NORTH ATLANTIC OSCILLATION Over the Atlantic there is a reversal of pressure called the **North Atlantic Oscillation (NAO)** that has an effect on the weather in Europe and along the east coast of North America. For example, in winter if the atmospheric pressure in the vicinity of the Icelandic low drops, and the pressure in the region of the Bermuda-Azores high rises, there is a corresponding large difference in atmospheric pressure between these two regions that strengthens the westerlies.* The strong westerlies in turn direct strong storms on a more northerly track into northern Europe, where winters tend to be wet and mild, and river flooding in spring is more likely. During this *positive phase* of the NAO, winters in the eastern United States tend to be wet and relatively mild, while northern Canada and Greenland are usually cold and dry (see Fig. 14.20a).

* Information on the average positions of high- and low-pressure areas throughout the world is given in Fig. 8.3a and Fig. 8.3b, on p. 212.

The *negative phase* of the NAO occurs when the atmospheric pressure in the vicinity of the Icelandic low rises, while the pressure drops in the region of the Bermuda high (see Fig. 14.20b). This pressure change results in a reduced pressure gradient and weaker westerlies that steer fewer and weaker winter storms across the Atlantic in a more westerly path. These storms bring wet weather to southern Europe and to the region around the Mediterranean Sea. Meanwhile, winters in Northern Europe are usually cold and dry, as are the winters along the east coast of North America. Greenland and northern Canada usually experience mild winters.

Although the NAO varies from year to year (and sometimes from month to month), it may exhibit a tendency to remain in one phase for several years. It is interesting to note that the NAO during the past 30 years or so has been trending toward a more positive phase, although in recent decades it has become more variable, which has led to drastic changes in weather patterns over Europe during this period.

ARCTIC OSCILLATION Closely related to the North Atlantic Oscillation is the **Arctic Oscillation (AO)**, where changes in atmospheric pressure between the Arctic and regions to the south cause changes in upper-level westerly winds. During the *positive warm phase* of the *AO* (see Fig. 14.21a) strong pressure differences produce strong westerly winds aloft that prevent cold arctic air from invading the United States, and so

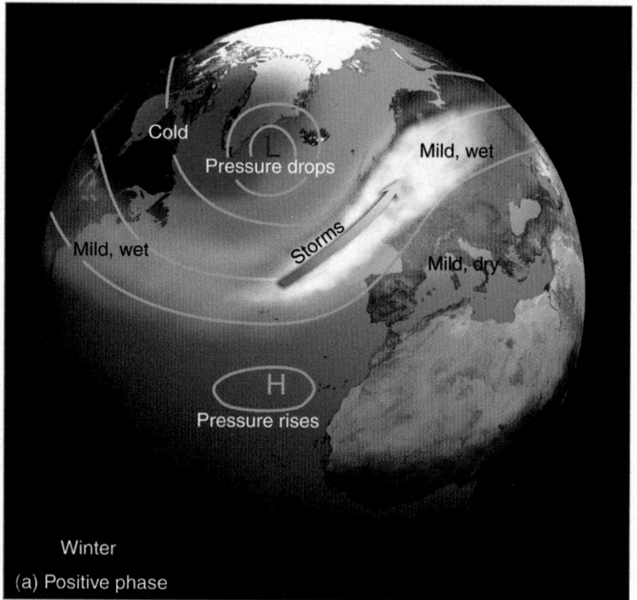

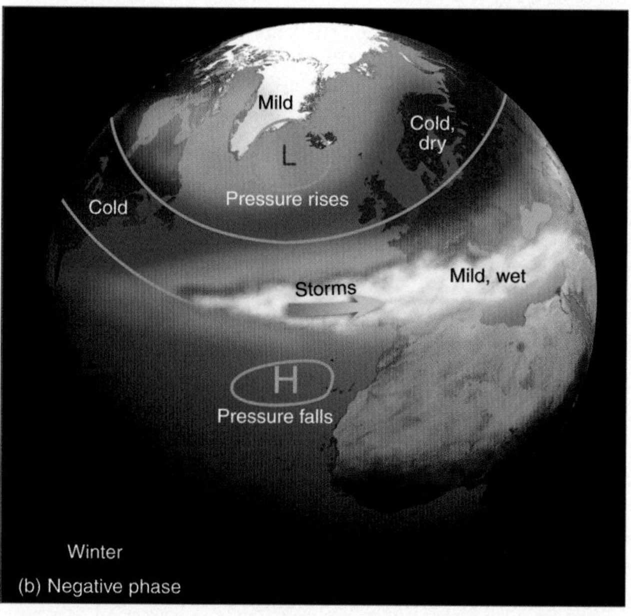

FIGURE 14.20 Change in surface atmospheric pressure and typical winter weather patterns associated with the North Atlantic Oscillation's (a) positive phase and (b) negative phase.

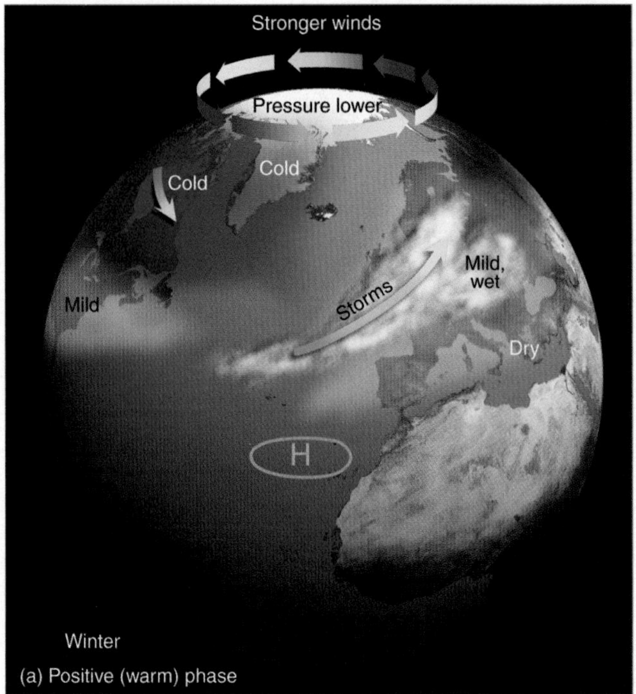

(a) Positive (warm) phase

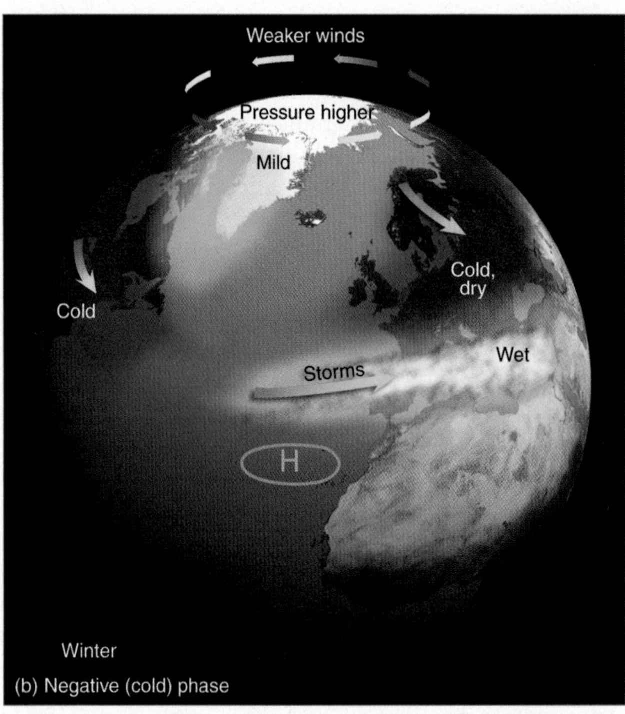

(b) Negative (cold) phase

FIGURE 14.21 Change in surface atmospheric pressure in polar regions and typical winter weather patterns associated with the Arctic Oscillation's (a) positive (warm) phase and (b) negative (cold) phase.

winters in this region tend to be warmer than normal. With cold arctic air in place to the north, winters over Newfoundland and Greenland tend to be very cold. Meanwhile, strong winds over the Atlantic direct storms into northern Europe, bringing with them wet, mild weather.

During the *negative cold phase* of the *AO* (Fig. 14.21b), small pressure differences between the arctic and regions to the south produce weaker westerly winds aloft. Cold arctic air is now able to penetrate farther south, producing colder than normal winters over much of the United States. Cold air also invades northern Europe and Asia, while Newfoundland and Greenland experience warmer than normal winters.

So when Greenland has mild winters, northern Europe has cold winters and vice versa. This seesaw in winter temperatures between Greenland and northern Europe has been known for many years. What was not known until recently is that, during the warm Arctic Oscillation phase, relatively warm, salty water from the Atlantic is able to move into the Arctic Ocean, where it melts sea ice, causing it to thin dramatically. During the cold phase, surface winds tend to keep warmer Atlantic water to the south, which promotes thicker sea ice. Although the Arctic Oscillation switches from one phase to another on an irregular basis, one phase may persist for several years in a row, bringing with it a succession of either cold or mild winters.

Summary

In this chapter, we looked at weather forecasting with emphasis placed upon the forecasting of extreme and hazardous weather. We saw how computer progs can be used to make weather forecasts. We learned that different computer progs are based upon different atmospheric models that describe the state of the atmosphere and how it will change with time. The atmosphere's chaotic behavior, along with flaws in the models and tiny errors (uncertainties) in the data, generally amplify as the computer tries to project weather further and further into the future.

At present, computer progs that predict the weather over a vast region are better at forecasting the position of mid-latitude highs and lows and their development than at forecasting local showers and thunderstorms. Presently, using a variety of techniques it is possible to forecast which areas are most likely to experience severe weather, but forecasting exactly where a severe thunderstorm will form, and which ones will produce tornadoes, is impossible at this time. As new information from atmospheric research programs is fed into the latest generation of computers, it is hoped that the progs will be able to show more skill in predicting the weather up to 10 days into the future.

Forecasting tomorrow's weather entails a variety of techniques and methods. Persistence, surface maps, satellite imagery, and Doppler radar are all useful when making a very short-range prediction. For short- and medium-range forecasts, the current analysis, satellite data, pattern recognition, meteorologist intuition, and experience, along with statistical information and guidance from the many computer progs supplied by the National Weather Service, all go into making a prediction. For monthly and seasonal long-range forecasts, meteorologists incorporate changes in sea surface temperature in the Pacific and Atlantic Oceans. The linkages between sea-surface temperature change and weather and climate conditions in some other region of the world are called *teleconnections.*

The large-scale interaction between the atmosphere and the ocean during El Niño, La Niña, and the Southern Oscillation affects global wind patterns, which in turn provide too much rain in some areas and not enough in others. Over the north-central Pacific, a reversal of surface water temperature—called the Pacific Decadal Oscillation—appears to influence winter weather along the west coast of North America. Over the Atlantic Ocean, a reversal of pressure, called the North Atlantic Oscillation, appears to influence weather and climate patterns in various regions of the world. And surface atmospheric pressure changes in the Arctic seem to influence weather and climate patterns over North America.

Key Terms

The following terms are listed (with page numbers) in the order they appear in the text. Define each. Doing so will aid you in reviewing the material covered in this chapter.

weather watch, 400
weather warning, 400
AWIPS, 402
meteogram, 402
soundings, 403
analysis, 404
numerical weather prediction, 404
atmospheric models, 404
prognostic chart (prog), 404
chaos, 407
ensemble forecasting, 407
persistence forecast, 408
steady-state (trend) forecast, 408
analogue forecasting method, 408
pattern recognition, 408
statistical forecast, 408
probability forecast, 408
weather type forecast, 409
climatological forecast, 410
very short-range forecast, 410
short-range forecast, 410
medium-range forecast, 410
long-range forecast, 410
teleconnections, 418
El Niño, 418
Southern Oscillation, 419
ENSO, 419
La Niña, 420
Pacific Decadal Oscillation (PDO), 423
North Atlantic Oscillation (NAO), 424
Arctic Oscillation (AO), 424

Review Questions

1. What is the function of the National Center for Environmental Prediction?
2. How does a *weather watch* differ from a *weather warning?*
3. List some of the tools a weather forecaster might use when making a short-range forecast.
4. In what ways has the computer assisted the meteorologist in making weather forecasts?
5. How does a prog differ from an analysis?
6. How are computer-generated weather forecasts prepared?
7. What are some of the problems associated with computer-model forecasts?
8. Make a persistence forecast for your area for this same time tomorrow. Did you use any skill in making this prediction? Explain.

9. Describe four methods of forecasting the weather and give an example for each one.
10. How does pattern recognition aid a forecaster in making a prediction?
11. How can ensemble forecasts improve medium-range weather forecasts?
12. Do all accurate forecasts show skill? Explain.
13. Do monthly and seasonal forecasts make specific predictions of rain or snow? Explain.
14. How can a sounding help in the prediction of severe weather and especially severe thunderstorms?
15. If the value of CAPE derived from a sounding over Atlanta, Georgia, is 3600 J/Kg, would you expect a forecast for Atlanta to be "fair and warmer?" Explain.
16. What atmospheric conditions cause the forecaster to issue a severe thunderstorm watch? What would be necessary for the same forecaster to issue a tornado warning?
17. What atmospheric conditions would the forecaster look for when predicting:
(a) strong Santa Ana winds in Southern California
(b) an air pollution alert in a mountain valley
(c) record-high temperatures in Arizona
(d) fog advisory in California's Central Valley
(e) blizzard conditions over Minnesota
(f) high-wind warning over Nebraska
(g) hard freeze for northern Texas
(h) thunderstorms for Tennessee
(i) drought over Florida
(j) ice storm in valleys of western North Carolina
(k) record-low temperatures in North Dakota
18. How does the thickness chart aid a forecaster in predicting whether falling precipitation will be in the form of rain or snow?
19. Explain how teleconnections can be used in making a long-range weather forecast.
20. What are the conditions over the tropical eastern and central Pacific Ocean during the phenomenon known as El Niño? La Niña?
21. What type of weather (cold/warm, wet/dry) would you expect over North America during a strong El Niño? During a strong La Niña?
22. Describe the ocean surface temperatures associated with the Pacific Decadal Oscillation. What climate patterns (cool/warm, wet/dry) tend to exist during the warm phase and the cool phase?
23. How does the positive phase of the North Atlantic Oscillation differ from the negative phase?
24. During the negative cold phase of the Arctic Oscillation, when Greenland is experiencing mild winters, what type of winters (cold or mild) is northern Europe usually experiencing?

Online Learning

STUDENT COMPANION WEBSITE: Visit this book's companion website at: www.cengage.com/ahrens/extreme1e and choose Chapter 14 for many study aids and ideas for further reading and research. These include flashcards, practice quizzing, and web links.

METEOROLOGY RESOURCE CENTER: For students with access, log on at www.cengage.com/login for more assets, including animations, videos, and more. If your textbook did not come with access, visit www.CengageBrain.com to purchase.

The Earth's Changing Climate

15

CONTENTS

RECONSTRUCTING PAST CLIMATES

CLIMATE THROUGHOUT THE AGES

CLIMATE CHANGE CAUSED BY NATURAL EVENTS

CLIMATE CHANGE CAUSED BY HUMAN (ANTHROPOGENIC) ACTIVITIES

GLOBAL WARMING

SUMMARY

KEY TERMS

REVIEW QUESTIONS

As our planet warms, glaciers over the Northern Hemisphere, such as the Meares Tidewater glacier in Alaska shown here, are melting at record rates, causing sea level to rise worldwide.

The climate is always changing. Evidence shows that climate has changed in the past, and nothing suggests that it will not continue to change. As the urban environment grows, its climate differs from that of the region around it. Sometimes the difference is striking, as when city nights are warmer than the nights of the outlying rural areas. Other times, the difference is subtle, as when a layer of smoke and haze covers the city. Climate change, in the form of a persistent drought or a delay in the annual monsoon rains, can adversely affect the lives of millions. Even small changes can have an adverse effect when averaged over many years, as when grasslands once used for grazing gradually become uninhabited deserts.

Climate change is taking place right now as the world is warming at an alarming rate. Consequently, in the Northern Hemisphere, polar sea ice in winter does not extend as far south as it once did, Greenland ice is melting rapidly, and sea level is rising worldwide. The main cause of this **global warming** appears to be human (anthropogenic) activities. We will, therefore, first look at the evidence for climate change in the past; then, we will investigate the causes of climate change due to both natural variations and human intervention. Near the end of the chapter we will examine how climate change might impact extreme weather and climate events on earth.

Reconstructing Past Climates

Not only is the earth's climate always changing, but a mere 18,000 years ago the earth was in the grip of a cold spell, with *alpine glaciers* extending their icy fingers down river valleys and huge ice sheets *(continental glaciers)* covering vast areas of North America and Europe (see ❱ Fig. 15.1). The ice at that time measured over one mile thick and extended as far south as New York and the Ohio River Valley. Perhaps the glaciers advanced 10 times during the last 2.5 million years, only to retreat. In the warmer periods, between glacier advances, average global temperatures were slightly higher than at present. Hence, some scientists feel that we are still in an ice age, but in the comparatively warmer part of it.

Presently, glaciers cover less than 10 percent of the earth's land surface. Most of this ice is in the Greenland and Antarctic ice sheets, and its accumulation over time has allowed scientists to measure past climatic changes. If global temperatures were to rise enough so that all of this ice melted, the level of the ocean would rise about 213 ft (65 m) (see ❱ Fig. 15.2). Imagine the catastrophic results: Many major cities (such as New York, Tokyo, and London) would be inundated. Even a rise in global temperature of several degrees Celsius might be enough to raise sea level by 3 ft or more, flooding coastal lowlands.

The study of the geological evidence left behind by advancing and retreating glaciers is one factor suggesting that global climate has undergone slow but

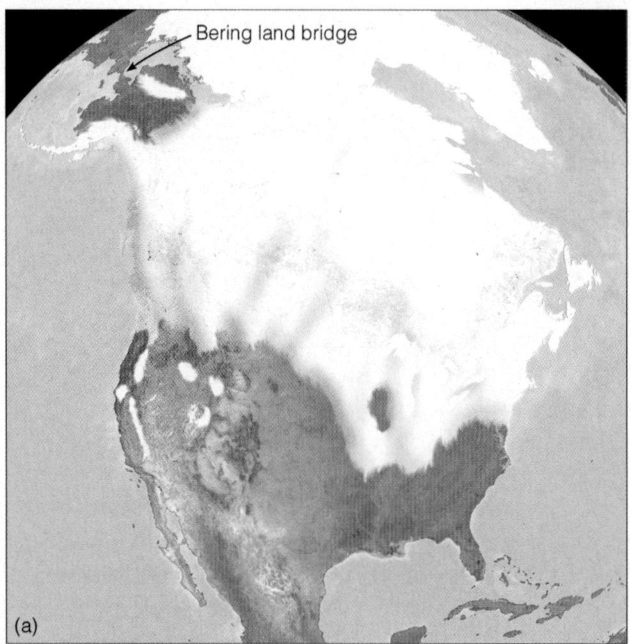

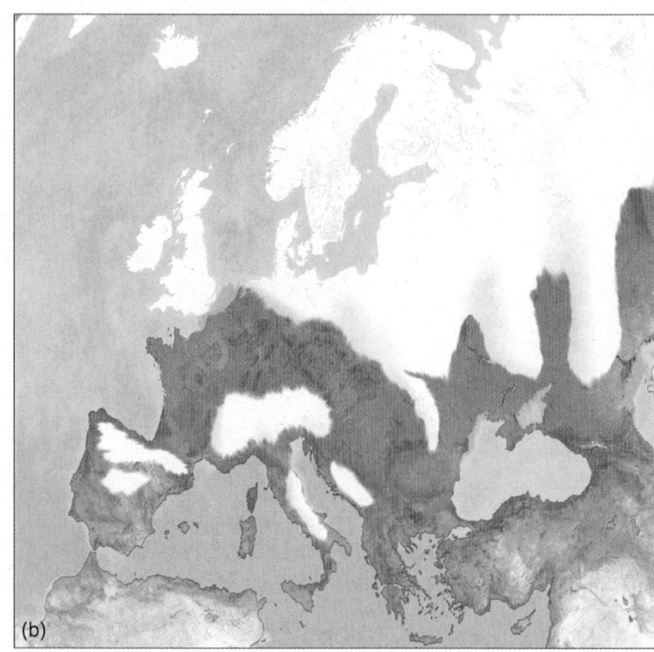

❱ FIGURE 15.1 Extent of glaciation about 18,000 years ago over (a) North America and over (b) western Europe.

© C. Donald Ahrens

FIGURE 15.2

If all the ice locked up in glaciers and ice sheets were to melt, estimates are that this coastal area of south Florida would be under 213 ft (65 m) of water. Even a relatively small three-foot rise in sea level would threaten half of the world's population with rising seas. In fact, latest research suggests sea level will rise 3 feet or more by the end of this century due to the rapid melting of ice in Greenland and Antarctica.

continuous changes. To reconstruct past climates, scientists must examine and then carefully piece together all the available evidence. Unfortunately, the evidence only gives a general understanding of what past climates were like. For example, fossil pollen of a tundra plant collected in a layer of sediment in New England and dated to be 12,000 years old suggests that the climate of that region was much colder than it is today.

Other evidence of global climatic change comes from core samples taken from ocean floor sediments and ice from Greenland and Antarctica. A multiuniversity research project known as CLIMAP (Climate: *l*ong-range *i*nvestigation *m*apping *a*nd *p*rediction) studied the past million years of global climate. Thousands of feet of ocean sediment obtained with a hollow-centered drill were analyzed. This sediment contained the remains of calcium carbonate shells of organisms that once lived near the surface. Because certain organisms can only live within a narrow range of temperature, the distribution and type of organisms within the sediment indicate the temperature of the surface water.

In addition, the oxygen-isotope* ratio of these shells provided information about the sequence of glacier advances. For example, most of the oxygen in sea water is composed of 8 protons and 8 neutrons in its nucleus, giving it an atomic weight of 16. However, about one out of every thousand oxygen atoms contains an extra 2 neutrons, giving it an atomic weight of 18. When ocean water evaporates, the heavy oxygen 18 tends to be left behind. Consequently, during periods of glacier advance, the oceans, which contain less water, have a higher concentration of oxygen 18. Since the shells of marine organisms are constructed from the oxygen atoms existing in ocean water, determining the ratio of oxygen 18 to oxygen 16 within these shells yields information about how the climate may have varied in the past. A higher ratio of oxygen 18 to oxygen 16 in the sediment record suggests a colder climate, whereas a lower ratio suggests a warmer climate. Using data such as these, the CLIMAP project was able to reconstruct the earth's surface ocean temperature for various times during the past (see Fig. 15.3).

Vertical ice cores extracted from ice sheets in Antarctica and Greenland provide additional information on past temperature patterns. Glaciers form over land where temperatures are sufficiently low so that, during the course of a year, more snow falls than will melt. Successive snow accumulations over many years compact the snow, which slowly re-crystallizes into ice. Since ice is composed of hydrogen and oxygen, examining the oxygen-isotope ratio in ancient cores provides a past record of temperature trends. Generally, the colder the air when the snow fell, the richer the concentration of oxygen 16 in the core. Moreover, bubbles of ancient air trapped in the ice can be analyzed to determine the past composition of the atmosphere (see Fig. 15.12, p. 441).

Ice cores also record the causes of climate changes. One such cause is deduced from layers of sulfuric acid in the ice. The sulfuric acid originally came from large volcanic explosions that injected huge quantities of sulfur into the stratosphere. The resulting sulfate aerosols eventually fell to the earth in polar regions as acid snow, which was preserved in the ice sheets. The Greenland ice cores also provide a continuous record of sulfur from

*Isotopes are atoms whose nuclei have the same number of protons but different numbers of neutrons.

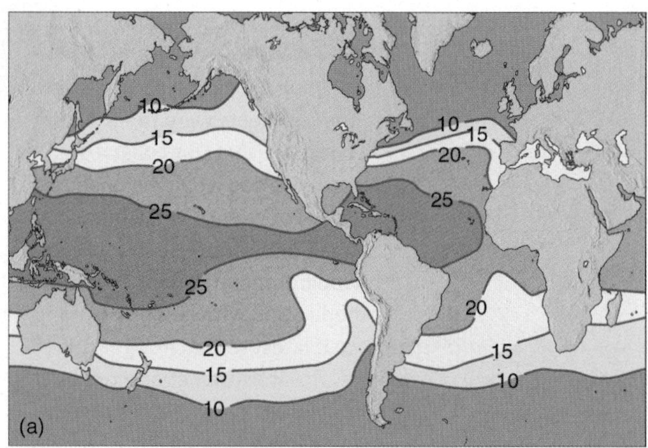

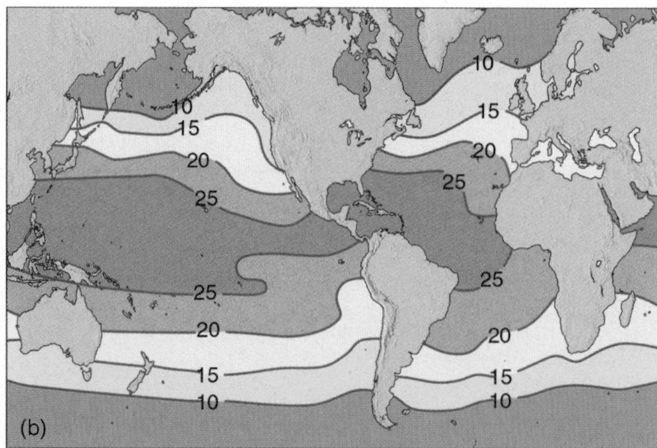

FIGURE 15.3 (a) Sea surface isotherms (°C) during August 18,000 years ago and (b) during August today. Apparently, during the Ice Age (diagram a) the Gulf Stream shifted to a more easterly direction, depriving northern Europe of its warmth and causing a rapid north-to-south ocean surface temperature gradient.

human sources. Moreover, ice cores at both poles are being analyzed for many chemicals that provide records of biological and physical changes in the climate system. Various types of dust collected in the cores indicate whether the climate was arid or wet.

Still other evidence of climatic change comes from the study of annual growth rings of trees, called **dendrochronology**. As a tree grows, it produces a layer of wood cells under its bark. Each year's growth appears as a ring. The changes in thickness of the rings indicate climatic changes that may have taken place from one year to the next. The density of late growth tree rings is an even better indication of changes in climate. The presence of frost rings during particularly cold periods and the chemistry of the wood itself provide additional information about a changing climate. Tree rings are only useful in regions that experience an annual cycle and in trees that are stressed by temperature or moisture during their growing season. The growth of tree rings has been correlated with precipitation and temperature patterns for hundreds of years into the past in various regions of the world.

Other data have been used to reconstruct past climates, such as:

1. records of natural lake-bottom sediment and soil deposits
2. the study of pollen in deep ice caves, soil deposits, and sea sediments
3. certain geologic evidence (ancient coal beds, sand dunes, and fossils), and the change in the water level of closed basin lakes
4. documents concerning droughts, floods, crop yields, rain, snow, and dates of lakes freezing
5. the study of oxygen-isotope ratios of corals
6. dating calcium carbonate layers of stalactites in caves
7. borehole temperature profiles, which can be inverted to give records of past temperature change at the surface
8. deuterium (heavy hydrogen) ratios in ice cores, which indicate temperature changes

Even with all of this knowledge, our picture of past climates is still incomplete. With this shortcoming in mind, we will examine what the information gained about past climates does reveal.

Climate Throughout the Ages

Throughout much of the earth's history, the global climate was probably much warmer than it is today. During most of this time, the polar regions were free of ice. These comparatively warm conditions, however, were interrupted by several periods of glaciation. Geologic evidence suggests that one glacial period occurred about 700 million years ago (m.y.a.) and another about 300 m.y.a. The most recent one—the *Pleistocene epoch* or, simply, the **Ice Age**—began about 2.5 m.y.a. Let's summarize the climatic conditions that led up to the Pleistocene.

About 65 m.y.a., the earth was warmer than it is now; polar ice caps did not exist. Beginning about 55 m.y.a., the earth entered a long cooling trend. After millions of years, polar ice appeared. As average temperatures continued to lower, the ice grew thicker, and by about 10 m.y.a. a deep blanket of ice covered the Antarctic. Meanwhile, snow and ice began to accumulate in high mountain valleys of the Northern Hemisphere, and alpine glaciers soon appeared.

About 2.5 m.y.a., continental glaciers appeared in the Northern Hemisphere, marking the beginning of the Pleistocene epoch. The Pleistocene, however, was not a period of continuous glaciation but a time when glaciers alternately advanced and retreated (melted back) over large portions of North America and Europe. Between the glacial advances were warmer periods called **interglacial periods**, which lasted for 10,000 years or more.

The most recent North American glaciers reached their maximum thickness and extent about 18,000–22,000 years ago (y.a.). At that time, average temperatures in Greenland were about 18°F (10°C) lower than at present and tropical average temperatures were about 7°F (4°C) lower than they are today. Because a great deal of water was in the form of ice over land, sea level was perhaps 395 ft (120 m) lower than it is now. The lower sea level exposed vast areas of land, such as the *Bering land bridge* (a strip of land that connected Siberia to Alaska as shown in Fig. 15.1a), which allowed human and animal migration from Asia to North America.

The ice began to retreat about 14,000 y.a. as surface temperatures slowly rose, producing a warm spell (see ❯ Fig. 15.4). Then, about 12,700 y.a., the average temperature suddenly dropped and northeastern North America and northern Europe reverted back to glacial conditions. About 1000 years later, the cold spell (known as the **Younger Dryas***) ended abruptly and temperatures rose rapidly in many areas. Beginning about 8000 y.a. the mean temperature dropped by as much as 3.6°F (2°C) over central Europe. During this cold period, which was not experienced worldwide, the European alpine timberline fell about 600 feet. The cold period ended, temperatures began to rise, and by about 6000 y.a. the continental ice sheets over North America were gone. This warm spell during the current interglacial period, or *Holocene epoch,* is sometimes called the **mid-Holocene maximum**, and because this warm period favored the development of plants, it is also known as the *climatic optimum.* About 5000 y.a, a cooling trend set in, during which extensive alpine glaciers returned, but not continental glaciers.

It is interesting to note that ice core data from Greenland reveal that rapid shifts in climate (from ice age conditions to a much warmer state) took place in as little as three years over central Greenland around the end of the Younger Dryas. The data also reveal that similar rapid shifts in climate occurred several times toward the end of the Ice Age. What could cause such rapid changes in temperature? One possible explanation is given in the Focus section on p. 434.

TEMPERATURE TRENDS DURING THE PAST 1000 YEARS ❯ Figure 15.5 shows how the average surface air temperature changed in the Northern Hemisphere during the last 1000 years. The data needed to reconstruct the temperature profile in Fig. 15.5 comes from a

*This exceptionally cold spell is named after the *Dryas*, an arctic flower.

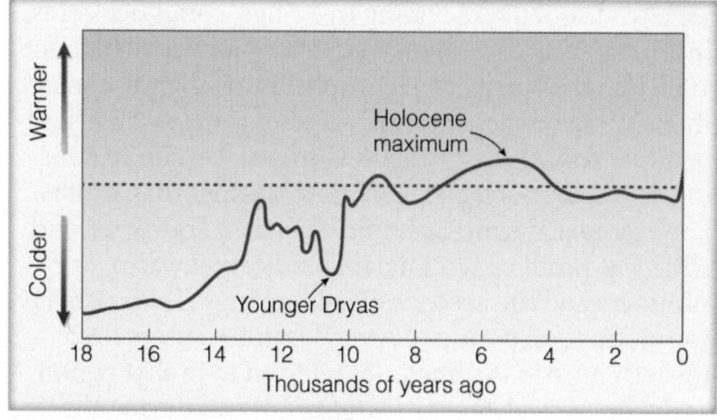

❯ **FIGURE 15.4** Relative air temperature variations (warmer and cooler periods) during the past 18,000 years. These data, which represent temperature records compiled from a variety of sources, only give an approximation of temperature changes. Some regions of the world experienced a cooling and other regions a warming that either preceded or lagged behind the temperature variations shown in the diagram.

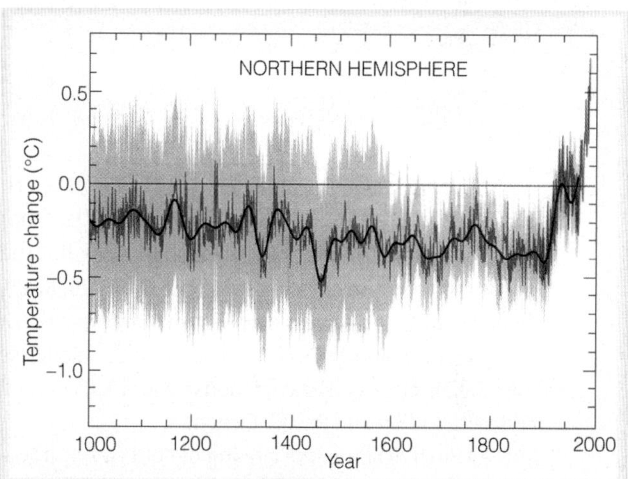

❯ **FIGURE 15.5** The average temperature variations over the Northern Hemisphere for the last 1000 years relative to the 1961 to 1990 average (zero line). Yearly temperature data from tree rings, corals, ice cores, and historical records are shown in blue. Yearly temperature data from thermometers are in red. The black line represents a smoothing of the data. (The gray shading represents a statistical 95 percent confidence range in the annual temperature data, after Mann, et al., 1999.) (*Source:* From Climate Change 2001: The Scientific Basis, 2001, by J.T. Houghton, et al. Copyright © 2001 Cambridge University Press. Reprinted with permission of the Intergovernmental Panel on Climate Change.)

variety of sources, including tree rings, corals, ice cores, historical records, and thermometers. Notice that about 1000 y.a., the Northern Hemisphere was slightly cooler than average (where average represents the average temperature from 1961 to 1990). However, certain regions in the Northern Hemisphere were warmer than others. For example, during this time vineyards flourished and wine was produced in England, indicating warm, dry summers and the absence of cold springs. This relatively warm, tranquil period of several hundred years over western Europe is sometimes referred to in that region as the *Medieval Climatic Optimum.* It was during the early part of the millennium that Vikings colonized Iceland and Greenland and traveled to North America.

Notice in Fig. 15.5 that the temperature curve shows a relatively warm period during the 11th to the 14th centuries—relatively warm, but still cooler than the 20th century. During this time, the relatively mild climate of Western Europe began to show large variations. For several hundred years the climate grew stormy. Both great floods and great droughts occurred. Extremely cold winters were followed by relatively warm ones. During the cold spells, the English vineyards and the Viking settlements suffered. Europe experienced several famines during the 1300s.

Again look at Fig. 15.5 and observe that the Northern Hemisphere experienced a slight cooling during the 15th to 19th centuries. This cooling was significant enough in certain areas to allow alpine glaciers to

FOCUS ON

EXTREME WEATHER

The Ocean's Influence on Rapid Climate Change

During the last glacial period, the climate around Greenland (and probably other areas of the world, such as northern Europe) underwent shifts, from ice-age temperatures to much warmer conditions in a matter of years. What could bring about such large fluctuations in temperature over such a short period of time? It now appears that a vast circulation of ocean water, known as the *conveyor belt,* plays a major role in the climate system.

Figure 1 illustrates the movement of the ocean conveyor belt, or *thermohaline circulation.** The conveyor-like circulation begins in the north Atlantic near Greenland and Iceland, where salty surface water is cooled through contact with cold Arctic air masses. The cold, dense water sinks and flows southward through the deep Atlantic Ocean, around Africa, and into the Indian and Pacific Oceans.

In the North Atlantic, the sinking of cold water draws warm water northward from lower latitudes. As this water flows northward, evaporation increases the water's salinity (dissolved salt content) and density. When this salty, dense water reaches the far regions of the North Atlantic, it gradually sinks to great depths. This warm part of the conveyor delivers an incredible amount of tropical heat to the northern Atlantic. During the winter, this heat is transferred to the overlying atmosphere, and evaporation moistens the air. Strong westerly winds then carry this warmth and moisture into northern and western Europe, where it causes winters to be much warmer and wetter than one would normally expect for this latitude.

Ocean sediment records along with ice-core records from Greenland suggest that the giant conveyor belt has switched on and off during the last glacial period. Such events have apparently coincided with rapid changes in climate. For example, when the conveyor belt is strong, winters in northern Europe tend to be wet and relatively mild. However, when the conveyor belt is weak or stops altogether, winters in northern Europe appear to turn much colder. This switching from a period of milder winters to one of severe cold shows up many times in the climate record. One such event—the Younger Dryas—illustrates how quickly climate can change and how western and northern Europe's climate can cool within a matter of decades, then quickly return back to milder conditions.

Apparently, one mechanism that can switch the conveyor belt off is a massive influx of freshwater. For example, about 11,000 years ago during the Younger Dryas event, freshwater from a huge glacial lake began to flow

*Thermohaline circulations are ocean circulations produced by differences in temperature and/or salinity. Changes in ocean water temperature or salinity create changes in water density.

increase in size and advance down river canyons. In many areas in Europe, winters were long and severe; summers, short and wet. The vineyards in England vanished, and farming became impossible in the more northern latitudes. Cut off from the rest of the world by an advancing ice pack, the Viking colony in Greenland perished. (Although climate change played a role in the demise of the Viking colony in northern Greenland, it was also their inability to adapt to the climate and learn hunting and farming techniques from the Eskimos that led to their downfall.)

There is no evidence that this cold spell existed worldwide. However, over Europe, this cold period has come to be known as the **Little Ice Age**. During these colder times, one particular year stands out: 1816. In Europe that year, bad weather contributed to a poor wheat crop, and famine spread across the land. In Northern America, unusual blasts of cold arctic air moved through Canada and the northeastern United States between May and September. The cold spells brought heavy snow in June and killing frosts in July and August. In the warmer days that followed each cold snap, farmers replanted, only to have another cold outbreak damage the planting. The year 1816 has come to be known as "the year without a summer" or "eighteen hundred and froze-to-death." The unusually cold summer was followed by a bitterly cold winter.*

*More information on this infamous year, 1816, is provided in Chapter 2 in the Focus section on p. 59.

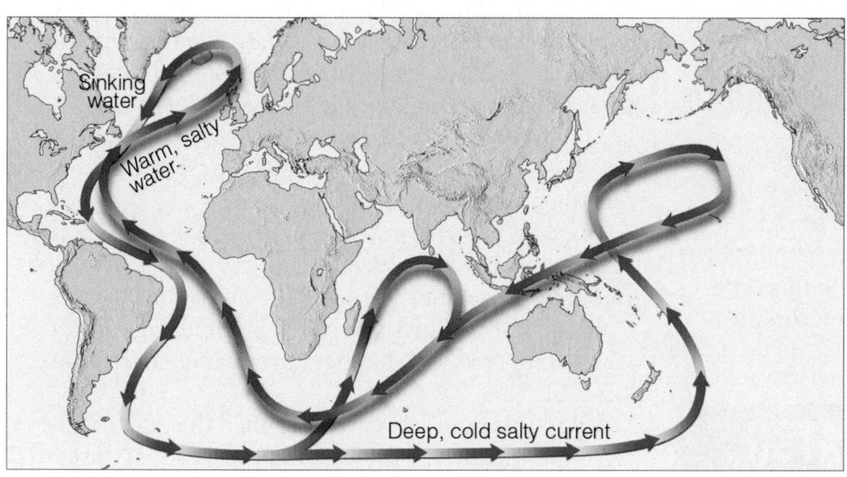

Figure 1 The ocean conveyor belt. In the North Atlantic, cold, salty water sinks, drawing warm water northward from lower latitudes. The warm water provides warmth and moisture for the air above, which is then swept into northern Europe by westerly winds that keep the climate of that region milder than one would normally expect. When the conveyor belt stops, winters apparently turn much colder over northern Europe.

down the St. Lawrence River and into the North Atlantic. This massive inflow of freshwater reduced the salinity (and, hence, density) of the surface water to the point that it stopped sinking. The conveyor shut down for about 1000 years during which time severe cold engulfed much of northern Europe. The conveyor belt started up again when freshwater began to drain down the Mississippi rather than into the North Atlantic. It was during this time that milder conditions returned to northern Europe.

Will increasing levels of CO_2 have an effect on the conveyor belt? Some climate models predict that as CO_2 levels increase, more precipitation will fall over the North Atlantic. This situation reduces the density of the sea water and slows down the conveyor belt. In fact, if CO_2 levels double (from their own current values), computer models predict that the conveyor belt will slow and that Europe will not warm as much as the rest of the world.

TEMPERATURE TREND DURING THE PAST 100-PLUS YEARS In the early 1900s, the average global surface temperature began to rise (see Fig. 15.6). Notice that, from about 1900 to 1945, the average temperature rose nearly 0.5°C, or 0.9°F. Following the warmer period, the earth began to cool slightly over the next 25 years or so. In the late 1960s and 1970s, the cooling trend ended over most of the Northern Hemisphere. In the mid-1970s, a warming trend set in that continued into the twenty-first century. In fact, over the Northern Hemisphere, the decade of the 1990s was the warmest of the 20th century, with 1998 and 2005 being the warmest years in over 1000 years. The exceptionally warm year of 1998 happened to coincide with a major El Niño warming of the tropical Pacific Ocean, whereas the warm year of 2005 did not. It appears that the increase in average temperature experienced over the Northern Hemisphere during the 20th century is likely to have been the largest increase in temperature of any century during the past 1000 years.

The average warming experienced over the globe, however, has not been uniform. The greatest warming has occurred in the arctic and over the mid-latitude continents in winter and spring, whereas a few areas have not warmed in recent decades, such as areas of the oceans in the Southern Hemisphere and parts of Antarctica. The United States has experienced less warming than the rest of the world. Moreover, most of the warming has occurred at night—a situation that has lengthened the frost-free seasons in many mid- and high-latitude regions, although, in recent decades, the warming has been equally distributed between day and night.

The changes in air temperature shown in Fig. 15.6 are derived from three main sources: air temperatures over land, air temperatures over ocean, and sea surface temperatures. There are, however, uncertainties in the temperature record. For example, during this time period recording stations have moved, and techniques for measuring temperature have varied. Also, marine observing stations are scarce. In addition, urbanization (especially in developed nations) tends to artificially raise average temperatures as cities grow (the urban heat island effect). When urban warming is taken into account, and improved sea-surface temperature information is incorporated into the data, the warming during the 20th century measures about 1°F (about 0.6°C). Over the past several decades this global warming trend has not only continued, but has increased to about 3.6°F (2.0°C) per century, with twelve of the warmest years on record occurring since 1995.

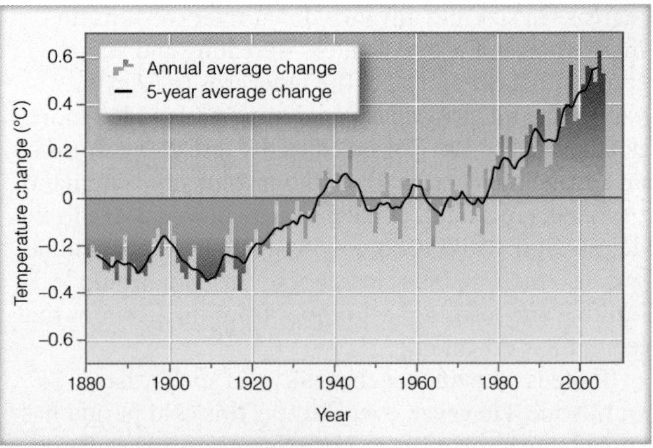

FIGURE 15.6 The red and blue bars represent the annual average temperature variations over the globe (land and sea) from 1880 through 2006. Temperature changes are compared to the average surface temperature from 1951 to 1980. The dark solid line shows the five-year average temperature change. (NASA)

A global increase in temperature of 1°F or 0.6°C may seem small, but global temperatures probably have not varied by more than 4°F during the past 10,000 years. Consequently, an increase of 1°F becomes significant when compared with temperature changes over thousands of years.

Up to this point we have examined the temperature record of the earth's surface and observed that the earth has been in a warming trend for more than 100 years. The main question regarding this global warming is whether the warming trend is due to natural variations in the climate system, or whether it is due to human activities. Or is it due to a combination of the two? As we will see later in this chapter, climate scientists believe that most of the recent warming is due to an enhanced greenhouse effect caused by increasing levels of greenhouse gases, such as CO_2. (The earth's atmospheric greenhouse effect is due mainly to the absorption and emission of infrared radiation by gases, such as water vapor, CO_2, methane, nitrous oxide, and chlorofluorocarbons. Refer back to Chapter 2, p. 45, for additional

EXTREME WEATHER WATCH

The prolonged warming period that affected earth from the 11th to the 14th century may have been responsible for the disappearance of the Anasazi, a group of Native Americans who inhabited what is now the southwestern United States. They abandoned their remarkable cliff dwellings, such as those at Mesa Verde (in southwestern Colorado) during the 13th century, and it is widely supposed that a prolonged drought was the major factor contributing to the demise of their culture.

information on this topic.) If human activities are at least partly responsible for this global warming, why has the earth undergone warming trends in the past, before humanity walked on the surface of the earth?

BRIEF REVIEW

Before going on to the next section, here is a brief review of some of the facts and concepts we covered so far:

- **The earth's climate is constantly undergoing change. Evidence suggests that throughout much of the earth's history, the earth's climate was much warmer than it is today.**
- **The most recent glacial period (or Ice Age) began about 2.5 million years ago. During this time, glacial advances were interrupted by warmer periods called *interglacial periods.* In North America, continental glaciers reached their maximum thickness and extent about 18,000 to 22,000 years ago and disappeared completely from North America by about 6000 years ago.**
- **The Younger Dryas event represents a time about 12,000 years ago when northeastern North America and northern Europe reverted back to glacier conditions.**
- **During the 20th century, the earth's surface temperature increased by about 0.6°C. This global warming has not only continued, but over the last several decades has increased to about 2°C per century (0.2°C/decade).**

Climate Change Caused by Natural Events

Why does the earth's climate change? There are three "external" causes of climate change. They are:

1. changes in incoming solar radiation
2. changes in the composition of the atmosphere
3. changes in the earth's surface

Natural phenomena can cause climate to change by all three mechanisms, whereas human activities can change climate by both the second and third mechanisms. In addition to these external causes, there are "internal" causes of climate change, such as changes in the circulation patterns of the ocean and atmosphere, which redistribute energy within the climate system.

Part of the complexity of the climate system is the intricate interrelationship of the elements involved. For example, if temperature changes, many other elements may be altered as well. The interactions among the atmosphere, the oceans, and the ice are extremely complex and the number of possible interactions among these systems is enormous. No climatic element within the system is isolated from the others, which is why the complete picture of the earth's changing climate is not totally understood. With this in mind, we will first investigate how feedback systems work; then we will consider some of the current theories as to why the earth's climate changes naturally.

CLIMATE CHANGE: FEEDBACK MECHANISMS In Chapter 2, we learned that the earth-atmosphere system is in a delicate balance between incoming and outgoing energy. If this balance is upset, even slightly, global climate can undergo a series of complicated changes.

Let's assume that the earth-atmosphere system has been disturbed to the point that the earth has entered a slow warming trend. Over the years the temperature slowly rises, and water from the oceans rapidly evaporates into the warmer air. The increased quantity of water vapor absorbs more of the earth's infrared energy, thus strengthening the atmospheric greenhouse effect.

This strengthening of the greenhouse effect raises the air temperature even more, which, in turn, allows more water vapor to evaporate into the atmosphere. The greenhouse effect becomes even stronger, and the air temperature rises even more. This situation is known as the **water vapor–greenhouse feedback**. It represents a **positive feedback mechanism** because the initial increase in temperature is reinforced by the other processes. If this feedback were left unchecked, the earth's temperature would increase until the oceans evaporated away. Such a chain reaction is called a *runaway greenhouse effect.* This water vapor–greenhouse positive feedback mechanism works in the case of a cooling planet also. For instance, if the earth's climate system was cooling, this positive feedback mechanism would amplify the cooling.

Another positive feedback mechanism is the **snow-albedo feedback**, in which an increase in global surface air temperature might cause snow and ice to melt in polar latitudes. This melting would reduce the albedo (reflectivity) of the surface, allowing more solar energy to reach the surface, which would further raise the temperature (see ❱ Fig. 15.7).

All feedback mechanisms work simultaneously and in both directions. Consequently, the snow-albedo feedback produces a positive feedback on a cooling planet as well. Suppose, for example, the earth were in a slow cooling trend. Lower temperatures might allow for a greater snow cover in middle and high latitudes, which would increase the albedo of the surface so that much of the incoming sunlight would be reflected back to space.

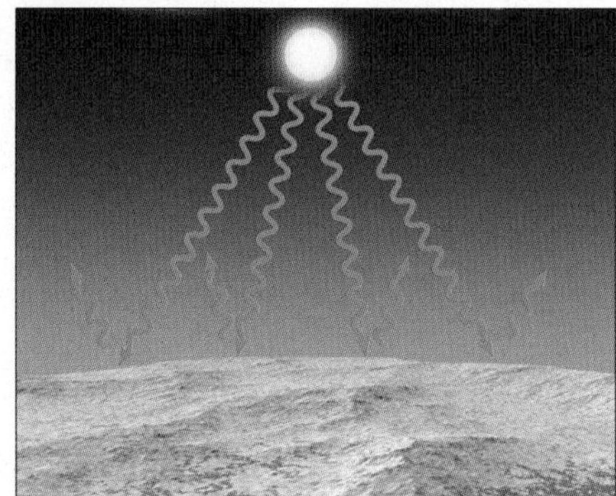

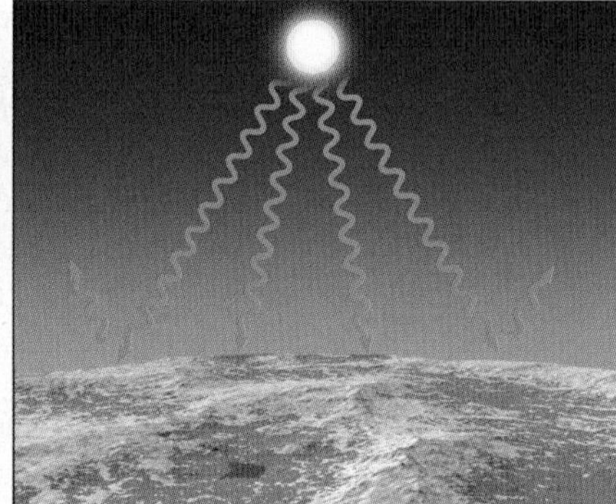

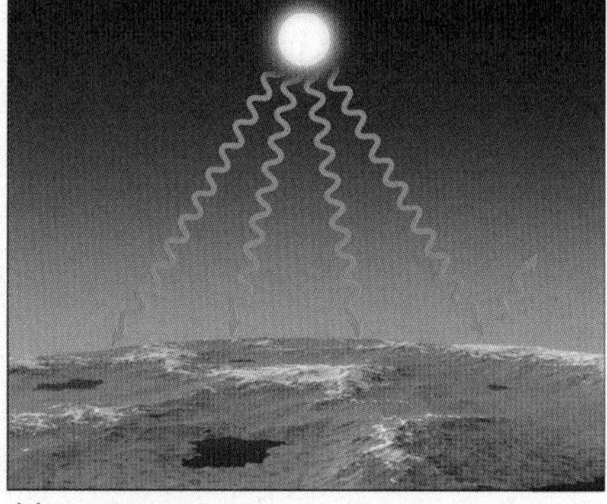

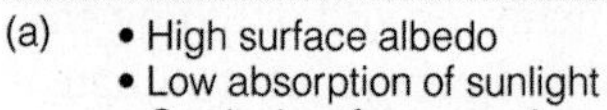
FIGURE 15.7 **On a warming planet, the snow-albedo positive feedback would enhance the warming. (a) In polar regions snow reflects much of the sun's energy back to space. (b) If the air temperature were to gradually increase, some of the snow would melt, less sunlight would be reflected, and more sunlight would reach the ground, warming it more quickly. (c) The warm surface would enhance the snow melt which, in turn, would accelerate the rise in temperature.**

Lower temperatures might further increase the snow cover, causing the air temperature to lower even more. If left unchecked, the snow-albedo positive feedback would produce a *runaway ice age,* which is highly unlikely on earth because other feedback mechanisms in the atmospheric system would be working to moderate the magnitude of the cooling.

To counteract the positive feedback mechanisms there are **negative feedback mechanisms**—those that tend to weaken the interactions among the variables rather than reinforce them. For example, a warming planet emits more infrared radiation. If the earth climate system were in a runaway greenhouse effect, the increase in radiant energy from the surface would greatly slow the rise in temperature and help to stabilize the climate. The increase in radiant energy from the surface as the planet warms is the strongest negative feedback in the climate system, and greatly lowers the possibility of a runaway greenhouse effect. Consequently, there is no evidence that a runaway greenhouse effect ever occurred on earth, and it is not very likely that it will occur in the future.

In summary, the earth-atmosphere system has a number of checks and balances called *feedback mechanisms* that help it counteract tendencies of climate change. Although we do not worry about a runaway greenhouse effect or an ice-covered earth anytime in the future, there is concern that large positive feedback mechanisms may be working in the climate system to produce accelerated melting of ice in polar regions, especially in Greenland.

CLIMATE CHANGE: PLATE TECTONICS AND MOUNTAIN BUILDING Earlier, we saw that one of the external causes of climate change is a change in the surface of the earth. During the geologic past, the earth's surface has undergone extensive modifications. One involves the slow shifting of the continents and the ocean floors. This motion is explained in the widely accepted **theory of plate tectonics**. According to this theory, the earth's outer shell is composed of huge plates that fit together like pieces of a jigsaw puzzle. The plates, which slide over a partially molten zone below them, move in relation to one another. Continents are embedded in the plates and move along like luggage riding piggyback on a conveyor belt. The rate of motion is extremely slow, only a few centimeters per year.

According to plate tectonics, the now existing continents were at one time joined together in a single huge continent, which broke apart. Its pieces slowly moved across the face of the earth, thus changing the distribution of continents and ocean basins, as illustrated in Fig. 15.8. Some scientists feel that, when landmasses are concentrated in middle and high latitudes (as they are today), ice sheets are more likely to form. During these times, there is a greater likelihood that more sunlight will be reflected back into space from the snow that falls over the continent in winter. Less sunlight absorbed by the surface lowers the air temperature, which allows for a greater snow cover, and, over thousands of years, the formation of continental glaciers. (The amplified cooling that takes place over the snow-covered land is the snow-albedo feedback mentioned earlier.)

Some scientists speculate that climatic change, taking place over millions of years, might be related to the

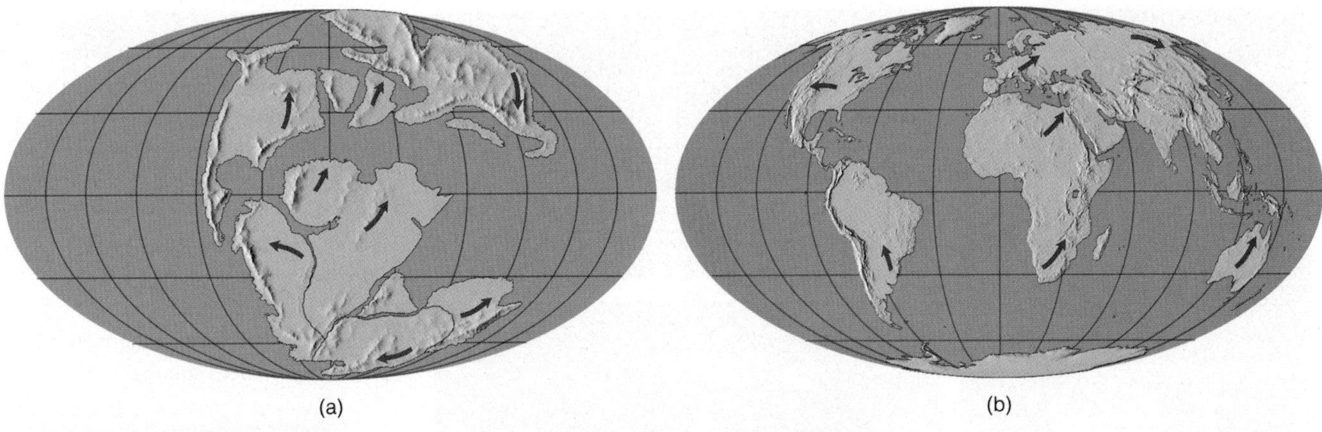

FIGURE 15.8 Geographical distribution of (a) landmasses about 150 million years ago, and (b) today. Arrows show the relative direction of continental movement.

rate at which the plates move and, hence, related to the amount of CO_2 in the air. For example, during times of rapid movement, an increase in volcanic activity vents large quantities of CO_2 into the atmosphere, which enhances the atmospheric greenhouse effect, causing global temperatures to rise.

Millions of years later, when movement decreases, less volcanic activity means less CO_2 is spewed into the atmosphere. A reduction in CO_2 levels weakens the greenhouse effect, which, in turn, causes global temperatures to drop. The accumulation of ice and snow over portions of the continents may promote additional cooling by reflecting more sunlight back to space.

A chain of volcanic mountains forming perpendicular to the mean wind flow may disrupt the airflow over them. By the same token, mountain building that occurs when two continental plates collide (like that which presumably formed the Himalayan mountains and Tibetan highlands) can have a marked influence on global circulation patterns and, hence, on the climate of an entire hemisphere.

Up to now, we have examined how climatic variations can take place over millions of years due to the movement of continents and the associated restructuring of landmasses. We will now turn our attention to variations in the earth's orbit that may account for climatic fluctuations that take place on a time scale of tens of thousands of years.

CLIMATE CHANGE: VARIATIONS IN THE EARTH'S ORBIT Another external cause of climate change involves a change in the amount of solar radiation that reaches the earth. A theory ascribing climatic changes to variations in the earth's orbit is the **Milankovitch theory**, named for the astronomer Milutin Milankovitch, who first proposed the idea in the 1930s. The basic premise of this theory is that, as the earth travels through space, three separate cyclic movements combine to produce variations in the amount of solar energy that falls on the earth.

The first cycle deals with changes in the shape (**eccentricity**) of the earth's orbit as the earth revolves about the sun. Notice in Fig. 15.9 that the earth's orbit changes from being elliptical (dashed line) to being nearly circular (solid line). To go from circular to elliptical and back again takes about 100,000 years. The greater the eccentricity of the orbit (that is, the more elliptical the orbit), the greater the variation in solar energy received by the earth between its closest and farthest approach to the sun.

Presently, we are in a period of low eccentricity, which means that our annual orbit around the sun is more circular. Moreover, the earth is closer to the sun in January and farther away in July (see Chapter 2, p. 52). The difference in distance (which only amounts to about

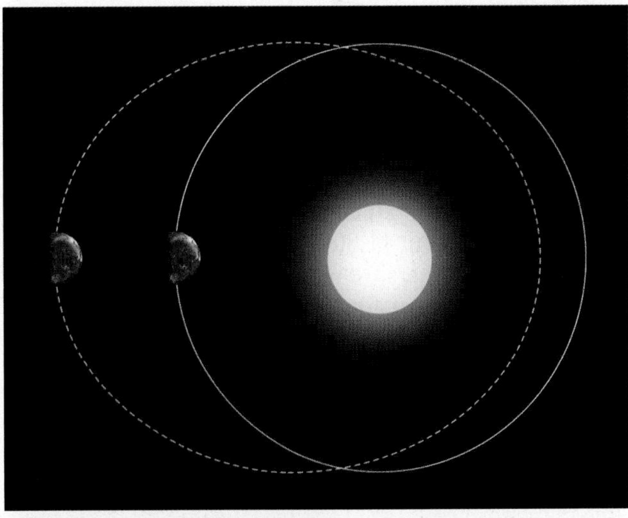

FIGURE 15.9 For the earth's orbit to stretch from nearly a circular (solid line) to an elliptical orbit (dashed line) and back again takes nearly 100,000 years. (Diagram is highly exaggerated and is not to scale.)

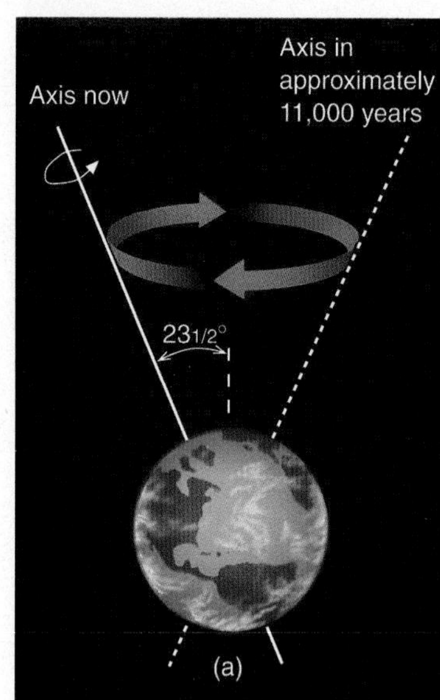

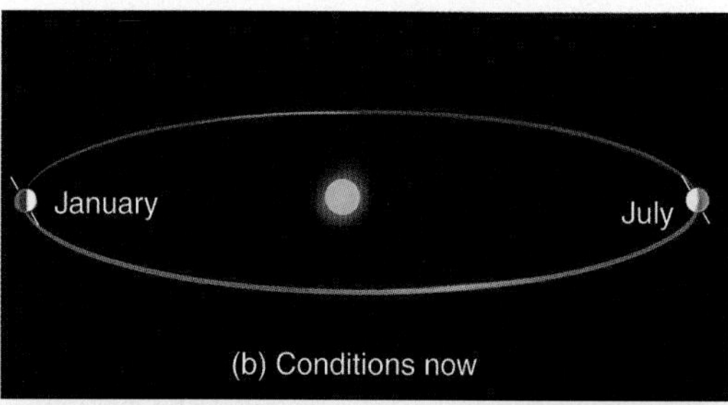

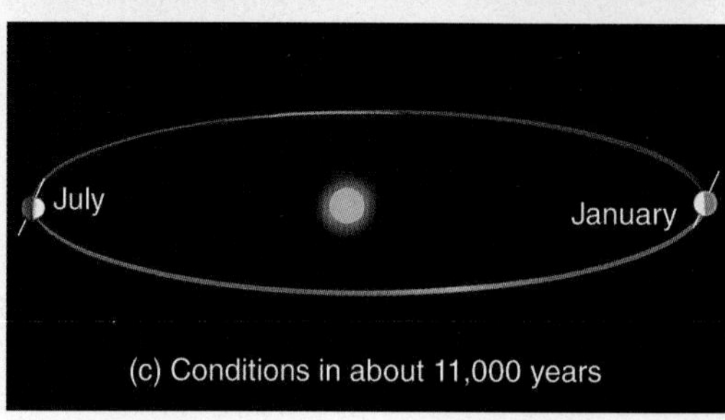

FIGURE 15.10

(a) Like a spinning top, the earth's axis of rotation slowly moves and traces out the path of a cone in space.
(b) Presently the earth is closer to the sun in January, when the Northern Hemisphere experiences winter.
(c) In about 11,000 years, due to precession, the earth will be closer to the sun in July, when the Northern Hemisphere experiences summer.

3 percent) is responsible for a nearly 7 percent increase in the solar energy received at the top of the atmosphere from July to January. When the difference in distance is 9 percent (a highly elliptical orbit), the difference in solar energy received between July and January will be on the order of 20 percent. In addition, the more eccentric orbit will change the length of seasons in each hemisphere by changing the length of time between the vernal and autumnal equinoxes. Although rather large percentage changes in solar energy can occur between summer and winter, the globally and annually averaged change in solar energy received by the earth (due to orbital changes) hardly varies at all. It is the distribution of incoming solar energy that changes, not the totals.

The second cycle takes into account the fact that, as the earth rotates on its axis, it wobbles like a spinning top. This wobble, known as the **precession** of the earth's axis, occurs in a cycle of about 23,000 years. Presently, the earth is closer to the sun in January and farther away in July. Due to precession, the reverse will be true in about 11,000 years (see Fig. 15.10). In about 23,000 years we will be back to where we are today. This means, of course, that if everything else remains the same, 11,000 years from now seasonal variations in the Northern Hemisphere should be greater than at present. The opposite would be true for the Southern Hemisphere.

The third cycle takes about 41,000 years to complete and relates to the changes in tilt (**obliquity**) with respect to the earth's orbit. Presently, the earth's orbital tilt is 23½°, but during the 41,000-year cycle the tilt varies from about 22° to 24½° (see Fig. 15.11). The smaller the tilt, the less seasonal variation there is between summer and winter in middle and high latitudes; thus, winters tend to be milder and summers cooler.

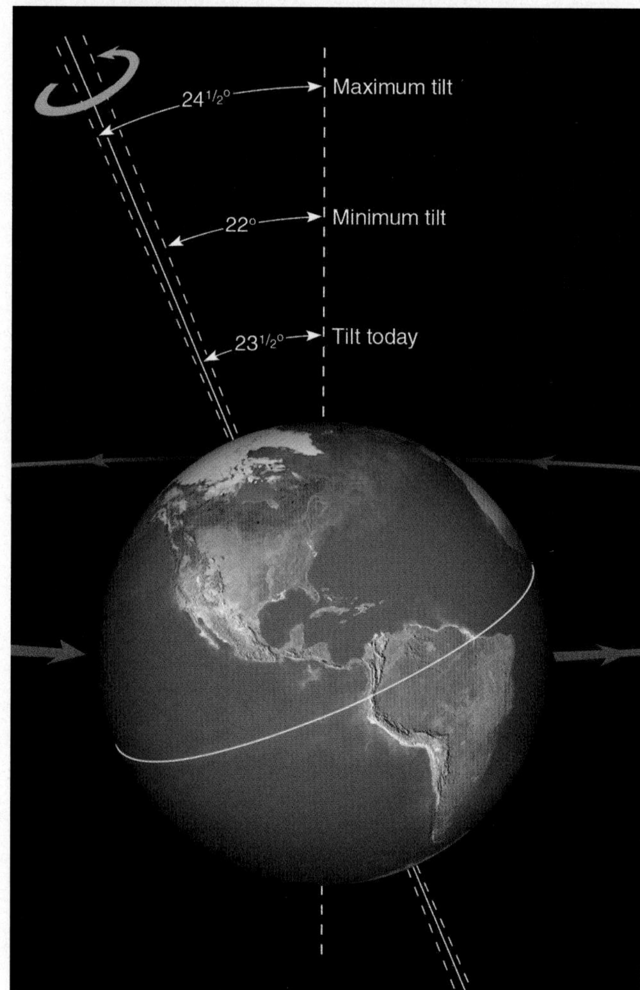

FIGURE 15.11 The earth currently revolves around the sun while tilted on its axis by an angle of 23½°. During a period of 41,000 years, this angle of tilt ranges from about 22° to 24½°.

Ice sheets over high latitudes of the Northern Hemisphere are more likely to form when less solar radiation reaches the surface in summer. Less sunlight promotes lower summer temperatures. During the cooler summer, snow from the previous winter may not totally melt. The accumulation of snow over many years increases the albedo of the surface. Less sunlight reaches the surface, summer temperatures continue to fall, more snow accumulates, and continental ice sheets gradually form. At this point, it is interesting to note that when all of the Milankovich cycles are taken into account, the present trend should be toward *cooler summers* over high latitudes of the Northern Hemisphere.

In summary, the Milankovitch cycles that combine to produce variations in solar radiation received at the earth's surface include:

1. changes in the shape (*eccentricity*) of the earth's orbit about the sun
2. *precession* of the earth's axis of rotation, or wobbling
3. changes in the tilt (*obliquity*) of the earth's axis

In the 1970s, scientists of the CLIMAP project found strong evidence in deep-ocean sediments that variations in climate during the past several hundred thousand years were closely associated with the Milankovitch cycles. More recent studies have strengthened this premise. For example, studies conclude that during the past 800,000 years, ice sheets have peaked about every 100,000 years. This conclusion corresponds naturally to variations in the earth's eccentricity. Superimposed on this situation are smaller ice advances that show up at intervals of about 41,000 years and 23,000 years. It appears, then, that eccentricity is the *forcing factor*—the external cause—for the frequency of glaciation, as it appears to control the severity of the climatic variation.

But orbital changes alone are probably not totally responsible for ice buildup and retreat. Evidence (from trapped air bubbles in the ice sheets of Greenland and Antarctica representing thousands of years of snow accumulation) reveals that CO_2 levels were about 30 percent lower during colder glacial periods than during warmer interglacial periods. Analysis of air bubbles in Antarctic ice cores reveals that methane follows a pattern similar to that of CO_2 (see ❱ Fig. 15.12). This

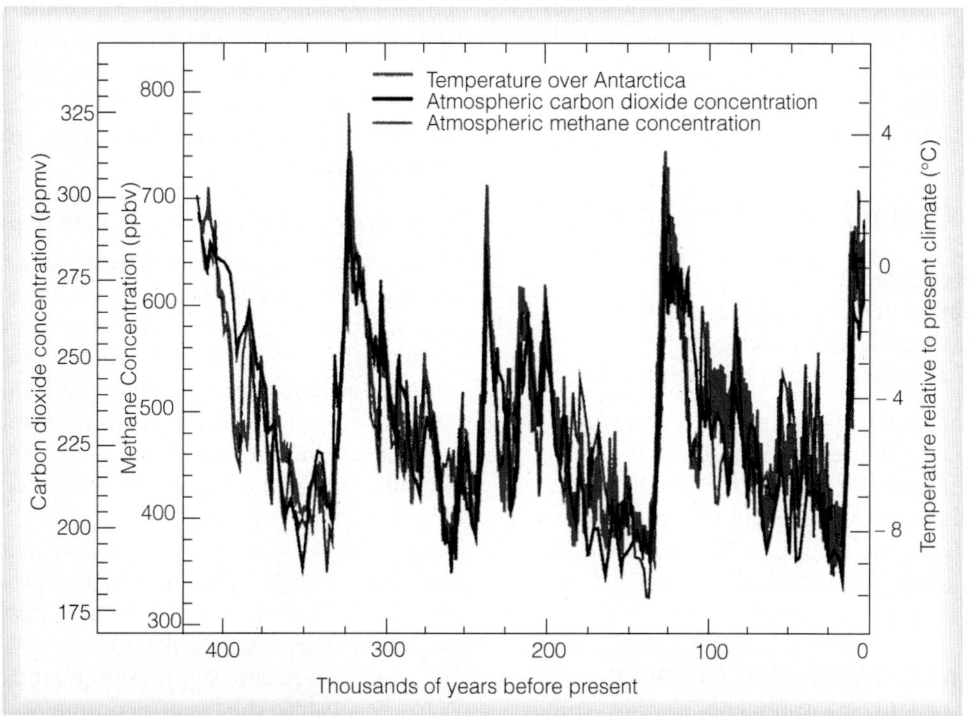

❱ **FIGURE 15.12** Variations of temperature (red line, °C), carbon dioxide (black line, ppmv), and methane (blue line, ppbv). Concentrations of gases are derived from air bubbles trapped within the ice sheets of Antarctica and extracted from ice cores. Temperatures are derived from the analysis of oxygen isotopes. (Note: ppmv represents parts per million by volume, and ppbv represents parts per billion by volume.) (*Source:* From Climate Change 2001: The Scientific Basis, 2001, by J.T. Houghton, et al. Copyright © 2001 Cambridge University Press. Reprinted with permission of the Intergovernmental Panel on Climate Change.)

knowledge suggests that lower atmospheric CO_2 levels may have had the effect of amplifying the cooling initiated by the orbital changes. Likewise, increasing CO_2 levels at the end of the glacial period may have accounted for the rapid melting of the ice sheets.*

The latest research shows that temperature changes thousands of years ago actually *preceded* the CO_2 changes. This observation indicates that CO_2 is a positive feedback in the climate system, where higher temperatures lead to higher CO_2 levels and lower temperatures to lower CO_2 levels. Consequently, CO_2 is an internal, natural part of the earth's climate system.

Just why atmospheric CO_2 levels have varied as glaciers expanded and contracted is not clear, but it appears to be due to changes in biological activity taking place in the oceans. Perhaps, also, changing levels of CO_2 indicate a shift in ocean circulation patterns. Such shifts, brought on by changes in precipitation and evaporation rates, may alter the distribution of heat energy around the world. Alteration wrought in this manner could, in turn, affect the global circulation of winds, which may explain why alpine glaciers in the Southern Hemisphere expanded and contracted in tune with Northern Hemisphere glaciers during the last ice age, even though the Southern Hemisphere (according to the Milankovitch cycles) was not in an orbital position for glaciation.

Still other factors may work in conjunction with the earth's orbital changes to explain the temperature variations between glacial and interglacial periods. Some of these are:

1. the amount of dust and other aerosols in the atmosphere
2. the reflectivity of the ice sheets
3. the concentration of other greenhouse gases
4. the changing characteristics of clouds
5. the rebounding of land, having been depressed by ice

Hence, the Milankovitch cycles, in association with other natural factors, may explain the advance and retreat of ice over periods of 10,000 to 100,000 years. But what caused the Ice Age to begin in the first place? And why have periods of glaciation been so infrequent during geologic time? The Milankovitch theory does not attempt to answer these questions.

CLIMATE CHANGE: VARIATIONS IN SOLAR OUTPUT Solar energy measurements made by sophisticated instruments aboard satellites suggest that the sun's energy output (called *brightness*) may vary slightly—by a fraction of one percent—with sunspot activity.

Sunspots are huge magnetic storms on the sun that show up as cooler (darker) regions on the sun's surface (Fig. 15.13). They occur in cycles, with the number and size reaching a maximum approximately every 11 years. During periods of maximum sunspots, the sun emits more energy (about 0.1 percent more) than during periods of sunspot minimums (see Fig. 15.14). Evidently, the greater number of bright areas (*faculae*) around the sunspots radiate more energy, which offsets the effect of the dark spots.

It appears that the 11-year sunspot cycle has not always prevailed. Apparently, between 1645 and 1715, during the period known as the **Maunder minimum**,* there were few, if any, sunspots. It is interesting to note that the minimum occurred during the "Little Ice Age," a cool spell in the temperature record experienced mainly over Europe. Some scientists suggest that a reduction in the sun's energy output was, in part, responsible for this cold spell.

Fluctuations in solar output may account for climatic changes over time scales of decades and centuries. Many theories have been proposed linking solar variations to climate change, but none have been proven. However, instruments aboard satellites and solar telescopes on the earth are monitoring the sun to observe how its energy output may vary. To date, these measurements show that solar output has only changed a fraction of one percent over several decades. Because many years of data are needed, it may be some time before we fully understand the relationship between solar activity and climate change on earth.

CLIMATE CHANGE: ATMOSPHERIC PARTICLES

Microscopic liquid and solid particles (*aerosols*) that enter the atmosphere from both human-induced and natural sources can have an effect on climate. The effect these particles have on the climate is exceedingly complex and depends upon a number of factors, such as the particle's size, shape, color, chemical composition, and vertical distribution above the surface. In this section, we will examine those particles that enter the atmosphere through natural means.

Particles can enter the atmosphere in a variety of ways. For example, wildfires can produce copious amounts of tiny smoke particles, and dust storms sweep tons of fine particles into the atmosphere. Smoldering volcanoes can release significant quantities of sulfur-rich aerosols into the lower atmosphere. And even the oceans are a major source of natural sulfur aerosols, as tiny drifting aquatic plants—phytoplankton—produce a form of sulfur that slowly diffuses into the atmosphere,

*It is interesting to note that during peak CO_2 levels, its concentration of about 325 ppm was still lower than its concentration of 385 ppm in today's atmosphere.

*This period is named after E. W. Maunder, the British solar astronomer who first discovered the low sunspot period sometime in the late 1880s.

© Carr Clifton/Minden Pictures

FIGURE 15.13 The sun plays a major role in the earth's climate system. Any significant change in the sun's energy output (brightness) can have a drastic effect on air temperatures at the earth's surface.

where it combines with oxygen to form sulfur dioxide, which in turn converts to *sulfate aerosols*. Although the effect these particles have on the climate system is complex, the overall effect they have is to *cool the surface* by preventing sunlight from reaching the surface.

Volcanic eruptions can have a definitive impact on climate. During volcanic eruptions, fine particles of ash and dust (as well as gases) can be ejected into the stratosphere (see Fig 15.15).* Scientists agree that the volcanic eruptions having the greatest impact on climate are those rich in sulfur gases. These gases, over a period of about two months, combine with water vapor in the presence of sunlight to produce tiny, reflective sulfuric acid particles that grow in size, forming a dense layer of haze. The haze may reside in the stratosphere for several years, absorbing and reflecting back to space a portion of the sun's incoming energy. The reflection of incoming sunlight by the haze tends to cool the air at the earth's surface, especially in the hemisphere where the eruption occurs.

*You may recall from Chapter 1 that the stratosphere is a stable layer of air above the troposphere, typically about 7 to 31 mi (11 to 50 km) above the earth's surface.

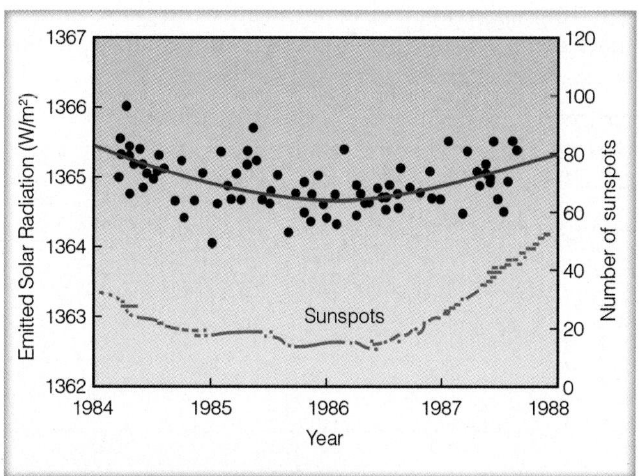

FIGURE 15.14 Changes in solar energy output (upper curve) in watts per square meter as measured by the *Earth Radiation Budget Satellite*. Bottom curve represents the yearly average number of sunspots. As sunspot activity increases from minimum to maximum, the sun's energy output increases by about 0.1 percent. (From V. Ramanathan, B. R. Barstrom, and E. F. Harrison, "Climate and earth's radiation budget," *Physics Today,* May, 1988, Fig. 5.)

FIGURE 15.15 Large volcanic eruptions rich in sulfur can affect climate. As sulfur gases in the stratosphere transform into tiny reflective sulfuric acid particles, they prevent a portion of the sun's energy from reaching the surface. Here, the Philippine volcano Mount Pinatubo erupts during June, 1991.

Two of the largest volcanic eruptions of the 20th century in terms of their sulfur-rich veil, were that of El Chichón in Mexico during April, 1982, and Mount Pinatubo in the Philippines during June, 1991. The eruption of Mount Pinatubo in 1991 was many times greater than that of Mount St. Helens in the Pacific Northwest in 1980. In fact, the largest eruption of Mount St. Helens was a lateral explosion that pulverized a portion of the volcano's north slope. The ensuing dust and ash (and very little sulfur) had virtually no effect on global climate as the volcanic material was confined mostly to the lower atmosphere and fell out quite rapidly over a large area of the northwestern United States.

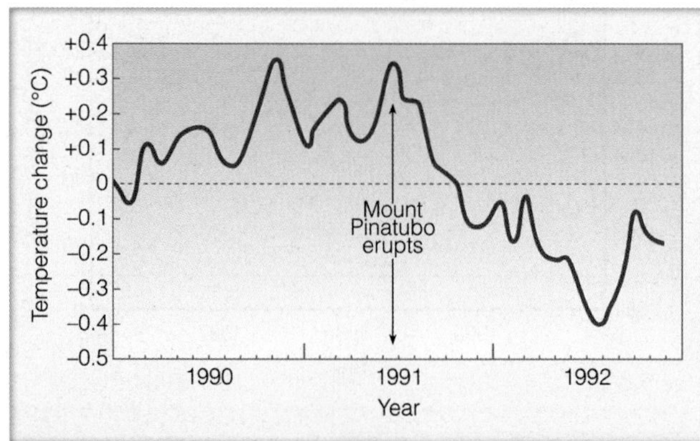

FIGURE 15.16 Changes in average global air temperature from 1990 to 1992. After the eruption of Mount Pinatubo in June, 1991, the average global temperature by July, 1992, decreased by almost 0.5°C (0.9°F) from the 1981 to 1990 average (dashed line). (Data courtesy of John Christy, University of Alabama, Huntsville, and R. Spencer, NASA Marshall Space Flight Center.)

Mount Pinatubo ejected an estimated 20 million tons of sulfur dioxide (more than twice that of El Chichón) that gradually worked its way around the globe. For major eruptions such as this one, mathematical models predict that average hemispheric temperatures can drop by about 0.2° to 0.5°C or more for one to three years after the eruption. Model predictions agreed with temperature changes brought on by the Pinatubo eruption, as in early 1992 the mean global surface temperature had decreased by about 0.5°C (see Fig. 15.16). The cooling might even have been greater had the eruption not coincided with a major El Niño event that began in 1990 and lasted until early 1995 (see Chapter 14, p. 418 for more information on El Niño). In spite of the El Niño, the eruption of Mount Pinatubo produced the two coolest years of the 1990s—1991 and 1992.

An infamous cold spell often linked to volcanic activity occurred during the year 1816, "the year without a summer" mentioned earlier. Apparently, a rather stable longwave pattern in the atmosphere produced unseasonably cold summer weather over eastern North America and western Europe. The cold weather followed the massive eruption in 1815 of Mount Tambora in Indonesia. In addition, a smaller volcanic eruption occurred in 1809, from which the climate system may not have fully recovered when Tambora erupted in 1815.

EXTREME WEATHER WATCH

The eruption of the volcano Toba on the Indonesian island of Sumatra about 74,000 years ago, ejected an estimated 720 cubic miles of material into the atmosphere (compared to an estimated 38 cubic miles ejected by the Tambora eruption of 1815). The Toba eruption may have lowered the earth's average temperature by 9°F (5°C) for several years following the event. Over middle-and-high latitudes the cooling may have been much greater. This change in the earth's climate system probably had a substantial impact on the distribution and migration of humans at that time.

In an attempt to correlate sulfur-rich volcanic eruptions with long-term trends in global climate, scientists are measuring the acidity of annual ice layers in Greenland and Antarctica. Generally, the greater the concentration of sulfuric acid particles in the atmosphere, the greater the acidity of ice layer. Relatively acidic ice has been uncovered from about A.D. 1350 to about 1700, a time that corresponds to a cooling trend over Europe referred to as the *Little Ice Age.* Such findings suggest that sulfur-rich volcanic eruptions may have played an important role in triggering this comparatively cool period and, perhaps, other cool periods during the geologic past. Moreover, recent core samples taken from the northern Pacific Ocean reveal that volcanic eruptions in the northern Pacific were at least 10 times larger 2.6 million years ago (a time when Northern Hemisphere glaciation began) than previous volcanic events recorded elsewhere in the sediment.

BRIEF REVIEW

Before going on to the next section, here is a brief review of some of the facts and concepts we covered so far:

- **The external causes of climate change include: (1) changes in incoming solar radiation; (2) changes in the composition of the atmosphere; (3) changes in the surface of the earth.**
- **The shifting of continents, along with volcanic activity and mountain building, are possible causes of climate change.**
- **The Milankovitch theory (in association with other natural forces) proposes that altering glacial and interglacial episodes during the past 2.5 million years are the result of small variations in the tilt of the earth's axis and in the geometry of the earth's orbit around the sun.**
- **Trapped air bubbles in the ice sheets of Greenland and Antarctica reveal that CO2 levels and methane levels were lower during colder glacial periods and higher during warmer interglacial periods. But even when the levels were higher, they still were much lower than they are today.**
- **Fluctuation in solar output (brightness) may account for climatic changes over time scales of decades and centuries.**
- **Volcanic eruptions rich in sulfur may be responsible for cooler periods that span years and decades in the geologic past.**

Climate Change Caused by Human (Anthropogenic) Activities

Earlier in this chapter we saw how increasing levels of carbon dioxide (CO_2) may have contributed to changes in global climate spanning thousands and even millions of years. Today, we are modifying the chemistry and characteristics of the atmosphere by injecting vast quantities of particles and greenhouse gases into the air without fully understanding the long-term consequences. Therefore, in this section, we will first look at how particles injected into the lower atmosphere by human activities may be affecting climate. Then we will examine how CO_2 and other trace gases appear to be enhancing the earth's greenhouse effect, producing global warming. Finally, we will explore how predicted changes in the earth's climate may induce changes in extreme weather events.

CLIMATE CHANGE: AEROSOLS INJECTED INTO THE LOWER ATMOSPHERE In the previous section we learned that tiny solid and liquid particles (aerosols) can enter the atmosphere from both human-induced and natural sources. The human-induced sources include emissions from factories, autos, trucks, aircraft, power plants, home furnaces and fireplaces, to name a few. Many aerosols are not injected directly into the atmosphere, but form when gases convert to particles. Some particles, such as sulfates and nitrates, mainly reflect incoming sunlight, whereas others, such as soot, readily absorb sunlight. Many of the particles that reduce the amount of sunlight reaching the earth's surface tend to cause a *net cooling* of the surface air during the day.

In recent years, the effect of highly reflective **sulfate aerosols** on climate has been extensively researched.

In the lower atmosphere (troposphere), the majority of these particles are related to human activities and come from the combustion of sulfur-containing fossil fuels. Sulfur pollution, which has more than doubled globally since preindustrial times, enters the atmosphere mainly as sulfur dioxide gas. There, it transforms into tiny sulfate droplets or particles. Since these aerosols usually remain in the lower atmosphere for only a few days, they do not have time to spread around the globe. Hence, they are not well mixed and their effect is felt mostly over the Northern Hemisphere, especially over polluted regions.

Sulfate aerosols not only reflect incoming sunlight back to space, but they also serve as cloud condensation nuclei, tiny particles on which cloud droplets form. Consequently, they have the potential for altering the physical characteristics of clouds. For example, if the number of sulfate aerosols and, hence, condensation nuclei inside a cloud should increase, the cloud would have to share its available moisture with the added nuclei, a situation that should produce many more (but smaller) cloud droplets. The greater number of droplets would reflect more sunlight and have the effect of brightening the cloud and reducing the amount of sunlight that reaches the surface.

In summary, sulfate aerosols reflect incoming sunlight, which tends to lower the earth's surface temperature during the day. Sulfate aerosols may also modify clouds by increasing their reflectivity. Because sulfate pollution has increased significantly over industrialized areas of eastern Europe, northeastern North America, and China, the cooling effect brought on by these particles may explain:

1. why the industrial regions of the Northern Hemisphere have warmed less than the Southern Hemisphere during the past several decades
2. why the United States has experienced less warming than the rest of the world
3. why up until the last few decades most of the global warming has occurred at night and not during the day, especially over polluted areas

The overall effect that aerosols in the lower atmosphere have on the climate system is not yet totally understood. Research is still being done. Information regarding the possible catastrophic effect on climate from particles injected into the atmosphere during nuclear war is given in the Focus section on p. 447.

CLIMATE CHANGE: INCREASING LEVELS OF GREENHOUSE GASES We learned in Chapter 2, p. 45, that carbon dioxide (CO_2) is a greenhouse gas that strongly absorbs infrared radiation and plays a major role in the warming of the lower atmosphere. Everything else being equal, the more CO_2 in the atmosphere, the warmer the surface air. We also know that CO_2 has been increasing steadily in the atmosphere, primarily due to human activities, such as the burning of fossil fuels (see Fig. 1.8, p. 10). However, deforestation—the reduction of forests—is also adding to this increase, through the process of photosynthesis, in which the leaves of trees remove CO_2 from the atmosphere. The CO_2 is then stored in leaves, branches, and roots. When the trees are cut and burned, or left to rot, the CO_2 goes back into the atmosphere.

In 2008, the annual average of CO_2 in the atmosphere was about 385 ppm. Present estimates are that if CO_2 levels continue to increase at the same rate that they have been (about 1.9 ppm per year), atmospheric concentrations will rise significantly by the end of this century, and the earth's surface will warm dramatically. To complicate the picture, increasing concentrations of other greenhouse gases—such as methane (CH_4), nitrous oxide (N_2O), and chlorofluorocarbons (CFCs)—all readily absorb infrared radiation and enhance the atmospheric greenhouse effect.* How will increasing levels of greenhouse gases influence climate in the future? Before we address this question, we will look at how humans may be affecting climate by changing the landscape.

CLIMATE CHANGE: LAND USE CHANGES All climate models predict that, as humanity continues to spew greenhouse gases into the air, the climate will change and the earth's surface will warm. But are humans changing the climate by other activities as well? Modification of the earth's surface taking place right now could potentially be influencing the immediate climate of certain regions. For example, studies show that about half the rainfall in the Amazon River Basin is returned to the atmosphere through evaporation and through transpiration from the leaves of trees. Consequently, clearing large areas of tropical rain forests in South America to create open areas for farms and cattle ranges will most likely cause a decrease in evaporative cooling. This decrease, in turn, could lead to a warming in that area of at least several degrees Celsius. In turn, the reflectivity of the deforested area will change. Similar changes in albedo result from the overgrazing and excessive cultivation of grasslands in semi-arid regions, causing an increase in desert conditions (a process known as **desertification**).

Currently, billions of acres of the world's range and cropland, along with the welfare of millions of people, are affected by desertification. Annually, millions of acres are reduced to a state of near or complete uselessness. The

*Refer back to Chapter 1 and Table 1.3 on p. 8 for additional information on the concentration of these gases.

main cause is overgrazing, although overcultivation, poor irrigation practices, and deforestation also play a role. The effect this will have on climate, as surface albedos increase and more dust is swept into the air, is uncertain.

It is interesting to note that some scientists feel that humans may have been altering climate way before modern civilizations came along. For example, retired Professor William Ruddiman of the University of Virginia suggests that humans have been influencing climate change for the past 8000 years. Although some climate scientists vehemently oppose his ideas, Ruddiman speculates that without pre-industrial farming, which produces methane and some carbon dioxide, we would have entered a naturally occurring ice age. He even suggests that the Little Ice Age of the 15th through the 19th centuries in Europe was human-induced because plagues, which killed millions of people, caused a reduction in farming.

The reasoning behind this idea goes something like this: As forests are cleared for farming, levels of CO_2 and methane increase, producing a strong greenhouse effect and a rise in surface air temperature. When catastrophic

FOCUS ON
EXTREME WEATHER

Catastrophic Climate Change Brought on by Nuclear War

A number of studies indicate that a nuclear war would drastically modify the earth's climate, instigating climate change unprecedented in recorded human history.

Researchers assume that a nuclear war would raise an enormous pall of thick, sooty smoke from massive fires that would burn for days, even weeks, following an attack. The smoke would drift higher into the atmosphere, where it would be caught in the upper-level westerlies and circle the middle latitudes of the Northern Hemisphere. Unlike soil dust, which mainly reflects incoming sunlight, soot particles readily absorb sunlight. Hence, for months, or perhaps years, after the war, sunlight would virtually be unable to penetrate the smoke layer, bringing darkness or, at best, twilight at midday.

Such reduction in solar energy would cause surface air temperatures over landmasses to drop below freezing, even during the summer, resulting in extensive damage to plants and crops and the death of millions (or possibly billions) of people. The dark, cold, and gloomy conditions that would be brought on by nuclear war are often referred to as *nuclear winter*.

As the lower troposphere cools, the solar energy absorbed by the smoke particles in the upper troposphere would cause this region to warm. The end result would be a strong, stable temperature inversion extending from the surface up into the higher atmosphere. A strong inversion would lead to a number of adverse effects, such as suppressing convection, altering precipitation processes, and causing major changes in the general wind patterns.

The heating of the upper part of the smoke cloud would cause it to rise upward into the stratosphere, where it would then drift around the world. Thus, about one-third of the smoke would remain in the atmosphere for up to a decade. The other two-thirds would be washed out in a month or so by precipitation. This smoke lofting, combined with persisting sea ice formed by the initial cooling, would produce climatic change that would remain for more than a decade.

Virtually all research on nuclear winter, including models and analog studies, confirms this gloomy scenario. Observations of forest fires show lower temperatures under the smoke, confirming part of the theory. A three-year study involving more than 300 scientists from more than 30 countries conducted by the Scientific Committee On Problems of the Environment (SCOPE) of the International Council of Scientific Unions has detailed the climatic, environmental, and agricultural effects of nuclear winter. The implications of nuclear winter are clear: A nuclear war would drastically alter global climate and would devastate our living environment.

Even with improved global superpower relations, and the end of the Cold War, the danger of nuclear winter remains a possibility. Presently, the current global nuclear arsenal is more than that needed to produce the effects of a nuclear winter. As other nations develop nuclear capability, the potential for nuclear winter remains with us. It will not disappear until the global nuclear weapons arsenal numbers in the hundreds, not in the thousands.

plagues strike—the bubonic plague, for instance—high mortality rates cause farms to be abandoned. As forests begin to take over the untended land, levels of CO_2 and methane drop, causing a reduction in the greenhouse effect and a corresponding drop in air temperature. When the plague abates, the farms return, forests are cleared, levels of greenhouse gases go up, and surface air temperatures rise.

Global Warming

We have previously pointed out several times in this chapter that the earth's atmosphere is in a warming trend that began around the beginning of the 20th century. Is there any evidence that this warming is due to increasing levels of greenhouse gases caused by human activities? If so, what is it?

Numerical climate models (mathematical models that simulate climate) predict that, by the end of this century, increasing concentrations of greenhouse gases could result in an additional global warming of at least several degrees Celsius. The newest, most sophisticated models take into account a number of important relationships, including the interactions between the oceans and the atmosphere, the processes by which CO_2 is removed from the atmosphere, and the cooling effect produced by sulfate aerosols in the lower atmosphere. The models also predict that, as the air warms, additional water will evaporate from the ocean's surface and enter the atmosphere as water vapor. The added water vapor (which is the most abundant greenhouse gas) will produce a feedback on the climate system by enhancing the atmospheric greenhouse effect and accelerating the temperature rise. (This phenomenon is the *water vapor-greenhouse feedback* described on p. 439.) Without this feedback produced by the added water vapor, the models predict that the warming will be much less.

EXTREME WEATHER WATCH

Be careful. Don't base climate change trends on a specific weather event. For example, a January cold wave in 2009 sent temperatures plummeting across North America. In Maine, the cold snap produced a state-record low of –50°F at Big Black River, and in Waterloo, Iowa, the temperature dropped to a record low –34°F. Yet, globally, January, 2009, was the seventh warmest January on record according to the Global Historic Climatology Network, whose data set began in 1880.

RECENT GLOBAL WARMING: PERSPECTIVE Since the beginning of the 20th century, the average global surface air temperature has risen by more than 1.4°F (0.8°C). Is this warming due to increasing greenhouse gases and an enhanced greenhouse effect? Before we can address this question, we need to review a few concepts we learned in Chapter 2.

Radiative Forcing Agents We know from Chapter 2 that our world without water vapor, CO_2, and other greenhouse gases would be a colder world—about 59°F (33°C) colder than at present. With an average surface temperature of about 0°F, much of the planet would be uninhabitable. In Chapter 2, we also learned that when the rate of the incoming solar energy balances the rate of outgoing infrared energy from the earth's surface and atmosphere, the earth-atmosphere system is in a state of *radiative equilibrium*. Increasing concentrations of greenhouse gases can disturb this equilibrium and are, therefore, referred to as **radiative forcing agents.** The **radiative forcing*** provided by extra CO_2 and other greenhouse gases increased by about 3 W/m^2 over the past several hundred years, with CO_2 contributing about 60 percent of the increase. So it is very likely that part of the warming during the last century is due to increasing levels of greenhouse gases. But what part does natural climate variability play in global warming? And with levels of CO_2 increasing by more than 25 percent since the early 1900s, why has the observed increase in global temperature been relatively small?

We know that the climate may change due to natural events. For example, changes in the sun's energy output (called *solar irradiance*) and volcanic eruptions rich in sulfur are two major natural radiative forcing agents. Studies show that since the middle 1700s, changes in the sun's energy output may have contributed a small positive forcing (about 0.12 W/m^2) on the climate system, most of which occurred during the first half of the 20th century. On the other hand, volcanic eruptions that inject sulfur-rich particles into the stratosphere produce a negative forcing, which lasts for a few years after the eruption. Because several major eruptions occurred between 1880 and 1920, as well as between 1960 and 1991, the combined change in radiative forcing due to both volcanic activity and solar activity over the past 25 to 45 years appears to be *negative*, which means that the net effect is that of *cooling* the earth's surface. Did this cooling in combination with the cooling produced by sulfur-rich aerosols in the lower troposphere reduce the overall warming of the earth's surface during the last century? The use of climate models can help answer this question.

*Radiative forcing is interpreted as an increase (positive) or a decrease (negative) in net radiant energy observed over an area at the tropopause. All factors being equal, an increase in *radiative forcing* may induce surface *warming*, whereas a *decrease* may induce surface *cooling*.

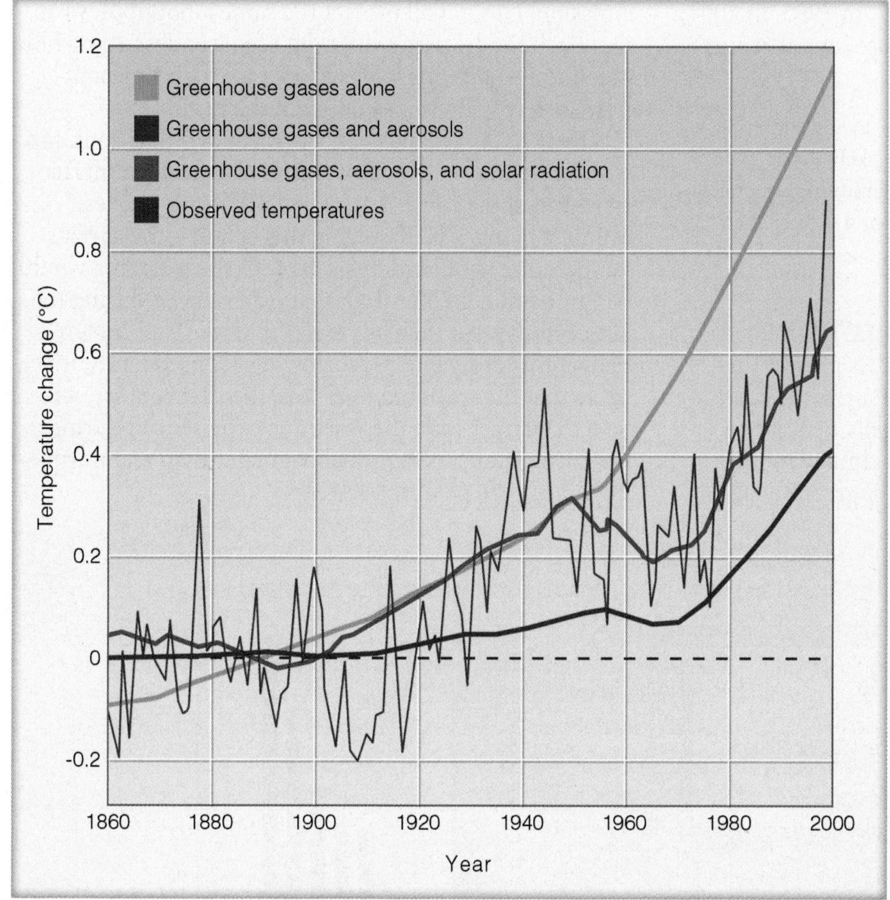

FIGURE 15.17

Projected surface air temperature changes from different climate models. Model input from greenhouse gases only is shown in yellow; input from greenhouse gases plus sulfate aerosols is shown in blue; input from greenhouse gases, sulfate aerosols, and solar energy changes is shown in red. The gray line shows observed surface temperature change. The dashed line is the 1880 to 1999 mean temperature. (Redrawn from "The Science of Climate Change" by Tom M. L. Wigley, published by the Pew Center of Global Climate Change.)

Climate Models and Recent Temperature Trends

The earth's average surface temperature increased by about 0.6°C from 1900 to 2000. How does this observed temperature change over the last century compare with temperature changes derived from climate models using different forcing agents? Before we look at what climate models reveal, it is important to realize that the interactions between the earth and its atmosphere are so complex that it is difficult to unequivocally *prove* that the earth's present warming trend is due entirely to increasing concentrations of greenhouse gases. The problem is that any human-induced signal of climate change is superimposed on a background of natural climatic variations ("noise"), such as the El Niño-Southern Oscillation (ENSO) phenomenon (discussed in Chapter 14). Moreover, in the temperature observations, it is difficult to separate a signal from the noise of natural climate variability. However, today's more sophisticated climate models are much better at filtering out this noise while at the same time taking into account those forcing agents that are both natural and human-induced.

Figure 15.17 shows the predicted changes in surface air temperature from 1860 to 2000 made by different climate models using various scenarios (different forcing agents). The gray line represents the actual changes in surface air temperature from 1860 to 2000. Notice that when only increasing levels of greenhouse gases are plugged into the model (yellow line), the model shows a surface temperature increase in excess of 1°C. When greenhouse gases and sulfate aerosols are both added to the model (blue line), the increase in surface temperature is much less; in fact, it is less than the temperature increases observed during the last century. However, when greenhouse gases, sulfate aerosols, and changes in solar radiation are *all* added to the model (red line), the projected temperature change and the observed temperature change closely match.

It is this match in projected and observed temperature trends that helps to explain why a global warming of 0.6°C measured during the last century was less than the warming projected by climate models that took into account only increasing levels of greenhouse gases. So without negative forcing agents acting on the climate system, global warming during the last century would have likely been greater than observed.

It is climate studies using computer models such as these that have led scientists to conclude that *most* of the warming during the latter decades of the 20th century is very likely due to increasing levels of greenhouse gases. In fact, the Intergovernmental Panel on Climate Change (IPCC), a committee of over 2000 leading earth scientists, considered the issues of climate change in a report published in 1990 and updated in 1992, in 1995, in 2001, and again in 2007. The 2007 Fourth Assessment report of the IPCC states that:

> Most of the observed increase in globally averaged temperatures since the mid-20th century is very likely due to the observed increase in anthropogenic greenhouse gas concentrations. (In the report "very likely" means a greater than 90 percent probability.)

FUTURE GLOBAL WARMING: PROJECTIONS As we can see in Fig. 15.18, climate models project that, due to increasing levels of greenhouse gases, the surface air temperature will increase substantially by the end of this century. Notice, however, that the climate models (or a single model using a wide range of greenhouse gas emissions) do not all project the same amount of warming. Each model uses a different scenario describing how greenhouse gas emissions will change with time and how society will utilize energy in the future.

The IPCC in its 2007 report concluded that doubling the concentration of CO_2 would likely produce surface warming in the range of 2°C to 4.5°C, with the best estimate being 3°C. If, during this century, the surface temperature should increase by 2°C, the warming would be three times greater than that experienced during the 20th century. An increase of 4.5°C would have potentially devastating effects worldwide. Consequently, it is likely that the warming over this, the 21st century, will be much larger than the warming experienced during the 20th century, and probably greater than any warming during the past 10,000 years.

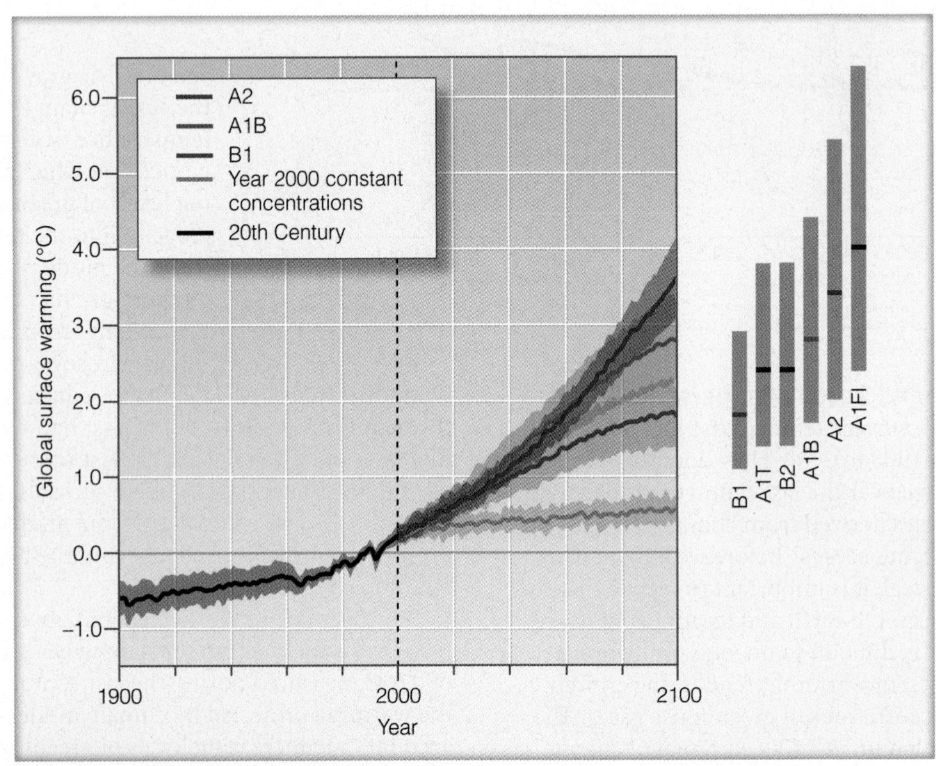

FIGURE 15.18 Global average projected surface air temperature changes (°C) above the 1980–1999 average (dark purple zero line) for the years 2000 to 2100. Temperature changes inside the graph and to the right of the graph are based on multi-climate models with different scenarios. Each scenario describes how the average temperature will change based on different concentrations of greenhouse gases and various forcing agents. The black line shows global temperature change during the 20th century. The orange line shows projected temperature change where greenhouse gas concentrations are held constant at the year 2000 level. The vertical gray bars on the right side of the figure indicate the likely range of temperature change for each scenario. The thick solid bar within each gray bar gives the best estimate for temperature change for each scenario. (*Source:* Climate Change 2007, *The Physical Science Basis*, by the Working Group 1 contribution to the Fourth Assessment Report to the IPCC © 2007. Reprinted by permission of the Intergovernmental Panel on Climate Change.)

Uncertainties about Greenhouse Gases There are, however, uncertainties in predicting the climate of the future. At this point in time, it is unclear how water and land will ultimately affect rising levels of CO_2. Currently, the oceans and the vegetation on land absorb about half of the CO_2 emitted by human sources. As a result, both oceans and landmasses play a major role in the climate system, yet the exact effect they will have on rising levels of CO_2 and global warming is not totally clear. For instance, the microscopic plants (phytoplankton) dwelling in the oceans extract CO_2 from the atmosphere during photosynthesis and store some of it below the oceans' surface, where they die. Will a warming earth trigger a large blooming of these microscopic plants, in effect reducing the rate at which atmospheric CO_2 is increasing?

Current models show that warming the earth tends to *reduce* both ocean and land intake of CO_2. Therefore, if levels of human-induced CO_2 emissions continue to increase at their present rate, more CO_2 should remain in the atmosphere to further enhance global warming. An example of how rising temperatures can play a role in altering the way landmasses absorb and emit CO_2 is found in the Alaskan tundra. There, temperatures in recent years have risen to the point where more frozen soil melts in summer than it used to. Accordingly, during the warmer months, deep layers of exposed decaying peat moss release CO_2 into the atmosphere. Until recently, this region absorbed more CO_2 than it released. Now, however, much of the tundra acts as a producing source of CO_2.

At present, deforestation accounts for about one-fifth of the observed increase in atmospheric CO_2. Hence, changes in land use could influence levels of CO_2 concentrations, especially if the practice of deforestation is replaced by reforestation. Furthermore, it is unknown what future steps countries will take in limiting the emissions of CO_2 from the burning of fossil fuels.

Currently it is not known how quickly greenhouse gases will increase in the future. We can see in Fig. 15.19 the dramatic rise in CO_2 levels during the 20th century. In the year 1990, carbon dioxide levels were increasing by about 1.5 ppm/year, whereas today they are increasing by about 1.9 ppm/year. If this trend continues, CO_2 concentrations could easily exceed 550 ppm by the end of this, the 21st century. In Fig 15.19 notice that the atmospheric concentration of methane has increased dramatically over the last 250 years, and it is still increasing. Also notice that atmospheric concentrations of nitrous oxide have risen quickly, and its concentration is still rising.

Since the mid-1990s, the atmospheric concentration of a group of greenhouse gases called *chlorofluorocarbons* (halocarbons) has been decreasing. However, the substitute compounds for chlorofluorocarbons, which are also greenhouse gases, have been increasing. Moreover,

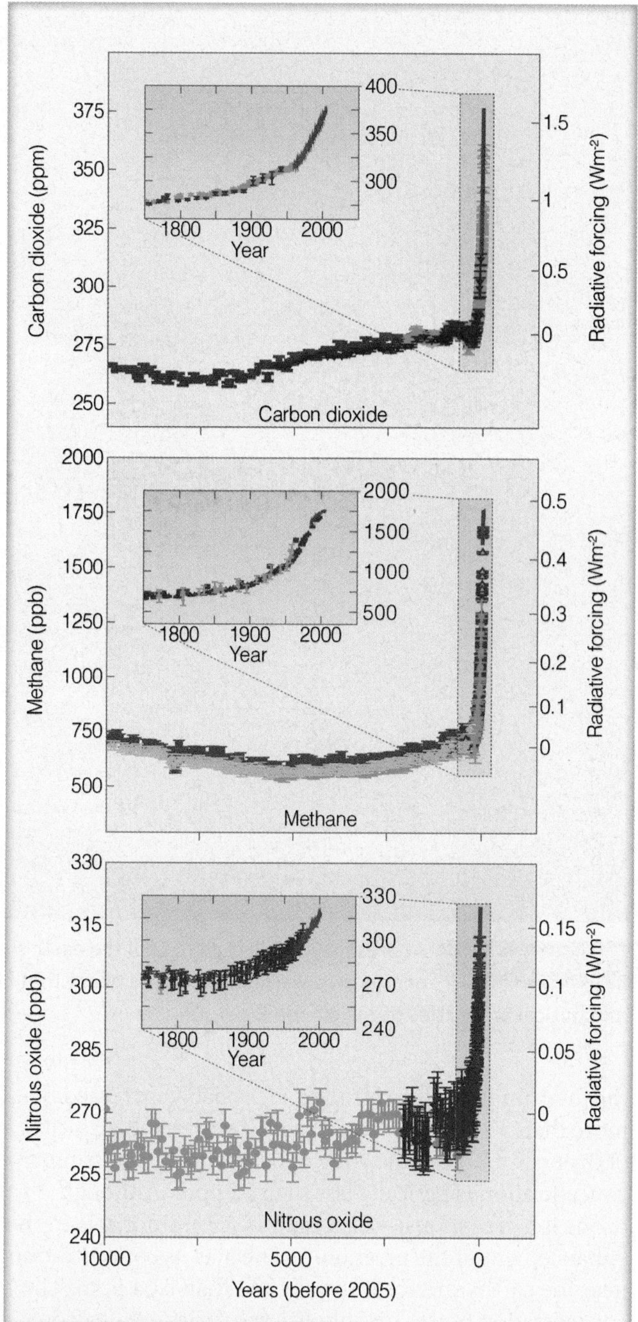

FIGURE 15.19 Changes in the greenhouse gases carbon dioxide, methane, and nitrous oxide indicated from ice core and modern data. (*Source:* Climate Change 2007, *The Physical Science Basis,* by the Working Group 1 contribution to the Fourth Assessment Report to the IPCC © 2007. Reprinted by permission of the Intergovernmental Panel on Climate Change.)

© C. Donald Ahrens

FIGURE 15.20 At present, clouds tend to cool the earth's surface by reflecting back to space much of the sun's incoming energy. In a warmer world, the effect that clouds will have on the climate system depends on the type, height, and optical properties of the clouds that form.

the total amount of surface ozone probably increased by more than 30 percent since 1750. However, the majority of ozone is found in the stratosphere where its maximum concentration is typically less than 12 ppm. Although ozone is a greenhouse gas, it plays a very minor role in the enhancement of the greenhouse effect as its concentration near the earth's surface is usually less than 0.04 ppm. The concentration of this greenhouse gas varies greatly from region to region, and depends upon the production of photochemical smog. The increase in surface ozone has probably led to a small increase in radiative forcing.

The Question of Clouds As the atmosphere warms and more water vapor is added to the air, global cloudiness might increase as well. How, then, would clouds—which come in a variety of shapes and sizes and form at different altitudes—affect the climate system? Clouds reflect incoming sunlight back to space, a process that tends to cool the climate, but clouds also emit infrared radiation to the earth, which tends to warm it. Just how the climate will respond to changes in cloudiness will probably depend on the type of clouds that form, their height above the surface, and their physical properties, such as liquid water (or ice) content, depth, and droplet size distribution. For example, high, thin cirriform clouds (composed mostly of ice) appear to promote a net warming effect: They allow a good deal of sunlight to pass through (which warms the earth's surface), yet because they are cold, they warm the atmosphere around them by absorbing more infrared radiation from the earth than they emit upward. Low stratified clouds, on the other hand, tend to promote a net cooling effect. Composed mostly of water droplets, they reflect much of the sun's incoming energy, which cools the earth's surface and, because their tops are relatively warm, they radiate to space much of the infrared energy they receive from the earth. Satellite data confirm that, overall, clouds presently have a *net cooling effect* on our planet, which means that, without clouds, our atmosphere would be warmer (see Fig. 15.20).

Additional clouds in a warmer world would not necessarily have a net cooling effect, however. Their influence on the average surface air temperature would depend on their extent and on whether low or high clouds dominate the climate scene. Consequently, the feedback from clouds could potentially enhance or reduce the warming produced by increasing greenhouse gases. Most models show that as the surface air warms, there will be more convection, more convective-type cumulus clouds, and an increase in cirrus clouds. This situation would tend to provide a small positive feedback on the climate system, and the effect of clouds on cooling the earth would be diminished.* (Information on how jet aircraft activity may be affecting clouds is given in the Focus section below.)

*In addition to the amount and distribution of clouds, the way in which climate models calculate the optical properties of a cloud (such as albedo) can have a large influence on the model's calculations. Also, there is much uncertainty as to how clouds will interact with aerosols, and what the net effect will be.

FOCUS ON
A SPECIAL TOPIC

Contrails and Climate Change

Human activities can have an effect on clouds that form in the upper troposphere. Jet aircraft flying at high altitudes often produce a cirrus-like trail of condensed vapor called a *condensation trail* or *contrail* (see Fig. 2). The condensation may come directly from the water vapor added to the air from the engine exhaust, or the exhaust may provide condensation nuclei, particles on which cloud droplets and ice crystals can form.

Some contrails disappear quickly after they form. If the relative humidity of the air is high, contrails may persist for some time, occasionally stretching across the sky as streamers of cirriform clouds that coalesce into a white canopy.*

Contrails can affect climate in a number of ways. For one thing, the added condensation nuclei can increase the number of ice crystals within cirriform clouds, which in turn can change the cloud's albedo. In addition, contrails can enhance cirriform cloudiness. Studies conducted during the mid-1960s showed that the occurrence of cirriform clouds above cities with heavy air traffic increased in tandem with commercial jet activity.

Contrails reflect sunlight and absorb infrared energy emitted from the earth's surface; hence, they have the ability to alter the temperature near the ground. Recent studies show that contrails have the potential to reduce the daily temperature range—the difference between the daily maximum and minimum temperature. As an example, after the terrorist attack on the World Trade Center in New York City on September 11, 2001, all commercial aircraft flights in the United States were cancelled for 3 days. During this period (with virtually no contrails), the average daily temperature range at the earth's surface across the United States increased by about 2°F above the climatological average.

The impact contrails will have on future climate is not known. Research is presently ongoing so that our understanding of the role contrails play in the climate system will improve.

*Cirriform clouds, which are mainly composed of ice, form at high altitudes, typically above about 20,000 feet in middle latitudes. Additional information on these clouds is given in Chapter 4, beginning on p. 105.

Figure 2 A contrail forming behind a jet aircraft.

The Ocean's Impact The oceans play a major role in the climate system, yet the exact effect they will have on climate change is not clear. For example, the oceans have a large capacity for storing heat energy. Thus, as they slowly warm, they should retard the rate at which the atmosphere warms. Overall, the response of ocean temperatures, ocean circulations, and sea ice to global warming will probably determine the global pattern and speed of climate change.

CONSEQUENCES OF GLOBAL WARMING: THE POSSIBILITIES If the world continues to warm as predicted by climate models, where will most of the warming take place? Climate models predict that land areas will warm more rapidly than the global average, particularly in the northern high latitudes in winter (see ❱ Fig. 15.21a). We can see in Fig. 15.21b that the greatest surface warming for the period 2001 to 2006 occurred over landmasses in the high latitudes of the Northern Hemisphere. These observations of global average temperature change suggest that climate models are on target with their warming projections.

As high-latitude regions of the Northern Hemisphere continue to warm, modification of the land may actually enhance the warming. For example, the dark green boreal forests* of the high latitudes absorb up to three times as much solar energy as does the snow-covered tundra. Consequently, the winter temperatures in subarctic regions are, on the average, much higher than they would be without trees. If warming allows the boreal forests to expand into the tundra, the forests may accelerate the warming in that region. As the temperature rises, organic matter in the soil should decompose at a faster rate, adding more CO_2 to the air, which might accelerate the warming even more. Trees that grow in a climate zone defined by temperature may become especially hard hit as rising temperatures place them in an inhospitable environment. In a weakened state, they may become more susceptible to insects and disease.

*The boreal forest consists of woodlands (northern part) and conifers and some hardwoods (southern part). Its northern boundary is next to the tundra along the Arctic tree line.

As the world warms, total rainfall must increase to balance the increase in evaporation. But precipitation will not be evenly distributed as some areas will get more precipitation, and others less (see ❱ Fig. 15.22). As you look at Fig. 15.22, keep in mind that the stippled areas represent regions where more than 90 percent of the models agree about whether precipitation will increase or decrease, and white areas represent those regions where less than 66 percent of the models agree about how precipitation will change. Notice in Fig. 15.22a that the models project an increase in winter precipitation over high latitudes of the Northern Hemisphere and a decrease in precipitation over areas of the subtropics. A decrease in precipitation in this region could have an adverse effect by placing added stress on agriculture. Some models even suggest that changes in global patterns of precipitation might cause more extreme rainfall events, such as floods and severe drought. In fact, it is interesting to note that during the warming of the 20th century, there appears to have been an increase in precipitation by as much as 10 percent over the middle- and

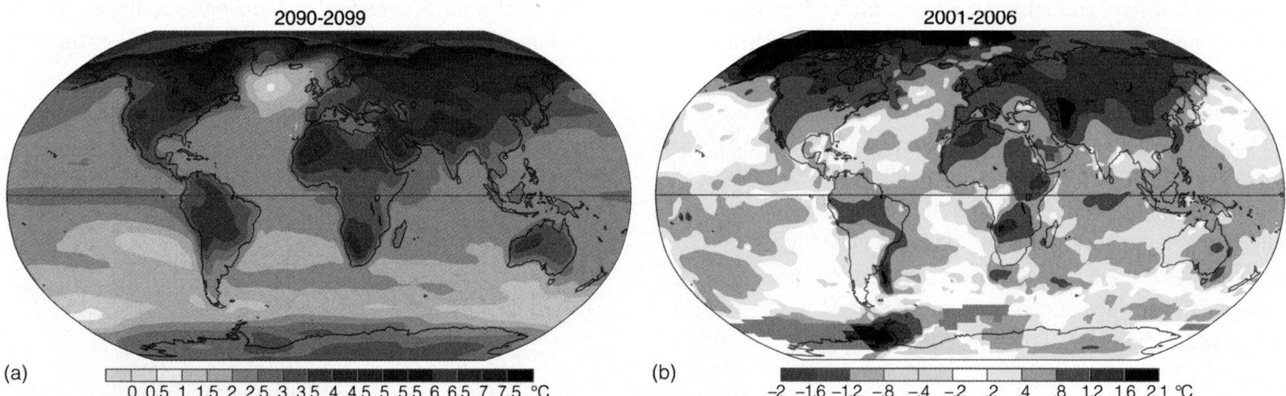

❱ FIGURE 15.21 (a) Projected surface air temperature changes averaged for the decade 2090–2099 (using the A1B scenario) compared to the average surface temperature for the period 1980–1999. The largest increase in air temperature is projected to be over landmasses and in the Arctic region. (b) The average change in surface air temperature for the period 2001–2006 compared to the average for the years 1951–1980. The greatest warming was over the Arctic region and the high-latitude landmasses of the Northern Hemisphere. (Diagram [a] *Source:* Climate Change 2007, *The Physical Science Basis* by the Working Group 1 contribution to the Fourth Assessment Report to the IPCC. Reprinted by permission of the Intergovernmental Panel on Climate Change. Diagram [b] Courtesy NASA.)

high-latitude land areas of the Northern Hemisphere. In contrast, it appears that over subtropical land areas, a decrease in precipitation has occurred. It also appears that there has been an increase in the frequency of heavy precipitation events during the last 50 years or so.

In mountainous regions of western North America, where much of the precipitation falls in winter, precipitation might fall mainly as rain, causing a decrease in snow-melt runoff that fills the reservoirs during the spring. In California, the reduction in water storage could threaten the state's agriculture.

Other consequences of global warming will likely be a rise in sea level as glaciers over land recede and the oceans continue to expand as they slowly warm. During the 20th century, sea level rose about 6 in. (15 cm), and today's improved climate models estimate that sea level will rise an additional 12 in. (30 cm) or more by the end of this century. The rise in sea level will depend on how much the temperature rises, and on how quickly the ice in Greenland and Antarctica melts. In fact, recent models suggest that sea level may rise more than 40 in. (100 cm) by the year 2100, as the ice in Greenland appears to be melting quite rapidly. Rising ocean levels could have a damaging influence on coastal ecosystems. In addition, coastal groundwater supplies might become contaminated with saltwater. And as we saw in Chapter 13, as sea surface temperatures increase (other factors being equal) the intensity of hurricanes will likely increase as well. (For more information on hurricanes and global warming, read the section about "Hurricanes in a Warmer World" on p. 393.)

In polar regions, as elsewhere around the globe, rising temperatures produce complex interactions among temperature, precipitation, and wind patterns. Hence, in polar areas more snow might actually fall in the warmer (but still cold) air, causing snow to build up or, at least, stabilize over the continent of Antarctica. Over Greenland, which is experiencing rapid melting of ice and snow, any increase in precipitation will likely be offset by rapid melting, and so the ice sheet is expected to continue to shrink. Presently, in the Arctic, warming has caused sea ice to shrink and thin. (Sea ice is formed by the freezing of sea water.) During 2005, the extent of Arctic sea ice was at a record minimum for every month except May. If the warming in this region continues at its present rate, polar sea ice in summer may be totally absent by the middle of this century (see ❯ Fig. 15.23).

Increasing levels of CO_2 in a warmer world might have additional consequences. For example, higher levels of CO_2 might act as a "fertilizer" for some plants, accelerating their growth. Increased plant growth consumes more CO_2, which might retard the increasing rate of CO_2 in the environment. On the other hand, the increased plant growth might force some insects to eat more, resulting in a net loss in vegetation. It is possible that a major increase in CO_2 might upset the balance of nature, with some plant species becoming so dominant that others are eliminated. In tropical areas, where many developing nations are located, the warming may actually decrease crop yield, whereas in cold

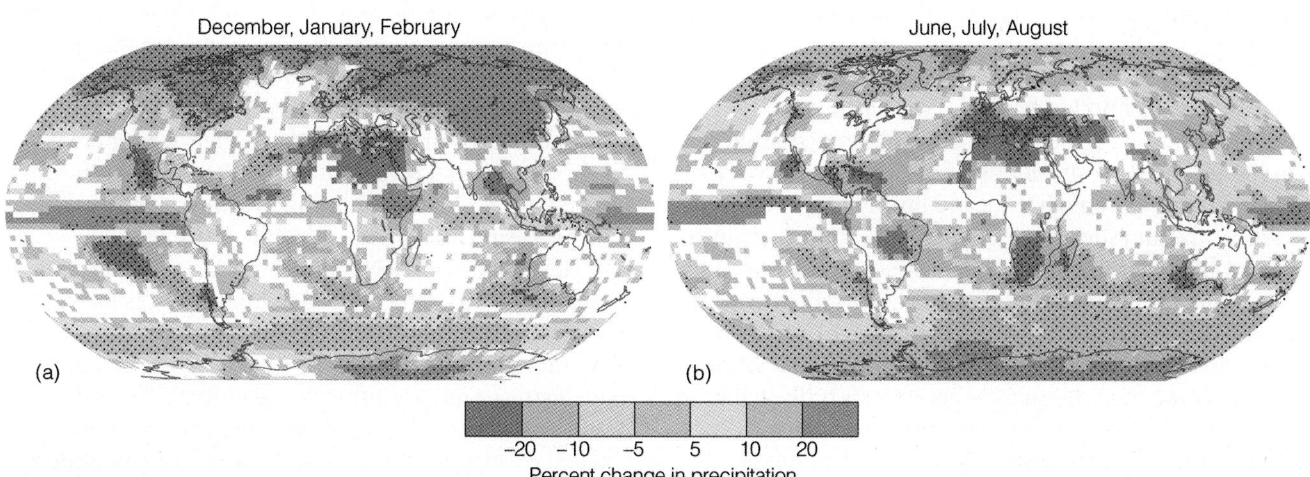

❯ FIGURE 15.22 Projected relative changes in precipitation (in percent) for the last decade of this century (2090–2099) compared to the average for the period 1980–1999. Values are multimodel averages for (a) December through February, and (b) June through August. The stippled areas represent regions where more than 90 percent of the models agree as to whether precipitation will increase or decrease; white regions show where less than 66 percent of the models agree about how precipitation will change. (*Source:* Climate Change 2007, *The Physical Science Basis,* by the Working Group 1 contribution to the Fourth Assessment Report to the IPCC. Reprinted by permission of the Intergovernmental Panel on Climate Change.)

FIGURE 15.23

The extent of Arctic sea ice in (a) March, 2005, when the ice cover was at or near its maximum and in (b) September, 2005, when the ice cover was near or at its minimum. The orange line represents the (a) median maximum and (b) median minimum extent of the ice cover for the period 1979–2000.

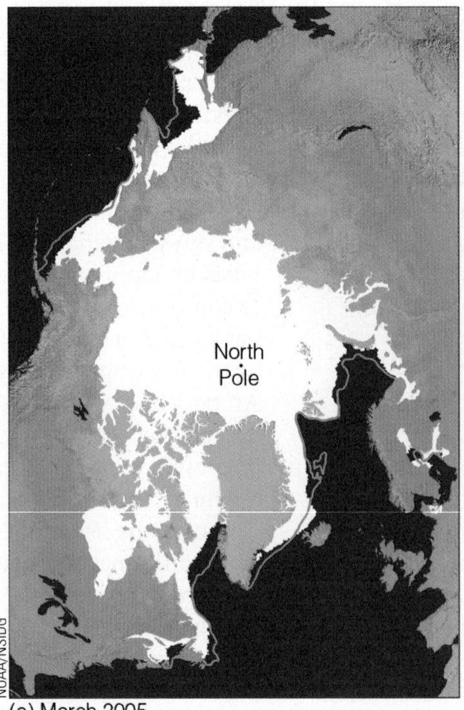

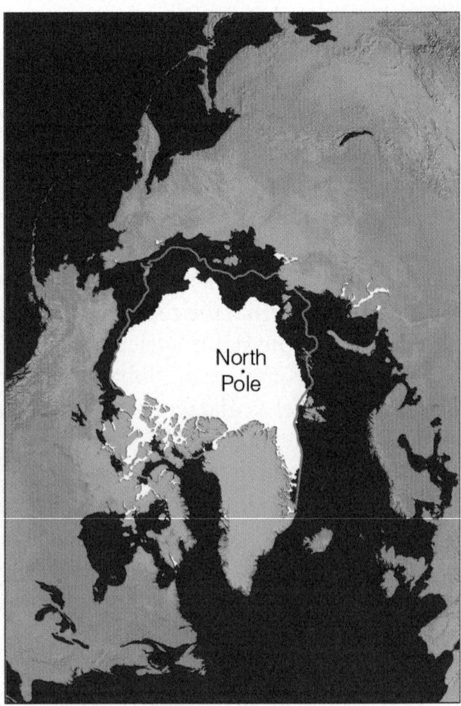

climates, where crops are now grown only marginally, the warming effect may actually increase crop yields. In a warmer world, higher latitudes might benefit from a longer growing season, and extremely cold winters might become less numerous with fewer bitter cold spells.

GLOBAL WARMING: POSSIBLE IMPACTS ON EXTREME WEATHER How will global warming influence extreme weather events? Will thunderstorms become more likely? What do the computer models suggest? In this section, we will address these questions by presenting possible scenarios for different types of extreme weather. Keep in mind that many of the possibilities suggested here are speculative and are based on computer simulations. The scenarios are designed to stimulate thought and discussion and are not intended to represent a forecast for the future.

Temperature and Humidity The earth is in a warming trend. If greenhouse gas emission continues to increase and the air temperature continues to rise, what will future temperature and humidity be like? To address this question we look to the past to project into the future. Over the past 50 years or so, global warming has resulted in widespread changes in extreme temperature. Cold days, cold nights, and frost have become less frequent, whereas hot days, hot nights, and heat waves have become more frequent. Therefore, it is likely that in the future there will be more summer heat waves with more record high temperatures. In fact, very hot days, by the end of this century, are projected to be about 10°F higher than at present. Nighttime temperatures are likely to be higher, too. The number of days each year with air temperature above 90°F is projected to increase throughout the United States. Some models predict that if CO_2 emissions continue to increase, it is likely that over parts of the southern United States, where a temperature of 90°F or above occurs on average about 60 days per year, by the end of this century the number of days over 90°F will increase to 150 days per year.

Recent modeling studies indicate that there will be an increased risk of more extreme, more frequent, and longer-lasting heat waves. Over much of the United States, heat wave events that now occur once every 20 years or so, by the end of this century will likely occur about every other year. The European heat wave of 2003 is an example of the type of extreme heat event that is likely to become much more common in the future. Some models indicate that by the 2040s more than half of European summers will be hotter than the summer of 2003.

In regions of the world that experience an increase in the air's water vapor content, average dew-point temperatures will rise. In humid regions, such as the southeastern United States and the Caribbean, the average Heat Index (HI), which combines air temperature and moisture to determine how hot it feels, is likely to rise also. This situation could make humid summer days more uncomfortable than they already are, putting added stress on both humans and animals. In addition, the increased demand for more energy to provide needed cooling might cause an increase in greenhouse gas emissions.

Precipitation—Drought and Floods As we saw earlier in this chapter, in a warmer world it is likely that, averaged over the entire earth, precipitation must increase to compensate for the added water vapor evaporated from the oceans. The models, however, project that the added precipitation will not be evenly distributed across the planet. Most models predict wetter winters and drier summers over most of the middle and high latitudes of the Northern Hemisphere, including North America.

The added water vapor in the air brings a risk of more intense precipitation events. In fact, where precipitation is likely to increase, the models predict more frequent and extreme flooding. In areas where precipitation is likely to decrease, more frequent and extreme drought is likely. In between the heavy rainfall events, the models project that there will be long periods with little or no precipitation. Over much of Europe, some models suggest that winters will be very wet, which, in turn, could produce an increased possibility for flooding.

One area of particular concern is the effect climate change might have on the monsoon of southeast Asia. More than half of the world's population depends upon the monsoon rains for much needed rain for agriculture and basic human needs. How the monsoon will respond to a warmer planet is not totally clear, but observations suggest that the monsoon over central India has intensified with the number of reported heavy rainfall events having doubled since the early 1950s.

Mid-Latitude Cyclonic Storms In Chapter 10, we learned that the polar jet stream is instrumental in the formation and growth of a mid-latitude cyclonic storm. The jet stream provides an area of diverging air aloft that promotes upward air motions and the lowering of surface air pressure. Everything else being equal, the stronger the winds of the jet stream, the stronger the area of upper-level diverging air, and the more intense the surface storm.

The polar jet stream forms, in part, due to horizontal variations in air temperature. Should the polar region warm dramatically as projected by climate models, what effect might this warming have on the polar jet stream? As polar regions warm, more than the rest of the world, the contrast in temperature between low and high latitude (the temperature gradient) should decrease. This decrease in temperature gradient could reduce the strength of the winds in the polar jet stream, which, in turn, might decrease the area of diverging air above the developing storm. Without strong upper-level support, surface storms might not become as intense as they are today.

Climate models project that, in the Northern Hemisphere, as polar regions warm, the polar jet stream will likely move northward, and that mid-latitude cyclonic storms will likely move northward, too. This shifting in the storm track might greatly affect winds, precipitation, and temperatures in the middle latitudes, especially in winter when mid-latitude cyclones are more strongly developed.

Lake-Effect Snowstorms Lake-effect snowstorms form when cold air passes over a warmer body of water. The body of water can be any relatively large lake, such as the Great Lakes or the Great Salt Lake. The cold air absorbs moisture evaporated from the lake, converts it into clouds, and eventually dumps the moisture as heavy snows on the downwind side of the lake. Heavy lake-effect snows occur over the Great Lakes during the fall and early winter before the lakes freeze over. The greater the contrast in temperature between the air and the water below, the greater the likelihood of a heavy snowstorm.

As the earth's surface warms, the temperature difference between the air and the water might diminish, which would reduce convection somewhat. This condition, in turn, might reduce the intensity of lake-effect snowstorms. On the other hand, a warmer lake would not freeze over as quickly as a colder one, so there could be more ice-free winter months, a situation that might lead to more lake-effect snowstorms. As a consequence, as the earth's surface warms, there may be fewer really heavy lake-effect snowstorm events, but a longer period of time in which these storms might form.

Thunderstorms and Tornadoes At present, climate models are not clear on the question of how the number of thunderstorms might change as the earth warms. Some models show that the intensity (updrafts) within thunderstorms might increase, suggesting that stronger storms might occur with greater frequency. We know that climate in the future is expected to be warmer than it is now. If the climate in summer turns out to be drier over the continental United States, there may be fewer thunderstorms due to the decrease in moisture that drives the storms. On the other hand, warmer conditions might initiate more convection and possibly more storms. So, at present, it is not clear whether a warmer world will enhance, diminish, or have no influence on thunderstorm activity over North America.

EXTREME WEATHER WATCH

An analysis of extreme temperature records at 300 weather stations in the United States selected for both their length of record and their geographic distribution indicates that, between the years 2000 and 2009, 114 of the sites recorded either their highest temperature ever or their warmest single month on record. Conversely, during this same period, only two stations reported either their lowest temperature ever or their coldest single month on record.

NASA

FIGURE 15.24 As the earth continues to warm, what effect will the warming have on the jet stream and the development of mid-latitude cyclonic storms similar to these two storms that have formed over the north Atlantic Ocean during November, 2006?

We saw in Chapter 12 that the number of reported tornadoes has been increasing in recent decades. This increase appears to be due mainly to advances in observational techniques, and not to global warming. So, how will tornadoes change as the climate warms? If thunderstorms become more intense with stronger updrafts, tornadoes might become more numerous if the proper wind shear is present. However, at the present time, it is unknown how tornado frequency and intensity will change (if at all) on a warmer planet.

Hurricanes As we learned in Chapter 13, hurricanes (tropical cyclones) form over warm tropical water where the surface water temperature is at least 80°F (26.5°C) and the winds above the surface are light. How will global warming influence both the strength and number of hurricanes that form each year?

Studies indicate that surface water temperatures in tropical regions have increased over the past several decades. As tropical oceans warm, they evaporate more water into the air. The added water vapor, along with the increase in surface water temperature, tends to increase the energy available for hurricane development. The added energy is likely to produce a more intense storm, and the stronger the storm, the higher its winds. Moreover, the added water vapor may increase the amount of rainfall that the storm deposits.

Climate models predict that, as the earth continues to warm, the surface of the sea in the tropics will also warm. Hurricanes that form over this warmer water should be more intense, with higher wind speeds, everything else being equal. Will hurricanes become more frequent as the earth warms? The answer to this question is not clear, as some climate models predict more hurricanes, whereas others predict fewer storms. However, as tropical water warms, it is likely that the temperature of surface water capable of fueling hurricanes will exist for a longer time. Hence, "hurricane season" may be longer than at present.

One problem with predicting hurricane frequency is that at this time it is impossible to forecast what other factors may arise as the world warms. For example, it is unclear if wind shear will be present over the tropical Atlantic and disrupt the formation of an organizing storm. We know that, as the tropical Pacific warms during El Niño years, there are fewer Atlantic Basin hurricanes due to stronger winds aloft. As the tropical Pacific Ocean becomes warmer than it is now, will stronger winds aloft be a more permanent fixture over the tropical Atlantic? And how will these stronger winds (if they should exist) influence both the frequency and intensity of hurricanes in this region? At this point, we have much to learn about hurricane formation as the earth's climate system changes.

Summary of Extreme Weather Events and Global Warming Many weather extremes have been changing over the past several decades. Most of North America has been experiencing more unusually hot days and nights, fewer unusually cold days and nights, and fewer frost days. Heavy downpours have become more frequent and intense. Droughts are becoming more severe in some regions. Climate models predict that, as the earth continues to warm due to increasing concentrations of greenhouse gases, these types of weather and climate changes will not only continue into the future, but will increase in both intensity and frequency.

Linking extreme weather events with global warming remains an area of vigorous research and debate. It is important to know what weather and climate will be like in the future, as it appears we are entering a period in which the impact of extreme weather events will likely produce a large potential for human suffering. What steps, then, are now being taken to curb global warming?

GLOBAL WARMING: EFFORTS TO CURB The most obvious way to curb global warming is to reduce greenhouse gas emissions by reducing the use of fossil fuels, such as oil and coal. Using alternative energy such as solar collectors and wind power—the world's two fastest growing energy sources—could also help with this endeavor.

In an attempt to mitigate the impact humans have on the climate system, representatives from 160 countries met at Kyoto, Japan, in 1997 to work out a formal agreement to limit greenhouse gas emissions in industrialized nations. The international agreement—called the *Kyoto Protocol*—was adopted in 1997, and was put into force in February, 2005.

The Protocol sets mandatory targets for reducing greenhouse gas emissions in countries that adopt the plan. Although the percent by which each country reduces its emissions varies, the overall goal is to reduce greenhouse gas emissions in developed countries by at least 5 percent below existing 1990 levels during the 5-year period of 2008 through 2012.

The agreement gives countries flexibility in meeting their emission-reduction goals. For example, a country that plants forests can receive "credit" for reducing greenhouse gases, because trees act as a "sink" and remove CO_2 from the atmosphere. Other types of "credits" may be given to industrialized countries that establish emission-reducing projects in developing countries. Although the plan has gained world wide acceptance, the United States has not signed the Protocol as of this writing. However, many large states such as California have implemented climate change policies. California's aggressive plan (adopted in 2006) sets targets for reducing greenhouse gas emissions to 1990 levels by the year 2020.

A study headed by Tom Wigley at the National Center for Atmospheric Research (NCAR) suggests that injecting sulfate aerosols into the stratosphere could slow down global warming. Using computer models to simulate climate, tons of sulfate aerosols—on the order of those lofted by Mount Pinatubo in 1991—were put into the stratosphere at various intervals.

The study concluded that injecting sulfate aerosols into the stratosphere every one to four years in conjunction with reducing greenhouse gases could provide a "grace period" of up to 20 years before major cutbacks in greenhouse gas emissions would be required. Of course, injecting sulfate particles into the stratosphere might have additional consequences, such as changing the temperature of the upper atmosphere and affecting the fragile ozone layer. Although the idea of injecting the stratosphere with sulfate particles has not been given much credence by scientists, the idea of reducing the impact of climate change through global scale technological fixes (called **geoengineering**) is intriguing. The science of geoengineering is fairly new, and typically poses costly technological challenges for the scientific community. For example, other geoengineering proposals might include fertilizing the oceans with plants that absorb CO_2, or putting reflective mirrors in space to reduce sunlight, or changing the reflective characteristics of clouds that would change the earth's albedo.

Cutting down on the emissions of greenhouse gases and pollutants has several potentially positive benefits. A reduction in greenhouse gas emissions could slow down the enhancement of the earth's greenhouse effect and reduce global warming while at the same time it would reduce a country's dependence on oil. A reduction in air pollutants might reduce acid rain, diminish haze, and slow the production of photochemical smog. Even if the greenhouse warming proves to be less than what modern climate models project, these measures would certainly benefit humanity.

Summary

In this chapter, we considered some of the many ways the earth's climate can be changed by both natural and human influences. First, we saw that the earth's climate has undergone considerable change during the geologic past. Some of the evidence for a changing climate comes from tree rings (dendrochronology), chemical analysis of oxygen isotopes in ice cores and fossil shells, and geologic evidence left behind by advancing and retreating glaciers. The evidence from these suggests that, throughout much of the geologic past (long before humanity arrived on the scene), the earth was warmer than it is today. There were cooler periods, however, during which glaciers advanced over large sections of North America and Europe.

We examined some of the possible causes of climate change, noting that the problem is extremely complex, as a change in one variable in the climate system almost immediately changes other variables. One theory of climate change suggests that the shifting of the continents, along with volcanic activity and mountain building, may account for variations in climate that take place over millions of years.

The Milankovitch theory proposes that alternating glacial and interglacial episodes during the past 2.5 million years are the result of small variations in the tilt of the earth's axis and in the geometry of the earth's orbit around the sun. Another theory suggests that certain cooler periods in the geologic past may have been caused by volcanic eruptions rich in sulfur. Still another theory postulates that climatic variations on earth might be due to variations in the sun's energy output.

We looked at temperature trends and found that, since the beginning of the last century, the earth's surface has warmed by more than 0.8°C. Scientific studies suggest that it is very likely that most of the warming during the last 50 years is due to increasing concentrations of greenhouse gases emitted by human activities. Sophisticated climate models project that, as levels of CO_2 and other greenhouse gases continue to increase, the earth will warm substantially by the end of this century. The average warming over the next several decades will likely be close to 0.2°C per decade. The models also predict that, as the earth warms, there will be a global increase in atmospheric water vapor, an increase in global precipitation, a more rapid melting of sea ice, and a rise in sea level. Extreme weather events will likely change, too. Although speculative, some studies indicate that drought, heavy downpours, excessive heat, and intense hurricanes are likely to become more commonplace in a warmer world.

Key Terms

The following terms are listed (with page numbers) in the order they appear in the text. Define each. Doing so will aid you in reviewing the material covered in this chapter.

global warming, 432
dendrochronology, 434
Ice Age, 434
interglacial period, 435
Younger Dryas (event), 435
mid-Holocene maximum, 435
Little Ice Age, 437
water vapor–greenhouse feedback, 439
positive feedback mechanism, 439
snow-albedo feedback, 439
negative feedback mechanism, 440
theory of plate tectonics, 440
Milankovitch theory, 441
eccentricity, 441
precession, 442
obliquity, 442
Maunder minimum, 444
sulfate aerosols, 447
desertification, 448
radiative forcing agents, 450
radiative forcing, 450
geoengineering, 461

Review Questions

1. What methods do scientists use to determine climate conditions that have occurred in the past?
2. Explain how the changing climate influenced the formation of the Bering land bridge.
3. How does today's average global temperature compare with the average temperature during most of the past 1000 years?
4. Explain how the ocean conveyor belt plays a role in the climate of northern Europe.
5. What is the Younger Dryas episode? When did it occur?
6. What are some of the possible causes for the Little Ice Age?
7. Describe how the earth's average surface temperature has changed since the early 1900s.
8. List some of the natural causes of climate change.
9. How does a positive feedback mechanism differ from a negative feedback mechanism? Is the water vapor-greenhouse feedback considered positive or negative? Explain.
10. How does the theory of plate tectonics explain climate change over periods of millions of years?
11. Describe the Milankovitch theory of climatic change by explaining how each of the three cycles alters the amount of solar energy reaching the earth.
12. Given the analysis of air bubbles trapped in polar ice during the past 400,000 years, were CO_2 levels generally higher or lower during colder glacial periods? Were methane levels higher or lower during these colder glacial periods?
13. How do the concentrations of CO_2 in today's atmosphere compare with maximum CO_2 levels thousands of years ago?
14. In climate change, what is a "forcing factor?"
15. How do sulfate aerosols in the lower atmosphere affect surface air temperatures during the day?
16. Describe the scenario of nuclear winter.
17. Do volcanic eruptions rich in sulfur tend to warm or cool the earth's surface? Explain.
18. Explain how variations in the sun's energy output might influence global climate.
19. List several ways in which human activities can cause climate change.
20. Climate models predict that increasing levels of CO_2 will cause the mean global surface air temperature to rise significantly by the year 2100. What other greenhouse gas *must* also increase in concentration in order for this condition to occur?
21. Describe some of the natural- and human-produced radiative forcing agents and their effect on climate.
22. (a) Describe how clouds influence the climate system.
    (b) Which clouds would tend to promote surface cooling—high clouds or low clouds?
23. In Fig. 15.17, p. 449, explain why the actual rise in surface air temperature (gray line) is much less than the projected rise in air temperature due to increasing levels of greenhouse gases (yellow line).
24. Why do climate scientists now believe that most of the warming experienced during the last 50 years was due to increasing levels of greenhouse gases, and not natural causes?
25. List some of the consequences that global warming might have on the atmosphere and its inhabitants.
26. Explain how global warming might influence the intensity and/or frequency of several extreme weather events.

Online Learning

STUDENT COMPANION WEBSITE: Visit this book's companion website at: www.cengage.com/ahrens/extreme1e and choose Chapter 15 for many study aids and ideas for further reading and research. These include flashcards, practice quizzing, and web links.

METEOROLOGY RESOURCE CENTER: For students with access, log on at www.cengage.com/login for more assets, including animations, videos, and more. If your textbook did not come with access, visit www.CengageBrain.com to purchase.

Units, Conversions, and Abbreviations

A

Length

1 kilometer (km)	=	1000 m
	=	3281 ft
	=	0.62 mi
1 mile (mi)	=	5280 ft
	=	1609 mi
	=	1.61 km
1 meter (m)	=	100 cm
	=	3.28 ft
	=	39.37 in.
1 foot (ft)	=	12 in.
	=	30.48 cm
	=	0.305 m
1 centimeter (cm)	=	0.39 in.
	=	0.01 m
	=	10 mm
1 inch (in.)	=	2.54 cm
	=	0.08 ft
1 millimeter (mm)	=	0.1 cm
	=	0.001 m
	=	0.039 in.
1 micrometer (μm)	=	0.0001 cm
	=	0.000001 m
1 degree latitude	=	111 km
	=	60 nautical mi
	=	69 statute mi

Area

1 square centimeter (cm^2)	=	0.15 $in.^2$
1 square inch ($in.^2$)	=	6.45 cm^2
1 square meter (m^2)	=	10.76 ft^2
1 square foot (ft^2)	=	0.09 m^2

Volume

1 cubic centimeter (cm^3)	=	0.06 $in.^3$
1 cubic inch ($in.^3$)	=	16.39 cm^3
1 liter (l)	=	1000 cm^3
	=	0.264 gallon (gal) U.S.

Speed

1 knot	=	1 nautical mi/hr
	=	1.15 statute mi/hr
	=	0.51 m/sec
	=	1.85 km/hr
1 mile per hour (mi/hr)	=	0.87 knot
	=	0.45 m/sec
	=	1.61 km/hr
1 kilometer per hour (km/hr)	=	0.54 knot
	=	0.62 mi/hr
	=	0.28 m/sec
1 meter per second (m/sec)	=	1.94 knots
	=	2.24 mi/hr
	=	3.60 km/hr

Force

1 dyne	=	1 gram centimeter per second per second
	=	2.2481×10^{-6} pound (lb)
1 newton (N)	=	1 kilogram meter per second per second
	=	10^5 dynes
	=	0.2248 lb

Mass

1 gram (g)	=	0.035 ounce
	=	0.002 lb
1 kilogram (kg)	=	1000 g
	=	2.2 lb

Energy

1 erg	=	1 dyne per cm
	=	2.388×10^{-8} cal
1 joule (J)	=	1 newton meter
	=	0.239 cal
	=	10^7 erg
1 calorie (cal)	=	4.186 J
	=	4.186×10^7 erg

Pressure

1 millibar (mb)	=	1000 dynes/cm^2
	=	0.75 millimeter of mercury (mm Hg)
	=	0.02953 inch of mercury (in. Hg)
	=	0.01450 pound per square inch (lb/in.^{2})
	=	100 pascals (Pa)
1 standard atmosphere	=	1013.25 mb
	=	760 mm Hg
	=	29.92 in. Hg
	=	14.7 lb/in.^{2}
1 inch of mercury	=	33.865 mb
1 millimeter of mercury	=	1.3332 mb
1 pascal	=	0.01 mb
	=	1 N/m^2
1 hectopascal (hPa)	=	1 mb
1 kilopascal (kPa)	=	10 mb

Power

1 watt (W)	=	1 J/sec
	=	14.3353 cal/min
1 cal/min	=	0.06973 W
1 horse power (hp)	=	746 W

Powers of Ten

PREFIX					
nano	one-billionth	=	10^{-9}	=	0.000000001
micro	one-millionth	=	10^{-6}	=	0.000001
milli	one-thousandth	=	10^{-3}	=	0.001
centi	one-hundredth	=	10^{-2}	=	0.01
deci	one-tenth	=	10^{-1}	=	0.1
hecto	one hundred	=	10^{2}	=	100
kilo	one thousand	=	10^{3}	=	1000
mega	one million	=	10^{6}	=	1,000,000
giga	one billion	=	10^{9}	=	1,000,000,000

Temperature

$$°C = \frac{5}{9}(°F - 32)$$

To convert degrees Fahrenheit (°F) to degrees Celsius (°C): Subtract 32 degrees from °F, then divide by 1.8.

To convert degrees Celsius (°C) to degrees Fahrenheit (°F): Multiply °C by 1.8, then add 32 degrees.

To convert degrees Celsius (°C) to Kelvins (K): Add 273 to Celsius temperature, as

$$K = °C + 273.$$

Table A.1

Temperature Conversions

°F	°C	°F	°C	°F	°C	°F	°C	°F	°C	°F	°C	°F	°C	°F	°C
-40	-40	-20	-28.9	0	-17.8	20	-6.7	40	4.4	60	15.6	80	26.7	100	37.8
-39	-39.4	-19	-28.3	1	-17.2	21	-6.1	41	5.0	61	16.1	81	27.2	101	38.3
-38	-38.9	-18	-27.8	2	-16.7	22	-5.6	42	5.6	62	16.8	82	27.8	102	38.9
-37	-38.3	-17	-27.2	3	-16.1	23	-5.0	43	6.1	63	17.2	83	28.3	103	39.4
-36	-37.8	-16	-26.7	4	-15.6	24	-4.4	44	6.7	64	17.8	84	28.9	104	40.0
-35	-37.2	-15	-26.1	5	-15.0	25	-3.9	45	7.2	65	18.3	85	29.4	105	40.6
-34	-36.7	-14	-25.6	6	-14.4	26	-3.3	46	7.8	66	18.9	86	30.0	106	41.1
-33	-36.1	-13	-25.0	7	-13.9	27	-2.8	47	8.3	67	19.4	87	30.6	107	41.7
-32	-35.6	-12	-24.4	8	-13.3	28	-2.2	48	8.9	68	20.0	88	31.1	108	42.2
-31	-35.0	-11	-23.9	9	-12.8	29	-1.7	49	9.4	69	20.6	89	31.7	109	42.8
-30	-34.4	-10	-23.3	10	-12.2	30	-1.1	50	10.0	70	21.1	90	32.2	110	43.3
-29	-33.9	-9	-22.8	11	-11.7	31	-0.6	51	10.6	71	21.7	91	32.8	111	43.9
-28	-33.3	-8	-22.2	12	-11.1	32	0.0	52	11.1	72	22.2	92	33.3	112	44.4
-27	-32.8	-7	-21.7	13	-10.6	33	0.6	53	11.7	73	22.8	93	33.9	113	45.0
-26	-32.2	-6	-21.1	14	-10.0	34	1.1	54	12.2	74	23.3	94	34.4	114	45.6
-25	-31.7	-5	-20.6	15	-9.4	35	1.7	55	12.8	75	23.9	95	35.0	115	46.1
-24	-31.1	-4	-20.0	16	-8.9	36	2.2	56	13.3	76	24.4	96	35.6	116	46.7
-23	-30.6	-3	-19.4	17	-8.3	37	2.8	57	13.9	77	25.0	97	36.1	117	47.2
-22	-30.0	-2	-18.9	18	-7.8	38	3.3	58	14.4	78	25.6	98	36.7	118	47.8
-21	-29.4	-1	-18.3	19	-7.2	39	3.9	59	15.0	79	26.1	99	37.2	119	48.3

Table A.2

SI Units* and Their Symbols

QUANTITY	NAME	UNITS	SYMBOL
length	meter	m	m
mass	kilogram	kg	kg
time	second	sec	sec
temperature	Kelvin	K	K
density	kilogram per cubic meter	kg/m^3	kg/m^3
speed	meter per second	m/sec	m/sec
force	newton	$m \cdot kg/sec^2$	N
pressure	pascal	N/m^2	Pa
energy	joule	$N \cdot m$	J
power	watt	J/sec	W

*SI stands for Système International, which is the international system of units and symbols.

B Equations and Constants

Gas Law (Equation of State)

The relationship among air pressure, air density, and air temperature can be expressed by

$$\text{Pressure} = \text{density} \times \text{temperature} \times \text{constant}.$$

This relationship, often called the gas law (or equation of state), can be expressed in symbolic form as:

$$p = \rho RT$$

where p is air pressure, ρ is air density, R is a constant, and T is air temperature.

UNITS/CONSTANTS

p = pressure in N/m^2 (SI)
ρ = density (kg/m^3)
T = temperature (K)
R = 287 J/kg • K (SI) or
$R = 2.87 \times 10^6$ erg/g • K

Stefan-Boltzmann Law

The Stefan-Boltzmann law is a law of radiation. It states that all objects with temperatures above absolute zero emit radiation at a rate proportional to the fourth power of their absolute temperature. It is expressed mathematically as:

$$E = \sigma T^4$$

where E is the maximum rate of radiation emitted each second per unit surface area, T is the object's surface temperature, and σ is a constant.

UNITS/CONSTANTS

E = radiation emitted in W/m^2 (SI)
$\sigma = 5.67 \times 10^{-8}$ W/m^2 • K^4 (SI) or
$\sigma = 5.67 \times 10^{-5}$ erg/cm^2 • K^4 • sec
T = temperature (K)

Wien's Law

Wien's law (or Wien's displacement law) relates an object's maximum emitted wavelength of radiation to the object's temperature. It states that the wavelength of maximum emitted radiation by an object is inversely proportional to the object's absolute temperature. In symbolic form, it is written as:

$$\lambda_{max} = \frac{w}{T}$$

where λ_{max} is the wavelength at which maximum radiation emission occurs, T is the object's temperature, and w is a constant.

UNITS/CONSTANTS

λ_{max} = wavelength (micrometers)
w = 0.2897 μm K
T = temperature (K)

Geostrophic Wind Equation

The geostrophic wind equation gives an approximation of the wind speed above the level of friction, where the wind blows parallel to the isobars or contours. The equation is expressed mathematically as:

$$V_g = \frac{1}{2\Omega \sin\phi \rho} \frac{\Delta p}{d}$$

where V_g is the geostrophic wind, Ω is a constant (twice the earth's angular spin), $\sin\phi$ is a trigonometric function that takes into account the variation of latitude (ϕ), ρ is the air density, Δp is the pressure difference between two places on the map some horizontal distance (d) apart.

UNITS/CONSTANTS

V_g = geostrophic wind (m/sec)
Ω = 7.29×10^{-5} radian*/sec
ϕ = latitude
ρ = air density (kg/m^3)
d = distance (m)
Δp = pressure difference (newton/m^2)

*2π radians equal 360°.

Hydrostatic Equation

The hydrostatic equation relates to how quickly the air pressure decreases in a column of air above the surface. The equation tells us that the rate at which the air pressure decreases with height is equal to the air density times the acceleration of gravity. In symbolic form, it is written as:

$$\frac{\Delta p}{\Delta z} = -\rho g$$

where Δp is the decrease in pressure along a small change in height Δz, ρ is the air density, and g is the force of gravity.

UNITS/CONSTANTS

Δp = pressure difference (newton/m^2)
Δz = change in height (m)
ρ = air density (kg/m^3)
g = force of gravity (9.8m/sec^2)

Relative Humidity

The relative humidity of the air can be expressed as:

$$RH = \frac{e}{e_s} \times 100\%.$$

UNITS/CONSTANTS
e = actual vapor pressure (millibars)
e_s = saturation vapor pressure (millibars)
RH = relative humidity (percent)

Computing Relative Humidity

To determine e and e_s, when the air temperature and dew-point temperature are known, consult Table B.1. Simply read the value adjacent to the air temperature and obtain e_s; read the value adjacent to the dew-point temperature and obtain e.

To compute the relative humidity, use Table B.1 and the expression:

$$RH = \frac{e}{e_s} \times 100\%$$

If the air temperature is 80°F (27°C), the value for e_s is 35 mb. If the dew-point temperature is 65°F (18°C), then the value for e is 21 mb, and the relative humidity is:

$$RH = \frac{21}{35} \times 100\%, \text{ or } 60\%$$

Table B.1

Saturation Vapor Pressure over Water for Various Air Temperatures

AIR TEMPERATURE		SATURATION VAPOR	AIR TEMPERATURE		SATURATION VAPOR
(°C)	(°F)	PRESSURE (MB)	(°C)	(°F)	PRESSURE (MB)
−18	(0)	1.5	18	(65)	21.0
−15	(5)	1.9	21	(70)	25.0
−12	(10)	2.4	24	(75)	29.6
−9	(15)	3.0	27	(80)	35.0
−7	(20)	3.7	29	(85)	41.0
−4	(25)	4.6	32	(90)	48.1
−1	(30)	5.6	35	(95)	56.2
2	(35)	6.9	38	(100)	65.6
4	(40)	8.4	41	(105)	76.2
7	(45)	10.2	43	(110)	87.8
10	(50)	12.3	46	(115)	101.4
13	(55)	14.8	49	(120)	116.8
16	(60)	17.7	52	(125)	134.2

Weather Symbols and the Station Model

C

Simplified Surface-Station Model

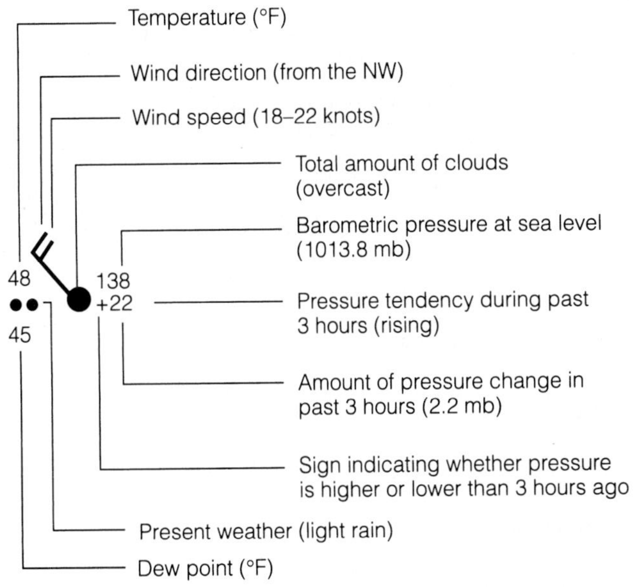

Cloud Coverage

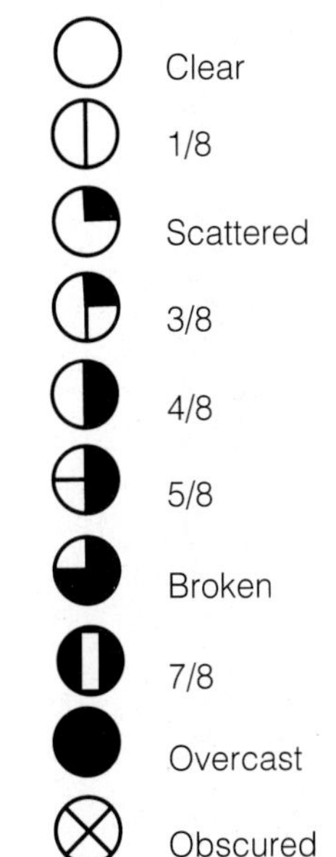

Clear
1/8
Scattered
3/8
4/8
5/8
Broken
7/8
Overcast
Obscured
Missing

Upper-Air Model (500 mb)

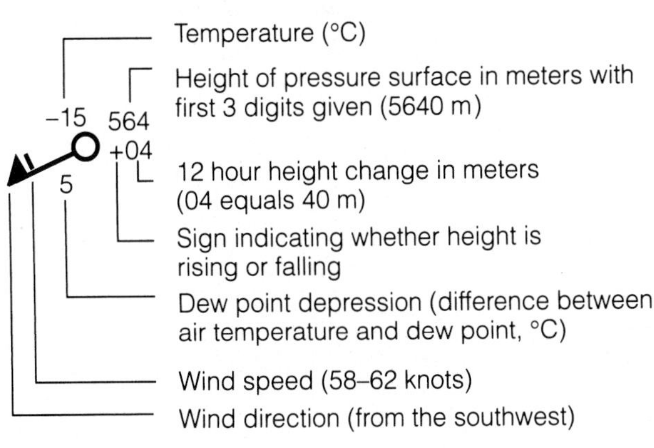

Common Weather Symbols

Light rain
Moderate rain
Heavy rain
Light snow
Moderate snow
Heavy snow
Light drizzle
Ice pellets (sleet)
Freezing rain
Freezing drizzle
Rain shower
Snow shower
Showers of hail
Drifting or blowing snow
Dust storm
Fog
Haze
Smoke
Thunderstorm
Hurricane

Wind Entries

	MILES (STATUTE) PER HOUR	KNOTS	KILOMETERS PER HOUR
	Calm	Calm	Calm
	1–2	1–2	1–3
	3–8	3–7	4–13
	9–14	8–12	14–19
	15–20	13–17	20–32
	21–25	18–22	33–40
	26–31	23–27	41–50
	32–37	28–32	51–60
	38–43	33–37	61–69
	44–49	38–42	70–79
	50–54	43–47	80–87
	55–60	48–52	88–96
	61–66	53–57	97–106
	67–71	58–62	107–114
	72–77	63–67	115–124
	78–83	68–72	125–134
	84–89	73–77	135–143
	119–123	103–107	144–198

Pressure Tendency

Rising, then falling

Barometer now higher than 3 hours ago:
- Rising, then steady; or rising, then rising more slowly
- Rising steadily or unsteadily
- Falling or steady, then rising; or rising, then rising more quickly

Steady, same as 3 hours ago

Falling, then rising, same or lower than 3 hours ago

Barometer now lower than 3 hours ago:
- Falling, then steady; or falling, then falling more slowly
- Falling steadily, or unsteadily
- Steady or rising, then falling; or falling, then falling more quickly

Front Symbols

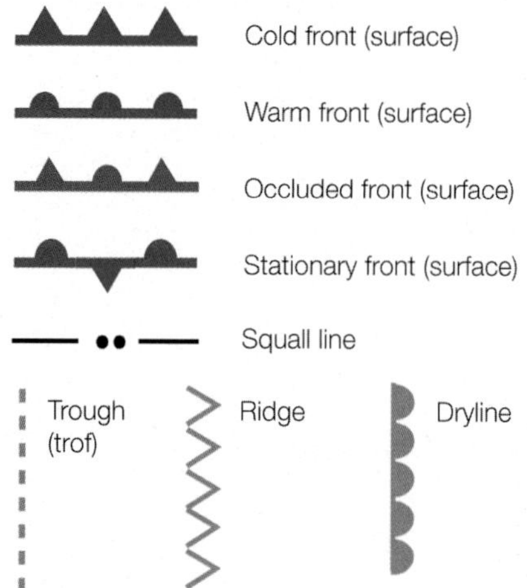

Humidity and Dew-Point Tables (Psychrometric Tables)

D

To obtain the dew point (or relative humidity), simply read down the temperature column and then over to the wet-bulb depression. For example, in Table D.1, a temperature of 10°C with a wet-bulb depression of 3°C produces a dew-point temperature of 4°C. (Dew-point temperature and relative humidity readings are appropriate for pressures near 1000 mb.)

Table D.1

Dew-Point Temperature (°C)

Air (Dry-Bulb) Temperature (°C)	WET-BULB DEPRESSION (DRY-BULB TEMPERATURE MINUS WET-BULB TEMPERATURE) (°C)															
	0.5	1.0	1.5	2.0	2.5	3.0	3.5	4.0	4.5	5.0	7.5	10.0	12.5	15.0	17.5	20.0
−20	−25	−33														
−17.5	−21	−27	−38													
−15	−19	−23	−28													
−12.5	−15	−18	−22	−29												
−10	−12	−14	−18	−21	−27	−36										
−7.5	−9	−11	−14	−17	−20	−26	−34									
−5	−7	−8	−10	−13	−16	−19	−24	−31								
−2.5	−4	−6	−7	−9	−11	−14	−17	−22	−28	−41						
0	−1	−3	−4	−6	−8	−10	−12	−15	−19	−24						
2.5	1	0	−1	−3	−4	−6	−8	−10	−13	−16						
5	4	3	2	0	−1	−3	−4	−6	−8	−10	−48					
7.5	6	6	4	3	2	1	−1	−2	−4	−6	−22					
10	9	8	7	6	5	4	2	1	0	−2	−13					
12.5	12	11	10	9	8	7	6	4	3	2	−7	−28				
15	14	13	12	12	11	10	9	8	7	5	−2	−14				
17.5	17	16	15	14	13	12	12	11	10	8	2	−7	−35			
20	19	18	18	17	16	15	14	14	13	12	6	−1	−15			
22.5	22	21	20	20	19	18	17	16	16	15	10	3	−6	−38		
25	24	24	23	22	21	21	20	19	18	18	13	7	0	−14		
27.5	27	26	26	25	24	23	23	22	21	20	16	11	5	−5	−32	
30	29	29	28	27	27	26	25	25	24	23	19	14	9	2	−11	
32.5	32	31	31	30	29	29	28	27	26	26	22	18	13	7	−2	
35	34	34	33	32	32	31	31	30	29	28	25	21	16	11	4	
37.5	37	36	36	35	34	34	33	32	32	31	28	24	20	15	9	0
40	39	39	38	38	37	36	36	35	34	34	30	27	23	18	13	6
42.5	42	41	41	40	40	39	38	38	37	36	33	30	26	22	17	11
45	44	44	43	43	42	42	41	40	40	39	36	33	29	25	21	15
47.5	47	46	46	45	45	44	44	43	42	42	39	35	32	28	24	19
50	49	49	48	48	47	47	46	45	45	44	41	38	35	31	28	23

Table D.2

Relative Humidity (Percent)

Air (Dry-Bulb) Temperature (°C)	WET-BULB DEPRESSION (DRY-BULB TEMPERATURE MINUS WET-BULB TEMPERATURE) (°C)																	
	0.5	1.0	1.5	2.0	2.5	3.0	3.5	4.0	4.5	5.0	7.5	10.0	12.5	15.0	17.5	20.0	22.5	25.0
−20	70	41	11															
−17.5	75	51	26	2														
−15	79	58	38	18														
−12.5	82	65	47	30	13													
−10	85	69	54	39	24	10												
−7.5	87	73	60	48	35	22	10											
−5	88	77	66	54	43	32	21	11	1									
−2.5	90	80	70	60	50	42	37	22	12	3								
0	91	82	73	65	56	47	39	31	23	15								
2.5	92	84	76	68	61	53	46	38	31	24								
5	93	86	78	71	65	58	51	45	38	32	1							
7.5	93	87	80	74	68	62	56	50	44	38	11							
0	94	88	82	76	71	65	60	54	49	44	19							
12.5	94	89	84	78	73	68	63	58	53	48	25	4						
15	95	90	85	80	75	70	66	61	57	52	31	12						
17.5	95	90	86	81	77	72	68	64	60	55	36	18	2					
20	95	91	87	82	78	74	70	66	62	58	40	24	8					
22.5	96	92	87	83	80	76	72	68	64	61	44	28	14	1				
25	96	92	88	84	81	77	73	70	66	63	47	32	19	7				
27.5	96	92	89	85	82	78	75	71	68	65	50	36	23	12	1			
30	96	93	89	86	82	79	76	73	70	67	52	39	27	16	6			
32.5	97	93	90	86	83	80	77	74	71	68	54	42	30	20	11	1		
35	97	93	90	87	84	81	78	75	72	69	56	44	33	23	14	6		
37.5	97	94	91	87	85	82	79	76	73	70	58	46	36	26	18	10	3	
40	97	94	91	88	85	82	79	77	74	72	59	48	38	29	21	13	6	
42.5	97	94	91	88	86	83	80	78	75	72	61	50	40	31	23	16	9	2
45	97	94	91	89	86	83	81	78	76	73	62	51	42	33	26	18	12	6
47.5	97	94	92	89	86	84	81	79	76	74	63	53	44	35	28	21	15	9
50	97	95	92	89	87	84	82	79	77	75	64	54	45	37	30	23	17	11

Table D.3

Dew-Point Temperature (°F)

Air (Dry-Bulb) Temperature (°F)	WET-BULB DEPRESSION (DRY-BULB TEMPERATURE MINUS WET-BULB TEMPERATURE) (°F)																							
	1	2	3	4	5	6	7	8	9	10	11	12	13	14	15	16	17	18	19	20	25	30	35	40
0	−7	−20																						
5	−1	−9	−24																					
10	5	−2	−10	−27																				
15	11	6	0	−9	−26																			
20	16	12	8	2	−7	−21																		
25	22	19	15	10	5	−3	−15	−51																
30	27	25	21	18	14	8	2	−7	−25															
35	33	30	28	25	21	17	13	7	0	−11														
40	38	35	33	30	28	25	21	18	13	7	−1	−14												
45	43	41	38	36	34	31	28	25	22	18	13	7	−1	−14										
50	48	46	44	42	40	37	34	32	29	26	22	18	13	8	0	−13								
55	53	51	50	48	45	43	41	38	36	33	30	27	24	20	15	9	1	−12						
60	58	57	55	53	51	49	47	45	43	40	38	35	32	29	25	21	17	11	4	−8				
65	63	62	60	59	57	55	53	51	49	47	45	42	40	37	34	31	27	24	19	14				
70	69	67	65	64	62	61	59	57	55	53	51	49	47	44	42	39	36	33	30	26	−11			
75	74	72	71	69	68	66	64	63	61	59	57	55	54	51	49	47	44	42	39	36	15			
80	79	77	76	74	73	72	70	68	67	65	63	62	60	58	56	54	52	50	47	44	28	−7		
85	84	82	81	80	78	77	75	74	72	71	69	68	66	64	62	61	59	57	54	52	39	19		
90	89	87	86	85	83	82	81	79	78	76	75	73	72	70	69	67	65	63	61	59	48	32		
95	94	93	91	90	89	87	86	85	83	81	80	79	78	76	74	73	71	70	68	66	56	43	24	
100	99	98	96	95	94	93	91	90	89	87	86	85	83	82	80	79	77	76	74	72	63	52	37	12
105	104	103	101	100	99	98	96	95	94	93	91	90	89	87	86	84	83	82	80	78	70	61	48	30
110	109	108	106	105	104	103	102	100	99	98	97	95	94	93	91	90	89	87	86	84	77	68	57	43
115	114	113	112	110	109	108	107	106	104	103	102	101	99	98	97	96	94	93	92	90	83	75	65	54
120	119	118	117	115	114	113	112	111	110	108	107	106	105	104	102	101	100	98	97	96	89	81	73	63

Table D.4

Relative Humidity (Percent)

Air (Dry-Bulb) Temperature (°F)	Wet-Bulb Depression (Dry-Bulb Temperature minus Wet-Bulb Temperature) (°F)																						
	1	2	3	4	5	6	7	8	9	10	11	12	13	14	15	16	17	18	19	20	25	30	35	40
0	67	31	1																					
5	73	46	20																					
10	78	56	34	13																				
15	82	64	46	29	11																			
20	85	70	55	40	26	12																		
25	87	74	62	49	37	25	13	1																
30	89	78	67	56	46	36	26	16	6															
35	91	81	72	63	54	45	36	27	19	10	2													
40	92	83	75	68	60	52	45	37	29	22	15	7												
45	93	86	78	71	64	57	51	44	38	31	25	18	12	6										
50	93	87	80	74	67	61	55	49	43	38	32	27	21	16	10	5								
55	94	88	82	76	70	65	59	54	49	43	38	33	28	23	19	14	9	5						
60	94	89	83	78	73	68	63	58	53	48	43	39	34	30	26	21	17	13	9	5				
65	95	90	85	80	75	70	66	61	56	52	48	44	39	35	31	27	24	20	16	12				
70	95	90	86	81	77	72	68	64	59	55	51	48	44	40	36	33	29	25	22	19	3			
75	96	91	86	82	78	74	70	66	62	58	54	51	47	44	40	37	34	30	27	24	9			
80	96	91	87	83	79	75	72	68	64	61	57	54	50	47	44	41	38	35	32	29	15	3		
85	96	92	88	84	80	76	73	69	66	62	59	56	52	49	46	43	41	38	35	32	20	8		
90	96	92	89	85	81	78	74	71	68	65	61	58	55	52	49	47	44	41	39	36	24	13	3	
95	96	93	89	85	82	79	75	72	69	66	63	60	57	54	51	49	46	43	41	38	27	17	7	1
100	96	93	89	86	83	80	77	73	70	68	65	62	59	56	54	51	49	46	44	41	30	21	12	4
105	97	93	90	87	83	80	77	74	71	69	66	63	60	58	55	53	50	48	46	43	33	23	15	7
110	97	93	90	87	84	81	78	75	73	70	67	65	62	60	57	55	52	50	48	46	36	26	18	11
115	97	94	91	88	85	82	79	76	74	71	68	66	63	61	58	56	54	52	49	47	37	28	21	13
120	97	94	91	88	85	82	80	77	74	72	69	67	65	62	60	58	55	53	51	49	40	31	23	17

Changing GMT and UTC to Local Time

E

The system of time used in meteorology is Greenwich Mean Time (GMT), which is also known as Coordinated Universal Time (UTC), and as Zulu (Z) Time. This is the time measured on the prime meridian (0° longitude) in Greenwich, England. Because Eastern Standard Time (EST) is 5 hours slower than GMT, to convert from GMT to EST simply requires subtracting 5 hours from GMT. Conversely, to change EST to GMT entails adding 5 hours to EST. Figure E.1 shows how to convert to GMT in various time zones of North America. Since in meteorology the time is given on a 24-clock, Table E.1 shows the relationship between the familiar two 12-hour periods of A.M. and P.M. and the 24-hour system.

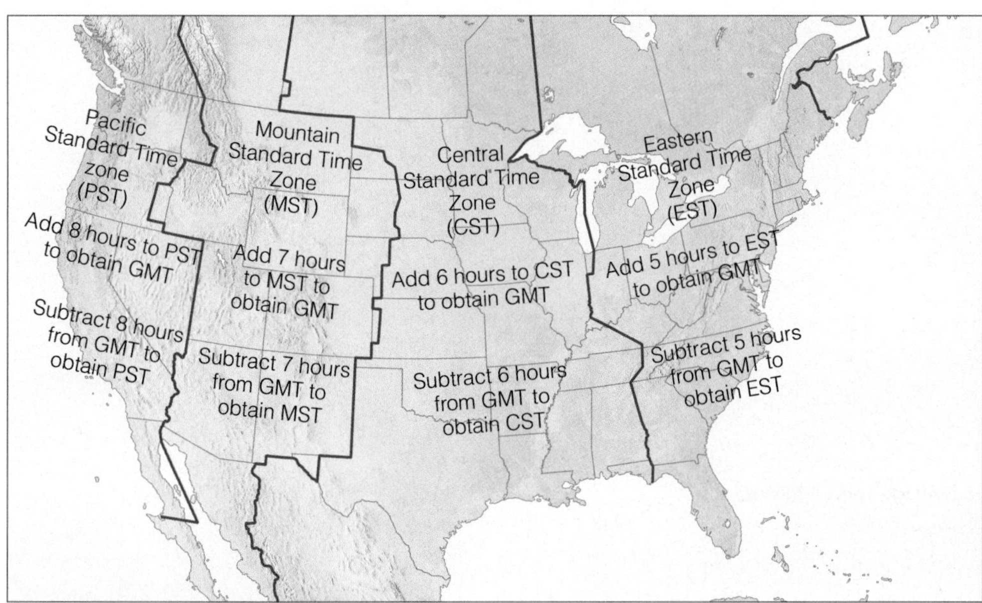

FIGURE E.1 The time zones of North America.

Table E.1

Conversion of A.M. and P.M. Time System to 24-Hour System

TIME	24-HR SYSTEM TIME	TIME	24-HR SYSTEM TIME	TIME	24-HR SYSTEM TIME	TIME	24-HR SYSTEM TIME
A.M.		A.M.		P.M.		P.M.	
12:00 (midnight)	0000	6:00	0600	12:00 (noon)	1200	6:00	1800
1:00	0100	7:00	0700	1:00	1300	7:00	1900
2:00	0200	8:00	0800	2:00	1400	8:00	2000
3:00	0300	9:00	0900	3:00	1500	9:00	2100
4:00	0400	10:00	1000	4:00	1600	10:00	2200
5:00	0500	11:00	1100	5:00	1700	11:00	2300

F

Average Annual Global Precipitation

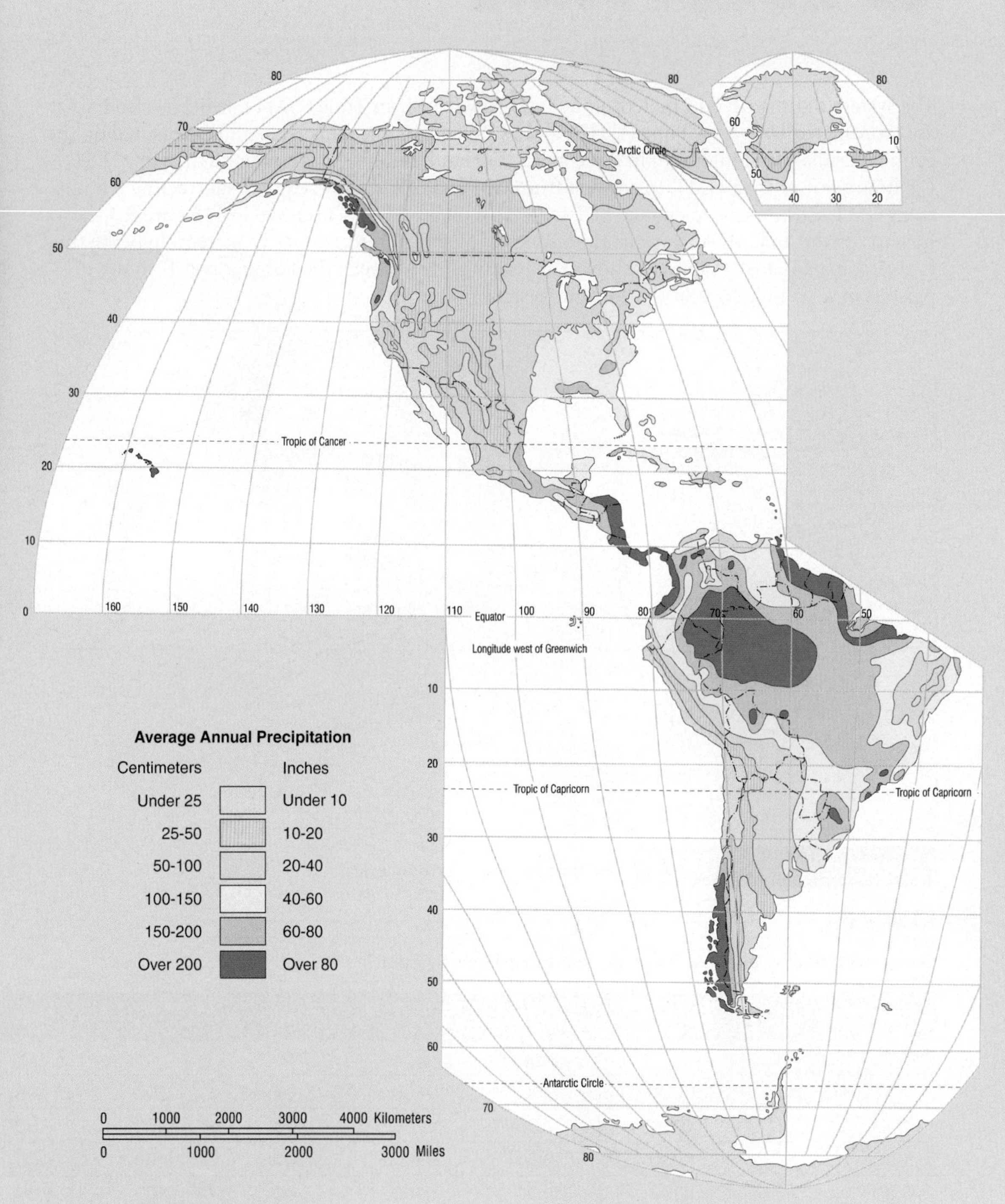

FIGURE F.1 World map of average annual precipitation.

A Western Paragraphic Projection developed at Western Illinois University

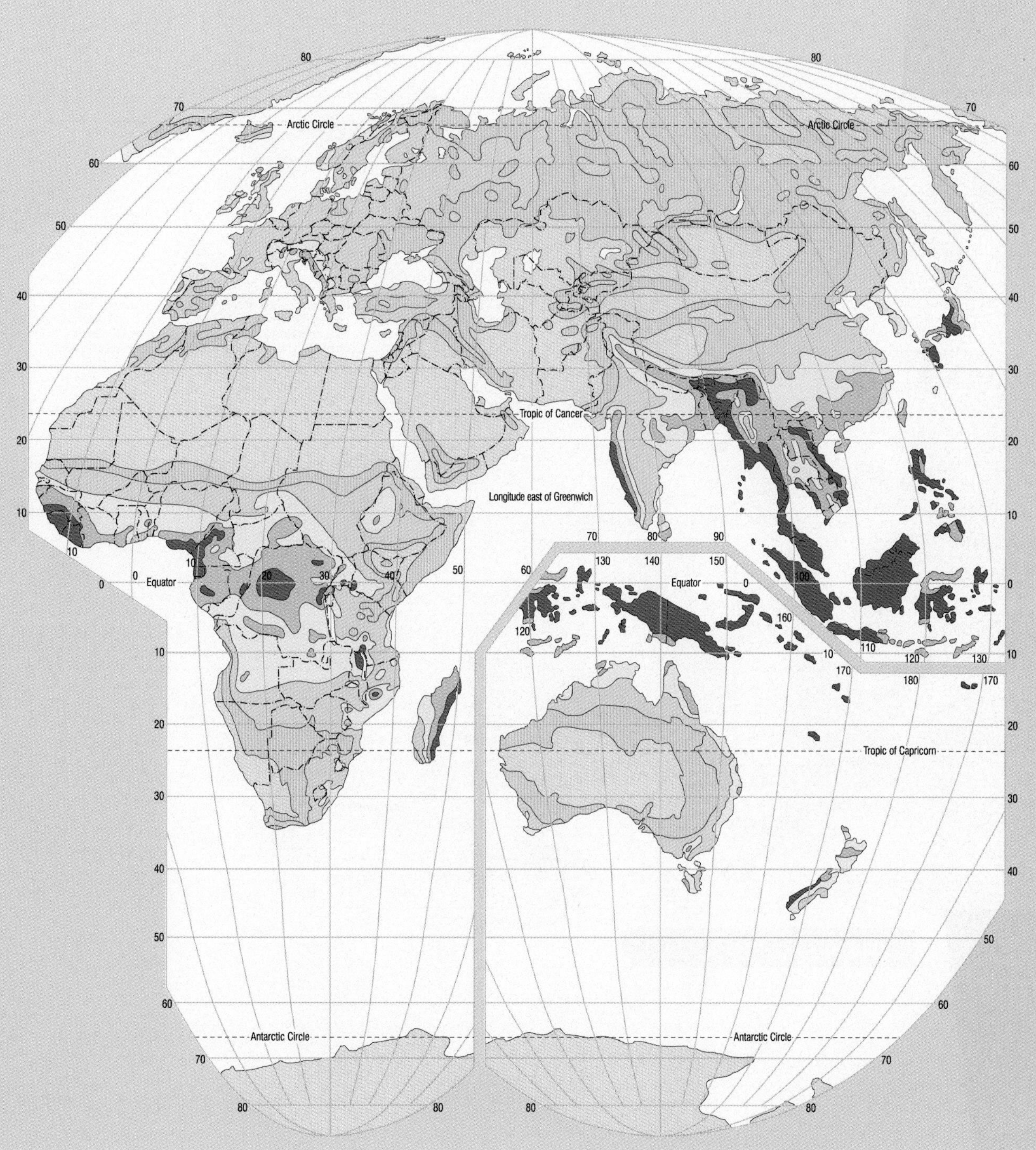

G Hurricane Tracking Chart

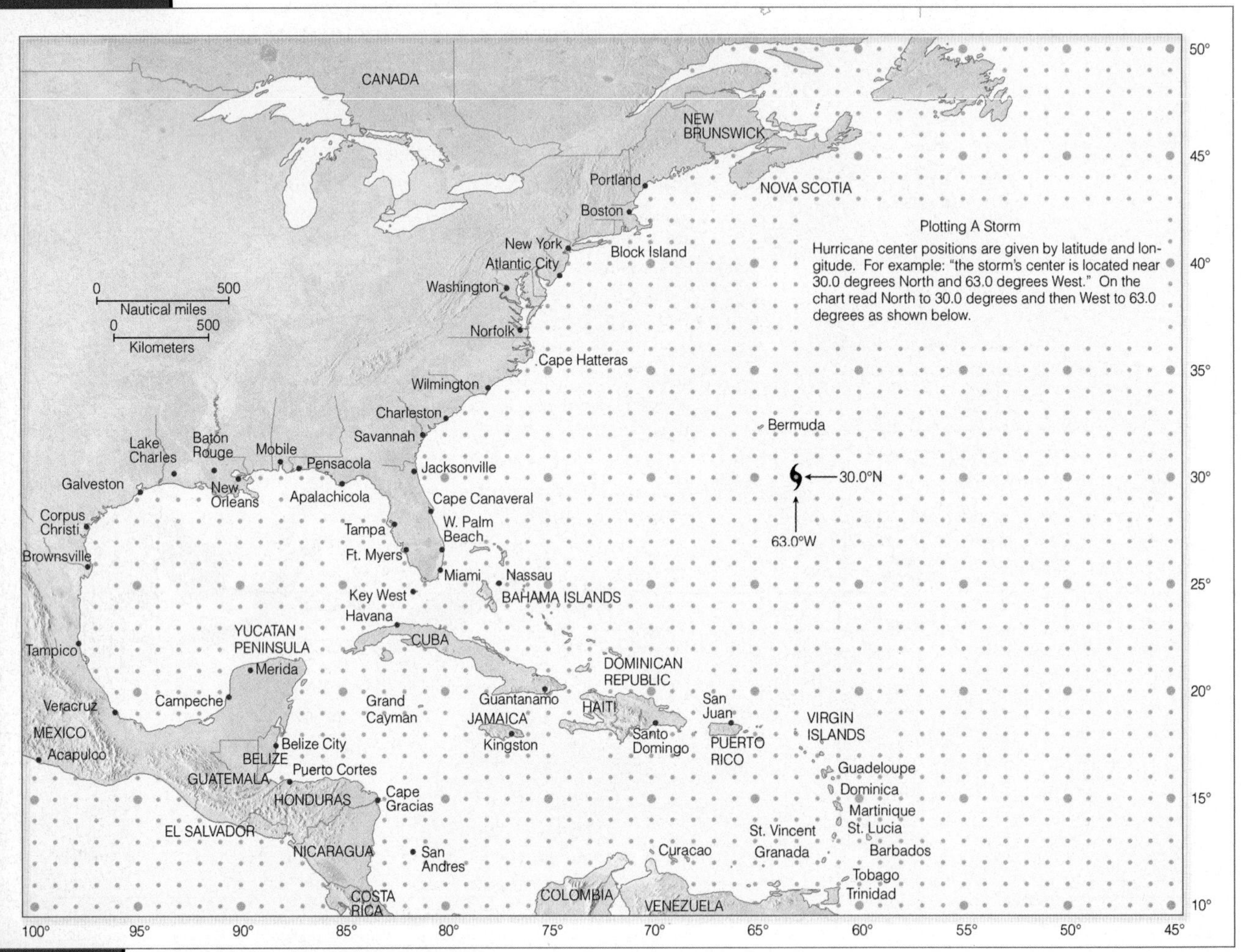

FIGURE G.1

A Western Paragraphic Projection developed at Western Illinois University

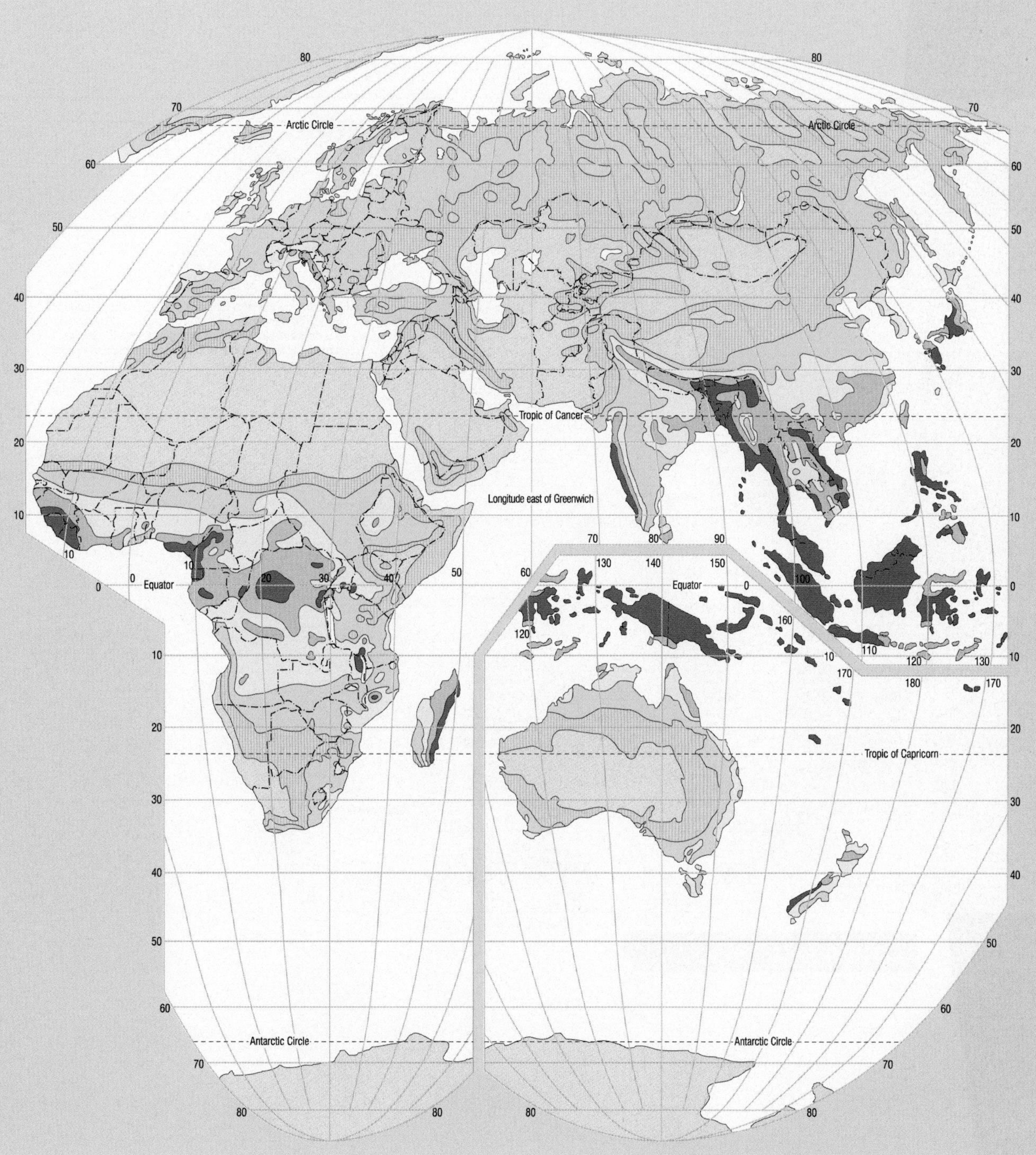

G

Hurricane Tracking Chart

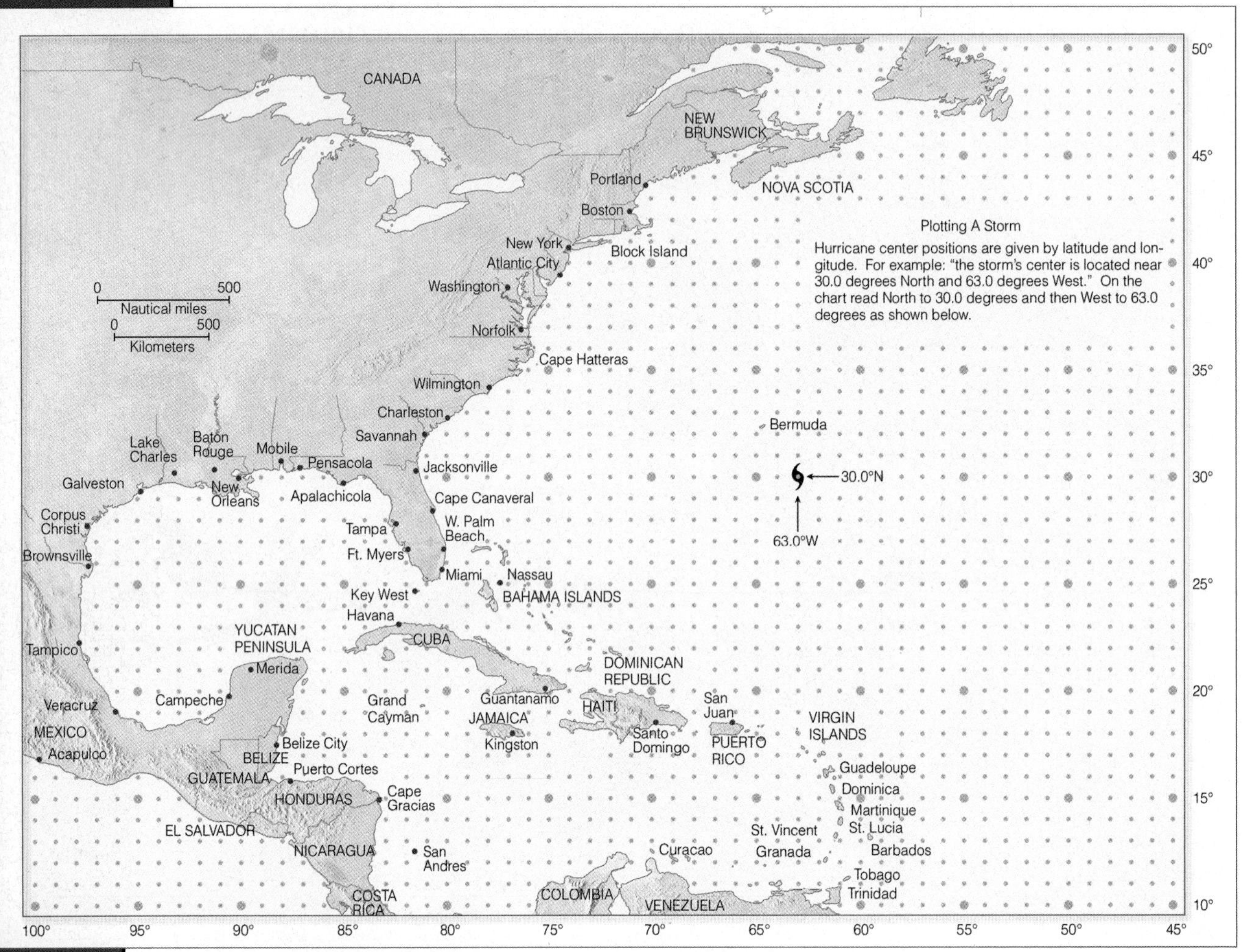

FIGURE G.1

Additional Reading Material

Periodicals

Selected nontechnical periodicals that contain articles on weather and climate.

Bulletin of the American Meteorological Society. Monthly. The American Meteorological Society, 45 Beacon St., Boston, MA 02108.

Meteorological Magazine. Monthly. British Meteorological Office, British Information Services, 845 Third Avenue, New York, NY.

National Weather Digest. Quarterly. National Weather Association, 4400 Stamp Road, Room 404, Marlow Heights, MD 20031. (Deals mainly with weather forecasting.)

Weather. Monthly. Royal Meteorological Society, James Glaisher House, Grenville Place, Bracknell, Berkshire, England.

Weatherwise. Bimonthly. Heldref Publications, 4000 Albermarle St., N.W., Washington, DC 20016.

Selected Technical Periodicals

EOS—Transaction of the American Geophysical Union. American Geophysical Union (AGU), Washington, DC.

Journal of Applied Meteorology. American Meteorological Society (AMS), Boston, MA.

Journal of Atmospheric and Oceanic Technology. AMS, Boston, MA.

Journal of Atmospheric Science. AMS, Boston, MA.

Journal of Climate. AMS, Boston, MA.

Journal of Geophysical Research. American Geophysical Union, Washington, DC.

Monthly Weather Review. AMS, Boston, MA.

Weather and Forecasting. AMS, Boston, MA.

Additional periodicals that frequently contain articles of meteorological interest.

American Scientist. Bimonthly. Sigma Xi, the Scientific Research Society, Inc., New Haven, CT.

Science. Weekly. American Association for the Advancement of Science, Washington, DC.

Scientific American. Monthly. Scientific American, Inc., New York, NY.

Smithsonian. Monthly. The Smithsonian Association, Washington, DC.

Books

The titles listed below may be drawn upon for additional information. Many are written at the introductory level. Those that are more advanced are marked with an asterisk.

Ahrens, C. Donald. *Essentials of Meteorology* (5th ed.), Thomson/Brooks/Cole, Belmont, CA, 2008.

Ahrens, C. Donald. Meterology *Today* (9th ed.), Cengage Learning/Brooks/Cole, Belmont, CA, 2009.

Anthes, R. A. *Tropical Cyclones: Their Evolution, Structure, and Effect,* American Meteorological Society, Boston, MA, 1982.

Arya, Pal S. *Air Pollution Meteorology,* Oxford University Press, New York, 1998.

Bigg, Grant R. *The Oceans and Climate,* Cambridge University Press, New York, 1996.

*Bluestein, Howard B. *Synoptic-Dynamic Meteorology in Midlatitudes. Vol. 1: Principles of Kinematics and Dynamics,* Oxford University Press, New York, 1992.

_____. *Synoptic-Dynamic Meteorology in Midlatitudes. Vol. II: Observations and Theory of Weather Systems,* Oxford University Press, New York, 1993.

_____. *Tornado Alley: Monster Storms of the Great Plains,* Oxford University Press, New York, 1999.

Bohren, Craig F. *Clouds in a Glass of Beer: Simple Experiments in Atmospheric Physics,* Wiley, New York, 1987.

_____. *What Light Through Yonder Window Breaks?,* Wiley, New York, 1991.

Boubel, Richard W., et al., *Fundamentals of Air Pollution* (3rd ed.), Academic Press, New York, 1994.

Burgess, Eric, and Douglass Torr. *Into the Thermosphere: The Atmosphere Explorers,* National Aeronautics and Space Administration, Washington, DC, 1987.

Burroughs, William J. *Watching the World's Weather,* Cambridge University Press, New York, 1991.

_____. *Climate Revealed,* Cambridge University Press, Cambridge, England, 1999.

Burt, Christopher C. *Extreme Weather, A Guide and Record Book,* W.W. Norton & Company, New York, 2007.

Carlson, Toby N. *Mid-Latitude Weather Systems,* American Meteorological Society, Boston, MA, 1998.

Climate Change 2007. The Physical Science Basis. Working Group 1 contribution to the Fourth Assessment Report of the IPCC, Cambridge University Press, New York, 2007.

*Cotton, W. R., and R. A. Anthes. *Storm and Cloud Dynamics,* Academic Press, New York, 1989.

Cotton, William R. *Storms,* ASTeR Press, Fort Collins, CO, 1990.

Cotton, William R., and Roger A. Pielke. *Human Impacts on Weather and Climate,* Cambridge University Press, New York, 1995.

Crowley, Thomas J. and Gerald R. North. *Paleoclimatology,* Oxford University Press, New York, 1991.

De Blij, H. J. *Nature on the Rampage,* Smithsonian Books, Washington, DC, 1994.

Doswell, Charles A. III, editor. *Severe Convection Storms,* American Meteorological Society, Boston, MA, 2001.

Elsner, James B., and A. Biral Kara. *Hurricanes of the North Atlantic,* Oxford University Press, New York, 1999.

Elsom, Derek M. *Atmospheric Pollution: A Global Problem* (2nd ed.), Blackwell Publishers, Oxford, England, 1992.

Emanuel, Kerry. *Divine Wind-The History and Science of Hurricanes,* Oxford University Press, Oxford, New York, 2005.

Encyclopedia of Climate and Weather, Vol. 1 and Vol. 2, Stephen H. Schneider, Ed., Oxford University Press, New York, 1996.

Energy and Climate Change. Report of the DOE Multi-Laboratory Climate Change Committee, Lewis Publishers, Chelsea, MI, 1991.

England, Gary A. *Weathering the Storm,* University of Oklahoma Press, Norman, OK, 1996.

Firor, John. *The Changing Atmosphere: A Global Challenge,* Yale University Press, New Haven, CT, 1990.

Fujita, T. T. *The Downburst–Microburst and Macroburst,* University of Chicago Press, Chicago, 1985.

Glossary of Meteorology. Todd S. Glickman, Managing Ed., American Meteorological Society, Boston, MA, 2000.

Glossary of Weather and Climate. Ira W. Geer, Ed., American Meteorological Society, Boston, MA, 1996.

*Graedel, T. E. and Paul J. Crutzen. *Atmospheric Change: An Earth System Perspective,* W. H. Freeman, New York, 1993.

Graedel, Thomas E., and Paul J. Crutzen. *Atmosphere, Climate, and Change,* W. H. Freeman, New York, 1995.

Greenler, Robert. *Rainbows, Halos and Glories,* Cambridge University Press, New York, 1980.

*Grotjahn, Richard. *Global Atmospheric Circulations: Observations and Theories,* Oxford University Press, Oxford, England, 1993.

Henson, Robert, *The Rough Guide to Climate Change,* (2nd ed.), Rough Guide, New York, 2008.

_____. *The Rough Guide to Weather,* Rough Guide, New York, 2007.

*Hobbs, Peter V. *Basic Physical Chemistry for Atmospheric Sciences,* Cambridge University Press, New York, 1995.

Hoyt, Douglas V. and Kenneth H. Schatten. *The Role of the Sun in Climate Change,* Oxford University Press, New York, 1997.

International Cloud Atlas. World Meteorological Organization, Geneva, Switzerland, 1987.

James, Bruce P., et al., Eds. *Climate Change 1995. Economic and Social Dimensions of Climate Change,* Cambridge University Press, Cambridge, England, 1996.

*Karoly, David J., and Dayton G. Vincent, Eds. *Meteorology of the Southern Hemisphere,* American Meteorological Society, Boston, MA, 1998.

Keen, Richard A. *Skywatch: The Western Weather Guide,* Fulcrum Incorporated, Golden, CO, 1987.

_____. *Skywatch East: A Weather Guide,* Fulcrum Incorporated, Golden, CO, 1992.

Kessler, Edwin. *Thunderstorm Morphology and Dynamics* (2nd ed.), University of Oklahoma Press, Norman, OK, 1986.

Kocin, Paul J., and L. W. Uccellini. *Northeast Snowstorms.* Vol. 1 and Vol. 2., American Meteorological Society, Boston, MA, 2004.

Kocin, Paul J., and L. W. Uccellini. *Snowstorms along the Northeastern Coast of the United States: 1955 to 1985,* American Meteorological Society, Boston, MA, 1990.

Laskin, David. *Braving the Elements: The Stormy History of American Weather,* Doubleday, New York, 1996.

Ludlum, D. M. *The Audubon Society Field Guide to North American Weather,* Alfred A. Knopf, New York, 1991.

Lynch, David K., and William Livingston. *Color and Light in Nature,* Cambridge University Press, New York, 1995.

Mason, B. J. *Acid Rain: Its Causes and Its Effects on Inland Waters,* Oxford University Press, New York, 1992.

Meinel, Aden, and Marjorie Meinel. *Sunsets, Twilights and Evening Skies.* Cambridge University Press, New York, 1983.

Nelson, Mike. *The Colorado Weather Book,* Westcliff Publishers, Englewood, CO, 1999.

Pretor-Pinney, Gavin. *The Cloud Spotters Guide,* Penguin Group, New York 2006.

Prospects for Future Climate. Special US/USSR Report on Climate and Climate Change, Lewis Publishers, Chelsea, MI, 1990.

Righter, Robert W. *Wind Energy in America,* University of Oklahoma Press, Norman, OK, 1996.

*Rogers, R. R. *A Short Course in Cloud Physics* (3rd ed.), Pergamon Press, Oxford, England, 1989.

Schaefer, Vincent J. *Peterson First Guide to Clouds and Weather,* Houghton Mifflin, Boston, MA, 1991.

Scorer, Richard S., and Arjen Verkaik. *Spacious Skies,* David and Charles Publishers, London, 1989.

Somerville, Richard C. *The Forgiving Air* (2nd Ed.), American Meteorological Society, Boston, MA, 2008.

*Stull, Roland B. *Meteorology Today for Scientists and Engineers* (2nd Ed.), Brooks/Cole Publishing Co., Pacific Grove, CA, 2000.

Van Andel, Tjeerd H. *New Views on an Old Planet—A History of Global Change,* Cambridge University Press, Cambridge, England, 1994.

Vasquez, Tim. *Storm Chasing Handbook,* Weather Graphics Technologies, Austin, TX, 2002.

Vasquez, Tim. *Weather Forecasting Handbook.* Weather Graphics Technologies, Austin, TX, 2002.

Vital Signs 2007-2008. The World Watch Institute, New York, 2007.

Wallace, John M. and Peter V. Hobbs. *Atmospheric Science: An Introductory Survey* (2nd ed.), Academic Press, Burlington, MA, 2006.

*Watson, Robert T., et al., Eds. *Climate Change 1995. Impacts, Adaptations and Mitigation of Climate Change: Scientific-Technical Analyses,* Cambridge University Press, Cambridge, England, 1996.

Weather. Smithsonian Field Guide, Harper Collins Publishers, New York, 2006.

Williams, Jack. *The AMS Weather Book: The Ultimate Guide to America's Weather.* American Meteorological Society, Boston, MA, 2009.

Glossary

A

Absolute humidity The mass of water vapor in a given volume of air. It represents the density of water vapor in the air.
Absolute zero A temperature reading of –273°C, –460°F, or 0°K. Theoretically, there is no molecular motion at this temperature.
Absolutely stable atmosphere An atmospheric condition that exists when the environmental lapse rate is less than the moist adiabatic rate. This results in a lifted parcel of air being colder than the air around it.
Absolutely unstable atmosphere An atmospheric condition that exists when the environmental lapse rate is greater than the dry adiabatic rate. This results in a lifted parcel of air being warmer than the air around it.
Accretion The growth of a precipitation particle by the collision of an ice crystal or snowflake with a supercooled liquid droplet that freezes upon impact.
Acid fog *See* Acid rain.
Acid rain Cloud droplets or raindrops combining with gaseous pollutants, such as oxides of sulfur and nitrogen, to make falling rain (or snow) acidic—pH less than 5.0. If fog droplets combine with such pollutants it becomes acid fog.
Actual Vapor Pressure *See* Vapor pressure.
Adiabatic process A process that takes place without a transfer of heat between the system (such as an air parcel) and its surroundings. In an adiabatic process, compression always results in warming, and expansion results in cooling.
Advection The horizontal transfer of any atmospheric property by the wind.
Advection fog Occurs when warm, moist air moves over a cold surface and the air cools to below its dew point.
Advection-radiation fog Fog that forms as relatively warm moist air moves over a colder surface that cooled mainly by radiational cooling.
Aerosols Tiny suspended solid particles (dust, smoke, etc.) or liquid droplets that enter the atmosphere from either natural or human (anthropogenic) sources, such as the burning of fossil fuels. Sulfur-containing fossil fuels, such as coal, produce sulfate aerosols.
Air density *See* Density.
Aggregation The clustering together of ice crystals to form snowflakes.
Air mass A large body of air that has similar horizontal temperature and moisture characteristics.
Air-mass thunderstorm *See* Ordinary thunderstorm.
Air parcel *See* Parcel of air.
Air pollutants Solid, liquid, or gaseous airborne substances that occur in concentrations high enough to threaten the health of people and animals, to harm vegetation and structures, or to toxify a given environment.
Air pressure (atmospheric pressure) The pressure exerted by the mass of air above a given point, usually expressed in millibars (mb), inches of mercury (Hg) or in hectopascals (hPa).
Albedo The percent of radiation returning from a surface compared to that which strikes it.
Aleutian low The subpolar low-pressure area that is centered near the Aleutian Islands on charts that show mean sea-level pressure.
Altimeter An instrument that indicates the altitude of an object above a fixed level. Pressure altimeters use an aneroid barometer with a scale graduated in altitude instead of pressure.
Altocumulus A middle cloud, usually white or gray. Often occurs in layers or patches with wavy, rounded masses or rolls.
Altostratus A middle cloud composed of gray or bluish sheets or layers of uniform appearance. In the thinner regions, the sun or moon usually appears dimly visible.
Analogue forecasting method A forecast made by comparison of past large-scale synoptic weather patterns that resemble a given (usually current) situation in its essential characteristics.
Analysis The drawing and interpretation of the patterns of various weather elements on a surface or upper-air chart.
Anemometer An instrument designed to measure wind speed.

Aneroid barometer An instrument designed to measure atmospheric pressure. It contains no liquid.
Annual range of temperature The difference between the warmest and coldest months at any given location.
Anticyclone An area of high atmospheric pressure around which the wind blows clockwise in the Northern Hemisphere and counterclockwise in the Southern Hemisphere. Also called a *high*.
Apparent temperature What the air temperature "feels like" for various combinations of air temperature and relative humidity.
Arctic Front In northern latitudes, the semi-permanent boundary that separates very cold, dense arctic air from the less-cold and less-dense polar air.
Arctic Oscillation (AO) A reversal of atmospheric pressure over the Arctic that produces changes in the upper-level westerly winds over northern latitudes. These changes in upper-level winds influence winter weather patterns over Northern America, Greenland, and Europe.
ASOS Acronym for Automated Surface Observing Systems. A system designed to provide continuous information of wind, temperature, pressure, cloud base height, and runway visibility at selected airports.
Atmosphere The envelope of gases that surround a planet and are held to it by the planet's gravitational attraction. The earth's atmosphere is mainly nitrogen and oxygen.
Atmospheric greenhouse effect The warming of an atmosphere by its absorbing and emitting infrared radiation while allowing shortwave radiation to pass on through. The gases mainly responsible for the earth's atmospheric greenhouse effect are water vapor and carbon dioxide. Also called the *greenhouse effect.*
Atmospheric models Simulation of the atmosphere's behavior by mathematical equations or by physical models.
Atmospheric stagnation A condition of light winds and poor vertical mixing that can lead to a high concentration of pollutants. Air stagnations are most often associated with fair weather, an inversion, and the sinking air of a high-pressure area.
Atmospheric window The wavelength range between about 8 and 11 µm in which little absorption of infrared radiation takes place.
Aurora Glowing light display in the nighttime sky caused by excited gases in the upper atmosphere giving off light. In the Northern Hemisphere it is called the *aurora borealis* (northern lights); in the Southern Hemisphere, the *aurora australis* (southern lights).
Autumnal equinox The equinox at which the sun approaches the Southern Hemisphere and passes directly over the equator. Occurs around September 23.
Avalanche A mass of snow moving rapidly downhill along a steep mountain slope.
AWIPS Acronym for Advanced Weather Interactive Processing System. New computerized system that integrates and processes data received at a Weather Forecasting Office from NEXRAD, ASOS, and analysis and guidance products prepared by NMC.

B

Back-door cold front A cold front moving south or southwest along the Atlantic seaboard of the United States.
Ball lightning A rare form of lightning that may consist of a reddish, luminous ball of electricity or charged air.
Banner cloud A cloud extending downwind from an isolated mountain peak, often in the shape of an elongated cloud plume on an otherwise cloud-free day.
Barograph A recording barometer.
Barometer An instrument that measures atmospheric pressure. The two most common barometers are the mercury barometer and the aneroid barometer.
Bergeron process *See* Ice-crystal process.
Bermuda high *See* Subtropical high.
Billow clouds Broad, nearly parallel lines of wavelike clouds oriented at right angles to the wind.
Black blizzard A term that describes the type of dust storm that blocked out the sun during the dust bowl years of the 1930s in the Great Plains of the United States.
Blackbody A hypothetical object that absorbs all of the radiation that strikes it. It also emits radiation at a maximum rate for its given temperature.
Black ice A thin layer of ice appearing dark in color.
Blizzard A severe weather condition characterized by low temperatures and strong winds (greater than 35 mi/hr) bearing a great amount of snow either falling or blowing. When these conditions continue after the falling snow has ended, it is termed a ground blizzard.
Boulder winds Fast-flowing, local downslope winds that may attain speeds of 100 knots or more. They are especially strong along the eastern foothills of the Rocky Mountains near Boulder, Colorado.
Bow echo A line of thunderstorms on a radar screen that appears in the shape of a bow. Bow echoes are often associated with damaging straight-line winds and small tornadoes.
Buoyant force (buoyancy) The upward force exerted upon an air parcel (or any object) by virtue of the density (mainly temperature) difference between the parcel and that of the surrounding air.
Buys-Ballot's law A law describing the relationship between the wind direction and the pressure distribution. In the Northern Hemisphere, if you stand with your back to the surface wind, then turn clockwise about 30°, lower pressure will be to your left. In the Southern Hemisphere, stand with your back to the surface wind, then turn counterclockwise about 30°, lower pressure will be to your right.

C

Cap cloud *See* Pileus cloud.
Carbon dioxide (CO_2) A colorless, odorless gas whose concentration is about 0.038 percent (385 ppm) in a volume of air near sea level. It is a selective absorber of infrared radiation and, consequently, it is important in the earth's atmospheric greenhouse effect. Solid CO_2 is called *dry ice.*
Carbon monoxide (CO) A colorless, odorless, toxic gas that forms during the incomplete combustion of carbon-containing fuels.
Celsius scale A temperature scale where zero is assigned to the temperature where water freezes and 100 to the temperature where water boils (at sea level).
Centripetal acceleration The inward-directed acceleration on a particle moving in a curved path.

Centripetal force The radial force required to keep an object moving in a circular path. It is directed toward the center of that curved path.

Chaos The property describing a system that exhibits erratic behavior in that very small changes in the initial state of the system rapidly lead to large and apparently unpredictable changes sometime in the future.

Chinook wall cloud A bank of clouds over the Rocky Mountains that signifies the approach of a chinook.

Chinook wind A warm, dry wind on the eastern side of the Rocky Mountains. In the Alps, the wind is called a *foehn.*

Chlorofluorocarbons (CFCs) Compounds consisting of methane (CH_4) or ethane (C_2H_6) with some or all of the hydrogen replaced by chlorine or fluorine. Used in fire extinguishers, as refrigerants, as solvents for cleaning electronic microcircuits, and as propellants. CFCs contribute to the atmospheric greenhouse effect and destroy ozone in the stratosphere.

Cirrocumulus A high cloud that appears as a white patch of clouds without shadows. It consists of very small elements in the form of grains or ripples.

Cirrostratus High, thin, sheetlike clouds, composed of ice crystals. They frequently cover the entire sky and often produce a halo.

Cirrus A high cloud composed of ice crystals in the form of thin, white, featherlike clouds in patches, filaments, or narrow bands.

Clear air turbulence (CAT) Turbulence encountered by aircraft flying through cloudless skies. Thermals, wind shear, and jet streams can each be a factor in producing CAT.

Clear ice A layer of ice that appears transparent because of its homogeneous structure and small number and size of air pockets.

Climate The accumulation of daily and seasonal weather events over a long period of time.

Climatic controls The relatively permanent factors that govern the general nature of the climate of a region.

Climatological forecast A weather forecast, usually a month or more in the future, which is based upon the climate of a region rather than upon current weather conditions.

Cloud A visible aggregate of tiny water droplets and/or ice crystals in the atmosphere above the earth's surface.

Cloudburst Any sudden and heavy rain shower.

Cloud seeding The introduction of artificial substances (usually silver iodide or dry ice) into a cloud for the purpose of either modifying its development or increasing its precipitation.

Coalescence The merging of cloud droplets into a single larger droplet.

Cold advection (or cold air advection) The transport of cold air by the wind from a region of lower temperatures to a region of higher temperatures.

Cold air damming A shallow layer of cold air that is trapped between the Atlantic coast and the Appalachian Mountains.

Cold air funnel A funnel cloud or tornado that descends from a convective cloud, most often a non-supercell thunderstorm. They often form along the west coast of North America, especially in later winter and early spring when a cold upper-level low moves into the region.

Cold front A transition zone where a cold air mass advances and replaces a warm air mass.

Cold occlusion *See* Occluded front.

Cold wave A rapid fall in temperature within 24 hours that often requires increased protection for agriculture, industry, commerce, and human activities.

Collision-coalescence process The process of producing precipitation by liquid particles (cloud droplets and raindrops) colliding and joining (coalescing).

Comma cloud A band of organized cumuliform clouds that looks like a comma on a satellite photograph.

Computer enhancement A process where the temperatures of radiating surfaces are assigned different shades of gray (or different colors) on an infrared picture. This technique allows specific features to be more clearly delineated.

Condensation The process by which water vapor becomes a liquid.

Condensation level The level above the surface marking the base of a cumuliform cloud.

Condensation nuclei Also called *cloud condensation nuclei.* Tiny particles upon whose surfaces condensation of water vapor begins in the atmosphere.

Conditionally unstable atmosphere An atmospheric condition that exists when the environmental lapse rate is less than the dry adiabatic rate but greater than the moist adiabatic rate. Also called *conditional instability.*

Conduction The transfer of heat by molecular activity from one substance to another, or through a substance. Transfer is always from warmer to colder regions.

Constant-height chart (constant-level chart) A chart showing variables, such as pressure, temperature, and wind, at a specific altitude above sea level. Variation in horizontal pressure is depicted by isobars. The most common constant-height chart is the surface chart, which is also called the *sea-level chart,* or surface weather map.

Constant-pressure chart (isobaric chart) A chart showing variables, such as temperature and wind, on a constant-pressure surface. Variations in height are usually shown by lines of equal height (contour lines).

Contact freezing The process by which contact with a nucleus such as an ice crystal causes supercooled liquid droplets to change into ice.

Continental arctic air mass An air mass characterized by extremely low temperatures and very dry air.

Continental polar air mass An air mass characterized by low temperatures and dry air. Not as cold as arctic air masses.

Continental tropical air mass An air mass characterized by high temperatures and low humidity.

Contour line A line that connects points of equal elevation above a reference level, most often sea level.

Contrail (condensation trail) A cloud-like streamer frequently seen forming behind aircraft flying in clear, cold, humid air.

Controls of temperature The main factors that cause variations in temperature from one place to another.

Convection Motions in a fluid that result in the transport and mixing of the fluid's properties. In meteorology, convection usually refers to atmospheric motions that are predominantly vertical, such as rising air currents due to surface heating. The rising of heated surface air and the sinking of cooler air aloft is often called *free convection.* (Compare with forced convection.)

Convective instability The state of an unsaturated layer of air where the upper part is dry and the lower part is humid and the entire layer is lifted to a point where it becomes saturated and unstable. Also called *potential instability* and *thermal instability.*
Convergence An atmospheric condition that exists when the winds cause a horizontal net inflow of air into a specified region.
Conveyor belt model (for middle-latitude storms) A three-dimensional picture of a mid-latitude cyclonic storm and the various air streams (called *conveyor belts*) that interact to produce the weather associated with the storm.
Coriolis force An apparent force observed on any free-moving object in a rotating system. On the earth, this deflective force results from the earth's rotation and causes moving particles (including the wind) to deflect to the right in the Northern Hemisphere and to the left in the Southern Hemisphere.
Cumulonimbus An exceptionally dense and vertically developed cloud, often with a top in the shape of an anvil. The cloud is frequently accompanied by heavy showers, lightning, thunder, and sometimes hail. It is also known as a thunderstorm cloud.
Cumulus A cloud in the form of individual, detached domes or towers that are usually dense and well defined. It has a flat base with a bulging upper part that often resembles cauliflower. Cumulus clouds of fair weather are called *cumulus humilis.* Those that exhibit much vertical growth are called *cumulus congestus* or towering cumulus.
Cumulus stage The initial stage in the development of an ordinary thunderstorm in which rising, warm, humid air develops into a cumulus cloud.
Cut-off low A cold upper-level low that has become displaced out of the basic westerly flow and lies to the south of this flow.
Cyclogenesis The development or strengthening of middle-latitude (extratropical) cyclones.
Cyclone An area of low pressure around which the winds blow counterclockwise in the Northern Hemisphere and clockwise in the Southern Hemisphere.

D

Daily range of temperature The difference between the maximum and minimum temperatures for any given day.
Dart leader The discharge of electrons that proceeds intermittently toward the ground along the same ionized channel taken by the initial lightning stroke.
Dendrochronology The analysis of the annual growth rings of trees as a means of interpreting past climatic conditions.
Density The ratio of the mass of a substance to the volume occupied by it. Air density is usually expressed either as g/cm^3 or kg/m^3.
Deposition A process that occurs in subfreezing air when water vapor changes directly to ice without becoming a liquid first.
Derecho Strong, damaging, straight-line winds associated with a cluster of severe thunderstorms that most often form in the evening or at night.
Desertification A general increase in the desert conditions of a region.
Dew Water that has condensed onto objects near the ground when their temperatures have fallen below the dew point of the surface air.
Dew point (dew-point temperature) The temperature to which air must be cooled (at constant pressure and constant water vapor content) for saturation to occur.
Dissipating stage The final stage in the development of an ordinary thunderstorm when downdrafts exist throughout the cumulonimbus cloud.
Divergence An atmospheric condition that exists when the winds cause a horizontal net outflow of air from a specific region.
Doldrums The region near the equator that is characterized by low pressure and light, shifting winds.
Doppler lidar The use of light beams to determine the velocity of objects such as dust and falling rain by taking into account the Doppler shift.
Doppler radar A radar that determines the velocity of falling precipitation either toward or away from the radar unit by taking into account the Doppler shift.
Doppler shift (effect) The change in the frequency of waves that occurs when the emitter or the observer is moving toward or away from the other.
Downburst A severe localized downdraft that can be experienced beneath a severe thunderstorm. (Compare Microburst and Macroburst.)
Downdraft An area of downward moving air within a cumulonimbus cloud.
Drizzle Small water drops between 0.2 and 0.5 mm in diameter that fall slowly and reduce visibility more than light rain.
Drought A period of abnormally dry weather sufficiently long enough to cause serious effects on agriculture and other activities in the affected area.
Dry adiabatic rate The rate of change of temperature in a rising or descending unsaturated air parcel. The rate of adiabatic cooling or warming is about 10°C per 1000 m (5.5°F per 1000 ft).
Dry-bulb temperature The air temperature measured by the dry-bulb thermometer of a psychrometer.
Dry lightning Lightning that occurs with thunderstorms that produce little, if any, appreciable precipitation that reaches the surface.
Dryline A boundary that separates warm, dry air from warm, moist air. It usually represents a zone of instability along which thunderstorms form.
Dry slot On a satellite image the dry slot represents the relatively clear region (or clear wedge) that appears just to the west of the tail of a comma cloud of a mid-latitude cyclonic storm.
Dust devil (or whirlwind) A small but rapidly rotating wind made visible by the dust, sand, and debris it picks up from the surface. It develops best on clear, dry, hot afternoons.
Dust storm A weather condition characterized by strong winds and dust-filled air extending over a large area.

E

Easterly wave A migratory wavelike disturbance in the tropical easterlies. Easterly waves occasionally intensify into tropical cyclones. They are also called *tropical waves.*

Eccentricity (of the earth's orbit) The deviation of the earth's orbit from elliptical to nearly circular.
Eddy A small volume of air (or any fluid) that behaves differently from the larger flow in which it exists.
EF (Enhanced Fujita) scale A modification of the original *Fujita Scale* that describes tornado intensity by observing damage caused by the tornado.
Ekman spiral An idealized description of the way the wind-driven ocean currents vary with depth. In the atmosphere it represents the way the winds vary from the surface up through the friction layer or planetary boundary layer.
Ekman transport Net surface water transport due to the Ekman spiral. In the Northern Hemisphere the transport is 90° to the right of the surface wind direction.
Electromagnetic waves Waves that have both electric and magnetic properties that travel at the speed of light. (*See also* radiant energy.)
El Niño An extensive ocean warming that begins along the coast of Peru and Ecuador and extends westward over the tropical Pacific. Major El Niño events, or strong El Niños, occur once every 2 to 7 years as a current of nutrient-poor tropical water moves southward along the west coast of South America.
Energy The property of a system that generally enables it to do work. Some forms of energy are kinetic, radiant, potential, chemical, electric, and magnetic.
Ensemble forecasting A forecasting technique that entails running several forecast models (or different versions of a single model), each beginning with slightly different weather information. The forecaster's level of confidence is based on how well the models agree (or disagree) at the end of some specified time.
ENSO (El Niño/southern oscillation) A condition in the tropical Pacific whereby the reversal of surface air pressure at opposite ends of the Pacific Ocean induces westerly winds, a strengthening of the equatorial countercurrent, and extensive ocean warming.
Entrainment The mixing of environmental air into a pre-existing air current or cloud so that the environmental air becomes part of the current or cloud.
Environmental lapse rate The rate of decrease of air temperature with elevation. It is most often measured with a radiosonde.
Evaporation The process by which a liquid changes into a gas.
Evaporation (mixing) fog Fog produced when sufficient water vapor is added to the air by evaporation, and the moist air mixes with relatively drier air. The two common types are steam fog, which forms when cold air moves over warm water, and frontal fog, which forms as warm raindrops evaporate in a cool air mass.
Exosphere The outermost portion of the atmosphere.
Extratropical cyclone A cyclonic storm that most often forms along a front in middle and high latitudes. Also called a *middle-latitude cyclonic storm,* a *depression,* and a *low*. It is not a tropical storm or hurricane.
Eye A region in the center of a hurricane (tropical cyclone) where the winds are light and skies are clear to partly cloudy.
Eyewall A wall of dense thunderstorms that surrounds the eye of a hurricane.
Eyewall replacement A situation within a hurricane (tropical cyclone) where the storm's original eyewall dissipates and a new eyewall forms outward, farther away from the center of the storm.

F

Fahrenheit scale A temperature scale where 32 is assigned to the temperature at which water freezes and 212 to the temperature at which water boils (at sea level).
Fallstreaks Falling ice crystals that evaporate before reaching the ground.
Fall wind A strong, cold katabatic wind that blows downslope off snow-covered plateaus.
Feedback mechanism A process whereby an initial change in an atmospheric process will tend to either reinforce the process (positive feedback) or weaken the process (negative feedback).
Flash flood A flood that rises and falls quite rapidly with little or no advance warning, usually as the result of intense rainfall over a relatively small area.
Foehn *See* Chinook wind.
Fog A cloud with its base at the earth's surface.
Forced convection On a small scale, a form of mechanical stirring taking place when twisting eddies of air are able to mix hot surface air with the cooler air above. On a larger scale, it can be induced by the lifting of warm air along a front (frontal uplift) or along a topographic barrier (orographic uplift).
Forked lightning The most common form of cloud-to-ground lightning where many branches of light extend from the main lightning channel.
Free convection *See* Convection.
Freeze A condition occurring over a widespread area when the surface air temperature remains below freezing for a sufficient time to damage certain agricultural crops. A freeze most often occurs as cold air is advected into a region, causing freezing conditions to exist in a deep layer of surface air. Also called *advection frost.*
Freezing rain and freezing drizzle Rain or drizzle that falls in liquid form and then freezes upon striking a cold object or ground. Both can produce a coating of ice on objects which is called *glaze.*
Friction layer The atmospheric layer near the surface usually extending up to about 1 km (3300 ft) where the wind is influenced by friction of the earth's surface and objects on it. Also called the *atmospheric boundary layer* and *planetary boundary layer.*
Front The transition zone between two distinct air masses.
Frontal thunderstorms Thunderstorms that form in response to forced convection (forced lifting) along a front. Most go through a cycle similar to those of ordinary thunderstorms.
Frontal wave A wavelike deformation along a front in the lower levels of the atmosphere. Those that develop into storms are termed unstable waves, whereas those that do not are called *stable waves.*
Frontogenesis The formation, strengthening, or regeneration of a front.
Frontolysis The weakening or dissipation of a front.
Frost A covering of ice produced by deposition on exposed surfaces when the air temperature falls below the frost point. Also called *hoarfrost.*
Frostbite The partial freezing of exposed parts of the body, causing injury to the skin and sometimes to deeper tissues.

Frost point The temperature at which the air becomes saturated with respect to ice when cooled at constant pressure and constant water vapor content.
Frozen dew The transformation of liquid dew into tiny beads of ice when the air temperature drops below freezing.
Fujita scale A scale developed by T. Theodore Fujita for classifying tornadoes according to the damage they cause and their rotational wind speed.
Fulgurite A rootlike tube (or several tubes) that forms when a lightning stroke fuses sand particles together.
Funnel cloud A tornado whose circulation has not reached the ground. Often appears as a rotating conelike cloud that extends downward from the base of a thunderstorm.

G

Gas law The thermodynamic law applied to a perfect gas that relates the pressure of the gas to its density and absolute temperature.
General circulation of the atmosphere Large-scale atmospheric motions over the entire earth.
Geoengineering Reducing the impact of climate change through global scale technological fixes, such as fertilizing the oceans with plants that absorb carbon dioxide.
Geostationary satellite A satellite that orbits the earth at the same rate that the earth rotates and thus remains over a fixed place above the equator.
Geostrophic wind A theoretical horizontal wind blowing in a straight path, parallel to the isobars or contours, at a constant speed. The geostrophic wind results when the Coriolis force exactly balances the horizontal pressure gradient force.
Glaciated cloud A cloud or portion of a cloud where only ice crystals exist.
Global scale The largest scale of atmospheric motion. Also called the *planetary scale.*
Global warming Increasing global surface air temperatures that show up in the climate record. The term global warming is usually attributed to human activities, such as increasing concentrations of greenhouse gases.
Gradient wind A theoretical wind that blows parallel to curved isobars or contours.
Graupel Ice particles between 2 and 5 mm in diameter that form in a cloud often by the process of accretion. Snowflakes that become rounded pellets due to riming are called *graupel* or *snow pellets.*
Greenhouse effect *See* Atmospheric greenhouse effect.
Greenhouse gases Gases in the earth's atmosphere, such as water vapor and carbon dioxide, that allow much of the sunlight to pass through, but are strong absorbers of infrared energy emitted by the earth and its atmosphere. Other greenhouse gases include methane, nitrous oxide, ozone, and chlorofluorocarbons.
Ground fog *See* Radiation fog.
Gulf stream A warm, swift, narrow ocean current flowing along the east coast of the United States.
Gust front A boundary that separates a cold downdraft of a thunderstorm from warm, humid surface air. On the surface its passage resembles that of a cold front.
Gustnado A relatively weak tornado associated with a thunderstorm's outflow. It most often forms along the gust front.

H

Haboob A dust or sandstorm that forms as cold downdrafts from a thunderstorm turbulently lift dust and sand into the air.
Hadley cell A thermal circulation proposed by George Hadley to explain the movement of the trade winds. It consists of rising air near the equator and sinking air near 30° latitude.
Hailstones Transparent or partially opaque particles of ice that range in size from that of a pea to that of golf balls.
Hailstreak The accumulation of hail at the earth's surface along a relatively long (10 km), narrow (2 km) band.
Halos Rings or arcs that encircle the sun or moon when seen through an ice crystal cloud or a sky filled with falling ice crystals. Halos are produced by refraction of light.
Haze Fine dry or wet dust or salt particles dispersed through a portion of the atmosphere. Individually these are not visible but cumulatively they will diminish visibility. Dry haze particles are very small, on the order of 0.1 μm. Wet haze particles are larger.
Heat A form of energy transferred between systems by virtue of their temperature differences.
Heat burst A sudden increase in surface air temperature often accompanied by extreme drying. A heat burst is associated with the downdraft of a thunderstorm, or a cluster of thunderstorms.
Heat capacity The ratio of the heat absorbed (or released) by a system to the corresponding temperature rise (or fall).
Heat index (HI) An index that combines air temperature and relative humidity to determine an apparent temperature—how hot it actually feels.
Heat lightning Distant lightning that illuminates the sky but is too far away for its thunder to be heard.
Heatstroke A physical condition induced by a person's overexposure to high air temperatures, especially when accompanied by high humidity.
Heat wave A period of abnormally hot weather often accompanied by high humidity.
Hectopascal Abbreviated hPa. One hectopascal is equal to 100 Newtons/m^2, or 1 millibar.
High *See* Anticyclone.
Hook echo The shape of an echo on a Doppler radar screen that indicates the possible presence of a tornado.
Horse latitudes The belt of latitude at about 30° to 35° where winds are predominantly light and the weather is hot and dry.
Humidity A general term that refers to the air's water vapor content. (*See* Relative humidity.)
Hurricane A tropical cyclone having wind speeds in excess of 74 mi/hr (64 knots).
Hurricane warning A warning given when it is likely that a hurricane will strike an area within 24 hours.
Hurricane watch A hurricane watch indicates that a hurricane poses a threat to an area (often within several days) and residents of the watch area should be prepared.
Hydrologic cycle A model that illustrates the movement and exchange of water among the earth, atmosphere, and oceans.
Hygrometer An instrument designed to measure the air's water vapor content. The sensing part of the instrument can be hair (hair hygrometer), a plate coated with carbon (electrical hygrometer), or an infrared sensor (infrared hygrometer).

Hygroscopic The ability to accelerate the condensation of water vapor. Usually used to describe condensation nuclei that have an affinity for water vapor.
Hypothermia The deterioration in one's mental and physical condition brought on by a rapid lowering of human body temperature.
Hypoxia A condition experienced by humans when the brain does not receive sufficient oxygen.

I

Ice Age *See* Pleistocene epoch.
Ice-crystal (Bergeron) process A process that produces precipitation. The process involves tiny ice crystals in a supercooled cloud growing larger at the expense of the surrounding liquid droplets. Also called the *Bergeron process.*
Ice fog A type of fog that forms at very low temperatures, composed of tiny suspended ice particles.
Ice jam flooding The break-up of river ice that gets caught in a narrow channel and produces local flooding, most often in spring.
Icelandic low The subpolar low-pressure area that is centered near Iceland on charts that show mean sea-level pressure.
Ice nuclei Particles that act as nuclei for the formation of ice crystals in the atmosphere.
Ice pellets *See* Sleet.
Ice storm A winter storm characterized by a substantial amount of precipitation in the form of freezing rain, freezing drizzle, or sleet.
Indian summer An unseasonably warm spell with clear skies near the middle of autumn. Usually follows a substantial period of cool weather.
Infrared radiation Electromagnetic radiation with wavelengths between about 0.7 and 1000 μm. This radiation is longer than visible radiation but shorter than microwave radiation.
Insolation The incoming solar radiation that reaches the earth and the atmosphere.
Instrument shelter A boxlike, often wooden, structure designed to protect weather instruments from direct sunshine and precipitation.
Interglacial period A time interval of relatively mild climate during the Ice Age when continental ice sheets were absent or limited in extent to Greenland and the Antarctic.
Intertropical convergence zone (ITCZ) The boundary zone separating the northeast trade winds of the Northern Hemisphere from the southeast trade winds of the Southern Hemisphere.
Inversion An increase in air temperature with height.
Ion An electrically charged atom, molecule, or particle.
Ionosphere An electrified region of the upper atmosphere where fairly large concentrations of ions and free electrons exist.
Isobar A line connecting points of equal pressure.
Isobaric map *See* Constant-pressure chart.
Isobaric surface A surface along which the atmospheric pressure is everywhere equal.
Isotach A line connecting points of equal wind speed.
Isotherm A line connecting points of equal temperature.
Isothermal layer A layer where the air temperature is constant with increasing altitude. In an isothermal layer, the air temperature lapse rate is zero.

J

Jet maximum *See* Jet streak.
Jet streak A region of high wind speed that moves through the axis of a jet stream. Also called *jet maximum.*
Jet stream Relatively strong winds concentrated within a narrow band in the atmosphere.

K

Katabatic (fall) wind Any wind blowing downslope. It is usually cold.
Kelvin A unit of temperature. A Kelvin is denoted by K and 1 K equals 1°C. Zero Kelvin is absolute zero, or −273.15°C.
Kelvin scale A temperature scale with zero degrees equal to the theoretical temperature at which all molecular motion ceases. Also called the *absolute scale.* The units are sometimes called *degrees Kelvin;* however, the correct SI terminology is *Kelvins,* abbreviated K.
Kinetic energy The energy within a body that is a result of its motion.
Knot A unit of speed equal to 1 nautical mile per hour. One knot equals 1.15 mi/hr.

L

Lake breeze A wind blowing onshore from the surface of a lake.
Lake-effect snows Localized snowstorms that form on the downwind side of a lake. Such storms are common in late fall and early winter near the Great Lakes as cold, dry air picks up moisture and warmth from the unfrozen bodies of water.
Land breeze A coastal breeze that blows from land to sea, usually at night.
Landspout Relatively weak nonsupercell tornado that originates with a cumuliform cloud in its growth stage and with a cloud that does not contain a mid-level mesocyclone. Its spin originates near the surface. Landspouts often look like waterspouts over land.
La Niña A condition where the central and eastern tropical Pacific Ocean turns cooler than normal.
Lapse rate The rate at which an atmospheric variable (usually temperature) decreases with height. (*See* Environmental lapse rate.)
Latent heat The heat that is either released or absorbed by a unit mass of a substance when it undergoes a change of state, such as during evaporation, condensation, or sublimation.
Lee-side low Storm systems (extratropical cyclones) that form on the downwind (lee) side of a mountain chain. In the United States lee-side lows frequently form on the eastern side of the Rockies and Sierra Nevada mountains.
Leeward The side of a mountain or mountain range that is on the downwind side of the main flow of air.
Lenticular cloud A cloud in the shape of a lens.
Level of free convection The level in the atmosphere at which a lifted air parcel becomes warmer than its surroundings in a conditionally unstable atmosphere.
Lightning A visible electrical discharge produced by thunderstorms.

Little Ice Age The period from about 1550 to 1850 when average temperatures over Europe were lower.
Local winds Winds that tend to blow over a relatively small area; often due to regional effects, such as mountain barriers, large bodies of water, local pressure differences, and other influences.
Long-range forecast Generally used to describe a weather forecast that extends beyond about 8.5 days into the future.
Longwave radiation A term most often used to describe the infrared energy emitted by the earth and the atmosphere.
Longwaves in the westerlies A wave in the upper level of the westerlies characterized by a long length (thousands of kilometers) and significant amplitude. Also called *Rossby waves.*
Low *See* Mid-latitude cyclonic storm.
Low-level jet streams Jet streams that typically form near the earth's surface below an altitude of about 2 km and usually attain speeds of less than 70 mi/hr.

M

Macroburst A strong downdraft (downburst) greater than 2.5 mi (4 km) wide that can occur beneath thunderstorms. A downburst less than 2.5 mi (4 km) across is called a *microburst.*
Macroscale The normal meteorological synoptic scale for obtaining weather information. It can cover an area ranging from the size of a continent to the entire globe.
Mammatus clouds Clouds that look like pouches hanging from the underside of a cloud.
Maritime air Moist air whose characteristics were developed over an extensive body of water.
Maritime polar air mass An air mass characterized by low temperatures and high humidity.
Maritime tropical air mass An air mass characterized by high temperatures and high humidity.
Mature cyclone A stage in the development of a mid-latitude cyclonic storm just before it begins to dissipate. This stage is often associated with the storm's lowest central pressure, highest wind speeds, and heaviest precipitation.
Mature thunderstorm The second stage in the three-stage cycle of an ordinary thunderstorm. This mature stage is characterized by heavy showers, lightning, thunder, and violent vertical motions inside cumulonimbus clouds.
Maunder minimum A period from about 1645 to 1715 when few, if any, sunspots were observed.
Mean annual temperature The average temperature at any given location for the entire year.
Mean daily temperature The average of the highest and lowest temperature for a 24-hour period.
Medium-range forecast Generally used to describe a weather forecast that extends from about 3 to 8.5 days into the future.
Mercury barometer A type of barometer that uses mercury to measure atmospheric pressure. The height of the mercury column is a measure of atmospheric pressure.
Meridional flow A type of atmospheric circulation pattern in which the north-south component of the wind is pronounced.
Mesocyclone A vertical column of cyclonically rotating air within a severe thunderstorm.
Mesohigh A relatively small area of high atmospheric pressure that forms beneath a thunderstorm.
Mesoscale convective complex (MCC) A large organized convective weather system comprised of a number of individual thunderstorms. The size of an MCC can be 1000 times larger than an individual air mass thunderstorm.
Mesoscale convective system (MCS) A large cloud system that represents an ensemble of thunderstorms that form by convection and produce precipitation over a wide area.
Mesoscale convective vortex A counterclockwise circulation about an area of low pressure that forms in the mid-levels of the atmosphere and is usually less than 150 miles in diameter.
Mesosphere The atmospheric layer between the stratosphere and the thermosphere. Located at an average elevation between 50 and 80 km above the earth's surface.
Meteogram A chart that shows how one or more weather variables has changed at a station over a given period of time or how the variables are likely to change with time.
Meteorology The study of the atmosphere and atmospheric phenomena as well as the atmosphere's interaction with the earth's surface, oceans, and life in general.
Microburst A strong localized downdraft (downburst) less than 2.5 mi (4 km) wide that occurs beneath thunderstorms. A strong downburst greater than 2.5 mi (4 km) across is called a *macroburst.*
Microclimate The climate structure of the air space near the surface of the earth.
Micrometer (μm) A unit of length equal to one-millionth of a meter.
Microscale The smallest scale of atmospheric motions.
Mid-Holocene maximum A warm period in geologic history (about 5000 to 6000 years ago) that favored the development of plants.
Middle latitudes The region of the world typically described as being between 30° and 50° latitude.
Mid-latitude cyclonic storm A cyclonic storm that most often forms along a front in middle and high latitudes. Also called an *extratropical cyclone,* a *depression,* and a *low.* It is not a tropical cyclone or hurricane.
Milankovitch theory A theory proposed by Milutin Milankovitch in the 1930s suggesting that changes in the earth's orbit were responsible for variations in solar energy reaching the earth's surface and climatic changes.
Millibar (mb) A unit for expressing atmospheric pressure. Sea-level pressure is normally close to 1013 mb.
Mixing ratio The ratio of the mass of water vapor in a given volume of air to the mass of dry air.
Moist adiabatic rate The rate of change of temperature in a rising or descending saturated air parcel. The rate of cooling or warming varies but a common value of 6°C per 1000 m (3.3°F per 1000 ft) is used.
Molecule A collection of atoms held together by chemical forces.
Monsoon wind system A wind system that reverses direction between winter and summer. Usually the wind blows from land to sea in winter and from sea to land in summer.

Mountain and valley breeze A local wind system of a mountain valley that blows downhill (mountain breeze) at night and uphill (valley breeze) during the day.
Multicell storms Thunderstorms often in a line, each of which may be in a different stage of its life cycle.
Multi-vortex tornado A single large tornado with multiple smaller whirls (vortices) rotating within the larger tornado.

N

Negative feedback mechanism *See* Feedback mechanism.
Neutral stability (neutrally stable atmosphere) An atmospheric condition that exists in dry air when the environmental lapse rate equals the dry adiabatic rate. In saturated air the environmental lapse rate equals the moist adiabatic rate.
NEXRAD An acronym for Next Generation Weather Radar. The main component of NEXRAD is the WSR 88-D Doppler radar.
Nimbostratus A dark, gray cloud characterized by more or less continuously falling precipitation. It is rarely accompanied by lightning, thunder, or hail.
Nitrogen (N_2) A colorless and odorless gas that occupies about 78 percent of dry air in the lower atmosphere.
Nocturnal inversion *See* Radiation inversion.
Nonsupercell tornado A tornado that occurs with a cloud that is often in its growing stage and one that does not contain a mid-level mesocyclone or wall cloud. Landspouts and gustnadoes are examples of nonsupercell tornadoes.
North Atlantic Oscillation (NAO) A reversal of atmospheric pressure over the Atlantic Ocean that influences the weather over Europe and over eastern North America.
Northeaster A name given to a strong, steady wind from the northeast that is accompanied by rain and inclement weather. It often develops when a storm system moves northeastward along the coast of North America. Also called *Nor'easter.*
Northern lights *See* Aurora.
Nowcasting Short-term weather forecasts varying from minutes up to a few hours.
Nuclear winter The dark, cold, and gloomy conditions that presumably would be brought on by nuclear war.
Numerical weather prediction (NWP) Forecasting the weather based upon the solutions of mathematical equations by high-speed computers.

O

Obliquity (of the earth's axis) The tilt of the earth's axis. It represents the angle from the perpendicular to the plane of the earth's orbit.
Occluded front (occlusion) A complex frontal system that ideally forms when a cold front overtakes a warm front. When the air behind the front is colder than the air ahead of it, the front is called a *cold occlusion.* When the air behind the front is milder than the air ahead of it, it is called a *warm occlusion.*
Offshore wind A breeze that blows from the land out over the water. Opposite of an onshore wind.
Onshore wind A breeze that blows from the water onto the land. Opposite of an offshore wind.
Open wave The stage of development of a wave cyclone (mid-latitude cyclonic storm) where a cold front and a warm front exist, but no occluded front. The center of lowest pressure in the wave is located at the junction of the two fronts.
Ordinary thunderstorm (formally called *air-mass thunderstorm*) A thunderstorm produced by local convection within a conditionally unstable air mass. It often forms in the afternoon and does not reach the intensity of a severe thunderstorm.
Orographic uplift The lifting of air over a topographic barrier. Clouds that form in this lifting process are called *orographic clouds.*
Outflow boundary A surface boundary separating cooler more-dense air from warmer less-dense air. Outflow boundaries form by the horizontal spreading of cool air that originated inside a thunderstorm.
Outgassing The release of gases dissolved in hot, molten rock.
Overrunning A condition that occurs when air moves up and over another layer of air.
Overshooting top A situation in an intense thunderstorm where rising air associated with strong convection penetrates into a stable layer (usually the stratosphere), forcing the upper part of the cloud to rise above its relatively flat anvil top.
Oxygen (O_2) A colorless and odorless gas that occupies about 21 percent of dry air in the lower atmosphere.
Ozone (O_3) An almost colorless gaseous form of oxygen with an odor similar to weak chlorine. The highest natural concentration is found in the stratosphere where it is known as stratospheric ozone. It also forms in polluted air near the surface, where it is the main ingredient of photochemical smog. Here, it is called *tropospheric ozone.*
Ozone hole A sharp drop in stratospheric ozone concentration observed over the Antarctic during the spring.

P

Pacific air Air from off the Pacific Ocean that, after moving over several mountain ranges, is relatively mild and fairly dry when it moves into the Great Plains of North America.
Pacific decadal oscillation (PDO) A reversal in ocean surface temperatures that occurs every 20 to 30 years over the northern Pacific Ocean.
Pacific high *See* Subtropical high.
Palmer Drought Severity Index (PDSI) An index that determines moisture deficiencies by comparing the precipitation received in an area during a given period of time with the average amount expected during the same period.
Parcel of air An imaginary small body of air a few meters wide that is used to explain the behavior of air.
Particulate matter Solid particles or liquid droplets that are small enough to remain suspended in the air. Also called *aerosols.*
Pattern recognition An analogue method of forecasting where the forecaster uses prior weather events (or similar weather map conditions) to make a forecast.

Permafrost A layer of soil beneath the earth's surface that remains frozen throughout the year.
Persistence forecast A forecast that the future weather condition will be the same as the present condition.
Photochemical smog *See* Smog.
Photon A discrete quantity of energy that can be thought of as a packet of electromagnetic radiation traveling at the speed of light.
Pileus cloud A smooth cloud in the form of a cap. Occurs above, or is attached to, the top of a cumuliform cloud. Also called a *cap cloud.*
Plate tectonics The theory that the earth's surface down to about 100 km is divided into a number of plates that move relative to one another across the surface of the earth. Once referred to as *continental drift.*
Pleistocene Epoch (or Ice Age) The most recent period of extensive continental glaciation that saw large portions of North America and Europe covered with ice. It began about 2 million years ago and ended about 10,000 years ago.
Polar easterlies A shallow body of easterly winds located at high latitudes poleward of the subpolar low.
Polar front A semipermanent, semicontinuous front that separates tropical air masses from polar air masses.
Polar front jet stream (polar jet) The jet stream that is associated with the polar front in middle and high latitudes. It is usually located at altitudes between 9 and 12 km.
Polar front theory A theory developed by a group of Scandinavian meteorologists that explains the formation, development, and overall life history of cyclonic storms that form along the polar front.
Polar low An area of low pressure that forms over polar water behind (poleward of) the main polar front.
Polar orbiting satellite A satellite whose orbit closely parallels the earth's meridian lines and thus crosses the polar regions on each orbit.
Pollutants Any gaseous, chemical, or organic matter that contaminates the atmosphere, soil, or water.
Positive feedback mechanism *See* Feedback mechanism.
Potential energy The energy that a body possesses by virtue of its position with respect to other bodies in the field of gravity.
Precession (of the earth's axis of rotation) The wobble of the earth's axis of rotation that traces out the path of a cone over a period of about 23,000 years.
Precipitation Any form of water particles—liquid or solid—that falls from the atmosphere and reaches the ground.
Pressure The force per unit area. *See also* Air pressure.
Pressure gradient The rate of decrease of pressure per unit of horizontal distance. On the same chart, when the isobars are close together, the pressure gradient is steep. When the isobars are far apart, the pressure gradient is weak.
Pressure gradient force (PGF) The force due to differences in pressure within the atmosphere that causes air to move and, hence, the wind to blow. It is directly proportional to the pressure gradient.
Pressure tendency The rate of change of atmospheric pressure within a specified period of time, most often three hours. Same as barometric tendency.
Prevailing westerlies The dominant westerly winds that blow in middle latitudes on the poleward side of the subtropical high-pressure areas. Also called *westerlies.*
Prevailing wind The wind direction most frequently observed during a given period.
Probability forecast A forecast of the probability of occurrence of one or more of a mutually exclusive set of weather conditions.
Prognostic chart (prog) A chart showing expected or forecasted conditions, such as pressure patterns, frontal positions, contour height patterns, and so on.
Psychrometer An instrument used to measure the water vapor content of the air. It consists of two thermometers (dry bulb and wet bulb). After whirling the instrument, the dew point and relative humidity can be obtained with the aid of tables.

R

Radar An electronic instrument used to detect objects (such as falling precipitation) by their ability to reflect and scatter microwaves back to a receiver. (*See also* Doppler radar.)
Radiant energy (radiation) Energy propagated in the form of electromagnetic waves. These waves do not need molecules to propagate them, and in a vacuum they travel at nearly 300,000 km per sec (186,000 mi per sec).
Radiational cooling The process by which the earth's surface and adjacent air cool by emitting infrared radiation.
Radiation fog Fog produced over land when radiational cooling reduces the air temperature to or below its dew point. It is also known as *ground fog* and *valley fog.*
Radiation inversion An increase in temperature with height due to radiational cooling of the earth's surface. Also called a *nocturnal inversion.*
Radiative equilibrium temperature The temperature achieved when an object, behaving as a blackbody, is absorbing and emitting radiation at equal rates.
Radiative forcing An increase (positive) or a decrease (negative) in net radiant energy observed over an area at the tropopause. An increase in radiative forcing may induce surface warming, whereas a decrease may induce surface cooling.
Radiative forcing agent Any factor (such as increasing greenhouse gases and variations in solar output) that can change the balance between incoming energy from the sun and outgoing energy from the earth and the atmosphere.
Radiosonde A balloon-borne instrument that measures and transmits pressure, temperature, and humidity to a ground-based receiving station.
Rain Precipitation in the form of liquid water drops that have diameters greater than that of drizzle.
Rain gauge An instrument designed to measure the amount of rain that falls during a given time interval.
Rain shadow The region on the leeside of a mountain where the precipitation is noticeably less than on the windward side.
Rawinsonde observation A radiosonde observation that includes wind data.
Reflection The process whereby a surface turns back a portion of the radiation that strikes it.
Refraction The bending of light as it passes from one medium to another.

Relative humidity The ratio of the amount of water vapor in the air compared to the amount required for saturation (at a particular temperature and pressure). The ratio of the air's actual vapor pressure to its saturation vapor pressure.
Return stroke The luminous lightning stroke that propagates upward from the earth to the base of a cloud.
Ribbon lightning Cloud-to-ground lightning that appears to be spread horizontally into a series of luminous streaks or ribbons.
Ridge An elongated area of high atmospheric pressure.
Rime A white or milky granular deposit of ice formed by the rapid freezing of supercooled water drops as they come in contact with an object in below-freezing air.
Riming *See* Accretion.
Roll cloud A dense, cylindrical, elongated cloud that appears to slowly spin about a horizontal axis behind the leading edge of a thunderstorm's gust front.
Rotor cloud A turbulent cumuliform type of cloud that forms on the leeward side of large mountain ranges. The air in the cloud rotates about an axis parallel to the range.
Rotors Turbulent eddies that form downwind of a mountain chain, creating hazardous flying conditions.

S

Saffir-Simpson scale A scale relating a hurricane's central pressure and winds to the possible damage it is capable of inflicting.
Saint (St.) Elmo's fire A bright electric discharge that is projected from objects (usually pointed) when they are in a strong electric field, such as during a thunderstorm.
Santa Ana wind A warm, dry wind that blows into southern California from the east off the elevated desert plateau. Its warmth is derived from compressional heating.
Saturation (of air) An atmospheric condition whereby the level of water vapor is the maximum possible at the existing temperature and pressure.
Saturation vapor pressure The maximum amount of water vapor necessary to keep moist air in equilibrium with a surface of pure water or ice. It represents the maximum amount of water vapor that the air can hold at any given temperature and pressure. (*See* Equilibrium vapor pressure.)
Scales of motion The hierarchy of atmospheric circulations from tiny gusts to giant storms.
Scattering The process by which small particles in the atmosphere deflect radiation from its path into different directions.
Sea breeze A coastal local wind that blows from the ocean onto the land. The leading edge of the breeze is termed a sea breeze front.
Sea-level pressure The atmospheric pressure at mean sea level.
Secondary low A low-pressure area (often an open wave) that forms near, or in association with, a main low-pressure area.
Seiches Standing waves that oscillate back and forth over an open body of water.
Selective absorbers Substances such as water vapor, carbon dioxide, clouds, and snow that absorb radiation only at particular wavelengths.
Semipermanent highs and lows Areas of high pressure (anticyclones) and low pressure (extratropical cyclones) that tend to persist at a particular latitude belt throughout the year. In the Northern Hemisphere, typically they shift slightly northward in summer and slightly southward in winter.
Sensible heat The heat we can feel and measure with a thermometer.
Sensible temperature The sensation of temperature that the human body feels in contrast to the actual temperature of the environment as measured with a thermometer.
Severe thunderstorms Intense thunderstorms capable of producing heavy showers, flash floods, hail, strong and gusty surface winds, and tornadoes. Severe thunderstorms are defined by the National Weather Service as having at least one of the following: large hail with a diameter of at least three-quarters of an inch and/or surface winds gusting to 58 mi/hr (50 knots) or greater, or produces a tornado.
Shear *See* Wind shear.
Sheet lightning Occurs when the lightning flash is not seen but the flash causes the cloud (or clouds) to appear as a diffuse luminous white sheet.
Shelf cloud A dense, arch-shaped, ominous-looking cloud that often forms along the leading edge of a thunderstorm's gust front, especially when stable air rises up and over cooler air at the surface. Also called an *arcus cloud*.
Short-range forecast Generally used to describe a weather forecast that extends from about 6 hours to a few days into the future.
Shortwave (in the atmosphere) A small wave that moves around longwaves in the same direction as the airflow in the middle and upper troposphere. Shortwaves are also called *shortwave troughs*.
Shortwave radiation A term most often used to describe the radiant energy emitted from the sun, in the visible and near ultraviolet wavelengths.
Shower Intermittent precipitation from a cumuliform cloud, usually of short duration but often heavy.
Siberian high A strong, shallow area of high pressure that forms over Siberia in winter.
Simoon A strong hot and very dry desert wind that blows over north Africa and the Middle East.
Sleet A type of precipitation consisting of transparent pellets of ice 5 mm or less in diameter. Same as ice pellets.
Smog Originally *smog* meant a mixture of smoke and fog. Today, *smog* means air that has restricted visibility due to pollution, or pollution formed in the presence of sunlight—*photochemical smog*.
Snow A solid form of precipitation composed of ice crystals in complex hexagonal form.
Snow-albedo feedback A positive feedback whereby increasing surface air temperatures enhance the melting of snow and ice in polar latitudes. This feedback reduces the earth's albedo and allows more sunlight to reach the surface, which causes the air temperature to rise even more.
Snowflake An aggregate of ice crystals that falls from a cloud.
Snow flurries Light showers of snow that fall intermittently.
Snow grains Precipitation in the form of very small, opaque grains of ice. The solid equivalent of drizzle.

Snow pellets White, opaque, approximately round ice particles between 2 and 5 mm in diameter that form in a cloud either from the sticking together of ice crystals or from the process of accretion. Also called *graupel.*

Snow rollers A cylindrical spiral of snow shaped somewhat like a child's muff. Snow rollers are produced by the wind.

Snow squall (shower) An intermittent heavy shower of snow that greatly reduces visibility.

Solar constant The rate at which solar energy is received on a surface at the outer edge of the atmosphere perpendicular to the sun's rays when the earth is at a mean distance from the sun. The value of the solar constant is about two calories per square centimeter per minute or about 1376 W/m^2 in the SI system of measurement.

Solar wind An outflow of charged particles from the sun that escapes the sun's outer atmosphere at high speed.

Sonic boom A loud explosive-like sound caused by a shock wave emanating from an aircraft (or any object) traveling at or above the speed of sound.

Sounding An upper-air observation, such as a radiosonde observation. A vertical profile of an atmospheric variable such as temperature or winds.

Source regions Regions where air masses originate and acquire their properties of temperature and moisture.

Southern Oscillation The reversal of surface air pressure at opposite ends of the tropical Pacific Ocean that occur during major El Niño events.

Specific heat The ratio of the heat absorbed (or released) by the unit mass of the system to the corresponding temperature rise (or fall).

Specific humidity The ratio of the mass of water vapor in a given parcel to the total mass of air in the parcel.

Squall line A line of thunderstorms that form along a cold front or out ahead of it.

Stable air See Absolutely stable atmosphere.

Standard atmosphere A hypothetical vertical distribution of atmospheric temperature, pressure, and density in which the air is assumed to obey the gas law and the hydrostatic equation. The lapse rate of temperature in the troposphere is taken as 6.5°C/1000 m or 3.6°F/1000 ft.

Standard atmospheric pressure A pressure of 1013.25 millibars (mb), 29.92 inches of mercury (Hg), 760 millimeters (mm) of mercury, 14.7 pounds per square inch (lb/in.^{2}), or 1013.25 hectopascals (hPa).

Stationary front A front that is nearly stationary with winds blowing almost parallel and from opposite directions on each side of the front.

Station pressure The actual air pressure computed at the observing station.

Statistical forecast A forecast based on a mathematical/statistical examination of data that represents the past observed behavior of the forecasted weather element.

Steady-state forecast A weather prediction based on the past movement of surface weather systems. It assumes that the systems will move in the same direction and at approximately the same speed as they have been moving. Also called *trend forecasting.*

Steam fog *See* Evaporation (mixing) fog.

Stefan-Boltzmann law A law of radiation which states that the amount of radiant energy emitted from a unit surface area of an object (ideally a blackbody) is proportional to the fourth power of the object's absolute temperature.

Stepped leader An initial discharge of electrons that proceeds intermittently toward the ground in a series of steps in a cloud-to-ground lightning stroke.

Storm surge An abnormal rise of the sea along a shore. Primarily due to the winds of a storm, especially a hurricane.

Straight-line wind Strong winds created by a thunderstorm's downdraft that flows outward, away from the storm in a straight line, more or less parallel to the ground.

Stratocumulus A low cloud, predominantly stratiform, with low, lumpy, rounded masses, often with blue sky between them.

Stratosphere The layer of the atmosphere above the troposphere and below the mesosphere (between 10 km and 50 km), generally characterized by an increase in temperature with height.

Stratus A low, gray cloud layer with a rather uniform base whose precipitation is most commonly drizzle.

Streamline A line that shows the wind-flow pattern.

Sublimation The process whereby ice changes directly into water vapor without melting.

Subpolar low A belt of low pressure located between 50° and 70° latitude. In the Northern Hemisphere, this "belt" consists of the Aleutian low in the North Pacific and the Icelandic low in the North Atlantic. In the Southern Hemisphere, it exists around the periphery of the Antarctic continent.

Subsidence The slow sinking of air, usually associated with high-pressure areas.

Subsidence inversion A temperature inversion produced by compressional warming—the adiabatic warming of a layer of sinking air.

Subtropical high A semipermanent high in the subtropical high-pressure belt centered near 30° latitude. The Bermuda High is located over the Atlantic Ocean off the east coast of North America. The Pacific High is located off the west coast of North America.

Subtropical jet stream The jet stream typically found between 20° and 30° latitude at altitudes between 12 and 14 km.

Suction vortices Small, rapidly rotating whirls perhaps 10 m in diameter that are found within large tornadoes.

Sulfate aerosols *See* Aerosols.

Sulfur dioxide (SO_2) A colorless gas that forms primarily in the burning of sulfur-containing fossil fuels.

Summer solstice Approximately June 21 in the Northern Hemisphere when the sun is highest in the sky and directly overhead at latitude 23½°N, the Tropic of Cancer.

Sunspots Relatively cooler areas on the sun's surface. They represent regions of an extremely high magnetic field.

Supercell storm An enormous severe thunderstorm that consists primarily of a single rotating updraft. Its organized internal structure allows that storm to maintain itself for several hours. The storm can produce large hail and dangerous tornadoes.

Supercell tornadoes Tornadoes that occur within supercell thunderstorms that contain well-developed mid-level mesocyclones.

Supercooled cloud (or cloud droplets) A cloud composed of liquid droplets at temperatures below 0°C (32°F). When the cloud is on the ground it is called *supercooled fog* or *cold fog.*

Supersaturation A condition whereby the atmosphere contains more water vapor than is needed to produce saturation with respect to a flat surface of pure water or ice, and the relative humidity is greater than 100 percent.
Super typhoon A tropical cyclone (typhoon) in the western Pacific that has sustained winds of 130 knots or greater.
Surface inversion *See* Radiation inversion.
Surface map A map that shows the distribution of sea-level pressure with isobars and weather phenomena. Also called a *surface chart.*
Synoptic scale The typical weather map scale that shows features such as high- and low-pressure areas and fronts over a distance spanning a continent. Also called the *cyclonic scale.*

T

Tcu An abbreviation sometimes used to denote a towering cumulus cloud (cumulus congestus).
Teleconnections A linkage between weather changes occurring in widely separated regions of the world.
Temperature The degree of hotness or coldness of a substance as measured by a thermometer. It is also a measure of the average motion or kinetic energy of the atoms and molecules in a substance.
Temperature inversion An increase in air temperature with height, often simply called an *inversion.*
Texas norther A strong, cold wind from between the northeast and northwest associated with a cold outbreak of polar or arctic air that brings a sudden drop in temperature. Sometimes called *blue norther.*
Thermal A small, rising parcel of warm air produced when the earth's surface is heated unevenly.
Thermal belts Horizontal zones of vegetation found along hillsides that are primarily the result of vertical temperature variations.
Thermal circulations Airflow resulting primarily from the heating and cooling of air.
Thermal lows and thermal highs Areas of low and high pressure that are shallow in vertical extent and are produced primarily by surface temperatures.
Thermograph An instrument that measures and records air temperature.
Thermometer An instrument for measuring temperature. The most common are liquid-in-glass, which have a sealed glass tube attached to a glass bulb filled with liquid.
Thermosphere The atmospheric layer above the mesosphere (above about 85 km) where the temperature increases rapidly with height.
Thunder The sound due to rapidly expanding gases along the channel of a lightning discharge.
Thunderstorm A local storm produced by cumulonimbus clouds. Always accompanied by lightning and thunder.
Tornado An intense, rotating column of air that often protrudes from a cumulonimbus cloud in the shape of a funnel or a rope whose circulation is present on the ground. (*See* Funnel cloud.)
Tornado alley A region in the Great Plains of the United States extending from Texas and Oklahoma northward into Kansas and Nebraska where tornadoes are most frequent.
Tornado outbreak A series of tornadoes that forms within a particular region—a region that may include several states. Often associated with widespread damage and destruction.
Tornado vortex signature (TVS) An image of a tornado on the Doppler radar screen that shows up as a small region of rapidly changing wind directions inside a mesocyclone.
Tornado warning A warning issued when a tornado has actually been observed either visually or on a radar screen. It is also issued when the formation of tornadoes is imminent.
Tornado watch A forecast issued to alert the public that tornadoes may develop within a specified area.
Trace (of precipitation) An amount of precipitation less than 0.01 in. (0.025 cm).
Trade wind inversion A temperature inversion frequently found in the subtropics over the eastern portions of the tropical oceans.
Trade winds The winds that occupy most of the tropics and blow from the subtropical highs to the equatorial low.
Training A situation where a series of individual thunderstorms keep moving over the same location.
Transpiration The process by which water in plants is transferred as water vapor to the atmosphere.
Tropical cyclone The general term for storms (cyclones) that form over warm tropical oceans.
Tropical depression A mass of thunderstorms and clouds generally with a cyclonic wind circulation of between 20 and 34 knots.
Tropical disturbance An organized mass of thunderstorms with a slight cyclonic wind circulation of less than 20 knots.
Tropical storm Organized thunderstorms with a cyclonic wind circulation between 35 and 64 knots.
Tropical wave A migratory wavelike disturbance in the tropical easterlies. Tropical waves occasionally intensify into tropical cyclones. They are also called *easterly waves.*
Tropopause The boundary between the troposphere and the stratosphere.
Troposphere The layer of the atmosphere extending from the earth's surface up to the tropopause (about 10 km above the ground).
Trough An elongated area of low atmospheric pressure.
Turbulence Any irregular or disturbed flow in the atmosphere that produces gusts and eddies.
Typhoon A tropical cyclone (hurricane) that forms in the western Pacific Ocean.

U

Ultraviolet (UV) radiation Electromagnetic radiation with wavelengths longer than X-rays but shorter than visible light.
Unstable air *See* Absolutely unstable atmosphere.
Upper-level front A front that is present aloft but usually does not extend down to the surface. Also called an *upper front* and *upper-air front.*
Upslope fog Fog formed as moist, stable air flows upward over a topographic barrier.
Upslope precipitation Precipitation that forms due to moist, stable air gradually rising along an elevated plain. Upslope precipitation is common over the western Great Plains, especially east of the Rocky Mountains.

Upwelling The rising of water (usually cold) toward the surface from the deeper regions of a body of water.
Urban heat island The increased air temperatures in urban areas as contrasted to the cooler surrounding rural areas.

V

Valley breeze *See* Mountain breeze.
Vapor pressure The pressure exerted by the water vapor molecules in a given volume of air.
Vernal equinox The equinox at which the sun approaches the Northern Hemisphere and passes directly over the equator. Occurs around March 20.
Very short-range forecast Generally used to describe a weather forecast that is made for up to a few hours (usually less than 6 hours) into the future.
Virga Precipitation that falls from a cloud but evaporates before reaching the ground. (*See* Fallstreaks.)
Visible radiation (light) Radiation with a wavelength between 0.4 and 0.7 μm. This region of the electromagnetic spectrum is called the *visible region.*
Visible region *See* Visible radiation.
Visibility The greatest distance an observer can see and identify prominent objects.

W

Wall cloud An area of rotating clouds that extends beneath a severe thunderstorm and from which a funnel cloud may appear. Also called a *collar cloud* or a *pedestal cloud.*
Warm advection (or warm air advection) The transport of warm air by the wind from a region of higher temperatures to regions of lower temperatures.
Warm-core low A low-pressure area that is warmer at its center than at its periphery. Tropical cyclones exhibit this temperature pattern.
Warm front A front that moves in such a way that warm air replaces cold air.
Warm occlusion *See* Occluded front.
Warm sector The region of warm air within a wave cyclone that lies between a retreating warm front and an advancing cold front.
Water equivalent The depth of water that would result from the melting of a snow sample. Typically about 10 inches of snow will melt to 1 inch of water, producing a water equivalent of 10 to 1.
Waterspout A column of rotating wind over water that has characteristics of a dust devil and tornado.
Water vapor Water in a vapor (gaseous) form. Also called *moisture.*
Water vapor–greenhouse effect feedback A positive feedback whereby increasing surface air temperatures cause an increase in the evaporation of water from the oceans. Increasing concentrations of atmospheric water vapor enhance the greenhouse effect, which causes the surface air temperature to rise even more.
Watt (W) The unit of power in SI units where 1 watt is equivalent to 1 joule per second.
Wave cyclone An extratropical cyclone that forms and moves along a front. The circulation of winds about the cyclone tends to produce a wavelike deformation on the front.
Wavelength The distance between successive crests, troughs, or identical parts of a wave.
Weather The condition of the atmosphere at any particular time and place.
Weather elements The elements of air temperature, air pressure, humidity, clouds, precipitation, visibility, and wind that determine the present state of the atmosphere, the weather.
Weather type forecasting A forecasting method where weather patterns are categorized into similar groups or types.
Weather warning A forecast indicating that hazardous weather is either imminent or actually occurring within the specified forecast area.
Weather watch A forecast indicating that atmospheric conditions are favorable for hazardous weather to occur over a particular region during a specified time period.
Westerlies The dominant westerly winds that blow in the middle latitudes on the poleward side of the subtropical high-pressure areas.
Wet-bulb depression The difference in degrees between the air temperature (dry-bulb temperature) and the wet-bulb temperature.
Wet-bulb temperature The lowest temperature that can be obtained by evaporating water into the air.
Whirlwinds *See* Dust devils.
Wind Air in motion relative to the earth's surface.
Wind-chill index The cooling effect of any combination of temperature and wind, expressed as the loss of body heat. Also called *wind-chill factor.*
Wind direction The direction from which the wind is blowing.
Wind profiler A Doppler radar capable of measuring the turbulent eddies that move with the wind. Because of this, it is able to provide a vertical picture of wind speed and wind direction.
Wind shear The rate of change of wind speed or wind direction, or both, over a given distance.
Wind speed The rate at which the air moves by a stationary object, usually measured in statute miles per hour (mi/hr), nautical miles per hour (knots), kilometers per hour (km/hr), or meters per second (m/sec).
Wind vane An instrument used to indicate wind direction.
Windward side The side of an object facing into the wind.
Wind waves Water waves that form due to the flow of air over the water's surface.
Winter solstice Approximately December 21 in the Northern Hemisphere when the sun is lowest in the sky and directly overhead at latitude 23½°S, the Tropic of Capricorn.
Wrap-around band A band of clouds and precipitation that wraps around the northwest side of a mid-latitude cyclone's central core of low pressure.

Y

Younger-Dryas event A cold episode that took place about 11,000 years ago, when average temperatures dropped suddenly and portions of the Northern Hemisphere reverted back to glacial conditions.

Z

Zonal wind flow A wind that has a predominant east-to-west component.

Index

A

Abercromby, 104
Absolute humidity, 82,85
Absolutely stable atmosphere, 124-125
Absolutely unstable atmosphere, 128-132
Absolute temperature scale, 14
Absolute zero, 14
Accretion, 149, 150, 151, 161
Acceleration, 184
Acid fog, 102-104
Acid rain, 12, 153
Acid snow, 429
Actual vapor pressure, 468
Additional reading material, 479-480
Adiabatic process
  defined, 123
  dry and moist rate, 123, 124, 126, 128, 129, 130, 131, 132
  reversible and irreversible, 123
Advanced Weather Interactive Processing System (AWIPS), 402
Advection
  cold and warm, 277
  defined, 40
  fog, 98, 99, 100
Aerosols, 11
  and climate change, 442, 445
  sulfate, 12, 445
Aggregation, 150, 153
Air (*see also* Atmosphere)
  rising and sinking, 23, 40
  saturated, 82
  weight of, 16-17
Aircraft icing, 160
  de-icing, 160
Air density
  changes with height, 17
  comparing moist air with dry air, 84
  defined, 16
  and watervapor content, 82
Air masses, 238-251 (*see also* specific types)
  affecting North America, 239-251
  characteristics, 239
  classification, 239
  defined, 238
  modification of, 243, 244
  source regions, 238, 240
Airmass weather, 250
Air parcel, 23, 81, 122, 298
Air pockets, 220
Air pollutants, defined, 11-12 (*see also* Air pollution)
Air pollution, 11-12
  and fog, 102-104
  and inversion, 103, 126
  smog, 103
Air pressure (*see* Pressure)
Air stagnation advisory, 414
Air temperature (*see* Temperature)
Albedo
  defined, 49
  of various substances, 49-50, 435
Alberta Clipper, 275
Aleutian low, 211, 212
Algorithms, 168, 354, 402
Altimeter, 18, 182, 183
Altitude corrections, 182-183
Altocumulus clouds,106, 107, 111, 139, 140
Altostratus clouds, 106, 107, 111
American Academy of Pediatrics, 68
American Meteorological Society (AMS), 31
Analysis, map, 404
Anemometer, 192, 202
Aneroid barometer, 18, 182
Aneroid cell, 18
Anticyclones
  formation of, 275-276
  movement of, 279-281
  semipermanent, 211
  structure, 273-281
  subtropical, 209, 210
  and vertical air movement, 276-277, 364
  winds around, 21, 182, 189-190, 364
Anticyclonic flow, 189
Anvil, of thunderstorm, 110, 111, 135, 299, 302, 307, 313
Aphelion, 52
Apparent temperature, 87
Arctic air masses, 239
Arthur M. Anderson, ship, 272-273
Arctic Oscillation (AO), 424-425
Arctic sea smoke, 101
Arcus cloud, 302, 304

Area conversions, 463
Argon, 8
Ash Wednesday Fire, 420
Atmosphere, 2-33
 absorption of radiation by gases in, 45-46
 carbon dioxide cycle in, 10
 composition of, 7-12
 compressibility, 16
 defined, 8
 general circulation of, 208-215
 heated from below, 48-49
 impurities in, 6, 8, 11-12
 layers, 15-16
 pressure, standard value, 17
 properties of, 13-25
 scales of motion, 19, 20
 standard, 14-16
 vertical structure of, 14-16
 viewed from space, 8
 warmed and cooled near the ground, 48-49
Atmospheric chaos, 407
Atmospheric density (*see* Air density)
Atmospheric greenhouse effect (*see* Greenhouse effect)
Atmospheric moisture (*see* Moisture)
Atmospheric pressure (*see* Pressure)
Atmospheric stability (*see* Stability)
Atmospheric stagnation, 414
Atmospheric window, 46
Aurora (borealis), l5, 54, 55
Automated Surface Observing System (ASOS), 74, 75, 90, 192, 402
Autumnal (fall) equinox, 53, 57
Avalanche, 155

B

Back door cold front (*see* Front)
Ball lightning, 324, 325
Bangladesh, killer cyclone of 1970, 390
Banner clouds, 112, 113
Barograph, 18
Barometers, 18, 182
 correction to sea level, 18, 19
Barometric pressure, 18
 highest recorded, 17
Bathythermograph, 369
Beaufort, Admiral Sir Francis, 195
Beaufort wind scale, 195, 196
Bel-Air fire, 229
Berg, 227 (*see also* Chinook wind)
Bergeron process, 147-149
Bergeron, Tor, 147, 270
Bering land bridge, 430, 433
Bermuda-Azores high, 211, 212, 213, 214
Billow clouds, 139, 220, 221
Bjerknes, Vilhelm, 270
Black blizzard, 232
Blackbody, 45
Black frost, 97
Black ice, 157
Blizzard, 231, 289-290
 of 1996, 411
 black, 232
 "Children's Blizzard," 290
 and deaths from, 6, 290
 defined, 154
 ground, 154
 and safety tips, 293
 warning, 401, 414
Blocking high, 422
Blowing snow, 154
Blue norther, 231
Boltzman, Ludwig, 43
Bora, 227 (*see also* Katabatic wind)
Boreal forest, 454
Boulder winds, 228 (*see also* Chinook wind)
Boundary layer (*see* Friction layer)
Bounded weak echo region (BWER), 349
Bow echo, 307-308, 209
Breeze
 lake, 223
 mountain and valley, 224-226
 sea and land, 223-224
Buoyant force, 129, 298
Burga, 231
Buys-Ballot, Christoph, 194
Buys-Ballot's law, 194

C

California norther, 231 (*see also* Santa Ana wind)
Calorie, defined, 37
Canadian high, 211, 212, 213
Cap cloud (*see* Pileus cloud)
CAPE (*see* Convective Available Potential Energy)
Captain Cosmo, fishing vessel, 275
Carbon dioxide
 amount in atmosphere, 8, 9-11, 451
 and climatic change, 47-48, 436, 446, 451
 and cloud seeding (dry ice), 149
 as a greenhouse gas, 9-11, 45-47, 436, 446, 451
 cycle, in atmosphere, 10
 measurements of, 10, 451
 role in absorbing infrared radiation, 45-47
 in Venus atmosphere, 47
Carbon monoxide, 12, 293
Castellanus clouds, 106, 139, 140
Cell
 convective, 40, 49
 Hadley, 208, 209, 210, 217
 polar, 210
 in thunderstorm, 301-306, 311
Celsius temperature scale, 14
 abbreviation, 7
 equation for, 14, 464
Centigrade temperature scale (*see* Celsius temperature scale)
Centrifugal force, 190
Centripetal acceleration, 190
Centripetal force, 190
Chaos, atmospheric, 407
Chaparral, 229
Chicago, the Windy City, 202
Chinook wall cloud, 227, 228
Chinook winds, 26, 227-229
 and human behavior, 26, 229
Chlorine in stratosphere, 11
Chlorofluorocarbons
 amount in atmosphere, 8
 and greenhouse effect, 11, 47, 436, 446, 451
 and ozone destruction, 11
Chromosphere, 54
Cirrocumulus clouds, 105, 106, 111
Cirrostratus clouds, 105, 106, 111
Cirrus clouds, 105, 111, 151
Clear air turbulence (CAT), 220, 221
Clear ice, 160
CLIMAP, 431
Climate
 defined, 13, 410
 extremes of, 25-31
 influence on humans, 25-31
 of the past, 13, 430-437
 urban, 436
Climate change, 428-460
 and aerosols, 442-446
 and atmospheric particles, 442-445
 causes of, 437-448
 and contrails, 453
 and ocean conveyor belt, 434-435
 and deforestation, 446
 evidence of, 430-432
 and feedback mechanism, 48, 437-438
 and global warming, 448-459
 and greenhouse gases, 47, 48, 446, 451
 by human (anthropogenic) activities, 445-448
 and increasing levels of CO2, 47-48, 436, 446, 451
 and land use changes, 446-448
 and the monsoon, 457

and mountain building, 438-439
by natural events, 437-445
and plate tectonics, 438-439
and radiative forcing agents, 448
and sea level changes, 11, 430, 431, 455
and sulfate pollution, 445-446
and surface modifications, 446
and variations in earth's orbit, 439-442
and variations in solar output, 442, 443
and sunspot activity, 442, 443
and volcanoes, 443-445
Climatic optimum, 433
Climate Prediction Center, 410, 422
Climatological forecast, 407, 410
Cloudburst, 23, 153
Cloud condensation nuclei (*see* Condensation nuclei)
Cloud droplets
growth of, 146, 147
size, 146, 147
Cloud seeding
different substances used, 149, 150
and hurricanes, 393
and lightning suppression, 329
natural, 151
overseeding, 150
and precipitation, 149-151
Clouds (*see also* specific types)
albedo of, 49-50
basic types, 111
and changing forms, 139-141
classification, 104
cold, 147
and cold front, 253-257
and convection, 133-135
defined, 104
development of, 24, 28, 132-141, 138
effect on daily temperatures, 46
electrification of, 321-322
and greenhouse effect, 48, 452-454
height of base, 105
high, 104, 105, 111, 453
identification of, 104-113
influence on climate change, 48, 452-454
low, 104, 105, 106-109, 111
major groups, 104
middle, 104, 105, 106, 111
orographic, 135
precipitation in, 111, 151-152
satellite observations of, 114-117
terms used in identifying, 104-113
and topography, 133, 135-139
unusual, 112-113
with vertical development, 109-111, 134
warm, 146, 147
and warm front, 257-259
Cloud streets, 244
Coalescence, 146, 147
Cold air damming, 256, 260
Cold-air funnels, 352, 353
Cold advection, 277
Cold front (*see* Fronts)
Cold core lows, 375
Cold occlusion, 262
Cold wave, 58
Collision and coalescence process, 146-147
Colorado low, 275
Columbia Gorge (coho) wind, 227
Comma cloud, 286
Compressional heating, 23, 122
and Chinook wind, 227-229
and continental polar air masses, 241
and eye of hurricane, 364
and Santa Ana wind, 229-231
Computer
and climate forecasts, 449-450
and cloud image enhancement, 47
and weather prediction, 402-403
Condensation, 9, 22, 37
latent heat of, 37
near the ground, 96-101
Condensation level, 129, 130, 136
Condensation nuclei, 22, 97, 146, 453
and growth of cloud droplets, 146
size, 146
Condensation trail (*see* Contrails)
Conditional instability (*see* Conditionally unstable atmosphere)
Conditionally unstable atmosphere, 128-130, 298
Conduction, 38-39, 40, 49
Confluence, 276
Conservation of angular momentum, 350, 368
Constant pressure charts (*see* Isobaric charts)
Constant pressure surface, flying along, 182-183
Continental arctic air masses, 239-244
and cold outbreaks, 241-242
and a record cold winter, 242
Continental drift (*see* Plate tectonics)
Continental polar air masses, 239-244, 248
Continental tropical air masses, 239, 249-251, 261
and record heat waves, 249, 250
Contour lines, defined, 180
Contrails, 453
Convection, 39-41, 49, 64, 122, 298
and cloud development, 49, 133-135
forced, 64
near the ground, 49
Convective Available Potential Energy (CAPE), 315, 415
Convective circulation, 40
Convective instability, 132, 314
Convergence, 275, 276
around jet streams, 279, 280, 281
cause of, 133, 275, 276
effect on surface pressure, 209-211, 275-281
with high-and-low-pressure areas, 275-281
in hurricanes, 362, 366
role in mid-latitude cyclone development, 276-277
with sea breeze, 223-224
Conveyor belt
model, mid-latitude cyclones, 282-283
model, ocean circulations, 434-435
Coriolis, Gaspard, 186
Coriolis effect (*see* Coriolis force)
Coriolis force, 21, 186-187, 348, 418
Corona discharge, 54, 325, 329
Counter current, in Pacific ocean, 418
Crop damage, due to temperature, 26, 27, 28, 69
Cumulonimbus clouds, 109, 110, 111, 135 (*see also* Thunderstorms)
distribution of ice and water in, 148
Cumulus clouds, 9, 84, 109, 111, 134, 135
and convection, 133-135
development of, 9, 109-111, 130, 133-135, 298-299
Cumulus congestus clouds, 109, 110, 135, 353
Cumulus fractus clouds, 109
Cupuric sulfide, 105
Cumulus humilis clouds, 109, 135
Cut-off low, 279
Cyclogenesis, 273, 275
Cyclones, middle latitude, 28, 182, 270-294
and blizzards, 289-290
compared with hurricanes, 374, 375
development of, 30, 270-271, 275-281
energy for development, 270, 281
families, 271, 272
regions of formation, 273-275
and global warming, 457
life cycle of, 30, 271
mature, 270, 271
movement of, 275-283
of November, 1975, 272-273
satellite image of, 30, 190, 200, 263, 283, 286, 291, 458
structure, 275-279
vertical air motions, 28, 30, 276-277
winds around, 21, 28, 182, 189-190, 275-277
Cyclones, tropical, 29, 363 (*see also* Hurricanes)
Bangladesh, 390
Bathurst Bay, 379
Great Bhala, 379
Cyclonic flow, 189
Cyclostrophic wind, 190

D

Daylight hours, for different latitudes and dates, 57
Deaths
  due to flooding, 6, 7
  and fog, 29
  and heat related, 5, 6, 27
  and hurricanes, 6, 29
  and lightning, 6, 7, 28
  and smog, 6
  and tornadoes, 6, 7, 27
  weather-related, 5
Deforestation, 446, 451
  and plagues, 448
Dendrite ice crystals, 153, 154
Dendrochronology, 432
Density (*see* Air density)
Deposition, 37, 96
Depression of wet bulb (*see* Wet bulb depression)
Depressions, middle latitude (*see* Cyclones, middle latitude)
Depressions, tropical, 369, 370
Derecho, 308, 309
Desertification, 446
Desert winds, 232-233
Dew, 96
Dew-point (temperature), 23, 24, 82
  average for January and July, 83
  extremes, 83-84
  and humid air, 82-88
  tables, 471, 473
Diablo wind, 220 (*see also* Santa Ana wind)
Diffluence, 276
Direct thermal circulation, 177
Discovery, space shuttle, 363
Disturbance, tropical, 369, 370
Divergence, 275-277
  around jet stream, 279-281
  cause of, 276
  effect on surface air pressure, 275-281
  with high-and-low-pressure areas, 275-281
  in hurricanes, 362
  in mid-latitude cyclones, 276-277
Dobson units, 11
Doldrums, 209, 210
Doppler lidar, 354
Doppler radar
  bounded weak echo region (BWER), 349
  defined, 168, 353
  determining winds, 354, 355
  and forecasting, 402
  hook echo, 349
  and hurricane and tropical storms, 381, 383, 384
  images, 169, 254, 306, 308, 309, 311, 349, 354, 402
  and precipitation, 168-169
  and severe weather, 305, 306, 308, 309
  target, 168
  and tornadoes, 349, 353-356
  on wheels, 354
  WSR-88D, 354
Doppler shift, 168, 353
Downbursts, 28, 303 (*see also* Thunderstorm, downdrafts)
Drizzle, 152
Drought, 169-172
  and the Dust Bowl years, 171-172
  and global warming, 456-457
  and the Palmer Index, 170
  some notable, 171-172
Dropsonde, 369, 392
Dry adiabatic rate of cooling and warming, 123, 124, 126, 128, 129, 130, 131, 132
Dryas, 433
Dry-bulb temperature, 90
Dry ice, in cloud seeding, 149
Dry lightning, 325
Dryline, 259-261, 315-316
Dry slot, 279, 283, 286
Dust clouds, 232, 303, 305
Dust devils, 232, 233
Dust storms, 206-207, 231
Dust wall, 232

E

Earth
  albedo of, 49-50
  annual energy balance, 7, 50-52
  atmosphere, overview of, 7-25
  average surface temperature, 7, 50-52
  changes in tilt, 440
  composite view from space, 4
  distance from sun, 7, 45, 52
  and its magnetic field, 54-55
  orbital variations, 53, 440
  radiation of, 7, 44
  rotation and revolution of, 52-60
  seasons, 52-60
  tilt of, 53, 440
Earth Radiation Budget Experiment, 48, 49, 443
Easterlies, polar, 210, 211, 213
Easterly wave, 362
Eccentricity of earth's orbit, 439, 441
Eddies, 19, 192, 220 (*see also* Cyclones; Anticyclones)
Edmund Fitzgerald, ship, 272-273
Ekman spiral, 377
Ekman transport, 377
El Chichón, volcano, 444
Electricity, and lightning, 321-322
Electrification of clouds, 321-322
Electromagnetic spectrum, 44
  visible region, 44
Electromagnetic waves, 41 (*see also* Radiation)
Elongated high (*see* Ridge)
Elongated low (*see* Trough)
El Niño, 367, 418-423, 436, 444
Embryo, hailstone, 161
Enhanced Fujita (EF) scale, 342-345
Energy,
  balance of earth and atmosphere, 45-52
  defined, 36
  heat, 36, 37
  kinetic, 36
  and latent heat, 36-38
  potential, 36
  radiant, 36
  in storms, 39
Energy conversions, 464
Ensemble forecasting, 392, 407
ENSO (El Niño/Southern Oscillation), 419
Entrainment, 134, 300
Environmental lapse rate, 124, 125, 128, 129, 130, 131
Environmental Protection Agency (EPA), 43
Equation of state, 466
Equatorial low, 210, 213
Equinox
  autumnal, 53, 57
  precession of, 440, 441
  vernal, 53, 58
Evaporation, 9, 22, 37
  factors that affect, 37-38
  latent heat of, 37
Evaporation (mixing) fog, 99
Exosphere, 16
Extended forecast, 410
Extratropical cyclone (*see* Cyclone, middle latitude)
Eye of hurricane (*see* Hurricanes)
Eyewall, 363, 365, 368
Eyewall replacement, 368, 386, 393

F

Faculae, 442
Fahrenheit, Daniel, 14
Fahrenheit temperature scale, 14
  abbreviation, 7
  equation for, 14, 464
Fall wind, 226-227
Fair weather cumulus (*see* cumulus humulis clouds)

Feedback mechanism
and climatic change, 48, 437-438
negative, 48, 438
positive, 48, 437
snow-albedo, 437, 438
water vapor-greenhouse, 48, 437, 448
Fetch, 198
Fine line, 261
Flares, solar, 54
Floods, 28, 165-168, 170, 292
in Big Thompson Canyon, 316, 317
in California, 1997, 246, 247
in China, 1931, 5
and deaths related to, 5, 6, 7, 28
flash, 28, 165-168
in Florence, Italy, 29
Great Flood of 1993, 167
and hurricanes, 29, 376-391
and ice jam, 167
and thunderstorms, 316
and tropical storm, 380-381
watches and warnings, 401
Flurries, of snow, 153
Foehn, 227 (*see also* Chinook wind)
Fog, 29, 94-95, 97-101
acid, 102-104
advisory, 401, 414
advection, 98, 99, 100
advection-radiation, 98
and auto accidents, 29, 102
and aviation , 102
and coast redwood trees, 98
and deaths related to, 29, 102
defined, 97
dissipation of, 98
effects of, 102
evaporation (mixing), 99
foggiest place in world, 102
frequency of heavy in United States, 102
frontal, 101
ground, 97
high fog, 98
ice, 99
number of days of, 102
pea soup, 103
in polluted air, 102-104
radiation, 69, 97, 98, 99, 102, 125
satellite images of, 102
steam, 100, 101, 136
upslope, 99, 100
valley, 98
visibility restrictions, 29, 97, 102
Fog drip, 98
Foggy weather, 101-104
Forced convection, 65
Forcing factor, of climate change, 441
Forecasting (*see* Weather forecasting)
Forecasts, accuracy of, 411-412
Free convection, 49
Freeze, 69, 97
Freezing drizzle, 157, 158
Freezing rain, 157, 158, 159 (*see also* Ice storms)
Friction, effect on wind, 192-194
Friction layer, 192
Frontal surface (zone), 251
Frontal wave, 270, 271
Frontogenesis, 255
Frontolysis, 255
Fronts, 28, 237, 251-264
arctic, 251
back door, 255-256
and clouds, 133, 256, 259
cold, 28, 30, 185, 252, 253-257, 263
defined, 28, 251
dew point, 259-260
dryline, 259
extraordinary cold front of 1836, 257
occluded (occlusion), 29, 30, 185, 252, 261-264
polar, 210, 213, 214, 21
quasi-stationary, 252
sea breeze, 224
smoke (smog), 224
stationary, 252-253, 271
symbols on weather map, 470
upper-level, 264
warm, 28, 30, 185, 252, 257-259, 263
Frost, 96, 97 (*see also* Freeze)
Frostbite, 25, 81
Frost point, 96
Fujita scale of damaging winds, 342-345 (*see also* EF scale)
Fujita, Theodore T., 342
Fulgurite, 325, 327
Funnel cloud, 122, 334, 350, 352, 353 (*see also* Tornadoes)

G

Gale warning, 401
Galileo, 18
Gas law, 466
General circulation, 208-215
and major pressure systems, 213
and precipitation patterns, 213-214
single-cell model, 208
three-cell model, 209-211
General Circulation Models (GCM), 208, 209-211, 412
Geoengineering, 459
Geostationary Operational Environmental Satellite (GOES), 114, 192, 406
Geostrophic wind, 188, 215
Geostrophic wind equation, 467
Glaciers, 13, 428-429, 430
Glaze, 157
Global Forecast System, 403, 405
Global Positioning System (GPS), 369
Global warming, 11, 47, 430, 448-459
and clouds, 452-453
consequences of, 454-456
and drought and floods, 456-457
and efforts to curb, 459
and El Niño-Southern Oscillation, 449
evidence for, 449
and extreme weather, 456-459
and hurricanes, 393-395, 458-459
and lake-effect snows, 457
and mid-latitude cyclonic storms, 457, 458
and oceans, 434-435, 454
and projected temperature changes, 11, 48, 450
and sea level rise, 11, 455
and temperature and humidity, 456
and thunderstorms, 457-458
and tornadoes, 457-458
uncertainties in, 451-452
Glossary, 481-494
Gradient, defined, 77
Gradient wind, 189
Graupel, 149, 154, 155, 159, 161, 321
Gravitational potential energy, 36
Gravity, 16
Gravity waves, 307
Gravity winds, 224
Great Basin High, 414
Greenhouse effect, 45-48, 436
and climate change, 47-48, 448-451
in earth's atmosphere, 9, 45-46, 436
enhancement of, 47-48
runaway, 437
and Venus, 47
Greenhouse gases, 9-11, 45, 46, 47-48
absorption of infrared radiation, 9, 45
and climate change, 47-48, 436, 446, 451-452
Greenland-Icelandic low, 211
Greenwich Mean Time (GMT), 475
Grid points, on chart, 404
Ground blizzard, 154
Ground fog, 97
Groundhog Day, and weather forecasting, 411
Gust front, 232, 300, 301-303, 304, 307, 313
Gustnadoes, 350, 352

H

Haboob, 232, 302
Hadley cell, 208, 209, 210, 217
Hadley, George, 208

Hail, 159-162, 339
 average number of days each year, 162, 319
 and deaths related to, 6
 formation, 161-162
 size, 159, 161
 soft, 321
 in thunderstorms, 27, 28, 298, 311, 313
 the white plague, 159
Hailstones, 159, 161, 321, 334
 embryo, 161
 wet and dry growth of, 162
Hailstreak, 162
Hair hygrometer, 90
Halos, 105, 106, 111
Hatteras low, 245, 275
Haze, 97, 125, 443
Heat
 advisories and warnings, 401
 and deaths related to, 5, 6, 64
 defined, 36
 extremes of, 65-67, 249, 250
 latent, 9, 36-38, 123, 364, 367
 and problems related to body, 26, 27, 86-87
 sensible, 37
 transfer, 38-40
Heat, balance of earth-atmosphere system, 45-52
Heat burst, 306
Heat capacity, 60
Heat conductivity, 40
Heat cramps, 86
Heat exhaustion, 26, 87
Heat index (HI), 87-88
Heat island (*see* Urban heat island)
Heat lightning, 324
Heat waves, 5, 27, 88, 250, 251
 in Chicago, July, 1995, 88-39
 and crop loss, 66
 and deaths related to, 5, 6, 64, 88
Heat stroke, 26, 87
 and car deaths relate to, 68
Hectopascal, 17
Helium, 8
High pressure areas (*see* Anticyclones)
Hildebrandsson, 104
Hoarfrost, 96
Hole in ozone layer, 11, 12
Holocene epoch, 433
Hook echo, 349
Horse latitudes, 210
Howard, Luke, 104
Human body, gain and loss of heat, 79-81
Humidity, 81-86 (*see* also specific types)
 absolute, 82, 85
 defined, 81
 and global warming, 456
 instruments, 90
 measurements of, 90
 specific, 82
 tables, 472, 474
Hurricane
 Allison, 167, 380, 381
 Andrew, 203, 379, 382-384, 392
 in Bangladesh, 390
 Beulah, 393
 Bill, 380
 Camille, 381, 382
 Carla, 382
 Charley, 389
 Dean, 393
 Debbie, 393
 Dennis, 389
 Donna, 382
 Dora, 369
 Elena, 363, 370, 372
 Emily, 389
 Emilia, 369
 Felix, 393
 Frances, 389
 in Galveston, Texas, 1900, 5, 387, 390
 Gordon, 372
 Great Hurricane of 1780, 391
 Hugo, 381-382, 384-385
 Humberto, 393
 Iniki, 373
 Isis, 372
 Ivan, 384-386, 389
 Iwa, 373
 Jeanne, 196, 379, 389
 John, 372
 Katrina, 29, 360, 361, 362, 365, 379, 382, 386-387, 388, 389, 392
 Mitch, 372, 391
 Nancy, 193
 Nora, 372, 373
 Olaf, 193
 Olivia, 203
 Ophelia, 117, 389
 Pauline, 372
 Rita, 30, 375, 386, 389
 during September, 1938, 389, 391
 Tico, 372
 Tina, 368
 Wilma, 363, 368, 381, 389
Hurricane damage-potential scale, 378
Hurricane hunters, 365, 369, 384
Hurricanes, 29, 361-395
 anatomy of, 362-365
 arctic, 375
 Atlantic season, 2004 and 2005, 389
 and climate change, 393-395, 458-459
 compared with mid-latitude cyclone, 374, 375
 and deaths related to, 5, 6, 29, 379-380, 381-391
 defined, 29, 362-363
 destruction, 381-391
 development of, 367-368
 Dopplar radar, images of, 369, 381, 383, 384
 over Eastern Pacific, 371-373
 and El Niño, 367
 erratic paths of, 372
 eye of, 29, 363, 364, 365, 369
 eyewall, 363, 365
 eyewall replacement, 368, 386, 393
 and flooding, 29, 376-391
 forecasts, 391-393
 formation and dissipation, 365-374
 as a heat engine, 368
 and the ITCZ, 366
 and latent heat, 364, 367
 landfall of, 378, 379
 and loop current, 386
 model of, 364
 modification of, 393
 most intense in United States, 382
 movement of, 370-374
 names, 374-376
 over North Atlantic, 374
 notable, 381-391
 and rain free area, 363
 record setting, 389
 satellite images of, 30, 370, 380, 382, 385, 387, 391
 and seeding of, 393
 in South Atlantic, March, 2004, 371, 372
 spiral rain bands in, 263
 stages of development, 369
 and the storm surge, 376-380
 structure of, 362-365
 and tornadoes, 336
 watches and warnings, 391-392, 401
Hydrocarbons, 12
Hydrogen, 8
Hydrogen sulfide, 418
Hydrologic cycle, 24-25
Hygrometers, 90
Hygroscopic particles, 97
 and cloud seeding, 149-151
Hydrostatic equation, 467
Hypothalamus gland, 86
Hypothermia, 25, 81
Hypoxia, 16

I

Ice
 clear and rime, 154, 157, 160
 vapor pressure over, 148-149
Ice ages, 432, 435
Ice cores, information from, 431-434
Ice Crystal (Bergeron) Process, 147-149

Ice crystals
in clouds, 148, 149, 151
forms of, 149, 150, 151, 153-154
and natural seeding, 151
Ice jam flooding, 167
Ice nuclei, 148, 150
Ice pellet (*see* Sleet)
Icelandic low, 211, 212
Ice storms, 26, 157, 158, 260, 417
Icing, aircraft, 160
Imager, in satellites, 114-115
Inches of mercury
atmospheric pressure, 19
defined, 17
Incipient cyclone, 270
Infrared (IR) radiation, wavelength of, 41, 44
Infrared, radiometer, 114-115
Insolation, 56
Instability, 128-132
absolute, 128
causes of, 130-132
conditional, 128-130
and thunderstorms, 134, 298, 313, 347
Instrument shelter, 74-75
Instruments (*see* specific types)
Interactive Forecast Preparation System, 402
Interglacial periods, 433
Intergovernmental Panel on Climate Change (IPCC) report from, 433, 441, 449-450, 451, 454, 455
Intertropical convergence zone (ITCZ), 210, 212
and hurricane formation, 366
movement of, 212, 213
and precipitation, 213-214
Inversion
and air pollution, 103, 126
and continental polar air mass, 239
defined, 15
nocturnal, 68
radiation, 68, 69
role in severe thunderstorm formation, 314-315
in the stratosphere, 15
subsidence, 126
surface, 68
trade wind, 367
Ionosphere, 16
Ions, in thunderstorms, 321-322
Irreversible pseudoadiabatic process, 123
Isobaric charts, 179, 180-182
Isobaric surface, 180-181
Isobars, defined, 179
sea level average, 216
Isotachs, defined, 218
Isothermal layer, defined, 68
Isotherms
defined, 75
January and July average, 75-77
Isotopes, of oxygen, 431

J

Jet streak (jet stream core), 218, 279-281, 313, 381
convergence and divergence around, 279, 280
and storm development, 279-281
Jet streams, 215-219, 279-281, 457
defined, 15, 215, 217
and developing mid-latitude cyclones, 279-281, 457, 458
formation of, 215-219
as heat transfer mechanism, 218
low-level jet, 218, 313
polar (front) jet, 217, 281, 313, 457
and thunderstorm development, 218, 307-308, 313-315, 337
rear-inflow in thunderstorms, 307, 308
in relation to tropopause, 15, 215, 217
subtropical jet, 217, 217
Joule, defined, 37, 315

K

Katabatic wind, 226-227
Kelvin, Lord, 14
Kelvin temperature scale (Kelvins), 14
Khamsin, 233
Kinetic energy, 36, 270, 281
Knot, defined, 15
Kocin, Paul, 284
Kyoto Protocol, 459

L

Lag in seasonal temperatures, 58, 64
Lake breeze, 223
Lake-effect (enhanced) snows, 135, 136-137, 166, 240
and global warming, 457
La Lombarde, 219
Lamarck, J., 104
Land breeze, 223
Land of the Midnight Sun, 58
Landspouts, 352, 353
Langmuir, Irving, 149
La Niña, 367, 418-423
Lapse rate, temperature, 14, 124
adiabatic, moist and dry, 123
defined, 14
environmental, defined, 124
standard, 14
superadiabatic, 128
Latent heat, 9, 36-38, 123
as source of energy in hurricanes, 364, 367
change of state, 36
Layers of atmosphere, 14-16
Lead iodide, 150
Lead time, of tornadoes, 412
Leaves, color changes, 57
Lee-side lows, 273
Lee wave clouds, 139
Length conversion, 463
Lenticular clouds, 112, 139, 140
Leste, 232, 233
Leveche, 233
Level of free convection, 129
Lidar, Doppler, 354
Lifting condensation level, 136
Light (*see* Radiation)
Lightning, 320-329
ball, 324, 325
bead, 324, 325
blue jet, 325, 326
"bolt from the blue", 320
cloud to ground (negative and positive), 322, 324, 328, 329
dart leader, 322
detection and suppression, 329
determining distance from, 320
development of, 323
different forms of, 324-324
dry, 325
electrification of clouds, 321-322
ELVES, 326
fatalities related to, 5, 7, 325-327
flash, 320, 329
forked, 323, 324
heat, 324
most prone areas, 325
protection from, 327-328
red sprite, 326
return stroke, 322, 323
ribbon, 324, 325
rods, 327, 328
sferics, 329
sheet, 324
shelter from, 327, 328
stepped leader, 322, 323
stroke, 29, 322-324
super bolts, 322
Lightning detection finder, 329
Little Ice Age, 435, 445
Long-range forecast, 410
Longwaves in atmosphere, 277
Loop current, 386

Low-level jet stream, 218
    and thunderstorm formation, 218, 313, 337
Low pressure area (*see* Cyclones, middle latitude)

M

Mackerel sky, 105, 111, 259
Macroscale, 19, 20
Magnetic storm, 54-55
Magnetosphere, 54, 55
Major El Niño event (*see* El Niño)
Mammatus clouds, 112, 113, 301, 302, 313, 350
Map analysis, 178-182
Mares' tails, 105
Maritime polar air masses, 239, 244-245, 246
Maritime tropical air masses, 239, 245-249, 261
Mass, defined, 16
    conversions, 464
Matter, states of, 22
Maunder, E.W., 442
Maunder minimum, 442
Max-min temperature shelter, 75
Medieval climatic optimum, 434
Medium-range forecast, 410
Melanin, and absorption of radiation, 42
Mercury barometer, 18
Meridional winds (flow), 191
Mesocyclone, 312, 313, 348
Mesohigh, 301
MesoNets, 355
Mesopause, 15
Mesoscale Convective System (MCS), 308
Mesoscale Convective Complexes (MCC), 308, 309, 310-311
Mesoscale Convective Vortex, 308, 309
Mesoscale winds, 19, 20, 219
Mesosphere, 15, 16
Meteogram, 402, 403
Meteorology, defined, 13
Meteors, 15
Methane
    amount in atmosphere, 8, 451
    and greenhouse effect, 11, 45, 46, 436, 446
Microburst, 303-306
Micrometers, defined, 41
Microscale winds, 19, 20
Middle latitude cyclonic storms (*see* Cyclones, middle latitude)
Mid-Holocene maximum, 433
Milankovitch, M., theory of climatic change, 439-442
Millibar, defined, 17, 178, 464
Mistral, 227 (*see also* Katabatic wind)
Mixing fog (*see* Evaporation (mixing) fog)
Mixing ratio, 82
Model Output Statistics (MOS), 408, 410
Models
    air motions in hurricane, 364
    atmospheric, 176, 177, 404
    climate, 448, 449-450
    general circulation, 209, 210
    Norwegian cyclone, 270
    station, 469
    used in weather forecasting, 404-408
Moist adiabatic rate of cooling or warming, 123, 124, 128, 129, 130, 131, 132
Moisture, 21-25 (*see also* Humidity)
Monomolecular film, and hurricane suppression, 393
Monsoon, 219-222
    summer, 219, 220, 222
    winter, 219, 222
Mountainadoes, 228
Mountain breeze, 224-226
Mountain building, and climate change, 438
Mountains
    and climatic change, 438-440
    hillside temperature variations, 69, 70
    influence on precipitation, 162-163
Mountain wave clouds, 139
Mount Pinatubo, volcano, 444, 459
Mount St. Helens, volcano, 444
Mount Tambora, volcano, 59, 444
Multicell storms, 302

N

National Center for Atmospheric Research (NCAR), 393, 459
National Centers for Environmental Prediction (NCEP), 400
National Hurricane Center (NHC), 203, 383, 384, 391
National Lightning Detection Network, 329
National Oceanic and Atmospheric Administration (NOAA), 31, 393, 422
NOAA reconnaissance (hurricane hunter) aircraft, 365, 384
National Weather Association (NWA), 31
National Weather Service, 31
    and forecasting, 288, 400, 401, 402, 406
    office in Buffalo, New York, 137
    and lightning probability forecasts, 329
    and naming hurricanes, 374
    tornado warnings by, 341, 345, 346
    UV forecast, 43
    and wind chill, 80
Negative feedback mechanism, 438
Neon, 8
Neutral stability, 126
Newton, Isaac, 183
Newton's laws of motion, 183-184
Next Generation Weather Radar (NEXRAD), 354
Nimbostratus clouds, 106, 107, 111, 151, 157
Nitrogen, in earth atmosphere, 8
Nitrogen dioxide, in earth atmosphere, 12
Nitrous oxide
    amount in atmosphere, 8, 451
    and greenhouse effect, 11, 45, 46, 436, 446
NOAA weather radio, 31, 341
Nocturnal drainage winds, 224
Nocturnal inversion (*see* Inversion)
Non-squall clusters, 362
Non-supercell tornadoes (*see* Tornadoes)
Norte, 231
North Atlantic Oscillation, 424
Northeasters, 245, 273, 274, 284
    with hurricane characteristics, 274
Northeast Snowfall Impact Scale (NESIS), 284
Northeast trades, 210, 213
Northern lights (*see* Aurora)
Norwegian cyclone model, 270
Nowcasting, 410
Nuclear (winter) war, 447
Numerical weather prediction, 404

O

Oakland hills fire, 230, 231
Obliquity, of earth's axis, 440, 441
Occluded fronts (*see* Fronts)
Occlusion, 262
Ocean currents, 78
    and climate change, 434-435
    as a control of temperature, 75, 78
    and conveyor belt model, 434-435
    effect on nearby land temperature, 78
    and fog formation, 98, 100
    thermohaline circulations, 434-435
Ocean-effect snows, 137
Ocean Niño Index, 421
Oceans
    influence on climate, 78
    influence on climate change, 434-435, 454
Open wave cyclone, 270, 271
Orographic, uplift, 135, 138, 163
    cloud development, 135, 138
Outflow boundary, 303, 305
Overrunning, 258, 270
Overshooting top, 301, 302, 313

Oxygen, in earth's atmosphere, 8
Oxygen-isotope ratios, 431
Ozone, 11
absorption of radiation, 45-47
amount in atmosphere, 8, 11
concentration over Antarctica, 11
destruction by chlorine, 11, 15
as a greenhouse gas, 42, 46, 452
hole, 11, 12
importance to life on earth, 11
maximum, 15
as a pollutant, 11
in stratosphere, 11, 15

P

Pacific air, 244-245
Pacific Decadal Oscillation (PDO), 423
Pacific high, 211, 212, 213, 214
Pacific Hurricane Center, 391
Palmer Drought Severity Index (PDSI), 170
Palmer Hydrological Drought Index (PHDI), 170
Parcel of air, 23, 81, 122
Pattern recognition forecast, 408
Perihelion, 52
Persistence forecast, 408
Photochemical smog (*see* Smog)
Photons, 41, 42
Photosynthesis, 9, 10, 446, 451
Phytoplankton, 10, 442, 451
Pileus cloud, 112, 113
Pineapple express, 246, 247
Plate tectonics, 13
and climatic change, 438-439
Pleistocene epoch, 432
Polar air masses, 210, 239-245
Polar cell, 210, 211
Polar easterlies, 210, 211, 213
Polar front, 210, 213, 214
Polar front theory of cyclones, 270-273
Polar high, 210, 213, 214
Polar (front) jet stream, 217, 457
Polar lows, 291-294, 375
Pollutants, 11-12
and the jet streams, 218
Pollution (*see* Air pollution)
Positive feedback mechanism, 437
Potential energy, 36
Power conversions, 464
Powers of Ten, 464
Precession, of earth's axis of rotation, 440, 441
Precipitation
acid, 12, 153
annual average, global, 163, 476-477
annual average, United States, 165, 167
in clouds, 151-152
and cloud seeding, 149-151
with cold fronts, 253-257
defined, 24, 146
and Doppler radar, 168-169
and driest cities in United States, 167
and extreme events, 162-172
and hydrologic cycle, 24
measuring, 168-169
and occluded front, 261-264
and records of, 163-165
related to general circulations, 213-214
and topography, 162-164
types, 152-162
wet and dry regions, 163-165, 213-214
wettest cities in United States, 165
with warm fronts, 257-259
Precipitation processes, 146-152
collision and coalescence, 146-147
ice-crystal (Bergeron), 147-149
Pre-frontal squall line, 306, 307
Pressure
associated with cold air and warm air aloft, 176-182
atmospheric, 16-19, 20-21, 176-178
average sea-level, 19, 211
barometric, 18, 182-183
causes of surface variations, 16-19, 176-178, 195-201
central, 270
changes aloft, causes of, 176-182
changes with height, 16, 17
conversions, 17, 464
defined, 16-19, 176-178
extremes, 195-203
in hectopascals, 17
horizontal changes and wind, 176-178
inches of mercury, 17, 19, 178
instruments, 18, 182-183
January and July average, aloft, 216
January and July average, sea level, 211, 212, 214-215, 216
mean sea-level, 19, 178
measurement of, 17, 179
in millibars, 17, 19, 178
reduction to sea level, 19, 178-179
relation to density and temperature, 176-178, 179, 466
standard, atmospheric, 17, 19, 178
station, 178
tendency, 253, 470
units, 17, 19, 178, 464
variation with altitude, 17-19, 176-178
world seasonal shift, 211-213
Pressure gradient, and winds, 184-185
Pressure gradient force (PGF), 177, 184-186
Prevailing westerlies, 210
Prevailing winds (*see* Winds, prevailing)
Probability forecast, 408
Profiler (*see* Wind profiler)
Prognostic chart (prog), defined, 404
Prominence on sun, 54
Psychromatic tables, 471-474
Psychrometer, 90
Purga, 231

Q

Queen Elizabeth II, ocean liner, 275
Quickscat satellite, 192, 193, 369

R

Radar (*see* Doppler radar)
Radiant energy (*see* Radiation)
Radiation (radiant energy), 7, 41-44
absorption and emission of, 45-52
effect on humans, 42
electromagnetic waves, 41
infrared, 44
intensity, 41
longwave (terrestrial), 44
radiative equilibrium, 45, 448
reflection of, 44, 49, 50, 51
scattering, 49, 50, 51
selective absorbers, 45-47
shortwave, 44
and skin cancer, 42
solar, 54-55, 56
and sunburning, 42, 49
of sun and earth, 44, 45-48, 49-52
and the Sun Protection Factor (SPF), 43
and tanning, 42
and temperature, 42-43, 46
ultraviolet (UV), 42, 44
variation with seasons, 57
visible light, 44
wavelength of, 41, 42
Radiational cooling, 67
Radiation fog, 97, 125
Radiation inversion, 68
Radiative equilibrium, 45, 448
Radiative equilibrium temperature, 45, 448
Radiative forcing, 448
Radiative forcing agents, 448
Radio, wavelengths of, 41, 44
Radiometers,114
Radiosonde, 75, 124, 179, 192, 392, 403
Rain, 24, 152-153, 159 (*see also* Precipitation)
acid, 12, 153
color of, 158
formation of, 146-152
intensity, 152-153, 162-168
*Rain*band and Intensity Change *Ex*periment (RAINEX), 368-369

Rainbows, 27, 334
Raindrop
  shape, 146
  size, 146, 147, 152
Rain-free base, of thunderstorms, 312, 313
Rain shadow, 135, 137, 138, 163
Rain streamers, 152
Rawinsonde observation, 192
Reading material, additional, 479-480
Rear-inflow jet, 307
Redwood trees, and fog, 98
Reflection (*see also*Albedo)
  and color of objects, 49
  of radiation, 44, 49, 50, 51
Relative humidity, 24, 82, 468
  comparing desert air with polar air, 85
  computation of, 24, 468
  daily variation of, 82
  defined, 24, 82
  and dew point, 82-88
  and dry air, 84-85
  effect on humans, 86-89
  equation for, 24, 82, 468
  extremes, 86, 88
  in the home, 86
  measuring, 90
  saturation, 82
  tables, 472, 474
Ridge, defined, 181 (*see also* Anticyclones)
  and upper-level charts, 277, 278
Rime, 154, 157, 160
Roll clouds, 302, 303, 304
Rossby, C. G., 277
Rossby waves (*see* Longwaves in atmosphere)
Rotor clouds, 112, 139
Ruddiman, William, 447
Runaway greenhouse effect, 437
Runaway ice age, 438

S

Saffir-Simpson scale, 287, 378
Saint Elmo's Fire, 325
Salt particles, as condensation nuclei, 22, 97, 146
Sandstorms, 231
Santa Ana wind, 26, 229-231
Satellite images
  computer enhancement of, 310
  infrared, 116, 117, 310, 382
  of middle latitude storms, 30, 116, 117, 247, 263
  of tropical storms, 247, 370, 380, 382, 391
  visible, 30, 116, 200, 370, 380, 385, 387, 391
  water vapor image, 117
Satellites
  Earth Radiation Budget Experiment (ERBE), 48, 49, 443
  and forecasting, 406
  geostationary, 114, 192, 406
  and identification of clouds, 114-117
  information from, 114-117, 192-192
  polar orbiting, 114
  QuickScat, 192, 193, 369
  Tiros 1, 114
  TRMM, 116, 117, 365
Saturation, 82
Saturation vapor pressure, 82
  for various air temperatures, 468
Scales of atmospheric motion, 19, 20 (*see also* Wind)
Scattering, of radiation, 49
Scatterometer, 192
Schaefer, Vincent, 149
Scientific Committee on Problems of the Environment (SCOPE), 447
Scud, 106, 107
Sea breeze, 223-224
Sea breeze convergence zone, 224
Sea breeze front, 224
Sea ice, 455
Sea level, and climate changes in (*see* Climate change)
Sea-level pressure, 17, 19, 178
Sea-level pressure chart, 179
Seasons, 52-60
  and apparent path of sun, 58
  defined, 52-53
  in Northern Hemisphere, 56-58
  in Southern Hemisphere, 56-58
Secondary low, 271
Seiches, 199
Selective absorbers (*see* Radiation)
Semipermanent highs and lows, 211
Sensible heat, 37
Sensible temperature, 79
Sferics, 233
Sharav, 233
Shelf cloud, 302, 304
Short-range forecast, 410
Shortwaves in atmosphere, 277
Shower, 153, 161
Siberian express, 242
Siberian high, 211, 212
Silver iodide, 150, 393
Simoom, 233
Sirocco, 233
Sky, color of, 49
Sleet, 157, 159
Small craft advisory, 401
Smog, 11, 103 (*see also* Air pollution)
  and deaths relate to, 6, 103
  London-type, 103
  Los Angeles-type, 11
  photochemical, 11
  visibility in, 103
Smog front, 224
Smoke front, 224
Snow, 62-63, 153-156, 159
  acid, 431
  advisory, 401
  albedo of, 49, 50
  blowing, 154
  and Chinook winds, 228-229
  and colors of, 155
  falling in above freezing air, 153
  flurries,153
  formation of, 147-149
  lake-effect (enhanced), 135, 136-137, 166
  melting level, 159
  record at Montague, New York, 166
  ocean-effect, 137
  squall, 153
  tapioca, 154-155
  upslope, 242
Snow-albedo feedback, 437, 438
Snow eaters, 229
Snowfall
  average annual in United States, 156
  effects of, 155-156
  records of, 156, 165, 166
  snowiest cities in United States, 156
Snowflake, 149, 153, 155, 157
  rimed, 154, 155
  shape of, 153, 154
Snow grains, 154
Snow pellets, 149, 154, 155
Snow rollers, 198
Snow squall, 153
Solar constant, 49
Solar irradiance, 448
Solar radiation, 49-50, 51, 56
  absorption by atmosphere, 49, 56-57
  absorption at earth's surface, 49, 56-57
  average annual incoming, 52
  changes in, 50, 439-441, 442, 443
Solar wind, 54
Solberg, Halvor, 270
Solstice
  summer, 53, 56
  winter 53, 57
Sonic boom, 320
Sound
  Doppler shift, 168, 353
  sonic boom, 320
  speed of, 320
  of thunder rumbling, 320
Sounder, in satellite, 114
Sounding, 75, 124, 314, 315
  and severe thunderstorms, 375
  and weather forecasting, 400, 403
Southeast trades, 210

Southern Oscillation, 418-423
Space weather, 54-55
Spaghetti plot, 407, 408
Specific humidity, 82
Spiral rain band, 363
Squall line, 255, 306-309, 362
Stable equilibrium, 122
Stability
atmospheric, 122-123, 124-127
and cloud development, 113, 132-141
changes due to lifting air, 124-127
and lake-effect snow storms, 136-137
neutral, 126
and pollution, 127
and subsidence inversion, 127
types of, 124-132
Standard atmospheric pressure, 17, 19, 178
Stationary front (*see* Fronts)
Station model, 469
Station pressure, 178
Statistical forecast, 408
Steady-state (trend) forecast, 408
Steam devils, 101
Steam fog, 136
Stefan-Boltzmann law, 43, 466
Stefan, Josef, 43
Stepped leader, 322, 323
StickNets, 355
STORMFURY, project, 393
Storm Based Warnings, 341
Storm Prediction Center, 341
Storms
Columbus Day, October, 1962, 288-289
comparing energy of, 39
convective, 298
December, 1992, 274
Great Blizzard of January 26, 1978, 289-290
magnetic, 54
March, 1962, 274
March Storm of 1993, 283-288
November 10, 1975, 272
pop-up, 298
ranking East Coast, 284
Storm of the Century, 283
top ten, Northeast United States, 284
Storm surge, 376-380
Storm warning, 401
Straight-line winds, 302, 307
Stratocumulus clouds, 106, 108, 111, 138, 141
Stratopause, 15
Stratosphere, 15, 16, 443
ozone in, 11
temperature of, 15
Stratus clouds, 98, 107, 108, 111
Stratus fractus clouds (*see* Scud)
Stream advisories, 401, 417
Streamlines, 362
Sublimation, 37, 96
Subpolar low, 210, 213
Subsidence inversions, 126
and stability, 127
Subtropical air, 246, 247, 248
Subtropical highs, and general circulation, 209, 210, 213, 214
Subtropical jet stream, 215, 217
Suction vortices, 340
Sulfates, and climate change, 442, 445, 459
Sulfur dioxide, 12, 443, 444
sulfuric acid, in ice cores, 431
Summer solstice, 53, 56
Sun
apparent path of, 58
changes in energy output, 442
changing position of rising and setting, 58
distance from earth, 7, 45, 52
electromagnetic spectrum of, 44
magnetic cycle of, 54-55, 442
protection from, 42-43
regions of, 54
and seasonal variations, 52-60
temperature of, 44
Sundowner wind, 231 (*see also* Santa Ann wind)
Sunspots, 54
and climate change, 442, 443
Sunstroke, 87
Superadiabatic lapse rate, 128
Supercell thunderstorms, 296-297, 306, 311-315
Classic (CL), 312
High precipitation (HP), 312
left-movers, 315
Low precipitation (LP), 312
right-movers, 315
and tornado formation, 347-350
Supercooled water, 148
Super typhoon (*see* Typhoons)
Surface inversion (*see* Inversion, radiation)
Surface charts, 178-182
Surface weather map, 179-180, 181, 185
Swells, 198, 377
Système International (SI) units, 49, 465

T

Tables
of constants and equations, 466-467
of dew point and humidity, 471-474
of wind chill, 80
Tapioca snow, 154-155
Teleconnections, 418, 423
Temperature, 13-16, 22, 36
absolute scale, 14
absolute zero, 14
advection (*see* Advection)
annual range, 77
apparent, 87
average change with height, 15, 64
and crop damage, 69
controls of, 75
conversions, 14, 464, 465
daily changes in, 64-79
daily range, 71-72, 73, 453
daytime warming, 64-67
defined, 13, 36
and density relationship, 176-178, 179, 466
and dew point, 23, 24, 82-88, 136
effects of humans, 79-89
effect on air's capacity for water vapor, 22-23, 82
extreme cold, 69-71, 239-243
extreme heat, 65-67, 249-250
global patterns of, 76
recent global warming, 11, 48, 448-450
and humidity, 72
horizontal variation between summer and winter, 75-77
and layers of atmosphere, 15
inside car, 67, 68
instruments, 74-75
inversion, 15, 68, 69, 127
lag in daily, 64
land versus water, 75, 77
lapse rate, average, 14, 124
lowest average value in atmosphere,15
maximum and minimum, 64-71
mean daily, 73
measurement near surface, 74-75
nighttime cooling, 67-69
normal, defined, 73
potential, 126, 239
radiative equilibrium, 45
rapid changes in, 228-229
record high, each state, 66
record high, world, 67
record low, each state, 71
record low, world, 72
regional variations, 75-79
of rising and sinking air, 23, 122-123
scales, 14
sensible, 79
in the thermosphere, 15, 16
trends over last 1000 years, 433-437, 449-450
wet bulb, 86, 90
world average, January, 76
world average, July, 76
vertical profile of, 14-16, 64, 69, 127, 315, 403
Texas norther, 231, 239
Thermal belts, 69, 70
Thermal circulations, 223

Thermal high and low pressure areas, 211, 212
  and general circulation, 211
Thermals, 40, 64, 133, 225
  smells of, 133
Thermistor, in radiosonde, 74
Thermocline, 419
Thermohaline circulation, 434
Thermometers, 74-75
Thermosphere, 15, 16
Thunder, 320, 321
Thunderstorm, watches and warnings, 401, 417
Thunderstorms, 28, 30, 110, 111, 120-121, 134, 135, 222, 224, 298-329
  air mass (ordinary) cell, 298-301
  and airline crashes, 303-305
  average number of days each year, 316, 318, 319
  cumulus (growth) stage, 298, 302
  defined, 28, 298
  development of, 298-316
  distribution, 316-320
  dissipating stage, 300, 301
  downdrafts and updrafts in, 28, 300, 302, 303, 313, 349
  and the dryline, 315-316
  electricity, 321-322
  and flash floods, 28, 145, 165-168, 316
  and forward flank downdraft, 349
  and global warming, 457-458
  and gust front, 301-303, 304, 307, 313
  and heat bursts, 306
  and hurricanes, 29, 298, 312, 313, 332-333, 336, 362-365
  life cycle of, 298-301
  and maritime tropical air masses, 245-249
  mature stage, 300, 302
  and microbursts, 303-306
  mini-supercell, 311
  multicell, 301-306
  non-squall clusters, 362
  ordinary cell (air mass), 298-301
  outflow boundary, 303, 305
  overshooting top, 301, 302
  and rear flank downdraft, 349
  and rear-inflow jet, 307, 308
  rotation of, 347-348
  severe, 28, 145, 298-316, 347-350
  stages of development, 298-301
  stratified region, 310
  supercell, 296-297, 306, 311-315, 332-333, 347-350
  squall line, 255, 306-309, 362
  training, 167, 316
Toba, volcano, 445
Tornado alley, 336, 337
Tornado outbreaks, 345-347
  April 3-4, 1974, 345-346
  May 3, 1999, 346-347
  Palm Sunday, 1965, 347
  Super Tuesday, 2008, 347
  Tri-state of 1925, 5, 335, 345
Tornadoes, 27, 332-357
  and deaths due to, 5, 6, 7, 335, 341, 345
  associated with hurricanes, 336, 379
  associated with thunderstorms, 298, 312, 313, 332-333, 347-352
  average number in United States, 338
  as cold-air funnel, 352, 353
  damage, 342-345
  decay stage, 336
  defined, 27, 334-335
  and Dixie Alley, 336
  and Doppler radar, 349, 353-356
  dust-whirl stage, 336
  EF scale, 342-345
  families, 345
  a field study of, 355
  and flying cows, 342
  formation of, 347-352
  Fujita scale, 342-345
  funnel clouds, 122, 334, 350, 352, 353
  and global warming, 457-458
  glow, 347
  incidence by state, 337
  lead time, 412
  life cycle of, 336
  mature stage, 336
  movement of, 334-336
  multi-vortex, 340, 351
  nonsupercell, 350-352, 353
  observations of, 339-357
  occurrence and distribution of, 336-339
  outbreaks of, 345-347
  records of, 337, 343
  seeking shelter from, 340-342
  stages of development, 336
  and suction vortices, 340, 341
  and VORTEX 2, 355
  supercell, 313, 347-350
  and vertical stretching, 350
  vortex, 350
  winds of, 337, 339-340, 353-356
Tornado vortex signature (TVS), 354
Tornado, watches and warnings, 341, 401
Torricelli, Evangelista, 18
Towering cumulus clouds (*see* Cumulus congestus clouds)
Trace gases, 8, 11
Trade wind inversion, 367
Trade winds, 210, 366-367
Training, 167, 316
Transpiration, and hydrologic cycle, 24, 25
Triple point, 271
Tropic of Cancer, 56, 58
Tropic of Capricorn, 57, 58, 59
Tropical air mass, 239, 245-251, 261
Tropical cyclone, 29, 363, 367 (*see also* Hurricanes)
  in Brazil, 372
  Nargis, 390, 391
  Sidr, 390
Tropical depression, 369, 370
Tropical disturbance, 369, 370
Tropical Rainfall Measuring Mission (TRMM) satellite, 116, 117, 365 (*see also* Satellites)
Tropical storms, 362, 366, 369, 370, 394 (*see also* Hurricanes)
  and flooding due to, 380
Tropical squall cluster, 362
Tropical wave, 362
Tropical weather, 362
Tropopause, 15
Troposphere, 15, 16, 443, 446
Trough
  along cold front, 253
  defined, 181
  on upper-level charts, 181, 277, 278
Tsunami, 199, 380
Turbulence, 218
  aircraft, 220-221
Twisters (*see* Tornadoes)
Typhoons, 363 (*see also* Hurricanes)
  Forrest, 368
  Freda, 288
  Morakot, 381
  super, 368, 378
  Tip, 363, 373

U

Uccellini, Louis, 284
UV Index, 43
Ultraviolet (UV) radiation, 41-44
Units and conversions, 463-465
Unstable equilibrium, 122
Unstable atmosphere (*see* Instability)
Upper air (level) charts, 178-182, 189, 191, 194, 195, 278, 405, 416
Upper-level front (*see* Fronts)
Upslope snow, 242
Upwelling, 79, 199, 418, 419
Urban heat island, 73, 436

V

Valley breeze, 224-226
Valley fog, 98
Vapor pressure
  defined, 82
  and computing relative humidity, 468
Venus, greenhouse effect on, 47

Verification of the Origins of Rotation in Tornadoes (VORTEX 2), 355
Vernal equinox, 53, 58
Vertical air motions
  in high- and low-pressure areas, 21, 277, 278, 279, 281
  in hurricanes, 364, 368
  in tornadoes, 342
Very short-range forecast, 410
Virga, 152, 306, 313
Visibility
  associated with warm fronts, 257-259
  reduction by fog, 29, 97
  reduction by pollutants, 103, 240
Visible light, 41, 44
Volcanoes (*see also* specific volcanoes)
  and climatic change, 443-445, 448
  and formation of atmosphere, 12
  gases emitted by, 12, 431, 443-445, 448
Volume conversions, 463
Vonnegut, Bernard, 150
Vortex (*see* Tornadoes)
Vortex chamber, 356
Vortex tubes, 347-348

W

Walker circulation, 419
Wall cloud, 312-313, 314, 350
Warm advection, 277
Warm core lows, 375
Warm fronts (*see* Fronts)
Warm occlusion, 262-263
Warm sector, 270
Water (*see also* Precipitation)
  change of state, 21-22, 37-38
  reflection of sunlight form, 49
  supercooled, 148
  temperature, contrasted with land, 75, 77
Water molecule, 21
  evaporating and condensing, 22
Waterspouts, 356-357
Water vapor, 9, 82
  air's capacity for, 22-23, 82
  amount in atmosphere, 8
  change of state, 21-22, 36
  as a greenhouse gas, 9, 45, 46, 48, 448
  properties of, 9
  role in absorbing infrared radiation, 9, 45-47
  role in climate change, 48, 436
Water vapor density (*see* Absolute humidity)
Water vapor-greenhouse feedback, 437, 448
Water vapor pressure, 82, 468
Watt, defined, 40
Wave clouds (*see* Billow clouds and Lenticular clouds)
Wave cyclone, 270, 272 (*see also* Cyclones, middle latitude)
Wave cyclone model, Norwegian, 270
Wavelength, defined, 41, 277
Waves, upper-level, 277
Waves, water, 198-200
Weather
  defined, 13
  cold and damp, 81
  disasters, 5
  elements of, 13
  foggy, 101-104
  influence on humans, 25-31, 79-89, 400
  record cold of 1983, December, 242
  record cold of 1989, 1990, December, 241-242
  record warmth of 1976, April, 246-247, 248
  record warmth of 1991, June, 250
  record warmth of 2005, July, 249
  in space, 54-55
  tropical, 362
Weather advisories, 400, 401
Weather Chanel, The, 31
Weather forecasting, 398-426
  accuracy and skill, 411-412
  and algorithms, 354
  analogue method, 408
  analysis, 404
  atmospheric models, 404, 405
  blizzards, 414-415
  and chaos, 407
  climatological, 410, 418-425
  cold weather, 414
  and computer progs, 404-406
  dense fog, 414
  ensemble, 392, 407
  extended, 410
  using the 500-mb chart, 405, 416
  flooding, 417
  hazardous and severe, 402, 412-418
  ice storms, 417
  improvement of, 404-408
  long-range, 410
  machine made, 404
  medium-range, 410
  meteogram, 402, 403
  methods of, 404-410
  numerical, 404-406
  outlooks, 410
  pattern recognition, 408
  persistence, 408
  probability, 408
  problems in, 400, 406-408
  prognostic (progs), 404
  record high temperatures, 412, 417
  using satellite information, 406
  Santa Ana winds, 414
  severe thunderstorms, 402, 415, 416
  short-range, 410
  snow storms, 414, 417
  and soundings, 400, 403
  statistical, 408
  steady state (trend) method, 408
  and thickness chart, 416
  tools, 402-404
  tornadoes, 415, 416
  types of, 410-411
  using surface charts, 412-417
  very short-range, 410
  weather type, 409
Weather information, acquisition of, 400
Weather symbols, 469-470
Weather type forecasting, 409
Weather watches and warnings, 400, 401
Weight, defined, 16
Westerlies, 210, 213, 277, 278
Wet-bulb depression, 90
Wet-bulb temperature, 86, 90
Whirlwinds (*see* Dust devils)
White Christmas, probability of, 408, 409
White Hurricane, 273
White frost, 96
Wien's (displacement) law, 42, 467
Wien, Wilhelm, 42
Wigley, Tom, 459
Willy-willy, 232
Wind, 19-21, 40, 176-204 (*see also* specific type)
  above friction layer, 188-189
  advisory, 401
  angle crosses isobars, 186-187, 191
  around high- and low-pressure areas, 20-21, 181-182, 189-190
  average surface, January and July, 211-215, 216
  and Beaufort scale, 195, 196
  blowing over water, 198-200
  and cold, 79-81
  cyclostrophic, 190
  defined, 19, 40, 208
  and deserts, 231-233
  direction, defined, 20, 182
  eddies, 19
  effect of friction on, 189, 192-194
  extreme, 195-200, 201-203, 219-233
  flow aloft, 188-189
  forces that influence, 184-187
  geostrophic, 188, 215
  global, 208-233
  gradient, 189
  gravity, 224
  and horizontal pressure changes, 175
  instruments, 192

(Wind, *continued*)
local systems, 219-233
and locating the center of storms, 194-195
measurements, 192
meridional, 191
monsoon, 219-222
nocturnal drainage, 224
in Northern Hemisphere, 195
onshore and offshore, 198, 223
and pressure patterns, 214-215
prevailing, 198
scales of motion, 19, 20
solar, 54, 55
in Southern Hemisphere, 190
speed, defined, 20
straight-line, 188-189, 302, 307, 308
in tornadoes, 353-356
trade, 210
units of speed, 463
on upper-level charts, 178-182, 189, 191, 194, 195
upslope and downslope, 224-225
and upwelling, 199
warnings, 401
and waves, 174-175
why it blows, 183-188
zonal, 191
Wind-chill, 25, 80
advisory, 401, 414
index, 80
record, 81
Wind profiler, 192, 193, 404
Wind sculptured trees, 198
Wind shear
and airline crashes, 28, 218, 220-221, 303-305
and air pockets, 220
and mid-latitude cyclonic storm development, 278
and severe thunderstorm development, 301, 313-315
and tornado development, 337
Wind sounding, 193
Wind vane, 192
Wind waves, 198
Windy afternoons, 225
Windy City, Chicago, 202
Windy places, 201
Winter solstice, 53, 57
Winter storm advisories and warnings, 401, 415, 417
Winter weather survival kit, 293
Witches of November, 273
World Meteorological Organization (WMO), 400
Wrap-around (cloud) band, 290, 291

Xenon, 8

Year without a summer, 59, 435, 444
Younger Dryas event, 433

Z

Zonal winds, 191
Zonda, 227 (*see also* Chinook wind)
Zulu time, 475